Linear
Operator
Theory

Linear
Operator
Theory

in
Engineering
and
Science

Arch W. Naylor
University of Michigan

George R. Sell
University of Minnesota

HOLT, RINEHART AND WINSTON, INC.
New York   Chicago   San Francisco
Atlanta   Dallas   Montreal
Toronto   London   Sydney

To
Andrée
and
Geraldine

Preface

The goal of this book is to present the basic facts of functional analysis in a form suitable for engineers, scientists, and applied mathematicians. Although the Definition–Theorem–Proof format of mathematics is used, careful attention is given to motivation of the material covered and many illustrative examples are presented.

The text can be used by students with various levels of preparation. However, the typical student is probably a first-year graduate student in engineering, one of the formal sciences, or mathematics. It is also possible to use this book as a text for a senior-level course. In order to facilitate students with varying backgrounds, a number of appendices covering useful mathematical topics have been included. Moreover, there has also been an attempt to make the pace in the beginning more gradual than that of later chapters.

The first five chapters are concerned with the "geometry" of normed linear spaces. The basic approach is to "disassemble" this geometric structure first, study the pieces, then reassemble and study the whole geometry. The pieces that result from this disassembly are set-theoretic, topological, and algebraic structures. Hence, Chapter 2 covers the appropriate set theory; Chapter 3 treats topological structure, in particular, metric spaces; and Chapter 4 handles algebraic structure, in particular, linear spaces. The reassembly takes place in Chapter 5 where normed linear spaces are studied. The main topic of this chapter is the geometry of Hilbert spaces.

The authors have found that the material covered in these first five chapters can be presented in a one-semester beginning graduate course. Indeed, the authors have done so a number of times in engineering and mathematics departments at a number of universities in the United States, Europe, and South America. Needless to say, the mode of presentation depends upon the audience. For certain audiences, motivation and examples are emphasized while proofs are only highlighted. For others, the converse is the case. An attempt has been made to make the book suitable for both modes of presentation. Moreover, there is material in the large collection of exercises appropriate for each type of audience.

Chapters 6 and 7 take the geometric structure developed in the first five chapters and apply it to the geometric analysis of linear operators. Chapter 6 covers the Spectral Theorem (the eigenvalue-eigenvector representation) for compact operators. Chapter 7 extends this material to certain discontinuous operators, in particular it treats those operators with compact resolvents. These two chapters also contain many illustrative examples.

Many chapters are divided into parts (Part A, Part B, and so forth). Part A contains basic introductory concepts. The subsequent parts of each chapter develop additional concepts and special topics. Thus, if a relatively quick introduction is desired, Part A can be covered first and material from the rest of the chapter can be added as needed.

For the person who is interested in getting to the spectral theory of linear operators as soon as possible it is recommended that he cover Part A of Chapters 3 and 4, Sections 1-8, 12-24 of Chapter 5, and then Chapters 6 and 7.

There is an important problem concerning integration theory. Although integration theory is not needed to understand the basic material covered, there are certain examples that do make reference to the Lebesgue integral and probability spaces. This problem can be handled in at least two ways. First, it can be more or less ignored. That is, the student can be told that there is such a thing as a Lebesgue integral and what its relation to the, presumably familiar, Riemann integral is. Probability spaces can be "glossed" over in the same way. The other way to approach the problem is to use the appendices. Appendix D gives an introduction to Lebesgue integration theory, and Appendix E presents the basic facts about probability spaces.

Each chapter is denoted by a numeral; that is, Chapter 3. The tenth section of the third chapter is denoted Section 3.10. However within Chapter 3, the 3 may be dropped and Section 10 used instead of Section 3.10. Theorem 5.5.4 (or Definition 5.5.4, Lemma 5.5.4, Corollary 5.5.4) refers to the fourth theorem in Section 5 of Chapter 5.

The notation "∎" is used to denote the end of proofs and examples. This allows the proof or examples to be skimmed on first reading.

The authors would like to thank a number of people who have aided in the development of this book. First, there are the students at various universities who have taken courses from one or the other of us based upon manuscript versions. Their suggestions have been invaluable. Next, we would like to thank colleagues who have aided us in various ways: H. Antosiewicz, M. Damborg, K. Irani, G. Kallianpur, W. Kaplan, W. Littman, W. Miller, R. Perret, W. Porter, T. Pitcher, P. Rejto, Y. Sibuya, H. van Nauta Lemke, and H. Weinberger. We especially want to thank F. Beutler for the many suggestions that arose out of his classroom use of the manuscript. Finally, we would like to thank the many secretaries at various universities who have helped in the preparation of the manuscript. In particular, we would like to thank the secretarial staffs of the Department of Electrical Engineering at the University of Michigan and the School of Mathematics at the University of Minnesota.

Arch W. Naylor
George R. Sell

Ann Arbor, Michigan
Minneapolis, Minnesota
July 1971

Contents

Introduction

1. BLACK BOXES

A great number of the mathematical problems of engineering and science can be fruitfully viewed as what are often referred to as "black box problems." One puts an "input" into a black box (Figure 1.1.1), the black box hums and

Figure 1.1.1.

whirls inside, and out comes an "output." Black box problems are questions about what black boxes do. The following are a few examples:

(1) If a black box is in fact an amplifier, questions can be asked about bandwidth, unit step response, distortion, and so on.

(2) Given an autonomous differential equation $\dot{x} = f(x)$, the initial state (that is, the initial conditions) may be viewed as the input and the resulting motion (or solution) may be viewed as the output. Many questions can be asked about the behavior of such equations; for example, questions about asymptotic growth, stability, periodicity, and so on.

(3) The input data to a digital computer is a string of symbols and its corresponding output is another string of symbols. The program determines what this black box does.

(4) Let $S = \{s_1, s_2, \ldots, s_n\}$ denote the state set for a Markov chain and $p(k), k = 0, 1, 2, \ldots$, denote the probability distribution over S at time k. Further let A denote the matrix of transition probabilities; that is,

$$p(k + 1) = Ap(k) \qquad k = 0, 1, 2, \ldots.$$

One can view the initial probability distribution $p(0)$ as an input and the resulting sequence $p(1), p(2), \ldots$ as the output. Or one can view $p(k)$ as an input and the resulting $p(k + 1)$ as an output.

(5) In the case of a plucked string, the initial stretched position of the string before release can be viewed as an input and the resulting string vibration can be viewed as an output.

(6) In a quantum-mechanical system, the wave function $\psi(x, t)$ may be viewed as an input and the integral $\int |\psi(x, t)|^2 \, dx$ or the partial derivative $\partial\psi/\partial t$ may be viewed as the output.

Needless to say, there is no end to the problems that can be formulated as black box problems.

As far as this book is concerned, the most important aspects of black box problems are that, once surface detail is removed, seemingly different problems become similar to one another and that certain patterns repeatedly appear in solution methods. For example, one does not treat linear time-invariant network problems as separate unrelated problems, rather one approaches them as a unified class of closely related problems. Similarly, it was noticed long ago that at a certain level of abstraction the matrix equation

$$\begin{bmatrix} y_1 \\ y_2 \\ \cdots \\ y_n \end{bmatrix} = \begin{bmatrix} a_{11} & & a_{1n} \\ & \cdots & \\ a_{n1} & & a_{nn} \end{bmatrix} \begin{bmatrix} x_1 \\ x_2 \\ \cdots \\ x_n \end{bmatrix}$$

and the integral equation

$$y(t) = \int_0^T k(t, \tau)x(\tau) \, d\tau \qquad t \in [0, T]$$

describe similar mathematical situations. Another way to say this is that these problems have similar mathematical structures.

The black box is thus an "operator" which transforms an input into an output. It is these operators that form the subject matter of our book. What we want to do, then, is recognize and study the essential mathematical structure of these operators. Although there are many kinds of operators, our goal here is to study those that can, once unessential details are removed, be viewed as transformations from normed linear spaces into normed linear spaces. This allows us to treat in a unified manner, matrix equations, integral equations, differential equations, difference equations, and random processes.

The real Euclidean plane is an example of a normed linear space. N-dimensional Euclidean space is another. Certain sequence spaces and function spaces are also examples. There are many other examples as we shall see later. For us the most important fact about normed linear spaces is that they all have a geometric structure that is very similar to ordinary two- or three-dimensional Euclidean geometry. This is particularly true for Hilbert spaces, a special subclass of normed linear spaces. *This geometric structure is the unifying theme of the material presented in this book.*

The first part (Chapters 2, 3, 4, and 5) of this book is devoted to a detailed study of this geometric structure. It turns out that the geometric structure of a normed linear space really involves three different kinds of structure: set-theoretic, topological, and algebraic. We illustrate this subdivision in the next section with the aid of a familiar example: the plane.

2.   STRUCTURE OF THE PLANE

The real Euclidean plane is a classic example of a normed linear space. As we have noted, it has set-theoretic, topological, and algebraic structure.

Set-Theoretic Structure

Before anything else, the plane is a set. In particular, it is the set of all ordered pairs of real numbers $x = (x_1, x_2)$. Denote this set by R^2, and note that $(7,1)$ and $(1,7)$ are different points in this set. We refer to this set as the underlying set.

Topological Structure

The type of topological structure that we are interested in here has to do with the concept of closeness. In particular, the Euclidean distance d between any two points $x = (x_1, x_2)$ and $y = (y_1, y_2)$ is

$$d(x,y) = \{ |x_1 - y_1|^2 + |x_2 - y_2|^2 \}^{1/2}.$$

The set R^2 equipped with this distance function is an example of what is called a metric space.

Algebraic Structure

The type of algebraic structure that we are interested in here is addition and scalar multiplication of points (vectors) in the plane. Thus, if $x = (x_1, x_2)$ and $y = (y_1, y_2)$, then

$$x + y = (x_1 + y_1, x_2 + y_2).$$

And if α is any real number,

$$\alpha x = (\alpha x_1, \alpha x_2).$$

With this structure on the set R^2 we have a linear space.

Combined Topological and Algebraic Structure

It is possible to have metric spaces that are not linear spaces and vice versa. As we have just seen here, it is also possible to have both a topological and an algebraic structure on the same underlying set. It happens very often that the topological and algebraic structure are blended together. In the case of the plane, and normed linear spaces in general, this blending is accomplished by means of the norm or length of vectors in the plane. If $x = (x_1, x_2)$, then the norm of x is given by

$$\|x\| = (x_1^2 + x_2^2)^{1/2}. \tag{1.2.1}$$

It follows that

$$d(x,y) = \|x - y\|$$

and

$$\|\alpha x\| = |\alpha| \, \|x\|,$$

where α is any scalar. Neither of the above two expressions would make sense if we did not have algebraic structure. We will see later that addition and scalar multiplication have continuity properties which are also a result of the blending of topological and algebraic structure.

Geometric Structure

When we put all the pieces together we are back to the plane with its familiar geometry. Some geometric facts are the result of the presence of topological structure only, some the result of the presence of algebraic structure only, and some involve both. We shall see which are which in the following four chapters.

Before we go on, it should be noted that the norm in (1.2.1) has some additional structure, namely that it is generated by an inner product. The inner product between the two vectors x and y in R^2 is given by

$$(x,y) = x_1 y_1 + x_2 y_2 .$$

Thus $\|x\| = (x,x)^{1/2}$. It should be noted here that there are other norms one can prescribe on R^2 that are not generated by inner products. We shall see in Chapter 5 that the geometric structure of spaces with inner products is much richer than those without.

3.   MATHEMATICAL MODELING

Since successful application of mathematics depends on successful mathematical modeling, it is worthwhile to say a few words about mathematical modeling. Roughly speaking, it is the formulation of a mathematical system whose mathematical behavior models certain aspects of a real system. For example, Ohm's law $e = Ri$ gives a mathematical model for the electrical behavior of a resistor.

The resistor can be used to illustrate the main problem of mathematical modeling. In order to formulate a mathematical model which models many aspects of a real system, one is usually led to a mathematical model of great complexity and such models are often mathematically intractable. For example, to model the high frequency as well as the high voltage behavior of our resistor could require a mathematical model involving nonlinear partial differential equations. Such equations are notoriously difficult. On the other hand, if one allows simple mathematical models only, one often ends up with a mathematical model which does not yield a sufficiently accurate or detailed description of the real system's behavior. For example, treating a long telephone or power line as a resistor without inductance and capacitance leads to a simple, yet usually inadequate mathematical model.

The formulation, then, of a mathematical model is a compromise between mathematical intractability and inadequate description of the system being modeled. There usually is a choice of mathematical models between these two extremes. For this reason, one usually talks about "a" mathematical model for a system not "the" mathematical model.

Another point to be made about mathematical modeling is that it is by no means a purely mathematical problem. It has a mathematical side, but it also has, for example, a physical or economical side. Indeed, mathematics alone would not allow us to arrive at Ohm's law. We need physics too. Mathematical modeling is the interface or bridge between pure mathematics and other disciplines.

4.   THE AXIOMATIC METHOD. THE PROCESS OF ABSTRACTION

The reader probably had his first encounter with the axiomatic method in the study of Euclidean geometry. Since all of mathematics and the subject matter of this book, in particular, is based on the axiomatic method, let us recall some of the features of axiomatic reasoning.

In every branch of mathematics one starts out with a collection of "undefinables." In the Euclidean geometry (of the plane) this includes "points" and "lines." Next, certain properties are stated. These properties (axioms, postulates) play the role of mathematical legislation and form the starting point of mathematical life, or reasoning. While these axioms usually have some basis in intuition, it should be emphasized that mathematical reasoning plays no role[1] in establishing these axioms. Some of the axioms of Euclidean geometry are: (a) the parallel postulate, and (b) if L is a line, then there exists a point not on L. Once the axioms have been chosen, one then tries to prove certain properties or theorems. For example, congruence or similarity of triangles is a question studied in Euclidean geometry.

The axiomatic method is *the* method of mathematics, in fact, it is mathematics. Even though there are many controversies in the mathematical community over the contents of sets of axioms, there is no question over the role of the axioms.

While the role of the axiomatic method in mathematics has been known for centuries, the emphasis of this role that one finds today is something which developed only recently. One can see this change by comparing the research papers of the last century with those published today. In the past it required very careful reading in order to determine the hypotheses needed in order to get a particular conclusion. Today, with most papers written in the definition-theorem-proof style, it is very easy to determine this.

Of course, axiomatic systems just do not happen. They must be formulated. As mentioned in Section 1, while working on seemingly diverse problems, one often finds that similar techniques are being employed. For example, the reader may be familiar with the z-transform and Laplace-transform techniques as applied to discrete-time and continuous-time systems, respectively. Another example would be the techniques used to study the harmonics of a vibrating string and the energy levels of the hydrogen atom. It is natural, then, to inquire into the essential features (or properties) of these techniques which allow them to be applied in different ways. By listing these properties (or axioms) as hypotheses and deriving results from them one thereby goes from a concrete problem to a more abstract

---

[1] We should note that a set of axioms should be consistent, that is, they should not lead to contradictory statements. This question of consistency is a very important question in mathematical logic, but we shall not go into it here. Instead, we refer the reader to Wilder [1].

problem. This process of abstraction plays a vital role in the development of mathematics because it allows one to gain insight into a larger class of problems.

This process of abstraction is very similar to the art of mathematical modeling discussed earlier. In modeling one seeks mathematical properties which describe (or model) a physical reality. In that realm there is a trade-off between finding a mathematically tractable model and deriving results sufficiently accurate for the physical reality. A similar trade-off occurs in the process of abstraction, namely, as one adds axioms (that is, becomes more concrete) one can derive sharper results; however, these results apply to a smaller class of objects. We shall experience this trade-off often in the following chapters. Indeed the next four chapters progress by adding axioms. In the next chapter we have the axioms of set theory only; consequently, there are few theorems to be proved. However, these theorems are widely applicable. Chapters 3 and 4 add more axioms and, hence, more theorems can be proven. However, these additional theorems are less widely applicable. We end Chapter 5 with a discussion of Hilbert spaces, and there we have a number of very sharp and important results.

5.   PROOFS OF THEOREMS

In the last section we discussed the role of axioms and hypotheses in the study of mathematics. However, there is another entity which plays a very important role, namely, the rules of logic.

In mathematical logic (metamathematics) the rules of logic are treated like axioms and they are studied as an axiomatic system. We shall not take this viewpoint here. Instead, we shall assume certain rules of logic and use them freely throughout. In this section we shall recall some of the more important rules and point out their role in the proofs of theorems.

The reader is undoubtedly familiar with the simple implication: "If A, then B." This means that if A is true or valid, then B is true or valid. It is sometimes written as "$A \Rightarrow B$." An example of this is:

If a is an even integer, then a^2 is an even integer.

The "if and only if" statement is a combination of two simple implications. That is, the statement "A if and only if B" is equivalent to *both* statements "If A, then B" and "If B, then A." We emphasize this point because beginning students sometimes misread the "if and only if" statements. Remember, in order to prove "A if and only if B" one must prove *two* things, namely "$A \Rightarrow B$" and "$B \Rightarrow A$." The "if and only if" statement is sometimes written as "$A \Leftrightarrow B$." We also say that B is a necessary and sufficient condition for A. An example of an "if and only if" statement is

$$\{x^2 + 2bx + c \geq 0 \quad \text{for all real} \quad x\} \Leftrightarrow \{b^2 - c \leq 0\},$$

where b, c, and x are real numbers.

It is customary in mathematics to give a definition in the following format:

An integer p is a prime number *if* the only divisors of p are p and 1.

Even though the word "if" is used in the defining clause it plays the role of an "if and only if" clause. In our notation, the definition above is equivalent to:

$$\{p \text{ is a prime}\} \Leftrightarrow \{p \text{ and } 1 \text{ are the only divisors of } p\}.$$

Let us now consider some of the actual techniques that are used in the proofs of theorems. The concept of the *direct proof* is the simplest. That is, in order to prove "$A \Rightarrow B$" one assumes A and then derives B. For example, if one wants to prove that

$$x \geq 0 \Rightarrow x < e^x, \tag{1.5.1}$$

one would assume that x is nonnegative and show directly that $x < e^x$. (See the Exercises.)

The *proof by contradiction* is different. For this one uses the "fact" that the statement "If A, then B" is equivalent to "If not B, then not A." Actually, the equivalence of these two statements is not a fact, but rather a rule of logic. In simplest terms, the rule of logic underlying this equivalence is the *Principle of Contradiction*, which reads:

"Either A or not A."

We shall accept this principle as one of the rules of logic. It is interesting to note that there is an entire school of mathematics, called the "institutionists" (Wilder [1; p. 243]), that does not accept the Principle of Contradiction.

Proof by contradiction goes as follows: In order to show that $A \Rightarrow B$, one assumes A and not B and then shows that this leads to a contradiction. In other words, if not B, then necessarily not A.

An example of a proof by contradiction occurs when one shows that $\sqrt{2}$ is irrational. The statement A is "$x = \sqrt{2}$" and the statement B is "x is irrational." Thus, A and not B becomes

"$x = \sqrt{2}$" and "x is rational."

If x is rational, then $x = p/q$, where p and q are integers without a common divisor. Statement A is equivalent to

$$p^2 = 2q^2.$$

This shows that p^2 is even. It follows then that p is even. So p^2 is divisible by 4. Hence q is even. Hence 2 is a common divisor for p and q. But this is a contradiction, for p and q have no common divisor. So if x is rational, $x \neq \sqrt{2}$. Or, if $x = \sqrt{2}$, then x is irrational.

Another type of proof which we shall use is proof by *Mathematical Induction*. Underlying this is the following rule of logic, called the Induction Principle.

Let $N = \{1, 2, \ldots\}$ denote the set of natural numbers (that is, the positive integers) and let M be a subset of N.

If the following two properties hold:

(1) 1 is in M, and
(2) if n is in M, then $(n + 1)$ is in M,

then $M = N$.

Let us give an application. Define S_n by

$$S_n = a + ar + \cdots + ar^{n-1}, \qquad n = 1, 2, \ldots.$$

We then claim that

$$S_n = a \frac{1 - r^n}{1 - r}, \qquad n = 1, 2, \ldots. \tag{1.5.2}$$

In order to prove (1.5.2) we shall let M denote those natural numbers for which (1.5.2) is true. If $n = 1$, then

$$S_1 = a = a \frac{1 - r}{1 - r};$$

hence, 1 is in M. Now assume that n is in M and consider S_{n+1}. Since

$$S_{n+1} = S_n + ar^n,$$

we have

$$S_{n+1} = a \frac{1 - r^n}{1 - r} + ar^n,$$

where the last equality follows from the assumption that n is in M. By use of simple arithmetic we see that

$$S_{n+1} = a \frac{1 - r^{n+1}}{1 - r},$$

so $(n + 1)$ is in M. It follows from the Induction Principle that $M = N$. Hence (1.5.2) has been proved.

EXERCISES

1. In order to prove (1.5.1) let $f(x) = x$ and $g(x) = e^x$. Show that $f(0) < g(0)$ and $f'(x) < g'(x)$ for $x > 0$. Now prove (1.5.1).

2. Show that $\sqrt{3}$ is irrational.

3. Show that the Induction Principle as stated in the text is equivalent to the following: Let M be a subset of the integers such that (a) M is not empty, and (b) $n \in M \Rightarrow (n + 1) \in M$. Then $k \in M \Rightarrow m \in M$ for all $m \geq k$.

4. Let $S_n = 1 + 2 + \cdots + n$. Show that $S_n = n(n + 1)/2$ for $n = 1, 2, \ldots.$

5. Let $S_n = 1^2 + 2^2 + \cdots + n^2$.   Show   that $S_n = n(n + 1)(2n + 1)/6$ for $n = 1, 2, \ldots$.

6. Let $h > 0$ and show that $(1 + h)^n > 1 + nh$ for $n = 2, 3, \ldots$.

SUGGESTED REFERENCES

Courant and Robbins [1].
Wilder [1].

2

Set-Theoretic Structure

1. INTRODUCTION

The purpose of this chapter is to present a brief review of certain basic set-theoretic concepts. The reader already familiar with these concepts may skim the chapter to develop familiarity with the notational conventions and then go on to Chapter 3.

We say that a set X is any well-defined collection of things. These things are referred to as the *members* or *elements* of X. In this book we are usually concerned with collections of numbers, sequences, functions, or, sometimes, collections of sets.

EXAMPLES

(1) The set R of all real numbers.

(2) The set of all sequences of the form

$$x = \{x_1, x_2, x_3, \ldots, x_k, \ldots\},$$

where x_k, $k = 1, 2, \ldots$, is a complex number.

(3) The set of all sequences of complex numbers $x = \{x_1, x_2, \ldots\}$ such that

$$\sum_{k=1}^{\infty} |x_k|^2 < \infty.$$

(4) The set $C[0,T]$ of all real-valued continuous functions x defined on the closed interval $0 \le t \le T$.

(5) The collection of all closed intervals on the real line.

We will usually denote sets by capital letters

$$A, B, X, \ldots.$$

We will use R and C to denote the sets of all real and complex numbers, respectively. The elements or members of a set will be denoted by lower case letters

$$a, b, x, \ldots.$$

If x is an element of a set A, we shall write this as $x \in A$. If x is not in A, we shall denote this by $x \notin A$.

One way of defining a set is by listing all of its elements. The set A consisting of the functions $f_1(t) = t, f_2(t) = t^2$, and $f_3(t) = t^3$ is denoted by

$$A = \{f_1, f_2, f_3\}.$$

Another way of defining a set B is (1) to assume that each element in B is an element in some well-defined *universal set*, say X, and (2) to list the properties that elements of the universal set must satisfy in order to be in B. For example, let X be the set of all sequences of complex numbers $x = \{x_1, x_2, x_3, \ldots\}$ and B be all elements of X possessing the property

$$\sum_{n=1}^{\infty} |x_n|^2 < \infty.$$

We shall use the following notation

$$B = \left\{ x \in X : \sum_{n=1}^{\infty} |x_n|^2 < \infty \right\},$$

which is read " B is the set of all elements of X such that (the colon stands for "such that") $\sum_{n=1}^{\infty} |x_n|^2 < \infty$. When there is no possibility of confusion we shall simply write

$$B = \left\{ x : \sum_{n=1}^{\infty} |x_n|^2 < \infty \right\}.$$

Abstracting this we see that if a set B is defined by a property P this can be written as

$$B = \{x \in X : P\},$$

which reads: "the set of all x in X such that P is true." Sometimes we define a set in terms of two properties P and Q. The set

$$B = \{x \in X : P \text{ and } Q\} = \{x \in X : P, Q\} \tag{2.1.1}$$

means "the set of all x in X such that both P and Q are true." [The comma in the second expression in (2.1.1) is to be read "and."] For example,

$$\{x \in R : x > 1 \text{ and } x < 2\} = \{x \in R : 1 < x < 2\},$$
$$\{x \in R : x > 1 \text{ and } x < 0\} = \varnothing,$$

where $\varnothing$ is used to denote the *empty* set; that is, the set with no elements.

The set

$$A = \{x \in X : P \text{ or } Q\} \tag{2.1.2}$$

means "the set of all x in X such that P, or Q, is true." In this case we use the so-called "inclusive or" which means P or Q or both. (The "exclusive or" is P or Q but not both.) For example,

$$\{x \in R : x > 2 \text{ or } x < 3\} = R.$$

A set A is said to be *finite* if it contains a finite number of elements. Otherwise, A is said to be an *infinite* set. A *countably infinite* set is one containing a countably infinite number of elements. A *countable* set is either finite or countably infinite An *uncountable* set is one that is infinite but not countably infinite. (See Appendix B on Cardinality.)

Two sets A and B are said to be *equal*, written $A = B$, if they both contain exactly the same elements. For example, the sets

$$A = \{1,2\} \text{ and } B = \{x: x^2 - 3x + 2 = 0\} \text{ are equal.}$$

We say that a set A is a *subset* of a set B if each element of A is also an element of B. We denote this by

$$A \subset B.$$

We also say that A is *contained* in B. Note that both B and $\varnothing$, the empty set, are always subsets of B. If x is an element of a set A, the subset of A containing exactly the element x is denoted by $\{x\}$. (Note that there is a difference between the element x and the set $\{x\}$.) If A is not a subset of B, we write

$$A \not\subset B$$

and say "A is not contained in B." A set A is a *proper subset* of a set B if $A \subset B$ and $A \neq B$. If $A \subset B$, we also sometimes say that B is a *superset* of A.

Two sets A and B are said to be *disjoint* if no element of A is in B and no element of B is in A. This is illustrated in Figure 2.1.1.

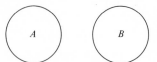

Figure 2.1.1.

2.   BASIC SET OPERATIONS

The *union* of sets A and B is the set made up of elements which belong to A or to B or to both. Refer to Figure 2.2.1. We denote the union of A and B by $A \cup B$ or $B \cup A$. In set-theoretic notation, we have

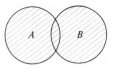

Figure 2.2.1. $A \cup B$ Shaded.

$$A \cup B = \{x: x \in A \text{ or } x \in B\}. \tag{2.2.1}$$

If $\{A_\alpha\}$ is an arbitrary collection[1] of sets, then the union of this collection, denoted $\bigcup_\alpha A_\alpha$, is the set made up of all elements x such that x belongs to at least one of the sets A_α.

---

[1] The collection $\{A_\alpha\}$ is indexed with the index α which ranges over some index set, and this index set may be finite, countably infinite, or uncountable.

EXAMPLE 1.   Let $A_1, A_2, A_3, \ldots$ be sets of real-valued functions defined on the interval $0 \leq t \leq 2\pi$. A_1 is the set of all functions of the form $x = a_1 \cos t + b_1 \sin t$, A_2 is the set of all functions of the form $x(t) = a_1 \cos t + b_1 \sin t + a_2 \cos 2t + b_2 \sin 2t$, and, A_k is the set of all functions of the form $x(t) = \sum_{i=1}^{k} (a_i \cos it + b_i \sin it)$, and so forth. It follows that

$$\bigcup_{n=0}^{\infty} A_n = \text{the set of all functions with a finite Fourier series expansion.} \quad \blacksquare$$

The *intersection* of a set A and a set B is the set made up of elements in both A and B. We denote the intersection by $A \cap B$ or $B \cap A$. See Figure 2.2.2. Equivalently,

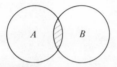

Figure 2.2.2. *$A \cap B$ Shaded.*

$$A \cap B = \{x: x \in A \text{ and } x \in B\}.$$

If $\{A_\alpha\}$ is any collection of sets, then the intersection of this collection, denoted $\bigcap_\alpha A_\alpha$, is the set made up of all elements x such that x belongs to every set A_α.

EXAMPLE 2.   Let $A_1, A_2, A_3, \ldots$ be sets of continuous functions defined by

$$A_k = \left\{ x \in C[0,T] : |x(t)| < 1 + \frac{1}{k} \text{ for all } t \right\}, \, k = 1, 2, \ldots.$$

Then

$$\bigcap_{k=1}^{\infty} A_k = \{x \in C[0,T] : |x(t)| \leq 1 \text{ for all } t\}. \quad \blacksquare$$

The *difference* of the sets A and B, denoted $A - B$, is the set made up of all elements of A that do not belong to B. See Figure 2.2.3. In other words,

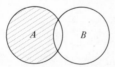

Figure 2.2.3. *$A - B$ Shaded.*

$$A - B = \{x: x \in A, x \notin B\}.$$

The *symmetric difference* of sets A and B is denoted by $A \triangle B$ and defined by (see Figure 2.2.4)

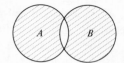

Figure 2.2.4. $A \triangle B$ Shaded.

$$A \triangle B = (A - B) \cup (B - A).$$

If X is a universal set and A is a set contained in X then the *complement* of A, denoted by A', is the set made up of all elements of X that do not belong to A. Equivalently,

$$A' = \{x \in X: x \notin A\} = X - A$$

(see Figure 2.2.5).

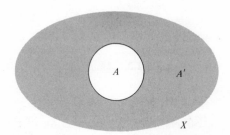

Figure 2.2.5. A' Shaded.

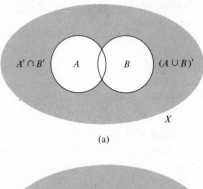

(a)

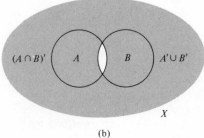

(b)

Figure 2.2.6. (a) $A' \cap B' = (A \cup B)'$
(b) $(A \cap B)' = A' \cup B'$.

Note that $X' = \emptyset$ and $\emptyset' = X$. Moreover, $X = A \cup A'$, $\emptyset = A \cap A'$, and $(A')' = A$ for arbitrary A.

The following two identities are often useful:

$$(A \cup B)' = A' \cap B',$$
$$(A \cap B)' = A' \cup B'.$$

(See Figure 2.2.6.)

De Morgan's Laws are generalizations of the last two identities and are stated as follows:

$$\left(\bigcup_\alpha A_\alpha\right)' = \bigcap_\alpha A_\alpha'$$

and

$$\left(\bigcap_\alpha A_\alpha\right)' = \bigcup_\alpha A_\alpha',$$

where $\{A_\alpha\}$ is any collection of sets contained in a universal set X.

3. CARTESIAN PRODUCTS

An *ordered pair* is a pair of objects x and y where one of the pair is designated as the first member of the pair and the other is designated as the second. We denote ordered-pairs by (x,y) with the obvious order.

EXAMPLE 1.   In the case of a system with two input channels, for example, the inputs to a stereoamplifier, denote the input on channel #1 by x_1 and channel #2 by x_2, then the system input is the ordered pair (x_1,x_2). Note that the system input $(e^{-t}, \sin t)$ is obviously different from the system input $(\sin t, e^{-t})$. ∎

If A and B are sets, then the set made up of all ordered pairs (a,b), where $a \in A$ and $b \in B$, is referred to as the *Cartesian product* of A and B. We write $A \times B$ for the Cartesian product and say "*A cross B*." Note that $A \times B$ is not the same thing as $B \times A$ unless $A = B$.

EXAMPLE 2.

(1) $R \times R$ or R^2 is the set of all ordered pairs of real numbers.

(2) $\{1,3\} \times \{7,5\} = \{(1,7), (1,5), (3,7), (3,5)\}$.

(3) $C[0,T] \times C[0,T]$ is the set of all ordered pairs of functions in $C[0,T]$. ∎

An *ordered n-tuple*, $(x_1,x_2, \ldots, x_n)$, is an *n*-tuple of objects, where one of them is designated as the first, one as the second, and so on until one is designated as the *n*th. When $n = 3$ we talk about *ordered triplets*.

If $X_1, X_2, \ldots, X_n$ are sets, we define the Cartesian product $X_1 \times X_2 \times \cdots \times X_n$ as the set of all ordered *n*-tuples $(x_1,x_2, \ldots, x_n)$, where $x_1 \in X_1, x_2 \in X_2, \ldots, x_n \in X_n$. Sometimes this is denoted by $\prod_{i=1}^n X_i$.

4.   SETS OF NUMBERS

There are a few sets of numbers which we will use throughout the text. These sets are the following:

The *natural numbers N*:

$$N = \{1,2,3,4,\ldots\}.$$

The *integers Z*:

$$Z = \{\ldots, -2,-1,0,1,2,\ldots\}.$$

The *rational numbers Q*:

$$Q = \{x: x = p/q, \text{ where } p \in Z, q \in Z, \text{ and } q \neq 0\}.$$

The *real numbers R*.
The *complex numbers C*.

In many ways the most interesting set of numbers is the *real number system R* or, as it is also called, the *real line*. Defining R exactly is not an issue here. Rather we shall assume R given and merely remark on certain properties of R.

We remark first that neither $+\infty$ nor $-\infty$ is a real number. When we adjoin $+\infty$ and $-\infty$ to R we have what is referred to as the *extended real numbers*.

A set $A \subset R$ is said to be *bounded from above* if there exists a real number u such that $a \leq u$ for all $a \in A$. The real number u is said to be an *upper bound* of A. We define *bounded from below* and *lower bound* in an analogous way. If a set A is both bounded from above and below, we say that A is *bounded*.

A real number M is said to be the *maximum* of a set $A \subset R$ if $M \in A$ and M is an upper bound for A. A real number m is said to be the *minimum* of a set $A \subset R$ if $m \in A$ and m is a lower bound for A. Needless to say, even a bounded set need have neither a maximum nor a minimum.

EXAMPLE 1.   The numbers 1 and 0 are, respectively, the maximum and minimum of the set $A = \{x \in R: 0 \leq x \leq 1\}$. The set $B = \{x \in R: 0 < x < 1\}$ has no maximum or minimum, however 1 and 0 are, respectively, upper bounds and lower bounds for B.   ∎

Let A be a nonempty set that is bounded from above, and let U denote the set of all upper bounds of A. For example, if A is the interval $0 < x < 1$, then $U = \{x \in R: 1 \leq x < \infty\}$. It is a fundamental fact about the real number system that the set U always has a minimum. This minimum of U is clearly the "least upper bound" of A. We denote it by sup A and say "the *supremum of A*." If A has a maximum, then clearly max A = sup A. If A is not bounded from above, we shall signal this fact by writing sup $A = \infty$. Further, if A is empty, we shall write sup $A = -\infty$. With these conventions the supremum exists for any subset of R. One sometimes sees l.u.b. A used in place of sup A.

Next let A be a set that is bounded from below, and let L denote the set of all lower bounds of A. Again, as long as A is nonempty L has a maximum. Obviously, this maximum is the "greatest lower bound" of A. We refer to it as "the *infimum* of A" and write inf A. Analogously to supremum, we say that inf $A = -\infty$ if A is not bounded from below and inf $A = \infty$ if A is empty. If A has a minimum, then min $A = $ inf A. One sometimes sees g.l.b. A used in place of inf A.

Sup A and inf A, then, are defined for any $A \subset R$. We remark in passing that a similar statement for the rational number system Q is not true.

EXAMPLE 2.   Let $A \subset Q$, the rational numbers, be the set

$$A = \{x \in Q : 0 < x < \sqrt{2}\}.$$

The set A has neither a maximum nor a supremum in Q.   ▌

Before we go on let us recall that a set A of complex numbers is said to be *bounded* if the set $\{|z| : z \in A\}$ is bounded in R.

5.   EQUIVALENCE RELATIONS AND PARTITIONS

Given a set X one is often interested in "cutting it up" into a family of disjoint subsets of X as illustrated in Figure 2.5.1. The technical term is "partition."

2.5.1 DEFINITION.   A family $\{A_\alpha\}$ of subsets of a set X is said to be a *partition* of X if

(1) $A_\alpha \cap A_\beta = \emptyset$ or $A_\alpha = A_\beta$, that is, the subsets A_α are pairwise disjoint or indexed more than once, and

(2) $\bigcup_\alpha A_\alpha = X$, that is, the union of the A_α's is all of X.

EXAMPLE 1.   Let $X = C[0,T]$ be the set of all continuous functions x defined on the interval $0 \le t \le T$. Let $\{A_\alpha\}$ be defined as follows: The index set is $R^+ = \{\alpha : 0 \le \alpha < \infty\}$ and

$$A_\alpha = \left\{ x \in C[0,T] : \int_0^T |x|^2\, dt = \alpha^2 \right\}.$$

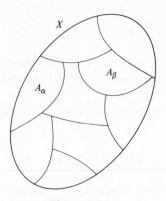

Figure 2.5.1.

It is clear that $A_\alpha \cap A_\beta = \varnothing$ whenever $\alpha \neq \beta$. It is also clear that one has

$$\int_0^T |x|^2 \, dt < \infty$$

for all $x \in C[0,T]$, so $X = \bigcup_{\alpha=0}^\infty A_\alpha$. ∎

The usual way that one characterizes a partition of a set is in terms of an equivalence relation. For this reason we now turn to relations and equivalence relations.

Given a set X, a *relation* on X can be defined as any subset of the Cartesian product $X \times X$. If R is a relation on X and the ordered pair (x,y) is in R, we say that "x is related to y under the relation R" and we write xRy. If the ordered pair (x,y) is not in R, we say that "x is not related to y under the relation R" and we write $x\not\!Ry$.

EXAMPLE 2.   Let X be a set of three people: Roger, aged 87, Waugh, aged 25, and Cuthburt, aged 104. Let R be the relation on X defined by xRy if and only if x is younger than y. The subset R of $X \times X$ will then be made up of the ordered pairs (Waugh,Roger), (Waugh,Cuthburt), (Roger,Cuthburt). ∎

A relation R on a set X is said to be *reflexive* if for each $x \in X$, one has xRx; that is, each x is related to itself.

EXAMPLE 3.   Let X be the real line. The relation R on X defined by

$$xRy \Leftrightarrow x \leq y$$

is reflexive. ∎

A relation R on a set X is said to be *symmetric* if xRy implies that yRx. That is, if x is related to y, then y is related to x.

EXAMPLE 4.   Let X be the real line, and let R be the relation on X defined by

$$xRy \Leftrightarrow |x| = |y|$$

R is symmetric and reflexive. ∎

A relation R on the set X is said to be *transitive* if xRy and yRz imply that xRz. In other words, if x is related to y and y is related to z, then x is related to z.

EXAMPLE 5.   Let X be the real line, and let the relation R on X be defined by

$$xRy \Leftrightarrow x < y.$$

R is transitive, but not symmetric nor reflexive. ∎

We are now ready to define equivalence relation.

2.5.2 DEFINITION.   A relation R on a set X is said to be an *equivalence relation* if it is reflexive, symmetric, and transitive.

If xRy, where R is an equivalence relation, we will say that "x is equivalent to y" and write $x \sim y$. Similarly, if $x\not{R}y$, we say that "x is not equivalent to y" and write $x \nsim y$. Needless to say, the notation "$x \sim y$" can lead to confusion if more than one equivalence relation is under consideration.

EXAMPLE 6.   We refer to Example 1 and note that the relation R on $X = C[0,T]$ defined by

$$xRy \Leftrightarrow \int_0^T |x(t)|^2 \, dt = \int_0^T |y(t)|^2 \, dt$$

is an equivalence relation.   ∎

We now come to the main point of this section: There is an intimate and natural connection between partitions and equivalence relations. In particular, any partition on a set X naturally determines an equivalence relation on X and vice versa.

Before stating the basic two theorems, we need the concept of an equivalence class. If R is an equivalence relation on a set X and $x \in X$, then the equivalence *class* determined by x is the subset

$$C_x = \{y \in X : y \sim x\};$$

that is, C_x is the set of elements of X that are equivalent to x. We sometimes denote C_x by $[x]$.

EXAMPLE 7.   (Continuing Example 6)

$$C_x = \left\{ y \in C[0,T] : \int_0^T |y(t)|^2 \, dt = \int_0^T |x(t)|^2 \, dt \right\}.   ∎$$

2.5.3 THEOREM.   *Let R be an equivalence relation on a set X. Then the family C of all equivalence classes is a partition of X.*

Proof: Let C_x and C_y be any two equivalence classes. We want to show that either $C_x = C_y$ or $C_x \cap C_y = \varnothing$. Suppose first that C_x and C_y are not disjoint. Let $w \in C_x \cap C_y$. By definition of equivalence class, wRx and wRy. Let z be any point in C_x, that is, zRx. We can then use the following line of argument:

$zRx, \ wRx, \ wRy \Rightarrow zRx, \ xRw \ wRy$ (by the symmetry of R)

$\qquad\qquad \Rightarrow zRw, \ wRy$ (by the transitivity of R)

$\qquad\qquad \Rightarrow zRy$ (by the transitivity of R)

$\qquad\qquad \Rightarrow z \in C_y \Rightarrow C_x \subset C_y.$

A similar argument shows that $C_y \subset C_x$, so $C_x = C_y$. Thus the only possibilities are $C_x = C_y$ or $C_x \cap C_y = \varnothing$. Since R is reflexive, we have $x \in C_x$ for all $x \in X$. It follows, then, that $X = \bigcup_x C_x$. ∎

EXAMPLE 8. Let $L_2[a,b]$ denote the set of all real-(or complex)- valued functions x such that

$$\int_a^b |x(t)|^2 \, dt < \infty.$$

We define the relation R by

$$xRy \Leftrightarrow \int_a^b |x(t) - y(t)|^2 \, dt = 0.$$

It follows that[2]

$x \sim y \Leftrightarrow \{$The set of points t, for which $x(t) \neq y(t)$, has measure zero$\}$.

The equivalence class C_x containing the function that is identically zero is a set made up of a huge number of functions each of which is zero almost everywhere. ∎

The next theorem provides the other half of the connection between partitions and equivalence relations.

2.5.4 THEOREM.   *Let $\{A_\alpha\}$ be a partition of a set X. Then the relation R defined by*

$$xRy \Leftrightarrow x \text{ is in the same partition subset as } y$$

is an equivalence relation on X.

The proof of this theorem follows immediately from the definition of a partition.

6.   FUNCTIONS

Most readers will have encountered real-valued functions of a real variable early in their training. Recall that such a function is a *rule f* which associates with each real number x another real number denoted by $f(x)$. However, the notion of a function is basically a set-theoretic concept and does not depend on the real numbers.

Suppose we have two sets X and Y. Suppose further that we have a rule f which assigns to *each* element in X *exactly one* element of Y. Then we say that f is a *Y-valued function defined on X*, or f is a *function defined on X with values in Y*. The terms mapping, transformation, and operator are sometimes used in place of function. We also say that f is a function which transforms, or maps, X into Y, and we will denote this by $f: X \rightarrow Y$.

An important point of notation: When we write f or $f(\cdot)$, we mean the function itself. When we write $f(x)$, we mean the element in Y assigned to the element x in X.

---

[2] The reader who is unfamiliar with the basic concepts of measure and integration theory can study these concepts in Appendix D. Roughly speaking, a set of measure zero is either finite or an infinite set whose "total length" (that is, measure) is zero.

If $f: X \to Y$, $g: X \to Y$, and $g(x) = f(x)$ for each $x \in X$, then we say that $f = g$, that is, they are one and the same function. We remind the reader not to confuse a function with its representation. That is, $f(x) = |x|$ and $g(x) = \exp(\log|x|)$ are *representations* of the same function.

If f is a function defined on a set X, we say that the set X is the *domain* of f and we write $\mathscr{D}(f) = X$. A function g such that $\mathscr{D}(f) \subset \mathscr{D}(g)$ and $g(x) = f(x)$ for each x in $\mathscr{D}(f)$ is said to be an *extension* of f. Next let f be a function with $\mathscr{D}(f) = X$, and let A be a subset of X. A function h such that $\mathscr{D}(h) = A$ and $h(x) = f(x)$ for each x in A is said to be *the restriction* of f to A. Sometimes h is denoted by $f|_A$.

Let $f: X \to Y$ be given. If $y = f(x)$, we say that y is *the image* of x or that y is *the value* of f at x. Also we say that x is a *pre-image* of y. We cannot say *the* pre-image, for there may be more than one x with y as its image. The set of all elements of Y that are images under f of elements of X is said to be the *range* of f and written $\mathscr{R}(f)$; that is,

$$\mathscr{R}(f) = \{y \in Y: y = f(x) \text{ for some } x \text{ in } X\}.$$

If $\mathscr{R}(f) \subset Y$, we say that f maps X *into* Y. If $\mathscr{R}(f) = Y$, we say that f maps X *onto* Y. If $\mathscr{R}(f)$ contains only one point, we then say that f is a *constant function*.

The mapping I of a set X onto itself given by $I(x) = x$ for all x in X is said to be the *identity mapping* on X.

If the domain of f does not contain two elements with the same image; that is, if

$$[f(x_1) = f(x_2)] \Rightarrow [x_1 = x_2],$$

then we say that f is a *one-to-one* mapping. Note that f is a one-to-one mapping if and only if every point y in $\mathscr{R}(f)$ has precisely one pre-image point in $\mathscr{D}(f)$.

The real-valued functions $y = x^2$ and $z = e^{-y}$ induce a mapping of x into z given by $z = e^{-x^2}$. This composition of two functions has an abstract formulation. Let $f: X \to Y$ and $g: Y \to Z$ be given. We define the composition of g and f by $(gf)(x) = g(f(x))$. This mapping is defined on X with values in Z. The composition of g and f is illustrated in Figure 2.6.1. We note that one can meaningfully write

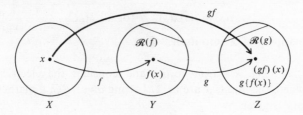

Figure 2.6.1.

gf or say that gf exists if and only if $\mathscr{D}(g) \supset \mathscr{R}(f)$. If this is not so, one says that gf does not exist. It may be that gf exists, but fg does not. Give an example.

If f_n maps a set X_n into set X_{n+1}, $n = 1, 2, \ldots, N$, the composition of $f_1, \ldots, f_N$ is defined by

$$(f_N f_{N-1} \cdots f_1)(x) = f_N\{\cdots f_2\{f_1(x)\}\ldots\},$$

for each $x \in X_1$. We leave it to the reader to show the associative law, that is, $(f_N f_{N-1} \cdots f_k)(f_{k-1} \cdots f_1) = f_N f_{N-1} \cdots f_1$ for arbitrary $1 < k < N$.

Let us recall what we mean by the graph of a real-valued function of a real variable. Consider the function $y = (x + 1)^2$. The graph of this function is a subset of the plane R^2. A point (x,y) of the plane lies in the graph if and only if $y = (x + 1)^2$. Now consider the general case. Let $f: X \to Y$. The *graph* of the function f is the subset of the product set $X \times Y$ made up of all ordered pairs (x,y) such that $y = f(x)$; that is,

$$\text{Gr}(f) = \text{graph}(f) = \{(x,y) \in X \times Y: y = f(x)\}.$$

Needless to say, it is not often that one can "plot" this graph.

Let us consider some illustrations of these concepts.

EXAMPLE 1.   Let $X = Y = R^n$, the set of all n-tuples of real numbers. Given an $n \times n$ real matrix A, the matrix equation

$$y = Ax \qquad x, y \in R^n$$

represents a mapping of X into itself. In Chapter 4 we will investigate the conditions under which this mapping is one-to-one and/or onto.   ∎

EXAMPLE 2.   Given a real number σ, let C_σ denote the vertical line in the complex plane defined by

$$C_\sigma = \{s: \text{Re}(s) = \sigma\}.$$

Let $L_{2,\sigma}(-i\infty,i\infty)$ be the set of all complex-valued functions $x(s)$ defined on C_σ such that

$$\frac{1}{2\pi i}\int_{C_\sigma} |x(s)|^2 \, ds < \infty.$$

Let $t(s)$ denote a bounded continuous complex-valued function defined on C_σ. Then

$$y(s) = t(s)x(s) \qquad \text{for all } s \in C_\sigma$$

represents a mapping of X into itself.   ∎

EXAMPLE 3.   Let $X = L_2(-\infty,\infty)$ and consider the electrical network of Figure 2.6.2 where the inputs x are in X. Letting $a = 1/RC$, a mathematical model for this network is given by

$$y(t) = \int_{-\infty}^{t} h(t - \tau)x(\tau) \, d\tau, \qquad (2.6.1)$$

where $h(t)$, the unit impulse response or weighting function, is given by

$$h(t) = \frac{1}{a} e^{-at}.$$

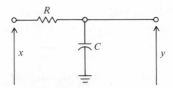

Figure 2.6.2.

We state without proof that (2.6.1) represents a mapping f of X into itself. It is shown in Exercise 1, Section 4.4, that f is one-to-one. ∎

EXAMPLE 4.   Let $X = N \times N$, the Cartesian product of the set of natural numbers with itself. Let $Y = \{0,1\}$. The mapping $\delta_{ij} : X \to Y$, $i, j \to 1, 2, \ldots$, defined by

$$\delta_{ij} = \begin{cases} 1, & \text{if } i = j \\ 0, & \text{if } i \neq j \end{cases}$$

is called the *Kronecker function*.

It should be noted that there is nothing sacred about the set N in this example. It could be replaced by any set, finite, infinite, even uncountably infinite. ∎

EXAMPLE 5. $C^n[a,b]$, where $n \geq 0$ is an integer, denotes the set of all real-valued (or complex-valued) functions x defined on the interval $a \leq t \leq b$, such that x is continuous, and the derivatives $d^k x/dt^k$ of order $k \leq n$ exist and are continuous on $[a,b]$. $C^\infty[a,b]$ denotes the set

$$C^\infty[a,b] = \bigcap_{n=0}^{\infty} C^n[a,b];$$

that is, $x \in C^\infty[a,b]$ if and only if x has continuous derivatives of all orders. Let $P(D)$ denote the differential operator

$$P(D) = \alpha_n \frac{d^n}{dt^n} + \alpha_{n-1} \frac{d^{n-1}}{dt^{n-1}} + \cdots + \alpha_1 \frac{d}{dt} + \alpha_0,$$

where the α's are constants.

It follows that $P(D)$ can be considered as a mapping of $C^\infty[a,b]$ into itself. It can also be considered as a mapping of $C^n[a,b]$ into $C^0[a,b] = C[a,b]$.

The interval $[a,b]$ in this example can be replaced by an arbitrary (even infinite) interval. For that matter we could replace it by a set Ω in R^m. In the latter case, $C^n(\Omega)$ would denote the collection of all functions $u = u(x_1, \ldots, x_m)$ defined on Ω with the property that all partial derivatives up to order n are continuous. Also, $C^\infty(\Omega) = \bigcap_{n=1}^{\infty} C^n(\Omega)$.

One sometimes refers to the fact that $u \in C^n(\Omega)$ by saying that *u is a C^n-function on Ω* or that *u is of class C^n*. ∎

EXAMPLE 6.   Sometimes functions are defined implicitly. For example let $k(t,s)$ be a continuous function defined for $0 \leq s \leq t \leq T$ and consider the Volterra integral equation

$$y(t) = x(t) + \int_0^t k(t, s)y(s)\, ds. \tag{2.6.2}$$

It is shown in Exercise 3.15.7 that there is a function $F: C^0[0,\alpha] \to C^0[0,\alpha]$ where $y = F(x)$ satisfies (2.6.2) provided α is a sufficiently small positive number. One can also show that F is one-to-one, see Exercise 2, Section 4.4.   ∎

Set Functions

If f maps a set X into a set Y, then there is a mapping $\mathscr{F}$ naturally associated with f which maps subsets of X into subsets of Y. Let $P(X)$ denote the set of all subsets of X. Then $\mathscr{F}: P(X) \to P(Y)$ is defined by

$$\mathscr{F}(A) = \{y \in Y: y = f(x) \text{ for some } x \text{ in } A\},$$

where $A \in P(X)$. Since $\mathscr{F}(A)$ is a subset of Y, it is an element of $P(Y)$. For instance, $\mathscr{F}(X) = \mathscr{R}(f)$, the range of f.

The mapping $\mathscr{F}$ maps $P(X)$ onto $P(Y)$ if and only if f maps X onto Y. $\mathscr{F}$ is one-to-one if and only if f is one-to-one. (Why?) Moreover,

$$\mathscr{F}(\varnothing) = \varnothing, \text{ the empty set}$$
$$\mathscr{F}(A_1 \cup A_2) = \mathscr{F}(A_1) \cup \mathscr{F}(A_2)$$
$$\mathscr{F}(A_1 \cap A_2) \subset \mathscr{F}(A_1) \cap \mathscr{F}(A_2).$$

EXAMPLE 7.   Let M be a mass initially at rest at a point A on a frictionless plane as shown in Figure 2.6.3. Assume that a force $x(t)$ is applied to M so that it moves along a straight line in the plane. The distance along this line measured from the point A denoted y, and the velocity of M is denoted by $v = dy/dt = \dot{y}$. If it is assumed that $C[0,\infty)$ is the set of allowable x's, then it is clear that

$$y(t) = \frac{1}{M} \int_0^t \int_0^\tau x(\theta)\, d\theta\, d\tau = \frac{1}{M} \int_0^t (t - \theta)x(\theta)\, d\theta,$$

where the last integral is obtained by a simple interchange of order of integration, and

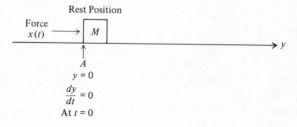

Figure 2.6.3.

$$v(t) = \frac{1}{M} \int_0^t x(\theta) \, d\theta$$

define a mapping $f: C[0,\infty) \to C[0,\infty) \times C[0,\infty)$ where an element in the range is the ordered pair of functions (y,v). That is, $(y,v) = f(x)$ where $y =$ displacement and $v =$ velocity.

The mapping $\mathscr{F}$, then, is a mapping of subsets of $C[0,\infty]$ into subsets of $C[0,\infty) \times C[0,\infty)$. For example, consider the set U of all inputs $x(t)$ such that $|x(t)| \leq u$, where u is a positive real number. It is not too difficult to show that $\mathscr{F}(U)$ is characterized by the set of all (y,v) in $C[0,\infty) \times C[0,\infty)$ such that

(1) $y, \dot{y}, \ddot{y}$ are in $C[0,\infty)$,

(2) v and $\dot{v}$ are in $C[0,\infty)$,

(3) $\dot{y} = v$,

(4) $|v(t)| \leq tu/M$,

(5) $|y(t)| \leq t^2 u/2M$. ∎

Now that the reader understands, we hope, the difference between $\mathscr{F}$ and f, we will henceforth economize on symbols and dispense with the symbol $\mathscr{F}$. It is customary to write $f(A)$ instead of $\mathscr{F}(A)$.

In addition to the function $\mathscr{F}$, there is another mapping of subsets into subsets which is naturally associated with f. It maps $P(Y)$ into $P(X)$. For the moment we will denote it $\mathscr{F}^{-1}$. We define $\mathscr{F}^{-1}$ as follows:

$$\mathscr{F}^{-1}(C) = \{x \in X: f(x) \in C\},$$

where $C \in P(Y)$. That is, $\mathscr{F}^{-1}(C)$ consists of all pre-image points of points in C.

The mapping $\mathscr{F}^{-1}$ is referred to as the *inverse set function* or *inverse image mapping* and the set $\mathscr{F}^{-1}(C)$ is said to be the *inverse image* of C. Note that

$$\mathscr{F}^{-1}(\varnothing) = \varnothing,$$
$$\mathscr{F}^{-1}(C_1 \cup C_2) = \mathscr{F}^{-1}(C_1) \cup \mathscr{F}^{-1}(C_2)$$
$$\mathscr{F}^{-1}(C_1 \cap C_2) = \mathscr{F}^{-1}(C_1) \cap \mathscr{F}^{-1}(C_2).$$

Also note that

$$\mathscr{F}^{-1}(\mathscr{R}(f) \cap C) = \mathscr{F}^{-1}(C)$$

for arbitrary C. Although the symbols might suggest it, $\mathscr{F}^{-1}$ is not necessarily the inverse[3] of $\mathscr{F}$. We note that $\mathscr{F}^{-1}$ always maps $P(Y)$ onto $P(X)$. $\mathscr{F}^{-1}$ is one-to-one if and only if f is one-to-one and $\mathscr{R}(f) = Y$. Again to economize on symbols, we will henceforth write $f^{-1}(C)$ instead of $\mathscr{F}^{-1}(C)$.

EXAMPLE 8.   Suppose that in Example 7 we let A be a set in $C[0,\infty) \times C[0,\infty)$ defined by

A term yet to be defined, but probably not new to the reader.

$$A = \begin{cases} \text{set of all motions of the mass such} \\ \text{that at time } T > 0, \, y(T) = b, \, v(T) = 0 \end{cases}.$$

It then follows that

$$f^{-1}(A) = \left\{ x \in C[0,\infty) : \frac{1}{M} \int_0^T (T - \theta)x(\theta) \, d\theta = b \text{ and } \frac{1}{M} \int_0^T x(\theta) \, d\theta = 0 \right\}. \quad \blacksquare$$

EXERCISES

1. Consider the space $C[0,T]$. In electronics a clipper circuit is a circuit whose output waveform is the input waveform (that is, voltage or current as a function of time) with the "top and bottom clipped off." More precisely, if $x(t)$ is the input waveform, then the output waveform $z(t)$ is given by

$$z(t) = \begin{cases} A, & \text{if } x(t) \geq A \\ x(t), & \text{if } -A < x(t) < A \\ -A, & \text{if } x(t) \leq -A, \end{cases} \quad (2.6.3)$$

where $A > 0$ is sometimes referred to as the clipping level. A circuit which would approximately realize this behavior is shown in Figure 2.6.4.

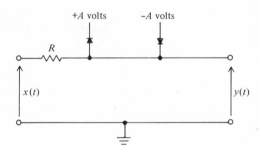

Figure 2.6.4.

(a) Does the statement (2.6.3) describe a mapping of $C[0,T]$ into $C[0,T]$?
(b) If so, characterize the range of the mapping.
(c) If so, is the mapping one-to-one?

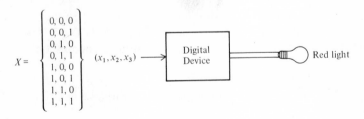

Figure 2.6.5.

2. Let X be the set of all ordered triplets of 1's and 0's as shown in Figure 2.6.5. Suppose we have a digital device into which we feed a point from X. The device is so constructed that if $x = (1,0,1)$ is the input, a red light lights; otherwise, the light remains off. Characterize this situation as a mapping of the set X into an appropriate set Y. Is this mapping one-to-one? Is this a mapping of X onto Y?

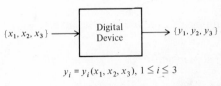

$$y_i = y_i(x_1, x_2, x_3), \ 1 \le i \le 3$$

Figure 2.6.6.

3. Let X be the same as in Exercise 2, and let $Y = X$. Now suppose that we have a proposed digital device which is to operate as follows (see Figure 2.6.6). At noon of day number one, x_1 is put into the machine and y_1 comes out of the machine almost instantaneously. At noon of the second day, x_2 goes in and y_2 comes out. Similarly, on the third day x_3 goes in and y_3 comes out. The following table characterizes the desired operation.

Input	Output
0,0,0	1,1,1
0,0,1	1,0,1
0,1,0	0,0,1
0,1,1	0,0,1
1,0,0	1,1,0
1,0,1	0,0,0
1,1,0	1,1,1
1,1,1	1,0,0

Have we characterized a mapping of X into itself. Obviously, yes. Is this mapping one-to-one? Is it onto? We will return to this exercise in Section 8.

7. INVERSES

The next topic, inverses, is simple yet extremely important. Roughly speaking, a mapping $F: X \to Y$ is invertible if there exists a mapping $G: Y \to X$ which unravels F. Let us make this idea precise.

2.7.1 DEFINITION. A mapping $F: X \to Y$ is said to be *invertible* if there exists a mapping $G: Y \to X$ such that GF and FG are the identity mappings on the sets X and Y, respectively. In this case G is said to be an *inverse* of F.

The next few lemmas introduce some important facts.

2.7.2 LEMMA.   *If a mapping $F: X \to Y$ is invertible, it has a unique inverse.*

Proof: Assume that G_1 and G_2 are inverses of F, and let y be any point in Y. Since FG_1 and $G_2 F$ represent the identity on Y and X, respectively, one has

$$G_2(y) = G_2(FG_1(y)) = G_2 F(G_1(y)) = G_1(y).$$

Hence $G_1 = G_2$.   ∎

Because of this lemma, we shall denote the inverse of F by F^{-1}.

2.7.3 LEMMA.   *If a mapping $F: X \to Y$ is invertible, then F^{-1} is invertible and $(F^{-1})^{-1} = F$.*

The validity of this lemma is obvious.

EXAMPLE 1.   Let X be the set of all real-valued C^1 functions x defined on the closed interval $0 \le t \le T$ such that $x(0) = 0$. Let Y be the set of all real-valued continuous functions x defined on $0 \le t \le T$. If F is defined by $y = dx/dt$, then

$$F^{-1}(y) = \int_0^t y(\tau)\, d\tau. \quad ∎$$

2.7.4 LEMMA.   *If a mapping $F: X \to Y$ is invertible, then F is one-to-one.*

Proof: Suppose F is invertible. Recall that F is one-to-one if every point y in Y has precisely one pre-image x in X, that is, there is precisely one point x in X such that $F(x) = y$. If x_1 and x_2 both satisfy $F(x_1) = F(x_2) = y$, then

$$x_1 = F^{-1}F(x_1) = F^{-1}F(x_2) = x_2.$$

Hence F is one-to-one.   ∎

2.7.5 LEMMA.   *If a mapping $F: X \to Y$ is invertible, then $\mathscr{R}(F) = Y$.*

Proof: If $y \in Y$, then $y = FF^{-1}(y) = F(x)$, where $x = F^{-1}(y)$. Hence $y \in \mathscr{R}(F)$, or $\mathscr{R}(F) = Y$.   ∎

Building upon Lemmas 2.7.4 and 2.7.5, we arrive at the following elegant characterization of invertible mappings.

2.7.6 THEOREM.   *A mapping $F: X \to Y$ is invertible if and only if it is one-to-one and $\mathscr{R}(F) = Y$.*

Proof: Lemmas 2.7.4 and 2.7.5 furnish the proof for the "only if" part of the theorem. We need to prove the "if" part. We assume that F is one-to-one and $\mathscr{R}(F) = Y$ and show it follows that F is invertible. We do so by exhibiting the inverse F^{-1}. Let y be an arbitrary element of Y. Since $\mathcal{R}(F) = Y$ and F is one-to-one, y has precisely one pre-image x in X. We define $G(y)$ to be x. We then have a mapping $G: Y \to X$ and it is clear that FG and GF are the identity mappings on Y and X, respectively. Hence F is invertible. ∎

An invertible mapping of X onto Y is sometimes called a *one-to-one correspondence* between X and Y, for it puts the elements of X and Y into a one-to-one correspondence.

Before the reader becomes too enamoured with Theorem 2.7.6, it should be noted that in practice determining whether or not a given mapping is one-to-one or onto is often a difficult task.

Suppose $F: X \to Y$ is one-to-one but $\mathcal{R}(F) \neq Y$. Then F is not invertible. However, if we replace Y with $\mathcal{R}(F)$, then $F: X \to \mathcal{R}(F)$ is one-to-one and a mapping of X onto $\mathcal{R}(F)$. In this case F is invertible, and this inverse is also denoted by F^{-1}. However, it should be emphasized that the domain of this inverse is $\mathcal{R}(F)$ and not Y.

In the last section we introduced a simplified notation for the *inverse set-function* associated with a mapping $F: X \to Y$. Recall that this was denoted by F^{-1}. Needless to say, this is apt to be confused with the *inverse* as defined in this section. However, these notational conventions are so standard it does not pay to tamper with them here. Furthermore, the usage will always be clear from context. Two points should be emphasized: (1) the inverse set-function is defined for *every* mapping $F: X \to Y$, (2) the inverse is defined only for certain mappings $F: X \to Y$.

EXAMPLE 2. Suppose that $X = Y = l_2$ is the set[4] of all infinite sequences $x = \{x_1, x_2, \ldots\}$ of real (or complex) numbers such that

$$\sum_{k=1}^{\infty} |x_k|^2 < \infty.$$

Consider the sequence $y = \{y_1, y_2, \ldots\}$, where $y_k = \lambda_k x_k$ and $|\lambda_k| \leq M$, a constant, for $k = 1, 2, \ldots$. The following infinite matrix equation is a convenient way to represent the dependence of y on x.

$$\begin{bmatrix} y_1 \\ y_2 \\ \cdots \end{bmatrix} = \begin{bmatrix} \lambda_1 & & 0 \\ & \lambda_2 & \\ 0 & & \cdots \end{bmatrix} \begin{bmatrix} x_1 \\ x_2 \\ \cdots \end{bmatrix}.$$

If the sequence y is in l_2 for arbitrary x in l_2, then a mapping of l_2 into itself has been described. But

$$\sum_{k=1}^{N} |y_k|^2 \leq M^2 \sum_{k=1}^{N} |x_k|^2 \leq M^2 \sum_{k=1}^{\infty} |x_k|^2.$$

Hence, y is in l_2. Denote this mapping of l_2 into itself by Λ.

---

[4] We will occasionally denote this set by $l_2(0, \infty)$. The symbol $l_2(-\infty, \infty)$ will denote the set of doubly infinite sequences $x = (\ldots, x_{-1}, x_0, x_1, x_2, \ldots)$ such that $\sum_{n=\infty}^{\infty} |x_n|^2 < \infty$, see Example 2.8.1.

It is easily shown that Λ is one-to-one if and only if $\lambda_k \neq 0$ for all k. Indeed, we want to show that $\Lambda(x) \neq \Lambda(z)$ whenever $x \neq z$. That is, the sequence $\{\lambda_k(x_k - z_k)\}$ is not a sequence of zeros whenever $x_j \neq z_j$ for at least one index j. The latter statement is now obviously the case if and only if $\lambda_k \neq 0$, $k = 1, 2, \ldots$.

If $|\lambda_k| \geq m > 0$ for all k, then $\mathcal{R}(\Lambda) = l_2$. Let $y = \{y_1, y_2, \ldots\}$ be an arbitrary sequence in l_2, and let $x = \{\lambda_1^{-1}y_1, \lambda_2^{-1}y_2, \ldots\}$. If x is in l_2, then it is clearly a pre-image of y and hence $\mathcal{R}(\Lambda) = l_2$. However,

$$\sum_{k=1}^{N} |x_k|^2 = \sum_{k=1}^{N} \left| \frac{1}{\lambda_k} y_k \right|^2 \leq \frac{1}{m^2} \sum_{k=1}^{N} |y_k|^2 \leq \frac{1}{m^2} \sum_{k=1}^{\infty} |y_k|^2,$$

which shows that x is in l_2.

On the other hand, we claim that if $|\lambda_k|$ is not bounded away from 0, then

$$\mathcal{R}(\Lambda) \neq l_2. \tag{2.7.1}$$

To prove this we first note that (2.7.1) is true if one or more λ_k's are zero. Thus we assume that $\lambda_k \neq 0$ for all k and $\lambda_{k_i} \to 0$ for some subsequence k_i. By taking a further subsequence we can choose the k_i so that $|\lambda_{k_i}| \leq 1/i$. We now claim that the sequence $y \in l_2$ defined by

$$y_k = \begin{cases} 1/i, & \text{if } k = k_i \\ 0, & \text{otherwise} \end{cases}$$

is not in $\mathcal{R}(\Lambda)$. Indeed, if $y \in \mathcal{R}(\Lambda)$, then its pre-image would have to be x where $|x_{k_i}| = |\lambda_{k_i}^{-1} y_{k_i}| \geq 1$. Since there are an infinite number of such k_i, one has

$$\sum_{k=1}^{\infty} |x_k|^2 = \infty.$$

Hence $x \notin l_2$, and (2.7.1) is valid.

We see, then, that $\Lambda: l_2 \to l_2$ can be one-to-one with range $\mathcal{R}(\Lambda) \neq l_2$. In that case, Λ is not invertible. However, if we restrict our attention to the range of Λ, the following matrix equation represents Λ^{-1}:

$$\begin{bmatrix} x_1 \\ x_2 \\ \cdot \\ \cdot \\ \cdot \end{bmatrix} = \begin{bmatrix} \dfrac{1}{\lambda_1} & & 0 \\ & \dfrac{1}{\lambda_2} & \\ 0 & & \cdots \end{bmatrix} \begin{bmatrix} y_1 \\ y_2 \\ \cdot \\ \cdot \\ \cdot \end{bmatrix}. \quad \blacksquare$$

We now turn to the concepts of left and right inverses.

2.7.7 DEFINITION.   A mapping $F: X \to Y$ is said to be *left invertible* if a mapping $G: Y \to X$ exists such that GF is the identity mapping on the set X.

2.7.8 THEOREM.   *A mapping $F: X \to Y$ is left invertible if and only if it is one-to-one.*

Proof: Consider the "if" first. Assume that $F: X \to Y$ is one-to-one. Then $F: X \to \mathscr{R}(F)$ is invertible, with $F^{-1}: \mathscr{R}(F) \to X$. Let $G: Y \to X$ be *any* extension of F^{-1}. It follows that GF is the identity on X.

The converse, or the "only if" part, is proved in the same manner that we proved Lemma 2.7.4. We ask the reader to check the details. ∎

If $F: X \to Y$ is left invertible, the mapping G given in Definition 2.7.7 is said to be a *left inverse* of F. The notation we shall use for this is $G = F_l^{-1}$. It should be clear from the proof that if $\mathscr{R}(F) \neq Y$, then G is not unique. That is, F has many left inverses. However, the restriction of any left inverse G to $\mathscr{R}(F)$ must be $F^{-1}: \mathscr{R}(F) \to X$, that is, $G|_{\mathscr{R}(F)} = F^{-1}$. This restriction is unique by Lemma 2.7.2.

EXAMPLE 3.   Returning to Example 2, one left inverse of Λ is as follows: Given $y \in l_2$

$$\Lambda_l^{-1}(y) = \begin{cases} \begin{bmatrix} \dfrac{1}{\lambda_1} & & 0 \\ & \dfrac{1}{\lambda_2} & \\ 0 & & \end{bmatrix} \begin{bmatrix} y_1 \\ y_2 \\ \cdot \\ \cdot \\ \cdot \end{bmatrix}, & \text{if } \sum_{k=1}^{\infty} \dfrac{1}{\lambda_k^2} y_k^2 < \infty, \\[4em] 0, & \text{otherwise.} \end{cases}$$

What are some other left inverses for Λ? ∎

2.7.9 DEFINITION.   A mapping $F: X \to Y$ is said to be *right invertible* if a mapping $H: Y \to X$ exists such that FH is the identity mapping on the set Y. The mapping H is called a *right inverse* of F and denoted F_r^{-1}.

2.7.10 THEOREM.   *A mapping $F: X \to Y$ is right invertible if and only if $\mathscr{R}(F) = Y$.*

Proof: Let us consider the "if" part of the theorem. Assume that $\mathscr{R}(F) = Y$ and let y_0 be an arbitrary element of Y. Since $y_0 \in \mathscr{R}(F)$, y_0 has at least one pre-image in X, that is, the set $F^{-1}(\{y_0\})$ is not empty. Moreover, if $y_1 \neq y_2$, the sets $F^{-1}(\{y_1\})$ and $F^{-1}(\{y_2\})$ in X are disjoint. We define a mapping $H: Y \to X$ by selecting an arbitrary representative (call it x) from each set $F^{-1}(\{y\})$, and letting $x = H(y)$. It is clear that FH is the identity on Y.

The converse, or the "only if" part is proved in exactly the same manner that we proved Lemma 2.7.5. We ask the reader to check the details. ∎

It should be noted that if $\mathscr{R}(F) = Y$ and if F is not one-to-one, then F will have more than one right inverse.

The results of this section are summarized in Figure 2.7.1.

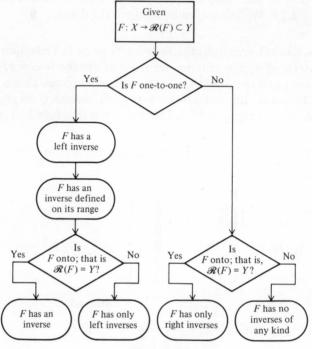

Figure 2.7.1.

EXAMPLE 4.   Let $X = R^4$ and $Y = R^2$ be the sets of all ordered quadruples and all ordered pairs of real numbers, respectively. The following matrix equation represents a mapping of X into Y.

$$\begin{bmatrix} y_1 \\ y_2 \end{bmatrix} = \begin{bmatrix} 1 & 1 & 1 & 1 \\ 1 & -1 & -1 & 1 \end{bmatrix} \begin{bmatrix} x_1 \\ x_2 \\ x_3 \\ x_4 \end{bmatrix}.$$

The range of the mapping is Y, and therefore, it has a right inverse. One right inverse can be represented as follows.

$$\begin{bmatrix} x_1 \\ x_2 \\ x_3 \\ x_4 \end{bmatrix} = \begin{bmatrix} \frac{1}{4} & \frac{1}{4} \\ \frac{1}{4} & -\frac{1}{4} \\ \frac{1}{4} & -\frac{1}{4} \\ \frac{1}{4} & \frac{1}{4} \end{bmatrix} \begin{bmatrix} y_1 \\ y_2 \end{bmatrix}.$$

Are there any other right inverses? ∎

EXAMPLE 5.   Let X and Y be intervals in the real line R. A function $f: X \to Y$ is said to be *monotone* if

(1) $s \le t \Rightarrow f(s) \le f(t)$, or

(2) $s \le t \Rightarrow f(s) \ge f(t)$.

If (1) holds one says that f is *monotone increasing*. If (2) holds f is *monotone decreasing*. If one has

(3) $s < t \Rightarrow f(s) < f(t)$, or

(4) $s < t \Rightarrow f(s) > f(t)$,

then f is said to be *strictly increasing*, or *strictly decreasing*, respectively. It is easy to see that if f is strictly increasing, or strictly decreasing, then f is one-to-one. When this happens, $f^{-1}: \mathcal{R}(f) \to X$ is defined and moreover f^{-1} is also strictly monotone. ∎

EXERCISES

1. Suppose[5] we have a " black box," T, as shown in Figure 2.7.2, whose input and output are the voltage waveforms $m(t)$ and $c(t)$, respectively. Assume that a mathematical model for this black box is the mapping of $L_2(-\infty,\infty)$ into itself represented by

$$c(t) = \int_{-\infty}^{t} e^{-(t-\tau)} m(\tau) \, d\tau.$$

$m(t)$ T $c(t)$

Figure 2.7.2.

Further, suppose we have another " black box," K, for which a suitable mathematical model is the mapping of $L_2(-\infty,\infty)$ into itself represented by

$$m(t) = ks(t), \qquad \text{for } -\infty < t < \infty,$$

where s and m are the input and output of K, respectively, and k is a real number.

Suppose we have K and T interconnected in a closed loop system as shown in Figure 2.7.3. One can view r and c as the system input and output, respectively. The function $s(t)$ is $r(t) - c(t)$ and is sometimes referred to as the error. It follows from the above diagram that given an error $s \in L_2(-\infty,\infty)$, the corresponding system input $r \in L_2(-\infty,\infty)$ is given by

$$r(t) = s(t) + k \int_{-\infty}^{t} e^{-(t-\tau)} s(\tau) \, d\tau.$$

Denote this mapping of $L_2(-\infty,\infty)$ into itself by F; that is, $r = F(s)$.

[5] The next four exercises have important application in stability theory, see Damborg and Naylor [1].

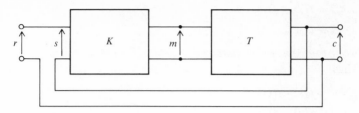

Figure 2.7.3.

Usually we are interested in determining how s depends on r; that is, we are interested in F^{-1}, if it exists.

Show by simple substitution that, for $1 + k > 0$, F^{-1} exists and can be represented[6] by

$$s(t) = r(t) - k \int_{-\infty}^{t} e^{-(1+k)(t-\tau)} r(\tau) \, d\tau.$$

Show again by substitution that for $1 + k < 0$, F^{-1} exists and can be represented by

$$s(t) = r(t) + k \int_{t}^{\infty} e^{-(1+k)(t-\tau)} r(\tau) \, d\tau.$$

2. (Continuation of Exercise 1.) Now consider the case $k = -1$; that is, F is represented by

$$r(t) = s(t) - \int_{-\infty}^{t} e^{-(t-\tau)} s(\tau) \, d\tau.$$

Show that $\mathscr{R}(F)$ is not $L_2(-\infty, \infty)$ by showing that if r is in the range of F, then

$$\lim_{N \to \infty} \int_{-N}^{N} r(t) \, dt = 0.$$

Since $\mathscr{R}(F)$ is not $L_2(-\infty, \infty)$, the mapping F cannot be invertible. However, if F is one-to-one, it will have a left inverse. Show that

$$s(t) = r(t) + \int_{-\infty}^{t} r(\tau) \, d\tau$$

represents a left inverse of F.

It happens, perhaps surprisingly, that

$$s(t) = r(t) - \int_{t}^{\infty} r(\tau) \, d\tau$$

also represents a left inverse F_l^{-1}. As a matter of fact,

$$s(t) = r(t) + \alpha \int_{-\infty}^{t} r(\tau) \, d\tau - \beta \int_{t}^{\infty} r(\tau) \, d\tau,$$

---

[6] Assume for the present that any required interchange of order of integration is allowable.

where α and β are any real numbers such that $\alpha + \beta = 1$, also represents F_l^{-1}. Show that this is indeed the case.

What can be said on the subject of a right inverse?

3. Given the "black boxes" in Exercise 1, we can usually model them using sets of functions other than $L_2(-\infty,\infty)$. For example, we can often model them using $L_2^\sigma(-\infty,\infty)$, where σ is a real number and $L_2^\sigma(-\infty,\infty)$ is the set of all real-valued functions x defined on $-\infty < t < \infty$ such that

$$\int_{-\infty}^{\infty} |x(t)|^2 e^{-2\sigma t}\, dt < \infty.$$

Assume that a mathematical model for T is given by the mapping of $L_2^\sigma(-\infty,\infty)$ into itself for $\sigma > -1$ represented by

$$c(t) = \int_{-\infty}^{t} e^{-(t-\tau)} m(\tau)\, d\tau.$$

Note that this mathematical model is the same as that used in Exercise 1 with $\sigma = 0$.

Similarly, assume that a mathematical model for K is the mapping of $L_2^\sigma(-\infty,\infty)$ into itself represented by

$$m(t) = ks(t).$$

Further, assume that K and T are interconnected as in Exercise 1. It then follows that a mathematical model characterizing the dependence of r on s is given by

$$r(t) = s(t) + k \int_{-\infty}^{t} e^{-(t-\tau)} s(\tau)\, d\tau,$$

where the above represents a mapping F of $L_2^\sigma(-\infty,\infty)$ into itself for $\sigma > -1$.

As before, we are interested in the inverse of F, if it exists. We now have two parameters to consider: k and σ.

Case 1. $\sigma > -(1 + k)$ and $\sigma > -1$.

Show that F^{-1} can be represented as follows:

$$s(t) = r(t) - k \int_{-\infty}^{t} e^{-(1+k)(t-\tau)} r(\tau)\, d\tau.$$

Case 2. $-1 < \sigma < -(1 + k)$.

Show that F^{-1} can be represented as follows:

$$s(t) = r(t) + k \int_{t}^{\infty} e^{-(1+k)(t-\tau)} r(\tau)\, d\tau.$$

4. Using Exercise 2 as a model, discuss the case $\sigma = -(1 + k)$ in Exercise 3.

5. Reconsider Exercises 1 and 2 on the set $L_2[0,\infty)$.

6. Show that if a mapping $T: X \to Y$ is both left and right invertible, then (i) T is invertible, (ii) the left and right inverses are unique, and that all three inverses $T^{-1}, T_l^{-1}, T_r^{-1}$ are the same.

7. Suppose that $F: X \to Y$ and $G: Y \to Z$ are invertible mappings. Show that the composition GF is invertible and that $(GF)^{-1} = F^{-1}G^{-1}$.

8.   SYSTEM TYPES

There are two concepts from system theory that are used in later examples, and it is convenient to introduce them here.

The first system concept is that of *time-invariance*. We shall discuss this concept first for discrete-time (sampled data) systems and then for continuous-time systems. In either event, the basic idea is the same. A system is time-invariant if the only effect of a translation in time of an input is a corresponding translation in time of the output. That is, if $y(t)$ is the output associated with an input $x(t)$, then $y(t + \tau)$, where τ is a constant, is the output associated with the input $x(t + \tau)$. Let us make this precise. First we consider the discrete-time case.

Let X denote a set of a doubly infinite sequences $x = \{\ldots, x_{-1}, x_0, x_1, \ldots\}$. The only requirement we place on X is that if $x = \{x_n\}$ is in X, then so is every translate of x. That is, if $\{x_n\} \in X$ and N is an integer (positive or negative), then $\{y_n\} \in X$, where $y_n = x_{n+N}, n = \ldots, -1, 0, 1, \ldots$. Briefly, X is closed under translations.

The sequences in X can be viewed as inputs and outputs to discrete-time systems, and time can be assumed to increase with the index; that is, x_1 comes before x_2.

Let $S_r: X \to X$ be the right shift (or translation) defined on X. In particular, if $x = \{\ldots, x_{-1}, x_0, x_1, \ldots\}$ and $S_r x = \{\ldots, y_{-1}, y_0, y_1, \ldots\}$, then $y_k = x_{k-1}$ for $k = \ldots, -1, 0, 1, \ldots$. The composition

$$\underbrace{S_r S_r S_r \cdots S_r}_{n \text{ times}}$$

is denoted S_r^n. The shift S_r is clearly invertible, so S_r^n is meaningful when n is a negative integer. Moreover, $S_r^0 = I$, the identity.

We are now ready to define time-invariance on X.

2.8.1 DEFINITION.   A mapping T of X into itself is said to be *time-invariant* if

$$S_r^n T = TS_r^n \tag{2.8.1}$$

for all integers n.

If x is an input and $y = Tx$ is the associated output, $S_r^n x$ is the input x shifted in time and $TS_r^n x$ is the output associated with this shifted version of x. $S_r^n Tx$ is a shifted version of y. Equation (2.8.1) states that this shifted version of y equals the output associated with the shifted input.

2.8.2 THEOREM. *A mapping $T: X \to X$ is time-invariant if and only if*

$$S_r T = T S_r.$$

Proof: If T is time-invariant, we clearly have $S_r T = T S_r$. Going the other way, if $S_r T = T S_r$, then

$$S_r^2 T = S_r(S_r T) = (S_r T)S_r = (T S_r)S_r = T S_r^2$$

and so forth. ∎

EXAMPLE 1. Let $X = l_2(-\infty, \infty)$, the set of all doubly infinite sequences $x = \{\ldots, x_k, x_{k+1}, \ldots, x_{j-1}, x_j, \ldots\}$ such that

$$\sum_{n=-\infty}^{\infty} |x_n|^2 < \infty.$$

Let T denote a mapping of X into itself that can be represented by

$$(Tx)_k = \sum_{j=-\infty}^{\infty} h_{(k-j)} x_j, \qquad k = \ldots, -1, 0, 1, 2, \ldots,$$

where

$$Tx = \{\ldots, (Tx)_k, (Tx)_{k+1}, \ldots, (Tx)_{j-1}, (Tx)_j, \ldots\}$$

and h is a sequence

$$h = \{\ldots, h_{-2}, h_{-1}, h_0, h_1, h_2, h_3, \ldots\}.$$

Then

$$(T S_r x)_k = \sum_{j=-\infty}^{\infty} h_{(k-j)} x_{j-1}$$

and

$$(S_r T x)_k = \sum_{j=-\infty}^{\infty} h_{(k-1-j)} x_j.$$

Letting $j = i - 1$ in the last equation shows that $T S_r = S_r T$. So T is time-invariant. ∎

Next we define time-invariance for the continuous time case.

Suppose Z is a set of functions defined on the real line, $(-\infty, \infty)$. Similarly to the discrete-time case we require that if $x \in Z$ and $y(t) = x(t + \tau)$, where τ is a constant, then $y \in Z$. The shift by τ, denoted S_τ, is the mapping of Z into itself defined by

$$(S_\tau x)(t) = x(t + \tau), \qquad t \in (-\infty, \infty).$$

2.8.3 DEFINITION. A mapping T of Z into itself is said to be *time-invariant* if

$$S_\tau T = T S_\tau$$

for all $-\infty < \tau < \infty$.

EXAMPLE 2.   Consider $Z = L_2(-\infty, \infty)$ and let T be a mapping of Z into itself represented by a convolution integral of the form

$$(Tx)(t) = \int_{-\infty}^{\infty} h(t-s)x(s)\, ds.$$

Then

$$(TS_\tau x)(t) = \int_{-\infty}^{\infty} h(t-s)x(s+\tau)\, ds$$

and

$$(S_\tau Tx)(t) = \int_{-\infty}^{\infty} h(t+\tau-s)x(s)\, ds.$$

By letting $s = \theta + \tau$ in the last equation shows that $S_\tau T = T S_\tau$ for arbitrary τ. Hence, T is time-invariant. ∎

A mapping T that is not time-invariant is said to be *time-varying*.

Causality is the next systems theory concept we consider. Roughly speaking, a system is causal if past output is independent of future inputs.

Let Y be a set of doubly infinite sequences similar to the set X described above. We no longer require, however, that Y be closed under translations or shifts.

2.8.4 DEFINITION.   A mapping T of Y into itself is said to be *causal* if for each integer N whenever two inputs $x = \{x_n\}$ and $y = \{y_n\}$ are such that $x_n = y_n$ for $n \le N$, it follows that $(Tx)_n = (Ty)_n$ for $n \le N$, where

$$Tx = \{\ldots, (Tx)_{-1}, (Tx)_0, (Tx)_1, (Tx)_2, \ldots\}$$

and

$$Ty = \{\ldots, (Ty)_{-1}, (Ty)_0, (Ty)_1, (Ty)_2, \ldots\}.$$

In other words, if the inputs x and y agree up to some time N, then the outputs Tx and Ty agree up to time N. In particular, Tx and Ty agree up to time N no matter what the inputs x and y are in the future beyond N.

EXAMPLE 3.   It is a simple matter to show that in Example 1 the operator T represents a causal system if and only if $h_{(k-j)} = 0$ for $j > k$. Indeed, in that case we have the Volterra equation

$$(Tx)_k = \sum_{j=-\infty}^{k} h_{(k-j)} x_j,$$

showing that $(Tx)_k$ is independent of x_{k+1}, x_{k+2}, and so on. ∎

Causality for continuous-time systems is defined in a similar manner. Let U be a set of functions defined on the real line.

2.8.5 DEFINITION. A mapping F of U into itself is said to be *causal* if for each time T, one has $(Fx)(t) = (Fy)(t)$ for $t \le T$, whenever $x(t) = y(t)$ for $t \le T$.

EXAMPLE 4. In the case of Example 2, the operator T represents a causal system if and only if $h(t - \tau) = 0$ for (almost) all $(t - \tau) < 0$. We can write T as a Volterra integral

$$(Tx)(t) = \int_{-\infty}^{t} h(t - s)x(s)\, ds,$$

showing that $(Tx)(t)$ is independent of $x(s)$ for $s > t$. ∎

There is a great deal more that can be said about the concepts of time-invariance and causality. Indeed, examples which appear later will do just that. However, for the moment we will let the matter rest.

EXERCISES

1. Is the digital system described in Exercise 3 of Section 6 causal?

2. A system is sometimes said to be *anticausal* if future output is independent of past input. For example, let $F: L_2(-\infty,\infty) \to L_2(-\infty,\infty)$. Then F is anticausal if for each time T one has $(Fx)(t) = (Fy)(t)$ for $t \ge T$ whenever $x(t) = y(t)$ for $t \ge T$. The definition is analogous for sampled data (or discrete-time) systems. Which of the following integral operators represents an anticausal mapping of $L_2(-\infty,\infty)$ into itself:

   (a) $y(t) = \int_{t}^{\infty} e^{(t-\tau)}x(\tau)\, d\tau.$

   (b) $y(t) = \int_{-\infty}^{\infty} e^{-(t-\tau)^2}x(\tau)\, d\tau.$

   (c) $y(t) = \int_{-\infty}^{t} e^{-(t-\tau)}x(\tau)\, d\tau.$

3. Consider the collection X of all mappings of $L_2(-\infty,\infty)$ into itself. Let $\mathscr{C} \subset X$ denote the causal mappings, and let $\mathscr{AC} \subset X$ denote the anticausal mappings. Show that $\mathscr{C} \cap \mathscr{AC}$ is not empty. Mappings in this intersection are referred to as *memoryless*.

4. Suppose that Y is a set of doubly infinite sequences $y = \{\ldots, y_{-1}, y_0, y_{+1}, y_2, \ldots\}$ and that T is an invertible mapping of Y onto itself. Show that the inverse of T is causal if and only if $Tx = \{\ldots, (Tx)_{-1}, (Tx)_0, (Tx)_1, \ldots\}$ and $Ty = \{\ldots, (Ty)_{-1}, (Ty)_0, (Ty)_1, \ldots\}$ are such that if $(Tx)_n = (Ty)_n$ for $n \le N$ then it follows that $x_n = y_n$ for $n \le N$, where $x = \{\ldots, x_{-1}, x_0, x_1, x_2, \ldots\}$ and $y = \{\ldots, y_{-1}, y_0, y_1, y_2, \ldots\}$. Carefully note that the causality of T^{-1} is independent of the causality of T.

5. Which of the mappings in Exercises 1, 2, 3, and 4 of Section 7 are causal?

SUGGESTED REFERENCES

Halmos [5]
Hausdorff [1]
Wilder [1]

3

Topological
Structure

1. INTRODUCTION

The word "topology" means literally "the study of places" or "the study of localities." This subject is a natural outgrowth of Euclid's study of plane geometry. Algebraic topology, combinational topology, differential topology, and point-set topology are all branches of this relatively young subject, with very few results dating before 1850.

The reader may have heard of some of the famous problems of topology, for example: the Four Color Problem, the Seven Bridges of Königsberg Problem, and the Cranky Neighbor Problem. The last two of these problems have been solved using topological methods, whereas the Four Color Problem is perhaps the most famous outstanding problem in topology.

In this chapter, and in the remainder of this book, we shall concentrate on one branch of topology, namely, point-set or general topology. By the end of this book, we hope the reader will appreciate the central role this branch of topology plays in modern analysis.

The basic geometric concept underlying the subject of point-set topology is the notion of "closeness" or "nearness." For most purposes the concept of closeness associated with a metric will be sufficient. Hence in this book and in this chapter, particularly, we shall study the topology of metric spaces.

This chapter is divided into two parts. The introductory metric space definitions and concepts are given in Part A, and a deeper discussion of metric space structure is given in Part B.

We recommend that the reader study Part A first and then go on to the rest of the book. The material from Part B can be filled in as needed.

Part A

Introduction
to Metric Spaces

2. METRIC SPACES: DEFINITION

A metric space is a pair of objects: a set, say X, and a metric—or distance function—$d(x,y)$. More precisely, we say that the pair (X,d) is a *metric space* if X is a set, called the *underlying set*, and $d(x,y)$ is a real-valued function, called the *metric*, defined for $x, y \in X$ and satisfying the following conditions or axioms:

(M1) (Positive) $d(x,y) \geq 0$ and $d(x,x) = 0$ (for all x and y in X).

(M2) (Strictly Positive) If $d(x,y) = 0$, then $x = y$ (for all x and y in X).

(M3) (Symmetry) $d(x,y) = d(y,x)$ (for all x and y in X).

(M4) (Triangle Inequality) $d(x,y) \leq d(x,z) + d(z,y)$ (for all x, y, and z in X).

All four of the above conditions are in harmony with our concept of distance in the Euclidean plane. Indeed, the distance function

$$d(x,y) = [(x_1 - y_1)^2 + (x_2 - y_2)^2]^{1/2}$$

discussed in Section 1.2 satisfies each of these properties. The reason for calling (M4) the triangle inequality is illustrated in Figure 3.2.1. The vertices of the

Figure 3.2.1.

triangle denote points x, y, and z in X, and the sides of the triangle have lengths $d(x,y)$, $d(x,z)$, and $d(z,y)$. The triangle inequality (M4) is, then, an abstract formulation of the triangle inequality of Euclidean geometry. It is very important to include (M4) as one of the defining properties for a metric. We invite the reader to watch for it as it is used time and time again in this chapter and later in the book.

It can happen that a nonempty set X may have more than one metric defined on it. For example, if $X = R^2$,

$$d_2(x,y) = [(x_1 - y_1)^2 + (x_2 - y_2)^2]^{1/2}$$

or

$$d_1(x,y) = |x_1 - y_1| + |x_2 - y_2|$$

are metrics on the set X. Hence, (X,d_2) and (X,d_1) are metric spaces. More importantly, they are *different* metric spaces even though they have the same underlying set X. In general, any nonempty set with more than one element can have an infinite number of metrics defined on it. Indeed, if d is a metric on X, then $d_\alpha(x,y) = \alpha d(x,y)$, $\alpha > 0$, is a metric on X. Other, less trivial examples abound.

A metric space (X,d) is, then, a set X with an additional structure defined by means of a metric function d. This additional structure is called the *topological structure*. The bulk of this chapter is devoted to a study of this structure.

When no confusion can arise we will often denote the metric space (X,d) by X.

If A is a nonempty subset of a metric space (X,d) and x is a point in (X,d), then we say that the *distance between x and the subset A*, denoted $d(x,A)$, is given by the real number

$$d(x,A) = \inf\{d(x,y): y \in A\}.$$

We say that the *diameter of the nonempty subset A*, denoted $\text{diam}(A)$, is given

$$\text{diam}(A) = \sup\{d(x,y): x, y \in A\}.$$

Note that $\text{diam}(A)$ may be $+\infty$. A set A is said to be *bounded* if its diameter is finite.

The following lemma is a simple consequence of the triangle inequality. We ask the reader to verify it.

3.2.1 LEMMA.   *Let (X,d) be a metric space. For any three points x, y, z in X one has*

$$|d(x,y) - d(y,z)| \le d(x,z).$$

EXERCISES

1. Let d be a metric on X and let $d_\alpha(x,y) = \alpha d(x,y)$, where $0 < \alpha < 1$. Show that $d_\alpha \not\equiv d$ if X has two or more points.

2. Show that $d(x,y) = |x - y|$ is a metric on the real numbers R; on the complex numbers C.

3. Let $d(x,y)$ be a metric on X. Show that

$$d_1(x,y) = \frac{d(x,y)}{1 + d(x,y)} \quad \text{and} \quad d_2(x,y) = \min(1, d(x,y))$$

are also metrics on X. Show that every set in the metric space (X,d_i) $(i = 1,2)$ is bounded.

4. A real-valued function $\rho(x,y)$ is said to be a *pseudometric* on X if it satisfies conditions (M1), (M3), and (M4).
   (a) Show that $\rho(x,y) \equiv 0$ is a pseudometric on any set X.
   (b) Show that $\rho\{(x_1,x_2), (y_1,y_2)\} = |x_1 - y_1|$ is a pseudometric in the plane R^2.

5. Show that if A is nonempty, in a metric space (X,d), then diam $A = 0$ if and only if A consists of a single point. Is this true in a pseudometric space?

3. EXAMPLES OF METRIC SPACES

The reader may wonder how metric spaces arise in practice. In some cases they arise quite naturally with the statement of a problem. For example, in surveying a flat piece of ground our set X is naturally chosen to be the set of all points in the area under consideration and our metric is naturally chosen to be the usual distance function. In other cases metric spaces are chosen on the basis of mathematical convenience. It often happens that the metric space which appears to be the most natural for a given problem leads to intractable mathematical difficulties. Consequently, we compromise and choose a metric space which is not as naturally related to the problem as we might desire but which does lead to a tractable mathematical setting. For example, the great use of square-root-integral-square and square-root-sum-square criteria, that is, criteria of the form

$$\left\{ \int |x(t) - y(t)|^2 \, dt \right\}^{1/2}$$

and

$$\left\{ \sum |x_i - y_i|^2 \right\}^{1/2}$$

throughout engineering and science is often more a result of a desire for mathematical convenience than of the universal naturalness of such criteria.

Let us consider some examples of metric spaces. Some of these examples are intended as illustrations of mathematical points while others are intended to illustrate the occurrence of metric spaces in applied mathematics.

Since some of these examples may be familiar, there is a possible source of confusion. In meaningful applications we seldom deal with sets which have a metric space structure only. Usually another structure is present. Furthermore, the additional structure may be so familiar that it may require a conscious effort to ignore it. Thus, it is possible to confuse the metric structure of an example with the other structure which may be present. We caution the reader to be on his guard against this source of confusion.

EXAMPLE 1.   We have already delved into the structure of the real Euclidean plane at some length in Section 1.2. Since it was our prototype, we are not surprised that it has enough structure—in fact, more than enough—to make it a metric space.

We mentioned the possibility that more than one metric could be defined on some sets. This is not only possible, it is quite common on R^2.

Let $x = (x_1, x_2)$ and $y = (y_1, y_2)$ be points in R^2. The Euclidean metric on R^2 is

$$d(x,y) = [(x_1 - y_1)^2 + (x_2 - y_2)^2]^{1/2}.$$

However, there are many other metrics for R^2. In particular, we claim that in each of the following cases the function $d(x,y)$ defined below is a metric on R^2:

(a) $d(x,y) = d_1(x,y) = |x_1 - y_1| + |x_2 - y_2|$,
(b) $d(x,y) = d_\infty(x,y) = \max\{|x_1 - y_1|, |x_2 - y_2|\}$,
(c) $d(x,y) = d_p(x,y) = \{|x_1 - y_1|^p + |x_2 - y_2|^p\}^{1/p}$,

where $1 \leq p < \infty$ ($p = 2$ corresponds to the Euclidean metric; $p = 1$ corresponds to (a); and (b) is sometimes called the $p = \infty$ metric).

(d) $d(x,y) = \{a|x_1 - y_1|^2 + b|x_1 - y_1||x_2 - y_2| + c|x_2 - y_2|^2\}^{1/2}$, where $a > 0$, $c > 0$, and $4ac - b^2 > 0$.

The proof that (a), (b), (c), and (d) are indeed metrics is left as an exercise. Although these examples yield a rich supply of metrics on R^2, it should not be imagined that we have come close to exhausting the supply. (What are some others?)

Each of the above metrics yields a different metric space even though the underlying set R^2 is always the same. We will see subsequently, when we discuss open sets, that these metric spaces are "equivalent" to one another in an important and useful sense. Although we are unable to discuss this concept of equivalence now, let us hasten to add that not all metric spaces with R^2 as the underlying set are equivalent to one another. For example, the function $d(x,y) = 0$, if $x = y$, and $d(x,y) = 1$, if $x \neq y$, defines a metric on R^2 which yields a metric space which is not equivalent to any of the preceding examples. ∎

EXAMPLE 2.   Let X be the set made up of all ordered n-tuples of real numbers, $x = (x_1, \ldots, x_n)$. We shall write this as $X = R^n$. The following are metrics on R^n.

(a) $d_p(x,y) = [|x_1 - y_1|^p + \cdots + |x_n - y_n|^p]^{1/p}$,   where $1 \leq p < \infty$.
(b) $d_\infty(x,y) = \max[|x_1 - y_1|, \ldots, |x_n - y_n|]$.

It is easy to show that (b) is a metric, and it is easy to show that (a) satisfies conditions (M1), (M2), and (M3). Showing that (a) satisfies the triangle inequality, condition (M4), follows from the Minkowski Inequality, see Appendix A. The function d_p is sometimes referred to as the l_p-metric on R^n. ∎

EXAMPLE 3.   Let X be the set made up of all ordered n-tuples of complex numbers, $x = (x_1, \ldots, x_n)$. We shall write this as $X = C^n$. What would be the analogs to (a) and (b) of Example 2 in this case? ∎

EXAMPLE 4.   Let $l_p = l_p(0,\infty)$ be the set made up of all infinite sequences of real (or complex) numbers, $x = (x_1,x_2,\ldots)$, such that the series $\sum_{i=1}^{\infty} |x_i|^p$ converges, where $1 \leq p < \infty$. We claim that the function

$$d_p(x,y) = \left[\sum_{i=1}^{\infty} |x_i - y_i|^p \right]^{1/p}$$

defines a metric on l_p. It follows from the Minkowski Inequality for infinite sums (Appendix A) that the series defining $d_p(x,y)$ always converges for x and y in l_p. Furthermore, we see that property (M4) for a metric is satisfied. The other three properties (M1), (M2), and (M3) are easily verified. Hence (l_p,d_p) is a metric space for each p, $1 \leq p < \infty$. Obviously, different p's yield different metric spaces.[1] (Do any two of these metric spaces have the same underlying set?)

Such sequence spaces occur throughout engineering and science. For example, the mathematical model for the set of all inputs to a sampled-data control system is often taken to be (l_2,d_2). Another example is to consider a typical point in (l_2,d_2) as the sequence of coefficients associated with the modes of vibration of a mechanical system which has an infinite number of modes (for example, a plucked string).

Since these spaces occur so often we shall refer to them as l_p "with the usual metric" or sometimes simply as l_p.   ∎

EXAMPLE 5.   Let l_∞ be the set of all bounded sequences of real (or complex) numbers. That is, a sequence $\{x_n\}$ is in l_∞ if and only if there exists a real number M such that $|x_n| \leq M$ for all n. For example, the sequence $\{\alpha,\alpha,\alpha,\ldots\}$, where α is a real number, is in l_∞; whereas the sequence $\{1,2,3,4,\ldots\}$ is not. Let

$$d_\infty(x,y) = \sup |x_n - y_n|.$$

It is easily shown that (l_∞,d_∞) is a metric space. This space is a generalization of (b) in Example 2. Note that here "sup" is used instead of "max."

This space is also referred to as l_∞ with the usual metric.

As in Example 4 we shall use $l_\infty(0,\infty)$ to denote the l_∞-space of ordinary sequences $x = (x_1,x_2,\ldots)$ and $l_\infty(-\infty,\infty)$ to denote the l_∞-space of doubly infinite sequences $x = (\ldots,x_{-1},x_0,x_1,\ldots)$.   ∎

EXAMPLE 6.   Let B be the set of all infinite sequences of positive integers $n = \{n_1,n_2,\ldots\}$, and let d be defined by

$$d(n,m) = \begin{cases} 0, & \text{if } n_i = m_i \text{ for } i = 1, 2, \ldots \\ \dfrac{1}{k}, & \text{where } k \text{ is the first index for which } n_i \neq m_i. \end{cases}$$

[1] We sometimes are interested in the space of doubly infinite sequences $x = (\ldots,x_{-1},x_0,x_1,\ldots)$ with $\sum_{n=-\infty}^{\infty} |x_n|^p < \infty$. We shall denote this space by $l_p(-\infty,\infty)$. The metric is then defined by $d_p(x,y) = (\sum_{n=-\infty}^{\infty} |x_n - y_n|^p)^{1/p}$.

(B,d) is a metric space called the Baire null-space. It is interesting to note that this metric satisfies a stronger triangle inequality, namely,

$$d(n,m) \leq \max\{d(n,p), d(p,m)\}.$$

Just to see that rather "weird looking" metrics can arise in practice, consider the following use of the Baire null-space as a mathematical model. Suppose that a function $s(t)$ is a signal to be sent through a communication system and that $s(t)$ is (1) sampled every second, (2) quantized, and (3) the quantization levels are encoded as positive integers. This processing of $s(t)$ is illustrated in Figure 3.3.1.

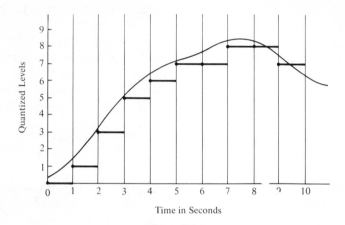

Figure 3.3.1.

The sequence of positive integers representing the signal $s(t)$ in Figure 3.3.1 would be

$$n_s = \{0,1,3,5,6,7,7,8,8,7,\ldots\}.$$

Further, suppose that in being processed by the remainder of the communication system, disturbances are introduced which tend to accumulate and eventually cause incorrect quantization levels to be received. If $n_r = \{n_{r_1}, n_{r_2}, \ldots\}$ denoted the received sequence associated with n_s, it is of interest to know how long the system runs before an error in quantization level occurs. Thus we can use the Baire null metric, $d(n_s, n_r)$, to characterize the distance or difference between the sent and the received sequence. The smaller $d(n_s, n_r)$ is, the longer the communication system runs before an error occurs. ∎

EXAMPLE 7. Let $X_1 = Q$, the set of all rational numbers, and let $d_1(x,y) = |x - y|$. Similarly, let $X_2 = R$, the set of all real numbers, and $d_2(x,y) = |x - y|$. The metric spaces (X_1,d_1) and (X_2,d_2) are different, and they differ from one another in an important way which will become clear when we discuss complete metric spaces (Sections 13 and 14). ∎

Now let us consider some metric spaces whose underlying sets are sets of functions.

EXAMPLE 8.   Let $X = C[0,T]$ be the set of all real-valued (or complex-valued) continuous functions defined on the interval $[0,T]$, where $T > 0$. We define a metric on $C[0,T]$ called the *sup-metric*, by

$$d_\infty(x,y) = d(x,y) = \sup\{\,|x(t) - y(t)| : 0 \le t \le T\}.$$

Let us show that the sup-metric satisfies the triangle inequality. Let x, y, and z be arbitrary elements of $C[0,T]$, then

$$
\begin{aligned}
d(x,y) &= \sup_{0 \le t \le T} |x(t) - z(t) + z(t) - y(t)| \\
&\le \sup_{0 \le t \le T} \{|x(t) - z(t)| + |z(t) - y(t)|\} \\
&\le \sup_{0 \le t \le T} |x(t) - z(t)| + \sup_{0 \le t \le T} |z(t) - y(t)| \\
&= d(x,z) + d(z,y).
\end{aligned}
$$

The remaining conditions on a metric are easily shown to be satisfied; hence, $d(x,y)$ is a metric and $(C[0,T], d)$ is a metric space. This metric space is often simply denoted by $C[0,T]$.   ∎

EXAMPLE 9.   Let $X = C[0,T]$ and now define

$$d_p(x,y) = \left[\int_0^T |x(t) - y(t)|^p \, dt\right]^{1/p},$$

where $1 \le p < \infty$. By using the Minkowski Inequality for integrals (Appendix A) it is easily seen that d_p defines a metric on $C[0,T]$. We note that when $p = 2$, the metric is the square-root-integral-square criterion referred to earlier.   ∎

EXAMPLE 10.   Let $X = L_p(I)$, $1 \le p < \infty$, be the set of all real- (or complex-) valued functions $x(\cdot)$ defined on the interval I such that

$$\int_I |x(t)|^p \, dt < \infty,$$

where the integral is the Lebesgue integral (see Appendix D). Define the metric on $L_p(I) = L_p$ by

$$d_p(x,y) = \left[\int_I |x(t) - y(t)|^p \, dt\right]^{1/p}.$$

This function d_p is referred to as the "usual metric" on $L_p(I)$.

There is a corresponding space $L_\infty(I)$. This consists of all real (or complex) valued functions $x(\cdot)$ defined on the interval I such that

$$\|x\|_\infty = \operatorname*{ess\,sup}_{t \in I} |x(t)| < \infty.$$

This means (see Section D.11) that a function $x(\cdot)$ belongs to $L_\infty(I)$ if and only if there is a real number N with the property that

$$|x(t)| \leq N \text{ (almost everywhere).}$$

In this case $\|x\|_\infty$ is then given by

$$\|x\|_\infty = \inf\{B: |x(t)| \leq B \text{ (almost everywhere)}\}.$$

The "usual metric" on $L_\infty(I)$ is then given by

$$d_\infty(x,y) = \|x - y\|_\infty.$$

This example raises a technical problem which occurs often with function spaces defined in terms of integrals. As it stands, d_p does not define a metric on L_p, for we can easily find an x and y in L_p such that $x \neq y$ but $d_p(x,y) = 0$. For example, let $x(t) = y(t)$ for all t except the point $t = 1$ and let $x(1) = y(1) + 1$. Clearly x and y are different functions and $d_p(x,y) = 0$. In other words, d_p is a pseudometric (compare with Exercise 4, Section 2) and not a metric. There are several ways of "changing" d_p into a metric. We adopt the following point of view:

We define a *new* equality between functions in L_p. We say that $x = y$ if and only if $d_p(x,y) = 0$. We see (Appendix D) that $x = y$ (in the new equality) if and only if $x(t) = y(t)$ everywhere except perhaps on a set of measure zero.

This new equality does turn d_p into a metric and (L_p, d_p) into a metric space. However there are several hidden defects that come with this point of view. First, we can no longer distinguish between functions that differ on finite sets, countable sets, or more generally, sets of measure zero. Secondly, a technical problem arises when one asks whether a given property, that is valid under the old equality, is now valid under the new equality. We shall deal with these difficulties as they arise.

The reader will see that they do not present serious obstacles. (See Example 8, Section 2.5). ∎

EXAMPLE 11.   Let $X = BC(I)$, where $BC(I)$ is the set of all real-valued (or complex-valued) continuous, bounded functions $x(t)$ defined on the finite or infinite interval I. Recall that a function x is bounded on I if there exists a real number M (the number M depends on the function x) such that $|x(t)| \leq M$ on I. Let the metric on $BC(I)$ be the sup-metric, which is given by

$$d(x,y) = \sup\{|x(t) - y(t)|: t \in I\}. \quad \blacksquare$$

EXAMPLE 12.   Let X be the set of all continuous functions $x(t,s)$ defined on some closed, bounded set D in the (t,s)-plane. Let the metric on X be the sup-metric:

$$d(x,y) = \sup\{|x(t,s) - y(t,s)|: (t,s) \in D\}. \quad \blacksquare$$

EXAMPLE 13.   Given a real number σ let C_σ denote the vertical line in the complex plane defined by

$$C_\sigma = \{s: \operatorname{Re}(s) = \sigma\}.$$

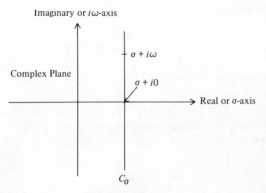

Figure 3.3.2.

Let $L_{2,\sigma}(-i\infty,i\infty)$ be the set of all complex-valued functions $x(s)$ defined on C_σ such that

$$\frac{1}{2\pi i}\int_{C_\sigma}|x(s)|^2\,ds < \infty.$$

Define the metric on $L_{2,\sigma}(-i\infty,i\infty)$ by

$$d_\sigma(x,y) = \left\{\frac{1}{2\pi i}\int_{C_\sigma}|x(s)-y(s)|^2\,ds\right\}^{1/2}.$$

The metric spaces $(L_{2,\sigma}(-i\infty,i\infty),\,d_\sigma)$ play an important role in Fourier and Laplace transform analysis.

The space $L_{2,\sigma}(-i\infty,i\infty)$ that arises when $\sigma = 0$ (that is, when the path of integration is the imaginary axis) will occur often in examples and exercises in this book. We shall denote this space simply by $L_2(-i\infty,i\infty)$. ∎

EXAMPLE 14.   Let C_σ be defined as in Example 13, and let Y_σ be the set of all continuous and bounded complex-valued functions $y(s)$ defined on C_σ. Define the metric on Y_σ by

$$d_\sigma(x,y) = \sup\{\,|x(s)-y(s)|: s\in C_\sigma\}.$$

The metric spaces (Y_σ,d_σ) also play an important role in Fourier and Laplace transform analysis. ∎

EXAMPLE 15.   Let $X = C(-\infty,\infty)$ be the set of all continuous, real- (or complex-) valued functions defined on the real line $R = (-\infty,\infty)$. Define

$$\rho_n(x,y) = \sup\{|x(t)-y(t)|: |t|\le n\},$$

$$\sigma_n(x,y) = \min\{1,\,\rho_n(x,y)\},$$

$$\sigma(x,y) = \sum_{n=1}^{\infty}\frac{1}{2^n}\,\sigma_n(x,y).$$

Then (X,σ) is a metric space. If we let

$$d(x,y) = \sup_{0<T}\left\{\min\left[\frac{1}{T}, \sup_{|t| \le T} |x(t) - y(t)|\right]\right\},$$

then (X,d) is another metric space.   ∎

EXAMPLE 16.   This example may appear trivial, yet it does have some mathematical importance and there are situations where it is an appropriate mathematical model. Let X be any nonempty set, and define $d(x,y)$ by

$$d(x,y) = \begin{cases} 0, & \text{if } x = y \\ 1, & \text{if } x \ne y. \end{cases}$$

It is a simple matter to show that (X,d) is a metric space. An example of a physical situation where (X,d) may serve as a mathematical model arises when X is a set of targets. If we miss the intended target, hitting some other one, we say that the miss distance or cost is 1, that is, all misses are equally weighted. If we hit the desired target, the miss distance or cost is zero.   ∎

EXAMPLE 17.   Let $X(n)$ be the set of all ordered n-tuples of "zeros" and "ones." For example,

$$X(3) = \{000,001,010,011,100,101,110,111\}.$$

For x and y in $X(n)$, let

$d(x,y) =$ number of places where x and y have different entries.

For example, in $X(3)$

$$d(110,110) = 0$$
$$d(010,110) = 1$$

and

$$d(101,010) = 3.$$

We leave it to the reader to show that $(X(n),d)$ is a metric space. This metric space occurs in switching and automata theory.   ∎

EXERCISES

1. For many of the examples of metric spaces given in this section, no proof was given that the claimed metric space was indeed a metric space. Supply the missing proofs.

2. Describe a mathematical model using each of the metric spaces given in the examples.

3. For each metric space given in Example 1, sketch the set of all points x such that $d(x,y_0) \le 1$, where $y_0 = (0,0)$.

4. A warehouse stores apples, rope, gasoline, sand, and plate glass. What are some of the ways that one might use a metric space as a mathematical model to characterize differences in warehouse content?

5. Let X be the set containing the cities Los Angeles, New York, Chicago, and San Francisco. How might an air traveler place a metric on this set? A rail traveler? An automobile traveler? A truck driver? A yacht owner?

6. Let $X = R^2$ and let

$$d(x,y) = \{ |x_1 - y_1|^{1/2} + |x_2 - y_2|^{1/2} \}^2.$$

Is $\{X,d\}$ a metric space?

7. In R^2 let $A = \{x = (x_1,x_2): (|x_1|^2 + |x_2|^2)^{1/2} < 1\}$, and

$$d(x,y) = \{ |x_1 - y_1|^2 + |x_2 - y_2|^2 \}^{1/2}.$$

Compute $d(x,A)$. Show that $d(x,A) = 0$ if and only if

$$|x_1|^2 + |x_2|^2 \le 1.$$

8. In Example 5, the metric d_∞ was defined with a "sup" instead of a "max." In order to see the necessity of this let $x = (x_1,x_2,\ldots)$, $y = (y_1,y_2,\ldots)$ be given by

$$x_n = \frac{1}{n+1} \quad \text{and} \quad y_n = \frac{n}{n+1}.$$

Show that $d_\infty(x,y) > |x_n - y_n|$ for all n.

On the other hand, in Example 8 "sup" could be replaced by "max." We will see why in Section 17. What about Example 11?

9. Sketch the set of points $x = (x_1,x_2)$ in R^2 for which

$$d_p(0,x) = 1,$$

where $0 = (0,0)$ and d_p is defined in Example 1. With x and y fixed show that $d_p(x,y)$ is decreasing in p. Hence

$$d_p(x,y) \ge d_q(x,y)$$

whenever $p \le q$.

10. Use the techniques of Exercise 9 to show that whenever $p \le q$ one has $d_p(x,y) \ge d_q(x,y)$ on R^n.

11. Let C be the unit circle in the complex plane, that is, $C = \{z: |z| = 1\}$. Let X denote all complex-valued functions $f(z)$ defined on C for which $\int_C |f(z)|^2 |dz| < \infty$. Show that

$$d(f,g) = \left(\int_C |f(z) - g(z)|^2 |dz| \right)^{1/2} = \left(\int_0^{2\pi} |f(e^{i\theta}) - g(e^{i\theta})|^2 \, d\theta \right)^{1/2}$$

is a metric on X.

12. Let X denote the class of all complex-valued functions $f(z)$ that are analytic for $|z| < 1$. Let $f(z) = \sum_{n=0}^{\infty} a_n z^n$, $g(z) = \sum_{n=0}^{\infty} b_n z^n$ be the power series expansion for f, g in X. Show that the following functions are metrics, or pseudo-metrics.
   (a) $d(f,g) = \sup\{|f(z) - g(z)| : |z| \leq \rho\}$, where $0 < \rho < 1$.
   (b) $d(f,g) = |a_0 - b_0|$.
   (c) $d(f,g) = |a_0 - b_0| + |a_1 - b_1|$.
   (d) $d(f,g) = \sum_{n=0}^{\infty} |a_n - b_n| \rho^n$, where $0 < \rho < 1$.

   (e) $d(f,g) = \sup\left\{\left|\int_{|\xi|=\rho} \dfrac{f(\xi) - g(\xi)}{z - \xi} \, d\xi\right| : |z| < \rho\right\}$, where $0 < \rho < 1$.

13. Let S be a nonempty set and let $X = B(S)$ denote the collection of all bounded, real-valued functions defined on S. Show that

$$d(f,g) = \sup\{|f(s) - g(s)| : s \in S\}$$

   is a metric on X.

14. Let $X(n)$ denote the collection of all differential operators of the form

$$P(D) = p_n D^n + p_{n-1} D^{n-1} + \cdots + p_1 D + p_0,$$

   where the coefficients p_i are real constants. Show that

$$d(P(D), Q(D)) = \sum_{i=0}^{n} |p_i - q_i|$$

   is a metric on $X(n)$, where $P(D)$ is given above and

$$Q(D) = \sum_{i=0}^{n} q_i D^i.$$

   Compare this metric space with the space (R^{n+1}, d_1) discussed in Example 2.

15. Let X denote the collection of all bounded closed intervals $[a,b]$ from the real line. Let $[a,b]$ and $[c,d]$ be two sets. Show that the symmetric difference $[a,b] \triangle [c,d]$ is the union of (at most) two bounded intervals. Define $\rho([a,b], [c,d])$ to be the sum of the lengths of these intervals. Show that ρ is a pseudometric on X. Discuss the relation $\rho([a,b], [c,d]) = 0$.

4.   SUBSPACES AND PRODUCT SPACES

In the last section we examined several examples of metric spaces. These examples by no means exhaust the supply of known metric spaces. If we are given some metric spaces, there are several ways of constructing new metric spaces. In this section we shall discuss two such methods, namely, subspaces and product spaces.

Subspaces

Let (X,d) be a metric space and let A be a subset of X. We can use the metric d to define a metric on A in a very natural way. If x and y are in A we simply let the distance between x and y be given by $d(x,y)$. Obviously, (A,d) is a metric space.

Since $A \subset X$, we say that (A,d) is a *subspace* of (X,d), or the topological structure on A is the structure *inherited from* (X,d). Needless to say, there are usually many metrics that can be placed on A; however, when we say that (A,d) is a subspace of (X,d) it is understood that the inherited metric is implied.

If (A,d) is a subspace of (X,d) and $A \neq X$, we sometimes emphasize the fact that A is not all of X by saying that (A,d) is a *proper* subspace of (X,d).

Let us consider a few examples of subspaces and their applications.

EXAMPLE 1. Suppose we are interested in a vibrating string as shown in Figure 3.4.1. Let $C[0,L]$ be the set of all real-valued, continuous functions z defined on the closed interval $0 \leq x \leq L$, and let

$$d(z_1,z_2) = \sup\{|z_1(t_1,x) - z_2(t_1,x)| : 0 \leq x \leq L\}.$$

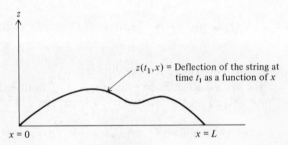

$z(t_1,x)$ = Deflection of the string at time t_1 as a function of x

$x = 0$ $x = L$

Figure 3.4.1.

Here we view the deflection of the vibrating string at time t_1 as a point $z(t_1, \cdot)$ in the metric space $\{C[0,L],d\}$. The motion of the vibrating string can be viewed as a curve $z(t, \cdot)$ in $\{C[0,L],d\}$ parameterized by t. However, not every point in $\{C[0,L], d\}$ can lie on this curve, for, as can be seen in Figure 3.4.1, the boundary conditions must be satisfied. In particular, the deflections at $x = 0$ and $x = L$ must always be zero. Let A be the subset of $C[0,L]$ made up of all real-valued, continuous functions z such that $z = 0$ at $x = 0$ and $x = L$. Obviously, (A,d) is a subspace of the metric space $\{C[0,L],d\}$, and all possible deflections of the vibrating string are contained in this subspace. ∎

EXAMPLE 2. Consider the following system of differential equations

$$\frac{dx_1}{dt} = a_{11}x_1 + a_{12}x_2,$$

$$\frac{dx_2}{dt} = a_{21}x_1 + a_{22}x_2. \tag{3.4.1}$$

This system of equations can be the mathematical model for many systems. For example, it can be used as a mathematical model for the electrical network shown in Figure 3.4.2. In this case, $a_{11} = -(R_1 + R_2)/C_1$, $a_{12} = R_2/C_1$, $a_{21} = R_2/C_2$,

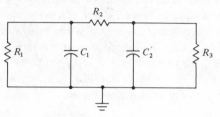

Figure 3.4.2.

and $a_{22} = -(R_2 + R_3)/C_2$. Moreover, x_1 and x_2 are the voltages across the capacitors C_1 and C_2, respectively.

Let us assume that $a_{11} = a_{22} = -2$ and $a_{12} = a_{21} = 1$. Then the general solution of (3.4.1) is

$$x_1(t) = \tfrac{1}{2}(e^{-t} + e^{-3t})x_1(0) + \tfrac{1}{2}(e^{-t} - e^{-3t})x_2(0),$$
$$x_2(t) = \tfrac{1}{2}(e^{-t} - e^{-3t})x_1(0) + \tfrac{1}{2}(e^{-t} + e^{-3t})x_2(0),$$ (3.4.2)

where $x_1(0)$ and $x_2(0)$ are the initial conditions.

Let X be the set of all ordered pairs (x_1,x_2) of bounded, continuous, real-valued functions defined on the interval $0 \le t < \infty$. Let

$$d(x,y) = \sup_t |x_1(t) - y_1(t)| + \sup_t |x_2(t) - y_2(t)|.$$

We leave it to the reader to show that (X,d) is a metric space.

For a given pair of initial conditions, $x_1(0)$ and $x_2(0)$, the ordered pair of functions given in (3.4.2) is a point in the metric space (X,d). Moreover, if A is the set of all solutions of (3.4.1), that is, all the ordered pairs given by (3.4.2), A is a set in (X,d).

Obviously, (A,d) is a subspace of (X,d). Is A a proper subset of X or does $A = X$? ∎

Product Spaces

The concept of product space is a little more complicated than that of a subspace. Let (X,d_x) and (Y,d_y) be two metric spaces. Recall (Section 2.3) that the product set $Z = X \times Y$ is defined as the collection of all ordered pairs (x,y) with $x \in X$ and $y \in Y$. Using the metrics d_x and d_y we can define a metric on Z in several ways. Letting $u = (x_1,y_1)$ and $v = (x_2,y_2)$ be elements of Z, the following functions are metrics on Z:

(a) $d_1(u,v) = d_x(x_1,x_2) + d_y(y_1,y_2)$.

(b) $d_2(u,v) = [d_x^2(x_1,x_2) + d_y^2(y_1,y_2)]^{1/2}$.

(c) $d_p(u,v) = [d_x^p(x_1,x_2) + d_y^p(y_1,y_2)]^{1/p}, \ 1 \le p < \infty$.

(d) $d_\infty(u,v) = \max\{d_x(x_1,x_2), d_y(y_1,y_2)\}$.

(e) $d(u,v) = [ad_x^2(x_1,x_2) + bd_x(x_1,x_2)d_y(y_1,y_2) + cd_y^2(y_1,y_2)]^{1/2}$, where $a > 0$, $c > 0$, $4ac - b^2 > 0$.

(Compare this with Example 1 of Section 3.)

The set $Z = X \times Y$ with any of the above metrics is referred to as the *product space* of (X,d_x) and (Y,d_y).

The reader may wonder if the above metrics are the only metrics that can be placed on $X \times Y$ to yield the product space of (X,d_x) and (Y,d_y). One may also wonder why we say *the* product space instead of *a* product space. After all, each different metric on Z yields a different metric space.

The answer to the first question is no. Except for the degenerate situation where X and Y each contain exactly one point, there is an infinite number of metrics which can be placed on Z to yield the product space. The metrics given above are merely examples from this infinite set. On the other hand, they are the ones most often employed.

The answer to the second question is that the metric spaces obtained are indeed different, yet in a sense they all turn out to be the same or equivalent. (We shall see this in Section 11.) An understanding of exactly in what sense they are equivalent must await the introduction of a few more topological concepts. Suffice it to say here that $Z = X \times Y$ with any of the above five metrics is the product space.

We will denote product spaces by $(X,d_x) \times (Y,d_y)$ or, where no confusion can arise, simply by $X \times Y$.

Let us now consider a few examples.

EXAMPLE 3.   Returning to Example 1 of this section, suppose that in addition to characterizing the deflection of the vibrating string we want to characterize the rate of change of the deflection. Thus, if $z = z(t,x)$ denotes the deflection of the string as a function of time and location,

$$v = v(t,x) = \frac{\partial}{\partial t} z(t,x)$$

will be the velocity of the string as a function of time and location. We will assume that $v(t, x)$ exists and that for each t it is a continuous function of x. We can then view the rate of change of deflection at time t as a point in the metric space $\{C[0,L], d\}$ discussed in Example 1. The evolution of this velocity can be viewed as a curve $v(t, \cdot)$ in $\{C[0,L], d\}$ parameterized by t. We can view the simultaneous deflection z and rate of change of deflection v as an ordered pair $\{z,v\}$, and we can view this ordered pair as a point s in the product space $(C[0,L], d) \times (C[0,L], d)$. The evolution of deflection and rate of change of deflection can be viewed as a curve $s(t, \cdot)$ in this product space parameterized by t, where

$$s(t,x) = (z(t,x), v(t,x)). \quad \blacksquare$$

EXAMPLE 4. Suppose that we are interested in comparing the economic performance of the United States economy from year to year. Further, suppose we do this by following certain economic indicators. For example, let the ordered N-tuple $x = \{x(n)\} = \{x(1),x(2),\ldots,x(N)\}$ be the daily Dow Jones industrial average, where N is the number of days during the year for which averages are given. (We assume that N does not change from year to year.) Letting X be the set of all possible N-tuples x, we assume that a good measure (for some purpose not stated here and not guaranteed) of differences in yearly performance is given by

$$d_X(x,y) = \frac{1}{N} \sum_{n=1}^{N} |x(n) - y(n)|.$$

We also might be interested in the monthly cost of living index. Let the 12-tuple $z = \{z(1),z(2),\ldots,z(12)\}$ be a year's record of this index. Let Z denote the set of all possible z's and assume that

$$d_Z(z,w) = \max_k |z(k) - w(k)|$$

is a "good" measure of differences in yearly performance of the cost of living index.

We can view the record x of the Dow Jones average for a given year as a point in the metric space (X,d_X) and the record z of the cost of living index as a point in the metric space (Z,d_Z). Finally, the behavior of the economy for a given year, say 1950, would be the ordered pair $p_0 = (x_0,z_0)$, where x_0 and z_0 are values for 1950. If we form the product $(P,d) = (X,d_X) \times (Z,d_Z)$, p_0 is obviously a point in it. Then, if p_1 is the behavior of the economy for 1961, the "distance" between p_1 and p_0 is $d(p_1,p_0)$. Of course, we can use any of the metrics (a), (b), (c), (d), or (e) for $d(p_1,p_0)$. As far as topological questions are concerned, our choice will turn out to be unimportant. On the other hand, our choice will obviously affect the way we interpret $d(p_1,p_0)$. However, the question of interpretation is a part of the mathematical modeling problem and not a purely mathematical question. ∎

Thus far we have discussed the product of two metric spaces. One can also define the product of more than two metric spaces. For example, assume that we are given n metric spaces $(X_1,d_1), (X_2,d_2), \ldots, (X_n,d_n)$. Let

$$X = X_1 \times X_2 \times \cdots \times X_n = \prod_{i=1}^{n} X_i,$$

where the ordered n-tuple $x = (x_1,\ldots,x_n)$, $x_i \in X_i$, $i = 1, 2, \ldots, n$, is an element of the set X. As before there are many ways of defining metrics on X; however, the three most common ways are

(a) $d_1(x,y) = d_1(x_1,y_1) + \cdots + d_n(x_n,y_n)$
(b) $d_2(x,y) = [d_1(x_1,y_1)^2 + \cdots + d_n(x_n,y_n)^2]^{1/2}$
(c) $d_\infty(x,y) = \max\{d_1(x_1,y_1), \ldots, d_n(x_n,y_n)\}.$

Later, after we define the concept of equivalence between metrics in Section 11, we shall show that the above metrics on X are equivalent to one another.

EXERCISES

1. Let $\xi = (\xi_1, \xi_2, \xi_3)$ be a R^3-valued random variable (see Appendix E) defined on a probability space $(\Omega, \mathscr{B}, P)$. Let Ξ denote those random variables ξ with $E(|\xi|^2) < \infty$, where E denotes the expectation and $|\xi|^2 = \xi_1{}^2 + \xi_2{}^2 + \xi_3{}^2$. Show that

$$d(\xi, v) = \{E(|\xi - v|^2)\}^{1/2}$$

is a metric on Ξ, where $\xi - v = (\xi_1 - v_1, \xi_2 - v_2, \xi_3 - v_3)$.

   (If B denotes a prescribed set in R^3—say a box—one sometimes is interested in the subspace A of Ξ given by

$$A = \{\xi \in \Xi : P(\xi \notin B) = 0\},$$

   that is, the probability ξ does not lie in B is zero.)

2. Show that the thrust vector $f(t) = (f_1(t), f_2(t), f_3(t))$ for a rocket engine can be modeled as a point in the product of metric spaces.

3. Let $N = \{1, 2, \ldots\}$ be the natural numbers with the usual metric $d(n, m) = |n - m|$. On $N \times N$ define a function by

$$\rho((n_1, n_2), (m_1, m_2)) = |n_1 m_2 - n_2 m_1| \cdot (n_2 m_2)^{-1}.$$

   Show that ρ is a pseudometric. Show that

$$\rho((n_1, 1), (m_1, 1)) = |n_1 - m_1|$$

$$\rho((1, n_2), (1, m_2)) = \left| \frac{1}{n_2} - \frac{1}{m_2} \right|.$$

4. Find some other examples of subspaces and product spaces.

5. CONTINUOUS FUNCTIONS

As has already been suggested, the introduction of abstract metric spaces allows the generalization of the concept of continuity. This generalization is one of the chief reasons for investigating metric spaces. An equally important reason, the generalization of the concept of convergence, is discussed in the next section.

Let us recall the definition of continuity for real-valued functions $F: R \to R$. We say that F is continuous at x_0 if for every real number $\varepsilon > 0$, there is a real number $\delta > 0$, such that $|F(x) - F(x_0)| < \varepsilon$, whenever $|x - x_0| < \delta$. The function F is said to be continuous if it is continuous at each point in the domain of definition. In this case the appropriate δ's are determined by x_0 as well as ε, so we sometimes write $\delta = \delta(\varepsilon, x_0)$. It can happen that a δ can be found which "works" for all x_0. More precisely, the following may hold: "For every $\varepsilon > 0$, there is a $\delta > 0$ such that for any x_0 in R one has $|F(x) - F(x_0)| < \varepsilon$ whenever $|x - x_0| < \delta$." Here δ depends on ε only, and F is said to be *uniformly continuous*.

Recall that a function $F: X \to Y$ is a set-theoretic concept. In order to define continuity we must require that the sets X and Y have some additional structure. In particular, the additional structure needed is a topological structure. More precisely, we will use the topological structure generated by metrics on X and Y.

3.5.1 DEFINITION.   Let $F: X \to Y$ be a mapping of the metric space (X, d_1) into the metric space (Y, d_2). The mapping F is said to be *continuous at the point* x_0 in X if for every real number $\varepsilon > 0$, there exists a real number $\delta > 0$ such that $d_2(F(x), F(x_0)) < \varepsilon$ whenever $d_1(x, x_0) < \delta$. The mapping F is said to be *continuous* if it is continuous at each point in its domain.

Note that this definition includes continuity of real-valued functions of a real variable as a special case. In particular, Definition 3.5.1 reduces to the familiar definition of continuity when $X = Y = R$ and $d_1(x,y) = d_2(x,y) = |x - y|$. Also note that, as before, if F is continuous at a point x_0, then the δ's are determined by x_0 as well as ε and we write $\delta = \delta(\varepsilon, x_0)$. The concept of uniform continuity is generalized in the obvious way.

3.5.2 DEFINITION.   A mapping F of a metric space (X, d_1) into a metric space (Y, d_2) is said to be *uniformly continuous* if for each $\varepsilon > 0$, there exists a $\delta = \delta(\varepsilon) > 0$ such that for any x_0 one has $d_2(F(x), F(x_0)) < \varepsilon$ whenever $d_1(x, x_0) < \delta$.

The function $\delta(\varepsilon, x_0)$ or $\delta(\varepsilon)$ is sometimes referred to as the *modulus of continuity of F*.

Let us consider a few examples of some mappings that are continuous and some that are not.

EXAMPLE 1.   Let $X = R^n$ and $Y = R^m$, and consider mappings F of X into Y that can be represented with the following matrix formulation.

$$
\begin{bmatrix} y_1 \\ \cdots \\ y_m \end{bmatrix} = \begin{bmatrix} a_{11} & a_{12} & \cdots & a_{1n} \\ a_{21} & & \cdots & \\ & & \cdots & \\ a_{m1} & & \cdots & a_{mn} \end{bmatrix} \begin{bmatrix} x_1 \\ \cdots \\ x_n \end{bmatrix},
$$

where the a_{ij}'s are real numbers. Consider the following metrics on X and Y.

For X:

$$
d_2(u,v) = \{ |u_1 - v_1|^2 + \cdots + |u_n - v_n|^2 \}^{1/2}.
$$

For Y:

$$
d_2(w,z) = \{ |w_1 - z_1|^2 + \cdots + |w_m - z_m|^2 \}^{1/2}.
$$

Let x_0 be an arbitrary element in X, that is,

$$x_0 = \begin{bmatrix} x_{01} \\ x_{02} \\ \cdots \\ x_{0n} \end{bmatrix}.$$

The image of x_0 under a mapping F is given by

$$y_0 = \begin{bmatrix} y_{01} \\ y_{02} \\ \cdots \\ y_{0m} \end{bmatrix} = \begin{bmatrix} a_{11}x_{01} + a_{12}x_{02} + \cdots + a_{1n}x_{0n} \\ a_{21}x_{01} + \qquad\qquad \cdots + a_{2n}x_{0n} \\ \cdots \\ a_{m1}x_{01} + \cdots \qquad\qquad + a_{m1}x_{mn} \end{bmatrix}.$$

It follows that if x is another arbitrary point in X and y is its image, then

$$d(y,y_0)^2 = \sum_{i=1}^{m} \left| \sum_{j=1}^{n} a_{ij}(x_j - x_{0j}) \right|^2.$$

Using the Schwarz Inequality (Appendix A), it follows that

$$d(y,y_0)^2 \le \sum_{i=1}^{m} \left(\sum_{j=1}^{n} |a_{ij}|^2 \right) \left(\sum_{j=1}^{n} |x_j - x_{0j}|^2 \right)$$

$$\le A^2 d(x,x_0)^2,$$

where $A = (\sum_{ij} |a_{ij}|^2)^{1/2}$. Then given an $\varepsilon > 0$ choose $\delta = \varepsilon/A$, provided $A \ne 0$. It is clear that $d(y,y_0) < \varepsilon$ whenever $d(x,x_0) < \delta$. Hence every such mapping F is continuous. (What happens if $A = 0$?) ▮

EXAMPLE 2.   Consider $Y = L_2(-\infty,\infty)$ with the usual metric (see Example 10, Section 3) and let X be the subspace of Y made up of all points x in Y such that

$$\int_{-\infty}^{\infty} \left| \int_{-\infty}^{t} x(\tau)\,d\tau \right|^2 dt < \infty.$$

It follows that

$$y(t) = \int_{-\infty}^{t} x(\tau)\,d\tau$$

represents a mapping of X into Y. Denote this mapping by F.

The mapping F is not continuous; in fact, it is not continuous at any point in X. In order to show this, we start with an arbitrary $x_0 \in X$ and we then seek an $\varepsilon > 0$, say ε_0, such that no matter which $\delta > 0$ we consider there is always at least one $x \in X$ with the property that $d(x,x_0) < \delta$ and $d(F(x),F(x_0)) \ge \varepsilon_0$.

Set $\varepsilon_0 = 1$. Let x be an arbitrary point in X, and let $y_0 = F(x_0)$ and $y = F(x)$. Then

$$y(t) - y_0(t) = \int_{-\infty}^{t} [x(\tau) - x_0(\tau)]\, d\tau.$$

Let x be chosen so that

$$x(t) - x_0(t) = \begin{cases} c, & 0 \le t \le 3T^2 \\ -c, & 3T^2 < t \le 6T^2 \\ 0, & \text{otherwise.} \end{cases}$$

Then $d(x, x_0) = \sqrt{6cT}$ and

$$y(t) - y_0(t) = \begin{cases} ct, & 0 \le t \le 3T^2 \\ c(6T^2 - t), & 3T^2 < t \le 6T^2 \\ 0, & \text{otherwise.} \end{cases}$$

Furthermore $d(y,y_0) = \sqrt{18cT^3}$. Let $\delta > 0$ be given and choose c and T so that

$$\sqrt{18cT^3} = 1 \qquad \text{and} \qquad \sqrt{6cT} < \delta.$$

One then has $d(x,x_0) < \delta$ and $d(y,y_0) \ge 1$, which proves that F is not continuous at x_0. Since x_0 is an arbitrary point in X, we see that F is nowhere continuous. ∎

EXAMPLE 3.   Let Y be the metric space $L_2(-i\infty,i\infty)$ defined in Example 13, Section 3 and let X be the subspace of (Y,d) made up of all functions such that

$$\int_{-\infty}^{\infty} \frac{|X(i\omega)|^2}{\omega^2}\, d\omega < \infty.$$

It then follows that if $Y = FX$ is given by

$$Y(i\omega) = \frac{X(i\omega)}{i\omega},$$

this represents a mapping F of X into Y. Using the same general approach as used in Example 2, it can be shown that F is not continuous at any point in X. ∎

The reader undoubtedly recognizes that Examples 2 and 3 are related to one another through the Fourier Transform, or the two-sided Laplace Transform.

EXAMPLE 4.   This example is similar to Example 2 except for the important difference that the functions considered are defined on a finite instead of an infinite interval. Let $X = Y = L_2[0,T]$, where $0 < T < \infty$, be given with the usual metric. Then

$$y(t) = \int_{0}^{t} x(\tau)\, d\tau$$

represents a mapping F of $L_2[0,T]$ into itself. Moreover, F is continuous. Let x_0 and x be arbitrary points in X, and let $y_0 = F(x_0)$ and $y = F(x)$. Then

$$|y(t) - y_0(t)| = \left|\int_0^t [x(\tau) - x_0(\tau)]\, d\tau\right| \le \int_0^t |x(\tau) - x_0(\tau)|\, d\tau$$

$$\le \left\{\int_0^T 1^2\, d\tau\right\}^{1/2} \left\{\int_0^T |x(\tau) - x_0(\tau)|^2\, d\tau\right\}^{1/2}$$

$$= \sqrt{T} \left\{\int_0^T |x(\tau) - x_0(\tau)|^2\, d\tau\right\}^{1/2}$$

where the Schwarz Inequality for integrals (Appendix A) is used in the second to the last step. Therefore,

$$\int_0^T |y(t) - y_0(t)|^2\, dt \le T^2 \int_0^T |x(\tau) - x_0(\tau)|^2\, d\tau$$

or

$$d(y,y_0) \le T d(x,x_0).$$

This last inequality implies that F is uniformly continuous. (Why?) ■

The reader should carefully compare Examples 2 and 4.

EXAMPLE 5.   Let the metric spaces X and Y be the same as those used in Example 4. Let $k(t,\tau)$ be a real-valued function defined on $[0,T] \times [0,T]$ such that

$$\int_0^T \int_0^T |k(t,\tau)|^2\, dt\, d\tau = M^2 < \infty.$$

Then

$$y(t) = \int_0^T k(t,\tau)x(\tau)\, d\tau$$

represents a continuous mapping K of $L_2[0,T]$ into itself.

Note that this is a generalization of Example 4 where

$$k(t,\tau) = \begin{cases} 1, & \text{for } t \ge \tau \\ 0, & \text{otherwise.} \end{cases}$$

Using essentially the same argument as used in Example 4, we find that

$$d(y,y_0) \le M d(x,x_0).$$

Hence, K is a continuous mapping of $L_2[0,T]$ into itself. ■

EXAMPLE 6.   If in Example 5 we use $L_2(-\infty,\infty)$, then we have a continuous mapping of $L_2(-\infty,\infty)$ into itself. However, the requirement that

$$\int_{-\infty}^{\infty} \int_{-\infty}^{\infty} |k(t,\tau)|^2\, d\tau\, dt = M^2 < \infty \tag{3.5.1}$$

is quite restrictive. That is, there are many integral operators with kernels that do not satisfy (3.5.1) but which do correspond to continuous mappings of $L_2(-\infty,\infty)$ into itself. For example,

$$k(t,\tau) = \begin{cases} e^{-(t-\tau)}, & \text{for } t \ge \tau \\ 0, & \text{otherwise} \end{cases}$$

corresponds (see Exercise 17, Section 5.6) to a continuous mapping of $L_2(-\infty,\infty)$ into itself, but

$$\int_{-\infty}^{\infty} \int_{-\infty}^{t} e^{-2(t-\tau)} \, d\tau \, dt = +\infty. \quad \blacksquare$$

EXAMPLE 7.   Let $X = Y = l_2(0,\infty)$ be given with the usual metric. (See Example 4, Section 3.)

Let A be an infinite matrix

$$\begin{bmatrix} a_{11}, & a_{12}, & \cdots \\ a_{21}, & & \cdots \\ & \cdots & \end{bmatrix}$$

such that

$$\sum_{j=1}^{\infty} \sum_{i=1}^{\infty} |a_{ij}|^2 = M^2 < \infty. \tag{3.5.2}$$

Using the Schwarz Inequality for infinite series the reader should be able to show that the following matrix formulation represents a continuous mapping of l_2 into itself:

$$\begin{bmatrix} y_1 \\ y_2 \\ \cdots \end{bmatrix} = \begin{bmatrix} a_{11}, & a_{12}, & \cdots \\ a_{21}, & & \cdots \\ & \cdots & \end{bmatrix} \begin{bmatrix} x_1 \\ x_2 \\ \cdots \end{bmatrix}. \tag{3.5.3}$$

It should be said that (3.5.2) is not a necessary condition for (3.5.3) to represent a continuous mapping of l_2 into itself. $\blacksquare$

A final yet important point concerns the continuity of inverses. If an invertible mapping F is continuous, the mere continuity of F does not say anything about the continuity of F^{-1}. Similarly, if F is not continuous, then this fact alone does not say anything about the continuity of F^{-1}.

EXERCISES

1. Let $C^\infty[0,T]$ denote the set of all real-valued infinitely differentiable functions x defined on $0 \le t \le T$. Let D be the mapping of $C^\infty[0,T]$ into itself defined by

$$Dx = \frac{dx}{dt}.$$

(a) Let d_1 be the sup-metric on $C^\infty[0,T]$. Is D a continuous mapping of $(C^\infty[0,T],d_1)$ into itself?

(b) Define a metric d_2 on $C^\infty[0,T]$ by

$$d_2(x,y) = d_1(x,y) + d_1(Dx,Dy).$$

Is D a continuous mapping of $(C^\infty[0,T],d_2)$ into $(C^\infty[0,T],d_1)$?

2. A delay line is a device whose output is ideally a delayed version of its input. That is, a mathematical model for a delay line as shown in Figure 3.5.1 is given by

$$y(t) = x(t - \tau).$$

Figure 3.5.1.

Suppose that $x, y \in L_2(-\infty,\infty)$, where $L_2(-\infty,\infty)$ has the usual metric. Is the mathematical model of the delay line a continuous mapping of $L_2(-\infty,\infty)$ into itself?

3. Let $Y = C[0,T]$ be given with the sup-metric $d(x,y)$. Let

$$X = (C[0,T],d) \times (C[0,T],d)$$

be the product space. Consider the mapping F of X into Y defined by $F(x) = x_1 x_2$, where $x = (x_1, x_2)$. (That is, F is a "multiplier.") Is F continuous? Is F uniformly continuous?

4. Define continuity and uniform continuity for functions defined on pseudo-metric spaces. Give examples of continuous and discontinuous functions on pseudometric spaces.

5. Let (X,d) be a metric space, where X is nonempty. Let $Y = BC(X,R)$ denote the collection of all bounded, continuous real-valued functions defined on X.

(a) Show that the functions

$$f_1 : x \to \frac{d(x,x_0)}{1 + d(x,x_0)}$$

$$f_2 : x \to d(x,x_1) - d(x,x_0)$$

$$f_3 : x \to 3$$

are in Y. (So Y is nonempty.)

(b) Show that

$$\sigma(f,g) = \sup\{|f(x) - g(x)|: x \in X\}$$

is a metric on Y.

6. Let (X,d_1) and (Y,d_2) be two metric spaces, where X and Y are nonempty. Assume that every set in Y is bounded. (See Exercise 3, Section 3.2.) Then also let $Z = C(X,Y)$ denote the space of all continuous functions from X into Y. Show that Z is nonempty. Show that

$$\sigma(f,g) = \sup\{d_2(f(x),g(x)): x \in X\}$$

is a metric on Z. What happens if one does not assume that every set in Y is bounded?

7. (Hölder continuity.) A real-valued function f defined on a closed, bounded interval I is said to satisfy an α-*Hölder condition* on I if there is a constant k (called the *Hölder coefficient*) such that $|f(t) - f(s)| \le k|t - s|^\alpha$ for $t, s \in I$. Let $C^\alpha(I)$ denote the collection of all such functions, where $0 < \alpha$.
   (a) Show that $f \in C^\alpha(I)$ for $1 < \alpha$, implies that $f(t) = $ constant.
   (b) Show that $0 < \beta \le \alpha \le 1$ implies that $C^\alpha(I) \subset C^\beta(I)$.
   (c) Show that for $0 < \alpha \le 1$, the function

$$d^\alpha(f,g) = \sup\{|f(t) - g(t)| : t \in I\} + \sup\left\{\frac{|f(t) - f(s)|}{|t - s|^\alpha} : t, s \in I, t \ne s\right\}$$

   is a metric on $C^\alpha(I)$.
   (d) Show that the mapping $f \to f$ of (C^α, d^α) into (C^β, d^β), where $0 < \beta \le \alpha \le 1$, is continuous.

8. Let (X,d) be a metric space and let $LC(X,R) = LC$ denote the space of real-valued (globally) *Lipschitz continuous* functions defined on X. That is, $f \in LC$ if there is a constant k such that

$$|f(x) - f(y)| \le kd(x,y) \qquad (x,y \in X). \tag{3.5.4}$$

Let $\|f\|$ denote the smallest number k that satisfies (3.5.4). (This is a pseudo-norm, see Exercise 3, Section 5.2.)
   (a) Show that if $f, g \in LC$ and $h(x) = f(x) + g(x)$, then $h \in LC$. Also $\alpha f \in LC$, where α is any real number.
   (b) Show that $\sigma(f,g) = \|f - g\|$ is a pseudometric on LC. Discuss the relationship $\sigma(f,g) = 0$.
   (c) Show that LC is nonempty. [*Hint*: Consider $f(x) = d(x,x_0)$ where x_0 is a fixed point in X.]
   (d) Let LC_{x_0} denote those f in LC that satisfy $f(x_0) = 0$. Show that σ is a metric on LC_{x_0}. Compare the space LC_{x_0} and LC_{x_1}, where $x_0 \ne x_1$. [*Note*: $\|f\|$ is a "norm" on LC_{x_0}, see Section 5.2.]

9. (Continuation of Exercise 8.) Let $LC_{x_0}{}^*$ denote the collection of all functions $l: LC_{x_0} \to R$ that satisfy the following conditions:
   (i) $l(f + g) = l(f) + l(g)$.
   (ii) $l(\alpha f) = \alpha l(f)$.

(iii) $\sup\{|l(f)|: \|f\| \leq 1\} < \infty$.

For $x \in X$, let δ_x be the *Dirac-function* $\delta_x(f) = f(x)$.

(a) Show that $\delta_x \in LC_{x_0}{}^*$ for every $x \in X$.

   For l_1, l_2 in $LC_{x_0}{}^*$, let

$$\|l_1 - l_2\| = \sup\{|l_1(f) - l_2(f)|: \|f\| \leq 1\}.$$

(b) Show that $\|\delta_x - \delta_y\| \leq d(x, y)$, for all $x, y \in X$.

(c) Show that $\sigma^*(l_1, l_2) = \|l_1 - l_2\|$ is a metric on $LC_{x_0}{}^*$.

10. (Continuation of Exercise 12, Section 3.) Let X denote the class of all complex-valued functions $f(z)$ that are analytic for $|z| < 1$. Let $D: X \to X$ be the differential operator $D: f \to df/dz$. By using the metrics, or pseudometrics, in Exercise 12, Section 3, discuss the continuity of D.

11. Let $f: X \to X$ be a continuous mapping, where X has a metric d. Let $\mathrm{Gr}(f)$ denote the graph (Section 2.6) of f in $X \times X$ and Δ the diagonal set

$$\Delta = \{(x,y): x = y\} = \{(x,x): x \in X\}.$$

Assume that $X \times X$ has the metric

$$d((x_1,y_1), (x_2,y_2)) = d(x_1,x_2) + d(y_1,y_2).$$

Define $g: \Delta \to \mathrm{Gr}(f)$ by

$$g(x,x) = (x, f(x)).$$

Show that g is continuous. Show that g is invertible. Is g^{-1} continuous?

12. In Definition 3.5.1 we used the strong inequalities "$d_2(F(x), F(x_0)) < \varepsilon$" and "$d_1(x,x_0) < \delta$" to define continuity at x_0. Show that one can replace either or both of these inequalities with the weak inequalities "$d_2(F(x), F(x_0)) \leq \varepsilon$" and "$d_1(x,x_0) \leq \delta$," without changing the concept of "continuity at x_0."

13. Show that the mapping $f: N \times N \to Q$, given by $(n_1,n_2) \to (n_1/n_2)$, is continuous when Q has the usual metric and $N \times N$ has the metric given in Exercise 3, of Section 4. Is f one-to-one?

6.   CONVERGENT SEQUENCES

In addition to the foregoing generalization of continuity, the introduction of metric spaces allows us to generalize the concept of convergent sequences. First, let us recall the definition of a convergent sequence of real numbers. We say that a sequence of real numbers $\{x_n\} = \{x_1,x_2,\ldots\}$ is convergent if there is a real number x_0 with the property that for each real number $\varepsilon > 0$ there is an integer N such that $|x_n - x_0| < \varepsilon$ whenever $n \geq N$. We say that x_0 is the limit of the sequence $\{x_n\}$. Since an appropriate integer N is determined by ε, we sometimes write $N = N(\varepsilon)$. Our intuitive picture is that N becomes larger as ε becomes smaller.

Suppose that instead of real numbers we have a sequence of points in a metric space (X,d). We again denote our sequence by $\{x_n\} = \{x_1,x_2,\ldots\}$. The generalization is straightforward.

3.6.1 DEFINITION.   A sequence $\{x_n\}$ of points in a metric space (X,d) is said to be *convergent* if there is point x_0 in (X,d) with the property that for each real number $\varepsilon > 0$ there is an integer N such that $d(x_n,x_0) < \varepsilon$ whenever $n \geq N$. The point x_0 is said to be the *limit* of the sequence $\{x_n\}$. This is sometimes written

$$\lim_{n \to \infty} x_n = \lim x_n = x_0,$$

or

$$x_n \to x_0 \text{ as } n \to \infty.$$

Note that x_0 is referred to as *the* limit and not as *a* limit. In doing so we are anticipating the next result which says that a convergent sequence in a metric space has precisely one limit.

3.6.2 LEMMA.   *Let $\{x_n\}$ be a convergent sequence in a metric space (X,d). If y_0 and x_0 in (X,d) are limits of the sequence $\{x_n\}$, then $y_0 = x_0$.*

Proof: If we can show that $d(x_0,y_0) < 2\varepsilon$ for every $\varepsilon > 0$, then it follows that $d(x_0,y_0) = 0$ and, from axiom (M2), that $x_0 = y_0$. Let $\varepsilon > 0$ be given. Since x_0 is a limit of $\{x_n\}$, there is an N_1 such that $n \geq N_1$ implies that $d(x_n,x_0) < \varepsilon$. Similarly, there is an N_2 such that $n \geq N_2$ implies that $d(x_n,y_0) < \varepsilon$. Let $M = \max[N_1 N_2]$. Then by the triangle inequality one has

$$d(x_0,y_0) \leq d(x_M,x_0) + d(x_M,y_0) < 2\varepsilon. \quad \blacksquare$$

Carefully note that the limit x_0 must be a point in (X,d). For example, if X is the open interval $0 < x < 1$ and $d(x,y) = |x - y|$, the sequence $\{\frac{1}{2},\frac{1}{3},\frac{1}{4},\ldots\}$ is not convergent because the apparent limit 0 is not a point in (X,d).

EXAMPLE 1.   Suppose that $X = C[0,1]$ is given with the metric

$$d_2(x,y) = \left\{\int_0^1 |x(t) - y(t)|^2 \, dt\right\}^{1/2}.$$

Consider the sequence $\{x_n\}$ in $(C[0,1],d_2)$, where

$$x_n(t) = \begin{cases} 1 - nt, & \text{for } 0 \leq t \leq 1/n \\ 0, & \text{for } 1/n < t \leq 1. \end{cases}$$

We claim that this sequence converges and the limit is $x_0(t) \equiv 0$. Indeed

$$d_2(x_n, x_0) = \left\{\int_0^1 |x_n(t) - x_0(t)|^2 \, dt\right\}^{1/2} = \left\{\int_0^{1/n} (1 - nt)^2 \, dt\right\}^{1/2} = (3n)^{-1/2}.$$

Given an $\varepsilon > 0$, if $n \geq N$, where N is an integer greater than ε^{-2}, then $d(x_n,x_0) < \varepsilon$. Hence, $x_0 = \lim_{n \to \infty} x_n$. $\quad \blacksquare$

EXAMPLE 2.   This example is the same as Example 1 except for a change of metric. Here we use the sup-metric d_∞. The sequence $\{x_n\}$ is now a sequence of points in the metric space $\{C[0,T],d_\infty\}$. Obviously the sequence does not converge to $x_0(t) \equiv 0$, for $d_\infty(x_n,x_0) = 1$ for each n. One might be tempted to say that $\{x_n\}$ converges to y_0, where

$$y_0(t) = \begin{cases} 1, & \text{for } t = 0 \\ 0, & \text{for } 0 < t \leq T. \end{cases}$$

But this would be nonsense, for y_0 is not a point in $\{C[0,T],d_\infty\}$. In fact the sequence $\{x_n\}$ just does not converge in this metric. We can see this by noting that for an arbitrary n we can find an $N > n$ such that $d_\infty(x_n,x_m) \geq \frac{1}{2}$ for all $m > N$. (Why does this observation show that $\{x_n\}$ cannot converge? See Exercise 8 below.) ∎

It should be mentioned that in practice we are not usually confronted with the problem of determining whether or not a point x_0 is the limit of sequence $\{x_n\}$. Usually we are primarily interested in determining whether or not a given sequence is convergent. Knowing this, we sometimes then seek the limit. Both of these problems are often difficult to solve.

EXERCISES

1. Let X denote the set of all bounded piecewise continuous[2] functions defined on $0 \leq t \leq T$, with the sup-metric d_∞.
   (a) Obviously $C[0,T]$ is a subspace of X. Let x_0 be an arbitrary point in $C[0,T]$. Suppose that x_0 is to be approximated by piecewise constant functions as shown in Figure 3.6.1. That is,

$$x_n(t) = x_0\left(j\frac{T}{n}\right) \quad \text{for} \quad j\frac{T}{n} \leq t < (j+1)\frac{T}{n} \quad \text{and} \quad j = 0,1,\dots,(n-1).$$

and

$$x_n(T) = x_0\left(\frac{n-1}{n}T\right).$$

Is it true that the sequence $\{x_n\}$ converges in (X,d)? If so, is it true that $x_0 = \lim_{n \to \infty} x_n$? [Hint: Use the fact that a real-valued continuous function, defined on a bounded closed interval is uniformly continuous, compare with Exercise 13, Section 17.]

---

[2] We say that a function $x(t)$ is piecewise continuous on $0 \leq t \leq T$ if (1) it is continuous at all but a finite number of points in $0 \leq t \leq T$, (2) if $x(t)$ is discontinuous at t_1, then left and right limits of $x(t)$ exist as t approaches t_1 from the left and right, and (3) if $x(t)$ is discontinuous at t_1, then $x(t_1)$ equals either the left or the right limit of $x(t)$. In other words, we allow only a finite number of "simple" discontinuities.

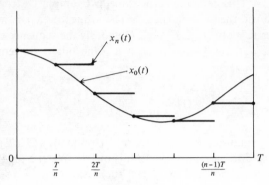

Figure 3.6.1.

(b) Suppose x_0 is not restricted to the subspace $C[0,T]$, and suppose that x_0 is approximated by functions x_n as in (a). Is it true that $x_0 = \lim_{n \to \infty} x_n$?

(c) Consider a different metric on X, namely,

$$d_2(x,y) = \left\{ \int_0^T |x(t) - y(t)|^2 \, dt \right\}^{1/2}.$$

Does the sequence $\{x_n\}$ converge to x_0 in (X,d_2)?

2. Suppose that in a metric space (X,d) a sequence $\{x_n\}$ converges to a point x_0. Does it follow that

$$d(x_1,x_0) \geq d(x_2,x_0) \geq d(x_3,x_0) \geq \cdots \geq d(x_n,x_0) \geq \cdots ?$$

Either prove that it does, or give a counterexample.

3. Consider the metric space (X,d) given in Example 16, Section 3. Characterize the collection of all convergent sequences in (X,d).

4. Consider the sequence $\{x_n\}$, where

$$x_n(t) = [\cos(n!\pi t)]^{2n}, \qquad n = 1, 2, \ldots$$

in the metric space $C[0,1]$ with the sup-metric d_∞. Is $\{x_n\}$ a convergent sequence?

5. If $\{x_n\}$ and $\{y_n\}$ are convergent sequences in metric space (X,d) show that the sequence of real numbers $\{d(x_n,y_n)\}$ converges to $d(x_0,y_0)$, where $x_0 = \lim_{n \to \infty} x_n$ and $y_0 = \lim_{n \to \infty} y_n$. (This exercise should be reconsidered after studying Section 7.)

6. Define a concept of sequential convergence for pseudometric spaces. Are the limits of sequences unique in pseudometric spaces? Give some examples.

7. (a) (Square Root Algorithm) Let a be a real number satisfying $0 \leq a \leq 1$ and set $b = 1 - a$. Let $y_0 = 0$ and

$$y_{n+1} = \tfrac{1}{2}(b + y_n^2).$$

Show that the sequence $\{y_n\}$ is bounded and monotone, and therefore convergent. Let $y = \lim y_n$, and $x = 1 - y$. Show that $x^2 = a$.

(b) Modify part (a) for the case $a > 1$.

(c) What happens when $a < 0$? When a is complex?

8. Let $\{x_n\}$ be a sequence in a metric space (X,d) with the property that for some $\varepsilon > 0$ one has $d(x_n, x_m) \geq \varepsilon$ for all n, m. Show that $\{x_n\}$ is not convergent.

9. Let $X = R^n$ be given with the metric $d(x,y) = \sum_{i=1}^{n} |x_i - y_i|$. Show that a sequence $\{x_n\}$ in X converges to x_0 if and only if $\lim x_{n,i} = x_{0,i}$ for each coordinate $x_{0,i}$, $1 \leq i \leq n$.

10. Let $x_n(t)$ be a sequence of continuous real-valued functions where $x_n(t)$ is periodic with period $\tau_n > 0$. Assume that $x_n(t) \to x(t)$ uniformly for t in R and $\tau_n \to \tau$. Show that $x(t)$ is periodic in t with period τ.

11. Let f and g be functions in $C[0,T]$. Define $x_0 = f$, $x_1 = g$, and

$$x_{n+1} = \tfrac{1}{2}(x_n + x_{n-1}), n = 1, 2, \ldots .$$

Show that $\lim x_n = \tfrac{1}{3}(f + 2g)$ when $C[0,T]$ has the sup-metric.

12. Consider the space $C[0,1]$ with the sup-metric d_∞. Let

$$g_n(t) = \frac{t^2}{t^2 + (1 - nt)^2}, \qquad n = 1, 2, \ldots .$$

Show that for each t, $g_n(t) \to 0$ as $n \to \infty$. Show that $d_\infty(g_n, 0) = 1$.

13. Consider the sequence in Exercise 12 in the space $L_1[0,1]$ with the usual metric

$$d_1(f,g) = \int_0^1 |f(t) - g(t)| \, dt.$$

Show that $d_1(g_n, 0) \to 0$ as $n \to \infty$. [Hint: Use the Lebesgue Dominated Convergence Theorem, Appendix D.]

14. Consider the space $C[0,T]$ with the sup-metric d_∞. Assume that $\{f_n\}$ is a sequence with $f_n \to f$ in $(C[0,T], d_\infty)$ and $\{t_n\}$ is a sequence with $t_n \to t$ in $[0,T]$. Show that $f_n(t_n) \to f(t)$ in R. (Use the usual metric on R and $[0,T]$.)

15. Show that

$$\lim_{n \to \infty} \sum_{i=0}^{n} \left(1 - 1 + \frac{(-1)^2}{2!} + \cdots + \frac{(-1)^n}{n!}\right)^i = \frac{e}{e - 1}.$$

[Hint: Apply Exercise 14 to $f_n(t) = \sum_{i=0}^{n} t^i$, $0 \leq t \leq 0.9$.]

16. Can a sequence of discontinuous functions converge uniformly to a continuous function?

17. Consider the following sequences of function $\{f_n\}$ defined for $0 \leq t < \infty$. Find all intervals on which these sequences converge uniformly.

(a) $\dfrac{t^n}{n}$

(b) $\dfrac{t^n}{n + t^n}$

(c) t^n

(d) $\dfrac{t^n}{n!}$

(e) $\dfrac{t}{n} e^{-(t/n)}$

(f) $\dfrac{\sin n^2 t}{nt}$.

7.   A CONNECTION BETWEEN CONTINUITY AND CONVERGENCE

Although it may appear that continuity and convergence are unrelated concepts, there is a very important connection between them. It is the purpose of this section to investigate this connection. In particular, we consider the problem of interchanging limits and functions; that is, when does $F(\lim_{n\to\infty} x_n)$ equal $\lim_{n\to\infty} F(x_n)$?

3.7.1 THEOREM.   *Let $F: (X,d_1) \to (Y,d_2)$ be given and let x_0 be a point in X. Then the following statements are equivalent:*

(a) *F is continuous at x_0.*
(b) $\lim F(x_n) = F(\lim x_n)$, *for every sequence $\{x_n\}$ with the property that* $\lim x_n = x_0$.

Statement (b) needs a word of explanation. It asserts two things: (i) The limit of $\{F(x_n)\}$ exists and (ii) this limit agrees with $F(x_0) = F(\lim x_n)$.

Proof: (a)$\Rightarrow$(b). Assume that F is continuous at x_0 and let $\{x_n\}$ be a sequence with the property that $\lim x_n = x_0$. We want to show that $\lim F(x_n) = F(x_0)$. Since F is continuous at x_0, for every $\varepsilon > 0$ one can find a $\delta > 0$ such that $d_2(F(x),F(x_0)) < \varepsilon$ whenever $d_1(x,x_0) < \delta$. Since $\lim x_n = x_0$, there is an N such that $d_1(x_n,x_0) < \delta$ whenever $n \geq N$. By combining these two statements we have $d_2(F(x_n),F(x_0)) < \varepsilon$ whenever $n \geq N$. Hence $\lim F(x_n) = F(x_0)$.

(b)$\Rightarrow$(a). Let $F(x_n) \to F(x_0)$ whenever $x_n \to x_0$. If F is not continuous at x_0, then there exists an $\varepsilon_0 > 0$ such that for each $\delta > 0$ there is an x with $d_1(x,x_0) < \delta$ and $d_2(F(x),F(x_0)) \geq \varepsilon_0$. Let x_1 be such an x for $\delta = 1$, x_2 for $\delta = \frac{1}{2}$, and in general, x_n for $\delta = 1/n$. We then have $d_1(x_n, x_0) < 1/n$ and $d_2[F(x_n),F(x_0)] \geq \varepsilon_0$, for all n. That is, the sequence $\{x_n\}$ converges to x_0, but quite clearly the sequence $\{F(x_n)\}$ does not converge to $F(x_0)$. But this contradicts our assumption that $F(x_n) \to F(x_0)$ whenever $x_n \to x_0$. Hence F must be continuous at x_0. ∎

The following theorem is an immediate consequence of Theorem 3.7.1.

3.7.2 THEOREM.   *Let F be a mapping of (X,d_1) into (Y,d_2). The mapping F is continuous if and only if*

$$F(\lim x_n) = \lim F(x_n) \qquad (3.7.1)$$

for every convergent sequence $\{x_n\}$ in (X,d_1). In other words, a mapping F is continuous if and only if it preserves convergent sequences.

If F is not continuous, it still may be true that $F(\lim_{n\to\infty} x_n) = \lim_{n\to\infty} F(x_n)$ for some convergent sequences; however, because of Theorem 3.7.2, this equality cannot hold for all convergent sequences.

The last two theorems seem simple and unobtrusive. However, they are very important. They will be used time and time again in this book. The reader should master them before proceeding.

EXAMPLE 1.   Let $X = C[0,T]$ be given with the sup-metric d_∞. Since

$$\left| \int_0^T x(t)\,dt - \int_0^T y(t)\,dt \right| \le \int_0^T |x(t) - y(t)|\,dt \le T d_\infty(x,\, y),$$

we see that the function $\int_0^T$ is a continuous function from X into the reals R. Let $\{x_n\}$ be any sequence of functions in X. We know that $x_n \to x_0$ in (X,d) if and only if the sequence of functions $\{x_n(t)\}$ converges uniformly to $x_0(t)$. It follows from Theorem 3.7.2 that

$$\lim \int_0^T x_n(t)\,dt = \int_0^T \lim x_n(t)\,dt \tag{3.7.2}$$

provided the sequence $\{x_n(t)\}$ converges uniformly.   ∎

EXERCISES

1. Are Theorems 3.7.1 and 3.7.2 true when we consider pseudometric spaces instead of metric spaces?

2. Let (X,d) be the metric space $C[0,T]$ with the sup-metric d_∞, and consider the mapping K of X into itself represented by $y = Kx$ where

$$y(t) = \int_0^T k(t,\tau)x(\tau)\,d\tau,$$

   where $k(t,\tau)$ is continuous on $[0,T] \times [0,T]$. Is it true that $x_n \to x_0$ implies that $K(x_n) \to K(x_0)$?

3. Let (Y,d_∞) be the metric space $\{C[0,T], d_\infty\}$ and let (X,d_∞) be the metric space $C^1[0,T]$ with the sup-metric d_∞, where $C^1[0,T]$ is the set of all continuous functions on $[0,T]$ with continuous first derivatives. Consider the differential mapping $y = Dx$ of X into Y given by $y = dx/dt$. Is it true that $x_n \to x_0$ implies that $D(x_n) \to D(x_0)$? If not, are there any convergent sequences $\{x_n\}$ such that $\lim_{n \to \infty} D(x_n) = D(\lim_{n \to \infty} x_n)$?

4. Let $\{x_n\}$ and $\{y_n\}$ be two sequences in a metric space (X,d) with $x = \lim x_n$ and $y = \lim y_n$. Show that if $d(x_n, y_n) \le k$ for all n, then $d(x,y) \le k$.

5. Use Theorem 3.7.2 to compute

$$\log\left(\lim_{N \to \infty} \prod_{n=1}^{N} 2^{1/n^2} e^{1/n!} \right).$$

6. Use Example 1 to prove the following: Let $\{x_n\}$ be a sequence in $C^1[0,T]$ and assume that $\{dx_n/dt\}$ converges uniformly on $[0,T]$ and that $\{x_n(0)\}$ converges. Then $\{x_n\}$ converges uniformly, say that $x = \lim x_n$, and moreover

$$dx/dt = \lim dx_n/dt.$$

7. Redo Exercise 5, Section 6, using the results of this section.

8. Let $f: (a,b) \to X$ be defined on an interval $a < t < b$ with range in a metric space (X,d). We say that $f(t) \to x_0$ as $t \to b^-$ if for every $\varepsilon > 0$ there is a $\delta > 0$ such that $d(f(t),x_0) \le \varepsilon$ whenever $0 < b - t \le \delta$.

   (a) Show that $f(t) \to x_0$ as $t \to b^-$ if and only if one has $f(t_n) \to x_0$ for every sequence t_n with $a < t_n < b$ and $t_n \to b$.

   (b) Show that $f(t) \to x_0$ as $t \to b^-$ if and only if there is an extension of f, $\tilde{f}: (a,b] \to X$, that is continuous at b.

9. Statement (a) in Theorem 3.7.1 assures us that the function F is continuous only at the point x_0.

   (a) Construct an example showing that F need not be continuous elsewhere.

   (b) Using your example explain why Equation (3.7.1) is not valid.

10. Let $\{x_n\}$ be a sequence of functions in $L_p(I)$ where I is a bounded interval, and $1 \le p \le \infty$. (See Example 10, Section 3.) Assume that there is an x in $L_p(I)$ such that $d_p(x_n, x) \to 0$ as $n \to \infty$, where d_p is the usual metric on $L_p(I)$.

   (a) Is it true that

$$\lim \int_I x_n \, dt = \int_I x \, dt?$$

   (b) What happens if I is an unbounded interval?

Part B

Some Deeper
Metric Space Concepts

8.   LOCAL NEIGHBORHOODS

The object of this section and the next is to show that the concepts of continuity and convergence are (in a sense) independent of the metric. The first step is to characterize these concepts in terms of "local neighborhoods." The importance of this characterization is that it is a natural stepping stone to a deeper understanding of topological structures.

3.8.1 DEFINITION.   Let (X,d) be a metric space and let x_0 be an arbitrary point in (X,d). The set

$$B_r(x_0) = \{x \in X: d(x,x_0) < r\},$$

where $0 < r < \infty$ is referred to as the *open ball* of radius r centered at x_0. The set

$$B_r[x_0] = \{x \in X: d(x,x_0) \le r\},$$

where $0 \le r < \infty$, is referred to as the *closed ball* of radius r centered at x_0. The set

$$S_r[x_0] = \{x \in X: d(x,x_0) = r\},$$

where $0 \le r < \infty$, is referred to as the *sphere* of radius r centered at x_0.

If one thinks of $X = R^3$ with

$$d(x,y) = \{(x_1 - y_1)^2 + (x_2 - y_2)^2 + (x_3 - y_3)^2\}^{1/2},$$

then if $x_0 = (0,0,0)$, $S_r[x_0]$ is the "boundary" of the ball, $B_r(x_0)$ is the "interior," and $B_r[x_0]$ is the "boundary plus the interior."[3]

We are now in a position to define local neighborhood.[4]

3.8.2 DEFINITION.   Let x_0 be an arbitrary point in a metric space (X,d). A subset N of (X,d) is said to be a *local neighborhood of x_0* if N is either $B_r(x_0)$ or $B_r[x_0]$, where $r \ne 0$. The positive number r is said to be the *radius* of the local neighborhood N. The open balls $B_r(x_0)$ are sometimes referred to as *open local neighborhoods*

---

[3] This terminology is used merely to help the reader understand the nature of the defined sets. Later, we shall give a precise definition of "boundary" and "interior." (See Exercise 24, Section 12.)

[4] We use the term "local" neighborhood, even though we shall not discuss other types of neighborhoods. The reason is that the term "neighborhood" has an accepted meaning which differs from our usage. (See Exercise 6.)

and the closed balls $B_r[x_0]$, $r > 0$, are sometimes referred to as *closed local neighborhoods*. The terms "open" and "closed" will be given technical meanings shortly. For the moment we shall use them only to distinguish between $B_r(x_0)$ and $B_r[x_0]$. We refer to the family of all local neighborhoods of a point x as the *local neighborhood system of x*. Thus, the local neighborhood system of x consists of all open balls and closed balls of nonzero radius centered at x. We shall denote this system by $\mathcal{N}(x)$. Let us consider some examples.

EXAMPLE 1. Let $X = C[0,T]$ be given with the sup-metric d_∞. Let x_0 be an arbitrary point in (X,d). Then the set of all continuous functions x such that $|x(t) - x_0(t)| < \frac{1}{2}$ for all $t \in [0,T]$ is an open local neighborhood of x_0. This set is illustrated in Figure 3.8.1. ∎

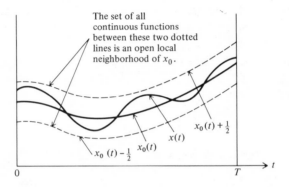

The set of all continuous functions between these two dotted lines is an open local neighborhood of x_0.

$x_0(t) + \frac{1}{2}$

$x(t)$

$x_0(t) - \frac{1}{2}$ $x_0(t)$

Figure 3.8.1.

EXAMPLE 2. Let $X = C[0,T]$ be given with the L_2-metric d_2, that is,

$$d_2(x,y)^2 = \int_0^T |x(t) - y(t)|^2 \, dt.$$

Let x_0 be an arbitrary point in (X,d_2). Then the set of all $x \in X$ such that

$$\left\{ \int_0^T |x(t) - x_0(t)|^2 \, dt \right\} < \frac{1}{2}$$

is an open local neighborhood of x_0. A few of the x's in this local neighborhood of x_0 are sketched in Figure 3.8.2. Note that in contrast to Example 1, we do not have a convenient way to represent this local neighborhood graphically. ∎

The following lemma states a key property of local neighborhood systems. Note that it is a consequence of the triangle inequality.

3.8.3 LEMMA. *Let $B_r(x_0)$ be any open local neighborhood. Then for every x in $B_r(x_0)$ there is a local neighborhood N_x of x with $N_x \subset B_r(x_0)$.*

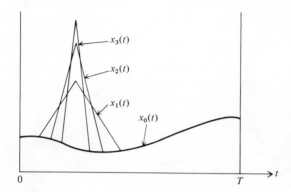

Figure 3.8.2.

Proof: Since $x \in B_r(x_0)$, we have $d(x_0,x) = \alpha < r$. By using the triangle inequality, we get

$$d(x_0,y) \le d(x_0,x) + d(x,y)$$

for any $y \in X$. In particular, if $d(x,y) < \beta$, where $\alpha + \beta = r$, then $d(x_0,y) < r$. In terms of local neighborhoods we have (see Figure 3.8.3)

$$N_x = B_\beta(x) \subset B_r(x_0). \quad \blacksquare$$

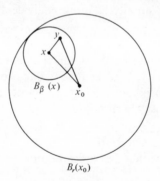

Figure 3.8.3.

Note that Lemma 3.8.3 is not true for closed local neighborhoods. (Why?)

We next turn to the characterization of the continuity of a function F in terms of local neighborhood systems and the inverse set-function F^{-1}.

3.8.4 THEOREM. *A function F mapping (X,d_1) into (Y,d_2) is continuous at a point x_0 in (X,d_1) if and only if the inverse image of every local neighborhood of $F(x_0)$ contains a local neighborhood of x_0. (Figure 3.8.4 illustrates this condition.)*

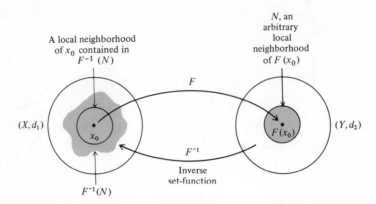

Figure 3.8.4.

Proof: First assume that F is continuous at x_0 and let N be any local neighbor-hood of $F(x_0)$. Let ε be the radius of N. By continuity of F there is a $\delta > 0$ such that $F(x) \in N$ whenever $d_1(x,x_0) < \delta$. In other words, $B_\delta(x_0) \subset F^{-1}(N)$.

Now assume that if N is any local neighborhood of $F(x_0)$, then $F^{-1}(N)$ con-tains a local neighborhood M of x_0. If we let ε be the radius of N and δ be the radius of M, we see that F is continuous at x_0. ∎

Since the conditions stated in Theorem 3.8.4 are necessary and sufficient for the continuity of F at x_0, an alternate but equivalent definition of continuity is possible.

3.8.5 DEFINITION (ALTERNATE).   A function $F: (X,d_1) \to (Y,d_2)$ is said to be *continuous* at a point x_0 in (X,d_1) if the inverse image of each local neighborhood of $F(x_0)$ contains a local neighborhood of x_0. (Note that this definition does not make use of the radius of any local neighborhood.)

We now turn to a characterization of convergent sequences in terms of the local neighborhood systems, but first a definition.

3.8.6 DEFINITION.   A sequence x_n of points in a set X is said to be *eventually in* a subset $A \subset X$ if there exists an integer N such that $x_n \in A$ for all $n \geq N$. In other words, the sequence gets into A and stays in A after a finite number of terms.

The next theorem is a direct consequence of the definitions of "convergent sequence" and "eventually in."

3.8.7 THEOREM.   *Let $\{x_n\}$ be a sequence in a metric space (X,d). The sequence $\{x_n\}$ converges to x_0 if and only if $\{x_n\}$ is eventually in every local neighborhood of x_0.*

As a result of this theorem, the following is obviously an alternate but equivalent definition for a convergent sequence.

3.8.8 DEFINITION (ALTERNATE).   A sequence $\{x_n\}$ of points in a metric space (X,d) is said to be *convergent* if there exists a point x_0 in (X,d) such that $\{x_n\}$ is eventually in each local neighborhood of x_0.

We see, then, that both continuity and convergence can be characterized in terms of local neighborhood systems. These facts are only a prelude to the characterizations given in the next section, characterizations which lead to certain deep and important concepts.

EXERCISES

1. Let $X = R^2$ be given with the metric $d_p(x,y)$ (Example 1, Section 3). Describe the local neighborhoods for d_1, d_2, and d_∞. Describe the corresponding local neighborhoods in R^3.

2. Let X be the set of all ordered n-tuples x of 1's and 0's; for example, $x = \{1,0,0,1,1,\ldots,0\}$. Let $d(x,y) =$ "number of places where x and y differ." For example, if $x = \{0,1,0,1\}$ and $y = \{1,1,1,1\}$, then $d(x,y) = 2$. Let x_0 be an arbitrary point in the metric space (X,d), and discuss the local neighborhood system of x_0.

3. Carry out a development similar to the one presented in this section for the pseudometric space environment.

4. Suppose that (A,d) is a subspace of a metric space (X,d). Let x_0 be an arbitrary point in A. How does the local neighborhood system of x_0 considered as a point in the metric space (A,d) compare with the local neighborhood system of x_0 considered as a point in the metric space (X,d)?

5. Let (X_1,d_1) and (X_2,d_2) be metric spaces, and let (X,d) be the product of (X_1,d_1) and (X_2,d_2). Let x_1 and x_2 be arbitrary points in X_1 and X_2, respectively. Then (x_1,x_2) is a point in X. What does the local neighborhood system of (x_1,x_2) "look like?" [*Note*: The local neighborhood system depends on how we put d_1 and d_2 together to form d (see Exercise 1).]

6. Let x_0 be an arbitrary point in a metric space (X,d). A subset A of (X,d) is said to be a *neighborhood* of x_0 if A contains a local neighborhood of x_0. (Note that the local neighborhoods are neighborhoods.) The set of all neighborhoods of a point x_0 is said to be the *neighborhood system* of x_0. (Note that the local neighborhood system of x_0 is contained in its neighborhood system.) Show that the following statements are true for arbitrary x_0:
   (a) The neighborhood system of x_0 is not empty, and x_0 is in each of its neighborhoods.
   (b) The intersection of two neighborhoods of x_0 is a neighborhood of x_0.
   (c) If A is a neighborhood of x_0, then each superset B of A (that is, $A \subset B$) is a neighborhood of x_0.

(d) Each neighborhood of x_0 contains a neighborhood of x_0 which in turn is a neighborhood of each of its points. That is, if A is a neighborhood of x_0, there is a neighborhood B of x_0 such that $B \subset A$ and if $x \in B$, then B is a neighborhood of x.

[*Remark*: Carefully note that if A is a neighborhood of x_0, it can easily happen that A is *not* a neighborhood of each of its points. Consider a closed local neighborhood, for example.]

7. Consider a nonempty set X with the metric

$$d(x,y) = \begin{cases} 1, & x \neq y \\ 0, & x = y. \end{cases}$$

Describe the local neighborhoods of a point in X.

8. Let d_1 and d_2 be metrics on a set X. Show that if there is a constant $k > 0$ such that $d_1(x,y) \leq k d_2(x,y)$ for all $x, y \in X$, then each d_1-local neighborhood contains a d_2-local neighborhood.

9. Let A be a Lebesgue measurable set in R with finite Lebesgue measure $m(A)$, and let $\chi_A(t)$ be the characteristic function of A.
(a) Show that

$$f(t) = \int_{-\infty}^{\infty} \chi_A(t + s)\chi_A(s)\, ds$$

is continuous and that $f(t) \leq m(A)$ for all t.
(b) Assume that $m(A) > 0$. Then show that the set

$$D(A) = \{x - y \colon x, y \in A\}$$

contains a neighborhood of the origin. [*Hint*: Show that there is a neighborhood U of the origin with the property that $f(t) > 0$ for $t \in U$. Then show that for any t with $f(t) > 0$ there is an s such that $t + s$ and s belong to A, and thus t is in $D(A)$.]

9.   OPEN SETS

In the last section we gave characterizations of continuity and convergence in terms of local neighborhood systems. To a limited extent these characterizations show that continuity and convergence are independent of the metric; that is, the metric was used to define the local neighborhood systems, but after that the formulations of continuity and convergence were essentially given in set-theoretic terminology. In particular, one could test for continuity, for instance, without knowing the specific value of the radius of each local neighborhood.

In this section we shall go one step further. We shall derive characterizations of continuity and convergence in terms of a distinguished family of sets, called open sets. This family is called the topology on the space. We shall see that the topology is generated by the metric, but in a very surprising way, it is independent of the metric. In fact, we shall show that many diverse looking metrics generate the same topology!

Let us see if we can obtain some insight into the topological structure of metric spaces by considering a special problem. Suppose we are given a set X with at least two points. We know that we can consider different metrics on X, and for each metric we obtain a different metric space. Let d_1 and d_2 be two distinct metrics on X. Our problem is to investigate the relation, if any, between the metric spaces (X,d_1) and (X,d_2). The purpose of this investigation is to show that different metric spaces can be essentially the same if we limit ourselves to questions of continuity of functions and convergence of sequences.

Given, then, that (X,d_1) and (X,d_2) are different metric spaces, are there any ways that they can be the same as far as continuity and convergence are concerned? One reasonable approach is to ask whether or not one, or both, of the following statements is true:

(a) Let F be a mapping of the set X into an arbitrary metric space (Y,d_3). The mapping $F: (X,d_1) \rightarrow (Y,d_3)$ is continuous if and only if the mapping $F: (X,d_2) \rightarrow (Y,d_3)$ is continuous. See Figure 3.9.1. (In other words, the class of all mappings defined on X that are continuous with respect to d_1 is the same as the class of all mappings that are continuous with respect to d_2.)

(b) A sequence $\{x_n\}$ converges to a point x_0 in (X,d_1) if and only if $\{x_n\}$ converges to x_0 in (X,d_2). (In other words, the class of sequences that converge with respect to d_1 is the same as the class convergent with respect to d_2 and the limit is independent of which metric is considered.)

It is easy to exhibit metric spaces (X,d_1) and (X,d_2) for which statements (a) and (b) are true. For example, if for each x the local neighborhood system generated by d_1 is exactly the same as the local neighborhood system generated by d_2, then (a) and (b) are true. This occurs if $d_1 = \alpha d_2$, where α is any positive constant. Again, the fact that the radius of a local neighborhood changes from d_1 to d_2 has no bearing on the issue.

A sufficient condition, then, for statements (a) and (b) to be true is that the local neighborhood systems be the same. However, this is not a necessary condition. In most interesting situations where (a) and (b) are true, the classes of local neighborhoods are not the same. For example, in the plane $X = R^2$, the metrics

$$d_1(x,y) = \max\{|x_1 - y_1|, |x_2 - y_2|\}$$

and

$$d_2(x,y) = [|x_1 - y_1|^2 + |x_2 - y_2|^2]^{1/2}$$

generate different local neighborhood systems, yet statements (a) and (b) are true for the metric spaces (R^2,d_1) and (R^2,d_2). Clearly we are not yet at the heart of the matter.

Let us give a name to the situation where statements (a) and (b) are true.

3.9.1 DEFINITION.   Let (X, d_1) and (X, d_2) be metric spaces with the same underlying set X. The metrics d_1 and d_2 are said to be *equivalent* if both statement (a) and statement (b) are true.

In the remainder of this section we shall show that there is a simple and elegant way to state necessary and sufficient conditions for the equivalence of two metrics.

We start with the identity mapping I of the set X onto itself. Considering the metric spaces (X,d_1) and (X,d_2), I can be viewed as a mapping of (X,d_1) onto (X,d_2). Since I is one-to-one and maps X onto X, it is invertible, and I^{-1} maps (X,d_2) onto (X,d_1). That is,

$$I: \quad (X,d_1) \to (X,d_2)$$
$$I^{-1}: \quad (X,d_2) \to (X,d_1).$$

Given the metric space structure, it makes sense to ask whether or not I or I^{-1} is continuous. Of course, it can easily happen that one or both are not continuous. The next theorem is one characterization of the equivalence of d_1 and d_2. It is not, however, the final word.

3.9.2 THEOREM.   *Let (X,d_1) and (X,d_2) be two metric spaces with the same underlying set. The following propositions are equivalent:*

(1) *The mappings $I: (X,d_1) \to (X,d_2)$ and $I^{-1}: (X,d_2) \to (X,d_1)$ are continuous.*

(2) *Statement (a) is true.*

(3) *Statement (b) is true.*

(4) *The metrics d_1 and d_2 are equivalent.*

Proof: Since (4) is equivalent to (2) and (3) taken together, it suffices to show that statements (1), (2), and (3) are equivalent.

(1) $\Rightarrow$ (2). Assume that (1) holds and let $F = F_1: (X,d_1) \to (Y,d_3)$ be continuous. We want to show that $F = F_2: (X,d_2) \to (Y,d_3)$ is continuous, see Figure 3.9.1. Let N be any local neighborhood of $F_1(x_0) = F_2(x_0)$ in (Y,d_3). Since F_1 is continuous, $F_1^{-1}(N)$ contains a local neighborhood M of x_0 in (X,d_1). Since I^{-1} is continuous,

$$(I^{-1})^{-1}(M) = M \subset F_2^{-1}(N)$$

contains a local neighborhood L of x_0 in (X,d_1). It follows from Theorem 3.8.4 that F_2 is continuous at x_0. Since x_0 is arbitrary, F_2 is continuous. We have shown that F_2 is continuous whenever F_1 is continuous. By using the continuity of I one can show, in the same way, that F_1 is continuous whenever F_2 is continuous.

(2) $\Rightarrow$ (1). This is trivial. Simply apply (2) twice. First to the case $(Y,d_3) = (X,d_2)$ and $F = I$, and then to the case $(Y,d_3) = (X,d_1)$ and $F = I^{-1}$.

(1) $\Leftrightarrow$ (3). Theorem 3.7.2 asserts that a function is continuous if and only if it preserves convergent sequences. Now observe that (3) says that the mappings I and I^{-1} preserve convergent sequences.   ∎

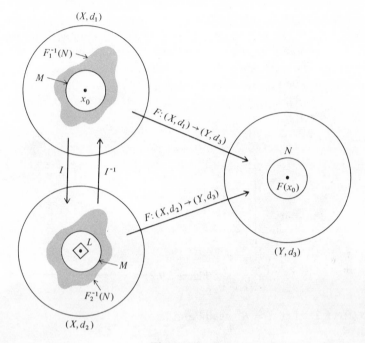

Figure 3.9.1.

We see, then, that continuity and convergence are preserved in the sense of statements (a) and (b) if and only if I and I^{-1} are continuous. It now behooves us to ask: When does this occur? The following lemma gives one answer in terms of the local neighborhood system.

3.9.3 LEMMA.   *The identity mappings $I: (X,d_1) \rightarrow (X,d_2)$ and*

$$I^{-1}: (X,d_2) \rightarrow (X,d_1)$$

are continuous at x if and only if

(1) *each local neighborhood of x in (X,d_1) contains a local neighborhood of x in (X,d_2), and*

(2) *each local neighborhood of x in (X,d_2) contains a local neighborhood of x in (X,d_1).*

The proof of this lemma involves merely noting that it is a restatement of the continuity of I and I^{-1} in the style of Theorem 3.8.4.

EXAMPLE 1.   Suppose $X = R^2$, the plane and

$$d_\infty(x,y) = \max\{ |x_1 - y_1|, |x_2 - y_2| \}$$
$$d_2(x,y) = [|x_1 - y_1|^2 + |x_2 - y_2|^2]^{1/2}.$$

Then local neighborhoods in (X, d_∞) "look like squares" and in (X, d_2) they "look like circles." Let x_0 be an arbitrary element of X. Obviously, each square centered at x_0 contains a circle centered at x_0 and vice versa. See Figure 3.9.2. Obviously, then, the metrics d_∞ and d_2 are equivalent. An analytic proof of this geometric fact will be given in Section 11. ∎

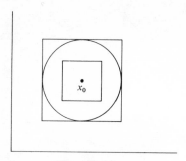

Figure 3.9.2.

EXAMPLE 2.   Let $X = R^2$ again, and let

$$d_0(x,y) = \begin{cases} 0, & \text{if } x = y \\ 1, & \text{if } x \neq y \end{cases}$$

$$d_2(x,y) = [\,|x_1 - y_1|^2 + |x_2 - y_2|^2]^{1/2}.$$

The local neighborhood system of each point x in (R^2, d_0) contains exactly two sets: the point set $\{x\}$ and the space R^2 itself. (Why?) Thus each local neighborhood of a point x in (X, d_2) contains a local neighborhood of x in (X, d_1), namely, $\{x\}$. However, the local neighborhood $\{x\}$ in (X, d_1) does not contain a local neighborhood of x in (X, d_2). (The reader should show that $I: (X, d_0) \to (X, d_2)$ is continuous but $I^{-1}: (X, d_2) \to (X, d_0)$ is not.) ∎

Lemma 3.8.3 says that each open local neighborhood in a metric space contains a local neighborhood of each of its points. Obviously there are sets other than open local neighborhoods that possess this property; for example, unions of open local neighborhoods. Heretofore we have used the term *open* in a relatively vague manner. Let us now give it a precise meaning.

3.9.4 DEFINITION.   A set A in a metric space (X, d) is said to be *open* if A contains a local neighborhood of each one of its points, that is, if for every x in A there is a local neighborhood N of x with $N \subset A$.

Note that the empty set $\varnothing$ and X itself are always open sets. Lemma 3.8.3 shows us that the open local neighborhoods are also open sets in the technical sense just defined.

And now for the most important definition of this section.

3.9.5 DEFINITION.   The class of all open sets in (X,d) is referred to as the *topology (generated by the metric d)* and it is denoted by $\mathcal{T}$.

Now let d_1 and d_2 be two metrics on a set X. Let $\mathcal{T}_1$ and $\mathcal{T}_2$ be the topologies generated by d_1 and d_2, respectively. The following theorem, which is the key result of this section, gives the promised characterization of the equivalence of two metrics.

3.9.6 THEOREM.   *Let (X,d_1) and (X,d_2) be two metric spaces with the same underlying set X. Then d_1 and d_2 are equivalent if and only if*

$$\mathcal{T}_1 = \mathcal{T}_2.$$

In other words, the metrics d_1 and d_2 are equivalent if and only if they generate the same class of open sets.

Proof: First assume that $\mathcal{T}_1 = \mathcal{T}_2$. We will show that $I: (X,d_1) \rightarrow (X,d_2)$ is continuous. A simple modification of the argument can be used to show the continuity of $I^{-1}: (X,d_2) \rightarrow (X,d_1)$.

Let x_0 be any point in (X,d_1), and let N_2 be any local neighborhood of x_0 with respect to the metric d_2. N_2 contains an open local neighborhood M_2 of x_0 with respect to d_2. Since $\mathcal{T}_1 = \mathcal{T}_2$, M_2 is also an open set with respect to d_1. Therefore $M_2 = I^{-1}(M_2)$ contains a local neighborhood of x_0 in (X,d_1). Thus, the inverse image of each local neighborhood of $I(x_0)$ contains a local neighborhood of x_0. Hence I is continuous at x_0 and therefore everywhere.

Let us now assume that I and I^{-1} are continuous. Let A be an open set with respect to d_2. Then for each x in A there is a local neighborhood N_2 of x in (X,d_2) with $N_2 \subset A$. Since I is continuous, the inverse image $N_2 = I^{-1}(N_2)$ contains a local neighborhood N_1 of x in (X,d_1). That is, $x \in N_1 \subset N_2 \subset A$. We have thus shown that every open set in $\mathcal{T}_2$ is also open in $\mathcal{T}_1$, that is, $\mathcal{T}_2 \subset \mathcal{T}_1$. By repeating this argument and using the continuity of I^{-1} we conclude that $\mathcal{T}_1 \subset \mathcal{T}_2$. Hence $\mathcal{T}_1 = \mathcal{T}_2$. ∎

We see, then, that open sets can be used to characterize the equivalence of metrics. It should not be surprising that this concept can also be used to characterize continuous functions and convergent sequences. As a matter of fact, these characterizations are so elegant that many authors use them as definitions.

3.9.7 THEOREM.   *A mapping $F: (X,d_1) \rightarrow (Y,d_2)$ is continuous if and only if the inverse image of each open set in (Y,d_2) is an open set in (X,d_1).*

Proof: First assume that F is continuous and let A be an open set in (Y,d_2). Let $x \in F^{-1}(A)$ be given and let $y = F(x)$. Then $y \in A$ and since A is open there is a local neighborhood M of y with $M \subset A$. By Theorem 3.8.4, $F^{-1}(M)$ contains a local neighborhood N of x. We then have $N \subset F^{-1}(M) \subset F^{-1}(A)$. Hence $F^{-1}(A)$ is open.

Now let us go the other way. Assume that for each open set A in (Y,d_2) the set $F^{-1}(A)$ is open in (X,d_1). Let x be an arbitrary point in (X,d_1), and let M be any open local neighborhood of $F(x)$. Then $F^{-1}(M)$ is an open set in (X,d_1) and $x \in F^{-1}(M)$. Hence there is a local neighborhood N of x with $N \subset F^{-1}(M)$. By Theorem 3.8.4, this shows that F is continuous, which completes the proof of the theorem. ▮

Using the above result, it is almost a triviality to prove that the composition of two continuous functions is continuous. We leave the proof as an exercise.

3.9.8 THEOREM.   *Let $f\colon (X,d_1) \to (Y,d_2)$ and $g\colon (Y,d_2) \to (Z,d_3)$ be continuous. Then the composition $h = gf$ is continuous.*

Before showing that we can also characterize convergence in terms of open sets, let us digress and consider the open mapping concept. Theorem 3.9.7 says that F is continuous if and only if the inverse image of each open set is an open set. A standard mistake is to turn this theorem around and say the wrong thing. It is *not true* that continuous functions necessarily map open sets onto open sets, that is, if A is open and F continuous this does not imply that $F(A)$ is open. For example, the constant mapping $F\colon X \to Y$, where $F(x) = y_0$ maps every nonempty subset of X into $\{y_0\}$. Very often the subset $\{y_0\}$ is not open.

However, functions that do map open sets onto open sets are important, so we introduce the following definition.

3.9.9 DEFINITION.   A mapping $F\colon (X,d_1) \to (Y,d_2)$ is said to be *an open mapping* if $F(A)$ is an open set in (Y,d_2) whenever A is an open set in (X,d_1).

We shall discuss this further in the exercises. Let us now look at the question of the convergence of sequences.

3.9.10 THEOREM.   *Let $\{x_n\}$ be a sequence in a metric (X,d). The sequence $\{x_n\}$ converges to a point x_0 in (X,d) if and only if the sequence is eventually in every open set containing x_0.*

Proof: First assume that $\lim x_n = x_0$. Then for every open set A containing x_0, there exists a local neighborhood N of x_0 contained in A. By Theorem 3.8.7 $\{x_n\}$ is eventually in N, hence $\{x_n\}$ is eventually in A.

On the other hand, if $\{x_n\}$ is eventually in each open set containing x_0, it is eventually in each open local neighborhood of x_0, and consequently $\{x_n\}$ is eventually in each local neighborhood of x_0. Hence $\lim x_n = x_0$, by Theorem 3.8.7. ▮

Needless to say, the conclusion of Theorem 3.9.10 can be and is often used as an alternative definition for sequential convergence in metric spaces.

What happens if the topologies $\mathcal{T}_1$ and $\mathcal{T}_2$ generated by metrics d_1 and d_2, respectively, are not the same? Obviously, the metrics d_1 and d_2 are not equivalent. Is there anything that we can say? In many situations there is. Suppose that $\mathcal{T}_1 \subset \mathcal{T}_2$, that is, each subset of X that is open with respect to $\mathcal{T}_1$ is also open with respect to $\mathcal{T}_2$. In this situation we say that $\mathcal{T}_2$ is *stronger than* (*finer than*) $\mathcal{T}_1$; or that $\mathcal{T}_1$ is *weaker than* (*coarser than*) $\mathcal{T}_2$. Needless to say, if $\mathcal{T}_1$ is both stronger and weaker than $\mathcal{T}_2$, then $\mathcal{T}_1 = \mathcal{T}_2$.

EXAMPLE 3.   Let X be a nonempty set and let

$$d(x,y) = \begin{cases} 0, & \text{if } x = y \\ 1, & \text{if } x \neq y. \end{cases}$$

Since the open ball $B_\alpha(x)$, $0 < \alpha < 1$, is simply the point set $\{x\}$ for each point x in X, every subset of X is an open set. Thus, the topology $\mathcal{T}$ generated by d is the class of all subsets of X. It follows that any function mapping (X,d) into some metric space is continuous. Moreover, $\mathcal{T}$ is obviously stronger than any other topology on X.

We mention without proof that at the other extreme is the topology generated by the pseudometric $\rho(x,y) \equiv 0$. It is made up of exactly two sets: X and $\varnothing$, the empty set. This is the weakest possible topology. (See Exercises 4 and 13.) ∎

The importance of $\mathcal{T}_1 \subset \mathcal{T}_2$ is as follows:

(a) Continuity with respect to $\mathcal{T}_1$ implies continuity with respect to $\mathcal{T}_2$. (Exercise 14.)

(b) Convergence with respect to $\mathcal{T}_2$ implies convergence with respect to $\mathcal{T}_1$. (Exercise 15.)

It can happen that neither $\mathcal{T}_1 \supset \mathcal{T}_2$ nor $\mathcal{T}_2 \supset \mathcal{T}_1$. In that case, we say that the topologies are *incommensurable*. Otherwise the topologies are *commensurable*.

EXERCISES

1. Let (A,d) be a subspace of a metric space (X,d). How is the topology of (A,d) related to the topology of (X,d)?

2. Let (X_1,d_1) and (X_2,d_2) be metric spaces and let (X,d) be the product of these two metric spaces. How is the topology of (X,d) related to the topologies of (X_1,d_1) and (X_2,d_2)?

3. Suppose that a precision cutting tool is to cut a piece in a form which can be represented by a curve x_0 in $C^1[0,T]$. The realized form of a given piece is represented by the curve x which is also in $C^1[0,T]$. The problem is to place a

metric on $C^1[0,T]$ which meaningfully characterizes the way in which x differs from x_0. Let the possible metrics be

$$d_\infty(x,y) = \text{sup-metric}$$

$$d_2(x,y) = \left\{\int_0^T [x(t) - y(t)]^2 \, dt\right\}^{1/2}$$

$$d_3(x,y) = \left\{\int_0^T |x(t) - y(t)|^3 \, dt\right\}^{1/3}$$

$$d(x,y) = d_\infty(x,y) + d_\infty(\dot{x},\dot{y}), \text{ where } \dot{x} = \frac{dx}{dt}.$$

(a) Is it true that as we make the error measured in terms of one of the metrics smaller and smaller that the error measured in terms of the others must become smaller and smaller?

(b) Let the topology generated on $C^1[0,T]$ by the metrics be denoted $\mathcal{T}_\infty$, $\mathcal{T}_2, \mathcal{T}_3, \mathcal{T}_0$. Which topologies are commensurable? If two topologies are commensurable, which is the stronger one? [*Hint*: See Exercise 8, Section 8.]

4. Carry out a development similar to the one of this section for the pseudo-metric spaces.

5. An important problem in modern control theory is the selection of an optimum input $u(t)$ from a set Ω, of allowable inputs. The set Ω is usually determined by a constraint on the inputs. For example, $u(t)$ may correspond to thrust, or force, and the amount of available thrust or force may be limited, that is, $|u(t)| \le M$. Then again u^2 may correspond to instantaneous power and total available energy may be limited, that is,

$$\int_0^T |u(t)|^2 \, dt \le N.$$

Or $u(t)$ may correspond to fuel rate and the total amount of fuel available may be limited, that is,

$$\int_0^T |u(t)| \, dt \le F.$$

It turns out that problems in this spirit can often be formulated within a metric space framework. It then happens that it is important to know whether or not the set of allowable inputs, Ω, is an open set. In what follows it is assumed that Ω is a subset of $X = C[0,T]$.

(a) Let X have the sup-metric d_∞. Which of the following are open sets?

    (1) $\Omega = \{u \in X : |u(t)| \le M\}$.
    (2) $\Omega = \{u \in X : |u(t)| < M\}$.
    (3) $\Omega = \{u \in X : \int_0^T |u(t)|^2 \, dt \le N\}$.
    (4) $\Omega = \{u \in X : \int_0^T |u(t)|^2 \, dt < N\}$.
    (5) $\Omega = \{u \in X : \int_0^T |u(t)| \, dt \le F\}$.
    (6) $\Omega = \{u \in X : \int_0^T |u(t)| \, dt < F\}$.

(b) Now let X have the metric d_2, where

$$d_2(x,y)^2 = \int_0^T |x(t) - y(t)|^2 \, dt.$$

Which of the Ω's from (a) is open?

(c) Now let X have the metric d_1, where

$$d_1(x, y) = \int_0^T |y(t) - x(t)| \, dt.$$

Which of the Ω's from (a) is open?

[*Remark*: Amusingly enough, "less than" or "less than or equal to" in the characterization of the above sets often lead to fundamentally different situations in optimum control theory. In other words, this is not a mere splitting of hairs, Neustadt [1].]

6. Show that a nonempty set A in a metric space (X,d) is an open set if and only if it is a union of open local neighborhoods.

7. Let $\mathscr{T}_1$ and $\mathscr{T}_2$ be the two topologies generated by metrics d_1 and d_2 on a set X. Let $I: (X,d_1) \to (X,d_2)$ and $I^{-1}: (X,d_2) \to (X,d_1)$ denote the identity maps.
   (a) Show that $\mathscr{T}_1 \subset \mathscr{T}_2$ if and only if I^{-1} is continuous.
   (b) Show that $\mathscr{T}_1 \subset \mathscr{T}_2$ if and only if every sequence that is convergent in (X,d_2) is also convergent in (X,d_1).

8. Prove Theorem 3.9.8.

9. Let d_2 be the Euclidean metric on the plane R^2, that is,

$$d_2(x,y) = \{ |x_1 - y_1|^2 + |x_2 - y_2|^2 \}^{1/2}.$$

   (a) Show that the mapping $(x_1,x_2) \to (y_1,y_2)$ given by

$$\begin{pmatrix} y_1 \\ y_2 \end{pmatrix} = \begin{pmatrix} 1 & 0 \\ 0 & 0 \end{pmatrix} \begin{pmatrix} x_1 \\ x_2 \end{pmatrix}$$

   is not an open mapping of (R^2,d_2) into (R^2,d_2).
   (b) Show that the mapping $(x_1,x_2) \to y_1$ given by $y_1 = x_1$ is an open mapping of (R^2,d_2) onto the real line R.

10. Let $X = C[0,1]$ and let $\mathscr{T}_p$, $1 \leq p \leq \infty$, be the topologies generated by the metrics

$$d_p(x,y) = \left(\int_0^1 |x(t) - y(t)|^p dt \right)^{1/p}, \qquad 1 \leq p < \infty$$

$$d_\infty(x,y) = \text{sup-metric}.$$

    Discuss the commensurability of these topologies. What happens if X is replaced by the Lebesgue space $L_\infty[0,1]$ or $L_p[0,1]$?

11. Let (X,d) be a metric space. Which of the following metrics are equivalent to d? (Prove your assertions.)

(a) $d_1(x,y) = \dfrac{d(x,y)}{1 + d(x,y)}$.

(b) $d_2(x,y) = \min(1, d(x,y))$.

(c) $d_3(x,y) = \sup_{t \in X} |d(x,t) - d(y,t)|$.

12. Let A be a finite set in a metric space (X,d). Show that the complement A' is open.

13. Let X be a nonempty set and let ρ be the pseudometric $\rho(x,y) \equiv 0$. What is the topology generated by ρ? Compare this with Example 3.

14. Let $\mathscr{T}_1$ and $\mathscr{T}_2$ be the two topologies generated by metrics d_1 and d_2 on a set X, and let (Y,d_3) a metric space. Denote the family of all continuous mappings of (X,d_i), $i = 1, 2$, into (Y,d_3) by C_i. Show that $C_1 \subset C_2$ if and only if $\mathscr{T}_1 \subset \mathscr{T}_2$.

15. (Continuation of Exercise 14.) Let CS_i, $i = 1, 2$, denote the family of all convergent sequences in (X,d_i). Show that $CS_2 \subset CS_1$ if and only if $\mathscr{T}_1 \subset \mathscr{T}_2$.

10. MORE ON OPEN SETS

In this section we shall investigate the general question of equivalence of metric spaces. For this study we shall not assume that the metric spaces have the same underlying set, as we did in the last section. Equivalence will be defined, as was done above, in terms of the concepts of continuity and convergence. We shall see that the topological structure of a metric space again plays a key role.

Let (X,d_1) and (Y,d_2) be metric spaces, and assume that the points in (X,d_1) can be placed into a one-to-one correspondence with the points in (Y,d_2). That is, assume there is an invertible mapping G of X onto Y.

If f maps (X,d_1) into a metric space (Z,d_3), then fG^{-1} maps (Y,d_2) into (Z,d_3). Conversely, if h maps (Y,d_2) into (Z,d_3), then hG maps (X,d_1) into (Z,d_3). (See Figure 3.10.1.) In fact, G puts the functions defined on (X,d_1) into one-to-one correspondence with the functions defined on (Y,d_2).

Similarly, if $\{x_n\}$ is a sequence in (X,d_1), then $\{y_n\} = \{G(x_n)\}$ is a sequence in (Y,d_2). Conversely, if $\{y_n\}$ is a sequence in (Y,d_2), then $\{x_n\} = \{G^{-1}(y_n)\}$ is a sequence in (X,d_1). Moreover, G puts the sequences in (X,d_1) into one-to-one correspondence with the sequences in (Y,d_2).

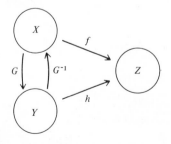

Figure 3.10.1.

The following two statements are the natural extensions of statements (a) and (b) of the foregoing section.

(a') *Let f be a mapping of* (X,d_1) *into an arbitrary metric space* (Z,d_3). *The mapping* $f: (X,d_1) \rightarrow (Z,d_3)$ *is continuous if and only if the mapping*

$$h = fG^{-1}: (Y,d_2) \rightarrow (Z,d_3)$$

is continuous.

(b') *A sequence* $\{x_n\}$ *in* (X,d_1) *converges to a point* x_0 *if and only if the sequence* $\{G(x_n)\}$ *in* (Y,d_2) *converges to* $G(x_0)$.

The mapping G is taking the role of the identity I in the preceding section. The reader will recall the importance of the continuity of I and I^{-1}. This fact motivates the following definition.

3.10.1 DEFINITION. A mapping $G: (X,d_1) \rightarrow (Y,d_2)$ is said to be a *homeomorphism* if (i) G is invertible and (ii) both G and G^{-1} are continuous. (Note that G is a homeomorphism if and only if G^{-1} is a homeomorphism.)

3.10.2 THEOREM. *Let* (X,d_1) *and* (Y,d_2) *be metric spaces, and let G be an invertible mapping of* (X,d_1) *onto* (Y,d_2). *Then the following statements are equivalent*:

(1) *The mapping G is a homeomorphism.*

(2) *Statement* (a') *is true.*

(3) *Statement* (b') *is true.*

The proof of this theorem is left to the reader.

What does this theorem say? The mapping G puts the points of (X,d_1) into one-to-one correspondence with the points in (Y,d_2). Moreover, and this is the important point, since G is a homeomorphism, *it also puts the open sets of* (X,d_1) *into one-to-one correspondence with the open sets of* (Y,d_2). If we view the mapping G as a renaming of points, we see that the metric spaces (X_1,d_1) and (Y,d_2) differ in the names given to their points but not in topological structure. Obviously, the "names given to points" is unimportant as far as continuity and convergence are concerned.

The next definition is the key concept introduced in this section.

3.10.3 DEFINITION. Two metric spaces (X,d_1) and (Y,d_2) are said to be *homeomorphic* (*to one another*) if there exists a homeomorphism mapping one of them onto the other.

Note that homeomorphic spaces (X,d_1) and (Y,d_2) can easily have many homeomorphisms mapping (X,d_1) onto (Y,d_2). Definition 3.10.3 requires only that there be at least one. A metric space (X,d) is always homeomorphic to itself, for the identity mapping I is always a homeomorphism. However, there are often other homeomorphisms of (X,d) onto itself.

There is, of course, a close connection between the concepts of homeomorphic metric spaces and equivalent metrics (Definition 3.9.1). In particular, two metrics d_1 and d_2 on a set X are equivalent if and only if I, the identity mapping, is a homeomorphism. Thus, d_1 and d_2 are equivalent if and only if the metric spaces (X,d_1) and (X,d_2) are homeomorphic in a special way, the special way being that $I: (X,d_1) \rightarrow (X,d_2)$ is a homeomorphism. Note that this does not rule out the possibility that (X,d_1) and (X,d_2) may be homeomorphic while I is not a homeomorphism.

If two metric spaces are homeomorphic to one another, it is clear that any property that can be characterized in terms of open set structure only is possessed by both or neither of them. We will see subsequently that separability and compactness are examples of such properties.

3.10.4 DEFINITION.   A property P is said to be a *topological property* (*topological invariant*) if whenever a metric space (X,d) possesses P every metric space homeomorphic to (X,d) possesses property P.

It should be noted that there is no requirement that distance be preserved by a homeomorphism. It can easily happen that $d_1(x,y) \neq d_2(Gx,Gy)$, where G is a homeomorphism. Since we are often interested in homeomorphisms which do preserve distance, we introduce the following concept:

3.10.5 DEFINITION.   A mapping G of (X,d_1) onto (Y,d_2) is said to be an *isometry* if $d_1(x,y) = d_2(Gx,Gy)$ for every pair x and y in (X,d_1). In this situation, we shall say that (X,d_1) and (Y,d_2) are *isometric* (to one another).

We leave it to the reader to show that an isometry is a homeomorphism.

If two metric spaces are isometric, they can be viewed as being essentially the same *metric* space. Again, the "names given to points" are different, but these names are unimportant for questions of distance as well as questions of continuity and convergence.

Let us end this section by delving a bit into the structure of topologies themselves. Suppose a set X and a class of subsets $\mathscr{A}$ of X are given. Is there a metric on X such that $\mathscr{A}$ is the topology generated by the metric? One can give an answer to this question, but the answer is beyond the scope of this book.[5] However, the following theorem presents an important part of the answer.

3.10.6 THEOREM.   *Let (X,d) be a metric space and let $\mathscr{T}$ be the topology generated by d. Then the following statements are valid:*

[5] An answer can be found in Kelley [1; pp. 124–130].

(1) $\emptyset$, the empty set, and X are in $\mathcal{T}$.

(2) If A_α is any collection of open sets, then $\bigcup_\alpha A_\alpha$ is an open set.

(3) If $A_1, \ldots, A_n$ is a finite collection of open sets, then $\bigcap_{i=1}^n A_i$ is an open set.

The proof is left as an exercise.

We see, then, that a topology on a metric space is closed under arbitrary unions and finite intersections. This may appear to be strangely asymmetrical. However, a simple example will illustrate what can happen. Consider the class of open local neighborhoods $B_{1/n}(x), n = 1, 2, \ldots$. These open local neighborhoods are open sets, and the intersection $\bigcap_n B_{1/n}(x)$ is simply the point set $\{x\}$, which is generally not an open set. (Give an example of a metric space where $\{x\}$ is not an open set.)

The converse of Theorem 3.10.6 is not true. That is, if we are given a collection of subsets $\mathcal{A}$ of X that is closed under arbitrary unions and finite intersections and that contains $\emptyset$ and X, it is not necessarily true that $\mathcal{A}$ is the topology generated by some metric on X. For a detailed discussion of a converse for Theorem 3.10.6 we again refer the reader to Kelley [1; pp. 124–130].

EXERCISES

1. Carry out a development for pseudometric spaces similar to the one carried out in this section for metric spaces.

2. Prove Theorem 3.10.6.

3. Give an example of two metric spaces with the same underlying set, say (X, d_1) and (X, d_2), that are homeomorphic to one another but for which the metrics d_1 and d_2 are not equivalent. [*Hint*: One approach is by way of product spaces.]

4. Sometimes a metric space is made up of "more than one piece." More precisely, a metric space (X, d), for that matter any topological space, is said to be *disconnected* if it is the union of two open, nonempty, disjoint subsets; that is, if there exist open subsets A and B of (X, d) such that $A, B \neq \emptyset$, $A \cap B = \emptyset$, and $X = A \cup B$. A metric space (X, d) is said to be *connected* if it is not disconnected.
   (a) Give an example of a disconnected metric space.
   (b) Give an example of a connected metric space.

5. Show that connectedness is a topological property.

6. Let $f: X \to Y$ be continuous, where (X, d_1) and (Y, d_2) are metric spaces. Let A be a connected set in X, that is, the metric space (A, d_1) is connected. Show that $f(A)$ is connected in Y, that is, $(f(A), d_2)$ is a connected space.

7. A set B in a metric space (X, d) is said to be *contractable* (*to a point* x_0) if there is a continuous mapping

$$F(x,t): B \times I \to B,$$

with the following properties:
(a) $F(x,0) = x$, for all x in B.
(b) $F(x,1) = x_0$, for all x in B.
(Here $I = \{t: 0 \leq t \leq 1\}$, and $B \times I$ has the metric

$$d'((x_1,t_1), (x_2,t_2)) = d(x_1,x_2) + |t_1 - t_2|.)$$

Show that contractibility is a topological property.

8. Use Exercise 7 to show that a circle in the plane is not homeomorphic with a line segment.

9. Suppose that F is a one-to-one mapping of a set X onto a metric space (Y,d_2).
   (a) Does $d_1(x_1,x_2) = d_2[F(x_1), F(x_2)]$ define a metric on X?
   (b) If so, is F a homeomorphism? Anything stronger?

10. Let (X,d) be a metric space, where X is nonempty, and let $Y = BC(X,R)$ denote the collection of all bounded, continuous real-valued functions defined on X. Assume that Y has the metric

$$\sigma(f,g) = \sup\{|f(x) - g(x)|: x \in X\},$$

see Exercise 6, Section 5. Let x_0 be a fixed point in X and define

$$f_y(x) = d(x,y) - d(x,x_0).$$

Show that the mapping $G: y \to f_y$ is an isometry from X onto a subspace of Y.

11. (Continuation of Exercise 10.) Assume that $X = R$ with the usual metric. Sketch a few of the functions f_y.

12. (Continuation of Exercise 9, Section 5.) Show that the mapping $x \to \delta_x$ of X into $LC_{x_0}{}^*$ is an isometry. [Hint: Compute $|\delta_x(f) - \delta_y(f)|$ and $\|f\|$ where $f(z) = \frac{1}{2}[d(x,z) - d(y,z)] + \frac{1}{2}[d(y,x_0) - d(x, x_0)]$.]

13. Let $f: (X_1,d_1) \to (X_2,d_2)$ be a homeomorphism of X_1 onto X_2, and let $g: (X_2,d_2) \to (X_3,d_3)$. Show that g is continuous if and only if

$$gf: (X_1,d_1) \to (X_3,d_3)$$

is continuous. What happens if f is not a homeomorphism?

14. Let $X = C(a,b)$ be the space of continuous real-valued functions defined on the open bounded interval (a,b). For $x, y \in X$ let

$$D(x,y) = \{t \in (a,b): x(t) \neq y(t)\}.$$

(a) Show that $D(x,y)$ is the union of disjoint open intervals.
(b) Let $d(x,y)$ denote the sum of the lengths of these intervals. Show that d is a metric on X.
(c) Find a sequence of bounded functions in X that converge to $x(t) = (t - a)^{-1}$.

11. EXAMPLES OF HOMEOMORPHIC METRIC SPACES

In this section we present several examples to illustrate the concepts of homeomorphism and isometry.

EXAMPLE 1.   On the set $X = R^2$, define the metrics

$$d_1(x,y) = |x_1 - y_1| + |x_2 - y_2|$$
$$d_2(x,y) = (|x_1 - y_1|^2 + |x_2 - y_2|^2)^{1/2}$$
$$d_\infty(x,y) = \max(|x_1 - y_1|, |x_2 - y_2|).$$

We will show that the three metrics d_1, d_2, and d_∞ are equivalent on R^2.

3.11.1 LEMMA.   *With d_1, d_2, and d_∞ defined as above one has*

(a) $d_\infty(x,y) \le d_2(x,y) \le d_1(x,y)$,

(b) $d_2(x,y) \le \sqrt{2} d_\infty(x,y)$,

(c) $d_1(x,y) \le \sqrt{2} d_2(x,y) \le 2d_\infty(x,y)$,

for all x, y in R^2.

Proof: Note that $d_2(x,y) \ge |x_1 - y_1|$ and similarly $d_2(x,y) \ge |x_2 - y_2|$. Consequently $d_2(x,y)$ is greater than the largest of these two numbers, that is, $d_2(x,y) \ge d_\infty(x,y)$. On the other hand, $d_2(x,y)^2 = |x_1 - y_1|^2 + |x_2 - y_2|^2 \le d_1(x,y)^2$, so $d_2(x,y) \le d_1(x,y)$. This completes the proof of (a).

To prove (b) we first note that $d_\infty(x,y)^2 = \max\{|x_1 - y_1|^2, |x_2 - y_2|^2\}$. Then $d_2(x,y)^2 \le 2\max\{|x_1 - y_1|^2, |x_2 - y_2|^2\}$ so $d_2(x,y) \le \sqrt{2}d_\infty(x,y)$.

To prove (c) we need an algebraic lemma[6] namely $2ab \le a^2 + b^2$ for all real numbers a and b. Applying it here we get

$$d_1(x,y)^2 = |x_1 - y_1|^2 + |x_2 - y_2|^2 + 2|x_1 - y_1||x_2 - y_2|$$
$$\le 2|x_1 - y_1|^2 + 2|x_2 - y_2|^2 = 2d_2(x,y)^2.$$

Hence, $d_1(x,y) \le \sqrt{2}\, d_2(x,y)$.   ∎

3.11.2 THEOREM.   *The metrics d_1, d_2, and d_∞ are equivalent on R^2. (Hence the metric spaces (R^2,d_1), (R^2,d_2), and (R^2,d_∞) are homeomorphic.)*

Proof: We shall use Theorem 3.9.2. Let $\{x_n\}$ be a sequence in R^2. By applying the last lemma to $d_1(x_0,x_n)$, $d_2(x_0,x_n)$, and $d_\infty(x_0,x_n)$, we see that

$$\lim x_n = x_0 \text{ in } (R^2,d_\infty) \Leftrightarrow \lim x_n = x_0 \text{ in } (R^2,d_1)$$
$$\Leftrightarrow \lim x_n = x_0 \text{ in } (R^2,d_2).   ∎$$

---

[6] The proof of this is simply to observe that $a^2 - 2ab + b^2 = (a - b)^2 \ge 0$.

EXAMPLE 2.   On the set $X = R^n$, define the metrics

$$d_1(x,y) = \sum_{i=1}^{n} |x_i - y_i|,$$

$$d_2(x,y) = \left\{ \sum_{i=1}^{n} |x_i - y_i|^2 \right\}^{1/2},$$

$$d_p(x,y) = \left\{ \sum_{i=1}^{n} |x_i - y_i|^p \right\}^{1/p}, \qquad 1 \le p < \infty,$$

$$d_\infty(x,y) = \max_{1 \le i \le n} |x_i - y_i|.$$

We ask the reader to modify the proof of Lemma 3.11.1 and Theorem 3.11.2 to establish the following facts. Further, see Exercise 1.

3.11.3 LEMMA.   *With d_1, d_2, and d_∞ defined on R^n as above, one has*

(a) $d_\infty(x,y) \le d_2(x,y) \le d_1(x,y)$
(b) $d_2(x,y) \le \sqrt{n}\, d_\infty(x,y)$
(c) $d_1(x,y) \le \sqrt{n}\, d_2(x,y)$

for all x, y in R^n.

3.11.4 THEOREM.   *The metrics d_1, d_2, and d_∞ are equivalent on R^n.* ∎

EXAMPLE 3.   Let R^n be given with the metric

$$d_2(x,y) = (|x_1 - y_1|^2 + \cdots + |x_n - y_n|^2)^{1/2}.$$

Let Y be the set made up of all functions y of the form

$$y(t) = a_1 \cos t + a_2 \cos 2t + \cdots + a_n \cos nt,$$

where $0 \le t \le 2\pi$ and $a_1, \ldots, a_n$ is any n-tuple of real numbers. Define a metric on the set Y by

$$\sigma_2(y_1, y_2) = \left\{ \int_0^{2\pi} |y_1(t) - y_2(t)|^2 \, dt \right\}^{1/2}.$$

Let G be the mapping of (X, d_2) onto (Y, σ_2) defined by

$$G\{x_1, \ldots, x_n\} = x_1 \cos t + \cdots + x_n \cos nt.$$

We leave it to the reader to show that G is invertible and that its inverse can be represented by

$$G^{-1}(y) = \left\{ \frac{1}{\pi} \int_0^{2\pi} y(t) \cos t \, dt, \frac{1}{\pi} \int_0^{2\pi} y(t) \cos 2t \, dt, \ldots, \frac{1}{\pi} \int_0^{2\pi} y(t) \cos nt \, dt \right\}.$$

Furthermore, we claim that G and G^{-1} are continuous. See Exercise 5. ∎

EXAMPLE 4.   Let $X = l_2(-\infty,\infty)$ and $Y = L_2[0,1]$ be given with the usual metrics. It will be shown in Example 2, Section 5.19, that these two metric spaces are homeomorphic. This is one of the most important examples of homeomorphic metric spaces. ∎

EXERCISES

1. Define $d_p(x,y)$ on R^2 by

$$d_p(x,y) = \{|x_1 - y_1|^p + |x_2 - y_2|^p\}^{1/p}, \qquad 1 \le p < \infty.$$

   Show that

$$d_\infty(x,y) \le d_p(x,y) \le d_1(x,y)$$
$$d_p(x,y) \le \sqrt{2}d_\infty(x,y)$$
$$d_1(x,y) \le \sqrt{2}d_p(x,y)$$

   for all x, y in R^2. [*Hint*: With x and y fixed, let $F(p) = d_p(x,y)$. Compute dF/dp and use Lemma 3.11.1.]

2. With d_p as given in Exercise 1 show that d_1 is equivalent to d_p on R^2.

3. Prove Lemma 3.11.3.

4. Prove Theorem 3.11.4.

5. The following questions refer to Example 3.
   (a) Show that G is a homeomorphism.
   (b) Is G the only homeomorphism between (X,d_2) and (Y,σ_2)?
   (c) Is G an isometry? If not, are there any isometries between (X,d_2) and (Y,σ_2)?
   (d) Is the mapping $H: X \to Y$ given by

$$H\{x_1,\ldots,x_n\} = x_1{}^3 \cos t + \cdots + x_n{}^3 \cos nt$$

   a homeomorphism?

6. Let X be the set of all ordered n-tuples of 1's and 0's with the metric given in Example 17, Section 3.

   Let W be a set containing n elements and let Y be the set of all subsets of W. Let A and B denote arbitrary elements in Y, and define a metric on Y by

$$d_2(A,B) = \text{number of elements of } W \text{ in the subset } A \triangle B,$$

   For example, in Figure 3.11.1, one has $d_2(A,B) = 3$.
   (a) Show that (X,d_1) and (Y,d_2) are isometric.
   (b) Which invertible mappings of (X,d_1) onto (Y,d_2) are isometries?
   (c) How many isometries are there of (X,d_1) onto (Y,d_2)? How many open sets are there in each of these topologies? If a metric space (Z,d_3) is homeomorphic to (X,d_1), how many open sets are there in the topology generated by d_3?

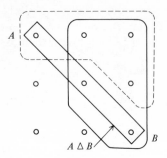

Figure 3.11.1.

7. Suppose that X and Y are finite sets containing exactly four elements, that is, $X = \{x_1, x_2, x_3, x_4\}$ and $Y = \{y_1, y_2, y_3, y_4\}$. Suppose we define metrics d_1 and d_2 on X and Y, respectively, with the aid of the two diagrams in Figure 3.11.2. In both diagrams the distance between adjacent nodes is 1. The distance between nonadjacent nodes is equal to the fewest number of branches that need be traversed in going from one node to the other. For example, $d_1(x_1, x_2) = 1$, $d_1(x_1, x_4) = 3$, $d_2(y_1, y_2) = 1$, $d(y_1, y_0) = 2$.

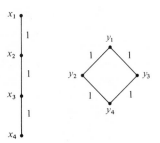

Figure 3.11.2.

   (a) Are the metric spaces (X, d_1) and (Y, d_2) homeomorphic to one another?
   (b) Are the metric spaces (X, d_1) and (Y, d_2) isometric to one another?
   (c) How many one-to-one mappings of (X, d_1) onto (Y, d_2) are there? How many, if any, of them are homeomorphisms? Isometries?
   (d) What are the topologies generated by d_1 and d_2?

8. Is it true that if two metric spaces have finite underlying sets with the same number of elements, then they are homeomorphic? Are they isometric?

9. Let (X, d_1) be the metric space

$$\{R^2, d_1(x, y) = |x_1 - y_1| + |x_2 - y_2|\}.$$

Let Y be the set of all functions y of the form $y(t) = a_1 + a_2 t$ where a_1 and a_2 are arbitrary real numbers and $0 \le t \le 1$. Define a metric on Y by

$$d_2(y_1,y_2) = \begin{cases} 0, & \text{if } y_1(t) \equiv y_2(t) \\ 1, & \text{otherwise.} \end{cases}$$

(a) What is the topology on Y generated by d_2?

(b) Which mappings of (Y,d_2) into (X,d_1) are continuous?

(c) Which mappings (X,d_1) into (Y,d_2) are continuous? [*Hint*: Consider constant mappings first.]

(d) Are any of the mappings in (c) invertible?

(e) Show that (X,d_1) and (Y,d_2) are not homeomorphic.

10. (Continuation of Exercise 3, Section 4.) Let Q^+ denote the positive rational numbers with the usual metric. Discuss the mapping $\xi: N \times N \to Q^+$ given by $\xi(n_1,n_2) = n_1 n_2^{-1}$.

11. Consider the interval $[0,1]$ with the usual metric. We say that two homeomorphisms f_0, f_1 from $[0,1]$ into $[0,1]$ are *homotopically equivalent* if there is a continuous mapping $F(t,s): [0,1] \times [0,1] \to [0,1]$ such that for each s, $F(\cdot,s)$ is a homeomorphism of $[0,1]$ into itself and $F(t,0) = f_0(t)$ and $F(t,1) = f_1(t)$. Show that every homeomorphism of $[0,1]$ into itself is homotopically equivalent to either $g(t) = t$ or $h(t) = 1 - t$. Show that g and h are not homotopically equivalent.

12. CLOSED SETS AND THE CLOSURE OPERATION

In this section we shall study the properties of closed sets, which we now define.

3.12.1 DEFINITION.   Let (X,d) be a metric space. A subset $A \subset X$ is said to be *closed* if its complement $A' = X - A$ is an open set.

Since $X' = \emptyset$ and $\emptyset' = X$ and since $\emptyset$ and X are both open sets, they are also closed sets. At first it may be surprising, but it is nevertheless true that sets can be both open and closed. (See Exercise 1.) Also note that it is possible for a set to be neither open nor closed.

EXAMPLE 1.   Let X be given by

$$X = \{x: 0 \le x \le 1 \quad \text{or} \quad 2 \le x \le 3\} = [0,1] \cup [2,3].$$

Let $d(x,y) = |x - y|$. It is easily shown that the set $A = \{x: 0 \le x \le 1\}$ is both an open and a closed subset of the metric space (X,d). Of course, A is not an open subset of (R,d), the real line with the usual metric. This latter fact shows that we have to be careful to state exactly which universe space X is being considered. ▮

In Section 11, we referred to "closed balls" and "closed local neighborhoods." Let us now prove that these "closed" sets are closed in the technical sense defined above.

3.12.2 LEMMA.   *Let (X,d) be a metric space and $B_r[x_0] = \{x \in X: d(x,x_0) \le r\}$, $S_r(x_0) = \{x \in X: d(x,x_0) = r\}$. Then $B_r[x_0]$ and $S_r(x_0)$ are closed sets for $0 \le r < \infty$.*

Proof: First consider $B_r[x_0]$. We must show that the complement $B_r[x_0]' = X - B_r[x_0]$ is open. Let $y \in B_r[x_0]'$. Then $d(y,x_0) = \alpha > r$. Now set $\beta = \alpha - r > 0$. Since

$$d(x_0,z) \ge d(y,x_0) - d(y,z) = \alpha - d(y,z),$$

we see that if $d(y,z) < \beta$, then $d(x_0,z) > \alpha - \beta = r$. In terms of local neighborhoods this means that $B_\beta(y) \subset B_r[x_0]'$, see Figure 3.12.1. Hence $B_r[x_0]'$ is open.

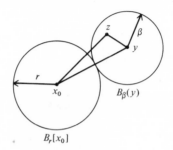

Figure 3.12.1.

Since $S_r(x_0)' = (B_r(x_0) \cup B_r[x_0]')$ is the union of two open sets, it is open and $S_r(x_0)$ is closed.   ∎

If $\mathscr{F}$ denotes the class of all closed sets in a metric space (X,d), then by using Theorem 3.10.6 and De Morgan's formulas for complementation we have the following theorem.

3.12.3 THEOREM.   *Let $\mathscr{F}$ denote the class of all closed sets in a metric space (X,d). Then*

(1) $\emptyset$, *the empty set, and X are in $\mathscr{F}$.*

(2) *If A_α is any collection of closed sets, then $\bigcap_\alpha A_\alpha$ is closed.*

(3) *If $A_1, \ldots, A_n$ is a finite collection of closed sets, then $\bigcup_i A_i$ is closed.*

Let us consider some examples.

EXAMPLE 2.   Let $X = R$ and d be the usual metric. It follows from Exercise 9.6, that each closed set in (X,d) is the complement of a union of open intervals. In this sense, then, it is "easy" to characterize the class of all closed sets in (X,d). Even so, the complement of a union of open intervals can be a rather complex object. One classic example is the Cantor set, which is constructed in Appendix D.

We invite the reader to review this construction. For now we merely make two observations:

(1) The Cantor set contains an uncountable number of points.

(2) The Cantor set does not contain a nontrivial interval.

It follows, then, that if we try to express a closed set as the union of intervals, we might need an uncountable number. This shows that closed sets can be quite pathological. ∎

EXAMPLE 3.   Let d_∞ be the sup-metric on $C[0,T]$. Then the set A of all $x \in C[0,T]$ such that $|x(t)| \leq 1$ for all t is a closed set. In fact, $A = B_1[0]$, that is, A is just the closed ball of radius 1 centered at $x(t) \equiv 0$. ∎

EXAMPLE 4.   Consider the metric space (l_2, d_2) (see Example 4, Section 3). Let A be the Hilbert cube, that is, the set of all points $x = \{x_1, x_2, \ldots\}$ in (l_2, d_2) such that $|x_n| \leq 1/n$. We leave it as an exercise to show that the Hilbert cube is closed. ∎

The next thing to do is investigate the structure of closed sets. The concept of point of adherence will be basic for this investigation.

3.12.4 DEFINITION.   Let A be a subset of a metric space (X,d). A point x in (X,d) is said to be a *point of adherence* of A if each open set of (X,d) containing x also contains a point y in A.

Notice that this is equivalent to saying that each local neighborhood N of x contains a point y in A. It should be noted that we do not ask that the point of adherence x be in A. We only ask that the local neighborhoods meet A. It should be clear that every point in a set A is a point of adherence of A.

We sometimes consider another kind of point that is close to a set A. A point x that is a point of adherence of $A - \{x\}$ is said to be a *point of accumulation* of A. It is clear that every point of accumulation is also a point of adherence.

EXAMPLE 5.   Let d be the usual metric on R and let $A = (0,1)$ and $B = \{0\}$. Clearly, each point in the interval $(0,1)$ along with 0 and 1 are points of adherence of A. In fact, these are all the points of adherence of A. Similarly $x = 0$ is the only point of adherence for B. (What about the points of accumulation?) ∎

EXAMPLE 6.   Let X be the set of all ordered n-tuples of 1's and 0's, and let

$$d(x,y) = \text{number of places where } x \text{ and } y \text{ disagree}.$$

Let A be an arbitrary set on the metric space (X,d). Since $B_{1/2}(x) = \{x\}$, every point of adherence of A lies in A. ∎

EXAMPLE 7.   Let $X = C[0,T]$ be given with the sup-metric d_∞ and let A be the set in (X,d_∞) made up of all functions x such that $x(0) = 0$ and $|x(t)| < 1$ for all t. (See Figure 3.12.2.) We note that the point $x_0(t) \equiv 1$ is not a point of adherence of A, since $d_\infty(x_0,x) = 1$ for each $x \in A$.   ∎

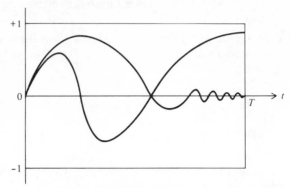

Figure 3.12.2.

EXAMPLE 8.   This example is a variation on the preceding one. Let $X = C[0,T]$ with the metric

$$d_2(x,y) = \left[\int_0^T |x(t) - y(t)|^2 \, dt\right]^{1/2}$$

and let the set A be defined as above. Now it is true that the point $x_0(t) \equiv 1$ is a point of adherence of A. (Why?)   ∎

If A is a set in a metric space (X,d), let $\bar{A}$ denote the set of all points of adherence of A. The set $\bar{A}$ is sometimes referred to as the *closure* of A. Let us see what can be said about $\bar{A}$.

3.12.5 THEOREM.   *A set A in a metric space (X,d) is closed if and only if $A = \bar{A}$.*

Proof: First assume that $A = \bar{A}$. We want to show that the complement A' is open. So let $x \in A'$. Since x is not a point of adherence of A, there is a local neighborhood N of x that does not meet A, that is, $N \cap A = \varnothing$. Hence A' is open, so A is closed.

Now assume that A is closed. Since the inclusion $A \subset \bar{A}$ is always true we want to show that $\bar{A} \subset A$. Let $x \in \bar{A}$ and assume that $x \notin A$. That is, x is in the open set A'. Hence there is a local neighborhood N of x with $N \cap A = \varnothing$. But this contradicts the fact that x is a point of adherence of A. Therefore one has $x \in A$.   ∎

3.12.6 THEOREM.   *Let A be a set in a metric space (X,d), then*

(1) $\bar{\bar{A}} = \bar{A}$, *and*
(2) $\bar{A}$ *is a closed set.*

Proof: (1) One always has the inclusion $\bar{A} \subset \bar{\bar{A}}$. Let $x \in \bar{\bar{A}}$ and let N be an arbitrary open local neighborhood of x. In order to show that $x \in \bar{A}$ we must show that there is a $z \in N \cap A$. Since x is a point of adherence of $\bar{A}$, there is a point y in $N \cap \bar{A}$. Let M be a local neighborhood of y (see Figure 3.12.3) satisfying $M \subset N$. (Here we use the fact that N is open.) Since y is a point of adherence of A there is a $z \in M \cap A \subset N \cap A$.

(2) This follows directly from (1) and Theorem 3.12.5. ∎

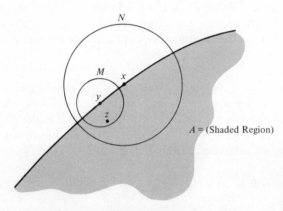

Figure 3.12.3.

The reader should be careful not to confuse the concepts of "point of adherence" and "limit of a sequence." Point of adherence is a concept attributable to *sets*. Limit is a concept attributable to *sequences*. Of course, the range of a sequence (that is, the set of all elements x_n in the sequence) is a set in (X,d), and it can have points of adherence.

Keeping the foregoing warning in mind, it is nevertheless still possible to characterize closed sets using the limit concept.

3.12.7 THEOREM (THE CLOSED SET THEOREM).   *A set A in a metric space (X,d) is a closed set if and only if every convergent sequence $\{x_n\}$ with $\{x_n\} \subset A$ has its limit in A.*

Proof: First assume that A is closed and let $\{x_n\}$ be a convergent sequence in A with $x_0 = \lim x_n$. We want to show that $x_0 \in A$. Since $x_0 = \lim x_n$, the sequence $\{x_n\}$ is eventually in each local neighborhood of x_0. But this means that x_0 is a point of adherence of A. Since A is closed, it follows from Theorem 3.12.5 that $x_0 \in A$.

Let us now show the converse. Assume that every convergent sequence in A has its limit in A. Let $x_0 \in \bar{A}$. Then there is at least one point x_n in $B_{1/n}(x_0) \cap A$ for each n. Since $x_0 = \lim x_n$, we see that $x_0 \in A$. Since A contains all of its points of adherence, it is closed. ∎

The two concepts of closed set and point of adherence lead to other concepts which are of use. We rather unimaginatively list them here and ask the reader to wade through them on the promise that they will indeed be useful.

3.12.8 DEFINITION.   A set A in a metric space (X,d) is said to be *dense* in (X,d) (or *everywhere dense*) if the closure of A is X, that is, $\bar{A} = X$.

Intuitively speaking, this says that for each point x in X, there are points in A arbitrarily close to x. Obviously, the set X is always dense in (X,d). However, there are many situations where a proper subset of a metric space is everywhere dense. For example, the rational numbers are dense in the real numbers, with the usual metric.

EXAMPLE 9.   Let d_∞ be the sup-metric on $C[0,T]$. Let $P[0,T]$ be the set of all polynomials, $p(t) = a_0 + a_1 t + \cdots + a_n t^n$, with real coefficients, defined on $[0,T]$, $n = 0, 1, 2, \ldots$. $P[0,T]$ is a subset of (X,d) and $P[0,T]$ is dense in $\{C[0,T], d\}$. The fact that this is the case is a consequence of the famous Weierstrass Approximation Theorem, see Exercises 20 and 21. ∎

3.12.9 DEFINITION.   A metric space (X,d) is said to be *separable* if it contains a dense set A that is countable.

For example, the set of real numbers with the usual metric forms a separable metric space since the set of rational numbers is dense and countable. Note that a separable metric space can easily contain more than one set which is dense and countable.

The next lemma follows from the definition of separable metric spaces.

3.12.10 LEMMA.   *A metric space (X,d) is separable if and only if there is a countable set $\{x_n\}$ with the following property: For each $\varepsilon > 0$ and each x in (X,d) there is at least one x_n with $d(x_n,x) < \varepsilon$.*

This lemma offers an interesting view of separable metric spaces. It says that an arbitrarily accurate approximation to any point in the separable metric space is contained in the countable set $\{x_n\}$.

Let us consider some examples of separable metric spaces.

EXAMPLE 10.   Let X be R^n or C^n with the metric

$$d_1(x,y) = |x_1 - y_1| + \cdots + |x_n - y_n|.$$

If we let A denote the set of all points, $x = (x_1,\ldots,x_n)$ in X with rational coordinates. A is clearly countable and dense. Hence (X,d) is separable. ∎

EXAMPLE 11.   Consider $l_2 = l_2(0,\infty)$ with the usual metric d_2. This metric space is separable. Let A be the subset of l_2 made up of all sequences $\{r_n\}$ which

(1) have rational entries only, and

(2) have only a finite number of nonzero entries.

First, let us show that A is countably infinite. Let A_k, $k = 1, 2, \ldots$, be the subset of A made up of sequences with $x_n = 0$ for $n \geq k$ and $x_{k-1} \neq 0$. Then $A = \bigcup_k A_k$. Each set A_k is countable, that is, A_k can be put into one-to-one correspondence with the positive integers. (Why?) [A_1 contains only one sequence: $(0,0,\ldots)$.] Let $a_k(j)$ denote the element of A_k corresponding to the positive integer j, that is, each $a_k(j)$ is a sequence. Form the following array:

$$a_1(1)$$

$$a_2(1), \rightarrow a_2(2), a_2(3), \rightarrow a_2(4), \ldots$$

$$a_3(1), a_3(2), a_3(3), \ldots$$

$$a_4(1), \ldots$$

This array contains every element in the set A exactly once, and using the diagonal counting pattern indicated by the arrows we can put A into one-to-one correspondence with the positive integers. Hence, A is countable.

Next let us show that $\bar{A} = l_2$. Let $x = \{x_1, x_2, \ldots\}$ be an arbitrary point in l_2. Since

$$\sum_{i=1}^{\infty} |x_i|^2 < \infty,$$

it follows that given any $\varepsilon > 0$ there exists an integer N such that

$$\sum_{i=N+1}^{\infty} |x_i|^2 < \frac{\varepsilon^2}{2}.$$

Obviously, there exists rational numbers $r_1, r_2, \ldots, r_N$ such that

$$\sum_{i=1}^{N} |x_i - r_i|^2 < \frac{\varepsilon^2}{2}.$$

Then $r = \{r_1, r_2, \ldots, r_N, 0, 0, \ldots\}$ is a point in A and $d(x,r) < \varepsilon$. By Lemma 3.12.10 we see that this space is separable. ∎

EXAMPLE 12.   The space $L_2(I)$ with the usual metric d_2 is also separable. The proof of this follows from the facts that separability is a topological concept (Exercise 15) and that $L_2(I)$ is isometric with l_2 (Section 5.19). ∎

EXERCISES

1. Show that a metric space (X,d) is disconnected if and only if it contains a set other than $\varnothing$ or X which is both open and closed. (For the definition of a disconnected metric space see Exercise 4, Section 10.)

2. What is the closure of the set A in Example 7?

3. What is the closure of the set A in Example 8?

4. Show that the space (l_p, d_p), $1 \le p < \infty$, is separable. (See Example 4, Section 3.) What happens for $p = \infty$?

5. We introduced the open mapping concept in Definition 3.9.9. There is also a closed mapping concept. A mapping F of a metric space (X, d_1) into a metric space (Y, d_2) is said to be a *closed mapping* if the image of each closed set in (X, d_1) is a closed set in (Y, d_2). Give an example of a closed mapping that is not continuous. Give an example of a continuous mapping that is not closed. Give an example of a mapping that is closed but not open. Give an example of a mapping that is open but not closed. What can be said about a one-to-one, continuous, closed mapping of (X, d_1) onto (Y, d_2)? (Unfortunately, the term closed is applied to transformations on normed linear spaces in a slightly different sense from the one used here. See Exercise 9, Section 5.6.)

6. Which of the sets in Exercise 5, Section 9 are closed?

7. Is Theorem 3.12.7 true for pseudometric spaces?

8. Show that $\text{diam}(A) = \text{diam}(\bar{A})$ for every subset A in a metric space (X, d).

9. Show that $\bar{A} = \bigcap B$ where the intersection is taken over all closed sets B with $A \subset B$. (Note that $\bar{A}$ is the smallest closed set containing A.)

10. Prove the following three statements about metric spaces:
    (a) Given any point x_0 in a metric space (X, d) and a closed set A with $x_0 \notin A$, there exist disjoint open sets 0_1 and 0_2 such that $x_0 \in 0_1$ and $A \subset 0_2$.
    (b) Given any pair of disjoint closed sets, A_1 and A_2, a metric space (X, d), there exist disjoint open sets 0_1 and 0_2 such that $A_1 \subset 0_1$ and $A_2 \subset 0_2$. (This shows that a metric space is normal, see Kelley [1; p. 112].)
    (c) Let U be an open set containing x. Show that there is an open set V containing x and such that the closure $\bar{V} \subset U$. (This shows that a metric space is regular, see Kelley [1; p. 113].)

11. Let X be a set with the metric

$$d(x,y) = \begin{cases} 0, & \text{if } x = y \\ 1, & \text{if } x \ne y. \end{cases}$$

Show that the metric space (X, d) is separable if and only if the set X is countable.

12. Show that the Hilbert cube (Example 4) is closed.

13. Let $\{x_n\}$ be a bounded sequence in (R, d) with infinitely many different values. Show the sequence converges if and only if the set $\{x \in X : x = x_n \text{ for some } n\}$ has precisely one point of accumulation. Describe $\lim x_n$.

14. Show that a set A in (X, d) is dense if and only if $U \cap A \ne \emptyset$ for every non-empty open set U.

15. Show that continuous mappings preserve separability. (Hence separability is a topological property.)

16. Show that the product of two separable metric spaces is a separable metric space. Show that any subspace of a separable metric space is separable.

17. In Example 10 it was shown that (R^n, d_1) and (C^n, d_1) are separable. What other metrics on these spaces would yield a separable metric space? (Use Exercise 15.)

18. Show that $d(x, A) = 0$ if and only if $x \in \bar{A}$.

19. Show that $\overline{B_r(x)} \subset B_r[x]$. What can one say about equality? [*Hint*: Consider $X = [0,1] \cup [2,3]$, $x = 1$, $r = 1$.]

20. (Bernstein Polynomials.) Let $f \in C[0,1]$ and define

$$B_n(x; f) = \sum_{k=0}^{n} f\left(\frac{k}{n}\right)\binom{n}{k} x^k(1-x)^{n-k},$$

where $\binom{n}{k} = \dfrac{n!}{k!(n-k)!}$ is the binomial coefficient.

(a) Show that

$$1 = \sum_{k=0}^{n}\binom{n}{k} x^k(1-x)^{n-k}.$$

(b) Show that

$$x = \sum_{k=0}^{n} \frac{k}{n}\binom{n}{k} x^k(1-x)^{n-k}$$

$$\left(1 - \frac{1}{n}\right)x^2 + \frac{1}{n}x = \sum_{k=0}^{n}\left(\frac{k}{n}\right)^2\binom{n}{k} x^k(1-x)^{n-k}$$

$$\frac{1}{n}x(1-x) = \sum_{k=0}^{n}\left(x - \frac{k}{n}\right)^2\binom{n}{k} x^k(1-x)^{n-k}$$

(c) Choose M and $\delta(\varepsilon)$ so that

$$|f(x)| \le M, \qquad 0 \le x \le 1$$
$$|f(x) - f(y)| \le \varepsilon, \quad \text{when } |x - y| \le \delta(\varepsilon).$$

(d) Let x be fixed and fix n so that

$$n \ge \max\{1/\delta(\varepsilon)^4; M^2/\varepsilon^2\}.$$

Now define

$$S = \left\{k: 0 \le k \le n \quad \text{and} \quad \left|x - \frac{k}{n}\right| < n^{-1/4}\right\}$$

$$L = \left\{k: 0 \le k \le n \quad \text{and} \quad \left(x - \frac{k}{n}\right)^2 \ge n^{-1/2}\right\}.$$

(e) Show that

$$\sum_{k \in S} \left| f(x) - f\left(\frac{k}{n}\right) \right| \binom{n}{k} x^k (1-x)^{n-k} \leq \varepsilon.$$

(f) Show that

$$\sum_{k \in L} \left| f(x) - f\left(\frac{k}{n}\right) \right| \binom{n}{k} x^k (1-x)^{n-k} \leq 2M \sum_{k \in L} \frac{\left(x - \frac{k}{n}\right)^2}{\left(x - \frac{k}{n}\right)^2} \binom{n}{k} x^k (1-x)^{n-k} \leq \varepsilon.$$

(g) Show that $B_n(x;f) \to f(x)$ uniformly for $0 \leq x \leq 1$.

21. Use Exercise 20 to prove the Weierstrass Theorem: Let f be a continuous real-valued function on a compact interval I. Then there are polynomials P_n on I such that $P_n \to f$ uniformly.

22. Show that the polynomials in Exercise 21 can be chosen to have rational coefficients. Hence the space $C(I)$, with the sup-metric, is separable.

23. Let X be the space of all complex-valued functions $x(t)$, $-\infty < t < \infty$, such that

$$\lim_{T \to \infty} \frac{1}{2T} \int_{-T}^{T} |x(t)|^2 \, dt < \infty.$$

Let the metric on X be given by

$$d(x,y) = \left\{ \lim_{T \to \infty} \frac{1}{2T} \int_{-T}^{T} |x(t) - y(t)|^2 \, dt \right\}^{1/2}.$$

Show that (X,d) is not separable. [*Hint*: Compute $d(e^{iat}, e^{ibt})$ where a, b are real numbers.]

24. Let A be a set in a metric space (X,d). A point x is said to be an *interior point* of A if A contains an open local neighborhood of x. Let Int A (interior of A) denote the collection of all interior points of A. Let Ext $A = $ Int A'. Let

$$\text{Bdy } A = (\text{Int } A \cup \text{Ext } A)'.$$

Ext A is called the *exterior* of A and Bdy A the *boundary* of A.
(a) Show that A is open if and only if $A = $ Int A.
(b) Show that $x \in $ Bdy A if and only if every open set containing x meets both A and A'.
(c) Show that A is closed if and only if $A = $ Int $A \cup $ Bdy A.
(d) Show that Bdy $(A) = $ Bdy (A').
(e) Show that for any set A, Int A and Ext A are open sets and Bdy A is a closed set.
(f) Is it possible to have $A = $ Bdy A?

25. Let $F: (X,d_1) \to (Y,d_2)$. Show that F is continuous if and only if the inverse image of every closed set in (Y,d_2) is a closed set in (X,d_1).

26. Let Q denote the set of rational numbers in R, where R has the usual metric. Discuss the interior, boundary, and exterior of Q and Q'.

27. Let C be the Cantor set in $[0,1]$, see Appendix D. Describe Int C, Ext C, and Bdy C.

28. Let $\{x_n\}$ be a sequence in a metric space (X,d) with $\lim x_n = x_0$. Assume that $d(x,x_n) \leq \varepsilon$ for all n. Show that $d(x,x_0) \leq \varepsilon$. [*Hint:* Use the fact that the interval $[0,\varepsilon]$ is closed in R.]

29. Let A be a nonempty set in a metric space (X,d) and define

$$\sigma(A;x) = \inf\{d(x,y): y \in A\}.$$

(a) Show that $\sigma(A;x) = 0$ if and only if $x \in \bar{A}$.

(b) Show that for A fixed $\sigma(A;x)$ is Lipschitz continuous in x.

   Let A and B be nonempty disjoint closed sets in (X,d) and define

$$f(x) = \frac{\sigma(A;x)}{\sigma(A;x) + \sigma(B;x)},$$

(c) Show that f is continuous and $0 \leq f(x) \leq 1$ for all x. (Is f Lipschitz continuous?)

(d) Show that $f(x) = 0$ precisely on A and $f(x) = 1$ precisely on B.

(e) What happens to f if A and B are not closed sets?

30. (Tietze Extension Theorem.) Let A be a closed set in a metric space (X,d) and let $f: A \to [0,1]$ be a continuous function. The following steps will lead to a proof of the fact that f has a continuous extension $g: X \to [0,1]$.

(a) Let $M = \sup\{f(x): x \in A\}$ and $A_1 = \{x \in A: f(x) \leq \frac{1}{3}M\}$ and $B_1 = \{x \in A: f(x) \geq \frac{2}{3}M\}$. Use Exercise 29 to show that there is a continuous function $\phi_1: X \to [0,1]$ with the property that $\frac{1}{3}M \leq \phi_1(x) \leq \frac{2}{3}M$ and $\phi_1(x) = \frac{1}{3}M$ on A_1 and $\phi_2(x) = \frac{2}{3}M$ on B_1.

(b) Show by induction that there is a sequence $\{\phi_n\}$ of continuous functions $\phi_n: X \to [0,1]$ such that

$$|f(x) - [\phi_1(x) + \cdots + \phi_n(x)]| \leq (\tfrac{2}{3})^n M, \quad (x \in A)$$
$$|\phi_n(x)| \leq (\tfrac{1}{3})(\tfrac{2}{3})^{n-1}M, \qquad\qquad (x \in X).$$

[*Hint:* Let

$$A_2 = \{x \in A: f(x) - \phi_1(x) \leq (\tfrac{1}{3})(\tfrac{2}{3})M\}$$

and

$$B_2 = \{x \in A: f(x) - \phi_1(x) \geq (\tfrac{2}{3})^2 M\}.]$$

(c) Show that $g_n = \phi_1 + \cdots + \phi_n$ converges uniformly on X, and the limit g is an extension of f.

31. (Continuation of Exercise 30.) What happens if the range of f in Exercise 30 is a bounded set in R? What happens if f is unbounded?

32. Let (X,d) be a metric space. Define a "distance" ρ between two bounded, nonempty subsets A and C of X by

$$\rho(A,C) = \inf\{\delta \colon A \subset B_\delta[C] \text{ and } C \subset B_\delta[A]\},$$

where $B_\delta[A] = \bigcup_{x \in A} B_\delta[x]$. Show that ρ defines a pseudometric on the collection X of all bounded, nonempty subsets of X. Show that $\rho(A,\overline{A}) = 0$. Show that $\rho(\{x\},\{y\}) = d(x,y)$, where $\{x\}$ and $\{y\}$ denotes the subsets containing exactly the points x and y, respectively.

33. Show that the interior of the Hilbert cube (Example 4) is empty.

13.   COMPLETENESS

Consider the metric space X consisting of all points in the half-open interval $(0,1]$—that is, $X = \{x \in R \colon 0 < x \leq 1\}$—with the usual metric. The sequence $\{1/n\} = \{1, \frac{1}{2}, \frac{1}{3}, \ldots\}$ lies in X. At first glance one might say that this sequence converges to 0. However, 0 is not a point in the space X. Therefore, it is nonsense to say that 0 is the limit. This particular sequence *in* X is just not convergent. On the other hand, we still feel that there is something special about this sequence and something special about the manner in which it fails to have a limit. What is special is that the sequence is a "Cauchy sequence" and the metric space X is not "complete." We investigate these concepts in this section.

First let us look at Cauchy sequences.

3.13.1 DEFINITION.   A sequence $\{x_n\}$ in a metric space (X,d) is said to be a *Cauchy sequence* if for each $\varepsilon > 0$ there exists an N such that $d(x_n,x_m) \leq \varepsilon$ for any choice of $n, m \geq N$. Notice that N depends on ε.

Another way of stating this definition is

$$\lim_{n,m \to \infty} d(x_n,x_m) = 0.$$

However, in this case it should be noted that this is a double limit and its precise meaning is really contained in the statement of Definition 3.13.1. For example, one could have $\lim_{n \to \infty} d(x_n,x_{2n}) = 0$, for a sequence $\{x_n\}$ that is not a Cauchy sequence.

It can be seen now that in the example given above $\{1/n\}$ is a Cauchy sequence. Let $N = 2/\varepsilon$. If $n, m \geq N$, then $1/n \leq \varepsilon/2$ and $1/m \leq \varepsilon/2$. Consequently, $|1/n - 1/m| \leq 1/n + 1/m \leq \varepsilon$ for all $n, m \geq N$.

The following lemma presents the key connection between convergent sequences and Cauchy sequences.

3.13.2 LEMMA.   *Let $\{x_n\}$ be a convergent sequence in a metric space (X,d). Then $\{x_n\}$ is a Cauchy sequence in (X,d).*

Proof: Let x_0 be the limit of the sequence $\{x_n\}$ in (X,d). Then for all n and m one has

$$d(x_n, x_m) \leq d(x_n, x_0) + d(x_0, x_m)$$

by the triangle inequality. Since x_0 is the limit of $\{x_n\}$, given $\varepsilon > 0$ there is an N such that $n, m \geq N$ implies that $d(x_n, x_0) < \varepsilon/2$ and $d(x_0, x_m) < \varepsilon/2$. But then by the above inequality one has $d(x_n, x_m) \leq \varepsilon$ whenever $n, m \geq N$. Hence, $\{x_n\}$ is a Cauchy sequence. ∎

We have thus seen that every convergent sequence is a Cauchy sequence, and we have seen an example of a Cauchy sequence that is *not* a convergent sequence. That is, being a Cauchy sequence does not imply that a sequence is convergent.

The example above suggests that the reason some Cauchy sequences fail to converge is a fault of the underlying set X. That is, in some sense a metric space (X,d) may have a "hole" in it. In the example cited above, the point 0 is "missing." If we were to "add" this point, then the sequence $\{1/n\}$ would converge. In the new space $[0,1] = X \cup \{0\}$ with the usual metric, it can be shown that every Cauchy sequence is convergent. That is, in the larger space a sequence is convergent if and only if it is a Cauchy sequence.

Metric spaces possessing this property—and many do—are so important that they are given a name.

3.13.3 DEFINITION.   A metric space (X,d) is said to be *complete* if each Cauchy sequence in (X,d) is a convergent sequence in (X,d).

It is difficult to overemphasize the importance of complete metric spaces. In many applications it is easier to show that a given sequence is a Cauchy sequence than to show that it is convergent. (The reason should be evident. The Cauchy test involves looking at the given sequence only, whereas the convergence test requires information outside the sequence, namely, the limit of the sequence.) However, if the underlying metric space is complete, showing that a sequence is a Cauchy sequence is enough. For example, many of the tests for convergence of sequences of real numbers are really tests for Cauchy sequences. In Section 15 we shall give one illustration of the importance of complete metric spaces, and this example by itself would be enough to justify giving attention to complete metric spaces. However, the primary significance of this concept, from our point of view, will be discussed in Chapter 5.

Let us consider some examples.

EXAMPLE 1.   The space of rational numbers with the usual absolute value metric is not a complete metric space. For example, the sequence $\{3, 3.1, 3.14, 3.141, 3.1415, \ldots\}$ is a Cauchy sequence but it is not convergent in the present metric space, for π is not a rational number. ∎

EXAMPLE 2.   The real numbers with the usual metric form a complete metric space. We assume that the reader has seen a development of the real number system. If not, we refer the reader to Rudin [1; pp. 1–47] for this development. ∎

EXAMPLE 3.   The complex numbers with the usual absolute value metric form a complete metric space. This fact follows directly from the completeness of the real numbers. ∎

3.13.4 LEMMA.   *Let $\{x_n\}$ be a Cauchy sequence in (X,d). Then the set $\{x_1,x_2,\ldots\}$ is bounded.*

Proof: With $\varepsilon = 1$ we can find an integer N such that $d(x_n,x_m) < 1$ whenever $n, m \geq N$. Now set

$$B = \max\{d(x_1,x_2), d(x_1,x_3),\ldots, d(x_1,x_N)\}.$$

Then it is easily verified that $d(x_1,x_n) \leq (B+1)$ for all n. It follows from the triangle inequality that $d(x_i,x_j) \leq 2(B+1)$ for all i and j. ∎

EXAMPLE 4.   We now show that the metric space (l_p,d_p), $1 \leq p < \infty$, is complete. (See Example 4, Section 3.) For each n let $x(n)$ denote an element of $l\hat{p}$, that is,

$$x(n) = \{x_1(n),x_2(n),\ldots\}.$$

Assume that $\{x(n)\}$, is a Cauchy sequence in (l_p,d_p). Given any $\varepsilon > 0$, there exists an $N(\varepsilon)$ such that

$$d_p(x(n), x(m)) = \left\{\sum_{k=1}^{\infty} |x_k(n) - x_k(m)|^p\right\}^{1/p} \leq \varepsilon \qquad \text{for } n, m \geq N(\varepsilon).$$

It follows that for $n, m \geq N(\varepsilon)$ one has

$$|x_k(n) - x_k(m)| \leq d_p(x(n), x(m)) \leq \varepsilon$$

for all k. But then, holding k fixed, the sequence of real or complex numbers $\{x_k(n)\}$ is a Cauchy sequence. Since the real numbers and the complex numbers are complete, the sequence $\{x_k(n)\}$ converges to a number $x_k(0)$. Let $x(0)$ denote the sequence $\{x_k(0)\}$. We show that $x(0)$ is in (l_p,d_p) and $\lim x(n) = x(0)$ in this metric space. Since $x(n)$ is a Cauchy sequence, Lemma 3.13.4 implies that there is a B such that

$$d_p(0,x(n)) = \left\{\sum_{k=1}^{\infty} |x_k(n)|^p\right\}^{1/p} \leq B < \infty$$

for all n. This implies that

$$\left\{\sum_{k=1}^{K} |x_k(n)|^p\right\}^{1/p} \leq B$$

for all n and each integer K. Since $\lim_{n\to\infty} x_k(n) = x_k(0)$, one has

$$\left\{\sum_{k=1}^{K} |x_k(0)|^p\right\}^{1/p} \leq B \tag{3.13.1}$$

for each integer K. By letting $K \to +\infty$ in (3.13.1), we see that $x(0) = \{x_k(0)\}$ is in (l_p, d_p).

Then, since $n, m \geq N(\varepsilon)$ implies that

$$\left\{\sum_{k=1}^{K} |x_k(n) - x_k(m)|^p\right\}^{1/p} \leq \varepsilon$$

for each integer K and $\lim_{m \to \infty} x_k(m) = x_k(0)$, it follows that

$$\left\{\sum_{k=1}^{K} |x_k(n) - x_k(0)|^p\right\}^{1/p} \leq \varepsilon \qquad (3.13.2)$$

for each integer K. By letting $K \to +\infty$ in (3.13.2) we get

$$d_p(x(n), x(0)) \leq \varepsilon$$

for all $n \geq N(\varepsilon)$. That is, $\lim_{n \to \infty} x(n) = x(0)$. Since $x(n)$ is an arbitrary Cauchy sequence and we have shown it to be convergent, the metric space (l_p, d_p) is complete. ∎

EXAMPLE 5. Let d_∞ be the sup-metric on $C[0,T]$. Let us show that $\{C[0,T], d_\infty\}$ is complete. Let $\{x_n\}$ be an arbitrary Cauchy sequence in $\{C[0,T], d_\infty\}$. Thus given any $\varepsilon > 0$, there is an $N(\varepsilon)$ such that $n, m \geq N(\varepsilon)$ implies that

$$|x_n(t) - x_m(t)| \leq d_\infty(x_n, x_m) \leq \varepsilon,$$

for all t. Then for fixed t, the sequence of numbers $\{x_n(t)\}$ converges to, say, $x_0(t)$. Since t is arbitrary, the sequence of functions $\{x_n(\cdot)\}$ converges pointwise to a function $x_0(\cdot)$. But $N(\varepsilon)$, being independent of t, implies that $\{x_n(\cdot)\}$ converges uniformly to $x_0(\cdot)$. But it is well known that if a sequence of continuous functions $\{x_n(\cdot)\}$ converges uniformly to a function $\{x_0(\cdot)\}$, then $x_0(\cdot)$ is continuous. See Exercise 19. Thus every Cauchy sequence in $\{C[0,T], d_\infty\}$ is convergent; hence, $\{C[0,T], d_\infty\}$ is complete. ∎

EXAMPLE 6. Let $1 \leq p < \infty$ and let R_p denote the space of all Riemann integrable functions x defined on $[a,b]$ with

$$\int_a^b |x(t)|^p \, dt < \infty$$

and define a metric by

$$d_p(x,y) = \left\{\int_a^b |x(t) - y(t)|^p \, dt\right\}^{1/p}.$$

It is shown in Appendix D that (R_p, d_p) is not complete. However, it is also shown in Appendix D that the Lebesgue space (L_p, d_p) (see Example 10, Section 3) is complete. ∎

EXAMPLE 7. Let (X,d) be a metric space and let $Y = BC(X,R)$ be the space of bounded, continuous, real-valued functions defined on X. Assume that Y has the sup-metric σ, see Exercise 10, Section 10. We claim that (Y,σ) is complete. Indeed if $\{f_n\}$ is a Cauchy sequence in (Y,σ), then

$$|f_n(x) - f_m(x)| \le \sigma(f_n, f_m)$$

for every $x \in X$. Therefore, for each $x \in X$, the sequence of real numbers $\{f_n(x)\}$ is a Cauchy sequence in R. Let $f(x) = \lim f_n(x)$. The rest of the argument now follows Example 5 and we omit the details. ∎

Let us now look at subspaces of a metric space. If $Y \subset X$, where (X,d) is a metric space, then (Y,d) is a metric space. We seek conditions under which the subspace (Y,d) is complete. It is important to note the meaning of this. That is, (Y,d) is complete if every Cauchy sequence in (Y,d) is convergent and (this is the important point) the limit is *in* Y. The following theorem gives the most useful result on this question.

3.13.5 THEOREM.   *Let (X,d) be a complete metric space and let (Y,d) be a subspace of (X,d). Then the subspace (Y,d) is complete if and only if Y is a closed set in (X,d).*

[Carefully note that Y is always a closed set in (Y,d), but this does not mean that Y is a closed set in (X,d).]

Proof: First assume that (Y,d) is complete. In order to show that Y is closed in (X,d), we will show that Y contains all of its points of adherence (Theorem 3.12.5). If y is a point of adherence of Y, then each open ball $B_{1/n}(y)$, $n = 1, 2, \ldots$, contains a point y_n in Y. Since $d(y_n, y) < 1/n$, $\{y_n\}$ is a convergent sequence in (X,d) converging to y. However, the sequence $\{y_n\}$ is a Cauchy sequence in the complete space (Y,d); therefore, $\{y_n\}$ converges to a point y_0 in (Y,d). Since the limit of a sequence is unique, $y = y_0$ or y is in Y. Hence, Y is a closed set in (X,d).

Now assume that Y is a closed set in (X,d). We want to show that (Y,d) is complete. Let $\{y_n\}$ be an arbitrary Cauchy sequence in (Y,d). Then $\{y_n\}$ is a Cauchy sequence in the complete metric space (X,d), so it converges to a limit y_0 in X. It follows from the Closed Set Theorem that $y_0 \in Y$. Hence (Y,d) is complete. ∎

EXAMPLE 8.   Let A be the Hilbert cube in (l_2, d_2), see Example 4, Section 12. Since (l_2, d_2) is complete (Example 4, Section 13) and the Hilbert cube is closed (Example 4, Section 12), the subspace (A,d) is complete. ∎

EXAMPLE 9.   Let d_∞ be the sup-metric on $C[0,T]$. Let $P[0,T]$ be the subset of $C[0,T]$ made up of all polynomials in t. The metric space $(C[0,T], d_\infty)$ is complete (Example 5, Section 13). However, the subset $P[0,T]$ is not closed. For example, the sequence in $P[0,T]$ given by

$$\{1, 1 + t, 1 + t + \frac{1}{2!} t^2, 1 + t + \frac{1}{2!} t^2 + \frac{1}{3!} t^3, \ldots\}$$

converges to e^t, which is not in $P[0,T]$. Since $P[0,T]$ is not closed, the subspace $(P[0,T],d_\infty)$ is not complete. ∎

We have said that if two metric spaces (X,d_1) and (Y,d_2) are isometric to one another, then they are, except for the names of the points in the underlying set, the same metric space. Thus the following theorem should not come as a surprise. We leave the proof as an exercise.

3.13.6 THEOREM. *Let (X,d_1) and (Y,d_2) be two metric spaces that are iso-metric to one another. Then (X,d_1) is complete if and only if (Y,d_2) is complete; that is, completeness is preserved by isometries.*

Although Theorem 3.13.6 is not a surprise, the next point may be somewhat of a shock. Let (X,d_1) and (Y,d_2) be homeomorphic to one another. It is *not* true that (X,d_1) is complete if and only if (Y,d_2) is complete. That is, homeomorphisms do not necessarily preserve completeness. It follows, then, that completeness is *not* a topological property. But, one may ask, how can this be? After all, complete-ness has something to do with convergence, and homeomorphisms preserve convergence (Section 7). True, but a homeomorphism does not necessarily pre-serve Cauchy sequences!

EXAMPLE 10. Let $X = (0,1]$ and d_1 be the usual metric, and let $Y = [1,\infty)$ and d_2 be the usual metric. The function $f: (X,d_1) \to (Y,d_2)$, where $y = f(x) = 1/x$, is a homeomorphism between (X,d_1) and (Y,d_2). The sequence $\{1/n\}$ in (X,d_1) is a Cauchy sequence. However, the corresponding sequence in (Y,d_2), that is, $\{f(1/n)\} = \{n\}$ is not a Cauchy sequence in (Y,d_2). On the other hand, a sequence $\{x_n\}$ in (X,d_1) is convergent if and only if $\{f(x_n)\}$ is convergent in (Y,d_2), for f is a homeomorphism. Finally note that since Y is a closed subset of R the space (Y,d_2) is complete, whereas (X,d_1) is not. ∎

Isometries, then, belong to the special class of homeomorphisms that preserve completeness; whereas, an arbitrary homeomorphism need not preserve complete-ness. On the other hand, it should not be surprising to find homeomorphisms, which are not isometries, that do preserve completeness. We can specify a very large class of such homeomorphisms.

3.13.7 DEFINITION. A homeomorphism f is said to be a *uniform homeo-morphism* if f and f^{-1} are uniformly continuous.

3.13.8 DEFINITION. Two metric spaces (X,d_1) and (Y,d_2) are said to be *uniformly homeomorphic (to one another)* if there exists a uniform homeomorphism mapping one of them onto the other.

3.13.9 THEOREM. *Let (X,d_1) and (Y,d_2) be uniformly homeomorphic. Then (X,d_1) is complete if and only if (Y,d_2) is complete.*

The proof of this theorem is outlined in the exercises.

Theorem 3.13.9 shows that uniform homeomorphisms preserve completeness. However, it is *not* true that if f is a homeomorphism between two complete metric spaces, then f is a uniform homeomorphism. For example, the real numbers with the usual metric, (R,d), is a complete metric space and $y = x^3$ is a nonuniform homeomorphism mapping (R,d) onto itself. Hence, uniform homeomorphisms are not the only homeomorphisms that preserve completeness.

There is another way to characterize complete metric spaces which is often useful. Before stating this characterization, we give a needed definition.

3.13.10 DEFINITION.   A sequence $\{A_n\}$ of subsets of a metric space (X,d) is said to be a *decreasing sequence* of subsets if

$$A_1 \supset A_2 \supset A_3 \supset \cdots .$$

3.13.11 THEOREM.   *Let (X,d) be a metric space. Then the following statements are equivalent:*

(a) *(X,d) is complete.*

(b) *$\bigcap_{n=1}^{\infty} A_n$ contains exactly one point, for every decreasing sequence of non-empty closed subsets with $\mathrm{diam}(A_n) \to 0$ as $n \to \infty$.*

Proof: (a) $\Rightarrow$ (b). Suppose (X,d) is complete and let $\{A_n\}$ be a decreasing sequence of nonempty subsets with $\mathrm{diam}(A_n) \to 0$ as $n \to \infty$. Let $A = \bigcap_{n=1}^{\infty} A_n$.

If x and y are in A, then $d(x,y) \le \mathrm{diam}(A) \le \mathrm{diam}(A_n)$ for every n. That is, $d(x,y) = 0$ or $x = y$. We must now show that A is not empty. Choose any sequence $\{x_n\}$ with $x_n \in A_n$. If $m \ge n$, then $x_m \in A_n$ and $d(x_n,x_m) \le \mathrm{diam}(A_n) \to 0$ as $n \to \infty$. Hence $\{x_n\}$ is a Cauchy sequence. Since (X,d) is complete, this sequence converges, call the limit x_0. Since A_n is closed, it follows from the Closed Set Theorem that $x_0 \in A_n$ for every n. Therefore $x_0 \in A$, which shows that $A = \bigcap_{n=1}^{\infty} A_n$ contains precisely one point.

(b) $\Rightarrow$ (a). Now suppose that for each decreasing sequence $\{A_n\}$ of nonempty closed sets with $\mathrm{diam}(A_n) \to 0$, the intersection $A = \bigcap_{n=1}^{\infty} A_n$ contains exactly one point. We want to show that (X,d) is complete. We shall outline the argument and ask the reader to check the details. Let $\{x_n\}$ be any Cauchy sequence in (X,d). Let $B_n = \{x_n, x_{n+1}, \ldots\}$ and $A_n = \bar{B}_n$. Then $\{B_n\}$ is a decreasing sequence and $\{A_n\}$ is decreasing by Exercise 9, Section 12. Also $\mathrm{diam}(A_n) = \mathrm{diam}(B_n)$ by Exercise 8, Section 12. Since $\{x_n\}$ is a Cauchy sequence one has

$$\mathrm{diam}(A_n) = \mathrm{diam}(B_n) \to 0 \text{ as } n \to \infty.$$

Since $A = \bigcap_{n=1}^{\infty} A_n$ contains one point x_0, one has $x_n \to x_0$ as $n \to \infty$. Indeed, $d(x_0,x_n) \le \mathrm{diam}(A_n) \to 0$ as $n \to \infty$.   ∎

EXERCISES

1. Carry out a development for pseudometric spaces analogous to the one carried out in this section for metric spaces.

2. Show that a Cauchy sequence $\{x_n\}$ in a metric space (X,d) is convergent if and only if it contains at least one convergent subsequence.

3. Let (B,d_2) be the subspace of (l_2,d_2) made up of all sequences with only a finite number of nonzero entries. Is (B,d_2) complete?

4. Prove Theorem 3.13.6.

5. Prove Theorem 3.13.9. [*Hint*: Let $f: X \to Y$ be a uniform homeomorphism between (X,d_1) and (Y,d_2). Show that $\{y_n\}$ is a Cauchy sequence in (Y,d_2) if and only if $\{f^{-1}(y_n)\}$ is a Cauchy sequence (X,d_1).]

6. Show that the space (l_∞,d_∞) (see Example 5, Section 3) is complete.

7. Let (X_1,d_1) and (X_2,d_2) be complete metric spaces. Show that the product space (X,d) is complete, where $X = X_1 \times X_2$ and $d(x,y) = d_1(x_1,y_1) + d_2(x_2,y_2)$.

8. Use Exercise 7 to show that the space (R^n,d_p), $1 \le p \le \infty$ (see Example 2, Section 3) is complete. Also show (C^n,d_p) is complete.

9. (Extension of Example 7.) Let (X,d_1) and (Y,d_2) be two metric spaces. Assume that (Y,d_2) is complete. Let $Z = BC(X,Y)$ denote the space of bounded continuous functions from X into Y. Assume that Z has the metric

$$\sigma(f,g) = \sup\{d_2(f(x), g(x)): x \in X\}.$$

Show that (Z,σ) is complete. [*Note*: (X,d_1) need not be complete.]

10. Show that the metric spaces in Examples 16 and 17 of Section 3 are complete.

11. Let I be an interval in R. Use the fact that R (with the usual metric) is complete to show that I is connected. Also prove the converse, namely, if A is a connected set in R, then A is an interval.

12. Let $f: [a,b] \to R$ be continuous with $f(a) < 0$ and $f(b) > 0$. Show that there is an x, $a < x < b$, with $f(x) = 0$. [*Hint*: Use Exercise 11 above and Exercise 6, Section 10.]

13. Let $f_1(x)$ and $f_2(x)$ be strictly monotone continuous functions defined for $0 \le x \le 1$ with $f_1(0) = f_2(0) = 0$. Show that there is a unique solution (r_0,x_1,x_2) for the equations

$$r_0 = f_1(x_1), \qquad r_0 = f_2(x_2) \qquad\qquad (3.13.3)$$
$$x_1 + x_2 = 1.$$

[*Hint*: Let r vary and let $x_1(r)$ and $x_2(r)$ be the solution of (3.13.3). Now apply Exercise 12 to

$$\phi(r) = x_1(r) + x_2(r) - 1.$$

This number r_0 arises as a critical feeding rate in a biological problem, see Sell and Weinberger [1].]

14. (Continuation of Exercise 9, Section 5.) Show that the space $LC_{x_0}^*$ is complete.

15. Let $g: X \to Y$ be uniformly continuous, where X and Y are metric spaces. Show that if $\{x_n\}$ is a Cauchy sequence in X, then $\{g(x_n)\}$ is a Cauchy sequence in Y.

16. Let $\{x_n\}$ and $\{y_n\}$ be two Cauchy sequences in a metric space (X,d). Show that $\{d(x_n,y_n)\}$ is a Cauchy sequence in R, where R has the usual metric.

17. (Baire Theorem.) Let (X,d) be a complete metric space where $X = \bigcup_{n=1}^{\infty} A_n$ and the sets A_n are closed. Show that at least one of the sets A_n contains a nonempty open local neighborhood. [*Hint*: Argue by contradiction and construct a decreasing sequence of open local neighborhoods $B_{1/2^n}(x_n)$ such that $B_{1/2^n}(x_n) \cap A_n = \emptyset$. Show that $\{x_n\}$ converges and that $x = \lim x_n$ is not in $\bigcup_{n=1}^{\infty} A_n$.]

18. Show that the space $C(-\infty,\infty)$ is complete with the metric

$$d(x,y) = \sum_{n=1}^{\infty} \frac{1}{2^n} \frac{\sup\{|x(t) - y(t)| : |t| \le n\}}{1 + \sup\{|x(t) - y(t)| : |t| \le n\}}.$$

What happens if we replace this metric with

$$d'(x,y) = \sum_{n=1}^{\infty} 2^{-n} \min(1, \sup\{|x(t) - y(t)| : |t| \le n\})?$$

19. Let $\{x_n\}$ be a sequence of real- or complex-valued functions defined on an interval I, and assume that $\{x_n\}$ converges uniformly to a limit x, that is, for every $\varepsilon > 0$ there is a N such that $|x_n(t) - x(t)| \le \varepsilon$ for all $n \ge N$ and all t in I. Show that $x(\cdot)$ is a continuous function. [*Hint*: Note that

$$|x(t) - x(s)| \le |x(t) - x_n(t)| + |x_n(t) - x_n(s)| + |x_n(s) - x(t)|.]$$

14.   COMPLETION OF METRIC SPACES

This section is devoted to explaining the following assertion: *Every metric space has a unique completion.*

Let us consider a simple case first. Suppose that (X,d) is a complete metric space and that (Y,d) is an arbitrary subspace of (X,d). As has been noted, (Y,d) is complete if and only if Y is a closed set in (X,d). In any event, the closure $\overline{Y}$ is a closed set in (X,d), and $(\overline{Y},d)$ is complete. Moreover, Y is dense in $(\overline{Y},d)$. Thus, in going from (Y,d) to $(\overline{Y},d)$, we fill in any "holes" that may exist in (Y,d). For example, let (X,d) be the real numbers with the usual metric, and let

$$Y = \{r: r \text{ is rational and } 0 < r < 1\}.$$

Then (Y,d) is not complete. However, the closure $\overline{Y} = [0,1]$, is complete, and Y is dense in $(\overline{Y},d)$.

Obviously the foregoing is a way to "complete" a metric space that is a subspace of a complete metric space. Many times, however, we would like to "complete" a metric space that is not specified as being a subspace of some complete metric space. A generally applicable concept of completion is given in the next definition.

3.14.1 DEFINITION. Let (X,d_1) be a metric space. A metric space (Y,d_2) is said to be *a completion of* (X,d_1) if

(1) (Y,d_2) is complete, and

(2) (X,d_1) is isometric with a dense subspace (Z,d_2) of (Y,d_2).

The situation is illustrated in Figure 3.14.1.

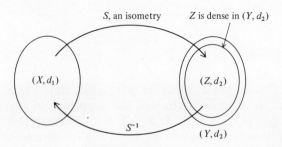

Figure 3.14.1.

Perhaps the first thing to notice about this concept is that a completion of (X,d_1) does not necessarily contain (X,d_1). This may seem bad, but do note that the completion does contain a dense subspace (Z,d_2) that is isometric to (X,d_1). We can think of (Z,d_2) as a mere renaming of the points of (X,d_1). Moreover, in certain cases (X,d_1) and (Z,d_2) are the same. This is the case in the example with which we started this section.

There are two questions which can be posed. First, which metric spaces have a completion? Second, how many "different" completions does a given space have? The answers are: (1) every metric space (X,d) has a completion and (2) all the completions of (X,d) are isometric with one another, that is, the completion is essentially unique.

3.14.2 THEOREM. *Let (X,d) be a metric space. Then (X,d) has a completion. Moreover, if (Y_1,d_1) and (Y_2,d_2) are two completions of (X,d), then (Y_1,d_1) and (Y_2,d_2) are isometric.*

The traditional method of proving this is outlined in the exercises. The reader should also note that the existence of a completion is a direct consequence of some earlier exercises, in particular, Exercises 8 and 9 of Section 5, Exercise 12, Section 10, Exercise 14, Section 13, and Theorem 3.13.5. We ask the reader to verify this. The fact that two completions are isometrically equivalent is a direct consequence of a more general result, which we now state.

3.14.3 THEOREM.   *Let $(\hat{X},\hat{d})$ be a completion of a metric space (X,d) and let $g:(X,d) \to (Y,\sigma)$ be a uniformly continuous function, where (Y,σ) is a complete metric space. Then there is a unique continuous "extension" $\hat{g}:(\hat{X},\hat{d}) \to (Y,\sigma)$ of g.* (See Figure 3.14.2.)

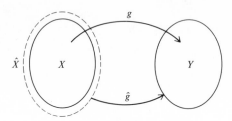

Figure 3.14.2.

We shall give the proof shortly. However, the term "extension" needs a word of explanation. It is customary to view the completion of a space as a process of adding ideal points. That is, if $S: X \to \hat{X}$ is an isometry that imbeds X into its completion $\hat{X}$, then instead of viewing X and $S(X)$ as different spaces one considers them to be the same. In this way the points in $\hat{X} - S(X)$ are "ideal" points and we complete $S(X)$ by "adding" them to $S(X)$.

This process is not as strange as it may sound. In fact, the generalized functions, such as the Dirac function, arising in the operational calculus, are examples of ideal points which one adds to a space of (ordinary) functions, as we now shall see.

EXAMPLE 1.[7]   Let X denote the collection of all functions x in $L_1(-1,1)$ with the property that $\int_{-1}^{1} |x(t)| \, dt = 1$. We define a metric on X as follows: Let $\{P_n\}$, $n = 1,2,\ldots$, denote the countable collection of all polynomials in t with rational coefficients. (Recall that $\{P_n\}$ is a dense set in $C[-1,1]$ with the sup-metric, see Exercise 22, Section 12.) Now for $x, y \in X$ let

$$\rho_n(x,y) = \min\left(1, \left|\int_{-1}^{1} [x - y]P_n \, dt\right|\right)$$

$$\rho(x,y) = \sum_{n=1}^{\infty} 2^{-n}\rho_n(x,y).$$

We claim that ρ is a metric on X. [It is easy to show that ρ is a pseudometric on X, and we define a new equality on X by saying $x = y$ if and only if $\rho(x,y) = 0$.]

---

[7] There are many (equivalent) ways of defining the Dirac function. The most common method is to use the theory of distributions, see A. Friedman [1]. Our primary purpose in this example is simply to show that the Dirac function can be viewed as an ideal point arising in the completion of a metric space.

Now consider the sequence

$$f_m(t) = \begin{cases} m, & |t| \le \dfrac{1}{2m} \\[2mm] 0, & \text{otherwise} \end{cases}$$

for $m = 1, 2, \ldots$. We claim that $\{f_m\}$ is a Cauchy sequence in this metric. To prove this we fix $\varepsilon > 0$. Now choose N so that

$$\sum_{n=N+1}^{\infty} 2^{-n} < \frac{\varepsilon}{2}.$$

Now consider the finite set of polynomials $\{P_1, P_2, \ldots, P_N\}$. Then for $n \ge m$ one has (see Figure 3.14.3)

$$\int_{-1}^{1} [f_n - f_m]P_i\, dt = -m\left[\int_{-1/2m}^{-1/2n} + \int_{1/2n}^{1/2m}\right]P_i\, dt + (n - m)\int_{-1/2n}^{1/2n} P_i\, dt.$$

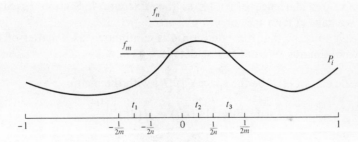

Figure 3.14.3.

By the Mean Value Theorem for integrals one has points t_1, t_2, t_3 in the intervals $[-1/2m, -1/2n]$, $[-1/2n, 1/2n]$, $[1/2n, 1/2m]$, respectively, such that

$$\int_{-1}^{1} [f_n - f_m]P_i\, dt = -m\left[\frac{1}{2m} - \frac{1}{2n}\right][P_i(t_1) + P_i(t_3)] + \frac{n-m}{n}P_i(t_2)$$

$$= \frac{n-m}{2n}[2P_i(t_2) - P_i(t_1) - P_i(t_3)]$$

$$\le \tfrac{1}{2}|2P_i(t_2) - P_i(t_1) - P_i(t_3)|.$$

Since P_i is continuous, we can choose M_i so that

$$\tfrac{1}{2}|2P_i(t_2) - P_i(t_1) - P_i(t_3)| < \frac{\varepsilon}{2N}$$

whenever $m \ge M_i$. (Why?) It follows then that $\rho_i(f_n, f_m) < \varepsilon/2N$ whenever $n \ge m \ge M_i$. Hence $\rho(f_n, f_m) < \varepsilon$ whenever $n \ge m \ge M$ where $M = \max(M_1, \ldots, M_N)$.

Thus $\{f_m\}$ is a Cauchy sequence, but the limit is not in X. By taking the completion of X we see that the limit of $\{f_m\}$ is the Dirac function δ_0. (δ_0 can be viewed as an operator on $C[-1,1]$ where $\delta_0(f) = f(0)$ for $f \in C[-1,1]$. Any function x in X can also be viewed in this way. Namely x is a mapping of $C[-1,1]$ into R given by $x: f \to \int_{-1}^{1} x(t)f(t)\, dt$. See Section 5.11 or Taylor [2,pp 33–35].) ∎

Proof of Theorem 3.14.3:   Since $g: X \to Y$ is uniformly continuous, it follows that g preserves Cauchy sequences. (See Exercise 15, Section 13.) Therefore, if $\{x_n\}$ is a Cauchy sequence in X, then $\{g(x_n)\}$ is a Cauchy sequence in Y. Since Y is complete, there is a point y in Y with $\lim g(x_n) = y$. If $x_n \to \hat{x}$ where $\hat{x} \in \hat{X}$, we then define $\hat{g}(\hat{x}) = y$. We must show that $\hat{g}(\hat{x})$ depends only on $\hat{x}$ and not on the sequence $\{x_n\}$. This follows from the uniform continuity of g. Let $\delta(\varepsilon)$ be the modulus of continuity of g. Let $\{x_n\}$ and $\{x_n'\}$ be two sequences with $\lim x_n = \lim x_n' = \hat{x}$, and define y and y' by $\lim g(x_n) = y$ and $\lim g(x_n') = y'$. Since

$$d(x_n, x_n') = \hat{d}(x_n, x_n') \le \hat{d}(x_n, \hat{x}) + \hat{d}(\hat{x}, x_n') \to 0,$$

we can find a N so that $d(x_n, x_n') < \delta(\varepsilon)$ whenever $n \ge N$. It follows that $\sigma(g(x_n), g(x_n')) \le \varepsilon$. Hence $\sigma(y, y') \le \varepsilon$. (See Exercise 4, Section 7.) Since ε is arbitrary we have $\sigma(y, y') = 0$, or $y = y'$.

It is easy to show that the extension $\hat{g}$ is continuous, as a matter of fact it is uniformly continuous. We omit these details. ∎

EXAMPLE 2.   Let X be the space $C[0,T]$ with the metric

$$d_2(x,y) = \left\{ \int_0^T |x(t) - y(t)|^2 \, dt \right\}^{1/2}.$$

Then the completion of X is $L_2[0,T]$. Define a mapping $K: X \to L_2[0,T]$ by $y = Kx$ where

$$y(t) = \int_0^T k(t,s)x(s)\, ds \tag{3.14.1}$$

and

$$\int_0^T \int_0^T |k(t,s)|^2 \, dt\, ds < \infty.$$

It is not difficult to show that K is uniformly continuous. (See Examples 4 and 5 of Section 5.) It follows from the last theorem that K has a unique extension to all of $L_2[0,T]$. The representation of this extension is given by (3.14.1) where the integral[8] is now the Lebesgue integral. ∎

EXERCISES

1. The following is another proof of Theorem 3.14.2. Let Y denote the collection of all Cauchy sequences $\{x_n\}$ from a metric space (X,d).
   (a) Show that $\sigma(\{x_n\}, \{x_n'\}) = \lim d(x_n, x_n')$ exists for $\{x_n\}$ and $\{x_n'\}$ in Y.

---

[8] One can view the operator K as being defined in terms of the Riemann integral. The extension of K would then require the Lebesgue integral.

(b) Define a "new" equality on Y by saying that $\{x_n\} = \{x_n'\}$ if

$$\lim d(x_n, x_n') = 0.$$

Show that σ is a metric on Y, in terms of this new equality.

(c) Show that (Y, σ) is complete.

(d) Find an isometry of X into Y. [*Hint*: Set $x_n = x$.]

2. Complete the proof of Theorem 3.14.2 as outlined in the text.

3. Let X denote the collection of all functions $f(z)$ analytic for $|z| < 2$. Define a metric on X by

$$d(f, g) = \left\{ \int_{|z| \leq 1} |f(z) - g(z)|^2 dx\, dy \right\}^{1/2}.$$

Show that (X, d) is not complete. What is the completion of (X, d)?

4. Theorem 3.14.3 can be used for defining functions. For example, find a continuous function $g(t)$, $0 \leq t \leq 1$, not identically zero such that

$$g(t + s) = g(t)g(s).$$

(a) Show that $g(0) = 1$.

(b) Let $g(1) = a$ and show that $g(r) = a^r$ for every rational r.

(c) Show that $g(t) = a^t$.

5. Find a continuous function $h(t)$, $1 \leq t$, that satisfies

$$h(ts) = h(t) + h(s).$$

(a) Show that $h(1) = 0$.

(b) Show that $h(t^r) = rh(t)$ for every rational r.

(c) Assume that h is not identically zero and show that $h(t) > 0$ for $t > 1$.

(d) Show that $h(t) = \log_b t$ for some $b > 1$.

6. Let X denote the space of all sequences of real numbers $x = \{x_1, x_2, \ldots\}$ with the property that only a finite number of the coordinates x_n are nonzero. Let d be the sup-metric on X. Show that (X, d) is not complete. Describe the completion of (X, d).

7. Discuss the completion of a pseudometric space. Show how Exercise 1 can be simplified in this case. What is the analogue of Theorem 3.14.3?

15.   CONTRACTION MAPPINGS

Given the concept of completeness introduced in the last two sections, we have the background for a study of contraction mappings. These mappings are extremely important, and they arise in a great number of applications.

3.15.1 DEFINITION.   Let (X, d) be a metric space and $f: X \to X$. We say that f is a *contraction*, or a *contraction mapping*, if there is a real number k, $0 \leq k < 1$, such that

$$d(f(x), f(y)) \leq kd(x, y)$$

for all x and y in X.

It follows immediately from the definition that a contraction mapping is uniformly continuous. The term k is sometimes called a *Lipschitz coefficient* for f. The reason contractions are important lies in the following fixed point theorem.

3.15.2 THEOREM. (CONTRACTION MAPPING THEOREM.)   *Let (X,d) be a complete metric space and let $f\colon X \to X$ be a contraction. Then there is one and only one point x_0 in X such that*

$$f(x_0) = x_0.$$

Moreover, if x is any point in X and x_n is defined inductively by $x_1 = f(x)$, $x_2 = f(x_1)$, ..., $x_n = f(x_{n-1})$, then $x_n \to x_0$ as $n \to \infty$. (That is, f has a unique fixed point x_0 and every sequence of iterations of f converges to this fixed point.)

Proof: Let us first show that every sequence of iterations of f converges to a fixed point. Next we show that f can have only one fixed point.

Let x be any point in X and define $x_1 = f(x)$, $x_2 = f(x_1)$, and in general, $x_n = f(x_{n-1})$. Thus $x_n = f^n(x)$. We will now show that the sequence $\{x_n\}$ is a Cauchy sequence. Assume that $n > m$, then

$$d(x_n, x_m) = d(f^n(x), f^m(x)) = d(f^m(x_{n-m}), f^m(x))$$
$$\leq kd(f^{m-1}(x_{n-m}), f^{m-1}(x)).$$

By induction, we get

$$d(x_n, x_m) \leq k^m d(x_{n-m}, x). \tag{3.15.1}$$

Using the triangle inequality, this becomes

$$d(x_n, x_m) \leq k^m[d(x_{n-m}, x_{n-m-1}) + \cdots + d(x_2, x_1) + d(x_1, x)].$$

By applying (3.15.1) we get

$$d(x_n, x_m) \leq k^m[k^{n-m-1} + \cdots + k + 1]\, d(x_1, x).$$

Since $0 \leq k < 1$, we get

$$d(x_n, x_m) \leq k^m \sum_{i=0}^{\infty} k^i\, d(x_1, x) = \frac{k^m}{1-k}\, d(x_1, x). \tag{3.15.2}$$

The right side of (3.15.2) can be made arbitrarily small by choosing m (and n) sufficiently large. That is, $d(x_n, x_m) \to 0$ as $n, m \to \infty$. Hence $\{x_n\}$ is a Cauchy sequence.

Since the space (X,d) is complete, $\{x_n\}$ converges. Let $x_0 = \lim x_n$. We now assert that x_0 is a fixed point of f. Since f is continuous, we know that

$$\lim_{n \to \infty} f(x_n) = f(\lim_{n \to \infty} x_n).$$

However, $f(\lim x_n) = f(x_0)$ and $\lim f(x_n) = \lim x_{n+1} = x_0$, so x_0 is a fixed point.

To show that the fixed point x_0 is unique we argue by contradiction. Assume that x_0 and y_0 are two distinct fixed points of f. We then get the contradiction

$$0 < d(x_0,y_0) = d(f(x_0),f(y_0)) \le kd(x_0,y_0) < d(x_0,y_0).$$

Hence f has only one fixed point. ∎

The fixed point theorem has several extensions. Let us consider one of the important ones.

3.15.3 COROLLARY. *Let (X,d) be a complete metric space and let f be a (not necessarily continuous) function, $f: X \to X$. If for some integer $p > 0$ the function f^p is a contraction, then f has a unique fixed point.*

Proof: Let $g = f^p$. By Theorem 3.15.2, g has a unique fixed point x_0. Let us show that x_0 is also a fixed point of f. (Notice that a fixed point of f is also a fixed point of g, so f can have at most one fixed point.)

Since $g = f^p$, it follows that $f(g(x)) = g(f(x))$ for all x in X. Since g is a contraction, there is a k, $0 \le k < 1$ such that

$$d(g(x),g(y)) \le kd(x,y)$$

for all x and y in X. If $f(x_0) \ne x_0$ we get the contradiction:

$$0 < d(f(x_0),x_0) = d(f(g(x_0)),g(x_0)) = d(g(f(x_0)),g(x_0))$$
$$\le kd(f(x_0),x_0), < d(f(x_0),x_0). \quad ∎$$

It is possible for f to be discontinuous while the composition $f \circ f$ is a contraction. For example, let $X = [0,1]$ and define

$$f(x) = \begin{cases} \frac{1}{4}, & \text{for } 0 \le x \le \frac{3}{4} \\ \frac{1}{2}, & \text{for } \frac{3}{4} < x \le 1. \end{cases}$$

Then $f \circ f(x) = \frac{1}{4}$ for all $0 \le x \le 1$, so $f \circ f$ is a contraction.

Let us now consider some examples of the use of contraction mappings.

EXAMPLE 1 (NONLINEAR FILTER). Consider the nonlinear[9] filter shown in Figure 3.15.1. (λ can be thought of as a "gain factor" of the linear element.) Assume that the initial conditions for the linear filter are 0. Then the (nonlinear) integral equation relating the input $u(t)$ to the output $z(t)$ is

$$z(t) = \lambda \int_0^t K(t,s)F(u(s),s)\,ds, \qquad 0 \le t \le T < \infty. \qquad (3.15.3)$$

We assume that $K(t,s)$ and $F(u,t)$ are continuous and that $F(0,t) = 0$ for all t. Further, assume that K and F are bounded, that is, $|K(t,s)| \le M$ for[10] $0 \le t,s \le T$

---

[9] Strictly speaking we are tacitly assuming an algebraic structure for discussing linearity or nonlinearity. This does not matter, since the precise assumptions of F and K are given below.
[10] This assumption for K is redundant, see Theorem 3.17.21.

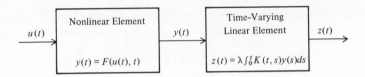

Figure 3.15.1.

and $|F(u,t)| \le N$ for $-\infty < u < \infty$ and $0 \le t \le T$. Finally, let us assume that F satisfies a global Lipschitz condition, that is, there is an $L > 0$ such that

$$|F(u,t) - F(v,t)| \le L|u - v|$$

for all u, v, and t.

Let us denote the mapping (3.15.3) of u into z by $z = f(u)$. Then with these assumptions, $f: C[0,T] \to C[0,T]$, that is, if u is a continuous function on $[0,T]$, then $z = f(u)$ is a continuous function on $[0,T]$.

Let us now show that the mapping $f: C \to C$ is continuous. Of course, to do this we must put a metric on C; we take the sup-metric d. Now

$$|f(u)(t) - f(v)(t)| = \left| \lambda \int_0^t K(t,s)\{F(u(s),s) - F(v(s),s)\}\, ds \right|$$

$$\le |\lambda| \int_0^t |K(t,s)| \cdot |F(u(s),s) - F(v(s),s)|\, ds$$

$$\le |\lambda|\, ML \int_0^t |u(s) - v(s)|\, ds \qquad (3.15.4)$$

or

$$|f(u)(t) - f(v)(t)| \le |\lambda|\, ML\, d(u,v)t. \qquad (3.15.5)$$

This implies that

$$d(f(u), f(v)) \le |\lambda|\, MLT\, d(u,v).$$

The last inequality shows that if $|\lambda| < (MLT)^{-1}$, then f is a contraction. Thus when λ is sufficiently small, we see that f has a unique fixed point. Since $F(0,t) = 0$ for all t, it follows that $u(t) \equiv 0$ is the fixed point.

We can actually show that for all λ, $u(t) \equiv 0$ is the only fixed point of f. We do this by applying the last corollary and show that for each λ there is a positive integer p such that f^p is a contraction.

First we assert that

$$|f^p(u)(t) - f^p(v)(t)| \le \frac{(|\lambda|\, MLt)^p}{p!}\, d(u,v). \qquad (3.15.6)$$

We prove this by induction. For $p = 1$, (3.15.6) reduces to (3.15.5). Now assume (3.15.6) is true for p and let us check $p + 1$. By (3.15.4) we get

$$|f^{p+1}(u)(t) - f^{p+1}(v)(t)| \le |\lambda|\, ML \int_0^t |f^p(u)(s) - f^p(v)(s)|\, ds.$$

By the induction hypothesis, this becomes

$$|f^{p+1}(u)(t) - f^{p+1}(v)(t)| \le |\lambda| \, ML \left\{ \int_0^t \frac{(|\lambda| \, MLs)^p}{p!} \, ds \right\} d(u,v).$$

By integrating we then get (3.15.6) for $p + 1$.

Now (3.15.6) implies that

$$d(f^p(u), f^p(v)) \le \frac{(|\lambda| \, MLT)^p}{p!} \, d(u,v).$$

With λ given, we can find an integer p such that

$$\frac{(|\lambda| \, MLT)^p}{p!} < 1.$$

Then, for this p, f^p is a contraction. Thus for every λ, f has a unique fixed point, namely $u(t) \equiv 0$.

This is a very interesting result, for it tells us that a fairly general class of filters has only the null function 0 as a fixed point. In other words, every nontrivial input is distorted. Moreover, one can modify this argument to show that the equation $f(u) = \alpha u$, where α is a nonzero constant, has only $u = 0$ as a solution. In other words, the mapping f has no eigenvectors. ∎

EXAMPLE 2. (EXISTENCE AND UNIQUENESS THEOREM FOR SOLUTIONS OF ORDINARY DIFFERENTIAL EQUATIONS.)   Let us consider the ordinary differential equation

$$\frac{dy}{dt} = f(y,t), \tag{3.15.7}$$

where f is a real-valued, continuous function defined on $R^1 \times R^1$. We shall seek a solution $y(t)$ for (3.15.7) which satisfies the initial condition

$$y(t_0) = y_0.$$

This is said to be the *initial value problem*. Since f is continuous, it is easily checked that a solution of the initial value problem is equivalent to a solution of the integral equation

$$y(t) = y_0 + \int_{t_0}^t f(y(s),s) \, ds. \tag{3.15.8}$$

Let us ask: When does (3.15.8) have a unique solution? In order to formulate an answer, let us consider the operator $z = F(y)$ where

$$z(t) = y_0 + \int_{t_0}^t f(y(s),s) \, ds.$$

F is then a mapping of one space of functions into another. By elementary calculus, we see that if y is continuous, then z is continuous. Thus $F: \mathscr{C} \to \mathscr{C}$, where $\mathscr{C}$ denotes the space of (real-valued) continuous functions defined on some interval I containing t_0.

Now $y(t)$ is a solution of (3.15.8) if and only if $y = F(y)$; that is, if and only if y is a fixed point of F. We now ask, under what conditions (on f) is the mapping F a contraction?

(In order to simplify the notation, let us set $y_0 = t_0 = 0$.)

Since f is continuous, it is bounded on the set $-1 \leq y \leq 1$, $-1 \leq t \leq 1$; that is,

$$|f(y,t)| \leq M$$

on this set. Now assume that f satisfies a Lipschitz condition in y. In other words, assume there is a constant K such that

$$|f(y,t) - f(x,t)| \leq K|y - x|$$

for all[11] t, x, y satisfying $-1 \leq t \leq 1$, $-1 \leq x \leq 1$, $-1 \leq y \leq 1$.

With M and K defined above, let X denote the set of all continuous real-valued functions $\phi(t)$ satisfying

$$|\phi(t)| \leq M|t|$$

on the interval $I = [-T,T]$ where $0 < T \leq 1$, $MT \leq 1$ and $KT < 1$. (See Figure 3.15.2.) X is then a subset of $\mathscr{C}(I)$. Moreover, if $\mathscr{C}(I)$ has the sup-metric d_∞, then X is a closed subset. To prove this, we want to show that the complement is open. So let ϕ be in X'. Then for some τ in I, $|\phi(\tau)| > M|\tau|$. Let $2\varepsilon = |\phi(\tau)| - M|\tau| > 0$. Now if ψ is in $B_\varepsilon(\phi)$, then $|\psi(\tau)| - M|\tau| > \varepsilon$. Hence ψ is in X', so X' is open.

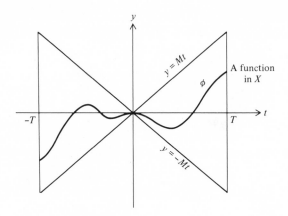

Figure 3.15.2.

Since $(\mathscr{C}, d_\infty)$ is complete and X is a closed subspace, it follows that (X, d_∞) is complete by Theorem 3.13.5. Let us now show that F maps X into X, and that F is a contraction.

---

[11] The expert will realize that we are being needlessly restrictive here. However, our purpose is not to prove the most general existence and uniqueness theorem, but rather to indicate an application of the Contraction Mapping Theorem.

Let $\phi \in X$, then $F(\phi)$ is in X since

$$|F(\phi)(t)| \leq \int_0^{|t|} |f(\phi(s), s)| \, ds \leq \int_0^{|t|} M \, ds \leq M \, |t|$$

for all t in I. (For this we have only used the fact that f is continuous. The Lipschitz condition will be used to show that F is a contraction.)

Let x and y be in X. Then

$$F(x)(t) - F(y)(t) = \int_0^t [f(x(s),s) - f(y(s),s)] \, ds.$$

Thus for $t \geq 0$ one gets

$$|F(x)(t) - F(y)(t)| \leq \int_0^t K \, |x(s) - y(s)| \, ds$$

$$\leq \int_0^t K \, d_\alpha(x,y) \, ds \leq Kt \, d_\infty(x,y).$$

For $t \leq 0$ one gets $|F(x)(t) - F(y)(t)| \leq K |t| \, d_\infty(x,y)$. For $|t| \leq T$ we get

$$d_\infty(F(x), F(y)) \leq KT \, d_\infty(x,y).$$

Since $KT < 1$, F is a contraction. Therefore, F has a unique fixed point and the initial value problem has a unique solution on the interval $[-T,T]$. ∎

EXAMPLE 3. (CLOSED LOOP FEEDBACK SYSTEM.)   Consider the closed loop feedback system illustrated in Figure 3.15.3. The equation for this system is determined as follows: Let r, ε, and c be real-valued functions of t and F a non-linear operator. Then $\varepsilon = r - c$ and $\varepsilon + F(\varepsilon) = (I + F)\varepsilon = r$.

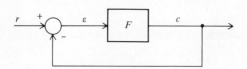

Figure 3.15.3.

Assume that r, ε, and c are points in the complete metric space BC, where BC is made up of all bounded continuous real-valued functions defined on R, with the sup-metric d_∞. Further, let F be a contraction mapping defined on BC. We would like to know several things about the mapping $(I + F)$:

(1) Is it one-to-one?

(2) Is the range BC?

(3) Is it invertible?

(4) If it is invertible, is the inverse continuous?

Using the fact that F is a contraction mapping, it is a simple matter to show that the answer to each of the above questions is yes.

Assume that $I + F$ is not one-to-one. Then there exists ε_1 and ε_2 such that $d_\infty(\varepsilon_1, \varepsilon_2), \neq 0$ and

$$(I + F)\varepsilon_1 = (I + F)\varepsilon_2$$

or

$$\varepsilon_1 - \varepsilon_2 = F\varepsilon_2 - F\varepsilon_1. \tag{3.15.9}$$

However (3.15.9) and the fact that F is a contraction implies that

$$0 < d_\infty(\varepsilon_1, \varepsilon_2) = d_\infty(F\varepsilon_1, F\varepsilon_2) < d_\infty(\varepsilon_1, \varepsilon_2),$$

which is a contradiction. Hence $I + F$ is one-to-one.

To show that the range of $(I + F)$ is BC, consider the equation

$$\varepsilon = r - F\varepsilon = G_r(\varepsilon).$$

We view the right-hand side as a mapping G_r of BC into itself parameterized by r. If we can show that G_r has a fixed point for each r, we will have shown that the range of $(I + F)$ is BC. Now

$$d_\infty(G_r\varepsilon_1, G_r\varepsilon_2) = \sup_t |r(t) - (F\varepsilon_1)(t) - r(t) + (F\varepsilon_2)(t)|$$

$$= \sup_t |(F\varepsilon_2)(t) - (F\varepsilon_1)(t)| = d_\infty(F\varepsilon_1, F\varepsilon_2).$$

Since F is a contraction mapping, it follows from the above equation that G_r is also a contraction mapping for arbitrary r. It follows from the Contraction Mapping Theorem that G_r has a fixed point for arbitrary r. Therefore $\mathscr{R}(I + F) = BC$.

Since $I + F$ is one-to-one and $\mathscr{R}(I + F) = BC$, it is invertible. Hence we can meaningfully write

$$\varepsilon = (I + F)^{-1}r.$$

Finally, let us consider the continuity of $(I + F)^{-1}$. Let r_1 and r_2 be arbitrary inputs and let

$$\varepsilon_1 = (I + F)^{-1}r_1,$$
$$\varepsilon_2 = (I + F)^{-1}r_2.$$

Then

$$\varepsilon_1 - \varepsilon_2 = r_1 - r_2 - F\varepsilon_1 + F\varepsilon_2.$$

This implies that

$$d_\infty(\varepsilon_1, \varepsilon_2) \leq d_\infty(r_1, r_2) + d_\infty(F\varepsilon_1, F\varepsilon_2).$$

Since F is a contraction mapping, there exists a constant k, $0 \leq k < 1$, such that $d_\infty(F\varepsilon_1, F\varepsilon_2) \leq k d_\infty(\varepsilon_1, \varepsilon_2)$. Hence

$$d_\infty(\varepsilon_1, \varepsilon_2) \leq \frac{1}{1 - k} d_\infty(r_1, r_2),$$

which shows that $(I + F)^{-1}$ is continuous. In fact it is uniformly continuous. Moreover, $(I + F)^{-1}$ maps bounded sets into bounded sets.

Let us define the *supremum of the incremental gain* of a transformation T of a metric space X into itself by

$$g(T) = \sup \left\{ \frac{d_\infty(Tx,Ty)}{d_\infty(x,y)} : x,y \in X \quad \text{and} \quad x \neq y \right\}.$$

Then we see that in the present example $g(F) < 1$ implies that

$$g((I + F)^{-1}) \leq 1/(1 - k).$$

Of course, the reader familiar with feedback theory will not be surprised that $I + F$ is " well-behaved " when the " loop gain " is less than unity. ∎

EXERCISES

1. In Example 3, nothing was said about F or $(I + F)^{-1}$ being causal, as in Section 2.8. If F is causal as well as a contraction, does it follow that $(I + F)^{-1}$ is causal?

2. Let $T(i\omega)$ denote the Fourier transform transfer function for a system that maps $L_2(-\infty,\infty)$ into itself. When does $T(i\omega)$ represent a contraction mapping?

3. Let $f(x,y)$ be a continuous real-valued function defined on a rectangle

$$\mathscr{R}: |x - x_0| \leq a, |y - y_0| \leq b,$$

and satisfying $y_0 = f(x_0,y_0)$. Assume there is a k, $0 \leq k < 1$, such that

$$|f(x,y) - f(x,y')| \leq k|y - y'|$$

for (x,y) and (x,y') in $\mathscr{R}$. Use the Contraction Mapping Theorem to show that there is an $\alpha > 0$ and a continuous function $y = g(x)$, defined for $|x - x_0| \leq \alpha$, such that $g(x) = f(x,g(x))$, for $|x - x_0| \leq \alpha$ and $y_0 = g(x_0)$.

4. (Continuation of Exercise 3.) Let $f(x,y)$ be continuous on $\mathscr{R}$ and satisfy $y_0 = f(x_0,y_0)$. Assume also that $\partial f/\partial y = f_y$ is continuous on $\mathscr{R}$ and that $f_y(x_0,y_0) = 0$. Show that there is an $\alpha > 0$ and a continuous function $y = g(x)$, defined for $|x - x_0| \leq \alpha$, such that $y_0 = g(x_0)$ and

$$g(x) = f(x,g(x)), \qquad |x - x_0| \leq \alpha.$$

5. (Implicit Function Theorem; continuation of Exercises 3 and 4.) Let $F(x,y)$ be a C^1 function on the rectangle $\mathscr{R}$ with $F(x_0,y_0) = 0$. Assume that

$$F_y(x_0,y_0) \neq 0.$$

Show that there is an $\alpha > 0$ and a continuous function $g(x)$, defined for $|x - x_0| \leq \alpha$, such that $y_0 = g(x_0)$ and

$$F(x,g(x)) = 0, \qquad |x - x_0| \leq \alpha.$$

[*Hint*: Apply Exercise 4 to $f(x,y) = y - F_y^{-1}(x_0,y_0)F(x,y)$.]

6. (Inverse Mapping Theorem.) Let $x = f(y)$ be a C^1 function defined for $|y - y_0| \le b$ with $x_0 = f(y_0)$ and $f'(y_0) \ne 0$. Show that there is an $\alpha > 0$ and a continuous function $y = g(x)$, defined for $|x - x_0| \le \alpha$, such that $x = f(g(x))$ for $|x - x_0| \le \alpha$ and $y_0 = g(x_0)$.

7. Consider the nonlinear Volterra integral equation

$$y(t) = h(t) + \int_0^t k(t,s) f(y(s),s) \, ds, \qquad (3.15.10)$$

where $h(t)$, $k(t,s)$, and $f(y,s)$ are continuous for $0 \le t$, $0 \le s$, and all y. Assume that there are positive constants A, B, and K such that

$$|f(y,s)| \le A + B|y|,$$
$$|f(y,s) - f(y',s)| \le K|y - y'|.$$

Show that there is an $\alpha > 0$ and a continuous function $y(t)$ defined for $0 \le t \le \alpha$, and satisfying (3.15.10). [*Hint*: Carefully examine Example 2.]

8. The condition that the Lipschitz coefficient k in Definition 3.15.1 satisfy $k < 1$ cannot be entirely omitted. Show that

$$f(x) = x - \tfrac{1}{2}e^x, \qquad x \le 0$$
$$= -\tfrac{1}{2} + \tfrac{1}{2}x, \qquad x \ge 0$$

has no fixed points but $|f(x) - f(y)| < |x - y|$ for all x and y in R. [*Remark*: One can show that certain nonexpansive mappings, that is, mappings satisfying $\|f(x) - f(y)\| \le \|x - y\|$, on Banach spaces have fixed points. See Browder [1].]

9. Is there a Contraction Mapping Theorem for pseudometric spaces? If so, what happens to uniqueness?

10. Show that the system

$$x_1 = \tfrac{1}{4}x_1 - \tfrac{1}{4}x_2 + \tfrac{2}{15}x_3 + 3,$$
$$x_2 = \tfrac{1}{4}x_1 + \tfrac{1}{5}x_2 + \tfrac{1}{2}x_3 - 1,$$
$$x_3 = -\tfrac{1}{4}x_1 + \tfrac{1}{3}x_2 - \tfrac{1}{3}x_3 + 2,$$

has a unique solution by using the Contraction Mapping Theorem.

16.   TOTAL BOUNDEDNESS AND APPROXIMATIONS

Compactness, the subject of the next section, is a property of metric spaces and subsets of metric spaces. We shall see that it is a topological property, that is, if the metric spaces X and Y are homeomorphic to one another, then X is compact if and only if Y is compact.

Before considering the concept of compactness itself, let us introduce the concept of total boundedness. Recall that a set A is said to be bounded if it has finite diameter.

EXAMPLE 1.   Let d_∞ be the sup-metric on $C[0,T]$. The set A, made up all functions x in $(C[0,T],d_\infty)$ such that $|x(t)| \leq 1$, is bounded, since $\text{diam}(A) = 2$. The set B made up of all x in $(C[0,T],d_\infty)$ such that

$$\int_0^T |x(t)|\, dt \leq 1$$

is not bounded. (Why?)   ∎

Now let A be a set contained in a metric space (X,d). Suppose that we are given an $\varepsilon > 0$ and that we want to find a distinguished subset of A, call it A_ε, with the property that for each point x in A there is a y in A_ε such that $d(x,y) \leq \varepsilon$. Figure 3.16.1 illustrates this idea.

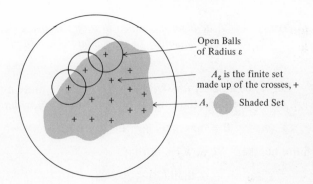

Figure 3.16.1.

This problem does not become interesting until one places further conditions on the distinguished set A_ε. For example, one may want A_ε to be finite. In fact, it is very convenient if a finite A_ε can be found for each $\varepsilon > 0$. Roughly speaking, it means that no matter how small ε is, a finite, albeit somewhat larger, set A_ε exists. For example, let (X,d) be the real Euclidean plane. Then the bounded rectangle shown in Figure 3.16.2(a) has this property, whereas the half-plane shown in Figure 3.16.2(b) does not.

(a)   (b)

Figure 3.16.2.

3.16.1 DEFINITION.   Let A be a set contained in a metric space (X,d). Given $\varepsilon > 0$, a subset A_ε of A is said to be an ε-*net of A* if (1) A_ε is finite and (2) for each $x \in A$ there is a $y \in A_\varepsilon$ such that $d(y,x) \leq \varepsilon$.

3.16.2 DEFINITION.   A set A in a metric space (X,d) is said to be *totally bounded* if for each $\varepsilon > 0$, A contains an ε-net.

For what it is worth we note that the empty set is always totally bounded. More important though, we note that any finite set is totally bounded. Of course, these examples are also examples of bounded sets.

However, total boundedness and boundedness are not the same thing, and it is important that the difference be understood. To begin with, total boundedness is a stronger property than boundedness.

3.16.3 LEMMA.   *Let A be a set contained in a metric space (X,d). If A has an ε-net for some $\varepsilon > 0$, then A is bounded. In particular, every totally bounded set is bounded.*

Proof: Let A_ε be an ε-set for A. Then A_ε contains a finite number of points $\{y_1, y_2, \ldots, y_n\}$. (If A_ε is empty, the conclusion is obvious.) Now let

$$B = \max\{d(y_i, y_j) : 1 \leq i, j \leq n\},$$

which is a finite number. We now claim that

$$\text{diam}(A) \leq B + 2\varepsilon.$$

Indeed, if x_1 and x_2 are any two points in A, then there are two points in A_ε—call them y_1 and y_2—such that

$$d(x_i, y_i) \leq \varepsilon, \qquad i = 1, 2.$$

Thus one has

$$d(x_1, x_2) \leq d(x_1, y_1) + d(y_1, y_2) + d(y_2, x_2) \leq B + 2\varepsilon;$$

hence $\text{diam}(A) \leq B + 2\varepsilon$, or A is bounded.   ∎

So total boundedness implies boundedness. However, an extremely important point is that boundedness does *not* imply total boundedness. This is illustrated in the examples below.

Before we turn to these examples, we ask the reader to verify the next assertion:

3.16.4 LEMMA.   *Let (X,d) be totally bounded. Then X is separable.*

EXAMPLE 2.   Consider the space (l_2, d_2) (see Example 4, Section 3) and let A be the set of all points in (l_2, d_2) such that $\sum_{n=1}^{\infty} |x_n|^2 \leq 1$. The set A is bounded. Since

$$d(x,y) = \left\{ \sum_{n=1}^{\infty} |x_n - y_n|^2 \right\}^{1/2} \leq \left\{ \sum_{n=1}^{\infty} |x_n|^2 \right\}^{1/2} + \left\{ \sum_{n=1}^{\infty} |y_n|^2 \right\}^{1/2}$$

by the Minkowski Inequality, we see that diam(A) ≤ 2. Yet the set A is not totally bounded. Consider the set $E = \{e_1, e_2, \ldots\}$ of points in A, where $e_1 = \{1,0,0,\ldots\}$, $e_2 = \{0,1,0,\ldots\}$, $e_3 = \{0,0,1,0,\ldots\}$, and so on. We see that $d_2(e_k, e_j) = \sqrt{2}$ for $k \neq j$. If an ε-net $A_{1/2}$ exists for $\varepsilon = \frac{1}{2}$, there must be an appropriate finite set in A. But the closed balls $B_{1/2}[e_k]$ and $B_{1/2}[e_j]$ are disjoint for $k \neq j$. Thus if the set $A_{1/2}$ contains a point within distance $\frac{1}{2}$ of each e_k, $A_{1/2}$ must have at least one of its points in each closed ball $B_{1/2}[e_k]$. It follows that $A_{1/2}$ must be at least countably infinite. Therefore, an ε-net for $\varepsilon = \frac{1}{2}$ does not exist, and A is not totally bounded. See Figure 3.16.3. ∎

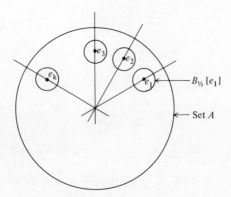

Figure 3.16.3.

EXAMPLE 3.   Consider the Lebesgue space $L_2(-\infty, \infty)$ with the usual metric d_2. Let A be the set of all points x in $L_2(-\infty, \infty)$ such that

$$\int_{-\infty}^{\infty} |x(t)|^2 \, dt \leq 1.$$

As in Example 2, it can be shown that the set A is bounded but not totally bounded. [*Hint*: Consider the Hermite functions

$$\Psi_n(t) = (2^n n! \sqrt{\pi})^{-1/2} H_n(t) e^{-t^2/2} \qquad n = 0, 1, 2, \ldots$$

where $H_n(t)$ is the Hermite polynomial

$$H_n(t) = (-1)^n e^{t^2} \left[\frac{d^n}{dt^n} e^{-t^2} \right] \qquad n = 0, 1, 2, \ldots$$

(See Section 7.14).] ∎

EXAMPLE 4.   Consider the metric d_2 on R^n, as in Example 2, Section 3, and let A be the set in (R^n, d_2) of all points $x = (x_1, x_2, \ldots, x_n)$ such that

$$\sum_{i=1}^{n} |x_i|^2 \leq 1.$$

The set A is, of course, bounded. Let us show that A contains an ε-net A_ε for each $\varepsilon > 0$. Let K be a positive integer such that $\sqrt{n} \leq \varepsilon K$. Let A_ε be the set of all n-tuples $(y_1, \ldots, y_n)$ such that $y_j \, (j = 1, 2, \ldots, n)$ can take only the values m/K (where m is an integer with $-K \leq m \leq K$) and

$$\sum_{j=1}^{n} |y_j|^2 \leq 1.$$

The set A_ε is finite and $A_\varepsilon \subset A$. It is apparent that for an arbitrary $x = (x_1, \ldots, x_n)$ in A there is a point $y = (y_1, \ldots, y_n)$ in A_ε such that $|x_i - y_i| \leq 1/K$ or

$$d(x,y) \leq \left\{ \sum_{j=1}^{n} \left(\frac{1}{K} \right)^2 \right\}^{1/2} = \frac{\sqrt{n}}{K} \leq \varepsilon.$$

Hence, A_ε is an ε-net and, ε being arbitrary, A is totally bounded. ∎

EXAMPLE 5.   Return to the metric space (l_2, d_2) of Example 2. Let A be the Hilbert cube, that is, the set of all points $x = \{x_n\}$ in (l_2, d_2) such that $|x_n| \leq 1/n$. It follows that

$$\lim_{N \to \infty} \sum_{n=N}^{\infty} |x_n|^2 = 0 \qquad \text{uniformly over } A,$$

that is, given an $\varepsilon > 0$ there exists an integer $N = N(\varepsilon)$ such that

$$\sum_{n=N}^{\infty} |x_n|^2 \leq \sum_{n=N}^{\infty} \frac{1}{n^2} \leq \left(\frac{\varepsilon}{2} \right)^2.$$

Carefully note that $N(\varepsilon)$ is independent of the point in A. Choose K to be an integer with $2\sqrt{N} \leq \varepsilon K$. Let A_ε be the set of points $y = \{y_n\}$ in A such that $y_n = 0$ for $n \geq N$, and such that the values of y_i, $i = 1, \ldots, N$, are restricted to the numbers $m/K \, (|m| \leq K)$, in the spirit of Example 4. It is then a simple matter to argue that for an arbitrary $x = \{x_1, x_2, \ldots\}$ in A there is a point y in A_ε such that $|x_i - y_i| \leq 1/K$, $i = 1, \ldots, N$, and

$$d_2^2(x,y) = \sum_{n=1}^{N} |x_n - y_n|^2 + \sum_{n=N+1}^{\infty} |x_n|^2 \leq \frac{N}{K^2} + \frac{\varepsilon^2}{4} \leq \varepsilon^2.$$

It follows immediately that A is totally bounded. ∎

This argument generalizes, and we have the following result.

3.16.5 THEOREM.   *A bounded set A in (l_2, d_2) is totally bounded if and only if for every $\varepsilon > 0$ there is a $N = N(\varepsilon)$ (independent of x) such that*

$$\sum_{n=N}^{\infty} |x_n|^2 \leq \varepsilon$$

for all $x = \{x_1, x_2, \ldots\}$ in A.

EXAMPLE 6.   Let d_∞ be the sup-metric on $C[0,T]$. The unit ball

$$B = \{x \in C[0,T] : d_\infty(0,x) \le 1\}$$

is a bounded set but not totally bounded. (We ask the reader to verify this.) We will show that given an $L \ge 0$, the subset of B given by

$$B_L = \{x \in B : |x(t) - x(s)| \le L|t - s|\}$$

is totally bounded.

Let $\varepsilon > 0$ be given. We construct an ε-net A_ε for B_L as follows: First choose an integer $K > 0$ such that $3 \le K\varepsilon$. Now choose N so that $3LT \le N\varepsilon$ and divide the interval $[0,T]$ into N equal parts $0 = t_0 < t_1 < \cdots < t_N = T$, where $t_i - t_{i-1} = h = TN^{-1}$. Now let A_ε be the collection of all continuous piecewise linear functions p defined on $[0,T]$ such that

$$p(t_i) = j/K \qquad \text{for some integer } j, \qquad |j| \le K, \qquad \text{and} \qquad i = 1, 2, \ldots, N,$$

and p is linear between t_i and t_{i+1} with $|p(t_{i+1}) - p(t_i)| \le Lh$. Clearly A_ε is finite. In fact A_ε contains no more than $(2K + 1)(KLh)^N$ elements. An illustration is given in Figure 3.16.4.

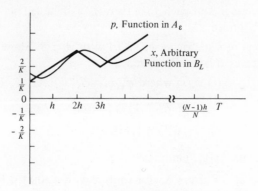

Figure 3.16.4.

Let $x \in B_L$. We claim that there is a p in A_ε that satisfies:

$$|x(t_i) - p(t_i)| \le \frac{1}{K}, \qquad 1 \le i \le N.$$

Indeed, since $|x(t_i) - x(t_{i-1})| \le Lh \le \varepsilon/3$, one then has for $t_i < t < t_{i+1}$

$$|x(t) - p(t)| \le |x(t) - x(t_i)| + |x(t_i) - p(t_i)| + |p(t_i) - p(t)|$$

$$< 2L|t - t_i| + \frac{\varepsilon}{3}$$

$$\le 2Lh + \frac{\varepsilon}{3} \le \varepsilon.$$

By our choice of K we get $d_\infty(x,p) \leq \varepsilon$, which shows that A_ε is an ε-net and that B_L is totally bounded. ∎

The reader can probably appreciate that total boundedness is a useful concept. Indeed the notion of approximating an arbitrary point by a special point from a pre-assigned finite collection has applications in many areas including numerical analysis. Going beyond this, we will show in the next section that total boundedness is intimately connected to the more general concept of compactness. In fact, we will eventually show (Theorem 3.17.13) that a metric space is compact if and only if it is (1) totally bounded and (2) complete.

EXERCISES

1. Let $A \subset R^n$, where R^n has the metric

$$d(x,y) = \sum_{i=1}^n |x_i - y_i|.$$

   Show that A is totally bounded if and only if A is bounded.

2. (Continuation of Exercise 1.) Find a metric on R^n that is equivalent to the metric in Exercise 1 and such that there is a bounded set $A \subset R^n$ where A is not totally bounded.

3. Let A, B be sets in a metric space (X,d) with $A \subset B$. Show that A is totally bounded whenever B is totally bounded.

4. Let $k(t,s) = \sum_{i=1}^n \phi_i(t)\, \psi_i(s)$ where $\phi_1, \ldots, \phi_n, \psi_1, \ldots, \psi_n$ belong to $L_2[0,T]$ where $0 < T \leq \infty$. Define $y = Kx$ by

$$y(t) = \int_0^T k(t,s)x(s)\, ds.$$

   Show that K maps $L_2[0,T]$ into itself. Let

$$A = \{\, y: y = Kx \text{ for some } x \text{ with } d_2(0,x) \leq 1 \},$$

   where d_2 is the usual metric on $L_2[0,T]$. Show that A is totally bounded.

5. Prove Lemma 3.16.4. [*Hint*: Consider a sequence of ε-nets for $\varepsilon = 2^{-n}$, $n = 1, 2, \ldots$.]

6. Discuss the concept of total boundedness in a pseudometric space.

7. Prove Theorem 3.16.5.

The following four exercises should be reconsidered after studying Section 3.17.

8. This exercise is the vector version of Example 6. Consider the space $X = C[I,R^n]$ of continuous functions $x(\cdot)$ defined on $I = [0,T]$ with values in R^n. Assume that R^n has the metric

$$d_1(x,y) = \sum_{i=1}^n |x_i - y_i|$$

and that X has the metric

$$d_\infty(x(\cdot),y(\cdot)) = \sup_{t \in I} d_1(x(t),y(t)).$$

Let B and B_L be defined by

$$B = \{x(\cdot) \in X: d_\infty(0,x(\cdot)) \leq 1\}$$
$$B_L = \{x(\cdot) \in B: d_1(x(t),x(s)) \leq L|t - s|\}.$$

Show that B is not totally bounded. Show that B_L is totally bounded.

9. Let $x(t) = e^{tA}x_0$ denote the solution of the linear differential equation $\dot{x}' = Ax$ with $x(0) = x_0 \in R^n$. Let d_1 be defined as in Exercise 8 and let S_K denote the collection of all such solutions of all such equations subject to the conditions

$$d_1(0,x_0) \leq 1,$$
$$\max_{i,j} |a_{ij}| \leq K, \tag{3.16.1}$$

where $A = (a_{ij})$. Show that S_K is a totally bounded subset of $C(I,R^n)$. (See Exercise 8.) [*Hint*: Observe that if A satisfies (3.16.1), then

$$d_1(0,Ax) \leq Kd_1(0,x)$$

for all $x \in R^n$.]

10. Consider the control differential equation

$$\dot{x}' = Ax + Bu, \tag{3.16.2}$$

where $x \in R^n$, $u \in R^m$, and A and B are fixed constant matrices of size $n \times n$ and $n \times m$, respectively. Let D denote the collection of all solutions $x(\cdot)$ of (3.16.2) subject to the conditions: $x(0) = 0$ and $d_1(0,u(t)) \leq 1$. Show that D is totally bounded in $C(I,R^n)$. (Compare with Exercise 8.)

11. It is known that if $x(t)$ is a solution of the Gronwall Inequality

$$|x(t)| \leq a(t) + \int_0^t |x(s)| \, b(s) \, ds, \qquad t \geq 0, \tag{3.16.3}$$

where $a(\cdot)$ and $b(\cdot)$ are continuous, nonnegative functions, then

$$|x(t)| \leq a_\infty(t)e^{B(t)},$$

where $a_\infty(t) = \max_{0 \leq s \leq t} a(s)$ and $B(t) = \int_0^t b(s) \, ds$.

   Let $y(t) = y_0 + \int_0^t x(s) \, ds$ where $x(\cdot)$ satisfies (3.16.3). Let $a(\cdot)$ come from a set A, $b(\cdot)$ from a set B, and let the corresponding $y(\cdot)$ range in a set Y. Find conditions on A, B and y_0 in order that Y be totally bounded in $C[0,T]$, where $C[0,T]$ has the sup-metric.

17.   COMPACTNESS

There are at least four equivalent ways of defining compactness in metric spaces. The definition we choose is based on the aspect of compactness that is used most in applications, namely, sequential compactness. We shall show in this section that sequential compactness in a metric space is equivalent to three other forms of compactness, Theorem 3.17.13.

Before we state the definition let us recall the concept of a subsequence. A sequence in a set X is a mapping x of the positive integers I^+ into X. The value of x at the point n is denoted by x_n. Now let $g: I^+ \to I^+$ be a strictly increasing mapping, that is, $g(m) < g(n)$ whenever $m < n$. A subsequence of $x: I^+ \to X$ is a sequence of the form $x \circ g: I^+ \to X$ where $g: I^+ \to I^+$ is strictly increasing. For example, let $x_n = 1/n$ and $g(n) = n^2$, then $x \circ g(n) = 1/n^2$. In other words, $\{1, \frac{1}{4}, \frac{1}{9}, \ldots\}$ is a subsequence of $\{1, \frac{1}{2}, \frac{1}{3}, \ldots\}$. In general if $\{x_1, x_2, \ldots\}$ is any sequence, and if $\{n_k\}$ are positive integers with $n_1 < n_2 < \ldots$, then $\{x_{n_1}, x_{n_2}, \ldots\}$ is a subsequence of $\{x_1, x_2, \ldots\}$.

We shall say that the sequence $\{x_1, x_2, \ldots\}$ *contains a convergent subsequence* if at least one of its subsequences is convergent. For example, the sequence $\{1, \frac{1}{2}, 3, \frac{1}{4}, 5, \frac{1}{6}, \ldots\}$ has $\{1, \frac{1}{2}, \frac{1}{4}, \frac{1}{6}, \ldots\}$ as a convergent subsequence. Note that the sequence itself is not convergent, and it contains many subsequences, such as $\{1, 3, 5, 7, \ldots\}$, which are not convergent.

We are now prepared to define sequential compactness.

3.17.1 DEFINITION.   A metric space (X, d) is said to be *sequentially compact* if every sequence in (X, d) contains a convergent subsequence. A set $A \subset X$ is said to be *sequentially compact* if the subspace (A, d) is sequentially compact. This means that every sequence in A contains a subsequence that converges to a point *in A*. For example, the set $(0, 1]$ is not sequentially compact in R since the sequence $\{1/n\}$ does not have a subsequence with a limit *in* $(0, 1]$.

Roughly speaking, a sequentially compact metric space is so " crowded " that no matter how hard one tries to choose a sequence, an infinite number of the elements will always " pile up " around at least one point in the metric space.

The following lemma is an immediate consequence of the Closed Set Theorem.

3.17.2 LEMMA.   *If A is a sequentially compact set in a metric space (X, d), then A is a closed set.*

3.17.3 THEOREM.   *Let (X, d) be a sequentially compact metric space. A set $A \subset X$ is sequentially compact if and only if A is closed.*

Proof: It follows from the last lemma that if A is sequentially compact, then it is closed. So now assume that A is closed. Let $\{x_n\}$ be a sequence in $A \subset X$. Since X is sequentially compact we can find a subsequence $\{x_{n_k}\}$ with limit x_0 in X. Since A is closed, it follows from the Closed Set Theorem that $x_0 \in A$. Hence A is sequentially compact.   ∎

It is important to note that we do assume (X, d) to be sequentially compact in the last theorem. The theorem is not true otherwise. That is, if (X, d) is not sequentially compact, then $A = X$ is a closed set that is not sequentially compact.

A set $A \subset X$ is said to have *compact closure* in (X,d) if the closure of A is a sequentially compact set in (X,d). In this case one sometimes says that A is *relatively compact* or *conditionally compact*.

We emphasized in the definition of sequential compactness that the limit of the subsequence had to be *in* the given set A. Not too surprisingly, when the limit of the subsequence exists but is not necessarily in A, we get compact closure.

3.17.4 LEMMA. *Let $A \subset X$ where (X,d) is a metric space. The following statements are equivalent:*

(a) *A has compact closure, that is, $\bar{A}$ is sequentially compact.*
(b) *Every sequence in A has a subsequence that converges (to a point in X).*

Proof: (a)$\Rightarrow$(b). If $\bar{A}$ is sequentially compact, then every sequence in $A \subset \bar{A}$ has a subsequence that converges to a point in $\bar{A} \subset X$.

(b)$\Rightarrow$(a). Let $\{x_n\}$ be a sequence in $\bar{A}$. It follows from the definition of the closure $\bar{A}$ that there is a sequence $\{y_n\}$ in A such that $d(x_n, y_n) \le 1/n$. From statement (b) we can choose a subsequence, call it $\{y_{n'}\}$, of $\{y_n\}$ such that $\{y_{n'}\}$ converges. Say that $z = \lim y_{n'}$. (It is clear that $z \in \bar{A}$.) Since

$$d(z, x_{n'}) \le d(z, y_{n'}) + d(y_{n'}, x_{n'}) \to 0,$$

we see that $z = \lim x_{n'}$. Hence $\bar{A}$ is sequentially compact. ∎

How is sequential compactness related to some of the other concepts introduced previously? More precisely, how is it related to completeness? To total boundedness? Before we answer this let us prove the following lemma.

3.17.5 LEMMA. *Let (X,d) be a sequentially compact metric space and let $\{M_n\}$ be a decreasing sequence (that is, $M_n \supset M_{n+1}$) of nonempty closed sets. Then $\bigcap_{n=1}^{\infty} M_n$ is nonempty.*

Proof: Choose $x_n \in M_n$ for $n = 1, 2, \ldots$. Since $\{M_n\}$ is decreasing one has $x_n \in M_N$ for all $n \ge N$ and all N. Since (X,d) is sequentially compact, there is a convergent subsequence $\{x_{n_k}\}$ with $x_0 = \lim x_{n_k}$ for some $x_0 \in X$. Since M_N is closed, it follows from the Closed Set Theorem that $x_0 \in M_N$ for every N. Hence $x_0 \in \bigcap_{N=1}^{\infty} M_N$. ∎

3.17.6 LEMMA. *If (X,d) is sequentially compact, then it is complete.*

Proof: Let $\{M_n\}$ be any decreasing sequence of nonempty closed sets with $\operatorname{diam}(M_n) \to 0$ as $n \to \infty$. It follows from Lemma 3.17.5 that $\bigcap_{n=1}^{\infty} M_n$ contains exactly one point. (Why?) It now follows from Theorem 3.13.11 that (X,d) is complete. ∎

3.17.7 LEMMA. *If (X,d) is sequentially compact, then it is totally bounded.*

Proof: We shall prove this by contradiction. If (X,d) does not contain an ε-net for some $\varepsilon > 0$, then we can find a sequence of points $\{x_n\}$ in X with the property that $d(x_n,x_m) \geq \varepsilon$ whenever $n \neq m$. But this implies that the sequence $\{x_n\}$ contains no convergent subsequence, compare with Exercise 8, Section 6. Hence (X,d) is not sequentially compact, and this is a contradiction.   ∎

One might paraphrase the last two lemmas as follows: Lemma 3.17.6 says that a sequentially compact metric space does not have any "holes" in it and Lemma 3.17.7 says that such a space is "cramped" or "crowded."

3.17.8 THEOREM.   *A metric space (X,d) is sequentially compact if and only if it is totally bounded and complete.*

Proof: Lemmas 3.17.6 and 3.17.7 show that a sequentially compact space is totally bounded and complete. Thus we must show here that total boundedness and completeness imply sequential compactness.

Let $S_1 = \{x_1(1),x_2(1),x_3(1),\ldots\}$ be an arbitrary sequence in (X,d). We denote the sequence with a one in parenthesis to distinguish it from subsequent sequences. Since (X,d) is totally bounded, there exists a finite collection of open balls, each with radius 2^{-1}, that covers (X,d). It follows that at least one of these open balls contains a subsequence of S_1. Denote this subsequence by $S_2 = \{x_1(2),x_2(2),\ldots\}$. Using the total boundedness of (X,d) again, there exists a subsequence of S_2 that is contained in an open ball of radius 2^{-2}. Denote this subsequence by $S_3 = \{x_1(3),x_2(3),\ldots\}$. We continue successively forming subsequences of subsequences in this manner so that the subsequence $S_n = \{x_1(n),x_2(n),\ldots\}$ lies in an open ball of radius 2^{-n}. We thus obtain the following infinite array:

$$S_1:   x_1(1), x_2(1), x_3(1), \ldots$$
$$S_2:   x_1(2), x_2(2), x_3(2), \ldots$$
$$S_3:   x_1(3), x_2(3), x_3(3), \ldots$$
$$\cdots$$

Next let S be the sequence made up of the diagonal entries in this array, that is, $S = \{x_1(1),x_2(2),x_3(3),\ldots\}$. Owing to our method of construction, S is a subsequence of S_1 and S is a Cauchy sequence. (Why?) Since (X,d) is complete, S converges. Thus, each sequence in (X,d) contains a convergent subsequence, and (X,d) is sequentially compact.   ∎

There is yet another way of characterizing sequentially compact metric spaces which is often useful.

3.17.9 DEFINITION.   A metric space (X,d) is said to possess the *Bolzano-Weierstrass property* if every infinite subset of (X,d) has at least one point of accumulation. A set A in (X,d) is said to possess the *Bolzano-Weierstrass property* if the space (A,d) has this property.

Note that if X or A is finite, it possesses the Bolzano-Weierstrass property because it has no infinite subsets. The intuitive idea behind the Bolzano-Weierstrass property is similar to that behind sequential compactness: No matter how hard one tries, one cannot select an infinite set that does not "pile up" around at least one point in the space. Not too surprisingly this property is equivalent to sequential compactness. We ask the reader to verify this.

There is still another form of compactness that is equivalent to sequential compactness. This is the so-called Heine-Borel compactness. To state this we need the following definition.

3.17.10 DEFINITION. Let A be a set in a metric space (X,d). A collection of sets $\{M_\alpha\}$ in (X,d) is said to be a *covering* of A if $A \subset \bigcup_\alpha M_\alpha$. A subcollection $\{M_\beta\}$ of a covering $\{M_\alpha\}$ with the property that $A \subset \bigcup_\beta M_\beta$ is said to be a *subcovering* of $\{M_\alpha\}$. Any covering or subcovering made up entirely of open sets is said to be an *open covering* or *open subcovering*.

Next the definition of compactness.

3.17.11 DEFINITION. A metric space (X,d) is said to be *compact* (*Heine-Borel compact*) if every open covering of (X,d) contains a finite open subcovering. A set A in a metric space (X,d) is said to be *compact* if the metric space (A,d) is compact; that is, if each open covering of A contains a finite open subcovering.

Let us emphasize that we ask that *each* open covering contains a finite open subcovering. We are *not* saying that a set A has a finite open covering, or that some open coverings have finite subcoverings. For example, the topology $\mathcal{T}$ of a metric space (X,d) is an open covering of (X,d). It contains a finite subcovering; namely, the collection consisting of X alone. In fact, any open covering that contains X contains a finite open subcovering. The interesting point about compact spaces is that even open coverings made up of "very small" open sets contain finite subcoverings.

This version of compactness is probably the hardest to understand. However, it is this version that is really the most fundamental. The reason for this is that Heine-Borel compactness can easily be generalized to topological spaces that are not metrizable.

For our purposes we have the following equivalence.

3.17.12 THEOREM. *A metric space (X,d) is sequentially compact if and only if it is compact.*

We shall outline a proof of this theorem in the exercises. It should be noted that it is not important, for the purpose of the book, that the reader master this proof. We shall only use this result as an excuse to use the phrase "compact" in place of "sequentially compact." All of our proofs will be based on the concept of sequential compactness and Theorem 3.17.8.

We have given four versions of compactness, all of them equivalent in metric spaces. Let us summarize this.

3.17.13 THEOREM. (COMPACTNESS THEOREM.) *Let* (X,d) *be a metric space. Then the following statements are equivalent:*

(a) (X,d) *is compact.*

(b) (X,d) *is sequentially compact.*

(c) (X,d) *is complete and totally bounded.*

(d) (X,d) *possesses the Bolzano-Weierstrass property.*

EXAMPLE 1. Let R be the real line with the usual metric, and let $A \subset R$. When is A compact? The following theorem, originally proved by Heine and Borel, characterizes compact subsets of R.

3.17.14 THEOREM. *A set* $A \subset R$ *is compact if and only if it is closed and bounded.*

Proof: If A is compact, then it is closed and totally bounded (Lemma 3.17.2 and Theorem 3.17.8). However, every totally bounded set is bounded (Lemma 3.16.3).

Now assume that A is closed and bounded. Since R is complete, it follows that A is complete (Theorem 3.13.5). Since A is bounded, it is totally bounded. (Why?) Therefore A is compact (Theorem 3.17.13). ∎

Is compactness a topological property? How does it behave under continuous mappings? Let us now look into these questions.

3.17.15 THEOREM. *Let* $f: X \to Y$ *be a continuous function, where* (X,d) *and* (Y,σ) *are metric spaces. If* (X,d) *is compact, then the range* $f(X)$ *is a compact set in* (Y,σ).

Proof: Let $\{y_n\}$ be a sequence in the range $f(X)$. Then there are corresponding points $\{x_n\}$ in X with $y_n = f(x_n)$. Since (X,d) is compact we can find a subsequence of $\{x_n\}$ that converges in X, say that $x_{n_k} \to x$. Since f is continuous one has $f(x_{n_k}) \to f(x)$ in $f(X)$ by Theorem 3.7.2. Hence $\{y_n\}$ has a convergent subsequence and $f(X)$ is compact. ∎

Since compactness is preserved under continuous mappings, it is obviously preserved by homeomorphisms.

3.17.16 COROLLARY. *Let* (X,d) *and* (Y,σ) *be homeomorphic metric spaces. Then* (X,d) *is compact if and only if* (Y,σ) *is compact.*

So compactness is a topological property, and compactness is equivalent to total boundedness with completeness. But wait! We have very carefully pointed out that completeness is not a topological property (Section 13). It also happens that total boundedness is not a topological property. For example, let $X = (0,1]$ and $Y = [1,\infty)$ where X and Y are equipped with the usual metric. Then the mapping $y = 1/x$ is a homeomorphism of X onto Y. But (X,d) is totally bounded and (Y,d) is not. Thus, completeness and total boundedness separately are not topological properties, whereas taken together they are!

We are often interested in products of compact metric spaces. Fortunately, products do not offer any problems.

3.17.17 THEOREM. *Let (X_1,d_1) and (X_2,d_2) be compact metric spaces. Then the product space (X,d), where $X = X_1 \times X_2$ and $d(x,y) = d_1(x_1,y_1) + d_2(x_2,y_2)$, is compact.*

Proof: Let $\{x_n\}$ be an arbitrary sequence in (X,d). Then $x_n = (x_1{}^n, x_2{}^n)$, where $\{x_1{}^n\}$ and $\{x_2{}^n\}$ are sequences in (X_1,d_1) and (X_2,d_2), respectively. Since (X_1,d_1) is sequentially compact, the sequence $\{x_1{}^n\}$ contains a convergent subsequence $\{x_1{}^{n_i}\}$. Similarly, the corresponding sequence $\{x_2{}^{n_i}\}$ in (X_2,d_2) contains a convergent subsequence $\{x_2{}^{n_{ij}}\}$. Since any subsequence of a convergent sequence is convergent, the subsequence $\{x_1{}^{n_{ij}}\}$ taken from $\{x_1{}^{n_i}\}$ is convergent. It follows that the sequence $(x_1{}^{n_{ij}}, x_2{}^{n_{ij}})$ in (X,d) is convergent. Hence, each sequence in (X, d) contains a convergent subsequence, and (X,d) is compact. ∎

3.17.18 COROLLARY. *Let (X_1,d_1) and (X_2,d_2) be compact metric spaces. Then the metric space (X,d'), where $X = X_1 \times X_2$ with d' any metric equivalent to $d(x,y) = d_1(x_1,y_1) + d_2(x_2,y_2)$, is compact.*

The proof of this corollary follows from the last theorem and the fact that compactness is a topological property.

3.17.19 COROLLARY. *Let $(X_1,d_1), \ldots, (X_n,d_n)$ be compact metric spaces. Then the metric space (X,d), where $X = X_1 \times X_2 \times \cdots \times X_n$ and*

$$d(x,y) = d_1(x_1,y_1) + \cdots + d_n(x_n,y_n)$$

or any equivalent metric, is compact.

A simple, though important, consequence of the above is the following characterization of compact sets in R^n, or C^n. We ask the reader to prove this result.

3.17.20 THEOREM. *Let R^n (or C^n) be given with the metric*

$$d(x,y) = \sum_{i=1}^{n} |x_i - y_i|.$$

Then a set $A \subset R^n$ (or C^n) is compact if and only if it is closed and bounded.

A useful observation concerns real-valued functions defined on compact sets.

3.17.21 THEOREM.   *Let f be a continuous real-valued function defined on a compact metric space (X,d). Then f is bounded, that is*

$$M = \sup\{f(x): x \in X\}$$

and

$$m = \inf\{f(x): x \in X\}$$

are finite. Moreover, there are points $x_{\min}$ and $x_{\max}$ in X such that $f(x_{\max}) = M$ and $f(x_{\min}) = m$.

Proof: It follows from Theorem 3.17.15 that $f(X)$ is a compact set in R, and Theorem 3.17.14 implies that $f(X)$ is closed and bounded. Therefore M and m are finite. Since $f(X)$ is closed, it is clear that M and m are in $f(X)$. Hence there are points $x_{\max}$ and $x_{\min}$ in X such that $f(x_{\max}) = M$ and $f(x_{\min}) = m$.   ∎

EXAMPLE 2. (ARZELA-ASCOLI'S THEOREM.)   Let (X,d_1) be a compact metric space and let (Y,d_2) be a complete metric space. Form the space $C = C(X,Y)$ of continuous functions defined on X with range in Y. If $f, g \in C$, define a metric ρ by

$$\rho(f,g) = \sup\{d_2(f(x), g(x)): x \in X\}.$$

It follows from Theorems 3.17.17 and 3.17.21, and the fact that

$$d_2(f(\cdot), g(\cdot)): X \to R$$

is continuous, that $\rho(f,g)$ is finite for all f and g.

A typical example occurs when $X = [a,b]$ is an interval, $Y = R$, the reals, and d_1 and d_2 are the usual metric. In this case (C,ρ) becomes the space $C[a,b]$ with the sup-metric.

Let us now show that (C,ρ) is a complete metric space. This is done as follows: Let $\{f_n\}$ be a Cauchy sequence in (C,ρ). Since $d_2(f_n(x),f_m(x)) \le \rho(f_n, f_m)$ for each x in X, we see that $\{f_n(x)\}$ is a Cauchy sequence in (Y,d_2) for each x in X. Since (Y,d_2) is complete this means that the function f defined by

$$f(x) = \lim_{n \to \infty} f_n(x)$$

maps X into Y. Now let us show that f is continuous. Let $\varepsilon > 0$ be given and choose an integer N so that $\rho(f_n,f_m) \le \varepsilon$ whenever $n, m \ge N$. Let $x \in X$. Since f_N is continuous we can find a $\delta > 0$ such that $d_2(f_N(x),f_N(x')) \le \varepsilon$, whenever $d_1(x,x') \le \delta$. Now choose $n \ge N$ and x' with $d_1(x,x') \le \delta$. One then has

$$d_2(f(x), f(x')) \le d_2(f(x), f_n(x)) + d_2(f_n(x), f_N(x)) + d_2(f_N(x), f_N(x'))$$
$$+ d_2(f_N(x'), f_n(x')) + d_2(f_n(x'), f(x'))$$
$$\le d_2(f(x), f_n(x)) + 3\varepsilon + d_2(f_n(x'), f(x')).$$

By taking the limit as $n \to \infty$ one gets

$$d_2(f(x), f(x')) \le 3\varepsilon$$

whenever $d_1(x, x') \le \delta$. Hence f is continuous.

Finally we must show that $\rho(f_n, f) \to 0$ as $n \to \infty$. However, since $\{f_n\}$ is a Cauchy sequence, for every $\varepsilon > 0$ there is a N such that

$$d_2(f_n(x), f_m(x)) \le \rho(f_n, f_m) \le \varepsilon, \qquad \text{for all } x \in X$$

and all $n, m \ge N$. If we let $m \to \infty$, then the above statement becomes

$$d_2(f_n(x), f(x)) \le \varepsilon, \qquad \text{for all } x \in X$$

and all $n \ge N$; that is, $\rho(f_n, f) \le \varepsilon$ whenever $n \ge N$. Hence $\rho(f_n, f) \to 0$ as $n \to \infty$.

Let A be a collection of functions from C. We seek conditions on A that will ensure that the closure $\bar{A}$ is a compact set in (C, ρ). For this we need two definitions:

3.17.22 DEFINITION. A family of functions A in C is said to be *pointwise compact* if for each $x \in X$ the set $\{f(x) : f \in A\}$ has compact closure in (Y, d_2).

3.17.23 DEFINITION. A family of functions A in C is said to be *equi-continuous* if for each $x \in X$ and $\varepsilon > 0$ there is a $\delta > 0$ such that $d_2(f(x), f(x')) \le \varepsilon$ for every f in A, whenever $d_1(x, x') \le \delta$. [*Note*: δ depends on ε and x but *not* on f. Hence the term "equi-continuity."] If δ can be chosen independent of x as well, the family is said to be *uniformly equi-continuous*.

3.17.24 THEOREM. (ARZELA-ASCOLI.) *Let A be a set in (C, ρ). Then the following statements are equivalent:*

(a) *The closure $\bar{A}$ is compact.*
(b) *The family A is pointwise compact and equi-continuous.*

Proof: (a) $\Rightarrow$ (b). In order to show that A is pointwise compact we fix x and choose a sequence $\{y_n\}$ in $\{f(x) : f \in A\}$. This means that $y_n = f_n(x)$ for some $f_n \in A \subset \bar{A}$. Since $\bar{A}$ is compact, this means that we can find a subsequence $\{f_{n'}\}$ that converges in (C, ρ), say that $f_{n'} \to f$ as $n' \to \infty$. If we let $y = f(x)$, then we see that $y_{n'} \to y$ as $n' \to \infty$. Hence A is pointwise compact. (The same argument shows that $\bar{A}$ is pointwise compact.)

Next we shall show that A is equi-continuous. Since $\bar{A}$ is compact, it is totally bounded, by Theorem 3.17.8. Let $\{f_1, \ldots, f_N\}$ be an ε-net for $\bar{A}$. If f is any point in $\bar{A}$ then there is an f_i in the ε-net such that

$$d_2(f(x), f_i(x)) \le \rho(f, f_i) \le \varepsilon.$$

It follows that

$$d_2(f(x_0), f(x')) \le 2\varepsilon + d_2(f_i(x_0), f_i(x')). \tag{3.17.1}$$

(Why?) Since each of the functions $\{f_1,\ldots,f_N\}$ is continuous at x_0, this means that we can find a $\delta = \delta(x_0,\varepsilon) > 0$ such that

$$d_2(f_i(x_0),f_i(x')) \leq \varepsilon \qquad (1 \leq i \leq N), \qquad (3.17.2)$$

whenever $d_1(x_0,x') \leq \delta$. By combining (3.17.1) and (3.17.2), we see that $\bar{A}$ is equi-continuous. Hence A is equi-continuous.

(b) $\Rightarrow$ (a). Since X is compact, we know that it is separable. (Theorem 3.17.8 and Lemma 3.16.4.) Let $D = \{x_n\}$ be a countable dense set in X. We shall use Lemma 3.17.4 to show that $\bar{A}$ is compact. Let $\{f_n\}$ be a sequence in A. Since $\{f_n(x_1)\}$ lies in a compact set in Y we can find a convergent subsequence, which we shall denote by $\{f_n^{(1)}\}$. Since $\{f_n^{(1)}(x_2)\}$ lies in a compact set in Y, we can find a convergent subsequence, which we shall denote by $\{f_n^{(2)}\}$. Continuing with x_3, x_4, and so on, we construct a family

$$\{f_n^{(1)}\}, \{f_n^{(2)}\}, \{f_n^{(3)}\}, \{f_n^{(4)}\}, \ldots$$

of sequences, each a subsequence of the preceding, and with the property that $\{f_n^{(k)}(x_i)\}$ converges for $1 \leq i \leq k$. It then follows that the diagonal sequence

$$\{f_n^{(n)}\} = \{f_1^{(1)}, f_2^{(2)}, f_3^{(3)}, \ldots\}$$

has the property that $\{f_n^{(n)}(x_i)\}$ converges for all x_i in D. Let $f(x_i) = \lim f_n^{(n)}(x_i)$, for $x_i \in D$.

To complete the proof we shall show three things: (i) $\{f_n^{(n)}(x)\}$ converges for each x in X. (We shall let $f(x) = \lim_{n \to \infty} f_n^{(n)}(x)$.) (ii) The limit function f is continuous. (iii) $\rho(f_n^{(n)},f) \to 0$ as $n \to \infty$.

In order to show that $\{f_n^{(n)}(x)\}$ converges for each x in X, we shall show that this sequence is a Cauchy sequence in Y. Let $\varepsilon > 0$ be given and choose $\delta > 0$ so that $d_2(f_n^{(n)}(x),f_n^{(n)}(x')) \leq \varepsilon$, for all n, whenever $d_1(x,x') \leq \delta$. Now choose x_i in D so that $d_1(x,x_i) \leq \delta$. Since $\{f_n^{(n)}(x_i)\}$ is a convergent sequence it is a Cauchy sequence. Hence, there is a N such that $d_2(f_n^{(n)}(x_i),f_m^{(m)}(x_i)) \leq \varepsilon$ whenever $n, m \geq N$. It follows that

$$d_2(f_n^{(n)}(x),f_m^{(m)}(x)) \leq d_2(f_n^{(n)}(x),f_n^{(n)}(x_i)) + d_2(f_n^{(n)}(x_i),f_m^{(m)}(x_i))$$
$$+ d_2(f_m^{(m)}(x_i),f_m^{(m)}(x)) \leq 3\varepsilon$$

whenever $n, m \geq N$. Hence $\{f_n^{(n)}(x)\}$ is a Cauchy sequence. Since Y is complete we see that $f(x) = \lim_{n \to \infty} f_n^{(n)}(x)$ is defined for all x in X.

The continuity of f follows from the equi-continuity, since

$$d_2(f(x),f(x')) = \lim_{n \to \infty} d_2(f_n^{(n)}(x),f_n^{(n)}(x')). \qquad (3.17.3)$$

That is if $\varepsilon > 0$ is given, we choose $\delta > 0$ so that by (3.17.3) and the equi-continuity of A we have $d_2(f(x),f(x')) \leq \varepsilon$, whenever $d_1(x,x') \leq \delta$.

The main step in showing that $\rho(f_n^{(n)},f) \to 0$ is to note that if $\{x_n\}$ is any sequence in X with $\lim x_n = x_0$, then $\lim f_n^{(n)}(x_n) = f(x_0)$. Indeed one does have

$$d_2(f_n^{(n)}(x_n),f(x_0)) \leq d_2(f_n^{(n)}(x_n),f_n^{(n)}(x_0)) + d_2(f_n^{(n)}(x_0),f(x_0))$$
$$\leq \varepsilon + d_2(f_n^{(n)}(x_0),f(x_0))$$

provided $d_1(x_n, x_0) \leq \delta$, where δ is given by the equi-continuity of A. By taking the limit as $n \to \infty$ we get

$$\limsup_{n \to \infty} d(f_n^{(n)}(x_n), f(x_0)) \leq \varepsilon.$$

Since ε is arbitrary we have $\lim f_n^{(n)}(x_n) = f(x_0)$. The final step, namely, $\rho(f_n^{(n)}, f) \to 0$, is now an easy exercise, see Exercise 15. ∎

EXAMPLE 3.   Let $u(t)$ be the downward thrust of a rocket of mass m. Suppose that $x(t)$ denotes the altitude of the rocket. Assume that $u(t) \geq 0$ and that the maximum thrust available is $M > 0$, that is, $u(t) \leq M$. Let the time interval over which the rocket burns be $[0,T]$, and assume that $u(t)$ is an element of $C[0,T]$.

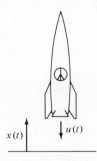

Figure 3.17.1.

Further, assume that changes in thrust are limited as follows:

$$|u(t) - u(s)| \leq V|t - s|, \qquad \text{for all } t, s \in [0,T],$$

where $V > 0$. Suppose the mathematical model for this system is given by

$$m \frac{d^2 x}{dt^2} = u - mg,$$

where

  (i) g is the gravitational constant.
 (ii) $mg < M$.
(iii) $u(0) = mg$.
 (iv) $x(0) = 0$ and $\dfrac{dx}{dt}(0) = 0$.
  (v) $\dfrac{dx}{dt}(T) = 0$.

One optimum control problem would be to select an input u so that $x(T)$ is maximized. Many techniques from optimum control theory are so-called indirect methods which start assuming that an optimum solution exists. For example, we might start by assuming u^* to be the optimum input and then build on this assumption. It is important, then, to be able to show that u^* does indeed exist. By using compactness, one can show that an optimum input u^* exists. Use Theorems 3.17.21 and 3.17.24. (Also compare with Example 6, Section 16.)   ∎

EXAMPLE 4.   Let Ω be a bounded open set in R^n and consider the space $L_p(\Omega)$, $1 \le p < \infty$, with the metric given by

$$d_p(f,g) = \left\{ \int_\Omega |f(x) - g(x)|^p dx \right\}^{1/p}.$$

Let $A = \{f\}$ denote a collection of functions from $L_p(\Omega)$ and assume that each function $f \in A$ is continuous on $\bar{\Omega}$. We claim that if A is pointwise compact (on $\bar{\Omega}$) and equi-continuous (on $\bar{\Omega}$), then the closure $\bar{A}$ is compact in $L_p(\Omega)$. The proof of this assertion is not difficult. It follows easily from the Arzela-Ascoli Theorem. Indeed, if $\{f_n\}$ is any sequence in A, then the Arzela-Ascoli Theorem assures us that there is a subsequence that converges uniformly on $\bar{\Omega}$, say that $f = \lim f_{n_k}$. It remains only to show that this subsequence $\{f_{n_k}\}$ converges to f in the metric d_p on $L_p(\Omega)$. However,

$$d_p(f, f_{n_k})^p = \int_\Omega |f(x) - f_{n_k}(x)|^p \, dx$$
$$\le \sup\{|f(x) - f_{n_k}(x)|^p : x \in \bar{\Omega}\} \cdot |\Omega|,$$

where $|\Omega|$ denotes the Lebesgue measure of Ω, which is finite. One then has $f_{n_k} \to f$ in (L_p, d_p).   ∎

EXAMPLE 5.   It is possible to characterize conditionally compact subsets of the space $L_p(-\infty, \infty)$ with the usual metric d_p when $1 \le p < \infty$. Specifically a set A in $L_p(-\infty, \infty)$ has compact closure if and only if

    (i) A is bounded, that is, there is a real number B such that $\int_{-\infty}^\infty |x(t)|^p dt \le B$ for all $x(\cdot)$ in A,

    (ii) $\lim_{T \to \infty} \left\{ \int_{-\infty}^{-T} |x(t)|^p \, dt + \int_T^\infty |x(t)|^p \, dt \right\} = 0$ uniformly over A, and

    (iii) $\lim_{\tau \to 0} \int_{-\infty}^\infty |x(\tau + t) - x(t)|^p \, dt = 0$ uniformly over A.

We will not prove this assertion here but instead we refer the reader to Dunford and Schwartz [1, pp. 298–301].   ∎

EXERCISES

1. Show that in the metric space (l_2, d_2) a set A has compact closure if and only if the following two conditions are satisfied:
   (a) A is bounded.
   (b) $\lim_{N \to \infty} \sum_{n=N}^\infty |x_n|^2 = 0$ uniformly over A.
   [*Hint*: See Theorem 3.16.5.]

2. What happens to Example 3 if we remove the restriction

$$|u(t) - u(s)| \leq V|t - s|?$$

3. Let d_∞ be the sup-metric on $C[0,T]$. Suppose that a set of output signals from a system (see Figure 3.17.2) are modeled by a set $A \subset C[0,T]$, where A is defined as the collection of all x in $C[0,T]$ such that

$$|x(t)| \leq 1, \qquad\qquad \text{for all } t \in [0,T],$$
$$|x(t_1) - x(t_2)| \leq V|t_1 - t_2|, \qquad \text{for } t_1, t_2 \in [0,T],$$

with $V > 0$. Assume that the output s of the system is not observed directly and that y, the output signal corrupted by additive noise $n \in C[0,T]$, is observed.

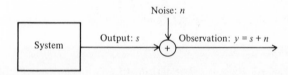

Figure 3.17.2.

Further assume that

$$|n(t)| \leq \frac{1}{10} \qquad \text{for all } t \in [0,T].$$

A set C of outputs, $C \subset A$, will be said to be distinguishable if $d(c_1,c_2) > 2/10$ for all $c_1, c_2 \in C$ such that $c_1 \neq c_2$. That is, the set of closed balls $\{B_{1/10}[c]|\, c \in C\}$ are pairwise disjoint. Using compactness, show that C can contain at most a finite number of output signals.

4. Discuss the concept of sequential compactness in a pseudometric space. Is the analog of Theorem 3.17.8 valid in this setting? What about Theorem 3.17.13?

5. Let A be a finite set in a metric space (X,d). Show that A is compact.

6. Let $\rho(x,y) \equiv 0$ be the trivial pseudometric on X. Characterize the (Heine-Borel) compact subsets of X.

7. Let $d(x,y) = 1 \ (x \neq y)$, $d(x,x) = 0$ be a metric on X. Characterize the compact subsets of X.

8. Choose sequences $\{a_n\}$ and $\{b_n\}$ in R so that $a_m \leq a_n \leq b_n \leq b_m$ whenever $m \leq n$. What is the set $\bigcap_{n=1}^{\infty} I_n$, where $I_n = [a_n, b_n]$?

9. Let $f: X \to Y$ be a one-to-one mapping of X onto Y. Assume that (X,d) and (Y,σ) are metric spaces and that (X,d) is compact. Assume that f is continuous. Show that f^{-1} is continuous. What happens if (X,d) is not compact?

10. A metric space (X,d) is said to be *locally compact* if every point x in X has compact local neighborhood. Which of the following spaces are locally compact?
    (a) R^n with $d(x,y) = \sum_{i=1}^{n} |x_i - y_i|$.
    (b) R with usual metric.
    (c) (l_2, d_2).
    (d) (l_p, d_p), $1 \le p \le \infty$.
    (e) (X,d) where $d(x,y) = 1$ $(x \ne y)$ and $d(x,x) = 0$.

11. (Continuation of Exercise 7, Section 5.) Let B^α be given by

$$B^\alpha = \{f \in C^\alpha(I): d^\alpha(0,f) \le 1\}.$$

Show that B^α is a compact set in $C_\beta(I)$ where $0 < \beta < \alpha \le 1$. [*Hint*: Apply the Arzela-Ascoli Theorem to the functions

$$g(t,s) = \frac{|f(t) - f(s)|}{|t - s|^\beta}$$

where $f \in B^\alpha$.]

12. (Continuation of Exercise 7, Section 15.) Consider the family of Volterra integral equations

$$x_n(t) = h_n(t) + \int_0^t k_n(t,s) f_n(x_n(s),s)\, ds, \qquad n = 1, 2, \ldots, \qquad (3.17.4)$$

for $0 \le t, s \le 1$. Assume that there are positive constants H, M, A, B, and K_n such that

$$|h_n(t)| \le H, |k_n(t,s)| \le M, |f_n(x,s)| \le A + B|x|$$

for all $0 \le t, s \le 1$, and all x, and

$$|f_n(x,s) - f_n(y,s)| \le K_n |x - y|$$

for all $0 \le s \le 1$, and all x, y. Assume further that h_n, k_n, and f_n are continuous.
    (a) Show that there are constants $\alpha > 0$ and $D > 0$ (independent of n) such that the (unique) solution $x_n(t)$ of (3.17.4) is defined for $0 \le t \le \alpha$ and satisfies $|x_n(t)| \le D$ for $0 \le t \le \alpha$.
    (b) Show that the sequence $\{x_n\}$ contains a uniformly convergent subsequence on $0 \le t \le \alpha$. [*Hint*: Apply the Arzela-Ascoli Theorem.]
    (c) Assume that $h_n \to h$, $k_n \to k$, and $f_n \to f$ where the convergence is uniform. Assume also that $x_n \to x$ uniformly for $0 \le t \le \alpha$. Show that $f_n(x(t),t) \to f(x(t),t)$ uniformly for $0 \le t \le \alpha$. Show that $x(t)$ satisfies

$$x(t) = h(t) + \int_0^t k(t,s) f(x(s),s)\, ds,$$

for $0 \le t \le \alpha$. [*Note*: The limit function f need not be Lipschitz continuous.]

13. Show that a continuous function defined on a compact metric space is uniformly continuous.

14. Use the result of Exercise 13 to show that "equi-continuity" and "uniform equi-continuity" are equivalent on compact metric spaces.

15. In Example 2, show that a sequence $\{f_n\}$ in (C,ρ) satisfies $\rho(f_n,f) \to 0$ as $n \to \infty$ if and only if $\lim f_n(x_n) = f(\lim x_n)$ for every convergent sequence $\{x_n\}$ in X.

16. In Example 2, show that a family A in (C,ρ) is equi-continuous if and only if its closure $\bar{A}$ is equi-continuous. Do the same for pointwise compactness.

17. Let $h \in C[0,1]$ and define $g_n(t) = \int_0^1 \cos^3(t + nu)\, h(u)\,du$. Show that $\{g_n\}$ has a subsequence that converges uniformly for $0 \le t \le 1$.

18. Let (X,d) be a complete metric space and $A \subset X$. Show that the closure $\bar{A}$ is compact if and only if A is totally bounded.

19. Let $A \subset X$ where (X,d) is a metric space. Show that the closure $\bar{A}$ is compact if and only if every sequence in A has a convergent subsequence. [*Note*: We do not ask that the limit of the convergent subsequence be *in* A.]

20. Complete the argument in Example 3.

21. Prove Theorem 3.17.20.

22. Show that sequential compactness is equivalent to the Bolzano-Weierstrass property.

23. The following steps will lead to a proof of the assertion that sequential compactness is equivalent to (Heine-Borel) compactness:
    (a) Show that if (X,d) is compact, then it is sequentially compact. [*Hint*: Let $\{x_n\}$ be a sequence in X. Show that for every $\varepsilon > 0$ there is an open ball B of radius ε that contains an infinite number of the $\{x_n\}$.]
    (b) Show that if (X,d) is sequentially compact, then it is separable. (Use Theorem 3.17.7.)
    (c) Show that if (X,d) is separable, then every open covering of (X,d) contains a countable subcovering.
    (d) Show that if (X,d) is sequentially compact, then it is compact. [*Hint*: Use Step (c) and argue by contradiction.]

24. (Extension of Arzela-Ascoli Theorem.) Let (X,d_1) be a compact metric space and let (Y,d_2) be a complete metric space. Let $A = \{f_1,f_2,\ldots\}$ be a sequence of continuous functions with the following properties:
    (a) A is pointwise compact.
    (b) There is a sequence of positive numbers $\{\varepsilon_n\}$ such that $\varepsilon_n \to 0$ as $n \to \infty$.
    (c) For each $x \in X$ and every $\varepsilon > 0$ there is a $\delta > 0$ such that

    $$d_2(f_n(x), f_n(x')) \le \varepsilon + \varepsilon_n, \qquad \text{for all } n,$$

    whenever $d_1(x,x') \le \delta$.
    Show that A has compact closure in (C,ρ). Show that A is equi-continuous.

25. Use Exercise 24 to solve the following problem: Let $\{g_n\}$ be a sequence of integrable (possibly unbounded) functions defined for $0 \le t \le 1$. Assume that there is a bounded function g with the property that $\int_0^1 |g_n(t) - g(t)| \, dt = \varepsilon_n \to 0$ as $n \to \infty$. Let

$$y_n(t) = \int_0^t g_n(s) \, ds, \qquad 0 \le t \le 1.$$

(a) Show that $\{y_n\}$ has a subsequence that converges uniformly for $0 \le t \le 1$.
(b) What is the limit of this subsequence?
(c) Is it true that $\lim y_n$ exists?

26. A sequence $\{f_n\}$ of real-valued continuous functions defined for $-\infty < t < \infty$ is said to *converge uniformly on compact sets* to a limit f if for every compact set $K \subset R$ one has

$$\sup_{t \in K} |f_n(t) - f(t)| \to 0$$

as $n \to \infty$. Show that $\{f_n\}$ converges to f in this sense if and only if $d(f_n, f) \to 0$ where d is the metric

$$d(f,g) = \sum_{n=1}^{\infty} 2^{-n} \min\left(1, \max_{|t| \le n} |f(t) - g(t)|\right).$$

27. Show that the Arzela-Ascoli Theorem extends to the space $(C(-\infty,\infty),d)$ where the metric d is defined in Exercise 26. [*Note*: This exercise is not a direct application of Theorem 3.17.24 because the domain $(-\infty,\infty)$ is not compact.]

28. (Continuation of Exercise 27.) Let $f \in C(-\infty,\infty)$ and let A denote the collection of all translates f_τ of f where $f_\tau(t) = f(\tau + t)$. Show that A has compact closure if and only if f is bounded and uniformly continuous.

29. Let (X,d) be a complete metric space, and let $\varepsilon_0 > 0$. For $0 \le \varepsilon \le \varepsilon_0$, let A_ε denote a subset of X. Assume that for $0 < \varepsilon < \varepsilon_0$, A_ε is totally bounded in X. Also assume that for $0 < \varepsilon < \varepsilon_0$, there is a mapping $\Phi_\varepsilon: A_0 \to A_\varepsilon$ such that $d(x,\Phi_\varepsilon x) \le \varepsilon$ for all $x \in A_0$. Show that $\bar{A}_0$ is compact. [*Note*: It is not necessary to assume that Φ_ε is continuous.]

30. Let (X,d) be the metric space Q made up of all the rational numbers with the usual absolute value metric. Show that a subset A of (X,d) is compact if and only if it is bounded and closed when considered as a subset of the real numbers. Hence the set $A = \{0,1/2,1/3,1/4,\ldots,1/k,\ldots\}$ is compact in Q. On the other hand, the subset $B = \{3,3.1,3.14,3.141,\ldots\}$ is not compact in Q.

31. Use the results of this section to reconsider
    (a) Exercise 8, Section 16.
    (b) Exercise 9, Section 16.
    (c) Exercise 10, Section 16.
    (d) Exercise 11, Section 16.

32. Let $\{B_n\}$ be a sequence of nonempty compact sets in a metric space (X,d). Assume that $\{B_n\}$ is decreasing, that is, $B_{n+1} \subset B_n$. (a) Show that $\bigcap_{n=1}^{\infty} B_n$ is nonempty and compact. (b) What happens if the sets B_n are not compact?

SUGGESTED REFERENCES

Bartle [1].
Boas [1].
Hardy, Littlewood, and Polya [1].
Hewitt [1].
Kelley [1].
Kolmogorov and Fomin [1].

Lee and Markus [1].
Maak [1].
Royden [1].
Rudin [1].
Simmons [1].

4

Algebraic Structure

1. INTRODUCTION

The study of algebraic structures is certainly one of the oldest endeavors in the world of mathematics. For example, with our current viewpoint many of the ancient problems of mathematics, such as the ruler-compass constructions of Euclidean geometry, are seen to be algebraic problems in disguise. However, it was not until rather recently that mathematicians began to observe the important unifying role of the concept of linear spaces. It is the algebraic structure of these spaces that interests us in this chapter.

In this chapter we leave aside topological considerations. Here, words and phrases such as limit, Cauchy sequence, open set, closed set, completeness, compact, closure, dense, isometry, homeomorphism, and, above all, continuity and convergence are suppressed. Recall that in Chapter 1 we divided the structure of the real Euclidean plane into three categories: set-theoretic, topological, and algebraic structure. In Chapter 2 we reviewed set-theoretic structure. In Chapter 3 we studied spaces with topological structure only, that is, metric and pseudometric spaces. Now we turn our attention to mathematical systems that have algebraic structure only, namely, linear spaces or, as sometimes referred to, "vector spaces."

We should note that there are other algebraic structures such as groups, rings, Boolean algebras, and so on. We will not discuss these structures here since they are not germane to our primary objective.

Part A

Introduction to Linear Spaces

2. LINEAR SPACES AND LINEAR SUBSPACES

A linear space consists of a set (the underlying set), a scalar field, and some structure. For present purposes the scalar field is always either the real numbers R or the complex numbers C. In cases where we do not wish to distinguish between R and C, we simply refer to the scalar field F. The structure of a linear space is based on operations of addition and scalar multiplication.

4.2.1 DEFINITION. A *linear space* over a scalar field F is a nonempty set X and

(1) A mapping of $X \times X$ into X, called addition and written $x_1 + x_2$

(2) A mapping of $F \times X$ into X, called scalar multiplication and written αx.

Addition and scalar multiplication must satisfy the following conditions:

(A1) $x_1 + x_2 = x_2 + x_1$ for all x_1, x_2 in X.

(A2) $x_1 + (x_2 + x_3) = (x_1 + x_2) + x_3$ for all x_1, x_2, x_3 in X.

(A3) There exists a (unique) element in X denoted by 0 and called the origin, such that $0 + x = x$ for every x in X.

(A4) Associated with each x in X is a (unique) point $-x$ in X such that $x + (-x) = 0$.

(SM1) $\alpha(\beta x) = (\alpha\beta)x$ for all α, $\beta \in F$ and all $x \in X$.

(SM2) $1x = x$ for all $x \in X$.

(SM3) $0x = 0$ for all $x \in X$.

(A & SM1) $\alpha(x_1 + x_2) = \alpha x_1 + \alpha x_2$ for all $\alpha \in F$ and all $x_1, x_2 \in X$.

(A & SM2) $(\alpha + \beta)x = \alpha x + \beta x$ for all α, $\beta \in F$ and all $x \in X$.

The somewhat strange numbering of the above conditions is employed to call attention to the nature of the conditions: (A1) for conditions on the addition operation, (SM1) for conditions on the scalar multiplication operation, and (A & SM1) for conditions on addition and scalar multiplication jointly.

It is not claimed that the above conditions are logically independent. For example, let $\alpha = 1$ and $\beta = 0$ in (A & SM2). Since $1 + 0 = 1$ in F, it follows that $1x = 1x + 0x$. From (SM2) $1x = x$; therefore, $x = x + 0x$ for all x. But from (A3),

161

the origin is the only point such that $x + y = x$ for all x. Therefore, $0x = 0$. Thus we have shown that (A & SM2), (SM2), and (A3) imply (SM3). We define linear spaces in terms of the above (logically dependent) set of conditions for the simple reason that this form of definition yields quick intuitive insight into the nature of linear spaces.

There are certain important conclusions which follow from the above conditions. These are: $-x = (-1)x; \alpha 0 = 0; -0 = 0; y + [x + (-y)] = x; \alpha x = 0 \Rightarrow \alpha = 0$ or $x = 0$ (or both); $x + y = x + z \Rightarrow y = z$; $\alpha x = \alpha y$ for $\alpha \neq 0 \Rightarrow x = y$; and $\alpha x = \beta x$ for $x \neq 0 \Rightarrow \alpha = \beta$.

Sometimes we refer to a linear space over the scalar field R as a *real linear space*, and a linear space over C as a *complex linear space*. Again, where no distinction is made we refer to the scalar field F. We shall consistently denote a linear space by the symbol denoting its underlying set, for example, X. One could write $(X, F, +, \cdot)$ to denote a linear space, but this would be far too cumbersome.

A subset Y of a linear space X is said to be a *linear subspace* if $x_1 + x_2 \in Y$ whenever $x_1, x_2 \in Y$ and $\alpha x \in Y$ whenever $\alpha \in F$ and $x \in Y$. The reader should verify that a linear subspace is itself a linear space, that is, the nine properties defining a linear space are satisfied. One should not confuse the concept of a linear subspace with the concept of a subspace of a metric space.

Next let us consider some examples. Not too surprisingly many of the examples of metric spaces given in Chapter 3 can also be used as examples of linear spaces. Of course, one has to shift his attention. In Chapter 3 topological structure was in the limelight and all other structure that happened to be present in an example was carefully ignored. Now this chapter brings linear space structure into the forefront.

We leave it to the reader to show that, considering the obvious definitions of addition and scalar multiplication, the following are examples of linear spaces: R, C, R^n, C^n, l_p with $1 \leq p \leq \infty$, and $C[0,T]$. Further, we invite the reader to find other linear spaces among the examples and exercises of Chapter 3.

EXAMPLE 1.   The plane: Let V^2 be the set of all vectors in a plane P emanating from a point 0 in this plane. Three such vectors are illustrated in Figure 4.2.1. Defining addition, scalar multiplication, the origin, and negation in the usual way we have a linear space. Of course, the critical reader may wonder if we have really

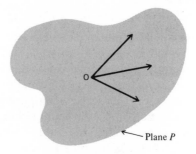

Plane P

Figure 4.2.1.

defined anything. We have not really said what a plane or a vector is. Nevertheless, this example does present an intuitive picture of a linear space.   ∎

EXAMPLE 2.   Many of the examples of linear spaces are function spaces and, as such, vector addition and scalar multiplication are defined in an obvious fashion. To explain, let S be any nonempty set and let X denote the collection of all scalar-valued functions defined on S. If $x, y \in X$, then we define $x + y$ by

$$(x + y)(s) = x(s) + y(s), \qquad \text{for all } s \in S,$$

which is certainly an element of X. Also if α is any scalar and $x \in X$, then αx is defined by

$$(\alpha x)(s) = \alpha(x(s)), \qquad \text{for all } s \in S,$$

which is also in X. In other words, we define these operations pointwise. It is a trivial matter, then, to show that X is a linear space.   ∎

EXAMPLE 3.   Let $X = L_p[0,T]$, $1 \leq p < \infty$, the set of all functions x defined on $[0,T]$ such that

$$\int_0^T |x(t)|^p \, dt < \infty. \qquad (4.2.1)$$

Addition, scalar multiplication, the origin, and negation are defined as in Example 2.   Of course, here one does have to show that $x + y$ and αx satisfy (4.2.1).

There is an important point to be made about the above linear space. If x and y are points in L_p and they differ only on a set of measure zero, then they are still different points in the linear space; and in most situations it is not desirable to distinguish between them. We confronted this situation before in Example 10 of Section 3.3 when we considered $L_p[0,T]$ as a metric space. We handle it here in exactly the same way. In particular, we use the "equality"

$$x = y \Leftrightarrow \int_0^T |x(t) - y(t)|^p \, dt = 0.$$

Given this equality, the linear space structure follows in a natural way.   ∎

EXAMPLE 4.   Let $(\Omega, \mathscr{F}, P)$ be a probability space, and let $L_2(\Omega, \mathscr{F}, P)$ denote the set of all complex-valued random variables X defined on $(\Omega, \mathscr{F}, P)$ such that

$$E\{|X|^2\} < \infty,$$

where E denotes the expectation operation. Addition and scalar multiplication are defined in the natural way. Indeed, this is a generalization of Example 3. We even have a "new equality" here too, namely

$$X = Y \Leftrightarrow E\{|X - Y|^2\} = 0.   ∎$$

EXERCISES

1. Let X be the linear space R^4. For what values of r, if any, is the set

$$A_r = \{x \in R^4 : x_1 + x_2 + x_3 + x_4 = r\},$$

where r is a real number, a linear subspace of X? For what values of r, if any, is the set

$$B_r = \{x \in R^4 : x_1{}^2 + x_2{}^2 + x_3{}^2 + x_4{}^2 = r^2\}$$

a linear subspace of X?

2. Let X be the linear space made up of all complex-valued functions $T(s)$ defined on the imaginary axis of the complex plane such that

$$\left| \int_{-i\infty}^{i\infty} |T(s)|^2 \, ds \right| < \infty,$$

where the integral is along the imaginary axis. That is, $X = L_2(-i\infty, i\infty)$. Let A be the set made up of all rational functions, that is, all functions of the form

$$T(s) = \frac{a_0 s^m + \cdots + a_m}{b_0 s^n + \cdots + b_n},$$

where $a_0 \neq 0$, $b_0 \neq 0$, m and n are integers. Is the set A a linear subspace of X? Next consider the subset B of A made up of all rational functions with $n > m$. Is B a linear subspace of X? What about the subset C of B made up of all functions with all their finite poles in the left-hand plane?

3. Show that the set of all $n \times m$ matrices can be viewed as a linear space.

4. Let X be the linear space $C[0,T]$. Which, if any, of the following subsets of X are linear subspaces?

$B_1 = \{x \in C[0,T] : x(0) = x(T)\}$,
$B_2 = \{x \in C[0,T] : x(0) = x(T) = 0\}$,
$B_3 = \{x \in C[0,T] : x(t_1) = x(t_2) \text{ for all } t_1, t_2 \text{ such that } t_1 + t_2 = T\}$,
$B_4 = \{x \in C[0,T] : x(0) = 1\}$,
$B_5 = \{x \in C[0,T] : \int_0^T x(\tau) \, d\tau = 1\}$,
$B_6 = \{x \in C[0,T] : |x(t_1) - x(t_2)| \leq 10|t_1 - t_2| \text{ for all } t_1, t_2 \in [0,T]\}$.

5. Show that if $\{B_\alpha\}$ is a family of linear subspaces of a linear space X, then $B = \bigcap_\alpha B_\alpha$ is a linear subspace of X. What about $\bigcup_\alpha B_\alpha$?

6. Let X be the linear space made up of all real-valued sequences. Show that A_1, the set of all sequences that have a finite number of nonzero entries only, is a linear subspace of X. Show that A_2, the set of all sequences that have an infinite number of nonzero entries, is not a linear subspace of X.

7. Often in systems theory the linear space $L_2(-\infty,\infty)$ is a good mathematical model for the set X of all inputs to a system as well as the set Y containing the range. Let $\mathscr{A}$ be the set of all mappings (linear and nonlinear) of X into Y. Show that $\mathscr{A}$ can be viewed as a linear space. Show that the subset $\mathscr{C} \subset \mathscr{A}$ of all mappings representing causal (Section 2.8) systems is a linear subspace of $\mathscr{A}$.

8. Let X be the linear space made up of all absolutely convergent sequences of real numbers. Show that B, the set of all absolutely convergent sequences of real numbers with limit zero, is a linear subspace of X.

9. Let X be the set of all convergent sequences of real numbers. Is X a linear space?

10. Let X denote the collection of all real-valued Lipschitz-continuous functions $x(t)$ defined for $-\infty < t < \infty$. That is, $x(t)$ satisfies $|x(t) - x(s)| \le k|t - s|$ for some constant k (which depends on x) and for all t, s. Show that X is a real linear space.

3.   LINEAR TRANSFORMATIONS

Continuous transformations play a central role in the case of metric spaces. Likewise there is a special class of transformations that plays a central role in the case of linear spaces, namely linear transformations.

4.3.1 DEFINITION.   A transformation L of a linear space X into a linear space Y, where X and Y have the same scalar field, is said to be a *linear transformation* if

(i) $L(\alpha x) = \alpha L(x)$ for all $x \in X$ and all scalars α, and

(ii) $L(x_1 + x_2) = L(x_1) + L(x_2)$ for all $x_1, x_2 \in X$.

Otherwise it is said to be a *nonlinear transformation*.

The scalar multiplication operations on the left- and right-hand sides of (i) above are, of course, those of X and Y, respectively. Similarly, the addition operations in (ii) are from X and Y, respectively. Please note carefully that (i) and (ii) must be satisfied for *all* x's and α's, not just some of them!

It might appear that one of the conditions (i) and (ii) implied the other; however, this is not so. For example, if $X = Y = C$, the complex numbers and L is the operation of complex conjugation, then $L(z_1 + z_2) = L(z_1) + L(z_2)$ for all $z_1, z_2 \in C$, but for complex scalars α with Im $\alpha \ne 0$ we have $L(\alpha z) \ne \alpha L(z)$. Going the other way, let $X = R^2$ and $Y = R$. Define a mapping $G: X \to Y$ by

$$G[(x_1,x_2)] = \begin{cases} x_1 + x_2, & \text{if } x_1 x_2 > 0 \\ 0, & \text{otherwise,} \end{cases}$$

where $x = (x_1,x_2) \in R^2$. The mapping G has the property that $G(\alpha x) = \alpha G(x)$ for all x and real α. On the other hand, $G(x + y)$ does not equal $G(x) + G(y)$ for all

$x, y \in R^2$. For example, if $x = (1,0)$ and $y = (0,1)$, then $G(x) = 0$, $G(y) = 0$, and $G(x + y) = 2$.

Note that it only makes sense to ask whether or not $L: X \to Y$ is linear if X and Y are both linear spaces and over the same scalar field. Otherwise, it is a meaningless question. Further, if we have a mapping S of a linear space X into a linear space Y, be it linear or not, it does not make sense to ask whether or not S is continuous. This question becomes meaningful only after we have added topological structure, as we shall see in the next chapter.

Up to this point nothing has been said that would guarantee that even one linear transformation exists. We circumvent the general existence problem by merely exhibiting some examples of linear transformations later in this section. It is possible to show that given any two nontrivial linear spaces X and Y over the same scalar field, there always exists a nontrivial linear transformation $L: X \to Y$. The proof of this theorem is outlined in Exercise 5 at the end of this section.

The *null space*, $\mathcal{N}(L)$, of a linear transformation $L: X \to Y$ is the subset of X defined by

$$\mathcal{N}(L) = \{x \in X: Lx = 0\}.$$

The origin of X is, of course, always in $\mathcal{N}(L)$, that is, $L(0) = 0$. A much more interesting fact is that $\mathcal{N}(L)$ is always a linear subspace of X. The *range space* $\mathcal{R}(L)$ of a linear transformation $L: X \to Y$ is

$$\mathcal{R}(L) = \{y = Lx: x \in X\}.$$

We note here that since L is linear, $\mathcal{R}(L)$ is a linear subspace of Y.

A reason for the importance of linear transformations is embodied in the following lemma.

4.3.2 LEMMA.   *A transformation L of X into Y, where X and Y are linear spaces over the same scalar field, is linear if and only if*

$$L(\alpha_1 x_1 + \cdots + \alpha_n x_n) = \alpha_1 L(x_1) + \cdots + \alpha_n L(x_n)$$

for all $x_1, x_2, \ldots, x_n \in X$, all scalars $\alpha_1, \alpha_2, \ldots, \alpha_n$, and all finite n.

The proof of this lemma follows immediately from Definition 4.3.1.

Lemma 4.3.2 shows that knowledge of how L transforms a finite number of points $\{x_1, \ldots, x_n\}$ allows a very simple characterization of how L transforms every point of the form $\alpha_1 x_1 + \cdots + \alpha_n x_n$. This fact, which is sometimes referred to as the *principle of superposition*, has far-reaching ramifications.

It should be carefully noted that nothing has been said in Lemma 4.3.2 about expressions of the form $\sum_{i=1}^{\infty} \alpha_i x_i$. Again, infinite series are not meaningful in the present linear space context, devoid of topological structure. Moreover, even if there were topological structure present, a transformation L being linear does not imply that

$$L\left(\sum_{i=1}^{\infty} \alpha_i x_i\right) = \sum_{i=1}^{\infty} \alpha_i L(x_i).$$

For example, differentiating an infinite series of functions term-by-term is not necessarily the same as differentiating the limit function. This is an important point, for the principle of superposition or linearity is sometimes mistakenly called upon to justify steps that just cannot be justified on the basis of linearity alone.[1]

Now let us consider some examples of linear transformations.

EXAMPLE 1. Consider the spring and mass system shown in Figure 4.3.1. With no applied force f the equilibrium or rest position is $x = 0$. Assume now that there is viscous friction between the mass and the surface it slides on, and that this viscous friction force is modeled by $-b(dx/dt)$. The combined restoring force of

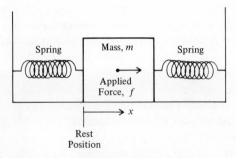

Figure 4.3.1.

the springs is modeled by $-kx$. We assume that we are interested in modeling the behavior of this system for times $t \geq 0$ and that $x(0) = 0$ and $(dx/dt)(0) = 0$. As is well-known, $f(t)$ and $x(t)$ are related by the differential equation

$$f = m \frac{d^2x}{dt^2} + b \frac{dx}{dt} + kx. \qquad (4.3.1)$$

We view f as a point in $C[0,\infty)$, the set of all real-valued continuous functions defined on $[0, \infty)$. Solving (4.3.1) with the initial conditions $x(0) = 0$ and $x'(0) = 0$ one gets

$$x(t) = \int_0^t h(t - s)f(s) \, ds, \qquad (4.3.2)$$

where

$$h(r) = \frac{1}{m(\lambda_1 - \lambda_2)} [e^{\lambda_1 r} - e^{\lambda_2 r}]$$

and λ_1 and λ_2 are roots of the equation

$$m\lambda^2 + b\lambda + k = 0.$$

---

[1] This problem is considered again in Chapter 5 (Theorem 5.6.2) where we have a topological as well as a linear space structure.

(We assume here that λ_1 and λ_2 are real and different. Otherwise $h(r)$ has a slightly different form.) Equation (4.3.2) then defines a mapping $L: C[0,\infty) \to C[0,\infty)$, where $x = Lf$. (Show that L is linear.)

Equation (4.3.1) really only describes the physical situation for relatively small motions x. One might construct a more globally applicable mathematical model by changing the mathematical model of the restoring force to include the effect of a spring being completely compressed. For example, instead of a linear spring $-kx$ one might use $-g(x)$, where $g(x)$ would model the abrupt increase in restoring force occurring when a spring is completely compressed. We then would obtain

$$f = m\frac{d^2x}{dt^2} + b\frac{dx}{dt} + g(x) = H(x), \tag{4.3.3}$$

where $H: C^2[0,\infty) \to C[0,\infty)$ and $C^2[0,\infty)$ is the linear subspace of $C[0,\infty)$ made up of all functions with continuous first and second derivatives.

Presumably (4.3.3) is a better mathematical model than (4.3.1) in that it more completely describes the physical system. On the other hand, H is a nonlinear transformation, and it is usually not possible to represent the inverse of H, if it exists, as simply as L is represented for (4.3.1). We have, then, a typical example of the often conflicting goals of mathematical modeling: (1) a complete description of the physical situation versus (2) a mathematically tractable model. ∎

EXAMPLE 2.   Let $L_2(-i\infty,i\infty)$ be the linear space defined in Example 13 of Section 3.3. Define a mapping L on $L_2(-i\infty,i\infty)$ by

$$Y(i\omega) = H(i\omega)X(i\omega), \qquad \text{for all } \omega,$$

where $H(i\omega)$ is a bounded, measurable function. It immediately follows that $\mathscr{R}(L) \subset L_2(-i\infty,i\infty)$. It is also easily shown that L is linear. Needless to say, this type of operation occurs very often in Fourier analysis. ∎

EXAMPLE 3.   Suppose that we are interested in the temperature of an infinite bar as a function of time, t, and position, x. Denote the temperature by $T(x,t)$. Further, assume that the bar is being heated along its length by a distributed heat source. The heat supplied per unit length at x and t is denoted $\phi(x,t)$. (See Figure 4.3.2.) Assume that at $t = 0$, $T(x,0) = 0$. Assume that ϕ is a point in the linear space X made up of all bounded continuous, real-valued functions defined on $(-\infty,\infty) \times [0,\infty)$. It is well-known that

$$T(x,t) = \int_0^\infty \int_{-\infty}^\infty H(x - x', t - t')\phi(x',t')\, dx'\, dt', \tag{4.3.4}$$

where

$$H(x,t) = \begin{cases} K\dfrac{e^{-x^2/4t}}{\sqrt{t}}, & \text{for } t > 0 \\[2mm] 0, & \text{for } t \le 0, \end{cases}$$

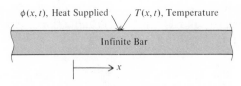

$\phi(x, t)$, Heat Supplied $T(x, t)$, Temperature

Infinite Bar

x

Figure 4.3.2.

and K is a constant. We leave it to the reader to show that (4.3.4) represents a linear transformation of X into itself. ∎

EXAMPLE 4.   In Section 2.8 we defined causal systems. Here we would like to show an important fact about causality in *linear* systems. Let X be a linear space made up of functions x defined on the real line. For each time T, let X_T denote the linear subspace of X made up of all functions x such that $x(t) = 0$ for[2] $t < T$. Further, let L be a linear mapping of X into itself. We claim that L is causal if and only if each linear subspace X_T is invariant under L, that is, if and only if $L(X_T) \subset X_T$ for all T.

First suppose that L is causal, and let T be any fixed time. Then let x be any point in X_T. We want to show that $Lx \in X_T$. Since L is causal and $x(t) = x_0(t)$ for $t < T$, where x_0 is the zero input, $(Lx)(t) = (Lx_0)(t)$ for $t < T$. But L is linear so $(Lx_0)(t) = 0$ for all t. Hence, $(Lx)(t) = 0$ for $t < T$ and $Lx \in X_T$.

Now suppose that for each T the linear subspace X_T is invariant under L. Let x_1 and x_2 be any two inputs such that $x_1(t) = x_2(t)$ for $t < T$. Letting $y_1 = Lx_1$ and $y_2 = Lx_2$, we want to show that $y_1(t) = y_2(t)$ for $t < T$. But $y_1 - y_2 = L(x_1 - x_2)$ and $(x_1 - x_2) \in X_T$. Since X_T is invariant under L, $(y_1 - y_2) \in X_T$ or $y_1(t) = y_2(t)$ for $t < T$.

It should be mentioned that if L is not linear, then the subspaces X_T being invariant under L does not imply causality. The next example shows this. ∎

EXAMPLE 5.   (This example is a continuation of the preceding one.) Let M denote the linear subspace of X defined by

$$M = \bigcup_T X_T,$$

that is, $x \in M$ if it vanishes to the left of some finite time. Let us assume that M is not all of X. For example, $M \neq X$ if $X = L_2(-\infty, \infty)$. We now define a mapping A of X into itself by defining it on M and $M' = X - M$. In particular, let

$$(Ax)(t) = x(t - 1) \qquad \text{for } x \in M$$

and

$$(Ax)(t) = x(t + 1) \qquad \text{for } x \in X - M.$$

---

[2] In spaces such as $L_2(-\infty, \infty)$ we say that $x(t) = 0$ almost everywhere in $(-\infty, T)$.

Clearly the linear subspaces X_T are invariant under A. Moreover, if there are just two distinct inputs x_1 and x_2 in $(X - M)$ and a time T such that $x_1(t) = x_2(t)$ for $t < T$, then A is not causal. ∎

EXERCISES

1. Let X and Y be linear spaces over the same scalar field. Show that $I: X \to X$, the identity transformation, and $0: X \to Y$, the zero transformation, are linear.

2. Let X and Y be linear spaces over the same scalar field. Show that $lt[X, Y]$, the set of all linear transformations of X into Y, is a linear space when addition of linear transformations and multiplication of linear transformations by scalars are defined as in Example 2, Section 2.

3. Let X be an arbitrary nonempty set and let Y be a linear space. Show $\mathscr{F}$, the set of all mappings of X into Y, is a linear space when addition of mappings and scalar multiples of mappings are defined as in Example 2, Section 2.

4. Let X, Y, and Z be linear spaces over the same scalar field, and let $L_1: X \to Y$ and $L_2: Y \to Z$ be linear. Show that the composition $L_2 L_1: X \to Z$ is linear.

5. Let X and Y be linear spaces over the same scalar field, and let M be a proper linear subspace of X. Further, let f be a linear transformation of M into Y. Show that there exists a linear extension F of f defined on X, that is, $F: X \to Y$ and $F(x) = f(x)$ for each $x \in M$. [*Hint*: Let x_0 be a point in X but not in M, and let M_0 be the linear subspace of X made up of all points $x + \alpha x_0$, where $x \in M$ and α is scalar. Show that there is a unique expression $x + \alpha x_0$ for each point in M_0. Define a linear transformation F_0 of M_0 into Y by

$$F_0(x + \alpha x_0) = f(x) + \alpha y_0,$$

where y_0 is an arbitrary point in Y. Then F_0 is an extension of f to M_0. Let E be the class of all linear transformations U with domain $\mathscr{D}(U) \subset X$, range $\mathscr{R}(U) \subset Y$, and which are linear extensions of f. Next introduce the following partial ordering on E: $U \leq V$ if $\mathscr{D}(U) \subset \mathscr{D}(V)$ and $U(x) = V(x)$ for all $x \in \mathscr{D}(U)$. Let C be an arbitrary totally ordered subset of E. Show that C has an upper bound. Finally apply Zorn's lemma.]

6. Suppose that we consider a system whose output is a delayed version of the input. That is, if $x(t)$ is the input, then the output $y(t) = x(t - \tau)$, where τ is a constant. Let X be the linear space $C(-\infty, \infty)$ of continuous real-valued functions defined on $(-\infty, \infty)$. Let D denote the system operation. Is D a linear transformation of X into itself? Suppose that instead of being constant the delay τ is given by $\tau = e^{-t}$. Do we have a linear transformation? Then suppose that $\tau = \exp\left[-\int_{-\infty}^{t} |x(\zeta)| d\zeta\right]$, where of course the linear space X must be selected so that the integral exists. Do we have a linear transformation?

7. Let $Y = C([0, \infty), R^n)$ be the linear space made up of all continuous mappings of $[0, \infty)$ into R^n, that is, each component is continuous. Let $X = C^1([0, \infty), R^n)$ be the linear subspace of Y made up of all elements of Y with continuous deriva-

tives, that is, each component has a continuous derivative. Does the expression $y = Tx$, where

$$Tx = \frac{dx}{dt} - Ax$$

and A is a real $n \times n$ matrix, represent a linear transformation of X into Y?

8. Let $Y = BC(-\infty,\infty)$ denote the space of all bounded real-valued continuous functions $y(t)$ defined for $-\infty < t < \infty$ and let X denote the space of all Lipschitz continuous functions; see Exercise 10, Section 2. Define $x = Ly$ by $x(t) = \int_0^t y(s)ds$. Show that L is a linear mapping of Y into X.

4. INVERSE TRANSFORMATIONS

The concept of the inverse mapping, which is a set-theoretic concept, was discussed in Section 2.7. Let X and Y be sets and let $G: X \to Y$ be a transformation. Recall (Theorem 2.7.6) that G is invertible if and only if G is a one-to-one mapping of X onto Y, that is, $\mathcal{R}(G) = Y$. Recall also that G has a left inverse on Y if and only if G is one-to-one, and that G has a right inverse on Y if and only if $\mathcal{R}(G) = Y$.

In this section we ask what more can be said when X and Y are linear spaces and the mapping G is a linear transformation. The first result is an elegant, although simple, statement about a linear transformation being one-to-one.

4.4.1 THEOREM. *Let $L: X \to Y$ be a linear transformation on two linear spaces X and Y. Then the transformation L is one-to-one if and only if the null space is trivial, that is $\mathcal{N}(L) = \{0\}$.*

Proof: Consider the "only if" part first, that is, assume that L is one-to-one, and let $x \in \mathcal{N}(L)$. Since

$$Lx = 0 \quad \text{and} \quad L0 = 0,$$

it follows from the one-to-one assumption that $x = 0$. Hence $\mathcal{N}(L) = \{0\}$.

Now consider the "if" part, that is, assume that $\mathcal{N}(L) = \{0\}$. Let x_1, x_2 be points in X. Since L is linear, one has

$$Lx_1 = Lx_2 \Leftrightarrow L(x_1 - x_2) = 0 \Leftrightarrow (x_1 - x_2) \in \mathcal{N}(L) = \{0\}$$
$$\Leftrightarrow x_1 = x_2.$$

Hence L is one-to-one. ∎

The next result tells us that the inverse of a linear transformation (when it exists) is necessarily linear.

4.4.2 THEOREM. *Let $L: X \to Y$ be an invertible linear transformation of X onto Y, where X and Y are linear spaces. Then the inverse $L^{-1}: Y \to X$ is linear.*

Proof: Let $y_1 = Lx_1$ and $y_2 = Lx_2$, where $y_1, y_2 \in Y$. Using the fact that L is linear and $L^{-1}L = $ identity on X we have

$$L^{-1}(y_1 + y_2) = L^{-1}(Lx_1 + Lx_2) = L^{-1}L(x_1 + x_2)$$
$$= x_1 + x_2 = L^{-1}y_1 + L^{-1}y_2.$$

Hence L^{-1} is additive. Similarly one has

$$L^{-1}(\alpha y_1) = L^{-1}(\alpha Lx_1) = L^{-1}L(\alpha x_1) = \alpha x_1 = \alpha L^{-1}y_1.$$

Hence L^{-1} is linear.   ∎

EXAMPLE 1.   Let us return to the operator L given by

$$Y(i\omega) = H(i\omega)X(i\omega), \tag{4.4.1}$$

which was discussed in Example 2, Section 3. But now assume that $H(i\omega)$ is bounded and continuous.

One can show that L is one-to-one if and only if the set

$$N = \{\omega : H(i\omega) = 0\}$$

has measure zero.

It is well-known that operators like L arise in the theory of Fourier transforms. In particular if one takes the Fourier transform of the convolution equation

$$y(t) = \int_0^t h(t - s)x(s)\, ds,$$

where $x, y \in L_2(-\infty, \infty)$ and $h \in L_1(-\infty, \infty)$ one arrives at (4.4.1) where Y, H, and X are the Fourier transforms of y, h, and x, respectively.   ∎

EXERCISES

1. Show that the linear transformation $y = Lx$ on $L_2(-\infty, \infty)$ given by

$$y(t) = \int_{-\infty}^t a^{-1}e^{-a(t-\tau)}x(\tau)\, d\tau$$

is one-to-one. [*Hint*: Show that $Lx = 0$ reduces to $\int_{-\infty}^t e^{a\tau}x(\tau)\, d\tau = 0$. Then differentiate and use Theorem D.13.3.]

2. Let $k(t,s)$ be continuous for $0 \le s \le t \le T$ and consider

$$y(t) = x(t) + \int_0^t k(t,s)y(s)\, ds. \tag{4.4.2}$$

The following steps will lead to a proof that the relationship $y = Fx$ implicitly given by (4.4.2) does define a linear mapping F on $C[0,\alpha]$ provided α is a sufficiently small positive number.

(a) Define G by $x = Gy$, where

$$x(t) = y(t) - \int_0^t k(t,s)y(s)\, ds.$$

Show that G is linear.

(b) Show that if G is one-to-one, then G^{-1} exists and is F, so F is linear.

(c) Let M satisfy $|k(t,s)| \leq M$ for $0 \leq s \leq t \leq T$, then show that $Gy = 0$ reduces to

$$|y(t)| \leq M \int_0^t |y(s)| \, ds.$$

Now use the Gronwall Inequality, see Cesari [1, p. 35], to show that G is one-to-one.

3. Show that if $L: X \to Y$ is left (right) invertible, then it has a linear left (right) inverse. Further, show that it is possible for a left (right) inverse of a linear transformation to be nonlinear.

5.   ISOMORPHISMS

One of the central concepts of mathematics is the concept that two mathematical structures are equivalent if they can be put into one-to-one correspondence in a way that preserves structure. We have already used it in several ways. For example, in Chapter 3 we introduced the concept of isometric metric spaces. Recall that (X,d_1) and (Y,d_2) are isometric if (i) there exists an invertible mapping F of (X,d_1) onto (Y,d_2) such that (ii) $d_2(F(x_1), F(x_2)) = d_1(x_1,x_2)$ for all x_1, $x_2 \in X$. Item (i) is the one-to-one correspondence, and (ii) is the structure preservation. This means that the only difference between two isometric metric spaces is in the names given to the elements in their underlying sets.

An analogous situation arises when we deal with linear spaces. Often two superficially different linear spaces are only different in the nature of the points in their underlying sets. Let us make this concept precise.

4.5.1 DEFINITION.   The linear spaces X and Y over the same scalar field F are said to be *isomorphic* if there exists a one-to-one linear mapping T of X onto Y. The mapping T is then said to be an *isomorphism* of X onto Y.

Obviously T puts X and Y into a one-to-one correspondence and it preserves linear space structure. It follows from Theorems 4.4.1 and 4.4.2 that $T: X \to Y$ is an isomorphism if and only if (i) T is one-to-one; (ii) T maps X onto Y; (iii) T is linear; and (iv) T^{-1} is linear.

Note there is no demand in Definition 4.5.1 that the isomorphism between two isomorphic linear spaces be unique. As a matter of fact, an infinite number of isomorphisms usually exist between two isomorphic linear spaces. This is one of the points illustrated in the following examples.

EXAMPLE 1.   Let $X = R^2$ and let V^2 be the plane discussed in Example 1, Section 2. The linear spaces R^2 and V^2 are isomorphic. One isomorphism, call it T_1, mapping R^2 onto V^2 can be defined as follows. Let v_1 and v_2 be any two non-collinear vectors in V^2 as illustrated in Figure 4.5.1. Then if $x = (x_1,x_2)$ is a point in R^2, define T_1 by

$$T_1(x) = x_1 v_1 + x_2 v_2,$$

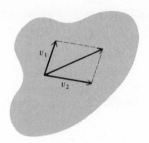

Figure 4.5.1.

where the operations on the right are the scalar multiplication and addition operations of V^2. Since the vectors v_1 and v_2 can be chosen almost arbitrarily, it is also clear that there is an infinite number of isomorphisms between R^2 and V^2. Another way of saying this is that there is an infinite number of different coordinate systems possible for V^2. ∎

EXAMPLE 2.   Let $X = R^2$, and let Z be the linear space made up of all functions z defined on $[0,T]$ of the form $z(t) = a_0 + a_1 t$, where a_0 and a_1 are arbitrary real numbers. The linear space structure on Z is, of course, defined in the obvious way. These two linear spaces are isomorphic. One isomorphism, call it T_1, mapping R^2 onto Z is defined by

$$T_1(x) = x_1 + x_2 t,$$

where $x = (x_1, x_2)$. We note in passing that the inverse of T_1 can be represented by

$$T_1^{-1}(z) = \left(z(0), \frac{dz}{dt} \bigg|_{t=0} \right).$$

It can also be shown that Z here and V^2 in Example 1 are isomorphic. (How?) ∎

EXAMPLE 3.   Let X denote the collection of all functions $\phi(t)$ that satisfy the differential equation

$$L(x) = a_0 x^{(n)} + a_1 x^{(n-1)} + \cdots + a_n x = 0, \tag{4.5.1}$$

where $x^{(j)} = d^j x / dt^j$ and the coefficients $\{a_0, \ldots, a_n\}$ are real constants and $a_0 \neq 0$. It is easily seen that X is a linear space. Furthermore, if we let $\phi_i(t)$, $i = 1, 2, \ldots, n$, denote the solution of (4.5.1) that satisfies $\phi_i^{(i-1)}(0) = 1$ and

$$\phi_i^{(j)}(0) = 0, \qquad j = 0, 1, \ldots, n-1 \text{ and } j \neq i-1,$$

then any solution $\phi(t)$ of (4.5.1) can be written as

$$\phi(t) = c_1 \phi_1(t) + \cdots + c_n \phi_n(t), \tag{4.5.2}$$

where the coefficients $\{c_1,\dots,c_n\}$ are real constants. In fact c_i is determined by $c_i = \phi^{(i-1)}(0)$, where $\phi^{(0)} = \phi$. We claim that X is isomorphic with R^n. The isomorphism $T: X \to R^n$ is given by

$$T(\phi) = (c_1,\dots,c_n),$$

where ϕ satisfies (4.5.2). The details are left as an exercise.   ∎

EXAMPLE 4.   Let $X = l_2(-\infty,\infty)$ and Z be the linear space made up of all complex-valued functions $f(z)$ defined on the unit circle of the complex plane such that

$$\frac{1}{2\pi i}\oint |f(z)|^2 \frac{dz}{z} < \infty.$$

The usual assumption is made that if $f_1(z)$ and $f_2(z)$ differ only on a set of measure zero, then f_1 and f_2 are considered to be the same function.

We claim that the following formula where

$$\mathscr{L}: (\dots,\xi_{-1},\xi_0,\xi_1,\dots) \to f(z)$$

$$f(z) = \sum_{n=-\infty}^{\infty} \xi_n z^{-n} \tag{4.5.3}$$

defines an isomorphism $\mathscr{L}$ of X onto Z. The reader probably recognizes that (4.5.3) is the *two-sided z-transform*. We remark that the convergence in (4.5.3) is convergence in the mean, that is,

$$\lim_{M,\,N\to\infty}\left\{\frac{1}{2\pi i}\oint \left| f(z) - \sum_{n=-N}^{M} \xi_n z^{-n}\right|^2 \frac{dz}{z}\right\} = 0.$$

We will delay the proof of the statement that $\mathscr{L}$ is an isomorphism until Chapter 5. Here we merely mention that the inverse of z can be represented by

$$\xi_k = \frac{1}{2\pi i}\oint z^k f(z)\frac{dz}{z}, \qquad k = \dots,\,-1,\,0,\,1,\,2,\,\dots.\quad ∎$$

EXERCISES

1. Let χ be the set of all linear spaces over a scalar field F. Show that the relation $X \sim Y \Leftrightarrow (X$ and Y are isomorphic) is an equivalence relation and, therefore, induces a partition on χ.

2. Referring to Example 2 above, under what conditions on real numbers c_{11}, c_{12}, c_{21}, c_{22} does

$$T_2(x) = (c_{11}x_1 + c_{12}x_2) + (c_{21}x_1 + c_{22}x_2)t,$$

where $x = (x_1,x_2)$, define an isomorphism from R^2 onto Z?

3. Show that the real linear space X made up of all functions of the form $x = a \cos(\omega t + \phi)$, where ω is fixed, is isomorphic to the complex numbers considered as a real linear space. (This fact is the cornerstone of the so-called phasor method of analyzing alternating current electrical networks.)

4. Let $L_2^{\sigma}(-\infty,\infty)$ denote the linear space made up of all complex-valued functions x defined on $(-\infty,\infty)$ such that

$$\int_{-\infty}^{\infty} |x|^2 e^{-2\sigma t}\, dt < \infty,$$

where σ is a real number. Show that $L_2^{\sigma}(-\infty,\infty)$ is isomorphic to $L_2^{\tau}(-\infty,\infty)$ for any σ and τ.

5. Show that $L_2(-\infty,\infty)$ is isomorphic to $L_2[0,\infty)$.

6. Show that $l_2(-\infty,\infty)$ is isomorphic with $l_2(0,\infty)$.

6.   LINEAR INDEPENDENCE AND DEPENDENCE

Linear independence is a property attributable to *sets* of points in a *linear space*. This section is devoted to presenting a precise formulation of this concept.

We shall use expressions of the form $\alpha_1 x_1 + \cdots + \alpha_n x_n$ often, so we introduce the following definition.

4.6.1 DEFINITION.   Let A be a set (perhaps infinite) in a linear space X. A point $x \in X$ is said to be a *linear combination of points in A* if there exists a finite set of points $\{x_1, x_2, \ldots, x_n\}$ in A and a finite set of scalars $\{\alpha_1, \alpha_2, \ldots, \alpha_n\}$ such that

$$x = \alpha_1 x_1 + \cdots + \alpha_n x_n. \tag{4.6.1}$$

If the set A is empty, we agree that the origin 0 is the unique point that is a linear combination of "points in A."

It should be noted that the expression for x in (4.6.1) may not be uniquely determined. Furthermore, it is important to note that no matter what the nature of the linear space X (that is, be it finite or infinite dimensional) or the set A, we consider *finite linear combinations only*.

EXAMPLE 1.   Let X be the linear space $C[0,T]$, and let A be the infinite set containing the continuous functions $\{1, t, t^2, t^3, \ldots\}$. The set of all linear combinations of points in A is the set of all polynomials in t, that is, all functions of the form

$$x(t) = a_0 + a_1 t + a_2 t^2 + \cdots + a_n t^n, \qquad t \in [0,T],$$

where $a_0, a_1, \ldots, a_n$ are scalars and $n = 0, 1, 2, 3, \ldots$. ∎

Let A be a set in a linear space X and let $V(A)$ be the set of all (finite) linear combinations of points in A. The important fact about the set $V(A)$ is that it is a linear subspace of X. We refer to it as the linear subspace *spanned* by A or simply the *span* of A. We leave it to the reader to show that $V(A)$ is the "smallest" linear subspace of X containing A; that is, if M is a linear subspace of X and $A \subset M$, then $V(A) \subset M$.

Now for the definitions of linear independence and linear dependence.

4.6.2 DEFINITION.   A set A in a linear space X is said to be *linearly independent* if for each point x in A, x is not a linear combination of points in the set $A - \{x\}$, that is, A with x removed. In other words, x is not in the linear subspace $V(A - \{x\})$. A set A in a linear space X is said to be *linearly dependent* if it is not linearly independent, that is if there exists at least one point x in A such that x is a linear combination of points in the set $A - \{x\}$.

EXAMPLE 2.   Let A be the set in a linear space X containing only the origin: $A = \{0\}$. Then the set $A - \{0\}$ is the empty set. But we have agreed to say that the origin is a linear combination of points in the empty set, so by Definition 4.6.2, the set A is linearly dependent. If B is an arbitrary set in X, then $\{0\} \cup B$ is linearly dependent. (Why?) If A is the empty set, then A is linearly independent. (Why?)   ∎

EXAMPLE 3.   Let A and X be the same as in Example 1. Let $x = t^k$ be an arbitrary point in A. Is x a linear combination of points in $A - \{x\}$? If it is, there exist integers $k_1, \ldots, k_n$, not equal to k, and nonzero scalars $a_1, \ldots, a_n$ such that

$$P(t) = t^k - a_1 t^{k_1} - \cdots - a_n t^{k_n} \equiv 0 \qquad t \in [0,T].$$

But this is clearly impossible since this polynomial can have only a finite number of zeros. Hence the set A is linearly independent.   ∎

The following theorem generalizes the method used in Example 3.

4.6.3 THEOREM.   *A set A in a linear space X is linearly independent if and only if for each nonempty finite subset of A, say $\{x_1, \ldots, x_n\}$, the only n-tuple of scalars satisfying the equation*

$$a_1 x_1 + a_2 x_2 + \cdots + a_n x_n = 0 \qquad (4.6.2)$$

is the trivial solution $a_1 = \cdots = a_n = 0$.

Proof: Let us do the "only if" part first, that is, assume that A is linearly independent. Let $\{x_1, \ldots, x_n\}$ be a finite collection of distinct points from A. Assume now that the equation

$$a_1 x_1 + \cdots + a_n x_n = 0 \qquad (4.6.3)$$

has a nontrivial solution for the scalars $a_1, \ldots, a_n$, that is, at least one of the scalars is nonzero. There is no loss in generality in assuming that $a_1 \neq 0$. If we let $b_i = -a_i/a_1$ for $i = 2, \ldots, n$, then one has

$$x_1 = b_2 x_2 + \cdots + b_n x_n, \qquad (4.6.4)$$

that is, x_1 is a linear combination of $\{x_2,\ldots,x_n\}$. This contradicts the fact that A is linearly independent. Hence (4.6.2) has the trivial solution $a_1 = \cdots = a_n = 0$ as its only solution.

Now for the "if" part of the theorem. Assume that the only solution of (4.6.2) is $a_1 = \cdots = a_n = 0$. If a point x_1 in A can be written as a linear combination of points in $A - \{x_1\}$, then (4.6.4) holds for some points $\{x_2,\ldots,x_n\}$ in A where $x_1 \neq x_i$, $2 \leq i \leq n$. But this implies that (4.6.2) has the nontrivial solution $a_1 = 1$, $a_2 = -b_2, \ldots, a_n = -b_n$, which is a contradiction. Hence A is linearly independent. ∎

The following is an obvious corollary to Theorem 4.6.3.

4.6.4 COROLLARY. *A nonempty set A in a linear space X is linearly dependent if and only if there is at least one nonempty finite subset of A, say $\{x_1,\ldots,x_n\}$, and scalars $a_1, \ldots, a_n$ where not all a_i's are zero, such that*

$$a_1 x_1 + \cdots + a_n x_n = 0.$$

Let us now turn to another way of characterizing linear independence. We motivate the next theorem with the following example.

EXAMPLE 4. Let X be the linear space R^3, and let A be the set $\{x_1,x_2,x_3\}$, where $x_1 = (1,0,0)$, $x_2 = (0,1,0)$, $x_3 = (0,0,1)$. The set A is linearly independent, and $V(A)$, the span of A, is R^3 itself. If $x = (a_1,a_2,a_3)$ is an arbitrary point in R^3, there is one and only one way to express x in the form

$$x = \alpha_1 x_1 + \alpha_2 x_2 + \alpha_3 x_3,$$

namely, $\alpha_1 = a_1$, $\alpha_2 = a_2$, $\alpha_3 = a_3$.

Continuing with the example, let C be the set $\{y_1,y_2,y_3\}$, where $y_1 = (1,0,0)$, $y_2 = (0,1,0)$, $y_3 = (1,1,0)$. The set C is linearly dependent, and $V(C)$ is the set of all points of the form $(a_1,a_2,0)$. Now if y is a point in $V(C)$, there is more than one way to express it in the form

$$y = \beta_1 y_1 + \beta_2 y_2 + \beta_3 y_3.$$

For example, if $y = (3,1,0)$, then one has

$$y = 2y_1 + y_2 = y_1 + y_3 = -y_2 + 2y_3. \quad ∎$$

The next theorem shows that the uniqueness of representation (or expression) concept illustrated by the preceding example characterizes linear independence.

4.6.5 THEOREM. *Let A be a nonempty set in a linear space X. The set A is linearly independent if and only if for each $x \neq 0$ in $V(A)$, there is one and only one finite subset of A, say $\{x_1,x_2,\ldots,x_n\}$, and a unique n-tuple of nonzero scalars, say $\{a_1, a_2, \ldots, a_n\}$, such that*

$$x = a_1 x_1 + \cdots + a_n x_n.$$

Proof: Let us consider the "only if" part first, that is, assume that A is linearly independent. Let $x \in V(A)$ where $x \neq 0$, and assume that

$$x = a_1 x_1 + \cdots + a_N x_N = b_1 y_1 + \cdots + b_M y_M,$$

where $\{x_1,\ldots,x_N\}$ and $\{y_1,\ldots,y_M\}$ are two sets in A and the coefficients $a_1,\ldots,a_N$, $b_1,\ldots,b_M$ are all nonzero. We then want to show two things: First that the sets $\{x_1,\ldots,x_N\}$ and $\{y_1,\ldots,y_M\}$ are the same. (From this fact we can, and do, assume that $x_1 = y_1, \ldots, x_N = y_N$ and $N = M$.) Secondly, we want to show that $a_1 = b_1, \ldots, a_N = b_N$.

First we note that

$$a_1 x_1 + \cdots + a_N x_N - b_1 y_1 - \cdots - b_M y_M = x - x = 0,$$

which is a special form of Equation (4.6.2). Since $a_1 \neq 0$ and since A is linearly independent, Theorem 4.6.3 assures us that x_1 must be included in the set $\{y_1,\ldots,y_M\}$, say that $x_1 = y_1$, and that $a_1 = b_1$. Similarly, since $a_2 \neq 0$, we see that x_2 lies in $\{y_1,\ldots,y_M\}$, say that $x_2 = y_2$, and that $a_2 = b_2$. By continuing in this fashion we see that the representation for x is unique.

Now consider the "if" part of the theorem. Assume that for each x in $V(A)$ the sets $\{x_1,\ldots,x_n\}$ and $\{a_1,\ldots,a_n\}$ are unique. We must show that A is linearly independent. Let x_0 be any point in A. Trivially $x_0 = x_0$, and by our assumption this is the only way to express x_0 as a linear combination of points in A. But then x_0 is not a linear combination of points in $A - \{x_0\}$. Hence, A is linearly independent. ∎

Yet another characterization of linear independence is given by the following theorem.

4.6.6 THEOREM.   *Let A be a set in a linear space X. The set A is linearly independent if and only if there is no proper subset A_0 of A such that $V(A_0) = V(A)$.*

We leave the proof of this theorem to the reader.

This is probably not the first time that the reader has confronted the concept of linear independence. Perhaps the only new aspects are the extension of this concept to linear spaces made up of arbitrary objects (for example, functions, sequences, random variables) and to infinite-dimensional linear spaces. Moreover, it should now be clear that linear independence is an algebraic concept and does not involve topological structure.

The following example illustrates the concept of a linear subspace being spanned by a set. Moreover, it is the basis for some later examples in this book.

EXAMPLE 5.   Let $X = l_2(-\infty,\infty)$. Let $S_r: X \to X$ be the right shift, that is, if $x = \{\ldots,x_{-1},x_0,x_1,x_2,\ldots\}$ and $y = S_r x = \{\ldots,y_{-1},y_0,y_1,y_2,\ldots\}$, then

$$y_k = x_{k-1} \qquad (k = \ldots,-1,0,1,2,\ldots).$$

In other words, S_r "shifts" the sequence x to the right by one position. As usual we denote the composition of S_r with itself n times by S_r^n. S_r is clearly invertible, so S_r^n with a negative n is meaningful. Moreover, $S_r^0 = I$.

Let x be any nonzero point in X, and consider the set A_x of all $y \in X$ that can be expressed in the form

$$y = \sum_{n=N}^{M} \alpha_n S_r^n x,$$

where $N, M = \ldots, -2, -1, 0, 1, 2, 3, \ldots$ and the α_n's are scalars. The set A_x is obviously a linear subspace of X. Indeed, it is the subspace spanned by the set

$$\{\ldots, S_r^{-2}x, S_r^{-1}x, x, S_r x, S_r^2 x, \ldots\}.$$

We can use the two-sided z-transform (Example 4, Section 5) to give a simple characterization of the subspace A_x. Let Z denote the linear space of functions used in Example 4, Section 5, and let $\mathscr{L}$ denote the two-sided z-transform, that is,

$$\mathscr{L}(\ldots, x_{-1}, x_0, x_1, \ldots) = \sum_{k=-\infty}^{\infty} x_k z^{-k},$$

where the convergence is convergence in the mean.

We remarked in Example 4, Section 5 that $\mathscr{L}$ is an isomorphism of X onto Z. Hence, if we characterize the linear subspace $\mathscr{L}(A_x)$ in Z that will be equivalent to characterizing A_x in X. But

$$\mathscr{L}(S_r x) = z^{-1} \mathscr{L}(x) \quad \text{and} \quad \mathscr{L}(S_r^n x) = z^{-n} \mathscr{L}(x)$$

so

$$\mathscr{L}\left(\sum_{n=N}^{M} \alpha_n S_r^n x\right) = \left\{\sum_{n=N}^{M} \alpha_n z^{-n}\right\} \mathscr{L}(x).$$

In other words, each point in the linear subspace $\mathscr{L}(A_x)$ of Z is of the form

$$p\mathscr{L}(x) \tag{4.6.5}$$

where p is a polynomial in z and $1/z$ given by

$$p(z) = \sum_{n=N}^{M} \alpha_n z^{-n}. \tag{4.6.6}$$

Or, a point in X is in the subspace A_x if and only if its z-transform is of the form (4.6.5).

We remark in passing that it should be obvious that $A_x \neq X$. In fact, if x and y are points in X such that the ratio $\mathscr{L}(x)/\mathscr{L}(y)$ is not of the form (4.6.6), then the subspaces generated by x and y are disjoint, that is, $A_x \cap A_y = \{0\}$. ∎

EXERCISES

1. Let T be an isomorphic mapping of a linear space X onto a linear space Y. Show that $A \subset X$ is linearly independent if and only if its image $T(A) \subset Y$ is linearly independent.

2. Let $X = L_2(\Omega, \mathscr{F}, P)$, the linear space made up of all complex-valued random variables x defined on a probability space $(\Omega, \mathscr{F}, P)$ such that

$$E\{x\bar{x}\} = \int_\Omega x\bar{x}\, dP = \int_\Omega |x|^2\, dP,$$

where $\bar{x}$ denotes the complex conjugate of x. Let $A \subset X$ be the set containing two random variables x_1 and x_2 such that

$$E\{x_1\bar{x}_1\} = 1, \qquad E\{x_2\bar{x}_2\} = 1, \qquad E\{x_1\bar{x}_2\} = 0.$$

Show that $y \in X$ is a linear combination of points in A if and only if

$$E\{y\bar{y}\} = |E\{y\bar{x}_1\}|^2 + |E\{y\bar{x}_2\}|^2.$$

Moreover, show that A is linearly independent.

3. Let X be the linear space $L_2[0,2\pi]$ and A the set of all functions $x_n(t) = e^{int}$, $n = 0, 1, 2, \ldots$. Show that A is linearly independent. [*Hint*: Assume that $\alpha_1 e^{in_1 t} + \cdots + \alpha_m e^{in_m t} \equiv 0$. Differentiate $(m-1)$ times.]

4. Show that a finite set $A = \{x_1,\ldots,x_n\}$ in a linear space X is linearly independent if and only if the only n-tuple of scalars satisfying the equation

$$a_1 x_1 + \cdots + a_n x_n = 0$$

is $a_1 = \cdots = a_n = 0$.

5. Let X be the linear space R^n, and let A be the set containing the n vectors

$$x_1 = \{x_{11}, x_{21}, \ldots, x_{n1}\},$$
$$x_2 = \{x_{12}, x_{22}, \ldots, x_{n2}\},$$
$$\cdots$$
$$x_n = \{x_{1n}, x_{2n}, \ldots, x_{nn}\}.$$

Show that this set is linearly independent if and only if

$$\det \begin{bmatrix} x_{11} & x_{12} & \cdots & x_{1n} \\ x_{21} & x_{22} & \cdots & x_{2n} \\ & & \cdots & \\ x_{n1} & x_{n2} & \cdots & x_{nn} \end{bmatrix} \neq 0.$$

6. Show that a set A in a linear space X is linearly independent if and only if every finite subset of A is linearly independent.

7. Prove Theorem 4.6.6.

8. Let the state of a dynamic system be a point in the linear space $X = R^n$. Let the state at time $k = 0, 1, 2, \ldots$ be denoted by x_k. Further suppose that the evolution of the system is characterized by a linear transformation $T: X \to X$; in particular, $x_k = Tx_{k-1}$. Let x_0 be an arbitrary initial state, and consider the set

$$A = \{x_0, Tx_0, T^2x_0, T^3x_0, \ldots\}.$$

Show that there exists an integer p such that for

$$A_p = \{x_0, Tx_0, \ldots, T^p x_0\}$$

one has $V(A) = V(A_p)$, that is, A and A_p span exactly the same linear subspace.

9. Let X be the linear space made up of all real-valued random variables defined on some probability space. Let A be a set of random variables in X. Show that if a random variable $z \in X$ is stochastically independent of each random variable in A, then z is not in $V(A)$.

10. Show that Theorem 4.6.5 remains true even when the set A is empty or when we consider the point $x = 0$ in $V(A)$.

11. Let X be a linear space. A set $K \subset X$ is said to be *convex* if

$$\lambda x + (1 - \lambda) y \in K \qquad (0 \le \lambda \le 1),$$

whenever $x, y \in K$. Let K_1 and K_2 be two convex sets in X.
(a) Show that $K_1 \cap K_2$ is convex.
(b) Is $K_1 \cup K_2$ convex?

12. Let $y = f(x)$ be a C^2 function defined for $-\infty < x < \infty$. Find a condition on d^2f/dx^2 in order that

$$K = \{(x,y) \in R^2 : y \ge f(x)\}$$

be convex.

13. Show that the polyhedron

$$P_n = \left\{ x = (x_1, \ldots, x_n) : x_i \ge 0 \text{ and } \sum_{i=1}^{n} x_i = 1 \right\}$$

is convex in R^n.

14. (Continuation of Exercise 13.) Let $(\lambda_1, \ldots, \lambda_n) \in P_n$.
(a) Show that

$$n^2 \le \sum_{i=1}^{n} \frac{1}{\lambda_i},$$

and show that equality holds if and only if $\lambda_i = 1/n$, $1 \le i \le n$. [*Hint:* Use mathematical induction.]
(b) Some stockbrokers recommend the method of "dollar cost averaging" for periodic investments, Engel [1; pp. 181 ff.]. Give a mathematical description of this method and use the inequality above to show that this

method results in a lower cost per share than the method of buying an equal number of shares at each investment period.

15. Let $A \subset X$ where X is a linear space.
   (a) Show that $V(A)$ is the smallest linear subspace of X that contains A. That is, show that if M is a linear subspace of X with $A \subset M$, then $V(A) \subset M$.
   (b) Show that A is a linear subspace of X if and only if $A = V(A)$.

7.   HAMEL BASES AND DIMENSION

We have used the terms "finite dimensional" and "infinite dimensional" somewhat casually up to this point. One thing we do in this section is give precise meanings to these terms. The other thing we do is introduce the concept of a Hamel basis for a linear space. A Hamel basis is important as we shall see, for several reasons. First it is the natural concept of basis for spaces that have linear structure only. Secondly, it allows one to distinguish between finite- and infinite-dimensional linear spaces. Indeed, we shall use exactly the same distinction in normed linear spaces.

On the other hand, Hamel basis is not the only concept of basis that arises in analysis. There are concepts of basis that involve topological as well as linear structure. Although these other bases reduce to Hamel bases on finite-dimensional linear spaces, they are usually quite different from Hamel bases on infinite-dimensional spaces. In fact, in applications involving infinite-dimensional spaces a useful basis, if one even exists, is usually something other than a Hamel basis. For example, a complete orthonormal set is far more useful in an infinite-dimensional Hilbert space than a Hamel basis.

A Hamel basis, then, is a purely algebraic concept that serves many important purposes, but it is not the last word on bases. Having said this, let us see what one is.

4.7.1 DEFINITION.   A set B in a linear space X is said to be a *Hamel basis* for X if (i) B is linearly independent and (ii) $V(B) = X$, that is, the span of B is X itself.

Of course, this definition is simply a generalization of the familiar concept of coordinate system.

EXAMPLE 1.   In the plane a set B containing any two noncollinear vectors is a Hamel basis or coordinate system for the plane.   ∎

EXAMPLE 2.   Let X be the real linear space made up of sequences $x = (x_1, x_2, \ldots)$ such that $\sum_{i=1}^{\infty} |x_i|^2 < \infty$. Let $A = \{e_1, e_2, \ldots\}$ where e_i is the sequence $e_i = (\delta_{i1}, \delta_{i2}, \ldots)$ and δ_{ij} is the Kronecker function. It is easily shown that A is linearly independent, so one suspects that A is a basis in the sense of Definition 4.7.1. *It is not!* Since we allow ourselves only finite sums, $a_1 x_1 + \cdots + a_n x_n$, we see that $V(A) \neq X$. In fact, $V(A)$ is the linear subspace of X made up of all sequences that are nonzero on only a *finite* number of entries. The reader may complain, saying he knows how to form infinite linear combinations of points in A to obtain arbitrary points in X, that is,

$$x = \sum_{n=1}^{\infty} x_n e_n .$$

While this is true, it is important to realize that the theory of infinite series requires more than just the linear space structure. We shall return to this point in Sections 5.17 and 5.18. ∎

So far we know only that some linear spaces have Hamel bases. The next theorem implies that every linear space has a Hamel basis.

   4.7.2 THEOREM. *If A is a linearly independent set in a linear space X, then there exists a Hamel basis B for X such that $A \subset B$.*

Since every linear space contains the empty set and the empty set is linearly independent, it follows from this theorem that every linear space has a Hamel basis. Of course, this theorem just says that a basis exists. It does not tell us how to find a basis. Since it is not crucial that the reader know the proof of this theorem, the proof is omitted here and outlined in the exercises. (Also see Appendix C.)

At this point a rather disconcerting thought should occur to the reader. If a linear space X has several different Hamel bases, do some Hamel bases have "fewer" points than others? Happily the answer to this question is no. In a meaningful sense all Hamel bases of a linear space X contain the same number of points.

   4.7.3 THEOREM. *If B_1 and B_2 are Hamel bases for a linear space X, then B_1 and B_2 have the same cardinal number.*

Recall that two sets have the same cardinal number if they can be put into a one-to-one correspondence with one another. The reader unfamiliar with the concept of cardinal numbers need merely note that intuitively this is a perfectly reasonable way to characterize the fact that two sets contain the same number of points. (See Appendix B.)

We shall not prove Theorem 4.7.3 here. One of the exercises at the end of this section sketches the proof for this finite-dimensional case and another problem sketches it for the infinite-dimensional case.

Theorems 4.7.2 and 4.7.3 furnish the foundation for a meaningful concept of dimension.

   4.7.4 DEFINITION.   The cardinal number of any Hamel basis of a linear space X is said to be the *dimension* of X. We denote the dimension of X by $\dim(X)$.

Again Theorem 4.7.2 shows that every linear space has a Hamel basis, and Theorem 4.7.3 shows that dimension is a property of the linear space in question and not dependent on the particular Hamel basis considered. If the dimension of X is finite, we say that X is a *finite-dimensional linear space*. Otherwise, we say that *X is an infinite-dimensional linear space. This then is the distinction between finite- and infinite-dimensional linear spaces no matter what additional structure (for example, a norm) may be present.*

The following theorem shows that for a given dimension there is essentially only one kind of linear space over a given scalar field.

4.7.5 THEOREM. *If X_1 and X_2 are linear spaces over the same scalar field, then X_1 and X_2 are isomorphic if and only if* dim $X_1 = $ dim X_2.

Proof: Suppose X_1 and X_2 are isomorphic, and let B_1 be a Hamel basis for X_1. Let $f: X_1 \rightarrow X_2$ be an isomorphism of X_1 onto X_2. We leave it to the reader to show that $B_2 = f(B_1)$ is a Hamel basis for X_2. Then since f is one-to-one, card$(B_1) = $ card(B_2); so dim$(X_1) = $ dim(X_2).

Now assume that dim$(X_1) = $ dim(X_2). Let B_1 and B_2 be Hamel bases for X_1 and X_2, respectively. Since dim$(X_1) = $ dim(X_2), there is a one-to-one mapping $\hat{f}$ of B_1 onto B_2. (That is what having the same cardinal number means.) Using this correspondence $\hat{f}$, we now define a linear mapping $f: X_1 \rightarrow X_2$: Let x be any point in X_1, $x \neq 0$. By Theorem 4.6.5, x can be expressed uniquely in the form

$$x = a_1 x_1 + \cdots + a_n x_n,$$

where $x_i \in B_1$ and $a_i \neq 0$, $i = 1, \ldots, n$. We let

$$f(x) = a_1 \hat{f}(x_1) + \cdots + a_n \hat{f}(x_n)$$

and

$$f(0) = 0.$$

It follows immediately from the fact that B_2 is a basis that f is one-to-one and maps X_1 onto X_2. Moreover, f is clearly linear. Hence, f is an isomorphism and X_1 and X_2 are isomorphic. ∎

4.7.6 COROLLARY. *If X is a finite-dimensional linear space over a scalar field F, where* dim$(X) = n$, *then X is isomorphic to F^n, the linear space made up of ordered n-tuples of scalars.*

In other words, all n-dimensional real linear spaces are isomorphic to R^n, and all complex ones to C^n.

Both the foregoing results deserve serious consideration. They raise the following question. Since all linear spaces of a given dimension over a given scalar field are essentially the same linear space, why bother to study and discuss linear spaces at an abstract level? The alternative would be to study a typical linear space for each dimension. For instance, with finite-dimensional spaces we could limit our study to F^n. There are two important reasons for not doing so.

First, each time we proved something about, say F^n, we would have to show that it held for all linear spaces isomorphic to F^n. Otherwise, we would not be making a statement about n-dimensional spaces in general. Needless to say, this would not be a saving of effort.

Secondly, in many applications, in fact in most, there is other structure (for example, topological) present, and this other structure is usually germane to the application. Attempting consistently to recast such problems in terms of "preferred linear space formulations" would be an extremely awkward practice.

The following theorem is an important statement about the dimensions of the range and null space of a linear transformation.

4.7.7 THEOREM.   *Let* $L\colon X \to Y$ *be a linear transformation where* X *and* Y *are linear spaces. Then the dimension of* $\mathcal{N}(L), \mathcal{R}(L)$, *and* X *are related by the formula*

$$\dim\{\mathcal{N}(L)\} + \dim\{\mathcal{R}(L)\} = \dim\{X\}.$$

This result is particularly useful when X is finite dimensional.

Proof: Let $B_{\mathcal{N}}$ be a Hamel basis for $\mathcal{N}(L)$. Then there exists (Theorem 4.7.2) a Hamel basis B_X for X such that $B_{\mathcal{N}} \subset B_X$. Let $B = \{y \in B_X : y \notin B_{\mathcal{N}}\}$. One clearly has $V(B) \cap \mathcal{N}(L) = \{0\}$. Furthermore, the set $L(B)$ in Y spans the range of L. (Why?) We assert that $L(B)$ is a linearly independent set in Y. To see this we consider the equation

$$\beta_1 L(x_1) + \cdots + \beta_n L(x_n) = 0, \tag{4.7.1}$$

where $\{x_1,\ldots,x_n\}$ is a finite set of distinct elements in B. Since L is linear (4.7.1) implies that

$$L(\beta_1 x_1 + \cdots + \beta_n x_n) = 0$$

or $(\beta_1 x_1 + \cdots + \beta_n x_n) \in \mathcal{N}(L)$. Since the point $(\beta_1 x_1 + \cdots + \beta_n x_n)$ is also in $V(B)$ one then has

$$\beta_1 x_1 + \cdots + \beta_n x_n = 0. \tag{4.7.2}$$

Since B is linearly independent, it follows from Theorem 4.6.3 that $\beta_1 = \cdots = \beta_n = 0$. By applying Theorem 4.6.3 again to Equation (4.7.1) we see that $L(B)$ is linearly independent in Y.

Since $L(B)$ is linearly independent and spans $\mathcal{R}(L)$ it follows that $L(B)$ is a Hamel basis for $\mathcal{R}(L)$. The conclusion now follows from standard cardinal arithmetic (see the exercises in Appendix B).   ∎

EXERCISES

1. Let X be a finite-dimensional space with $\dim(X) = n$. Show that every set containing $n + 1$ points is linearly dependent.

2. Let A be a linear subspace of a linear space X. Show that $\dim(A) \leq \dim(X)$. Moreover, if X is finite dimensional and A is a proper linear subspace of X, show that $\dim(A) < \dim(X)$.

3. Consider the following differential equation defined on $C^2[0,\infty)$

$$\frac{d^2x}{dt^2} + b\frac{dx}{dt} + cx = 0. \qquad (4.7.3)$$

If X denotes the set of all solutions of (4.7.3) show that X is a linear subspace of $C^2[0,\infty)$ and that $\dim(X) = 2$.

4. Show that if A is a set in a linear space X with $V(A) = X$, then A contains a Hamel basis of X.

5. Let X be the real linear space made up of all functions of the form $x(t) = a\cos(\omega t + \phi)$, where ω is fixed. Show that $B = \{\cos \omega t, \sin \omega t\}$ is a basis for X.

6. Prove Theorem 4.7.2. [*Hint*: Let P be the class made up of all linear independent sets in the linear space X, that is, an element of P is a linearly independent set in X. Let P be partially ordered by inclusion. Then use Zorn's lemma. Compare with Exercise 5, Section 3.]

7. Prove Theorem 4.7.3 for the finite-dimensional case. [*Hint*: Assume that $B_1 = \{x_1,\ldots,x_n\}$ is a finite Hamel basis for X, and let B_2 be any other Hamel basis of X. Let x_i be a point B_1, and let $B_2(x_i)$ be the unique finite set (Theorem 4.6.5) of points in B_2 needed to express x. Show that

$$B_2 = B_2(x_1) \cup B_2(x_2) \cup \cdots \cup B_2(x_n)$$

and that B_2 is finite. Let $B_2 = \{y_1, \ldots, y_m\}$. The point y_1 is a linear combination of points in B_1, that is, $y_1 = a_1x_1 + \cdots + a_nx_n$. Argue that at least one of the coefficients $a_1, \ldots, a_n$ is nonzero; for example, $a_1 \neq 0$. Then $x_1 = (1/a_1)y_1 - (a_2/a_1)x_2 - \cdots - (a_n/a_1)x_n$. Deleting x_1, the set $y_1, x_2, \ldots, x_n$ spans X. Continue along this line of argument to show that $n \geq m$. Then reverse the argument and show that $m \geq n$.]

8. Prove Theorem 4.7.3 for the infinite-dimensional case. [*Hint*: For each $x \in B_1$, let $B_2(x)$ be the unique finite set of points in B_2 needed to express x. Show that $y \in B_2$ implies that $y \in B_2(x)$ for some x. Then show that

$$B_2 = \bigcup_{x \in B_1} B_2(x).$$

Then show that $\operatorname{card}(B_2) \leq \operatorname{card}(B_1)$. Now reverse the roles of B_1 and B_2 to get $\operatorname{card}(B_1) \leq \operatorname{card}(B_2)$.]

9. Show that $C[0,T]$ is infinite dimensional. [*Hint*: Construct a linearly independent set of dimension n, where n is an arbitrary integer.]

10. Let L be a linear transformation of X into Y where X and Y are both finite dimensional.
    (a) Show that L maps X onto Y if and only if $\dim \mathscr{R}(L) = \dim Y$.
    (b) Show that L is one-to-one if and only if $\dim \mathscr{R}(L) = \dim X$.
    (c) Show that L is invertible if and only if $\dim X = \dim Y = \dim \mathscr{R}(L)$.
    (d) What can one say about infinite-dimensional spaces?

8.   THE USE OF MATRICES TO REPRESENT LINEAR TRANSFORMATIONS

Let us now turn to the important topic of the representation of linear transformations by matrices. Let X and Y be finite dimensional, and let $T: X \to Y$ be linear. Roughly speaking, if $Tx = y$; x is expanded in terms of a basis of X, y is expanded in terms of a basis of Y, then a matrix is used to relate the coefficients in these two expansions.

Let $B_1 = \{x_1, x_2, \ldots, x_n\}$ and $B_2 = \{y_1, y_2, \ldots, y_m\}$ be bases for X and Y, respectively. We know from Section 7 that $x \in X$ and any $y \in Y$ can be expressed uniquely in the form

$$x = \alpha_1 x_1 + \cdots + \alpha_n x_n$$

and

$$y = \beta_1 y_1 + \cdots + \beta_m y_m,$$

respectively. Thus for any $x \in X$, $T(x)$ can uniquely be expressed

$$T(x) = \beta_1 y_1 + \cdots + \beta_m y_m$$

and from the linearity of T

$$T(x) = \alpha_1 T(x_1) + \cdots + \alpha_n T(x_n).$$

But $T(x_1), \ldots, T(x_n)$ are points in Y, so they can be uniquely expressed

$$T(x_1) = t_{11} y_1 + \cdots + t_{m1} y_m,$$
$$T(x_2) = t_{12} y_1 + \cdots + t_{m2} y_m,$$
$$\cdots$$
$$T(x_n) = t_{1n} y_1 + \cdots + t_{mn} y_m,$$

where the t_{ij}'s are scalars. Therefore, we have

$$T(x) = \beta_1 y_1 + \cdots + \beta_m y_m$$

and

$$T(x) = \alpha_1(t_{11} y_1 + \cdots + t_{m1} y_m) + \cdots + \alpha_n(t_{1n} y_1 + \cdots + t_{mn} y_m)$$
$$= (t_{11}\alpha_1 + \cdots + t_{1n}\alpha_n) y_1 + \cdots + (t_{m1}\alpha_1 + \cdots + t_{mn}\alpha_n) y_m.$$

Again the expansion of $T(x)$ in terms of the basis B_2 is unique. Therefore,

$$\beta_1 = t_{11}\alpha_1 + t_{12}\alpha_2 + \cdots + t_{1n}\alpha_n$$
$$\beta_2 = t_{21}\alpha_1 + t_{22}\alpha_2 + \cdots + t_{2n}\alpha_n$$
$$\cdots$$
$$\beta_m = t_{m1}\alpha_1 + t_{m2}\alpha_2 + \cdots + t_{mn}\alpha_n.$$

Or in terms of matrices

$$\begin{bmatrix} \beta_1 \\ \beta_2 \\ \cdots \\ \beta_m \end{bmatrix} = \begin{bmatrix} t_{11} & t_{12} & \cdots & t_{1n} \\ t_{21} & t_{22} & \cdots & t_{2n} \\ & \cdots & & \\ t_{m1} & \cdots & & t_{mn} \end{bmatrix} \begin{bmatrix} \alpha_1 \\ \alpha_2 \\ \cdots \\ \alpha_n \end{bmatrix}.$$

We say that the matrix

$$[T] = \begin{bmatrix} t_{11} & \cdots & t_{1n} \\ t_{m1} & & t_{mn} \end{bmatrix}$$

represents the linear transformation $T: X \to Y$. Carefully note that T and $[T]$ are not the same thing. T is a rule for assigning y's to x's, while $[T]$ is an $m \times n$ array of scalars. The matrix $[T]$ represents the linear transformation T. More precisely, $[T]$ represents T relative to the bases B_1 and B_2, in the sense that $[T]$ together with B_1 and B_2 can be used to solve the equation $y = Tx$.

EXAMPLE 1.   Let X be the linear space made up of all third degree polynomials, that is, all $x(t)$ of the form

$$x(t) = \alpha_1 + \alpha_2 t + \alpha_3 t^2 + \alpha_4 t^3, \quad -\infty < t < \infty.$$

Let Y be the linear space made up of all second degree polynomials, that is,

$$y(t) = \beta_1 + \beta_2 t + \beta_3 t^2.$$

Let D be the derivative operation restricted to X. Clearly $\mathcal{R}(D) = Y$ and D is linear. One basis for X is

$$B_1 = \{x_1, x_2, x_3, x_4\},$$

where $x_1 = 1$, $x_2 = t$, $x_3 = t^2$, $x_4 = t^3$. Similarly, a basis for Y is

$$B_2 = \{y_1, y_2, y_3\},$$

where $y_1 = 1$, $y_2 = t$, $y_3 = t^2$. Then

$$D(x_1) = 0,$$
$$D(x_2) = y_1,$$
$$D(x_3) = 2y_2,$$
$$D(x_4) = 3y_3.$$

Thus

$$\begin{bmatrix} \beta_1 \\ \beta_2 \\ \beta_3 \end{bmatrix} = \begin{bmatrix} 0 & 1 & 0 & 0 \\ 0 & 0 & 2 & 0 \\ 0 & 0 & 0 & 3 \end{bmatrix} \begin{bmatrix} \alpha_1 \\ \alpha_2 \\ \alpha_3 \\ \alpha_4 \end{bmatrix}$$

and relative to the bases B_1 and B_2 the linear transformation D is represented by the matrix

$$[D] = \begin{bmatrix} 0 & 1 & 0 & 0 \\ 0 & 0 & 2 & 0 \\ 0 & 0 & 0 & 3 \end{bmatrix}.$$

Of course, if one changes bases, the matrix that represents D will probably change. For example, if instead of B_1 we let $B_3 = \{1 + t, t + t^2, t^2 + t^3, 1 + t^3\}$ be the basis for X, then

$$D(x_1) = y_1,$$
$$D(x_2) = y_1 + 2y_2,$$
$$D(x_3) = 2y_2 + 3y_3,$$
$$D(x_4) = 3y_3,$$

and relative to B_3 and B_2 the transformation D is now represented by

$$[D] = \begin{bmatrix} 1 & 1 & 0 & 0 \\ 0 & 2 & 2 & 0 \\ 0 & 0 & 3 & 3 \end{bmatrix}.$$

Carefully note that in each case the same transformation, $D: X \to Y$, is being represented. Roughly speaking, the matrices change because we change our coordinate systems.   ∎

EXERCISES

1. Let $[T]$ be an $m \times n$ matrix of scalars, and let X and Y be linear spaces where $\dim(X) = n$ and $\dim(Y) = m$. Show that there exists a linear transformation $T: X \to Y$ such that $[T]$ represents T relative to some Hamel basis.

2. Let X and Y be finite-dimensional linear spaces with $\dim X = n$ and $\dim Y = m$. Consider the linear space $lt[X,Y]$ of all linear transformation $T: X \to Y$ and the linear space M_{mn} of all $m \times n$ matrices of scalars. The relation "$[T]$ represents T relative to B_1 and B_2" is a mapping of $lt[X,Y]$ into M_{mn}. Show that it is an isomorphism. In other words, $lt[X,Y]$ and M_{mn} are isomorphic.

3. Let X, Y, and Z be linear spaces over the same scalar field, and let B_1, B_2, and B_3 be bases for X, Y, and Z, respectively. Let $T_1: X \to Y$ and $T_2: Y \to Z$ be linear. Show that the matrix $[T_2 T_1]$ that represents the composition $T_2 T_1: X \to Z$ relative to B_1 and B_3 is $[T_2 T_1] = [T_2][T_1]$ where $[T_1]$ represents T_1 relative to B_1 and B_2, $[T_2]$ represents T_2 relative to B_2 and B_3, and $[T_2][T_1]$ denotes the usual matrix product.

4. Let X be a finite-dimensional linear space and let $B_1 = \{x_1,\dots,x_n\}$ and $B_2 = \{y_1,\dots,y_n\}$ be two bases for X. Thus any $x \in X$ can be expressed uniquely in the form

$$x = \alpha_1 x_1 + \cdots + \alpha_n x_n$$

or

$$x = \beta_1 y_1 + \cdots + \beta_n y_n.$$

Show that an $n \times n$ matrix can be used to represent the transformation from the B_1 to B_2 coordinate system. Moreover, show that the same $n \times n$ matrix is the representation of the identity transformation $I: X \to X$ relative to B_1 and B_2.

5. Let X and Y be linear spaces over the same scalar field, and let B_1 and B_2 be countable Hamel bases for X and Y, respectively. Further, let $T: X \to Y$ be linear. Show that an infinite matrix can be used to represent T.

6. Discuss the use of matrices to represent (in the sense of this section) the linear transformations that model linear (time-invariant and time-varying) sampled data systems.

7. Let X be the linear space made up of all functions $\Phi(t,x)$ of the form

$$\Phi(t,x) = \alpha_1 + (\alpha_{21}t + \alpha_{22}x) + (\alpha_{31}t^2 + \alpha_{32}tx + \alpha_{33}x^2) + \cdots$$
$$+ (\alpha_{n1}t^{(n-1)} + \alpha_{n2}t^{(n-2)}x + \cdots + \alpha_{nn}x^{(n-1)}),$$

where $(t,x) \in [0,T] \times [0,L]$, n is a fixed positive integer, and the α's are scalars. What is the dimension of X? Show that the equation

$$\frac{\partial \Phi}{\partial t} - k \frac{\partial^2 \Phi}{\partial x^2} = \Psi$$

(where k is a constant) represents a linear transformation of X into itself. Represent this linear transformation with a matrix. Is the transformation one-to-one? Does it map X onto itself?

8. Consider the linear operator $K: X \to X$ given by $y = Kx$, where

$$y(t) = \int_0^{2\pi} k(t,s)x(s)ds, \qquad k(t,s) = 4 \cos 2(t-s),$$

and X is the linear space spanned by $\{1, \cos s, \cos 2s, \sin s, \sin 2s\}$.
(a) Express K as a matrix.
(b) Is K one-to-one?
(c) Does it map X onto itself?

9. Let L be a linear transformation of X *onto* Y where X is finite dimensional. Let $[L]$ be a matrix representing L. Show that if L is one-to-one, then $[L]$ is a square matrix and det $[L] \neq 0$.

10. Let $X = Y$ denote the space of all fourth degree polynomials in t and define $L: X \to Y$ by $L = D^2 + 2D + I$ where D is the differential operator, that is,

$$Lx = \frac{d^2x}{dt^2} + 2\frac{dx}{dt} + x.$$

(a) Represent L by a matrix L in terms of the basis $\{1,t,t^2,t^3,t^4\}$ on X and Y.
(b) Represent $L^2 = LL$ in terms of this basis. Show that $[L^2] = [L][L]$ in terms of the usual matrix product.
(c) Repeat steps (a) and (b) for $M = \alpha D^2 + \beta D + \gamma I$.

11. Let X_1 be the linear space of all functions of the form $\{\alpha_0 + \alpha_1 \cos t + \beta_1 \sin t\}$ for $0 \leq t \leq 2\pi$ and define $L: X_1 \to X_1$ by $y = Lx$ where

$$y(t) = \int_0^{2\pi} [1 + \cos(t-s)]x(s)\,ds.$$

(a) Represent L as a matrix operator.

(b) Do the same where X_1 is replaced by X_2, the collection of all functions of the form $\{\alpha_0 + \alpha_1 \cos t + \alpha_2 \cos 2t + \beta_1 \sin t + \beta_2 \sin 2t\}$ and L maps X_2 into itself.

(c) Do the same for the operator $M: X_2 \to X_2$ given by

$$y(t) = \int_0^{2\pi} [1 + \cos(t - s) + \sin 2(t + s)]x(s)\,ds.$$

12. Let X_n denote the space of all polynomials in t of degree $< n$.

(a) Consider the operator

$$y = Lx \leftrightarrow y(t) = \frac{d}{dt}(t^2 - 1)\frac{dx}{dt}$$

on X_4. Find a representation of $L: X_4 \to X_4$ with respect to the basis

$$A = \left\{1, t, \frac{3}{2}t^2 - \frac{1}{2}, \frac{5}{2}t^3 - \frac{3}{2}t\right\}.$$

(b) Find a basis B for X_4 such that the operator

$$y = Hx \leftrightarrow y(t) = \frac{d^2x}{dt^2} - 2t\frac{dx}{dt}$$

can be represented by a diagonal matrix.

(c) Can the operators L and H be represented by diagonal matrices on X_5? (The operators L and H generate the Legendre polynomials and the Hermite polynomials, respectively. See Section 7.14 and Exercise 4, Section 7.5.)

9.   EQUIVALENT LINEAR TRANSFORMATIONS

Two linear spaces are viewed as being essentially the same linear space if they are isomorphic to one another. In more or less the same spirit two linear transformations can be essentially the same. The idea is to connect the two by means of isomorphic transformations of their domains and ranges. Let $T: X \to Y$ and $\mathcal{T}: \mathcal{X} \to \mathcal{Y}$ be linear transformations. Further suppose that X and $\mathcal{X}$ are isomorphic and that Y and $\mathcal{Y}$ are also isomorphic. Let $U: X \to \mathcal{X}$ and $W: Y \to \mathcal{Y}$, be the isomorphisms. The situation is illustrated in Figure 4.9.1. We see that we

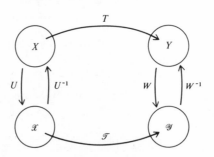

Figure 4.9.1.

have two mappings of X into Y; namely, T and $T' = W^{-1}\mathcal{T}U$. Similarly, $\mathcal{T}$ and $\mathcal{T}' = WTU^{-1}$ are two mappings of $\mathcal{X}$ into $\mathcal{Y}$. One does not expect T and T' or $\mathcal{T}$ and $\mathcal{T}'$ to be the same. However, it can happen—and this is very interesting when it does—that one has $T = T'$, or equivalently that $\mathcal{T} = \mathcal{T}'$.

EXAMPLE 1.   Let $X = Y = L_2(-\infty,\infty)$ and $\mathcal{X} = \mathcal{Y} = L_2(-i\infty,i\infty)$ and let $T: X \to Y$ be

$$(Tx)(t) = \int_{-\infty}^{t} e^{-(t-\tau)}x(\tau)\,d\tau, \qquad -\infty < t < \infty,$$

and $\mathcal{T}: \mathcal{X} \to \mathcal{Y}$ be

$$(\mathcal{T}x)(i\omega) = \frac{1}{i\omega + 1}\,x(i\omega), \qquad -\infty < \omega < \infty.$$

It is well-known that $\mathcal{F}$, the Fourier transform, is an isomorphic mapping of $L_2(-\infty, \infty)$ onto $L_2(-i\infty, i\infty)$ (Section 5.19) and that $T = \mathcal{F}^{-1}\mathcal{T}\mathcal{F}$ and $\mathcal{T} = \mathcal{F}T\mathcal{F}^{-1}$. ∎

Let us formalize the above remarks in a definition.

4.9.1 DEFINITION.   Let $X, \mathcal{X}, Y, \mathcal{Y}$ be linear spaces over the same scalar field where X and $\mathcal{X}$ are isomorphic and Y and $\mathcal{Y}$ are isomorphic. The linear transformations $T: X \to Y$ and $\mathcal{T}: \mathcal{X} \to \mathcal{Y}$ are said to be *isomorphically equivalent* (*to one another*) if there exist isomorphisms $U: X \to \mathcal{X}$ and $W: Y \to \mathcal{Y}$ such that

$$T = W^{-1}\mathcal{T}U$$

and

$$\mathcal{T} = WTU^{-1}.$$

It is trivial to show that $T = W^{-1}\mathcal{T}U$ if and only if $\mathcal{T} = WTU^{-1}$.

There is a special case of two linear transformations being isomorphically equivalent which is particularly important; namely, similarity.

4.9.2 DEFINITION.   Let X and $\mathcal{X}$ be isomorphic linear spaces. The linear transformations $T: X \to X$ and $\mathcal{T}: \mathcal{X} \to \mathcal{X}$ are said to be *similar* if there exists an isomorphism $U: X \to \mathcal{X}$ such that

$$T = U^{-1}\mathcal{T}U$$

and

$$\mathcal{T} = UTU^{-1}.$$

The situation described in this definition is illustrated in Figure 4.9.2.

It is difficult to overemphasize the importance of the two concepts presented in Definitions 4.9.1 and 4.9.2. In fact, these concepts are basic to the entire theory of linear operators. Given a linear transformation $T: X \to Y$ (or $T: X \to X$) one tries to find an isomorphically equivalent (or similar) transformation $\mathcal{T}: \mathcal{X} \to \mathcal{Y}$

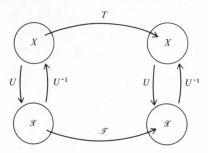

Figure 4.9.2.

(or $\mathcal{T}: \mathcal{X} \to \mathcal{X}$) which is somehow easier to work with than T. For example, it is usually easier to work with the transfer functions obtained by Fourier or Laplace transforms than to work with the corresponding differential or integral operators.

We end this section with a warning. The meanings of the terms "isomorphically equivalent" and "similar" are those given in Definitions 4.9.1 and 4.9.2 and absolutely nothing more. They are algebraic concepts. Two linear transformations T and $\mathcal{T}$ that are isomorphically equivalent or similar are essentially the same *linear* transformation from an algebraic point of view. Yet in specific cases two isomorphically equivalent transformations can differ in a world of ways not covered by these definitions. To name just one, there is the whole question of topological structure that may be present. We shall study this further in the next chapter.

EXERCISES

1. Show that the entire construction given in Section 8 can be viewed as a special case of linear transformations being isomorphically equivalent. [*Hint*: Given bases B_1 and B_2 for X and Y, respectively, let $U: X \to F^n$ and $W: Y \to F^m$ denote the operations of finding the "coordinates" of points in X and Y, respectively. Then show that $T: X \to Y$ is isomorphically equivalent to a linear matrix transformation $\mathcal{T}: F^n \to F^m$.]

2. Let X be the sequence space $l_2(-\infty, \infty)$ and let Z be the same as the linear space Z in Example 4, Section 5. Let $S_r: X \to X$ be the right shift operator. Let $\mathcal{T}: z \to z$ be defined by $(\mathcal{T}y)(z) = z^{-1}y(z)$ for $|z| = 1$. Show that $\mathcal{T}$ and S_r are similar. [*Hint*: Use the z-transform.]

3. Let X be a linear space, and let $lt[X,X]$ be the linear space of linear transformations of X into itself. Show that the relation $L_1 \sim L_2$ defined on $lt[X,X]$ by $L_1 \sim L_2 \Leftrightarrow \{L_1 \text{ and } L_2 \text{ are similar}\}$ is an equivalence relation. Let $lt[X,Y]$ denote the linear space of all linear transformations of X into Y. Show that $\{L_1 \text{ and } L_2 \text{ are isomorphically equivalent}\}$ is an equivalence relation on $lt[X,Y]$.

4. Let $X, Y, \mathcal{X}, \mathcal{Y}$ be finite-dimensional linear spaces over the same scalar field, where X and $\mathcal{X}$ are isomorphic and Y and $\mathcal{Y}$ are isomorphic. Let $T: X \to Y$ and

$\mathcal{T} : \mathcal{X} \rightarrow \mathcal{Y}$ be linear. Show that T and $\mathcal{T}$ are isomorphically equivalent if and only if $\dim[\mathcal{R}(T)] = \dim[\mathcal{R}(\mathcal{T})]$, where $\mathcal{R}(\cdot)$ denotes the range.

5. Let S and T be linear operators on R^2 and assume that there is a basis for R^2 in which S and T have the following representations.

$$S = \begin{bmatrix} 1 & 1 \\ 0 & 1 \end{bmatrix}, \quad T = \begin{bmatrix} \lambda & 0 \\ 0 & \mu \end{bmatrix}.$$

(a) Show that S and T are not similar.
(b) Are S and T ever isomorphically equivalent?

6. Define the 2×2 matrices σ_k, $k = 1, 2, 3$, by

$$\sigma_1 = \begin{bmatrix} 0 & 1 \\ 1 & 0 \end{bmatrix}, \quad \sigma_2 = \begin{bmatrix} 0 & -i \\ i & 0 \end{bmatrix}, \quad \sigma_3 = \begin{bmatrix} 1 & 0 \\ 0 & -1 \end{bmatrix}.$$

(These are referred to as the *Pauli spin matrices*.)
(a) Show that:

$$\begin{cases} \sigma_k^2 = I & k = 1, 2, 3 \\ \sigma_1 \sigma_2 = -\sigma_2 \sigma_1 = i\sigma_3 \\ \sigma_2 \sigma_3 = -\sigma_3 \sigma_2 = i\sigma_1 \\ \sigma_3 \sigma_1 = -\sigma_1 \sigma_3 = i\sigma_2. \end{cases} \qquad (4.9.1)$$

(b) Let S be a nonsingular 2×2 matrix and define τ_k by $\tau_k = S\sigma_k S^{-1}$, $k = 1, 2, 3$. Show that τ_1, τ_2, and τ_3 satisfy (4.9.1).
(c) Let τ_1, τ_2, and τ_3 be three 2×2 matrices that satisfy (4.9.1). Show that there is a nonsingular matrix S with the property that $\tau_k = S\sigma_k S^{-1}$, $k = 1, 2, 3$.

7. Let M and N be linear operators on C^n that can be represented as diagonal matrices

$$[M] = \mathrm{diag}\{\mu_1, \ldots, \mu_n\},$$
$$[N] = \mathrm{diag}\{v_1, \ldots, v_n\},$$

in terms of some basis. Show that M and N are similar if and only if the two sets $\{\mu_1, \ldots, \mu_n\}$ and $\{v_1, \ldots, v_n\}$ are the same.

Part B

Further Topics

10. DIRECT SUMS AND SUMS

In Chapter 3 we saw how two metric spaces can be put together to form a new metric space called the product space. In this section we shall do a similar thing for linear spaces. Suppose that X_1 and X_2 are linear spaces over the same scalar field. The linear spaces X_1 and X_2 can be combined to form a new linear space referred to as the *direct sum of X_1 and X_2* and denoted by $X_1 \oplus X_2$. The underlying set of $X_1 \oplus X_2$ is the Cartesian product, $X_1 \times X_2$, of the underlying sets of X_1 and X_2. Thus a point in $X_1 \oplus X_2$ is an ordered pair (x_1,x_2), where $x_1 \in X_1$ and $x_2 \in X_2$. Addition is defined by $(x_1,x_2) + (y_1,y_2) = (x_1 + y_1, x_2 + y_2)$. Note that " $+$ " on the left is defined in terms of the addition operations of X_1 and X_2 appearing on the right. Scalar multiplication is defined by $\alpha(x_1, x_2) = (\alpha x_1, \alpha x_2)$. The origin is $(0,0)$ and $-(x_1,x_2) = (-x_1, -x_2)$.

Let us consider some examples.

EXAMPLE 1.   Consider a system with two input channels (for example, settings of valves #1 and #2) and two output channels (for example, temperature #1 and flow rate #2) as shown in Figure 4.10.1. Assume that the function at

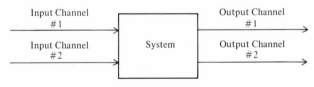

Figure 4.10.1.

each channel is a point in the linear space $C[0,T]$. Then an input to the system— as opposed to one of the input channels—is an ordered pair of continuous functions, the input on channel #1 and the input on channel #2. Similarly assume that the system output is an ordered pair of continuous functions. Thus the mathematical model for the system would be a transformation of $C[0,T] \oplus C[0,T]$ into itself. ∎

EXAMPLE 2.   Let us see that the operation of multiplication can represent either a linear or nonlinear system. Suppose we have some device, see Figure 4.10.2, that multiplies continuous functions of time together on a pointwise basis. If we let $x_2(t) = f(t)$, where f is a fixed point in $C[0,T]$, and we consider the resulting

196

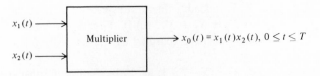

Figure 4.10.2.

mapping $x_1(t) \to f(t)x_1(t)$ of $C[0,T]$ into itself, then this mapping is linear. On the other hand, if we consider the mapping of $C[0,T] \oplus C[0,T]$ into $C[0,T]$ represented by

$$(x_1,x_2) \to x_0 = x_1 x_2,$$

this mapping is nonlinear. Note that whether or not the multiplier is "linear" depends on how it is being interpreted. ∎

EXAMPLE 3. Similarly, addition can represent either a linear or nonlinear system. Suppose we have some device, see Figure 4.10.3, that adds continuous

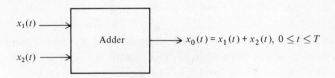

Figure 4.10.3.

functions together on a pointwise basis. If we let $x_1(t) = f(t), f \in C[0,T]$, then the mapping

$$x_0(t) = x_1(t) + f(t) \qquad 0 \le t \le T$$

of $C[0,T]$ into itself is linear if and only if $f(t) \equiv 0$. But the transformation

$$(x_1,x_2) \to x_0$$

given by

$$x_0(t) = x_1(t) + x_2(t)$$

is a linear mapping of $C[0,T] \oplus C[0,T]$ into $C[0,T]$. ∎

Needless to say, we form the direct sum $X_1 \oplus X_2 \oplus \cdots \oplus X_n$ of n linear spaces over the same scalar field in the obvious way.

Often when we are discussing two linear spaces, X_1 and X_2, over the same scalar field, they are in fact both linear subspaces of some containing linear space X. In that situation we have two ways to put X_1 and X_2 together to form a new linear space: *direct sum* and *(inner) sum*. The sum of X_1 and X_2, denoted $X_1 + X_2$, is the linear subspace of X made up of all points $x = x_1 + x_2$, where $x_1 \in X_1$ and $x_2 \in X_2$. That $X_1 + X_2$ is indeed a linear subspace of X is trivial. Note also that (a) $X_1 \subset X_1 + X_2$, (b) $X_2 \subset X_1 + X_2$, (c) if $X_1 \subset X_2$, then $X_1 + X_2 = X_2$, and (d) $X_1 + X_2$ is the linear subspace spanned by the set $X_1 \cup X_2$.

EXAMPLE 4.   Let $X = C[-T,T]$; let X_1 be the linear subspace of X made up of all even functions (that is, $x(t) = x(-t)$ for all $t \in [-T,T]$); and let X_2 be the linear subspace of X made up of all odd functions (that is, $x(t) = -x(-t)$ for all $t \in [-T,T]$). In this case it turns out that $X_1 + X_2 = X$. Indeed, if x is any point in X, then

$$x(t) = \frac{x(t) + x(-t)}{2} + \frac{x(t) - x(-t)}{2} = x_e(t) + x_0(t),$$

where $x_e \in X_1$ and $x_0 \in X_2$. ∎

We say that two linear subspaces X_1 and X_2 of a linear space X are *disjoint*[3] if $X_1 \cap X_2 = \{0\}$, that is, their intersection contains exactly one point, the origin.

4.10.1 LEMMA.   *Let X_1 and X_2 be linear subspaces of a linear space X. Then for each x in $X_1 + X_2$ there is a unique $x_1 \in X_1$ and a unique $x_2 \in X_2$ such that $x = x_1 + x_2$ if and only if X_1 and X_2 are disjoint.*

Proof: Consider the "if" part. Assume that X_1 and X_2 are disjoint. Let $x = x_1 + x_2 = y_1 + y_2$, where $x_1, y_1 \in X_1$ and $x_2, y_2 \in X_2$. Then $x_1 - y_1 = y_2 - x_2$. But $(x_1 - y_1) \in X_1$ and $(y_2 - x_2) \in X_2$ and $X_1 \cap X_2 = \{0\}$; therefore, $x_1 - y_1 = 0$ and $x_2 - y_2 = 0$.

Now consider the "only if" part. Suppose that x_1 and x_2 are uniquely determined for each x in $X_1 + X_2$. We must show that it follows that X_1 and X_2 are disjoint. If X_1 and X_2 are not disjoint, there exists an x^*, $x^* \neq 0$, in $X_1 \cap X_2$. Then if $x = x_1 + x_2$, it follows that $x = (x_1 + \alpha x^*) + (x_2 - \alpha x^*)$ for any scalar α. But then x_1 and x_2 are not unique, which is a contradiction. ∎

Although the sum $X_1 + X_2$ and the direct sum $X_1 \oplus X_2$ are different linear spaces, it is possible to compare them. In fact there is a *natural mapping*, call it ϕ, of $X_1 \oplus X_2$ into $X_1 + X_2$ defined by

$$\phi[(x_1,x_2)] = x_1 + x_2.$$

We immediately note that ϕ is a linear mapping of $X_1 \oplus X_2$ onto $X_1 + X_2$. However, ϕ need not be invertible, for it need not be one-to-one. But when ϕ is invertible, it follows that ϕ is an isomorphism and that $X_1 + X_2$ and $X_1 \oplus X_2$ are isomorphic. We now seek conditions under which ϕ is an isomorphism, that is, conditions under which $X_1 + X_2$ and $X_1 \oplus X_2$ are algebraically equivalent.

---

[3] This is an example of a minor ambiguity one sometimes finds in mathematical terminology, that is, "disjoint" linear subspaces are not "disjoint" as sets since they must contain the origin. However, the dual usage of "disjoint" is so natural that it is now universally accepted.

4.10.2 THEOREM.   *Let X_1 and X_2 be linear subspaces of a linear space X. The natural mapping ϕ of $X_1 \oplus X_2$ onto $X_1 + X_2$ is an isomorphism if and only if X_1 and X_2 are disjoint.*

The proof of this theorem involves a simple application of Lemma 4.10.1, and we ask the reader to check this.

Since $X_1 + X_2$ and $X_1 \oplus X_2$ are isomorphic when $X_1 \cap X_2 = \{0\}$, it is common practice when X_1 and X_2 are disjoint to refer to $X_1 + X_2$ as the direct sum of X_1 and X_2 and denote it by $X_1 \oplus X_2$. Needless to say, this can be confusing and sometimes misleading. This practice is particularly dangerous when the containing linear space has topological structure or algebraic structure in addition to its linear space structure.[4]

If $X = X_1 + X_2$, where X_1 and X_2 are disjoint linear spaces, then we shall say that X_2 is an *algebraic complement* of X_1. The next result is simple, but we caution the reader to observe that it is an algebraic fact and *not* a topological fact.

4.10.3 THEOREM.   *Let X_1 be a linear subspace of a linear space X. Then X_1 has an algebraic complement.*

Proof: Let B_1 be a Hamel basis for X_1 and choose B_2 so that $B_1 \cap B_2 = \varnothing$ and $B = B_1 \cup B_2$ is a Hamel basis for X. (See Theorem 4.7.2.) The linear space X_2 generated by B_2 is then an algebraic complement of X_1.   ∎

One can show (see Exercise 5) that every algebraic complement of a linear subspace X_1 has the same dimension. Indeed, this follows directly from the last result in the case of a finite-dimensional space X. In other words, the dimension of the algebraic complements of X_1 is a property of X_1. We refer to this dimension as the *co-dimension* of X_1. Roughly speaking, as the dimension goes up, the co-dimension goes down and vice versa.

EXAMPLE 5.   In this example we show that two linear time-invariant mappings do not necessarily commute. In particular, we exhibit two noncommuting linear time-invariant mappings T_1 and T_2 defined on a linear subspace X of $l_2(-\infty, \infty)$.

Let $x \in l_2(-\infty, \infty)$, $x \neq 0$, and let A_x denote the linear subspace generated by x and all shifted versions of x as in Example 5, Section 6. We know from this example that we can find a $y \in l_2(-\infty, \infty)$, $y \neq 0$, such that A_y, the subspace generated by y, is disjoint from A_x. Given x and y it is not difficult to show that one can also always find a z, $z \neq 0$, such that the subspaces A_x, A_y, and A_z are mutually disjoint. Let $X = A_x + A_y + A_z$. Since each of the A-subspaces is invariant under shifting, it follows that X is also.

We define the linear time-invariant operator T_1 on X by setting

$$T_1 x = y, \quad T_1 y = z, \quad \text{and} \quad T_1 z = x.$$

----

[4] This issue arises again in Section 5.20.

It follows from linearity and time-invariance that the above three conditions uniquely determine T_1 on all of X. Similarly, we define the linear time-invariant operator T_2 on X by setting

$$T_2 x = z, \quad T_2 y = y, \quad \text{and} \quad T_2 z = x.$$

It then follows that $T_2 T_1 x = T_2(y) = y$ and $T_1 T_2 x = T_1(z) = x$, and this shows that T_1 and T_2 do not commute.

We see then that linearity and time-invariance alone are not enough to guarantee that two operators commute. Let us note though that if such operators are also continuous (see Section 5.19), then they must commute. We will not prove this fact here, but the idea of the proof is simply to show (using transform techniques) that if T_1 and T_2 are two *continuous*, linear, time-invariant operators, then there is an invertible operator S such that

$$ST_1 S^{-1} = H_1 \quad \text{and} \quad ST_2 S^{-1} = H_2,$$

where H_1 and H_2 denote multiplication operators. [For example, if the base space for the operators T_1 and T_2 was $L_2(-\infty,\infty)$ instead of $l_2(-\infty,\infty)$, then S would be the Fourier transform, and H_1 and H_2 would have the form given in Example 2, Section 3.] Since H_1 and H_2 commute, it is easy to see that T_1 and T_2 commute. ▌

EXERCISES

1. Let $X = L_2[-\pi,\pi]$, and let $X_1 = V(A_1)$ and $X_2 = V(A_2)$, where

$$A_1 = \{1, \cos t, \cos 2t, \ldots\}$$

   and

$$A_2 = \{\sin t, \sin 2t, \ldots\}.$$

   Show that $X_1 \oplus X_2$ and $X_1 + X_2$ are isomorphic.

2. If $X_1, X_2, \ldots, X_n$ are linear subspaces of a linear space X, then there is a natural mapping ϕ of $X_1 \oplus X_2 \oplus \cdots \oplus X_n$ into $X_1 + X_2 + \cdots + X_n$. When is ϕ an isomorphism?

3. Show that $\dim(X_1 \oplus X_2) = \dim(X_1) + \dim(X_2)$.

4. Let X_1 and X_2 be linear subspaces of a linear space X dim $X < \infty$. Show where

$$\dim(X_1 + X_2) = \dim(X_1) + \dim(X_2) - \dim(X_1 \cap X_2).$$

5. Let M be a linear subspace of a linear space X. Show that if N_1 and N_2 are two algebraic complements of M, then dim $N_1 = $ dim N_2. (That is, the co-dimension of M depends only on M and X, but not on the choice of algebraic complement.)

6. Let X be a linear space and assume that $X = X_1 + \cdots + X_N$ where $X_1, \ldots, X_N$ are linear subspaces of X with $X_i \cap X_j = \{0\}$ for $i \neq j$. Let B_i be a Hamel basis for X_i. Show that $B = B_1 \cup \cdots \cup B_N$ is a Hamel basis for X.

11.   PROJECTIONS

Consider the plane X shown in Figure 4.11.1 with the two designated one-dimensional linear subspaces M and N. Since M and N are disjoint, we know from Section 10 that each point x in the plane can be uniquely expressed in the form

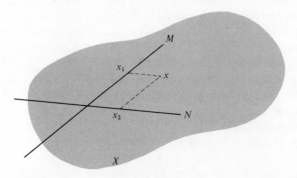

Figure 4.11.1.

$x = x_1 + x_2$, where $x_1 \in M$ and $x_2 \in N$. Suppose we consider the mapping P which maps the plane X into itself and which is defined by $P(x_1 + x_2) = x_1$. Geometrically P projects the plane X onto the subspace M along the subspace N, that is, $\mathscr{R}(P) = M$ and $\mathscr{N}(P) = N$. It is easily shown that P is linear. We see also that $P^2 = P$.

We want to extract the essence of the above situation and define a general notion of projection. It turns out that what we are interested in hinges on the fact that P is linear and that $P^2 = P$.

4.11.1 DEFINITION. A linear transformation P of a linear space X into itself is said to be a *projection* if $P^2 = P$.

We note that $P^2 = P$ does not imply that P is linear. For example, the non-linear mapping f of R into itself defined by $f(x) = +1$ for $x \geq 1$; $f(x) = 0$ for $-1 < x < 1$; and $f(x) = -1$ for $x \leq -1$ is such that $f \circ f = f$.

The following three theorems show that the Definition 4.11.1 does lead to a concept of projection which agrees with our intuitive concept of projection.

4.11.2 THEOREM. *Let P be a projection defined on a linear space X. Then the range and the null space, $\mathscr{R}(P)$ and $\mathscr{N}(P)$, are disjoint linear subspaces of X such that $X = \mathscr{R}(P) + \mathscr{N}(P)$. That is, $\mathscr{R}(P)$ and $\mathscr{N}(P)$ are algebraic complements of one another.*

Proof: We already know, of course, that the range and null space of a linear transformation are linear subspaces. Let us show that $\mathscr{R}(P)$ and $\mathscr{N}(P)$ are disjoint. Let $x \in \mathscr{R}(P) \cap \mathscr{N}(P)$. Since $x \in \mathscr{R}(P)$, there is a y such that $Py = x$, and $P^2y = Px = x$. Since $x \in \mathscr{N}(P), Px = 0$ or $x = 0$. Hence, $\mathscr{R}(P) \cap \mathscr{N}(P) = \{0\}$. Now let us show that $\mathcal{R}(P) + \mathcal{N}(P) = X$. Let x be an arbitrary point in X. Define $y = Px$ and $z = x - y$. One then has $x = y + z$ where $y \in \mathcal{R}(P)$. However, $Pz = P(x - y) = Px - P^2x = 0$, that is, $z \in \mathcal{N}(P)$. ∎

Since $\mathcal{R}(P) + \mathcal{N}(P) = X$ and $\mathcal{R}(P) \cap \mathcal{N}(P) = \{0\}$, we see from Lemma 4.10.1 that each $x \in X$ can be uniquely expressed in the form $x = x_1 + x_2$, where $x_1 \in \mathcal{R}(P)$ and $x_2 \in \mathcal{N}(P)$. Moreover, $Px = P(x_1 + x_2) = x_1$. Referring to Figure 4.11.1, we see that it makes sense in general to say that "P is the projection onto the subspace $\mathcal{R}(P)$ along the subspace $\mathcal{N}(P)$."

We note that if P is a projection, then so is $I - P$, with $\mathcal{R}(I - P) = \mathcal{N}(P)$ and $\mathcal{N}(I - P) = \mathcal{R}(P)$. Hence, if $x = x_1 + x_2$, where $x_1 = Px$, then $x_2 = (I - P)x$.

Suppose that instead of starting with a projection as in Theorem 4.11.2, we start with two disjoint linear subspaces, M and N with $X = M + N$. Does there exist a projection P with $\mathcal{R}(P) = M$ and $\mathcal{N}(P) = N$? The answer is yes.

4.11.3 THEOREM. *Let M and N be two disjoint linear subspaces of a linear space such that $M + N = X$. Then there exists a projection P defined on X such that $\mathcal{R}(P) = M$ and $\mathcal{N}(P) = N$.*

Proof: Again from Lemma 4.10.1 each $x \in X$ can be uniquely expressed $x = x_1 + x_2$, where $x_1 \in M$ and $x_2 \in N$. Let P be defined by $P(x) = x_1$. P is obviously the desired projection. ∎

Now let us start with one subspace.

4.11.4 THEOREM. *Let M be a linear space X. Then there exists a projection P such that $\mathcal{R}(P) = M$.*

Proof: This follows directly from Theorems 4.10.3 and 4.11.3. ∎

Let us consider some examples of projections.

EXAMPLE 1. Let $X(i\omega)$ be the Fourier transform of a signal $x(t)$, where $X(i\omega) \in L_2(-i\infty, i\infty)$. Let P be the transformation of $L_2(-i\infty, i\infty)$ into itself defined by

$$P[X(i\omega)] = \begin{cases} X(i\omega), & \text{for } -\omega_0 \leq \omega \leq \omega_0, \, \omega_0 > 0 \\ 0, & \text{otherwise.} \end{cases}$$

Those familiar with linear filter theory will recognize that P corresponds to an ideal low pass filter with unit gain in the passband $(-\omega_0 \leq \omega \leq \omega_0)$. They will also recall that P does not correspond to a causal system. We ask the reader to show that P is a projection. ∎

EXAMPLE 2. Let X be a linear space, and let $X_1, \ldots, X_n$ be linear subspaces of X such that X_i and X_j $(i, j = 1, 2, \ldots, n)$ are disjoint for $i \neq j$ and such that

$X = X_1 + \cdots + X_n$. Let P_j, $j = 1, 2, \ldots, n$, be the projection on X for which $\mathscr{R}(P_j) = X_j$ and $\mathscr{N}(P_j) = X_1 + \cdots + X_{j-1} + X_{j+1} + \cdots + X_n$. Then

$$T = \lambda_1 P_1 + \lambda_2 P_2 + \cdots + \lambda_n P_n, \tag{4.11.1}$$

where $\lambda_1, \ldots, \lambda_n$ are scalars, defines a linear transformation of X into itself. The restriction T_j of T to X_j is a mapping X_j into X. The range of T_j is X_j. Considering T_j as a mapping of X_j into itself, one has $T_j = \lambda_j I_j$, where I_j is the identity operator on X_j. Show that if X is finite dimensional, then T can be represented by a diagonal matrix of the form

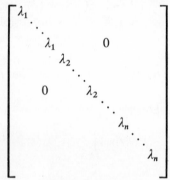

Show that $P_i P_j = 0$ for $i \neq j$. Further, show that T is a projection if and only if λ_j is either 0 or 1 for $j = 1, \ldots, n$. (By the end of this book the reader will recognize that the construction given in this example is extremely important.) ∎

EXAMPLE 3. Let X be the linear space $L_2(-\infty, \infty)$, and for each time T let the transformation P_T of X into itself be defined by

$$(P_T x)(t) = \begin{cases} x(t), & \text{for } -\infty < t \leq T \\ 0, & \text{for } T < t < \infty. \end{cases}$$

Each P_T is obviously a projection. ∎

EXERCISES

1. Let $X = L_2(-\infty, \infty)$ and consider the class of linear transformations of X into itself that can be represented in the form $y = Hx$ where $H \neq 0$ and

$$y(t) = \int_{-\infty}^{t} h(t - \tau) x(\tau) \, d\tau \qquad t \in (-\infty, \infty),$$

where h is in $L_1(-\infty, \infty)$ and $h(t) = 0$ for $t < 0$. (The latter requirement guarantees that the operator is causal.) Show that no nonzero transformation in the class is a projection. [*Hint*: Use Fourier transform methods.[5] Let

$$R(i\omega) + iX(i\omega) = \int_{-\infty}^{\infty} h(t) e^{-i\omega t} \, dt.$$

[5] The fact that Fourier transform methods are indeed rigorously applicable here is a fact that can be assumed for the purposes of this problem.

Show that

$$R(i\omega) = \int_{-\infty}^{\infty} \left[\frac{h(t) + \overline{h(-t)}}{2} \right] e^{-i\omega t}\, dt,$$

$$iX(i\omega) = \int_{-\infty}^{\infty} \left[\frac{h(t) - \overline{h(-t)}}{2} \right] e^{-i\omega t}\, dt,$$

where $\overline{h}$ denotes the complex conjugate of h. Further, show that $R + iX$ corresponds to a projection if and only if $X(i\omega) = 0$ and $R(i\omega) = 1$ or 0 for all $\omega \in (-\infty,\infty)$. Then conclude that this is impossible under the assumption that $h(t) = 0$ for $t < 0$, except for the trivial case $h(t) \equiv 0$. See Example 8, Section 5.9.]

2. Let P be a projection on a linear space X. Show that the range of P is given by $\mathscr{R}(P) = \{x \in X : Px = x\}$.

3. Let $X = L_2[-\pi,\pi]$ and show that

$$(Px)(t) = \int_{-\pi}^{\pi} K(t,\tau)x(\tau)\, d\tau,$$

where

$$K(t,\tau) = \frac{1}{2\pi} \sum_{n=-10}^{n=+10} e^{in(t-\tau)}$$

represents a projection on X.

4. Let $X = C[0,T]$ and define P by

$$(Px)(t) = x(0)(1 - t), \qquad \text{for } 0 \le t \le T.$$

Show that P is a projection.

5. Complete the argument of Example 2.

6. Let $P(t)$ be an $n \times m$ projection matrix defined for $-\infty < t < \infty$, that is, $[P(t)]^2 = P(t)$. Assume that the coefficients in P are C^1 functions and consider the matrix differential equation

$$X' = (P'P - PP')X,$$

where $P' = dP/dt$. Let $X(t)$ be a solution of the above equation. Show that $P(t)X(t)$ is also a solution of this equation. [*Hint*: Show that $PP'P = 0$.]

12. LINEAR FUNCTIONALS AND THE ALGEBRAIC CONJUGATE OF A LINEAR SPACE

There is a kind of linear transformation that is so important and used so often that it is given a special name; namely, a linear functional.

4.12.1 DEFINITION. Let X be a linear space over the scalar field F. A linear transformation l of X into its scalar field F is said to be a *linear functional* on X.

We use some special notation for linear functionals. Instead of $l(x)$, we write $\langle x, l \rangle$; and sometimes we write $\langle \cdot, l \rangle$ instead of l.

EXAMPLE 1.   Suppose that we have a tank Ω filled with a substance whose mass per unit volume at the point (x,y,z) is denoted $\rho(x,y,z)$. Assume that ρ is a point in the linear space X made up of all real-valued continuous functions defined throughout Ω. The total mass contained in the tank is given by

$$M = \langle \rho, l \rangle = \int_{\Omega} \rho \, dv,$$

where dv denotes a differential volume. Obviously, l, the operation of mapping the density ρ into total mass M, is a linear functional.   ∎

One of the main applications of linear functionals is their use in the characterization of subsets of linear spaces. For example, the null space of a linear functional, that is, $\mathcal{N}(l) = \{x \in X : \langle x, l \rangle = 0\}$, is a linear subspace of X. If l is a nontrivial ($l \neq 0$) linear functional, then $\mathcal{N}(l)$ has some interesting properties. In particular, $\mathcal{N}(l)$ is a very large proper linear subspace of X. In fact, we will now show that $\mathcal{N}(l)$ is maximal in the following sense: If A is any linear subspace of X such that $\mathcal{N}(l) \subset A$ and $\mathcal{N}(l) \neq A$, then $A = X$. Another way to say the same thing is that the co-dimension of $\mathcal{N}(l)$ is one.

4.12.2 THEOREM.   *Let l be a nontrivial linear functional on a linear space X and let M be an algebraic complement of the null space $\mathcal{N}(l)$. Then dim $M = 1$. Moreover, if A is any linear subspace of X with $\mathcal{N}(l) \subset A$ and $\mathcal{N}(l) \neq A$, then $A = X$.*

Proof: Let x_0 be a point in X such that $l(x_0) \neq 0$. Then let x be any point in X and let

$$z = x - \frac{l(x)}{l(x_0)} x_0.$$

It follows that $l(z) = 0$ and

$$x = z + \frac{l(x)}{l(x_0)} x_0.$$

So each point $x \in X$ can be expressed as the sum of a point in $\mathcal{N}(l)$ and a point in the one-dimensional linear subspace M spanned by x_0. Therefore, $X = \mathcal{N}(l) + M$ and $\mathcal{N}(l) \cap M = \{0\}$. Hence, M is an algebraic complement of $\mathcal{N}(l)$ and codim $\mathcal{N}(l) = 1$. Since $\mathcal{N}(l) \neq A$, we can also choose x_0 so that it is in $A - \mathcal{N}(l)$. The rest of the theorem follows immediately.   ∎

Since the intersection of linear subspaces is a linear subspace, we see that if $\{l_\alpha\}$ is a set of linear functionals on X, then $S = \bigcap_\alpha \mathcal{N}(l_\alpha)$ is a linear subspace of X. Thus we can use linear functionals and sets of linear functionals to characterize linear subspaces.

We can also use linear functionals to introduce a generalized plane concept. Recall that in R^3 a plane is characterized as being the set of all points $x = (x_1, x_2, x_3)$ such that

$$\alpha_1 x_1 + \alpha_2 x_2 + \alpha_3 x_3 = \alpha, \qquad (4.12.1)$$

where $\alpha_1, \alpha_2, \alpha_3$, and α are real numbers and $(\alpha_1, \alpha_2, \alpha_3) \neq (0,0,0)$. Now the left side of (4.12.1) defines a linear functional, call it l, on R^3. So this plane can also be viewed as the set of all points in R^3 such that $\langle x, l \rangle = \alpha$. Of course, different linear functionals l and different constants α can yield different planes in R^3. In general, we make the following definition.

4.12.3 DEFINITION. Let X be a linear space over a scalar field F. Given a linear functional l on X and a scalar α, the set

$$H = \{x \in X : \langle x, l \rangle = \alpha\}$$

is said to be the *hyperplane* in X determined by l and α. When $F = R$, the sets $\{x \in X : \langle x, l \rangle \geq \alpha\}$, $\{x \in X : \langle x, l \rangle > \alpha\}$, $\{x \in X : \langle x, l \rangle \leq \alpha\}$, and $\{x \in X : \langle x, l \rangle < \alpha\}$ are referred to as *half-spaces* determined by the hyperplane H. A set $A \subset X$ is said to be on *one side of the hyperplane H* if A is contained in one of the half-spaces. The set A lies *strictly on one side of H* if in addition A does not intersect H.

EXAMPLE 2. Consider the real space $X = L_2[0,T]$, and let l_1, l_2 be the linear functionals on X defined by

$$\langle x, l_1 \rangle = \int_0^T y_1(t)x(t)\, dt,$$

$$\langle x, l_2 \rangle = \int_0^T y_2(t)x(t)\, dt,$$

where y_1 and y_2 are given points in $L_2[0,T]$. The sets $\{x \in X : \langle x, l_1 \rangle = 1.2\}$, $\{x \in X : \langle x, l_1 \rangle = 0.8\}$, $\{x \in X : \langle x, l_2 \rangle = 1.1\}$, and $\{x \in X : \langle x, l_2 \rangle = 0.9\}$ are hyperplanes in X. The set

$$A = \{x \in X : 0.8 \leq \langle x, l_1 \rangle \leq 1.2\} \cap \{x \in X : 0.9 \leq \langle x, l_2 \rangle \leq 1.1\}$$

is the intersection of four half-spaces or—if you will—the intersection of two "slabs." In application, it might be that x was some input valve position as a function of time and the numbers $\langle x, l_1 \rangle$ and $\langle x, l_2 \rangle$ were final temperatures reached at time $t = T$ at two points in the system. The set A would be the set of all inputs x that yield final temperatures sufficiently close to certain "ideal" valves. ∎

So we can use linear functionals to characterize certain kinds of sets in a linear space. The question now arises, how rich is our supply of linear functionals? For example, a linear functional l on X characterizes a linear subspace of X with co-dimension one, namely, $\mathcal{N}(l)$. But what about the other way around? Suppose that M is any linear subspace of X with co-dimension one. Does there exist a linear functional l on X such that $\mathscr{N}(l) = M$? Fortunately, the answer is yes. The same question also has a geometric facet.

A hyperplane can be viewed as a translated linear subspace of co-dimension one. Indeed, consider the hyperplane $H = \{x \in X : \langle x, l \rangle = \alpha\}$ and let x_0 be an arbitrary point in H. It is easily shown that the set $M = \{x \in X : x + x_0 \in H\}$ is a linear subspace of co-dimension one. Indeed M is the null space of the linear functional l, that is, $\mathscr{N}(l) = M$. The situation is illustrated in Figure 4.12.1. Again one

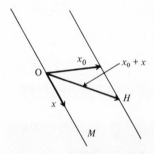

Figure 4.12.1.

can wonder about the supply of linear functionals on X. Suppose we are given a linear subspace M of co-dimension one and a point x_0. Let $H = \{x \in X : x - x_0 \in M\}$. Does there exist a linear functional l and a scalar α such that $\langle x, l \rangle = \alpha$ if and only if $x \in H$? The answer here is also yes.

The next two theorems contain the answers to the preceding questions.

4.12.4 THEOREM.   *Let M be a linear subspace of a linear space X, and let l_M be a linear functional defined on M. Then there exists a linear functional l defined on X such that l is an extension of l_M, that is, l is defined on X and $\langle x, l \rangle = \langle x, l_M \rangle$ for all $x \in M$.*

4.12.5 THEOREM.   *Let M_0 be a proper subspace of a linear space X, and let x_0 be a point in $X - M_0$. Then there exists a linear functional l on X such that $\langle x, l \rangle = 0$ if $x \in M_0$ and $\langle x_0, l \rangle = 1$.*

Proof of Theorem 4.12.4 is contained in the solution of Exercise 5, Section 3 So assuming Theorem 4.12.4, let us go on to the proof of Theorem 4.12.5. Let $[x_0]$ denote the one-dimensional linear subspace of X spanned by x_0. Let $M = M_0 + [x_0]$. Since $M_0 \cap [x_0] = \{0\}$, a point $x \in M$ can be expressed uniquely in the form $x = x' + \alpha x_0$, where $x' \in M_0$ and α is a scalar. Let l_M be the linear functional defined on M such that $\langle x' + \alpha x_0, l_M \rangle = \alpha$. Clearly $\langle x, l_M \rangle = 0$ for all $x \in M_0$ and $\langle x_0, l_M \rangle = 1$. Then using Theorem 4.12.4 extend l_M to X. This extension, call it l, is a linear functional on X of the desired form.   ∎

Let X^f denote the set of all linear functionals defined on a linear space X. We have just seen that in a meaningful sense there are many linear functionals in X^f. It also happens—with surprisingly widespread ramifications—that X^f can be viewed as a linear space over the same scalar field F. If l_1 and l_2 are in X^f, then $l_1 + l_2$ is defined by

$$\langle x, l_1 + l_2 \rangle = \langle x, l_1 \rangle + \langle x, l_2 \rangle, \qquad \text{for all } x \in X.$$

Similarly, we define scalar multiplication by

$$\langle x, \alpha l \rangle = \alpha \langle x, l \rangle, \qquad \text{for all } x \text{ in } X.$$

The origin in X^f is the zero functional. Moreover, one has

$$\langle x, -l \rangle = -\langle x, l \rangle, \qquad \text{for all } x \text{ in } X.$$

We refer to the linear space X^f as the *algebraic conjugate* of X.

EXERCISES

1. Show that if $\langle x, x' \rangle = 0$ for all $x' \in X^f$, then $x = 0$.

2. Let B be a Hamel basis for a finite-dimensional space X. If $B = \{x_1, \ldots, x_n\}$, then we know that each $x \in X$ can uniquely be written in the form

$$x = \alpha_1 x_1 + \cdots + \alpha_n x_n,$$

where the α's are scalars. Show that if z' is any point in X^f, there exist fixed scalars $\beta_1, \ldots, \beta_n$ such that

$$\langle x, z' \rangle = \alpha_1 \beta_1 + \cdots + \alpha_n \beta_n, \qquad \text{for all } x \in X.$$

3. Show that if X is finite dimensional, then $\dim(X^f) = \dim(X)$. (In the infinite-dimensional case this is not true.)

4. Let B_1 be a Hamel basis for a linear space X. If x_α is an element of B_1, let $x_\alpha' \in X^f$ be such that $\langle x_\alpha, x_\alpha' \rangle = 1$ and $\langle x_\beta, x_\alpha' \rangle = 0$, where x_β is any other element of B_1. Show that the mapping T of X into itself defined by

$$T(x) = \langle x, x_{\alpha_1}' \rangle x_{\alpha_1} + \cdots + \langle x, x_{\alpha_n}' \rangle x_{\alpha_n}$$

in a projection onto the subspace spanned by the set $\{x_{\alpha_1}, x_{\alpha_2}, \ldots, x_{\alpha_n}\}$. What is the null space of T?

13.   TRANSPOSE OF A LINEAR TRANSFORMATION

Suppose that $L: X \to Y$ is a linear transformation and let X^f and Y^f denote the algebraic conjugates of X and Y, respectively. Let y' be any point in Y^f and consider a hyperplane

$$H_Y = \{y \in Y : \langle y, y' \rangle = \alpha\}.$$

One is often interested, either directly or indirectly, in the inverse image of H_Y under the linear mapping L, that is,

$$L^{-1}(H_Y) = \{x \in X: \langle Lx,y' \rangle = \alpha\}.$$

The situation is sketched in Figure 4.13.1. The first thing to note is that the ex-

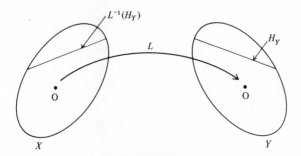

Figure 4.13.1.

pression $\langle L(\cdot),y' \rangle$ represents the composition $y'L$ of the linear transformation L and y'. But then $y'L$ is a linear mapping of X into the scalar field F; therefore, the expression $\langle L(\cdot),y' \rangle$ represents a linear functional on X. Denoting this functional by x', we have

$$\langle x,x' \rangle = \langle Lx,y' \rangle, \qquad \text{for all } x \in X.$$

As long as $x' \neq 0$, $L^{-1}(H)_Y$ is the hyperplane $H_X = \{x \in X: \langle x, x' \rangle = \alpha\}$. If $x' = 0$ and $\alpha = 0$, then $L^{-1}(H_Y) = X$. If $x' = 0$ and $\alpha \neq 0$, then $L^{-1}(H_Y)$ is empty. (Why?)

EXAMPLE 1.   Let $X = R^2$, $Y = R^3$, and L be the linear transformation represented by the following matrix equation

$$\begin{bmatrix} y_1 \\ y_2 \\ y_3 \end{bmatrix} = \begin{bmatrix} 7 & 2 \\ -1 & 4 \\ 2 & 0 \end{bmatrix} \begin{bmatrix} x_1 \\ x_2 \end{bmatrix}.$$

Further, let

$$H_Y = \{y \in Y: y_1 + y_2 + y_3 = 10\}.$$

We claim that the inverse image of H_Y under L is the hyperplane

$$H_X = \{x \in X: 8x_1 + 6x_2 = 10\}. \quad \blacksquare$$

Of course what we need to do next is find an orderly way to determine the x' that is associated with a y'. First note that implicit in the foregoing discussion is a mapping of Y^f into X^f. Indeed, given any $y' \in Y^f$, its image under this new mapping is $x' \in X^f$, where x' is the composition $y'L$. Let us denote this mapping by L^T, that is, $x' = L^T y'$. In other words, L^T is defined so that

$$\langle x, L^T y' \rangle = \langle Lx, y' \rangle, \tag{4.13.1}$$

for all $x \in X$ and all $y' \in Y^f$. It is easy to show that L^T is a linear mapping. Indeed if $x_i' = L^T y_i'$ for $i = 1, 2$, then

$$\langle x, x_1' + x_2' \rangle = \langle x, x_1' \rangle + \langle x, x_2' \rangle = \langle x, L^T y_1' \rangle + \langle x, L^T y_2' \rangle$$
$$= \langle Lx, y_1' \rangle + \langle Lx, y_2' \rangle = \langle Lx, y_1' + y_2' \rangle.$$

Hence $(x_1' + x_2') = L^T(y_1' + y_2')$. Similarly one has, $\alpha x' = L^T(\alpha y')$. Figure 4.13.2 illustrates (4.13.1). Let us now formalize what has just been said in a definition.

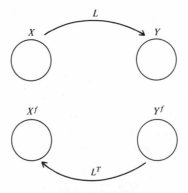

Figure 4.13.2.

4.13.1 DEFINITION. Let $L: X \to Y$ be a linear transformation. A linear transformation $L^T: Y^f \to X^f$ such that

$$\langle x, L^T y' \rangle = \langle Lx, y' \rangle,$$

for all $x \in X$ and all $y' \in Y^f$ is said to be the (*algebraic*) *transpose* of L.

Of course, the development leading up to the definition above shows that every L has a unique transpose, L^T.

The next example shows why we refer to L^T as the transpose of L.

EXAMPLE 2. (This example is a continuation of Example 1.) Since $X = R^2$ and $Y = R^3$ and since the dimension of a linear space and its algebraic conjugate are the same in the finite-dimensional case (Exercise 3, Section 12), X^f and Y^f are, respectively, two- and three-dimensional real linear spaces. The transpose L^T, then, is a linear transformation of a three- into a two-dimensional linear space. In the spirit of Section 8, we can represent L^T using a real matrix with two rows and three columns. Of course, the actual 2×3 matrix depends on the bases chosen for X^f and Y^f. We choose two very special ones. Let the basis of X^f be $\{\eta_1',\eta_2'\}$, where the linear functional η_i' on X is defined by $\langle x,\eta_i'\rangle = x_i$, $i = 1, 2$. We leave it to the reader to show that $\{\eta_1',\eta_2'\}$ is linearly independent and spans X^f. Similarly, let the basis of Y^f be $\{v_1',v_2',v_3'\}$ where $\langle y,v_i'\rangle = y_i$, $i = 1, 2, 3$. Then for any $x' \in X^f$ and $y' \in Y^f$, we have $x' = \alpha_1\eta_1' + \alpha_2\eta_2'$ and $y' = \beta_1 v_1' + \beta_2 v_2' + \beta_3 v_3'$. It then follows (after a little thought) that the matrix representation of L^T is

$$\begin{bmatrix} \alpha_1 \\ \alpha_2 \end{bmatrix} = \begin{bmatrix} 7 & -1 & 2 \\ 2 & 4 & 0 \end{bmatrix} \begin{bmatrix} \beta_1 \\ \beta_2 \\ \beta_3 \end{bmatrix}.$$

Carefully note that the above matrix is the transpose of the matrix in Example 1. Needless to say, other bases could be chosen for X^f and Y^f such that the matrix representing L^T would not be the transpose of the matrix representing L. ∎

The concepts of the transpose and the inverse of a linear operator can be put together to get further information about each. We note the following facts. The proofs are outlined in the exercises.

4.13.2 THEOREM. *Let L be a linear transformation of X into Y. Then $\mathscr{R}(L) = Y$ if and only if L^T is one-to-one.*

4.13.3 THEOREM. *Let L be a linear transformation of X into Y. Then L is one-to-one if L^T maps Y^f onto X^f.*

Carefully note the lack of symmetry between these two theorems. In Theorem 4.13.2 the statement is "if and only if" whereas in Theorem 4.13.3 it is merely "if." There are linear spaces X for which the latter theorem becomes an "if and only if" statement, among these linear spaces are all those of finite dimension.

EXERCISES

The first four exercises are concerned with the proof of Theorems 4.13.2 and 4.13.3.

1. Show that if L is a linear transformation of X into Y, then

$$\mathscr{N}(L) = \{x \in X : \langle x,x'\rangle = 0 \text{ for all } x' \in \mathscr{R}(L^T)\}.$$

[*Hint*: Denoting the set on the right by A, first show that $\mathscr{N}(L) \subset A$ as follows: Let x be any point in $\mathscr{N}(L)$ and show that $x \in A$ by using

$$\langle Lx,y'\rangle = \langle x,L^T y'\rangle,$$

for all $x \in X$ and $y' \in Y^f$. Next show that $\mathscr{N}(L) \supset A$ as follows: Let x be any point in A and show that $\langle Lx,y'\rangle = 0$, for all $y' \in Y^f$. Then use the result of Exercise 1, Section 12 to show that $x \in \mathscr{N}(L)$.]

2. Prove Theorem 4.13.3. [*Hint*: It follows from the foregoing exercise that L is one-to-one if and only if $A = \{0\}$. Therefore, show that[6] $\mathscr{R}(L^T) = X^f$ implies that $A = \{0\}$. Use Exercise 1, Section 12 again.]

3. Let L be a linear transformation of X into Y. Show that

$$\mathscr{R}(L) = \{y \in Y: \langle y, y' \rangle = 0 \text{ for all } y' \in \mathscr{N}(L^T)\}.$$

[*Hint*: Denote the set on the right by B. First show that $\mathscr{R}(L) \subset B$ as follows: Let y be any point in $\mathscr{R}(L)$. Letting x denote a pre-image of y, use

$$\langle Lx, y' \rangle = \langle x, L^T y' \rangle$$

to conclude that $\langle y, y' \rangle = 0$, for all $y' \in \mathscr{N}(L^T)$. Next show that $\mathscr{R}(L) \supset B$ by arguing by contradiction. That is, assume that there exists a point $y_0 \in B$ which is not in $\mathscr{R}(L)$. Use Theorem 4.12.5 to show there exists a functional $y_0' \in Y^f$ such that $\langle y, y_0' \rangle = 0$, for all $y \in \mathscr{R}(L)$ and $\langle y_0, y_0' \rangle = 1$. Then show that $y_0' \in \mathscr{N}(L^T)$. Since $y_0 \in B$, it then follows that $\langle y_0, y_0' \rangle = 0$, which is a contradiction.]

4. Prove Theorem 4.13.2. [*Hint*: Use the result of Exercise 3. Show that $B = Y$ if and only if $\mathscr{N}(L^T) = \{0\}$. First assume $\mathscr{N}(L^T) = \{0\}$. It follows almost trivially that $B = Y$. Next assume that $B = Y$. Let y_0' be any point $\mathscr{N}(L^T)$. Since, by assumption $B = Y$, $\langle y, y_0' \rangle = 0$, for all $y \in Y$.]

5. Generalize Example 2 to linear mappings of the real linear spaces R^n into R^m. That is assume that $L: R^n \to R^m$ is given by the matrix equation

$$\begin{bmatrix} y_1 \\ \vdots \\ y_m \end{bmatrix} = \begin{bmatrix} a_{11} \cdots a_{1n} \\ \cdots \\ a_{m1} \cdots a_{mn} \end{bmatrix} \begin{bmatrix} x_1 \\ \vdots \\ x_n \end{bmatrix}.$$

Show that with respect to appropriate bases on R^{nf} and R^{mf}, L^T can be represented by the matrix operator

$$\begin{bmatrix} a_{11} \cdots a_{m1} \\ \cdots \\ a_{1n} \cdots a_{mn} \end{bmatrix}.$$

6. What happens in Exercise 5 for linear mappings of the complex linear spaces C^n and C^m ?

SUGGESTED REFERENCES

Halmos [4].
Indritz [1].
Nering [1].

---

[6] The reason that $\mathscr{R}(L^T) = X^f$ is not a necessary condition for L to be one-to-one is that in certain infinite-dimensional linear spaces X there exist proper "total" subspaces M of X^f. That is, M has the property that if $\langle x, x' \rangle = 0$ for all $x' \in M$ then $x = 0$. One could have $\mathscr{R}(L^T) = M$ when L is one-to-one Roughly speaking, M contains a "rich supply" of linear functionals without being all of X^f.

5

Combined Topological and Algebraic Structure

1. INTRODUCTION

The concept of a continuous mapping was introduced in Chapter 3 (Section 5). This is a topological concept. A linear mapping was defined in Chapter 4 (Section 3). This is an algebraic concept. Our purpose in this chapter is to introduce a new class of spaces which, among other things, allows one to combine the two concepts of continuity and linearity. That is, we shall introduce the important concept of a continuous linear transformation.

The underlying space which allows one to combine the concepts of continuity and linearity is the normed linear space. This space is formed by suitably combining, by means of a norm, the topological structure of metric spaces and the algebraic structure of linear spaces. This combination is performed in such a manner that the two structures, topological and algebraic, are compatible. That is, addition and scalar multiplication are continuous, and the implied metric has a certain algebraic structure.

Part A of this chapter is devoted to a discussion of a number of elementary facts about normed linear spaces and Banach spaces. We ask the reader to note carefully the geometric nature of the structure of these spaces.

Part B of this chapter treats inner product spaces and Hilbert spaces. These are normed linear spaces with some very important additional structure. In particular an inner product and the concept of orthogonality are present. Because of this additional structure, the geometry of Hilbert spaces is relatively simple to understand. Indeed, it is more or less a generalization of Euclidean geometry to infinite-dimensional spaces.

It is difficult to overemphasize the importance of Hilbert spaces. A truly amazing number of problems in engineering and science can be fruitfully treated with geometric methods in Hilbert spaces. We shall illustrate some of them in the examples of this and the following chapters.

Part A

Banach
Spaces

2. DEFINITIONS

Let $x = (x_1, x_2)$ denote a point or vector in the real Euclidean plane. The Euclidean length of this vector is given by

$$\|x\| = (x_1{}^2 + x_2{}^2)^{1/2}.$$

The Euclidean distance between two points $x = (x_1, x_2)$ and $y = (y_1, y_2)$ is given by $\|x - y\|$, that is,

$$\|x - y\| = [(x_1 - y_1)^2 + (x_2 - y_2)^2]^{1/2}. \tag{5.2.1}$$

In this way one can view $\|x\|$ as a real-valued function defined on the real Euclidean plane. Moreover, this function generates a distance function or metric by means of (5.2.1). This function on the plane is an example, in fact, the archetypal example, of a norm on a linear space.

It is our desire to extend the foregoing concepts of length and distance to linear spaces (other than the plane) which leads us to seek a conception of "norm" which incorporates the essential features of length and distance. Experience has shown that the following definition is what we seek:

5.2.1 DEFINITION.   A real-valued function $\|x\|$ defined on a linear space X, where $x \in X$, is said to be a *norm on X* if

(N1) $\|x\| \geq 0$ (Positivity),

(N2) $\|x + y\| \leq \|x\| + \|y\|$ (Triangle inequality),

(N3) $\|\alpha x\| = |\alpha| \, \|x\|$, α an arbitrary scalar   (Homogeneity),

(N4) $\|x\| = 0$ if and only if $x = 0$ (Positive definiteness),

where x and y are arbitrary points in X. The number $\|x\|$ is referred to as the *norm of x*, or *length of x*.

Axiom (N1) says that the length of a vector is nonnegative, and Axiom (N4) says that only the origin (or zero vector) has length 0. Axiom (N2) is a type of triangle inequality, and, as we shall see, it is related to the triangle inequality for a metric. The Homogeneity Axiom (N3) says that scalar multiplication results in a stretching (or shrinking) of the length of x by a factor $|\alpha|$.

5.2.2 DEFINITION. A *normed linear space* is a pair $(X, \|\cdot\|)$, where X is a linear space and $\|\cdot\|$ is a norm defined on X. When no confusion is likely we will denote $(X, \|\cdot\|)$ by X.

We have asserted that a normed linear space combines the topological structure of a metric space with the algebraic structure of a linear space. Although the asserted algebraic structure is evident from the definition, the topological structure requires a few remarks.

We claim that the function

$$d(x,y) = \|x - y\|,$$

where x and y are two points in the normed linear space $(X, \|\cdot\|)$, is a metric defined on X. Let x, y, and z be arbitrary points in X. Axiom (N1) asserts that $\|x - y\| \geq 0$ and Axiom (N4) implies that $\|x - y\| = 0$ when $x = y$. Hence $\|x - y\|$ satisfies Axiom (M1) in the definition of a metric in Section 3.2. The "only if" part of Axiom (N4) shows that if $\|x - y\| = 0$, then $x = y$, which is Axiom (M2) for a metric. Axiom (M3) for a metric follows from (N3) by setting $\alpha = -1$. Finally from (N2) one gets

$$\|x - y\| = \|x - z + z - y\| \leq \|x - z\| + \|z - y\|,$$

which is the triangle inequality (M4) for a metric. Thus, $\|x - y\|$ is indeed a metric on X. Since $\|x\| = \|x - 0\|$ we see that the norm of a vector is equal to its distance from the origin.

The norm, then, generates a metric on X. We shall show in the exercises that there are metrics that are not generated by norms in the sense defined above. The concept of a norm does restrict the class of metrics.

Since a normed linear space is a metric space, one can ask questions about continuity and convergence utilizing the mathematical apparatus developed in Chapter 3. A convention which is universally accepted is the following:

> *Whenever one discusses the topological (or metrical) properties of a normed linear space $(X, \|\cdot\|)$, the metric is defined in terms of the given norm by $d(x,y) = \|x - y\|$.*

It is important to emphasize that, just as there are many metrics which can be defined on a given set, there are also many norms which can be defined on a given linear space X. Each norm gives a new normed linear space, and each norm defines a different metric on X. Different norms may define equivalent metrics, in which case, it is reasonable to say that the norms are *equivalent*. This concept is discussed again in Section 9.

Once and for all we answer a few elementary questions of continuity. The norm considered as a mapping of the normed linear space X into the reals R, is continuous. The addition operation in X (considered as a mapping of $X \times X$ into X) and the scalar multiplication operation (considered as a mapping of $F \times X$ into X, where F is the scalar field) are continuous. We leave the proofs of these statements as exercises.

One metric space concept is worthy of special note at this point, and that is completeness. It turns out that normed linear spaces with this property play a crucial role, and that prompts the following definition.

5.2.3 DEFINITION.   A normed linear space is said to be a *Banach space* if it is complete.

EXERCISES

1. Show that the norm $\|\cdot\|$ considered as a mapping of a normed linear space X into the reals is continuous. [*Hint*: Use the triangle inequality.]

2. Show that addition and scalar multiplication are continuous.

3. A function $\|x\|$ on a linear space X that satisfies conditions (N1), (N2), and (N3) is said to be a *pseudonorm*. Let $\|x\|$ be a pseudonorm on X. Show that $\rho(x,y) = \|x - y\|$ is a pseudometric on X.

4. Characterize all possible norms on the real line R, where R is considered a real linear space. On the complex plane C, where C is considered as a complex linear space. Show that C may have other norms when it is considered as a real linear space.

5. Let $(X, \|\cdot\|)$ be a normed linear space and let $S_r = \{x \colon \|x\| = r\}$ where $r > 0$. Assume that $X \neq \{0\}$. Show that $(X, \|\cdot\|)$ is a Banach space if and only if the metric space $\{S_r, \|\cdot\|\}$ is complete for some $r > 0$.

6. Let p satisfy $0 < p < 1$ and consider the space $L_p[0,1]$ of all functions with

$$\|x\| = \int_0^1 |x(t)|^p \, dt < \infty.$$

Show that $\|x\|$ is not a norm on $L_p[0,1]$. Show that $d(x,y) = \|x - y\|$ is a metric on $L_p[0,1]$. [*Hint*: Note that if $0 \le \alpha \le 1$, then $\alpha \le \alpha^p \le 1$.]

7. Define

$$\sigma_n(f) = \sup\{|f(t)| \colon |t| \le n\},$$

$$\rho_n(f) = \min(1, \sigma_n(f)),$$

$$\|f\| = \sum_{n=1}^{\infty} 2^{-n} \rho_n(f),$$

where $f \in C(-\infty, \infty)$.

(a) Show that $\sigma_n(f)$ is a pseudonorm on $C(-\infty, \infty)$.

(b) Show that $\rho_n(f)$ and $\|f\|$ are not norms.

(c) Show that $d(f,g) = \|f - g\|$ is a metric on $C(-\infty, \infty)$.

(d) Show that $d(f_n, f) \to 0$ as $n \to \infty$ if and only if $f_n(t) \to f(t)$ uniformly on compact sets in $-\infty < t < \infty$.

8. (Generalization on Exercises 6 and 7.) Let X be a linear space and let $\|x\|$ be a real-valued function defined on X. Show that $d(x,y) = \|x - y\|$ is a metric if and only if $\|x\|$ satisfies (N1), (N2), (N4), and $\|x\| = \|-x\|$ for all x in X.

9. Let $(X, \|\cdot\|)$ be a normed linear space and let $\{x_n\}$ be a sequence in X with $x = \lim_{n \to \infty} x_n$. Assume that $\|x_n - y\| \le a$ for all n. Show that $\|x - y\| \le a$.

10. Show that a Hamel basis for a Banach space is either finite or uncountably infinite. (Completeness is important for this. Use Exercise 17, Section 3.13.)

11. Let X denote the collection of all sequences $x = (x_1, x_2, \ldots)$ of complex numbers and define

$$\|x\| = \sum_{n=1}^{\infty} 2^{-n} \min(1, |x_n|).$$

(a) Is $\|\cdot\|$ a norm on X?
(b) Does $\|x - y\|$ define a metric on X? If so, explain the meaning of $\|x^{(n)} - x\| \to 0$ as $n \to \infty$.

12. Employ a normed linear space in the construction of a mathematical model.

13. Give an example of a metric d on a linear space such that $d(x, 0)$ is *not* a norm.

14. A real-valued function f defined on a linear space X is said to be *convex* if $f(\alpha x + \beta y) \le \alpha f(x) + \beta f(y)$ for all $x, y \in X$ and all real numbers α and β such that $0 \le \alpha, \beta \le 1$, and $\alpha + \beta = 1$. Show that a norm is a convex function.

3. EXAMPLES OF NORMED LINEAR SPACES

Many of the examples of metric spaces presented in Chapter 3 are also normed linear spaces, that is, the metric is generated by a norm. Some of these will be discussed in the exercises. In many of the examples below, we shall leave the proof that a certain function is a norm as an exercise.

EXAMPLE 1. Let $x = (x_1, \ldots, x_n)$ be a point in R^n. We define a norm on R^n by

$$\|x\|_p = \left[\sum_{i=1}^{n} |x_i|^p \right]^{1/p}, \qquad (5.3.1a)$$

for $1 \le p < \infty$, and

$$\|x\|_\infty = \max\{|x_1|, \ldots, |x_n|\}, \qquad (5.3.1b)$$

for $p = \infty$. It is easily seen that for each $1 \le p \le \infty$, $\|x\|_p$ is a norm on R^n. The only difficult step is the proof of the triangle inequality, but this is a direct consequence of the Minkowski Inequality. (See Appendix A.)

Equation (5.3.1) also defines a norm on the complex linear space C^n.

It follows (see Exercise 8, Section 3.13) that $(R^n, \|\cdot\|_p)$ and $(C^n, \|\cdot\|_p)$ are Banach spaces, for $1 \le p \le \infty$. ∎

EXAMPLE 2. Consider the set l_p, $1 \le p < \infty$, of all sequences $x = (x_1, x_2, \ldots)$ of scalars with the property that

$$\sum_{i=1}^{\infty} |x_i|^p < \infty:$$

It was shown in Chapter 4 that l_p is a linear space. If we define

$$\|x\|_p = \left[\sum_{i=1}^{\infty} |x_i|^p \right]^{1/p}, \tag{5.3.2}$$

we see that $\|x\|_p$ is a norm on l_p. The triangle inequality for $\|x\|_p$ follows from the Minkowski Inequality for infinite sums. (See Appendix A.)

It was shown in Section 3.13 that l_p with the metric determined from (5.3.2) is complete; therefore, the space $(l_p, \|\cdot\|_p)$ is a Banach space. The norm $\|x\|_p$ is referred to as the *usual norm* on l_p, $1 \le p < \infty$. ∎

EXAMPLE 3.   We can define a norm $\|x\|_\infty$ on the space l_∞ of all bounded sequences $x = (x_1, x_2, \ldots)$ of scalars as

$$\|x\|_\infty = \sup\{|x_i| : 1 \le i < \infty\}.$$

We leave it as an exercise to show that $(l_\infty, \|\cdot\|_\infty)$ is a Banach space. ∎

EXAMPLE 4.   Let (T,d) be a metric space and let $X = BC(T,R)$ denote the space of all real-valued, bounded, continuous functions defined on T. Thus $x \in X$ if and only if $x(t)$ is a real-valued, continuous function defined on T with

$$\|x\|_\infty = \sup\{|x(t)| : t \in T\}$$

being finite. It is easily shown that X is a linear space. We ask the reader to show that $\|\cdot\|_\infty$ is a norm on X. This is referred to as the *sup-norm*.

It is shown in Example 7, Section 3.13, that $BC(T,R)$ is complete; therefore, it is a Banach space.

Several commonly employed cases of this example occur when T is an interval on the real line, such that $T = [0,1]$, $T = [0,\infty)$, or $T = (-\infty,\infty)$. (In this case, d is the usual metric on T.) Other examples occur when $T = R^n$ or $T = C^n$ with one of the metrics given in Example 1. ∎

EXAMPLE 5.   Let (T,d) be a compact metric space. Then every real-valued continuous function defined on T is bounded, by Theorem 3.17.21. In other words, $BC(T,R)$ is precisely $C(T,R)$, the space of real-valued, continuous functions defined on T. It follows from that last example that $(C(T,R), \|\cdot\|_\infty)$ is a Banach space. ∎

EXAMPLE 6.   Consider the Lebesgue space $L_p[0,1]$, $1 \le p < \infty$, consisting of all scalar-valued measurable functions $x(t)$ defined for $0 \le t \le 1$ such that

$$\|x\|_p = \left[\int_0^1 |x(t)|^p \, dt \right]^{1/p}$$

is finite. The Minkowski Inequality for integrals (see Appendix A) shows that $\|x\|_p$ satisfies the triangle inequality. The other properties for a norm are easily verified and we see that $\|x\|_p$ is a norm[1] and $(L_p, \|\cdot\|_p)$ is a normed linear space. It follows from Theorem D.11.2 that this is a Banach space.

---

[1] Strictly speaking, $\|x\|_p$ is only a pseudonorm. However, we can change it to a norm by introducing a new equality on $L_p[0,1]$ (see Example 10, Section 3.3). That is, we say that $x = 0$ (in the new equality) if $\|x\|_p = 0$. It follows from Exercise 3, Section D.7, that $x = 0$ (in the new equality) if and only if $x(t) = 0$ almost everywhere.

We could replace [0,1] with any interval (finite or infinite) I or, more generally, by any measurable set A. In the latter case the norm would be defined by

$$\|x\|_p = \left(\int_A |x(t)|^p \, dt\right)^{1/p}.$$

Even more generally, if $(\Omega, \mathcal{M}, \mu)$ is any measure space, where μ is a positive measure, and $L_p(\Omega)$ denotes the collection of all scalar-valued measurable functions x defined on Ω with

$$\|x\|_p = \left(\int_\Omega |x(\omega)|^p \, \mu(d\omega)\right)^{1/p}$$

finite, then $\|x\|_p$ is a norm on $L_p(\Omega)$ and $(L_p(\Omega), \|\cdot\|_p)$ is a Banach space. Needless to say, an extremely important example of this Banach space occurs when $(\Omega, \mathcal{M}, \mu)$ is a probability space $(\Omega, \mathcal{F}, P)$ (see Appendix E) and the x's are random variables on $(\Omega, \mathcal{F}, P)$.

In any event, the norm $\|x\|_p$ is referred to as the *usual norm* on L_p. There are, of course, other norms that can be placed on L_p. ∎

EXAMPLE 7.   The Lebesgue space L_∞ is defined as follows: Let I be an interval (finite or infinite) and let $L_\infty(I)$ denote the collection of all scalar-valued measurable functions x defined on I with the property that there is a B, $0 \le B < \infty$, such that

$$|x(t)| \le B \qquad \text{(almost everywhere on } I\text{)}.$$

One can define a norm on $L_\infty(I)$ by

$$\|x\|_\infty = \text{ess sup}\{|x(t)| : t \in I\}$$
$$= \inf\{B : |x(t)| \le B \text{ almost everywhere on } I\}.$$

The norm $\|x\|_\infty$ is called the *essential supremum* of the function x. It is referred to as the *usual norm* on L_∞. This norm is sometimes referred to (somewhat inaccurately) as the *sup norm*. It is shown in Theorem D.11.4 that $(L_\infty(I), \|\cdot\|_\infty)$ is a Banach space. ∎

EXAMPLE 8.   Let $I = [a,b]$ be a bounded interval and let

$$P : a = t_0 < t_1 < \cdots < t_n = b$$

be a partition of I. Any scalar-valued function f, for which

$$V(f) = \sup\left\{\sum_{i=1}^n |f(t_i) - f(t_{i-1})| : P \text{ is a partition of } I\right\} \tag{5.3.4}$$

is finite, is said to be a *function of bounded variation* and $V(f)$ is said to be the *total variation* of f. One can define a norm on the collection $BV(I)$ of all functions of bounded variation by setting

$$\|f\| = |f(a+)| + V(f), \tag{5.3.5}$$

where $f(a+) = \lim_{t \to a} f(t)$. We leave it as an exercise to show that this is indeed a norm. ∎

EXAMPLE 9.   Let Ω be a nonempty set and let $\mathscr{M}$ be a σ-algebra of sets in Ω. Let X denote the collection of all real-valued measures μ on Ω that can be decomposed into its positive and negative parts, $\mu = \mu^+ - \mu^-$, where μ^+ and μ^- are positive measures (see Example 2, Section D.10). Let Y denote the sub-collection of those measures μ for which $\mu^+(\Omega) < \infty$ and $\mu^-(\Omega) < \infty$. A norm is given on Y by setting

$$\|\mu\| = \mu^+(\Omega) + \mu^-(\Omega). \quad \blacksquare$$

EXAMPLE 10.   (THE HÖLDER SPACES.)   Let α satisfy $0 < \alpha \leq 1$ and define $C^\alpha[0,1]$ to the space of all scalar-valued functions x that satisfy

$$|x(t) - x(s)| \leq K|t - s|^\alpha \qquad (0 \leq t,s \leq 1) \tag{5.3.6}$$

for some finite K. It follows that if x satisfies (5.3.6), then x is continuous. Let

$$N_\alpha(x) = \inf\{K : (5.3.6) \text{ is satisfied}\}.$$

For example, if $x(t) = \cos \pi t$, then $N_1(x) = \pi$.

Now define a norm on $C^\alpha[0,1]$ by

$$\|x\|_\alpha = \|x\|_\infty + N_\alpha(x),$$

where $\|x\|_\infty$ is the sup norm. This space is discussed further in the exercises. $\blacksquare$

EXAMPLE 11.   Let $I = (a,b)$ be an open interval in R and let $C^n(I)$ denote the collection of all scalar-valued functions defined on I with n continuous derivatives. As usual, let $C^\infty(I) = \bigcap_{n=0}^\infty C^n(I)$, and let $C_0^\infty(I)$ denote those functions in $C^\infty(I)$ that have compact support[2] in I. If $u(t)$ is any differentiable function let $Du = du/dt$, $D^2 u = D(Du)$, and so on. We ask the reader to show that each of the following is a norm on $C_0^\infty(I)$. For $1 \leq p < \infty$, $n = 0, 1, \ldots$, and $u \in C_0^\infty(I)$ let

$$\|u\|_{n,\,p} = \left[\int_I \sum_{i=0}^n |D^i u(t)|^p \, dt \right]^{1/p}.$$

This also defines a norm on the collection of all functions u in $C^n(I)$ for which $\|u\|_{n,p}$ is finite. $\blacksquare$

EXAMPLE 12.   This is an extension of Example 11. Replace I by Ω, where Ω is an open set in R^m. Define $C^n(\Omega)$, $C^\infty(\Omega)$, and $C_0^\infty(\Omega)$ as above, but now in terms of partial derivatives. Let $\alpha = (\alpha_1,\ldots,\alpha_m)$ be a vector with integral entries α_i where $\alpha_i \geq 0$. Let $|\alpha| = \sum_{i=1}^m \alpha_i$. Define the differential operator D^α by

$$D^\alpha u = \frac{\partial^{|\alpha|} u}{\partial x_1^{\alpha_1} \cdots \partial x_m^{\alpha_m}},$$

---

[2] A function u is said to have compact support if there exists a compact set M in I such that $u(t) = 0$ in $I - M$.

where $u(x) = u(x_1,\ldots,x_m)$. We ask the reader to show that each of the following is a norm on $C_0^\infty(\Omega)$. For $1 \le p < \infty$ and $u \in C_0^\infty(\Omega)$ let

$$\|u\|_{n,p} = \left[\int_\Omega \sum_{|\alpha| \le n} |D^\alpha u(x)|^p \, dx \right]^{1/p}.$$

These spaces are used in generalized function theory or distribution theory. ∎

EXAMPLE 13.   Let X denote the linear space $L_2[0,T]$ with the usual norm. Let Y denote the set of all linear mappings l of X into its scalar field that can be represented by

$$l(x) = l_y(x) = \int_0^T y(t)x(t)\, dt, \qquad \text{for all } x \in X,$$

where $y \in C[0,T]$, that is, each y determines an l_y. By defining addition and scalar multiplication in the natural way, we see that Y is a linear space. Furthermore, we claim that

$$\|l_y\| = \sup_{\|x\| \le 1} |l_y(x)|$$

defines a norm on Y. (Does it?) We also remark that $\|l_y\|$ can be computed from

$$\|l_y\| = \left\{ \int_0^T |y(t)|^2 \, dt \right\}^{1/2}.$$

EXERCISES

1. Show that $(l_\infty, \|\cdot\|_\infty)$ is a Banach space.

2. Let c_0 denote all sequences $x = (x_1, x_2, \ldots)$ of scalars with the property that $\lim x_n = 0$.
   (a) Show that $\|x\|_\infty = \sup\{|x_i| : 1 \le i < \infty\}$ is a norm on c_0.
   (b) Show that c_0 is a linear subspace of l_∞.
   (c) Show that c_0 is a closed subset of $(l_\infty, \|\cdot\|_\infty)$.

3. (Continuation of Exercise 2.) Let c denote the collection of all convergent sequences $x = (x_1, x_2, \ldots)$ of scalars.
   (a) Show that $\|x\|_\infty$ is a norm on c.
   (b) Show that $(c, \|\cdot\|_\infty)$ is a closed subset of $(l_\infty, \|\cdot\|_\infty)$.

4. Show that $\|\cdot\|_\infty$ is a norm on $BC(T,R)$ (see Example 4).

5. Let $C_0 = C_0(-\infty,\infty)$ denote the space of real-valued continuous functions, defined for $-\infty < t < \infty$, with *compact support*.
   (a) Show that $\|x\| = \sup\{|x(t)| : t \in R\}$ is a norm on C_0.
   (b) Show that $(C_0, \|\cdot\|_\infty)$ is not complete.
   (c) Show that C_0 is a linear subspace of $X = BC(R,R)$ (see Example 4).
   (d) Is C_0 a closed subset of $(X, \|\cdot\|_\infty)$?

6. In Example 7, show that the essential supremum defines a norm on $L_\infty(I)$. Show that $(L_\infty, \|\cdot\|_\infty)$ is a Banach space.

7. Let I be a finite interval and $1 \le p \le p' \le \infty$. Show that $L_{p'}(I) \subset L_p(I)$. Are these spaces equal? What happens if I is an infinite interval?

8. (Generalization of Exercise 7.) Let $(\Omega, \mathscr{F}, P)$ be a probability space. Show that $L_\infty(\Omega) \subset L_2(\Omega) \subset L_1(\Omega)$.

9. This exercise refers to Example 8.
   (a) Show that $V(f)$ is a pseudonorm on $BV(I)$.
   (b) Show that (5.3.5) defines a norm on $BV(I)$.

10. (Continuation of Exercise 9.) Relabel $V(f) = V(f; a,b)$. This is the total variation of f on the interval $[a,b]$. For $a \le t \le b$, let $V(f; a,t)$ denote the total variation of f on the interval $[a,t]$.
    (a) Show that $V(f; a,t)$ is an increasing function of t.
    (b) Show that $g(t) = V(f; a,t) - f(t)$ is an increasing function of t.
    (c) Show that a function f on I has bounded variation if and only if it can be written as the difference of two monotone functions.

11. (Continuation of Exercise 9.) A function f of bounded variation on $I = [a,b]$ is said to be *normalized* if $f(a) = 0$ and f is continuous from the right, that is, $f(t + 0) = f(t)$ for $a \le t \le b$. Let $NBV(I)$ denote the collection of all normalized functions of bounded variation on I. Show that $V(f)$ is a norm on $NBV(I)$.

12. (This exercise refers to Example 10.) Consider the unit ball in $C^\alpha[0,1]$, where $0 < \alpha \le 1$.

$$B_\alpha = \{x \in C^\alpha[0,1]: \|x\|_\alpha \le 1\}.$$

Show that B is a compact set in $(C[0,1], \|\cdot\|_\infty)$. [*Hint*: Use Ascoli's Theorem.]

13. (Continuation of Exercise 12.) Let $0 < \alpha < \beta \le 1$. Show that the unit ball B_β is a compact set in $(C^\alpha[0,1], \|\cdot\|_\alpha)$. [*Hint*: Note that

$$N_\alpha(x) = \sup\{|x(t) - x(s)|\, |t - s|^{-\alpha}: t \ne s\}.$$

Now we will apply Ascoli's Theorem to the function of two variables $y(t,s) = |x(t) - x(s)|\, |t - s|^{-\alpha}$, when $x \in B_\beta$.]

14. (Orlicz Spaces.[3]) Let $p(t)$ be a right continuous real-valued function defined for $t \ge 0$ such that $p(0) = 0$, $p(t) > 0$ for $t > 0$ and $p(\infty) = \lim_{t \to \infty} p(t) = \infty$. Let $M(u) = \int_0^{|u|} p(t)\, dt$ and assume that there are constants $\alpha > 0$, $u_0 > 0$ such that

$$up(u) \le \alpha M(u), \qquad u \ge u_0.$$

Let $q(s) = \sup\{t: p(t) \le s\}$ and $N(v) = \int_0^{|v|} q(s)\, ds$ and assume that there are constants $\beta > 0$ and $v_0 > 0$ such that

$$vq(v) \le \beta N(v), \qquad v \ge v_0.$$

Let $I = [a,b]$ be a finite interval, and let $L_M(I)$ be the collection of all real-valued functions u defined on I for which

$$\rho(u; M) = \int_a^b M(u(x))\, dx < \infty.$$

---

[3] For more details see Krasnosel'skii and Rutickii [1].

Define $L_N(I)$ similarly. For $u \in L_M(I)$ let

$$\|u\|_M = \sup\left\{\left|\int_a^b u(x)v(x)\,dx\right| : v \in L_N(I) \text{ and } \rho(v; N) \leq 1\right\}.$$

(a) Show that $\|u\|_M$ is a norm on $L_M(I)$.

(b) Show that $p(t) = t^{\alpha-1}$, for $\alpha > 1$, satisfies the conditions above. Compute $q(s)$, $\rho(u,M)$, and $\rho(v; N)$. Compute $\|u\|_M$ when $\alpha = 2$.

15. Consider the space M^n for all $n \times n$ matrices with real coefficients. Since M^n can be identified with R^{n^2} we can use any of the norms in Example 1 on M^n. However, another norm is often-times used. Let $\|\cdot\|$ be a fixed norm on R^n. Define the norm of a matrix $[L]$ in M^n by

$$\|[L]\| = \sup\{\|Lx\| : \|x\| \leq 1\}.$$

(a) Show that $\|[L]\|$ is a norm on M^n.

(b) Show that $\|[L][M]\| \leq \|[L]\| \cdot \|[M]\|$, where $[L][M]$ denotes the matrix product of the two matrices $[L]$ and $[M]$.

(c) Let $[L]$ be the 2×2 matrix $\begin{pmatrix} a & b \\ b & c \end{pmatrix}$ with $a > 0$, $c > 0$ and assume that R^2 has the Euclidean norm $\|x\|_2 = (x_1{}^2 + x_2{}^2)^{1/2}$. Show that

$$\|L\| = \tfrac{1}{2}[a + c + \sqrt{(a-c)^2 + 4b^2}].$$

(d) What happens if the matrices have complex coefficients? (The concept of norm introduced in this exercise is generalized in Section 8.)

16. Let H_p, $1 \leq p < \infty$, denote the collection of all functions $f(z)$ that are analytic for $|z| < 1$ with the property that

$$\|f\|_p = \sup_{r<1}\left\{\int_0^{2\pi} |f(re^{i\theta})|^p\,d\theta\right\}^{1/p} < \infty.$$

Show that $\|f\|_p$ is a norm and that $(H_p, \|\cdot\|_p)$ is a Banach space.

17. In Example 1, show that $\|x\|_\infty = \lim_{p\to\infty} \|x\|_p$.

18. In Examples 11 and 12, show that $\|u\|_{n,p}$ defines a norm on $C_0{}^\infty(I)$ and $C_0{}^\infty(\Omega)$, respectively.

4.   SEQUENCES AND SERIES

The first offspring of the wedding of topological and algebraic structure is the concept of an infinite series. Let $(X, \|\cdot\|)$ be a normed linear space and let $\{x_n\}$ be a sequence in X. Since X is a linear space, one can consider finite sums of the form

$$y_m = x_1 + x_2 + \cdots + x_m = \sum_{n=1}^m x_n$$

which generate a new sequence $\{y_m\}$, the sequence of *partial sums*. Since X is also a metric space, we can test whether the sequence $\{y_m\}$ converges to a limit y, which means that $y_m \to y$ as $m \to \infty$ if and only if $\|y_m - y\| \to 0$ as $m \to \infty$. If this limit y exists, we say that the infinite series $\sum_{n=1}^{\infty} x_n$ *converges* and we write

$$y = \sum_{n=1}^{\infty} x_n.$$

The infinite series is said to *diverge* if the sequence of partial sums $\{y_m\}$ fails to have a limit. It is important to emphasize that, just as in the case of metrics, convergence and divergence depend on the norm used on the underlying linear space X.

Let us now turn to Banach spaces, which the reader will recall are *complete* normed linear spaces. The story of convergence of infinite series in Banach space is simpler because one can use the Cauchy test for convergence without knowing the limit of the sequence of partial sums.

5.4.1 LEMMA. (THE CAUCHY TEST.)   *Let X be a Banach space. An infinite series $\sum_{n=1}^{\infty} x_n$ converges in X if and only if for each $\varepsilon > 0$ there is an integer N such that*

$$\left\| \sum_{i=m}^{n} x_i \right\| \le \varepsilon$$

whenever $n \ge m > N$.

The proof of this lemma is trivial, since it merely states that the series converges if and only if the sequence of partial sums is a Cauchy sequence.

Although the last lemma is useful, a somewhat stronger statement is more practical and widely used. A series $\sum_{n=1}^{\infty} x_n$ in a normed linear space X is said to *converge absolutely* if the series of absolute values $\sum_{n=1}^{\infty} \|x_n\|$ is convergent.

The series $\sum_{n=1}^{\infty} \|x_n\|$ is a series of real numbers and it converges or diverges on the real line. Since the real line is complete, it follows from Lemma 5.4.1 that the series $\sum_{n=1}^{\infty} \|x_n\|$ converges if and only if for each $\varepsilon > 0$ there is an integer N such that

$$\sum_{i=m}^{n} \|x_i\| < \varepsilon$$

whenever $n \ge m > N$.

How are absolute convergence and convergence related? In a Banach space we have the following answer.

5.4.2 THEOREM.   *Let X be a Banach space. If the series $\sum_{n=1}^{\infty} x_n$ is absolutely convergent, then it is convergent.*

Proof: It follows from the triangle inequality (N2) that

$$\left\| \sum_{i=m}^{n} x_i \right\| \le \sum_{i=m}^{n} \|x_i\|.$$

Using the Cauchy Test it follows that the series $\sum_{n=1}^{\infty} x_n$ is convergent.   ∎

EXAMPLE 1.   Let X be the Banach space $C[-\pi,\pi]$ with the sup norm $\|\cdot\|_\infty$. The series

$$\sum_{n=1}^{\infty} \frac{3^n}{n!} \cos nt$$

is absolute convergent since

$$\sum_{n=1}^{\infty} \left\| \frac{3^n}{n!} \cos nt \right\| = \sum_{n=1}^{\infty} \frac{3^n}{n!} = e^3 - 1.$$

Therefore, this series is convergent.   ∎

This next example shows that in an incomplete space an absolutely convergent series need not converge.

EXAMPLE 2.   Let X be the normed linear space $C[-\pi,\pi]$ with the norm

$$\|x\| = \left(\int_{-\pi}^{\pi} |x|^2 \, dt \right)^{1/2}.$$

This space is not complete. The series

$$\sum_{n=1}^{\infty} \frac{1}{n} \sin nt$$

is absolutely convergent; however, it is not convergent. Indeed, in the Banach space $L_2[-\pi,\pi]$ one has

$$\sum_{n=1}^{\infty} \frac{1}{n} \sin nt = y(t),$$

where

$$y(t) = \begin{cases} -\dfrac{\pi}{4}, & -\pi \le t < 0 \\[2mm] \dfrac{\pi}{4}, & 0 \le t \le \pi. \end{cases}$$

But $y(t)$, not being continuous, is not in X. However, the series does converge in $L_2[-\pi,\pi]$.   ∎

The reader familiar with series of real numbers may recall that the nice thing about absolutely convergent series of real numbers is that no matter how we rearrange the terms in the series, the rearranged series (i) converges and (ii) the limit is the same. Almost the same thing is true in this more general setting.

We will say that a series $\sum x_n$ is *unconditionally convergent* if it and each of its rearrangements (i) are convergent and (ii) have the same limit. Recall that a rearrangement is a reordering of the terms in an infinite series.

5.4.3 THEOREM.   *Let X be a Banach space. If a series $\sum_{n=1}^{\infty} x_n$ is absolutely convergent, then it is unconditionally convergent.*

In the case of series of real numbers it can be shown that unconditional convergence implies absolute convergence. *This is not the case* for Banach spaces in general. Dvoretzky and Rogers [1] have shown that absolute convergence is equivalent to unconditional convergence if and only if the Banach space is finite dimensional.

EXERCISES

1. Prove Theorem 5.4.3. [*Hint*: Consider $\sum x_n$ and $\sum x_n'$, where one is a rearrangement of the other. Given any $\varepsilon > 0$, there exists an integer N such that $m \geq n > N$ implies $\sum_{i=n}^{m} \|x_i\| < \varepsilon$. Choose an integer p such that $x_1, x_2, \ldots, x_N$ are contained in the set $\{x_1', x_2', x_3', \ldots, x_p'\}$.]

2. Let $\{y_n\}$ be a sequence in a normed linear space $(X, \|\cdot\|)$. Show that this sequence converges if and only if the series $\sum_{n=1}^{\infty} x_n$ converges where $x_n = y_n - y_{n-1}$ and $y_0 = 0$.

3. (Continuation of Exercise 15, Section 3.) We will let A be an $n \times n$ matrix with $\|A\| = a < 1$.
   (a) Show that $\|A^n\| \leq a^n$ for $n = 1, 2, \ldots$,
   (b) Let $B_0 = (I + A)(I - A), B_1 = (I + A^2)B_0$, and $B_n = (I + A^{2^n})B_{n-1}$, where I denotes the identity matrix. Show that $\|B_n - I\| \leq a^{2^{n+2}}$.
   (c) Let $C_n = B_0 + \sum_{i=1}^{n} \{B_i - I\}$. Show that the infinite series converges absolutely.
   (d) What is $\lim C_n$?
   (e) Let $D_n = (I + A^{2^n})(I + A^{2^{n-2}}) \cdots (I + A)$. What is $\lim D_n$?

4. (Continuation of Exercise 15, Section 3.) Show that

$$\exp(A) = e^A = \sum_{n=0}^{\infty} \frac{A^n}{n!}$$

is well-defined, where A is an $n \times n$ matrix. Compute e^A where A is a diagonal matrix.

5. (Continuation of Exercise 15, Section 3.) Let A be an $n \times n$ matrix with $\|A\| < 1$.
   (a) Show that

$$\log(I - A) = \sum_{n=1}^{\infty} \frac{1}{n} A^n$$

is well-defined.
   (b) Show that $\exp[\log(I - A)] = I - A$.

6. (Continuation of Exercise 4.) We will let $f(z)$ be an entire function, that is, $f(z) = \sum_{n=0}^{\infty} c_n z^n$ for all z in the complex plane C.
   (a) Show that for any $n \times n$ matrix

$$f(A) = \sum_{n=0}^{\infty} c_n A^n$$

is well-defined.
   (b) What is $f(A)$ where A is a projection matrix, that is, when $A^2 = A$?

7. (Continuation of Exercise 6.) Consider the power series $\sum_{n=0}^{\infty} c_n A^n$, where A is an $n \times n$ matrix with complex coefficients and the c_n are complex numbers.
   (a) Show that there is a number r, $0 \le r \le +\infty$, such that this series is absolutely convergent if $\|A\| < r$ and for any $s > r$, there is a matrix A with $\|A\| = s$ such that the series is divergent. (The number r is called the radius of convergence.)
   (b) Show that r satisfies

$$r^{-1} = \limsup_{n \to \infty} \sqrt[n]{|c_n|}.$$

8. (Integral Test.) Let $\sum_{n=1}^{\infty} x_n$ be an infinite series in a Banach space $(X, \|\cdot\|)$ and assume that there is a decreasing positive function $f(t)$ defined for $0 \le t < \infty$ such that $\|x_n\| < f(n)$. Show that if $f \in L_1[0,\infty)$, then the series $\sum_{n=1}^{\infty} x_n$ is absolutely convergent.

9. (Weierstrass M-Test.) Let $\sum_{n=1}^{\infty} x_n$ be an infinite series in a Banach space $(X, \|\cdot\|)$ and assume that $\|x_n\| \le M_n$. Show that if $\sum_{n=1}^{\infty} M_n < \infty$, then the series $\sum_{n=1}^{\infty} x_n$ is absolutely convergent.

10. Consider the norms $\|\cdot\|_p$, $1 \le p \le \infty$, on $C[0,T]$.
    (a) Show that if $\|x_n - x\|_{\infty} \to 0$ as $n \to \infty$, then for every p, $1 \le p < \infty$, one has $\|x_n - x\|_p \to 0$ as $n \to \infty$.
    (b) Show that if $1 \le r \le s \le \infty$, then $\|x\|_r \le T^{s-r/sr}\|x\|_s$.
    (c) Show that if $1 \le r \le s \le \infty$ and $\|x_n - x\|_s \to 0$, then $\|x_n - x\|_r \to 0$.
    (d) Find an example of a sequence $\{x_n\}$ in $C[0,T]$ and a point x in $C[0,T]$ with $\|x_n - x\|_1 \to 0$ but $\|x_n - x\|_{\infty} \nrightarrow 0$.

11. Let l_p $(p > 0)$ be the collection of all sequences $x = \{x_1, x_2, \ldots\}$ of scalars with the property that $\sum_{n=1}^{\infty} |x_n|^p < \infty$.
    (a) Show that $\|x\|_{\infty} = \sup\{|x_n| : n = 1, 2, \ldots\}$ is a norm on l_p.
    (b) Show that for $0 < p < \infty$, $(l_p, \|\cdot\|_{\infty})$ is not a Banach space.

12. Let A be a $n \times n$ matrix and let t be a real number. Then e^{tA} is given by

$$e^{tA} = \sum_{n=0}^{\infty} \frac{t^n A^n}{n!}.$$

(See Exercise 4.)
   (a) Show that $x(t) = e^{tA} x_0$ is a solution of

$$\frac{dx}{dt} = Ax$$

that satisfies $x(0) = x_0$.
   (b) Let $\|\cdot\|_2$ be the Euclidean norm on R^n. For σ real, let Y_σ denote the collection of all continuous functions $x : [0,\infty) \to R^n$ such that

$$\int_0^{\infty} \|x(t)\|_2^2 \, e^{-2\sigma t} \, dt < \infty.$$

Define a norm $\|\cdot\|_\sigma$ on Y_σ by

$$\|x(\,\cdot\,)\|_\sigma = \left(\int_0^\infty \|x(t)\|_2^{\,2}\, e^{-2\sigma t}\, dt \right)^{1/2}.$$

Show that there is a σ such that the mapping $T: x_0 \to e^{tA}x_0$ of R^n into Y_σ is continuous.

5.   LINEAR SUBSPACES

A *linear subspace* M of a normed linear space X is, as one would expect, a linear subspace in the algebraic sense equipped with the norm of X, that is, the normed linear space $(M,\|\cdot\|)$ is a linear subspace of $(X,\|\cdot\|)$. We have already seen some examples of this in the exercises following Section 3. In this section we present some of the fundamental facts about subspaces. The first lemma shows that open linear subspaces are rather uninteresting.

5.5.1 LEMMA.   *Let M be a linear subspace of a normed linear space X. If M is open (as a subset of X), then $M = X$.*

Proof: Let x be any point in X. We want to show that $x \in M$. Since M is a linear subspace, the origin 0 lies in M. So we can assume that $x \neq 0$. Since M is open, there is a local neighborhood U of 0 that lies in M. That is, there is an $\varepsilon > 0$ such that if $z \in X$ and $\|z\| < \varepsilon$, then $z \in M$. It follows, then, that for any $x \in X$

$$z = \frac{\varepsilon}{2\,\|x\|}\, x$$

lies in M. Since M is a linear subspace, the point $x = 2\|x\|\varepsilon^{-1} z$ lies in M.   ∎

The case where M is closed is far more interesting.

Using the fact that addition, scalar multiplication and norm are continuous operations, one can easily establish the following result.

5.5.2 THEOREM.   *The closure of a linear subspace is a closed linear subspace.*

Needless to say, the fact that we get a closed set is immediate. What has to be shown is that this closed set is also a linear subspace.

Now let M be a linear subspace in a Banach space X. We know already that M is a normed linear space, and we may ask whether M is a Banach space. That is, is M complete? The answer is simple, elegant, and not too surprising.

5.5.3 THEOREM.   *Let M be a linear subspace of a Banach space X. Then M is a Banach space if and only if M is closed.*

EXAMPLE 1.   Exercises 2 and 3 of Section 3 gave two examples of closed linear subspaces of the Banach space $(l_\infty, \|\cdot\|_\infty)$. As a matter of fact, the space $(c_0, \|\cdot\|_\infty)$

is a closed linear subspace of $(c, \|\cdot\|_\infty)$. Exercise 5, Section 3 gives an example of a linear subspace of a Banach space that is not closed. ∎

EXAMPLE 2.   Consider the real space $C[0,1]$ with the norm

$$\|x\|_2 = \left\{ \int_0^1 |x(t)|^2 \, dt \right\}^{1/2}.$$

This space is not complete. Let M denote the collection of all functions $x \in C[0,1]$ of the form $x(t) = a \sin 2\pi t + b \cos 2\pi t$, where a and b are scalars. Then

$$\|x\|_2{}^2 = \int_0^1 (a^2 \sin^2 2\pi t + 2ab \sin 2\pi t \cdot \cos 2\pi t + b^2 \cos^2 2\pi t) \, dt$$

$$= \tfrac{1}{2}(a^2 + b^2).$$

We ask now, is the space M complete? We can show that it is by showing that it is isometrically equivalent to a complete metric space, namely C, the complex numbers, considered as a two-dimensional real normed linear space, with the norm

$$\|z\| = \|a + ib\| = \frac{1}{\sqrt{2}} (a^2 + b^2)^{1/2}.$$

The identification is simply $x(\cdot) \to a + ib$. Since C is complete, it follows that M is complete by Theorem 3.13.6.

In this case M is a two-dimensional subspace of the given normed linear space. In Section 10 we shall show that every finite-dimensional linear subspace, of a normed linear space, is complete. ∎

If we have a proper linear subspace M in a normed linear space X, a little reflection shows that it is possible for M to be dense in X. For example, if $X = l_2$ and M is the linear subspace made up of all sequences with at most a finite number of nonzero terms, then M is a proper subspace of X and dense in X. On the other hand, if M is closed, the only way it can be dense in X is to be equal to X. In other words, if M is a proper closed subspace of X, then there are points a nonzero distance from M; and this brings us to an important geometric concept.

In ordinary three-dimensional Euclidean space, a vector x is orthogonal to a plane M if and only if $d(x,M) = \|x\|$. Refer to Figure 5.5.1. In normed linear

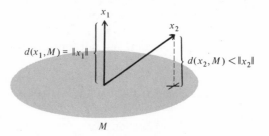

Figure 5.5.1.

spaces in general there is no concept orthogonality (because there is no inner product). However, if M is a proper linear subspace of a normed linear space X, we can still ask if $d(x,M) = \|x\|$. It is tempting to argue on the basis of geometric intuition that if M is closed (so M will not be dense in X) and X is complete, then we can always find an x in $X - M$ such that $d(x,M) = \|x\|$. Unfortunately, this is one of those places where geometric intuition can go wrong. All we can say in general is stated in the next theorem.

5.5.4 THEOREM. (RIESZ THEOREM.[4])   *Let M be a proper closed linear subspace of a normed linear space X, and let $\varepsilon > 0$. Then there exists an $x \in X$ with $\|x\| = 1$ such that $d(x,M) \geq 1 - \varepsilon$.*

Carefully note that this theorem does not claim that there exists a unit vector x such that $d(x,M) = 1$. Sometimes there is, but then again sometimes there is not (see Example 3 below). We shall see subsequently that this pathology cannot occur in Hilbert spaces.

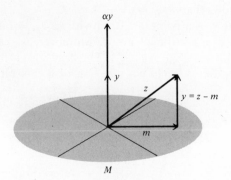

Figure 5.5.2.

Proof: It goes without saying that $0 \leq d(x,M) \leq \|x\|$ for all $x \in X$. Since M is closed and not all of X, there are x's such that $0 < d(x,M) \leq \|x\|$ [that is, $d(x,M)$ is nonzero]. The issue is to find an x such that $(1 - \varepsilon) < d(x,M) \leq \|x\|$. In other words, we want to find an x that is "almost orthogonal" to M. If $M = \{0\}$, the proof is trivial, so we assume M contains some nonzero vectors. Since M is closed, there is a point $z \in X - M$ such that $d(z,M) = \delta > 0$. See Figure 5.5.2. Since

$$d(z,M) = \inf\{\|z - m'\| : m' \in M\},$$

there is an $m \in M$ such that $\|z - m\| < \delta(1 + \varepsilon)$. Moreover, if $y = z - m$, then $d(y,M) = d(z,M) = \delta$. Thus, if α is any scalar, we have $\|\alpha y\| < \alpha\delta(1 + \varepsilon)$ and $d(\alpha y,M) = \alpha\delta$. (Why?) Finally, if we let $x = (1/\|y\|)y$, then $\|x\| = 1$ and $d(x,M) > 1/(1 + \varepsilon) > 1 - \varepsilon$.   ∎

---

[4] This theorem should not be confused with the Riesz Representation Theorem (Theorem 5.21.1).

The next example shows that, in general, one cannot set $\varepsilon = 0$ in the above theorem.

EXAMPLE 3.   Let X denote the linear space made up of all real-valued continuous functions $x(t)$ defined on the interval $0 \leq t \leq 1$ and that satisfy $x(0) = 0$. Let $\|\cdot\|_\infty$ denote the sup-norm. Then $(X, \|\cdot\|_\infty)$ is a Banach space. [It is a closed linear subspace of the Banach space $(C[0,1], \|\cdot\|_\infty)$.] Let M denote the linear subspace of X consisting of those functions $x(t)$ that satisfy

$$l(x) = \int_0^1 x(t)\, dt = 0.$$

We leave it to the reader to show that M is closed. (See Theorem 5.6.6.)

We want to show that if $x_0 \in X$ with $\|x_0\|_\infty = 1$, then $d(x_0, M) < 1$. That is, even though X is complete and M is closed, there is no x_0 that is "orthogonal" to M.

Suppose for a moment that we consider M as a linear subspace of $(C[0,1], \|\cdot\|_\infty)$. Suppose further that x_0 is a point in $C[0,1]$ with $\|x_0\|_\infty = 1$ and $|x_0(0)| = 1$. Since $y(0) = 0$ for each $y \in M$, it follows that $\|x_0 - y\|_\infty \geq 1$ for all y in M. Then since $\|x_0 - 0\|_\infty = 1$, it follows that $d(x_0, M) = 1$.

Next suppose that x_1 is a point in $C[0,1]$ with $\|x_1\|_\infty = 1$ and $|x_1(0)| = \alpha < 1$. It follows that $\|x_1 - y\|_\infty \geq \alpha$ for all $y \in M$. Moreover, we can show that there exists a $\hat{y} \in M$ such that $\alpha \leq \|x_1 - \hat{y}\|_\infty < 1$. The basic idea is illustrated by Figure 5.5.3. That is, $\alpha \leq d(x_1, M) < 1$. The construction of a $\hat{y}$ is rather tedious, so we

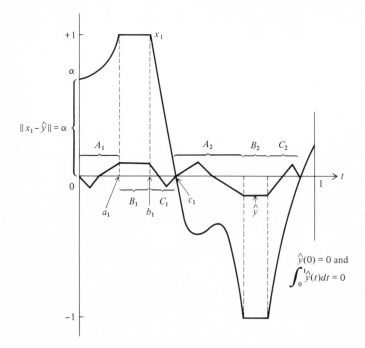

Figure 5.5.3.

shall merely sketch it here. Assume that $x_1(0) = \alpha \geq 0$. Since x_1 is continuous, there exists a maximal nontrivial interval[5] A_1 starting at $t = 0$ such that $-1 < x_1(t) < 1$ for all $t \in A_1$. Then since $\|x_1\|_\infty = 1$ and x_1 is continuous, one has $|x_1(a_1)| = 1$ with $a_1 = \sup A_1$. Note that $a_1 \notin A_1$. Next we have the maximal closed interval $B_1 = [a_1, b_1]$ upon which $x_0(t) = x_0(a_1)$. It can happen that $a_1 = b_1$. Then if $b_1 \neq 1$, we have the nontrivial interval C_1 where $0 \leq x_1(t) < 1$, if $x_1(b_1) = 1$, or $-1 < x_1(t) \leq 0$, if $x_1(b_1) = -1$. We construct $\hat{y}$ on $I_1 = A_1 \cup B_1 \cup C_1$ so that $\hat{y}(0) = 0$,

$$\int_{I_1} \hat{y}(t)\, dt = 0 \qquad \text{and} \qquad \sup_{t \in I_1} |x_1(t) - \hat{y}(t)| < 1.$$

This construction is continued until we have defined $\hat{y}$ on all of $[0,1]$.

Hence, for $\|x\|_\infty = 1$ we have that $d(x,M) = 1$ if and only if $|x(0)| = 1$.

Return now to the original problem where X is the Banach space containing the closed linear subspace M. Since all the $x \in X$ satisfy $x(0) = 0$, it follows that there is no x_0 in X with $\|x_0\| = 1$ such that $d(x_0, M) = 1$.

We leave it to the reader to show that if $z_n \in X$, where

$$z_n(t) = \begin{cases} nt, & \text{for } 0 \leq t < \dfrac{1}{n} \\[2mm] 1, & \text{for } \dfrac{1}{n} \leq t \leq 1 \end{cases} \qquad n = 1, 2, 3, 4, \ldots,$$

then $\|z_n\| = 1$, and $\lim_{n \to \infty} d(z_n, M) = 1$. (See Figure 5.5.4.) ∎

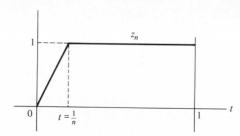

Figure 5.5.4.

EXERCISES

1. Prove Theorem 5.5.2.

2. (Continuation of Exercise 11, Section 3.) Show that $NBV(I)$ is a closed linear subspace of $BV(I)$.

3. Consider $L_1(I)$ with the usual norm $\|\cdot\|_1$. Let $z \in L_\infty(I)$. Now show that $M = \{x \in L_1(I): \int_I xz\,dt = 0\}$ is a closed linear subspace of $L_1(I)$. When is M a proper subspace?

---

[5] If $A_1 = [0,1]$, then set $\hat{y} = 0$. Otherwise continue with the rest of the construction.

4. Consider $L_p(I)$ with the usual norm $\|\cdot\|_p$ where $1 \leq p < \infty$. Let $z \in L_q(I)$ where $p^{-1} + q^{-1} = 1$. Show that $M = \{x \in L_p(I): \int_I xz\, dt = 0\}$ is a closed linear subspace of $L_p(I)$.

5. Consider $L_\infty(I)$ with the usual norm $\|\cdot\|_\infty$. Let $z \in L_1(I)$. Now show that $M = \{x \in L_\infty(I): \int_I xz\, dt = 0\}$ is a closed linear subspace of $L_\infty(I)$. What happens if we replace $L_\infty(I)$ with $BC(I)$?

6. Let $y_1, y_2, \ldots, y_n$ be a finite collection of functions in $L_2(I)$, where $L_2(I)$ has the usual norm $\|\cdot\|_2$. Show that

$$M = \left\{x \in L_2(I): \int_I xy_i\, dt = 0, 1 \leq i \leq n\right\}$$

is a closed linear subspace of $L_2(I)$.

7. Show that one can choose $\varepsilon = 0$ in Theorem 5.5.4 when M is of finite dimension.

8. Let X be a linear subspace in a Banach space B and assume that

$$\inf_{x \in X} \|y - x\| = d(y,X) \leq C\,\|y\|$$

for all $y \in B$. Show that if $C < 1$, then X is dense in B. [*Hint*: Use the Riesz Theorem.]

9. Show that the trivial subspace $\{0\}$ is closed.

6.   CONTINUOUS LINEAR TRANSFORMATIONS

The most important offspring of the wedding between topological and algebraic structure is the concept of a continuous linear transformation.

5.6.1 DEFINITION.   Let X and Y be normed linear spaces. A mapping $L: X \to Y$ is said to be a *continuous linear transformation* if it is (a) linear and (b) continuous.

As the reader probably suspects, many of the examples of continuous transformation in Chapter 3 are also linear. Likewise, many of the examples of linear transformations in Chapter 4 are also continuous. In particular, Examples 1, 4, 5, 6, 7 of Section 5 in Chapter 3 and Examples 2 and 3 of Section 3 in Chapter 4 are of this nature. Let us consider some other examples.

EXAMPLE 1.   Let $X = BC[0,\infty)$, the normed linear space made up of all bounded continuous functions defined on $[0,\infty)$ with the sup-norm $\|x\|_\infty$. Let T denote the operation of evaluating the "running average," that is,

$$(Tx)(t) = \frac{1}{t}\int_0^t x(\tau)\, d\tau.$$

It can be seen (from L'Hospital's Rule, for example) that

$$\lim_{t \to 0}\left\{\frac{1}{t}\int_0^t x(\tau)\,d\tau\right\} = x(0).$$

Moreover, Tx is clearly a continuous function of t, and

$$\left|\frac{1}{t}\int_0^t x(\tau)\,d\tau\right| \le \frac{1}{t}\int_0^t |x(\tau)|\,d\tau \le \|x\|_\infty,$$

for all t, so Tx is also bounded. Hence, T is a mapping of $BC[0,\infty)$ into itself. T is obviously linear. Let us show that T is continuous. Suppose that $x_0, x \in BC[0,\infty)$ and that $\varepsilon > 0$ is given. Then

$$\|Tx - Tx_0\| = \sup_t \left|\frac{1}{t}\int_0^t x(\tau)\,d\tau - \frac{1}{t}\int_0^t x_0(\tau)\,d\tau\right|$$

$$\le \sup_t \left\{\frac{1}{t}\int_0^t |x(\tau) - x_0(\tau)|\,d\tau\right\}$$

$$\le \|x - x_0\|_\infty.$$

So $\|x - x_0\| < \delta = \varepsilon$ implies $\|Tx - Tx_0\| < \varepsilon$. It follows that T is continuous at x_0. Since x_0 is arbitrary, T is continuous. ∎

EXAMPLE 2. Let X be the normed linear space made up of all bounded analytic functions $f(z)$ defined inside the unit circle, that is, $|z| < 1$. We define the norm by

$$\|f\| = \sup_{|z| < 1} |f(z)|.$$

Since f is analytic, it has a unique Taylor power series expansion

$$f(z) = a_0 + a_1 z + a_2 z^2 + \cdots.$$

Let T be the mapping of X into itself defined by

$$(Tf)(z) = a_0 + a_1 z, \qquad \text{for } |z| < 1.$$

Now $a_0 = f(0)$ and $a_1 = (df/dz)(0)$ so T is linear. Indeed T is a projection. Let us show that T is continuous.

We know from Cauchy's integral formula for analytic functions that

$$\frac{d^n f}{dz^n}(z_0) = \frac{n!}{2\pi i}\int_C \frac{f(z)\,dz}{(z - z_0)^{n+1}}, \qquad n = 0, 1, 2, \ldots,$$

where C is a closed curve within and on which f is analytic. Thus $|a_0| \le \|f\|$ and $|a_1| \le \|f\|$. Therefore, $|(Tf)(z)| = |a_0 + a_1 z| \le 2\|f\|$ and so $\|Tf\| \le 2\|f\|$. It follows that if $f, f_0 \in X$, then $\|Tf - Tf_0\| \le 2\|f - f_0\|$ and T is continuous.

In summary, then, T is a continuous projection. ∎

EXAMPLE 3.   Consider the integral operator $y = Kx$, where

$$y(t) = \int_I k(t,\tau)x(\tau)\, d\tau$$

and t, τ belong to some interval I. Assume that

$$B = \int_I \int_I |k(t,\tau)|^p \, d\tau \, dt < \infty, \tag{5.6.1}$$

where $1 \leq p < \infty$. Then K represents a continuous linear mapping of $L_q(I)$ into $L_p(I)$, where $p^{-1} + q^{-1} = 1$.

   In order to show this we shall assume that $1 < p < \infty$. (The case $p = 1$ is left as an exercise.) It follows from the Hölder Inequality for integrals (see Appendix A) that

$$\left| \int_I k(t,s)x(s)\, ds \right| \leq \left[\int_I |k(t,s)|^p \, ds \right]^{1/p} \left[\int_I |x(s)|^q \, ds \right]^{1/q}.$$

Therefore,

$$\int_I |y(t)|^p \, dt \leq \int_I \left[\int_I |k(t,s)|^p \, ds \right] \left[\int_I |x(s)|^q \, ds \right]^{p/q} dt$$

$$\leq \left[\int_I \int_I |k(t,s)|^p \, ds \, dt \right] \left[\int_I |x(s)|^q \, ds \right]^{p/q}.$$

In other words, $\|y\|_p \leq B^{1/p}\|x\|_q$, where B is given by (5.6.1). It follows then from $\|y - y_0\|_p \leq B^{1/p}\|x - x_0\|_q$ that K is a continuous linear mapping of $L_q(I)$ into $L_p(I)$. Of particular importance is the case $p = q = 2$.   ∎

EXAMPLE 4.   Consider the integral operator $y = Kx$, where

$$y(t) = \int_{-\infty}^{\infty} k(t - \tau)x(\tau)\, d\tau, \qquad t \in (-\infty,\infty). \tag{5.6.2}$$

This integral represents a time-invariant operator. We assume that $k \in L_1(-\infty,\infty)$. Let us now show that K is a bounded linear transformation of $L_1(-\infty,\infty)$ into itself. Let $y_i = Kx_i$, $i = 1, 2$. Then

$$\int_{-\infty}^{\infty} |y_1(t) - y_2(t)| \, dt = \int_{-\infty}^{\infty} \left| \int_{-\infty}^{\infty} k(t - \tau)[x_1(\tau) - x_2(\tau)] \, d\tau \right| dt$$

$$\leq \int_{-\infty}^{\infty} \int_{-\infty}^{\infty} |k(t - \tau)| \, |x_1(\tau) - x_2(\tau)| \, d\tau \, dt.$$

By interchanging the order of integration (see Appendix D) and using the fact that for every τ one has $\int_{-\infty}^{\infty} |k(t - \tau)| \, dt = \|k\|_1$, one gets

$$\|y_1 - y_2\|_1 \leq \|k\|_1 \|x_1 - x_2\|_1.$$

Hence, K is continuous.

   More generally, if $k \in L_1(-\infty,\infty)$, then (5.6.2) represents a continuous linear transformation of $L_p(-\infty,\infty)$ into itself. (See Exercise 17.)   ∎

EXAMPLE 5.   In Example 2 of Section 3.5 we gave an example of a discontinuous operator which is also linear. Let us consider here another discontinuous linear operator.

Let $X = C^1[-1,1]$ and $Y = C[-1,1]$ be given with the sup-norm $\|\cdot\|_\infty$. Consider the differential operator $D: x \to dx/dt$. Let us show that D is discontinuous at the origin. Indeed if $x_n(t) = n^{-1} \sin nt$, then $\|x_n\|_\infty = n^{-1}$. However $Dx_n = \cos nt$, so $\|Dx_n\|_\infty = 1$. Hence D is discontinuous at $x = 0$. One can easily show that D is discontinuous everywhere. ∎

So much for examples. Let us now turn to some of the properties of continuous linear transformations.

Recall that in Lemma 4.3.2 we showed that a transformation $L: X \to Y$ is linear if and only if

$$L(\alpha_1 x_1 + \cdots + \alpha_n x_n) = \alpha_1 L(x_1) + \cdots + \alpha_n L(x_n),$$

for all $x_1, \ldots, x_n$ in X and all scalars $\alpha_1, \ldots, \alpha_n$. There we carefully pointed out that we were only considering finite linear combinations. Now that we have topological structure present we can consider the infinite case.

5.6.2 THEOREM.   *Let X and Y be normed linear spaces, and let L be a transformation of X into Y. The transformation L is a continuous linear transformation if and only if*

$$L\left(\sum_{i=1}^{\infty} \alpha_i x_i \right) = \sum_{i=1}^{\infty} \alpha_i L(x_i), \tag{5.6.3}$$

for every convergent series $\sum_{i=1}^{\infty} \alpha_i x_i$.

Carefully note that we are only considering series $\sum_{i=1}^{\infty} \alpha_i x_i$ that converge. The reason should be obvious. Further note that if we choose to call the conclusion of this theorem the *principle of superposition*, then the principle of superposition is a characterization of *continuous* linear transformations.

Proof: The proof of this theorem is a direct consequence of Theorem 3.7.2 and Lemma 4.3.2. The former asserts that L is continuous if and only if

$$L\left(\lim_{n \to \infty} z_n \right) = \lim_{n \to \infty} L(z_n),$$

for every convergent sequence $\{z_n\}$.

First assume that L is linear and continuous, and let $\sum \alpha_i x_i$ be any convergent series in X. Then the sequence of partial sums

$$z_n = \sum_{i=1}^{n} \alpha_i x_i$$

is convergent, and from Theorem 3.7.2 we have

$$L\left(\lim_{n \to \infty} z_n \right) = \lim_{n \to \infty} L(z_n).$$

By definition $\lim_{n \to \infty} z_n = \sum_{i=1}^{\infty} \alpha_i x_i$, so

$$L\left(\sum_{i=1}^{\infty} \alpha_i x_i\right) = \lim_{n \to \infty} L(z_n)$$

$$= \lim_{n \to \infty} \sum_{i=1}^{n} \alpha_i L(x_i) = \sum_{i=1}^{\infty} \alpha_i L(x_i).$$

Now assume that (5.6.3) holds for every convergent series $\sum \alpha_i x_i$. We want to show that L is linear and continuous. Using series with only a finite number of nonzero terms and Lemma 4.3.2, we see that L is linear. So let us turn to showing that L is continuous. Let $\{z_n\}$ be any convergent sequence in X, and let $x_n = z_n - z_{n-1}$, where $z_0 = 0$. It follows that $z_n = \sum_{i=1}^{n} x_i$ and that the series $\sum_{i=1}^{\infty} x_i$ is convergent with $\lim_{n \to \infty} z_n = \sum_{i=1}^{\infty} x_i$. It follows from (5.6.3) that

$$L\left(\sum_{i=1}^{\infty} x_i\right) = \sum_{i=1}^{\infty} L(x_i)$$

or

$$L\left(\lim_{n \to \infty} z_n\right) = \lim_{n \to \infty} \sum_{i=1}^{n} L(x_i).$$

Then using the linearity of L, or (5.6.3) again, we have

$$L\left(\lim_{n \to \infty} z_n\right) = \lim_{n \to \infty} L\left(\sum_{i=1}^{n} x_i\right)$$

$$= \lim_{n \to \infty} L(z_n).$$

Since $\{z_n\}$ was an arbitrary convergent sequence, it follows from Theorem 3.7.2 that L is continuous. ∎

EXAMPLE 6. Let us define $J: L_2[-\pi,\pi] \to L_2[-\pi,\pi]$ by $y = Jx$, where $y(t) = \int_0^t x(s)\, ds$. This mapping is a continuous linear mapping. Since the series $\sum_{n=1}^{\infty} (1/n) \sin nt$ converges in L_2, with the usual metric, one has

$$\int_0^t \sum_{n=1}^{\infty} \frac{1}{n} \sin ns\, ds = \sum_{n=1}^{\infty} \frac{1}{n} \int_0^t \sin ns\, ds$$

$$= \sum_{n=1}^{\infty} \frac{1}{n^2} (\cos nt - 1). \quad \blacksquare$$

EXAMPLE 7. Let $\{\phi_1, \phi_2, \ldots\}$ be a sequence of functions in the real space $L_2(I)$ with the usual norm. Assume that $\int_I \phi_n(t)\phi_m(t)\, dt = \delta_{nm}$ where δ_{nm} is the Kronecker function. Let $a = \{a_1, a_2, \ldots\}$ be a sequence of real numbers in l_2. Then form the series

$$\phi = \sum_{i=1}^{\infty} a_i \phi_i.$$

One can show that this series converges in $L_2(I)$ by using the Cauchy Test (Lemma 5.4.1). Indeed, one has for $n \geq m$

$$\left\| \sum_{i=m}^{n} a_i \phi_i \right\|^2 = \int_I (a_n \phi_n + \cdots + a_m \phi_m)^2 \, dt$$

$$= \sum_{i,\, j=m}^{n} \int_I a_i a_j \phi_i \phi_j \, dt = \sum_{i,\, j=m}^{n} a_i a_j \delta_{ij}$$

$$= \sum_{i=m}^{n} |a_i|^2.$$

Since this sequence $\{a_i\}$ is in l_2, we know that $\lim_{n,m \to \infty} \sum_{i=m}^{n} |a_i|^2 = 0$. Hence, the original series converges in $L_2(I)$.

Let $K: L_2(I) \to R$ be given by $x \to \int_I k(t)x(t) \, dt$, where k is a fixed element in $L_2(I)$. K is linear and by the Hölder Inequality for integrals, from Appendix A, one has

$$\left| \int_I k(t)[x_1(t) - x_2(t)] \, dt \right| \leq \left\{ \int_I |k(t)|^2 \, dt \right\}^{1/2} \left\{ \int_I |x_1(t) - x_2(t)|^2 \, dt \right\}^{1/2}$$

$$\leq \|k\| \, \|x_1 - x_2\|.$$

It follows that K is continuous. Therefore,

$$K\phi = \int_I k(t) \left\{ \sum_{n=1}^{\infty} a_n \phi_n(t) \right\} dt = \sum_{n=1}^{\infty} a_n \left\{ \int_I k(t)\phi_n(t) \, dt \right\}. \quad \blacksquare$$

At this point let us note a very important distinction between continuous and discontinuous linear transformations. So far in each of the examples in this section, the ratio

$$\frac{\|Lx\|}{\|x\|} \qquad (\|x\| \neq 0)$$

is bounded for continuous linear transformations and unbounded for discontinuous linear transformations. We shall now show that this is, in fact, always the case. First, we need the concept of a bounded linear transformation.

5.6.3 DEFINITION.   Let $L: X \to Y$ be a linear transformation, where X and Y are normed linear spaces. We shall say that L is *bounded* if there is a real number $M \geq 0$ such that

$$\|Lx\| \leq M\|x\|,$$

for all x in X.

Before we proceed, a word concerning the notation is needed. Since X and Y are allowed to be different normed linear spaces, one sometimes uses different notation for the norm on each space. For example, one might use $\|\cdot\|_X$ and $\|\cdot\|_Y$ to denote these norms. However, where no confusion will arise, this distinction is dropped. In any event, in the above inequality $\|x\|$ denotes the norm in X (since x is in X) and $\|Lx\|$ denotes the norm in Y (since $y = Lx$ is in Y).

The ratio $\|Lx\|/\|x\|$, where $\|x\| \neq 0$, can be viewed as the "gain" or "amplification" of the operator L for the "input" x. It is not difficult to see that this gain is bounded if and only if L is bounded. Indeed, if L is bounded, then $\|Lx\|/\|x\| \leq M$; and if $\|Lx\|/\|x\|$ is bounded, then $\|Lx\| \leq \{\sup_{x \neq 0} \|Lx\|/\|x\|\} \|x\| = M\|x\|$.

The next theorem—and it is a very important one—shows how boundedness is related to continuity for linear transformations.

5.6.4 THEOREM.   *A linear transformation $L: X \to Y$, where X and Y are normed linear spaces, is continuous if and only if it is bounded.*

Proof: Let us prove the "if" part first, that is, assume that L is bounded. It follows from the linearity of L and the definition of bounded that one has

$$\|Lx - Lx_0\| = \|L(x - x_0)\| \leq M\|x - x_0\|.$$

By setting $\delta = \varepsilon/M$ we see that L is continuous at x_0. Since x_0 is arbitrary we see that L is continuous.

Now for the "only if" part. Assume that L is continuous. Then L is continuous at 0. Thus for $\varepsilon = 1$, there is a $\delta > 0$ such that $\|Lx'\| \leq 1$ whenever $\|x'\| \leq \delta$. If x is any point in X, $x \neq 0$, let $x' = \beta x$ where $\beta = \delta\|x\|^{-1}$. Since $\|x'\| = \delta$, one has $\|Lx'\| \leq 1$. Therefore,

$$1 \geq \|Lx'\| = \|L(\beta x)\| = |\beta| \, \|Lx\|$$

which implies that

$$\|Lx\| \leq \frac{1}{\beta} = \frac{1}{\delta} \|x\| . \tag{5.6.4}$$

That is, (5.6.4) holds for every point x in X with $x \neq 0$. But it also holds for $x = 0$, so we see that L is bounded. (Carefully note that δ is independent of x.) ∎

The last theorem has a rather interesting mathematical interpretation. Since L is linear one has $L(0) = 0$. Now if one examines the definition of bounded, one sees that L is bounded if and only if L is continuous at $x = 0$. The last theorem, then, asserts that a linear transformation is continuous (everywhere) if and only if it is continuous at a single point, namely, 0. It is a simple exercise to show that 0 can be replaced by any other point x in X. Hence, we have the following result.

5.6.5 LEMMA.   *Let $L: X \to Y$ be a linear transformation, where X and Y are normed linear spaces. If L is continuous at one point x in X, then L is continuous everywhere.*

Finally, it is interesting to note that if a linear transformation is continuous, it is uniformly continuous. This fact follows from the boundedness of the transformation.

Theorems 5.6.2 and 5.6.4 are fundamental statements about continuous linear transformations. We urge the reader to master them before continuing. Moreover, we remark that we shall henceforth use the terms "continuous linear transformation" and "bounded linear transformation" interchangeably.

We end this section by noting a simple but important fact about continuous linear transformations.

5.6.6 THEOREM. *If $L: X \to Y$ is a continuous linear transformation, then $\mathcal{N}(L)$, the null space of L, is a closed linear subspace of X.*

The proof of this theorem is left as an exercise.

EXERCISES

1. Show that the mapping K in Example 3 is continuous when $p = 1$.

2. Show that a linear transformation is continuous if and only if it is uniformly continuous.

3. Let B_1 and B_2 be two Banach spaces and let X be a linear subspace of B_1. Let $L: X \to B_2$ be a continuous linear transformation,
   (a) Show that L has a continuous extension $\tilde{L}$ defined on the closure $\bar{X}$.
   (b) Show that $\tilde{L}$ is necessarily linear.
   (c) Show that $\tilde{L}$ is unique. [*Hint*: Use Theorem 3.14.3.]

4. Use the result of Exercise 3 to characterize the Lebesgue integral as an extension of the Riemann integral. (Compare with Appendix D.)

5. Let A, B be two continuous linear mappings of X into Y, where X and Y are normed linear spaces.
   (a) Show that $A + B$ is a continuous linear mapping of X into Y.
   (b) Show that for every scalar α, the mapping αA is a continuous linear mapping of X into Y. (Thus, in other words, the collection of continuous linear mapping of X into Y is a linear space. We shall show in Section 8 that this is a normed linear space.)

6. Show by means of specific examples that the Principle of Superposition fails for discontinuous linear transformations.

7. Show by examples that Theorem 5.6.4 and Lemma 5.6.5 fail for nonlinear mappings.

8. (a) Prove Lemma 5.6.5.
   (b) Prove Theorem 5.6.6.

9. (Closed operators.) Let B_1 and B_2 be two Banach spaces and let X be a linear subspace of B_1. A linear operator $L: X \to B_2$ is said to be *closed* if whenever $x_n \to x$ in B_1 and $y_n = Lx_n \to y$ in B_2 one has $x \in X$ and $y = Lx$. Show that every continuous linear operator is closed. [*Remark*: There are closed linear operators that are not continuous as is shown in Exercise 11. Also, do not confuse this concept with that of a closed mapping given in Exercise 5, Section 3.12. We shall return to this concept in Section 7.10.]

10. Let $Gr(L)$ be the graph of L, that is,

$$Gr(L) = \{(x,Lx): x \in X\} \subset B_1 \times B_2.$$

(a) Show that $Gr(L)$ is a linear subspace of $B_1 \times B_2$.
(b) Show that L is closed if and only if $Gr(L)$ is a closed linear subspace of $B_1 \times B_2$, where $B_1 \times B_2$ has the norm $\|(x_1,x_2)\| = \|x_1\| + \|x_2\|$.

11. Let $B_1 = B_2 = C[0,1]$ be given with the sup norm and let $X = C^1[0,1]$. Let $D: X \to C[0,1]$ be the differential operator $D: x \to dx/dt$. [Note: D is linear and discontinuous.] Show that D is closed.

12. Let $X = C^n[0,1]$ and let $P(z)$ be a polynomial in z of degree n. Now we let $P(D): X \to C[0,1]$ be the associated polynomial differential operator. Show that $P(D)$ is closed, where X and $C[0,1]$ have the sup norm.

13. (Continuation of Exercise 9.) Let $B_1 = B_2 = L_2(-i\infty,i\infty)$ be given with usual norm. Let $f(i\omega)$ be any measurable, scalar-valued function defined on the imaginary axis $(-i\infty,i\infty)$. Let $X = \{x \in L_2(-i\infty,i\infty): fx \in L_2(-i\infty,i\infty)\}$ and define $F: X \to L_2(-i\infty,i\infty)$ by $y = Fx$, where $y(i\omega) = f(i\omega)x(i\omega)$.

(a) Show that F is closed.
(b) Find conditions on f in order that $X = L_2(-i\infty,i\infty)$. Is F continuous in this case? (A special case of this problem occurs when f is continuous.)

14. (Continuation of Exercise 9.) Let $B_1 = B_2 = C(D)$, where D is the unit disk$\{(r,\theta): 0 \le r \le 1\}$ in the plane, given with the sup-norm. Let X denote the collection of C^2 functions $u(r,\theta)$ such that $u(1,\theta) = 0$ for $0 \le \theta \le 2\pi$, and let $\nabla^2: X \to C(D)$ be the Laplacian operator, that is,

$$\nabla^2 u = \partial^2 u/\partial r^2 + 1/r \, \partial u/\partial r + (1/r^2) \, \partial^2 u/\partial \theta^2.$$

Show that ∇^2 is closed.

15. Let $0 < \alpha < 1$ and define $f_1(t) = t$

$$f_n(t) = \begin{cases} t, & 0 \le t \le 1/n \\ t^{-\alpha} - (n-1)t, & 1/n < t \le 1/(n-1) \\ 0, & 1/(n-1) < t \le 1, \end{cases}$$

for $n = 2, 3, \ldots$. In which spaces $L_p[0,1]$ does the infinite series $\sum_{n=1}^{\infty} f_n$ converge? Compute

$$\int_0^t \left[\sum_{n=1}^{\infty} f_n(t) \right] dt.$$

16. Let $\rho(x)$ be a nonnegative real-valued C^∞-function on R^m with $\rho(x) = 0$ for $x \ge 1$ and $\int_{R^m} \rho(x) \, dx = 1$. (For an appropriate choice of the constant C the function

$$\rho(x) = \begin{cases} 0, & |x| \ge 1 \\ C \exp\left\{\dfrac{1}{|x|^2 - 1}\right\}, & |x| \le 1 \end{cases}$$

satisfies these conditions.) For $\varepsilon > 0$ define the *mollifier operator* J_ε by

$$J_\varepsilon u(x) = \varepsilon^{-m} \int_\Omega \rho\left(\frac{x-y}{\varepsilon}\right) u(y)\, dy,$$

where $u(y)$ is defined in a bounded open domain Ω in R^m.

(a) Show that if $u \in L_p(\Omega)$ the $J_\varepsilon(u) \in C_0^\infty(R^m)$, that is, $J_\varepsilon(u)$ is a C^∞-function with compact support in R^m.

(b) Show that $J_\varepsilon: L_p(\Omega) \to L_\infty(\Omega)$ is a bounded linear transformation.

(c) Show that $\|J_\varepsilon u - u\|_p \to 0$ as $\varepsilon \to 0$ for each $u \in L_p(\Omega)$. [*Hint*: First show that

$$J_\varepsilon u(x) - u(x) = \int_{|\xi| \leq 1} \rho(\xi)[u(x - \varepsilon\xi) - u(x)]\, d\xi.$$

Then show that (c) holds when u is continuous. Finally, approximate an arbitrary u in L_p with a continuous function to get the general result.]

(d) Show that if $u \in C_0^\infty(\Omega)$, then

$$D^\alpha(J_\varepsilon u) = (-1)^{|\alpha|} J_\varepsilon(D^\alpha u).$$

17. Let $k \in L_1(-\infty, \infty)$. Show that the integral operator K given by

$$y(t) = \int_{-\infty}^{\infty} k(t - \tau) x(\tau)\, d\tau$$

is a continuous linear mapping of $L_p(-\infty, \infty)$ into itself, when $1 \leq p \leq \infty$. [*Hint*: Note that if $x(\cdot) \in L_p(-\infty, \infty)$, and $z(\cdot) \in L_q(-\infty, \infty)$, where $p^{-1} + q^{-1} = 1$, then

$$\left| \int_{-\infty}^{\infty} \int_{-\infty}^{\infty} k(t-\tau) x(\tau)\, d\tau\, z(t)\, dt \right| \leq \|k\|_1 \|x\|_p \|z\|_q.$$

Now apply Exercise 5, Section D.11.]

18. Define $L: L_2(-\infty, \infty) \to L_2(-\infty, \infty)$ by $L: x(t) \to x(-t)$. Is L linear and continuous?

7.   INVERSES AND CONTINUOUS INVERSES

Let $L: X \to Y$ be a mapping of X into Y. In Section 2.7, we saw that L has an inverse defined on its range if and only if L is one-to-one. In Section 4.4, we saw that if L is linear, then it is one-to-one if and only if $\mathcal{N}(L) = \{0\}$, that is, the null space is trivial. These considerations require only algebraic structure. However, when X and Y are normed linear spaces, one can inquire into the continuity of L and L^{-1}.

The first problem we would like to solve is: Find conditions on L that guarantee that $L^{-1}: \mathcal{R}(L) \to X$ exists and is continuous.

5.7.1 THEOREM. *Let* $L: X \to Y$ *be a linear transformation, where X and Y are normed linear spaces. If there is a constant $m > 0$ such that*

$$m\|x\| \le \|Lx\|, \qquad x \in X, \tag{5.7.1}$$

then L has a continuous inverse L^{-1} defined on its range $\mathscr{R}(L)$,

$$L^{-1}: \mathscr{R}(L) \to X$$

and

$$\|L^{-1}y\| \le \frac{1}{m}\|y\|, \tag{5.7.2}$$

for all y in $\mathscr{R}(L)$. Conversely, if there is a continuous inverse $L^{-1}: \mathscr{R}(L) \to X$, then there is a positive constant m such that (5.7.1) *holds for all x in X.*

Note that we do not require the operator L to be continuous. Also, this theorem says nothing about the range of L; that is, (5.7.1) being satisfied does not allow us to say that L is or is not a mapping of X *onto* Y. In general, this question must be handled separately.

A linear operator $L: X \to Y$ that satisfies (5.7.1) is said to be *bounded below*.

Proof: First assume that (5.7.1) holds for some $m > 0$. In order to show that $L^{-1}: \mathscr{R}(L) \to X$ exists, we must show that the null space $\mathscr{N}(L)$ contains only the origin. This fact follows immediately from (5.7.1). Indeed, if $x \ne 0$, then $Lx \ne 0$ since $\|Lx\| \ge m\|x\| > 0$. We now want to show that $L^{-1}: \mathscr{R}(L) \to X$ is continuous, or equivalently (Theorem 5.6.4), bounded. If $y \in \mathscr{R}(L)$, then there is an $x \in X$ such that $y = Lx$ and $x = L^{-1}y$. From (5.7.1), we get

$$\|L^{-1}y\| = \|x\| \le \frac{1}{m}\|Lx\| = \frac{1}{m}\|y\|, \tag{5.7.3}$$

for all y in $\mathscr{R}(L)$. Hence, L^{-1} is bounded.

Now assume that $L^{-1}: \mathscr{R}(L) \to X$ exists and is continuous. Then there is an $m > 0$ such that (5.7.2) holds. By reversing the reasoning of (5.7.3), one easily verifies that (5.7.1) holds. ∎

EXAMPLE 1. Let T be a continuous linear mapping of a Banach space B into itself, and consider the linear mapping $\lambda I - T$, where λ is a scalar. It follows from the triangle inequality that

$$\|(\lambda I - T)x\| \ge \big|\|\lambda x\| - \|Tx\|\big|,$$

for all $x \in B$.

Since T is bounded, there is an $M \ge 0$ such that $\|Tx\| \le M\|x\|$, so for $|\lambda| > M$ one has

$$\|(\lambda I - T)x\| \ge (|\lambda| - M)\|x\|.$$

Hence, by Theorem 5.7.1, the mapping $\lambda I - T$ has a continuous inverse defined on its range for sufficiently large $|\lambda|$, in particular, for $|\lambda| > M$. Moreover, the mapping

$$K_y(x) = \frac{1}{\lambda} y + \frac{1}{\lambda} Tx,$$

where y is a point in B, is a contraction mapping for sufficiently large $|\lambda|$. (Why?) So from the Contraction Mapping Theorem (Theorem 3.15.2) we know that K_y has a unique fixed point x. Since y is arbitrary, we have that given any $y \in B$ there exists a unique $x \in B$ such that

$$(\lambda I - T)x = y.$$

In other words, the range of $(\lambda I - T)$ is B for sufficiently large $|\lambda|$. ∎

EXAMPLE 2.   Let $T(i\omega)$ be a continuous complex-valued function defined on the imaginary axis of the complex plane. It may be viewed as the transfer function of some physical system. We define an operator $y = Tx$ formally by

$$y(i\omega) = T(i\omega)x(i\omega), \qquad \omega \in (-\infty, \infty).$$

More precisely, we choose the domain of T to be

$$\mathscr{D}(T) = \{x \in L_2(-i\infty, i\infty) : Tx \in L_2(-i\infty, i\infty)\}.$$

It follows that $\mathscr{R}(T) \subset L_2(-i\infty, i\infty)$. In this example we investigate some of the properties of the operator T.

If $|T(i\omega)|$ is bounded above, that is, $|T(i\omega)| \leq B$ for all ω, then

$$\|Tx\|^2 = \frac{1}{2\pi} \int_{-\infty}^{\infty} |T(i\omega)|^2 |x(i\omega)|^2 \, d\omega \leq \frac{B^2}{2\pi} \int_{-\infty}^{\infty} |x(i\omega)|^2 \, d\omega = B^2 \|x\|^2.$$

We see, then, that if $|T(i\omega)|$ is bounded above, then the operator T is bounded; moreover, we see that $\mathscr{D}(T) = L_2(-i\infty, i\infty)$.

If $|T(i\omega)| \geq b > 0$ for all ω, then

$$\|Tx\|^2 = \frac{1}{2\pi} \int_{-\infty}^{\infty} |T(i\omega)|^2 |x(i\omega)|^2 \, d\omega \geq b^2 \|x\|^2,$$

in other words, T is bounded below. In this case it follows from Theorem 5.7.1 that T^{-1} exists, is continuous and satisfies $\|T^{-1}y\| \leq b^{-1}\|x\|$ for all $y \in \mathscr{R}(T)$. In fact T^{-1} is given by $x = T^{-1}y$, where

$$x(i\omega) = \frac{1}{T(i\omega)} y(i\omega).$$

If $|T(i\omega)|$ satisfies $0 < b \leq |T(i\omega)| \leq B < \infty$ for all ω, then T and T^{-1} are both linear and continuous with $\mathscr{D}(T) = \mathscr{R}(T) = L_2(-i\infty, i\infty)$.

What happens to T if $|T(i\omega)|$ vanishes on some infinite interval $\omega_1 \leq \omega \leq \omega_2$ where $\omega_1 < \omega_2$? If we set

$$x(i\omega) = \begin{cases} 1, & \omega_1 \leq \omega \leq \omega_2 \\ 0, & \text{otherwise}, \end{cases}$$

that is, $x = \mathcal{X}_{[\omega_1,\omega_2]}$, where $\mathcal{X}_{[\omega_1,\omega_2]}$ denotes the characteristic function of $[\omega_1,\omega_2]$, then $Tx = 0$. Hence $x \in \mathcal{N}(T)$ and T is not one-to-one. More generally if $|T(i\omega)| = 0$ on a set A with Lebesgue measure $0 < m(A) < \infty$, then $T\mathcal{X}_A = 0$ where $\mathcal{X}_A$ is the characteristic function of A. Hence, $\mathcal{X}_A \in \mathcal{N}(T)$. Since $\|\mathcal{X}_A\|^2 \neq 0$ we see that $\mathcal{N}(T)$ is nontrivial.

The behavior of T, where $|T(i\omega)|$ vanishes only on a set of measure zero, is discussed in the exercises.   ∎

EXERCISES

1. This exercise refers to Example 2. Assume that $|T(i\omega)| = 0$ at $\omega = 0$. Let $x_n(i\omega) = n$ for $|\omega| \leq n^{-2}$ and $x_n(i\omega) = 0$ otherwise.
   (a) Show that x_n and Tx_n are in $L_2(-i\infty, i\infty)$.
   (b) Compute $\|x_n\|$.
   (c) Show that $\|Tx_n\| \to 0$ as $n \to \infty$, thereby showing that T is not bounded below.

2. (Continuation of Exercise 1.) Let $A = \{\omega : T(i\omega) = 0\}$. Assume that A is finite.
   (a) Show that $\mathcal{R}(T) \neq L_2(-\infty, \infty)$.
   (b) Show that $T^{-1} : \mathcal{R}(T) \to \mathcal{D}(T)$ is well-defined but not continuous.
   (c) What happens if A is countable?
   (d) What happens if A is a general set of measure zero?
   (e) Assume that A is empty but that $|T(i\omega)|$ is not bounded below. Discuss the behavior of T and T^{-1}.

3. What happens in Example 2 if we do not assume $T(i\omega)$ to be continuous?

4. Consider the space $X = C^1[0,T]$ with two norms:

$$\|x\|_1 = \|x\|_\infty = \text{sup-norm},$$
$$\|x\|_2 = \|x\|_\infty + \|x'\|_\infty,$$

where $x' = dx/dt$.
   (a) Show that the identity mapping $I : (X, \|\cdot\|_1) \to (X, \|\cdot\|_2)$ is bounded below, but not continuous.
   (b) Use this fact to compare the topologies on these spaces, referring to Section 3.9.

5. In Exercise 2, Section 7.7 we will show that if $f \in C^2[a,b]$, then

$$\int_a^b |f'(t)|^2 \, dt \leq 54 \left[\frac{1}{(b-a)^2} \int_a^b |f(t)|^2 \, dt + (b-a)^2 \int_a^b |f''(t)|^2 \, dt \right].$$

Furthermore we note that if

$$f \in C^1[a,b] \quad \text{with} \quad f(a) = 0,$$

then

$$\int_a^b |f(t)|^2 \, dt \leq (b-a)^2 \int_a^b |f'(t)|^2 \, dt.$$

Consider the operator $D^2: u \to u''$ on the domain

$$\mathscr{D}(D^2) = \{u \in C^2[0,1]: u(0) = u(1) = 0\}.$$

(a) Show that the range of D^2 lies in $C[0,1]$.
(b) Show that the equation $u'' = f$ has a unique solution u in $\mathscr{D}(D^2)$ when $f \in C[0,1]$.
(c) Let

$$\|f\|_{(0)} = \left(\int_0^1 |f(t)|^2 \, dt \right)^{1/2}$$

and

$$\|u\|_{(2)} = (\|u\|_{(0)}^2 + \|u'\|_{(0)}^2 + \|u''\|_{(0)}^2)^{1/2}$$

be the norms on $C[0,1]$ and $\mathscr{D}(D^2)$, respectively. Show that there is a real constant K such that

$$\|u\|_{(2)} \leq K\|D^2 u\|_{(0)},$$

where u is the solution given by part (b).
(d) What can one say about the inverse of D^2?

6. (Continuation of Exercise 5.) Consider the operator $D^4: u \to u^{(iv)}$ on the domain

$$\mathscr{D}(D^4) = \{u \in C^4[0,1]: u, u' \text{ vanish at } 0, 1\}.$$

(a) Discuss the equation $u^{(iv)} = f$ where $f \in C[0,1]$, $C[0,1]$ has the norm $\|f\|_{(0)}$, and $\mathscr{D}(D^4)$ has the norm

$$\|u\|_{(4)} = \left(\sum_{i=0}^4 \|u^{(i)}\|_{(0)}^2 \right)^{1/2}.$$

(b) What can one say about the inverse of D^4?

8.   OPERATOR TOPOLOGIES

The space $Blt[X, Y]$ of all bounded linear transformations of X into Y, where X and Y are normed linear spaces, is itself a linear space. (See Exercise 5, Section 6.) We shall show here that it is also a normed linear space. However, before doing this let us consider an example.

EXAMPLE 1.   Consider the (complex) normed linear space $L_2(I)$ with the usual norm $\|\cdot\|$. Let $\{\phi_1, \phi_2, \ldots\}$ be a sequence in $L_2(I)$ with the property that $\|\phi_n\| = 1$ and $\int_I \phi_n(t)\overline{\phi_m(t)} \, dt = 0$ when $n \neq m$. Let $\{\alpha_1, \alpha_2, \ldots\}$ be a sequence of complex numbers with $\sum_{n=1}^{\infty} |\alpha_n|^2 < \infty$. Now define

$$k_N(t,s) = \sum_{n=1}^N \alpha_n \phi_n(t)\overline{\phi_n(s)}, \qquad N = 1, 2, \ldots.$$

The function $k_N(t,s)$, $N = 1, 2, \ldots$, is a point in the Banach space $L_2(I \times I)$. Since

$$\|k_N - k_M\|^2 = \int_I \int_I |k_N(t,s) - k_M(t,s)|^2 \, dt \, ds$$

$$= \int_I \int_I \left| \sum_{n=M+1}^{N} \alpha_n \phi_n(t)\overline{\phi_n(s)} \right|^2 \, dt \, ds$$

$$= \sum_{n=M+1}^{N} |\alpha_n|^2,$$

it follows that $\{k_N\}$ is a Cauchy sequence in the complete space $L_2(I \times I)$. Hence, the sequence $\{k_N\}$ is convergent in $L_2(I \times I)$. Let

$$k(t,s) = \sum_{n=1}^{\infty} \alpha_n \phi_n(t)\overline{\phi_n(s)}.$$

Let K_N and K be the integral operators

$$(K_N x)(t) = \int_I k_N(t,s)x(s) \, ds, \quad t \in I,$$

$$(Kx)(t) = \int_I k(t,s)x(s) \, ds, \quad t \in I.$$

It was shown in Example 3, Section 6 that both K_N and K are bounded linear transformations of $L_2(I)$ into itself and

$$\|Kx\| \le \|k\| \cdot \|x\|, \qquad \|K_N x\| \le \|k_N\| \cdot \|x\|.$$

Since $K - K_N$ is an integral operator with kernel (or "unit impulse" response) $k(t,s) - k_N(t,s)$, one also has

$$\|(K - K_N)x\| \le \|k - k_N\| \cdot \|x\| \tag{5.8.1}$$

and

$$\lim_{N \to \infty} \|k - k_N\| = 0.$$

In other words, the sequence of operators $\{K_N\}$ converge to K in some sense. If we restrict x to satisfy $\|x\| \le 1$, we see from (5.8.1) that $\|(K - K_N)x\|$ converges to zero uniformly. Let us now make this precise. ∎

5.8.1 DEFINITION.   Let X and Y be normed linear spaces and let T be a bounded linear transformation of X into Y. We define the *norm of T* to be

$$\|T\| = \inf\{M : \|Tx\| \le M\|x\| \text{ for all } x \in X\}. \tag{5.8.2}$$

It follows from the definition of boundedness that $\|T\|$ is finite. Furthermore, one obviously has

$$\|Tx\| \le \|T\| \cdot \|x\|, \qquad x \in X. \tag{5.8.3}$$

We want to show that (5.8.2) defines a norm in the precise sense of Definition 5.2.1. However, before doing this it is convenient to note some alternate formulation of $\|T\|$.

5.8.2 LEMMA.   *Let X and Y be normed linear spaces and let $T \in Blt[X, Y]$. Then the norm $\|T\|$ agrees with all of the following:*

$$\|T\| = \sup\{\|Tx\| : \|x\| \leq 1\}, \tag{5.8.4a}$$

$$\|T\| = \sup\{\|Tx\| : \|x\| = 1\}, \tag{5.8.4b}$$

$$\|T\| = \sup\left\{\frac{\|Tx\|}{\|x\|} : x \neq 0\right\}. \tag{5.8.4c}$$

We leave the proof of this as an exercise. It should be noted that (5.8.4b) and (5.8.4c) are valid only when the space X is nontrivial.

Before we show that $\|T\|$ is a norm, let us give a physical interpretation of this number. Picture T as representing a physical system. Then the ratio $\|Tx\|/\|x\|$ can be viewed as the "gain" or "amplification" of T at the point x. Roughly speaking, we see by (5.8.4c) that $\|T\|$ is the "maximum" gain, or better, the least upper bound of the gain.

5.8.3 THEOREM.   *Let X and Y be normed linear spaces. Then Equation (5.8.2) defines a norm on $Blt[X, Y]$.*

Proof: Referring to the definition of a norm (Definition 5.2.1) we see that properties (N1) and (N4) obviously hold. We can prove (N3) by using Equation (5.8.4a).

$$\|\alpha T\| = \sup\{\|\alpha Tx\| : \|x\| \leq 1\} = \sup\{|\alpha| \|Tx\| : \|x\| \leq 1\}$$
$$= |\alpha| \sup\{\|Tx\| : \|x\| \leq 1\} = |\alpha| \|T\|.$$

Let us now prove (N2). Since we know that $\|Tx\| \leq \|T\| \cdot \|x\|$ and $\|Sx\| \leq \|S\| \cdot \|x\|$ for $T, S \in Blt[X, Y]$ one has

$$\|(T + S)x\| = \|Tx + Sx\| \leq \|Tx\| + \|Sx\| \leq (\|T\| + \|S\|)\|x\|.$$

It follows from (5.8.2) that $\|T + S\| \leq \|T\| + \|S\|$.   ∎

We have thus shown that $Blt[X, Y]$ is a normed linear space. It is important to note that the norm of T does depend on X and Y and the norms on these spaces. Furthermore, it should also be mentioned that the norm is generally not easy to compute. Oftentimes one has to be satisfied with only a crude estimate.

The next theorem shows that operator norms always have an additional property that norms in general do not have.

5.8.4 THEOREM.   *Let $T \in Blt[X, Y]$ and $S \in Blt[Y, Z]$ where X, Y and Z are normed linear spaces. Then the composition ST is in $Blt[X, Z]$ and*

$$\|ST\| \leq \|S\| \cdot \|T\|. \tag{5.8.5}$$

Proof: ST is obviously a linear transformation of X into Z. Since

$$\|STx\| \leq \|S\| \cdot \|Tx\| \leq \|S\| \cdot \|T\| \cdot \|x\|$$

we see that ST is bounded and that (5.8.5) holds. ∎

It is possible to define other norms on $Blt[X,Y]$. One can sometimes choose these norms so that both (5.8.3) and (5.8.5) are satisfied. (See Exercise 3.) In this book we shall always use (5.8.2), or one of its equivalent formulations, as the definition of the norm $\|T\|$. This is referred to as the *usual norm* on $Blt[X,Y]$.

Sequential convergence in $Blt[X,Y]$ with respect to the given norm is referred to by a special term in the literature.

5.8.5 DEFINITION. Let X and Y be normed linear spaces. A sequence $\{T_n\}$ in $Blt[X,Y]$ is said to *converge uniformly* to T in $Blt[X,Y]$ if $\lim_{n \to \infty} \|T - T_n\| = 0$. This is also referred to as *convergence in the uniform topology* or *convergence in the operator norm topology*.

The " uniformity " means that the convergence of $\{T_n\}$ to T is uniform over the unit ball $\{x \in X : \|x\| \leq 1\}$. Needless to say, Example 1 exhibits this kind of convergence.

One may ask whether $Blt[X,Y]$ is a Banach space. The answer, which we state here for reference, is proved in the exercises. (See Exercises 4 and 16.)

5.8.6. THEOREM. *Let X and Y be normed linear spaces. Then $Blt[X,Y]$, with the usual norm, is a Banach space (that is, complete) if Y is a Banach space.*

Referring to Theorem 3.14.3, the next result should not be too surprising.

5.8.7 THEOREM. *Let D be a dense linear subspace of a normed linear space X, and let $L: D \to Y$ be a bounded linear mapping of D into a Banach space Y. Then there is a unique bounded linear extension L_e of L defined everywhere on X, moreover, $\|L\| = \|L_e\|$.*

Proof: The first task is to define L_e. Let x be any point in X, and let $\{x_n\}$ be a sequence in D converging to x. We know one exists because D is dense in X. Then $\|Lx_n - Lx_m\| \leq \|L(x_n - x_m)\| \leq \|L\| \|x_n - x_m\|$, so $\{Lx_n\}$ is a Cauchy sequence in Y. Since Y is complete, let $y \in Y$ be the limit of $\{Lx_n\}$. It is not difficult to show that $y = L_e x$ defines a continuous linear mapping of X into Y such that $L_e x = Lx$ for all $x \in D$. We leave the remainder of the proof as an exercise. ∎

There is another form of convergence of operators which we will need later. Before defining it though let us consider an example.

EXAMPLE 2. Let $X = Y = L_2(-i\infty, i\infty)$ with the usual norm. If $x \in X$, we define $y = P_n x$, $n = 1, 2, 3, \ldots$ by

$$(P_n x)(i\omega) = \begin{cases} x(i\omega), & \text{for } |i\omega| \leq n \\ 0, & \text{otherwise,} \end{cases}$$

then for any $x \in X$, we have

$$\|Ix - P_n x\|^2 = \frac{1}{2\pi} \int_{|i\omega| > n} |x(i\omega)|^2 \, d\omega \to 0, \text{ as } n \to \infty, \qquad (5.8.6)$$

where I is the identity operator. So we see that $P_n x \to Ix$ for all $x \in X$. In some sense, then, the sequence $\{P_n\}$ converges to I. Let us now compute the operator norm $\|I - P_n\|$. Let x_n be any point in X such that $\|x_n\| = 1$ and $x_n(i\omega) = 0$ for $|i\omega| \leq n$. Then $P_n x_n = 0$, so $(I - P_n)x_n = x_n$. Since $\|(I - P_n)x_n\| = \|x_n\| = 1$, we have $\|I - P_n\| \geq 1$ for all n. (One can actually show that $\|I - P_n\| = 1$.) Clearly, it is impossible for $\{P_n\}$ to converge uniformly to I.

We see, then, that even though $\{P_n\}$ converges to I in the sense of (5.8.6), this sequence does not converge to I in the norm on $Blt[X,X]$. On the other hand, convergence in the sense of (5.8.6) is important, so we give it a name. ∎

5.8.8 DEFINITION.   Let X and Y be normed linear spaces. We shall say that a sequence $\{T_n\}$ in $Blt[X,Y]$ *converges strongly* to $T \in Blt[X,Y]$ if

$$\lim_{n \to \infty} \|(T_n - T)x\| = 0$$

for each x in X.

In other words, for each x in X the sequence $\{T_n x\}$ in Y converges (in Y) to Tx. This is obviously a direct generalization of the familiar concept of pointwise convergence of functions. The notation used to signify this type of convergence is

$$\lim_{n \to \infty} T_n = {}_s T, \qquad (5.8.7)$$

the "s" standing for "strong convergence."

In Example 2 we showed that $\lim_{n \to \infty} P_n = {}_s I$, but that $\{P_n\}$ did not converge to I in the operator norm topology. We see, then, that a sequence may converge strongly but not uniformly.

The next lemma shows that the converse is not possible, that is, uniform convergence always implies strong convergence. (See Figure 5.8.1.)

5.8.9 LEMMA.   *If the sequence* $\{T_n\}$ *in* $Blt[X,Y]$ *converges uniformly to* T *in* $Blt[X,Y]$, *then it converges strongly to* T.

Proof: Since $\|(T - T_n)x\| \leq \|T - T_n\| \cdot \|x\|$, we see that $\lim \|T - T_n\| = 0$ implies that $\lim \|(T - T_n)x\| = 0$ for each x in X. ∎

Needless to say, when we turn to infinite series of linear operators, we also have uniform and strong convergence. If the sequence of partial sums is strongly convergent, the series is strongly convergent. Similarly, uniform convergence of a series means uniform convergence of the sequence of partial sums. If a series $\sum_{n=1}^{\infty} T_n$

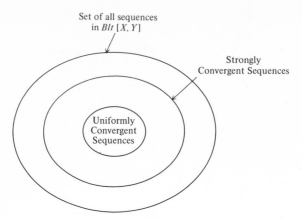

Figure 5.8.1.

converges strongly to T we shall write this as $\sum_{n=1}^{\infty} T_n =_s T$. Uniform convergence of a series shall be denoted in the usual fashion, namely $\sum_{n=1}^{\infty} T_n = T$.

We end this section with a few more examples.

EXAMPLE 3.   Let $H = C^n$, the finite-dimensional space made up of all ordered n-tuples of complex numbers $x = \{\xi_1, \xi_2, \ldots, \xi_n\}$ with the usual inner product

$$(x,y) = \xi_1 \bar{v}_1 + \xi_2 \bar{v}_2 + \cdots + \xi_n \bar{v}_n,$$

where $y = \{v_1, v_2, \ldots, v_n\}$ and $\bar{v}_i$ denotes the complex conjugate of v_i. The norm is given by $\|x\| = \sqrt{(x,x)}$.

Let L be the bounded linear transformation of H into itself represented by the matrix equation

$$y = Lx,$$

where we use the symbol L to denote an $n \times n$ matrix as well as the linear transformation itself. Then

$$\|Lx\|^2 = \bar{x}^{\mathrm{T}} \bar{L}^{\mathrm{T}} Lx,$$

where $\bar{L}^{\mathrm{T}}$ denotes the complex conjugate of the transpose of L. Since $\bar{L}^{\mathrm{T}}L$ is a positive (semidefinite) Hermitian matrix, the maximum of (5.8.2) over the unit ball exists and is equal to λ_1^2, where $\lambda_1^2 \geq \lambda_2^2 \geq \cdots \geq \lambda_n^2$ are the eigenvalues of the matrix $\bar{L}^{\mathrm{T}}L$. (The validity of this statement is shown in Chapter 6.) It follows that $\|L\| = |\lambda_1|$.

It should be mentioned that some authors define the norm of an $n \times n$ matrix by

$$\hat{\|L\|} = \left\{ \sum_{i,j=1}^{n} |l_{ij}|^2 \right\}^{1/2}.$$

It can be shown that this norm can be expressed in terms of the eigenvalues of $L^T L$ as

$$\hat{\|L\|} = \{\lambda_1{}^2 + \lambda_2{}^2 + \cdots + \lambda_n{}^2\}^{1/2}.$$

Obviously, $\hat{\|L\|}$ is rarely equal to $\|L\|$. ∎

EXAMPLE 4.   Suppose that we wish to approximate a pure time delay by a low pass filter whose transfer function is a ratio of polynomials, that is, a lumped parameter system. In particular, let $B = L_2(-\infty,\infty)$ and let S_τ be the delay by τ seconds operator defined by

$$(S_\tau x)(t) = x(t - \tau), \qquad \text{for } t \in (-\infty,\infty).$$

If $\mathscr{F}$ denotes the Fourier transform, then it is well-known that

$$S_\tau = \mathscr{F}^{-1} \Delta \mathscr{F},$$

where Δ denotes the mapping of $L_2(-i\infty,i\infty)$ into itself defined by

$$(\Delta X)(i\omega) = e^{-i\omega\tau} X(i\omega),$$

where $X = \mathscr{F}x$, that is, Δ is the operation of multiplication by the (transfer) function $e^{-i\omega\tau}$.

Now assume that the nth approximation to S_τ has a transfer function of the form

$$\Delta_n(i\omega) = k_n \frac{(i\omega + z_1) \cdots (i\omega + z_m)}{(i\omega + p_1) \cdots (i\omega + p_n)},$$

where $n > m$ are integers; k_n is a real number; $z_1, \ldots, z_m$ are complex numbers and zeros of $\Delta_n(s)$; $p_1, p_2, \ldots, p_n$ are complex numbers and poles of $\Delta_n(s)$. Assume further that the poles are in the left-hand plane and that the poles and zeros with nonzero imaginary parts occur in complex conjugate pairs, that is, if p is a pole (or zero), then $\bar{p}$ is a pole (or zero) also. Since $n > m$, it follows that

$$\lim_{\omega \to \pm\infty} |\Delta_n(i\omega)| = 0,$$

that is, $\Delta_n(i\omega)$ is a low pass filter. Then because

$$\text{ess sup } |e^{-i\omega\tau} - \Delta_n(i\omega)| \geq 1$$

it follows that

$$\|\Delta - \Delta_n\| \geq 1,$$

for $n = 1, 2, 3, \ldots$. Thus, no matter how we select a sequence of approximations $\{\Delta_n\}$, it cannot converge uniformly to Δ. On the other hand, there are many sequences $\{\Delta_n\}$ that converge strongly to Δ. ∎

EXAMPLE 5. (NEUMAN SERIES.)   Suppose B is a Banach space and L is a continuous linear mapping of B into itself. Further suppose that we are interested in the inverse, if one exists, of the transformation $\lambda I - L$, where λ is a complex number. We can carry out division formally as follows (see Example 1, Section 7):

$$\frac{1}{\lambda} + \frac{1}{\lambda^2}L + \frac{1}{\lambda^3}L^2 + \cdots + \frac{1}{\lambda^k}L^{k-1} + \cdots$$

$$\lambda I - L\overline{\big)I}$$

$$I - \frac{1}{\lambda}L$$

$$\overline{\frac{1}{\lambda}L}$$

$$\frac{1}{\lambda}L - \frac{1}{\lambda^2}L^2$$

$$\overline{\frac{1}{\lambda^2}L^2} \quad \text{and so on.}$$

Of course, this is just formal manipulation, and one has to show that it is meaningful in some sense or another. One thing we can do is show that the series converges to the inverse of $(\lambda I - L)$ if $|\lambda| > \|L\|$. In this case the series

$$\frac{1}{\lambda}\sum_{k=0}^{\infty}\frac{1}{\lambda^k}L^k,$$

which is called the *Neuman series*, is absolutely convergent since

$$\|L^k\|/|\lambda^k| \le \|L\|^k/|\lambda|^k < 1.$$

But B is complete, so $Blt[B,B]$ is also. It follows from Theorem 5.4.2 that the above series is convergent in $Blt[B,B]$. Let T denote the limit. We want to show that $T = (\lambda I - L)^{-1}$. Let

$$T_N = \frac{1}{\lambda}\sum_{k=0}^{N}\frac{1}{\lambda^k}L^k, \quad N = 1, 2, 3, \ldots.$$

Then

$$Tx = \lim_{N\to\infty} T_N x \quad \text{or} \quad T = {}_s\lim_{N\to\infty} T_N.$$

Since $(\lambda I - L)$ is continuous, one has

$$(\lambda I - L)Tx = \lim_{N\to\infty}(\lambda I - L)T_N x$$

$$= \lim_{N\to\infty}\left(I - \frac{1}{\lambda^{N+1}}L^{N+1}\right)x = x,$$

for all $x \in B$.

We have shown that $(\lambda I - L)T = I$. Similarly, we can show that $T(\lambda I - L) = I$, so $T = (\lambda I - L)^{-1}$.

We hasten to add that $|\lambda| > \|L\|$ is only a sufficient condition for $(\lambda I - L)^{-1}$ to exist and be continuous. It is not difficult to find cases where $\|L\| \geq |\lambda|$ and $(\lambda I - L)^{-1}$ is in $Blt[B,B]$. ∎

EXERCISES

1. Prove Lemma 5.8.2.

2. Consider R^2 with the norm $\|\cdot\|_p$, as in Example 1, Section 3. Let T be a matrix operator

$$T = \begin{pmatrix} a & b \\ c & d \end{pmatrix},$$

mapping $(R^2, \|\cdot\|_p)$ into $(R^2, \|\cdot\|_p)$.
(a) Compute $\|T\|$ when $b = c$ and $p = 2$.
(b) Compute $\|T\|$ in general.

3. Let M_n denote the space of real $n \times n$ matrices. For $A = (a_{ij}) \in M_n$ define

$$n(A) = \sum_{i,j} |a_{ij}|.$$

(a) Show here that $\|Ax\|_1 \leq n(A)\|x\|_1$, where $x \in R^n$ and R^n have the norm $\|x\|_1 = \sum_{i=1}^{n} |x_i|$.
(b) Let $A, B \in M_n$. Show that $n(AB) \leq n(A)n(B)$.
(c) Let $\|A\|$ be given by Definition 5.8.1 where $X = Y = R^n$ with the norm $\|x\|_1$. Compare $n(A)$ and $\|A\|$. When are they equal?

4. Prove Theorem 5.8.6. [*Hint*: Note that $\|(T_n - T_m)x\| \leq \|T_n - T_m\| \|x\|$. Use this to define T by $Tx = \lim T_n x$, where $\{T_n\}$ is a Cauchy sequence. Show that $T \in Blt[X,Y]$ and that $\|T_n - T\| \to 0$ as $n \to \infty$. See Exercise 16, below.]

5. In Example 2 show that for $N \neq M$ one has $\|P_N - P_M\| = 1$. (Thus $\{P_N\}$ is not a Cauchy sequence in the norm on $Blt[X,X]$.)

6. Let $\{T_n\}$ be a Cauchy sequence in the usual norm on $Blt[X,Y]$. Assume that $T \in Blt[X,Y]$ and that $\lim T_n = {}_s T$. Show that $\|T_n - T\| \to 0$ as $n \to \infty$, in other words, $\lim T_n = T$.

7. (Uniform Boundedness Principle.) Let $\{T_n\}$ be a sequence in $Blt[X,Y]$ where X is a Banach space. Assume that $\{\|T_n x\|\}$ is bounded for each x in X, that is,

$$\sup_n \|T_n x\| \leq B(x) < \infty,$$

where B depends on x. Now show that $\{\|T_n\|\}$ is bounded. [*Hint*: Let $A_m = \{x \in X : \|T_n x\| \leq m \text{ for all } n\}$. Show that A_m is closed and $X = \bigcup_{m=1}^{\infty} A_m$. Now apply Exercise 17, Section 3.13.]

8. Let $\{T_n\}$ be a sequence in $Blt[X,Y]$ where X is a Banach space. Assume that for each x in X, the limit $Tx = \lim_{n \to \infty} T_n x$ exists. Show that $T \in Blt[X,Y]$. [*Hint*: Use Exercise 7.]

9. Show that the Uniform Boundedness Principle (Exercise 7) fails if the domain X is not a Banach space. [*Hint*: Let $T_n = T$ be an appropriate unbounded operator.]

10. Explain why in Exercise 7 one does not have to assume that the space Y is a Banach space.

11. (Open Mapping Theorem.) The object of this exercise is to prove the following: Let $T \in Blt[X,Y]$, where X and Y are Banach spaces. If $T(X) = Y$, then T is an *open mapping*, that is, T maps open sets onto open sets. Let $B = \{x \in X : \|x\| < 1\}$ be the open unit ball in X, and $nB = \{nx : \|x\| \le 1\}$, where $n = 1, 2, \ldots$.
    (a) Show that if $T(nB)$ contains a nonempty open sphere, then T is an open mapping.
    (b) Show that if the closure $\overline{T(nB)}$ contains a nonempty open sphere, then T is an open mapping.
    (c) Show that $Y = \bigcup_{n=1}^{\infty} T(nB) = \bigcup_{n=1}^{\infty} \overline{T(nB)}$ and apply Exercise 17, Section 3.13.

12. (Continuation of Exercise 11.) Show that if $T \in Blt[X,Y]$ where X and Y are Banach spaces and if T is one-to-one with $T(X) = Y$, then T^{-1} is a bounded linear transformation.

13. (Closed Graph Theorem.) Let X and Y be Banach spaces and let $T: \mathcal{D}(T) \to Y$ be a linear transformation, where $\mathcal{D}(T)$ is a linear subspace of X. Let $Gr(T)$ be the graph of T in $X \times Y$. Show that if both $\mathcal{D}(T)$ and $Gr(T)$ are closed, then T is bounded. [*Hint*: Apply Exercises 11 and 12 to the mapping $S: (x, Tx) \to x$.]

14. Use the Closed Graph Theorem to show that if an unbounded linear operator is closed, its domain is not a Banach space.

15. Let $X = Y = L_2(-\infty, \infty)$. Show that the set C of all causal bounded linear operators (see Section 2.8) is a closed linear subspace $Blt[X,Y]$ with the usual norm. Similarly, show that the set (TI) of all time-invariant bounded linear operators is a closed linear subspace of $Blt[X,Y]$. Do the same thing for $X = Y = l_2$.

16. Prove the converse of Theorem 5.8.6 that is, show that if Y is not complete and $X \ne \{0\}$, then $Blt[X,Y]$ is not complete. [*Hint*: Let $\{y_n\}$ be a Cauchy sequence in Y that is not convergent. Show that there exists a Cauchy sequence $\{T_n\}$ in $Blt[X,Y]$ and a point x in X such that $T_n x = y_n$ for all n. Then show that the sequence $\{T_n\}$ is not uniformly convergent in $Blt[X,Y]$.]

17. Let X and Y be normed linear spaces, and let L_n, $n = 1, 2, \ldots$, be continuous linear mappings of X into Y. Assume that L is a mapping of X into Y. Further assume that for each $n = 1, 2, \ldots$ there exists a $M_n \ge 0$ such that

$$\|Lx - L_n x\| \le M_n \|x\|, \qquad \text{for all } x \in X.$$

Finally assume that $M_n \to 0$ as $n \to \infty$. Show that it follows that L is a continuous linear mapping of X into Y. (This exercise shows that if the sequence $\{L_n\}$ converges "uniformly" its limit must be continuous and linear.)

18. (Continuation of Exercise 17.) Assume that

$$\lim_{n \to \infty} \|L_n x - Lx\| = 0, \qquad \text{for all } x \in X,$$

that is, $\{L_n\}$ converges strongly to L. Show that L is linear.

19. Let B be the Banach space $L_2(-\infty,\infty)$, and let $L: B \to B$ be defined by $(Lx)(t) = kx(t + \tau)$, where $\tau > 0$ and $|k| < 1$. Consider the difference equation

$$(I + L)x = y.$$

Show that the solution x for a given y is

$$x(t) = y(t) - ky(t + \tau) + k^2 y(t + 2\tau) - k^3 y(t + 3\tau) + \cdots.$$

(It is interesting to compare this exercise with Example 3, Section 3.15.)

20. Extend Theorem 5.8.7 to the case where $L: D \to Y$ is a bounded linear mapping but Y is not assumed to be complete.

21. In Example 4, it was stated that there do exist sequences $\{\Delta_n\}$ that converge strongly to Δ. Construct such a sequence.

22. Show that the Neuman series in Example 5 converges uniformly provided $\|L\| < |\lambda|$.

23. Let $\mathcal{S}$ denote the collection of all operators $S: L_2(-\infty,\infty) \to L_2(-\infty,\infty)$ of the form

$$S = \alpha_1 S_{\tau_1} + \cdots + \alpha_n S_{\tau_n},$$

where $S_\tau: x(t) \to x(t - \tau)$ is the shift operator. Show that there are bounded linear time-invariant operators on $L_2(-\infty,\infty)$ that *cannot* be written as the uniform limit of operators in $\mathcal{S}$. [*Hint*: Apply the Fourier transform to $\mathcal{S}$ and show that $\mathcal{S}$ is mapped into the almost periodic functions in $L_\infty(-i\infty,i\infty)$. Now use the following two facts: (1) The collection of almost periodic functions is closed under uniform limits, compare with Besicovitch [1], and (2) there is a bounded linear time-invariant operator whose Fourier transform is not an almost periodic function, compare with Bochner [1, p. 144].]

9. EQUIVALENCE OF NORMED LINEAR SPACES

The reader should now be able to guess how one would define the concept of equivalence between normed linear spaces. Basically one wants a mapping that preserves both the algebraic and the topological structure. Let us make this precise.

5.9.1 DEFINITION. Two normed linear spaces X and Y are said to be *topologically isomorphic* if there exists a continuous linear transformation ϕ of X onto Y such that the inverse ϕ^{-1} exists and is continuous. In this case, the mapping ϕ is said to be a *topological isomorphism* of X onto Y.

A stronger form of equivalence is sometimes employed with normed linear spaces, and almost always employed with Hilbert spaces.

5.9.2 DEFINITION.   Two normed linear spaces X and Y are said to be *isometrically isomorphic* if there exists a linear transformation ϕ of X onto Y such that

$$\|\phi x\| = \|x\| \tag{5.9.1}$$

for all x in X. In this case, the mapping ϕ is said to be an *isometric isomorphism*.

Note that if (5.9.1) holds, then ϕ is bounded (that is, continuous), and ϕ^{-1} exists and is also an isometric isomorphism.

It is important that the reader not read too much into these equivalences. "Topologically isomorphic" signals the fact that algebraic and topological structures are essentially the same, but it says *nothing* about other structures that X and Y may have. Similarly, "isometrically isomorphic" says *nothing* about structures that X and Y may have in addition to their normed linear space structure. For example, one can consider the space of complex numbers as a real normed linear space. Then it is simple to show that C and R^2, where R^2 is a two-dimensional Euclidean space are isometrically isomorphic. One isometric isomorphism ϕ would be the mapping which takes the complex number z into the ordered pair (Re z, Im z). However, C has an operation of multiplication defined on it, whereas R^2 does not. Another example would be the normed linear space $\mathcal{M}^n$ of real-valued $n \times n$ matrices, where $M = \{m_{jk}\}$ denotes a point in $\mathcal{M}^n$ and

$$\|M\| = \left(\sum_{j,k} |m_{jk}|^2 \right)^{1/2}.$$

It is easily shown that $\mathcal{M}^n$ and R^{n^2}, where R^{n^2} is an n^2-dimensional Euclidean space, are isometrically isomorphic. However, there is an operation of matrix multiplication defined on $\mathcal{M}^n$ but not on R^{n^2}.

It is possible to give a necessary and sufficient condition for two normed linear spaces X and Y to be topologically isomorphic. Quite properly this condition is a condition on a linear transformation between X and Y.

5.9.3 THEOREM.   *Let* $(X, \|\cdot\|_X)$ *and* $(Y, \|\cdot\|_Y)$ *be two normed linear spaces. X and Y are topologically isomorphic if and only if there exists a linear transformation ϕ with domain X and range Y and two positive constants m and M such that*

$$m\|x\|_X \leq \|\phi x\|_Y \leq M\|x\|_X, \qquad x \in X. \tag{5.9.2}$$

Proof: If X and Y are topologically isomorphic, then there exists a continuous linear transformation ϕ of X onto Y with a continuous inverse $\phi^{-1} \colon Y \to X$. This means that

$$x = \phi^{-1}\phi x \quad \text{and} \quad y = \phi\phi^{-1}y$$

for all x in X and all y in Y, and there exist positive constants m and M such that

$$\|\phi x\|_Y \leq M \|x\|_X, \qquad x \in X$$

$$\|\phi^{-1}y\|_X \leq \frac{1}{m} \|y\|_Y, \qquad y \in Y.$$

Hence (5.9.2) holds.

Conversely, if there exists a linear transformation ϕ of X onto Y that satisfies (5.9.2), then ϕ is continuous because it is bounded. Furthermore, Theorem 5.7.1 guarantees that ϕ^{-1} exists and is continuous. Thus, X and Y are topologically isomorphic. ∎

EXAMPLE 1.   In Example 2, Section 7 we constructed a mapping T by $y = Tx$ where

$$y(i\omega) = T(i\omega)x(i\omega).$$

We showed that if $T(i\omega)$ satisfies

$$0 < b \le |T(i\omega)| \le B < \infty$$

for all ω, then T was a topological isomorphism of $L_2(-i\infty, i\infty)$ onto itself. ∎

EXAMPLE 2.   Let X denote the collection of all real solutions $\phi(t), 0 \le t \le 1$, of the differential equation $x'' - x = 0$. Thus, $\phi(t) = c_1 e^t + c_2 e^{-t}$ where c_1 and c_2 are real constants. Assume that X has the sup-norm $\|\cdot\|_\infty$. We define a mapping $T: X \to R^2$ by

$$T(c_1 e^t + c_2 e^{-t}) = (c_1, c_2).$$

It follows from the standard theory of differential equations that T is an isomorphism of X onto R^2. Let us now show that T is a topological mapping where R^2 has the norm $\|(c_1, c_2)\|_1 = |c_1| + |c_2|$. Since

$$\sup |c_1 e^t + c_2 e^{-t}| \le \sup |c_1| e^t + \sup |c_2| e^{-t} \le e(|c_1| + |c_2|),$$

where the sup is taken over $0 \le t \le 1$, we have

$$e^{-1} \|\phi\|_\infty \le \|T\phi\|_1,$$

so T is bounded below. In order to show that T is bounded above we solve the equations

$$\phi(0) = c_1 + c_2,$$
$$\phi(1) = c_1 e + c_2 e^{-1},$$

for c_1 and c_2. Using the fact that $|\phi(0)| \le \|\phi\|_\infty$ and $|\phi(1)| \le \|\phi\|_\infty$ this leads to the estimate

$$\|T\phi\|_1 = |c_1| + |c_2| \le \frac{4e}{e - e^{-1}} \|\phi\|_\infty.$$

Hence T is a topological isomorphism.

In this example, T is an isomorphism between two finite-dimensional normed linear spaces of the same dimension. It turns out that such mappings are always topological (see Theorem 5.10.5). ∎

EXAMPLE 3.   Let Y denote the complex linear space of all complex-valued polynomial functions of a real variable, that is, each y in Y can be written as

$$y(t) = a_0 + a_1 t + \cdots + a_n t^n,$$

where $n = 0, 1, 2, 3, \ldots$, the coefficients $\{a_0, \ldots, a_n\}$ are complex numbers and t is real. Define a norm on Y by

$$\|y\| = |a_0| + \cdots + |a_n|. \tag{5.9.3}$$

Let X be the linear subspace of Y consisting of all polynomials $x(t)$ that satisfy $x(0) = 0$, that is, the coefficient a_0 vanishes. Define the mapping $L: X \to Y$ by

$$Lx = L(a_1 t + \cdots + a_n t^n) = a_1 + 2a_2 t + \cdots + na_n t^{n-1},$$

that is, L is the differentiation operator.

Let us show that L is not continuous under the norm given by (5.9.3). For example, define the sequence of unit vectors $\{x_N\}$ by

$$x_N = \frac{1}{N} \{t + t^2 + \cdots + t^N\}, \qquad N = 1, 2, 3, \ldots .$$

Then

$$\|Lx_N\| = \left\| \frac{1}{N} (1 + 2t + \cdots + Nt^{N-1}) \right\| = \frac{1}{2} (N + 1) \to \infty, \qquad \text{as } N \to \infty,$$

so L is unbounded. L is obviously one-to-one. Let us now show that L maps X onto Y. Let $y = b_0 + b_1 t + \cdots + b_n t^n$ be an arbitrary point in Y. Since

$$L\left(b_0 t + \frac{b_1}{2} t^2 + \cdots + \frac{b_n}{n+1} t^{n+1} \right) = y,$$

we see that the range of L is all of Y. It follows that $L^{-1}: Y \to X$ exists. Since

$$L^{-1}y = L^{-1}(a_0 + \cdots + a_n t^n) = a_0 t + \cdots + \frac{a_n}{n+1} t^{n+1},$$

we have $\|L^{-1}y\| \le \|y\|$, so L^{-1} is continuous. Although L is an (algebraic) isomorphism of the linear space X onto the linear space Y, it is not a topological isomorphism, since L is not continuous. Nevertheless, X and Y are topologically isomorphic. We just need to consider another mapping. Indeed, the mapping ϕ defined by

$$\phi(a_1 t + \cdots + a_n t^n) = a_1 + a_2 t + \cdots + a_n t^{n-1}$$

with

$$\phi^{-1}(b_0 + b_1 t + \cdots + b_n t^n) = b_0 t + b_1 t^2 + \cdots + b_n t^{n+1}$$

is a topological isomorphism of X onto Y. As a matter of fact, ϕ is an isometric isomorphism.

Let us show yet another normed linear space that is isometrically isomorphic to Y. Let us consider l_1, that is, the normed linear space made up of all infinite sequences $z = \{z_1, z_2, z_3, \ldots\}$ of complex numbers such that $\sum_{i=1}^{\infty} |z_i| < \infty$ and with $\|z\| = \sum_{i=1}^{\infty} |z_i|$. Let W be the linear subspace of l_1 made up of all sequences which contain only a finite number of nonzero entries. We remark in passing that W is dense in l_1. We ask the reader to verify the mapping Ψ of Y onto W defined by

$$\Psi(a_0 + a_1 t + \cdots + a_n t^n) = \{a_0, a_1, \ldots, a_n, 0, 0, \ldots\}$$

is an isometric isomorphism of Y onto W. ∎

In Section 3.9 we studied families of metrics defined on the same underlying set, that is, (X, d_1) and (X, d_2). There we introduced the concept of equivalent metrics. Recall that two metrics are equivalent if and only if they generate the same topology. We shall say that two norms, $\|\cdot\|_a$ and $\|\cdot\|_b$, on a linear space X are *equivalent* if the metrics generated by these norms are equivalent. The following corollary shows that this equivalence can be characterized in a straightforward manner.

5.9.4 COROLLARY. *Let X be a linear space and let $\|\cdot\|_a$ and $\|\cdot\|_b$ be norms defined on X. These norms are equivalent if and only if there exist positive constants m and M such that*

$$m\|x\|_a \le \|x\|_b \le M\|x\|_a \tag{5.9.4}$$

for all x in X.

Proof: Let I be the identity mapping of $(X, \|\cdot\|_a)$ onto $(X, \|\cdot\|_b)$. Obviously, I^{-1} exists. The mappings I and I^{-1} are both continuous if and only if the topologies generated by $\|\cdot\|_a$ and $\|\cdot\|_b$ are the same. (Why?) But I and I^{-1} are also both continuous if and only if they are bounded, that is, if and only if (5.9.4) holds. (See Theorems 5.6.4 and 5.7.1.) ∎

EXAMPLE 4.   This is a continuation of Example 2. Let X denote the collection of all solutions $\phi(t)$, $0 \le t \le 1$, of the differential equation $x'' - x = 0$. Let $\|\phi\|_\infty$ denote the sup-norm and

$$\|\phi\|_2 = \left(\int_0^1 |\phi(t)|^2 \, dt \right)^{1/2}.$$

First we note that

$$\|\phi\|_2{}^2 = \int_0^1 |\phi(t)|^2 \, dt \le \int_0^1 \|\phi\|_\infty{}^2 \, dt = \|\phi\|_\infty{}^2.$$

If $\phi(t) = c_1 e^t + c_2 e^{-t}$, then it is easily shown that

$$\|\phi\|_2{}^2 = \tfrac{1}{2}(e^2 - 1)c_1{}^2 + 2c_1 c_2 + \tfrac{1}{2}(1 - e^{-2})c_2{}^2 \tag{5.9.5}$$
$$= f(c_1, c_2) \ge 0.$$

By standard analytic geometry we see that the level curves $f(c_1,c_2) = $ constant in the $c_1 c_2$-plane are ellipses centered at the origin. In particular we see that $f(c_1,c_2) > 0$ when $(c_1 c_2) \neq (0,0)$. [Another way to see the last fact is that if $f(c_1,c_2) = 0$ for some $(c_1,c_2) \neq (0,0)$, then e^t and e^{-t} would be linearly dependent over $0 \leq t \leq 1$. This is absurd.] Let

$$m^2 = \min\{f(c_1, c_2) \colon |c_1| + |c_2| = 1\}.$$

Since f is continuous on a compact set, this minimum exists and is positive. We then have

$$\|\phi\|_2{}^2 \geq m^2 = m^2(|c_1| + |c_2|)^2, \tag{5.9.6}$$

provided $|c_1| + |c_2| = 1$. We now claim that (5.9.6) holds for all (c_1,c_2). It clearly holds for $(c_1,c_2) = (0,0)$. So if $(c_1,c_2) \neq (0,0)$ and $\alpha = |c_1| + |c_2|$, then

$$\|\phi\|_2{}^2 = \int_0^1 |c_1 e^t + c_2 e^{-t}|^2 \, dt = \alpha^2 \int_0^1 \left| \frac{c_1}{\alpha} e^t + \frac{c_2}{\alpha} e^{-t} \right|^2 dt$$

$$= \alpha^2 f\left(\frac{c_1}{\alpha}, \frac{c_2}{\alpha} \right) \geq \alpha^2 m^2 = m^2(|c_1| + |c_2|)^2.$$

Finally, it follows from Example 2 that

$$\|\phi\|_\infty \leq e(|c_1| + |c_2|) \leq \frac{e}{m} \|\phi\|_2.$$

Hence $\|\phi\|_2$ and $\|\phi\|_\infty$ are equivalent on X. ∎

EXERCISES

1. Let X denote the collection of all real solutions ϕ of the differential equation

   $$x'' + bx' + cx = 0,$$

   where b and c are real constants. Assume that X has the sup-norm $\|\phi\|_\infty$. Show that X is topologically equivalent with R^2 with the norm $\|(c_1,c_2)\| = |c_1| + |c_2|$. [*Hint*: Distinguish between the three cases where the roots of $r^2 + br + c = 0$ are real and different, real and the same, and complex conjugates.]

2. (Continuation of Exercise 1.) Assume that coefficients b and c are complex and that X is the space of all complex solutions. Show that X is topologically equivalent with C^2.

3. (Continuation of Exercise 1.)
   (a) Show that the norm $\|\phi\|_2 = (\int_0^1 |\phi(t)|^2 \, dt)^{1/2}$ is equivalent to $\|\phi\|_\infty$ on X.
   (b) Show that these norms are *not* equivalent on the larger space $C[0,1]$.

4. Show that the norms $\|\cdot\|_p$, $1 \leq p \leq \infty$, are equivalent on R^n, see Example 1, Section 3. [*Hint*: See Section 3.11.]

5. (a) Show that $(R^2, \|\cdot\|_1)$ and $(R^2, \|\cdot\|_\infty)$ are isometrically isomorphic.
   (b) Show that $(R^3, \|\cdot\|_1)$ and $(R^3, \|\cdot\|_\infty)$ are topologically isomorphic but not isometrically isomorphic. [*Hint*: Sketch the unit ball $\{x: \|x\| \leq 1\}$ in each of these spaces.]

6. A Banach space B is said to be the *completion* of a normed linear space X if there is an isometric isomorphism $\phi: X \to B$ with the property that the range $\phi(X)$ is dense in B. Show that any two completions of a given normed linear space X are isometrically isomorphic. [*Hint*: Use Exercise 3, Section 6.]

7. (Continuation of Exercise 1, Section 3.14.) The following steps will lead to a proof that every normed linear space $(X, \|\cdot\|)$ has a completion. Let Y denote the collection of all Cauchy sequences $\{x_n\}$ from X.
   (a) Show that $n(\{x_n\}) = \lim \|x_n\|$ exists for every $\{x_n\}$ in Y.
   (b) Define a "new" equality on Y by saying that $\{x_n\} = \{x_n'\}$ if $n(\{x_n - x_n'\}) = 0$. Show that Y is a linear space in terms of this new equality and that $n(\cdot)$ is a norm on Y.
   (c) Show that $(Y, n(\cdot))$ is complete.
   (d) Show that the mapping $\phi: x \to (x, x, x, \dots)$ is an isometric isomorphism of X into Y and that $\phi(X)$ is dense in Y.

8. (a) Show that $l_2(0, \infty)$ and $l_2(-\infty, \infty)$ are isometrically isomorphic.
   (b) Show that $l_p(0, \infty)$ and $l_p(-\infty, \infty)$, $1 \leq p \leq \infty$, are isometrically isomorphic.

9. (a) Show that $L_2[0, 1]$ and $L_2[0, \infty]$ are topologically isomorphic. Are they isometrically isomorphic?
   (b) Let I and J be two nontrivial intervals. Show that $L_p(I)$ and $L_p(J)$, $1 \leq p \leq \infty$, are topologically isomorphic.

10. Define the mapping

$$U_\tau: f(t) \to f(\tau + t)$$

for $\tau \in R$.
   (a) Show that U_τ is an isometric isomorphism of $L_p(-\infty, \infty)$ onto $L_p(-\infty, \infty)$ for $1 \leq p < \infty$.
   (b) Is U_τ a topological isomorphism of $L_\infty(-\infty, \infty)$ onto $L_\infty(-\infty, \infty)$?

11. Let A be a bounded linear operator on $L_2(-\infty, \infty)$ and define U_τ by

$$U_\tau = e^{i\tau A} = \sum_{n=0}^{\infty} \frac{(i\tau)^n A^n}{n!}$$

as in Exercise 4, Section 4,
   (a) Show that U_τ is an isometric isomorphism on $L_2(-\infty, \infty)$.
   (b) Show that U_τ is an isometric isomorphism of $L_p(-\infty, \infty)$ onto $L_p(-\infty, \infty)$ for $1 \leq p < \infty$.
   (c) Show that if $\{\tau_n\}$ is a sequence in R with $\lim \tau_n = 0$, then $\lim U_{\tau_n} = {}_sI$, the identity.
   (d) What happens to the above for $p = \infty$?

12. Let Ω be a compact metric space and let $\pi: \Omega \times R \to \Omega$ be a continuous flow on Ω, that is, $\pi(x,0) = x$, $\pi(x,t+s) = \pi(\pi(x,t),s)$ and π is continuous. Assume there is a probability measure μ on Ω with the property that $\mu(A) = \mu(\pi(A,t))$ for all t and every measurable set A. Let $f: \Omega \to R$ be given and define the mapping U_τ, $-\infty < \tau < \infty$, by

$$(U_\tau f)(x) = f(\pi(x,\tau)).$$

(a) Show U_τ is an isometric isomorphism of $L_p(\Omega,\mu)$ onto itself for $1 \le p < \infty$.

(b) Show that if $\{\tau_n\}$ is a sequence in R with $\lim \tau_n = 0$, then $\lim U_{\tau_n} = {}_sI$.

(c) What happens to the above for $p = \infty$?

(d) What happens on $C(\Omega,R)$ with the sup-norm?

13. Let X and Y be normed linear spaces. Show that there may exist an isometry $\phi: X \to Y$ that is not linear. [*Hint*: See Exercise 11, Section 3.10.]

14. Let X and Y be topologically isomorphic normed linear spaces. Show that X is complete if and only if Y is complete.

15. (a) Characterize the family of all topological isomorphisms of the real space R^2 onto itself.

(b) Show that if R^2 has the Euclidean norm $\|x\|_2 = (|x_1|^2 + |x_2|^2)^{1/2}$, then every isometric isomorphism of R^2 onto itself is a matrix operator of the form

$$\begin{pmatrix} \cos\theta & \sin\theta \\ -\sin\theta & \cos\theta \end{pmatrix}$$

for some θ.

10.   FINITE-DIMENSIONAL SPACES

One should not be surprised to learn that the theory of finite-dimensional normed linear spaces is much simpler than that of general normed linear spaces. In this section we shall explore this simplification. We shall show, among other things, that all finite-dimensional normed linear spaces are Banach spaces (that is, complete), every linear subspace is closed, boundedness implies total boundedness, and all linear transformations are continuous. First, however, we need to prove a lemma regarding the representation of a vector in terms of a basis.

If $\{x_1,\dots,x_n\}$ is a (Hamel) basis[6] in a finite-dimensional normed linear space X, we know from Section 4.7 that each x in X can be expressed uniquely in the form

$$x = \alpha_1 x_1 + \cdots + \alpha_n x_n,$$

where the α_i's are scalars. The following lemma shows that each coefficient α_i is a continuous linear function of x. We shall use this fact repeatedly below.

---

[6] The algebraic concept of a Hamel basis for a linear space was introduced in Section 4.7. As long as we are discussing finite-dimensional normed linear spaces this is usually the concept of basis employed. Although every infinite-dimensional normed linear space has a Hamel basis, this purely algebraic concept is of limited usefulness. There are concepts of basis for infinite-dimensional normed linear spaces that do combine topological and algebraic structure; however, with one extremely important exception, they are not discussed in this book. The exception, which is discussed in Section 17, is the concept of an orthonormal basis for a Hilbert space.

5.10.1 LEMMA. *Let X be a finite-dimensional normed linear space and let $\{x_1, \ldots, x_n\}$ be a basis for X. Then each coefficient α_i, $1 \leq i \leq n$, in the expansion*

$$x = \alpha_1 x_1 + \cdots + a_n x_n \tag{5.10.1}$$

is a continuous linear function of x. In particular, there is a constant M such that $|\alpha_i| \leq M\|x\|$ for $1 \leq i \leq n$ and all $x \in X$.

Proof: Let $l_i \colon X \to F$ be the mapping of X into the scalar field F given by $\alpha_i = l_i(x)$, $1 \leq i \leq n$. It is easy to see that l_i is linear. We shall now show that such a constant M exists.

If we can show that there is an $m > 0$ such that

$$m(|\alpha_1| + \cdots + |\alpha_n|) \leq \|x\| = \|\alpha_1 x_1 + \cdots + \alpha_n x_n\|, \tag{5.10.2}$$

then it follows that $|\alpha_i| \leq m^{-1} \|x\|$.

Let us first prove (5.10.2) for sets of coefficients $\{\alpha_1, \ldots, \alpha_n\}$ that satisfy the condition $|\alpha_1| + \cdots + |\alpha_n| = 1$. (See Figure 5.10.1.) Let

$$A = \{(\alpha_1, \ldots, \alpha_n) \in F^n \colon |\alpha_1| + \cdots + |\alpha_n| = 1\}.$$

Vectors $x = \alpha_1 x_1 + \alpha_2 x_2$ on
the lines connecting the
four vectors $x_1, x_2, -x_1, -x_2$ are
those such that $|\alpha_1| + |\alpha_2| = 1$

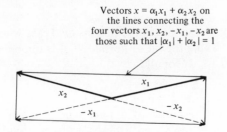

Figure 5.10.1.

It follows from Theorem 3.17.20 that A is a compact set in F^n, where F^n has the norm $\|\alpha\|_1 = \sum_{i=1}^{n} |\alpha_i|$.

Let $f \colon A \to R$ be given by[7]

$$f(\alpha_1, \ldots, \alpha_n) = \|\alpha_1 x_1 + \cdots + \alpha_n x_n\|.$$

This mapping f is continuous (see Exercise 3) and $f \geq 0$. Let

$$m = \inf\{f(\alpha_1, \ldots, \alpha_n) \colon (\alpha_1, \ldots, \alpha_n) \in A\}.$$

It follows from Theorem 3.17.21 that there is a point $(\alpha_1{}^0, \ldots, \alpha_n{}^0)$ in A with $f(\alpha_1{}^0, \ldots, \alpha_n{}^0) = m$. (See Figure 5.10.2.) If $m = 0$, then $\alpha_1{}^0 x_1 + \cdots + \alpha_n{}^0 x_n = 0$ and this contradicts the fact that $\{x_1, \ldots, x_n\}$ is a basis. It then follows that $m > 0$ and (5.10.2) holds for this value of m and for $(\alpha_1, \ldots, \alpha_n)$ in A.

[7] Compare this argument with Example 4, Section 9.

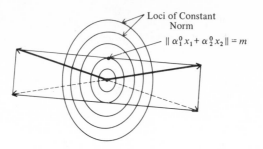

Figure 5.10.2.

For an arbitrary set of coefficients $(\alpha_1,\ldots,\alpha_n)$, set $\beta = |\alpha_1| + \cdots + |\alpha_n|$. If $\beta = 0$, (5.10.2) obviously holds. If $\beta \neq 0$, then

$$\|\alpha_1 x_1 + \cdots + \alpha_n x_n\| = \beta \left\| \frac{\alpha_1}{\beta} x_1 + \cdots + \frac{\alpha_n}{\beta} x_n \right\|$$

$$= \beta f\left(\frac{\alpha_1}{\beta}, \ldots, \frac{\alpha_n}{\beta}\right) \geq m\beta$$

$$\geq m(|\alpha_1| + \cdots + |\alpha_n|). \quad \blacksquare$$

The next example shows an infinite-dimensional case where all the functionals l_i are continuous but where the set of norms $\{\|l_i\|\}$ is unbounded. That is, there is no $M \geq 0$ such that $\|l_i\| \leq M$ for all i.

EXAMPLE 1. Suppose a sampled-data random process $\{x(1),x(2),x(3),\ldots\}$ can be modeled as the response of a filter whose input is a random process $\{n(0),n(1),n(2),\ldots\}$, where the n's are stochastically independent, $E\{n(i)\} = 0$, and $E\{|n(i)|^2\} = 1$ for all i. In particular, we will now suppose that $x(1) = n(0)$ and $x(k + 1) = x(k) + a(k)n(k)$ for $k = 1, 2, \ldots$, where $\{a(k)\}$ is a sequence of real numbers such that $0 < a(k) \leq 1$, $a(k + 1) < a(k)$ for all k and $\lim_{k \to \infty} a(k) = 0$. We can view this random process as a sequence in the normed linear space X made up of all complex-valued random variables x such that $E\{|x|^2\} < \infty$ and $E\{x\} = 0$ with $\|x\| = \sqrt{E\{|x|^2\}}$.

Let M denote the linear subspace of X made up of all finite linear combinations of vectors in the set $B = \{x(1),x(2),\ldots\}$. It should be clear that B is a Hamel basis for M and, consequently, M is infinite dimensional. The linear functional $\{l_k\}$ can be defined by $l_k(x(j)) = \delta_{kj}$, where δ_{kj} is the Kronecker function. One then has

$$l_k(n(j)) = \begin{cases} 0, & k \neq j \text{ or } k \neq j + 1, \\ -1/a(k), & k = j, \\ 1/a(k), & k = j + 1. \end{cases}$$

It can be shown that each l_k is continuous with $\|l_k\| = 1/a(k - 1)$. However, $\|l_k\| \to \infty$ as $k \to \infty$, because $a(k) \to 0$ as $k \to \infty$. $\quad \blacksquare$

The last example should not be misinterpreted. There are plenty of infinite-dimensional examples in which one has $\|l_k\| \leq M$ for all k. It simply does not occur in Example 1.

EXAMPLE 2. (CONTINUATION OF EXAMPLE 1.)   Let $\{y(0),y(1),y(2),\ldots\}$ be the set of random variables, where $y(0) = n(0)$, $y(k) = \alpha_k n(0) + \beta_k n(k)$, and $\alpha_k{}^2 + \beta_k{}^2 = 1$. Moreover, $\beta_k > \beta_{k+1} > 0$ and $\lim_{k \to \infty} \beta_k = 0$. It should be clear that the set $\{y(k)\}$ is linearly independent.

Let M be a subspace defined as above in Example 1. Then the linear functional l_1 such that $l_1(y(0)) = 1$ and $l_1(y(k)) = 0$ for $k = 1, 2, \ldots$ is not continuous. This follows from the fact that $\lim_k y(k) = y(0)$. ∎

The first thing to be said about finite-dimensional normed linear spaces is that they are always complete.

5.10.2 THEOREM.   *If a normed linear space X is finite dimensional, it is a Banach space.*

Proof: Let $\{x_1,x_2,\ldots,x_n\}$ be any basis for X and let $\{z_k\}$ be any Cauchy sequence in X. We must show that the sequence $\{z_k\}$ is convergent. Let

$$z_k = \alpha_{k1}x_1 + \cdots + \alpha_{kn}x_n$$

for $k = 1, 2, \ldots$. It follows from Lemma 5.10.1 that there is a constant M such that

$$|\alpha_{kj} - \alpha_{lj}| \leq M\|z_k - z_l\|, \qquad 1 \leq j \leq n.$$

Hence each sequence of scalars $\{\alpha_{kj}\}$ is a Cauchy sequence. Since the scalar field (the real or complex numbers) is complete, the sequence $\{\alpha_{kj}\}$ is convergent. Let $\alpha_{0j} = \lim_{k \to \infty} \alpha_{kj}, j = 1, 2, \ldots, n$. If we let $z_0 = \alpha_{01}x_1 + \cdots + \alpha_{0n}x_n$, then

$$\|z_k - z_0\| = \|(\alpha_{k1} - \alpha_{01})x_1 + \cdots + (\alpha_{kn} - \alpha_{0n})x_n\|$$
$$\leq |\alpha_{k1} - \alpha_{01}|\,\|x_1\| + \cdots + |\alpha_{kn} - \alpha_{0n}|\,\|x_n\|.$$

It quickly follows that $\{z_k\}$ converges to z_0. ∎

5.10.3 THEOREM.   *Let M be a finite-dimensional linear subspace of a normed linear space X. Then M is closed.*

In particular, then, a linear subspace M of a finite-dimensional normed linear space X is always closed.

Proof: It follows from Theorem 5.10.2 that every Cauchy sequence in M converges and has its limit in M. Since every convergent sequence is a Cauchy sequence, it follows that every convergent sequence in M has its limit *in M*. By applying the Closed Set Theorem, we see that M is closed. ∎

It is undoubtedly true that one of the major simplifications arising from finite dimensionality is that *all* linear transformations are continuous.

5.10.4 THEOREM.   *Let $L: X \to Y$ be a linear operator where X and Y are normed linear spaces. If X is finite dimensional, then L is continuous.*

Note that Y need not be finite dimensional.

Proof: Let $\{x_1, \ldots, x_n\}$ be a Hamel basis for X. Then any point in x in X can be uniquely expressed as $x = \alpha_1 x_1 + \cdots + \alpha_n x_n$, where the α's are scalars. Let $D = \max\{\|Lx_i\| : 1 \leq i \leq n\}$. Then

$$\|Lx\| = \|\alpha_1 L x_1 + \cdots + \alpha_n L x_n\| \leq |\alpha_1| \|L x_1\| + \cdots + |\alpha_n| \|L x_n\|$$
$$\leq D\{|\alpha_1| + \cdots + |\alpha_n|\}.$$

But from Lemma 5.10.1 there exists a constant M such that

$$|\alpha_1| + \cdots + |\alpha_n| \leq M\|x\|,$$

which implies that

$$\|Lx\| \leq DM\|x\|.$$

Hence L is bounded or, equivalently, continuous. (Carefully note how this proof uses the fact that X is finite dimensional.)   ∎

We know from Theorem 4.7.5 that two linear spaces (over the same vector field) are isomorphic if and only if they have the same (algebraic) dimension. In Section 9 we introduced the two concepts of topological and isometric isomorphisms. In general it is not true that those latter two forms of equivalence are determined simply by dimension. However, one can say the following for the finite-dimensional case.

5.10.5 THEOREM.   *Let $L: X \to Y$ be an isomorphism of X onto Y where X is finite dimensional. Then L is a homeomorphism, that is, L is a topological isomorphism. In particular, two finite-dimensional normed linear spaces (over the same scalar field) are topologically isomorphic if and only if they have the same dimension.*

This is a direct consequence of Theorem 5.10.4. Note that we say "topologically isomorphic." Two finite-dimensional spaces of the same dimension are not necessarily isometrically isomorphic. (See the Exercise 1.)

The following is a very easily proven theorem.

5.10.6 THEOREM.   *Let $\|\cdot\|_1$ and $\|\cdot\|_2$ be two norms on a finite-dimensional linear space X. Then $\|\cdot\|_1$ and $\|\cdot\|_2$ are equivalent.*

Let us now turn to a topological characterization of the algebraic concept of finite dimensionality.

5.10.7 THEOREM. *Let X be a normed linear space and let $D = \{x \in X \colon \|x\| \leq 1\}$. Then X is finite dimensional if and only if D is compact.*

Proof: First assume that X is finite dimensional and let $D = \{x \colon \|x\| \leq 1\}$. Let $\{z_k\}$ be a sequence in D. We want to show that there is a convergent subsequence with limit in D. Let $\{x_1, \ldots, x_n\}$ be a basis for X and let

$$z_k = \alpha_{k1} x_1 + \cdots + \alpha_{kn} x_n.$$

It follows from Lemma 5.10.1 that there is a constant M such that

$$|\alpha_{k1}| + \cdots + |\alpha_{kn}| \leq M \|z_k\| \leq M.$$

Since the coefficients $(\alpha_{k1}, \ldots, \alpha_{kn})$ lie in a closed bounded set in F^n, we can find a subsequence that converges in F^n, say that

$$(\alpha_{k'1}, \ldots, \alpha_{k'n}) \to (\alpha_{01}, \ldots, \alpha_{0n}).$$

One then has $z_{k'} \to z_0 = \alpha_{01} x_1 + \cdots + \alpha_{0n} x_n$. Since $\|z_0\| = \lim \|z_{k'}\| \leq 1$, we see that $z_0 \in D$. Hence D is compact.

Now assume that D is compact. We shall show that X has finite dimension by contradiction. Let $x_1 \in X$ with $\|x_1\| = 1$. Let M_1 be the linear space generated by $\{x_1\}$. If $\dim X \geq 2$, then by the Riesz Theorem (5.5.4) we can find a vector $\{x_2\}$ in X such that $\|x_2\| = 1$ and $\|x_1 - x_2\| \geq \frac{1}{2}$. We now proceed by induction. Assume that we have chosen vectors $\{x_1, \ldots, x_n\}$ with the property that $\|x_i\| = 1$, $1 \leq i \leq n$, and $\|x_i - x_j\| \geq \frac{1}{2}$ for $i \neq j$. Let M_n be the linear space generated by $\{x_1, \ldots, x_n\}$. If $\dim X \geq n + 1$, then we can find a vector x_{n+1} in X such that $\|x_{n+1}\| = 1$ and $\|x_{n+1} - x_i\| \geq \frac{1}{2}$ for $1 \leq i \leq n$. In this way, if $\dim X$ is not finite, we can construct a sequence $\{x_1, x_2, \ldots\}$ in D with the property that $\|x_i - x_j\| \geq \frac{1}{2}$ for $i \neq j$. But such a sequence has no convergent subsequence and we have contradicted the fact that D is compact. ∎

Recall that a metric space is said to be separable if it contains a countable set that is dense in the metric space. It is easy to see that any finite-dimensional normed linear space is separable. Indeed, if $\{x_1, x_2, \ldots, x_n\}$ is a basis, the set made up of all x's of the form $x = r_1 x_1 + \cdots + r_n x_n$, where the r's are rational or have rational real and imaginary parts, is a countable dense set.

EXAMPLE 3. Let X and Y be normed linear spaces, and let $L \colon X \to Y$ be a continuous linear transformation. Suppose that the range of L, $\mathscr{R}(L)$, is finite dimensional. Y itself may be infinite dimensional. Then L has an interesting property, namely, it maps bounded sets in X into compact sets in Y. In order to see this let A be any bounded set in X. It follows that there is an $r > 0$ such that $A \subset B_r[0]$, the closed ball of radius r. Since L is continuous, $L(A) \subset B_{r'}[0]$, where $r' = \|L\| r$. But $L(A) \subset \mathscr{R}(L) \cap B_{r'}[0]$, and from Theorem 5.10.7 the set $\mathscr{R}(L) \cap B_{r'}[0]$ is compact. Thus L maps bounded sets into compact sets. ∎

We can conclude this section by observing that things are better with finite-dimensional normed linear spaces(!)

EXERCISES

1. Let X and Y be finite-dimensional normed linear spaces with the same dimension. It follows from Theorem 5.10.5 that they are topologically isomorphic. This does not mean that they are isometrically isomorphic, as you are now asked to show. Let $X = Y = R^2$ and define the two norms

$$\|x\|_1 = |x_1| + |x_2| \quad \text{and} \quad \|x\|_2 = (|x_1|^2 + |x_2|^2)^{1/2}.$$

Show that $(X, \|\cdot\|_1)$ is not isometrically isomorphic with $(X, \|\cdot\|_2)$.

2. Show that a normed linear space X is finite dimensional if and only if it has the property that every closed, bounded set is compact.

3. Show that the function $f(\alpha_1, \ldots, \alpha_n)$, defined in the proof of Lemma 5.10.1, is continuous. [*Hint*: Show that

$$|f(\alpha_1, \ldots, \alpha_n) - f(\beta_1, \ldots, \beta_n)| \leq M(|\alpha_1 - \beta_1| + \cdots + |\alpha_n - \beta_n|)$$

for some M.]

4. Prove Theorem 5.10.6.

5. Let M and N be linear subspaces of a normed linear space X and define

$$\delta(M,N) = \inf\{\|x - y\| : x \in M, y \in N, \|x\| = \|y\| = 1\}.$$

Show that if X is finite dimensional, then $\delta(M,N) > 0$ if and only if $M \cap N = \{0\}$.

6. Show that the conclusion of Exercise 5 is false in infinite-dimensional spaces.

7. Let X be a normed linear space with the property that every linear mapping $L: X \to Y$ is continuous. Show that X is finite dimensional.

11.   NORMED CONJUGATE SPACE AND CONJUGATE OPERATOR

In Section 4.12 we discussed the algebraic conjugate of a linear space. Recall that the algebraic conjugate X^f of a linear space X is the linear space made up of all linear functionals on X. In the case of normed linear spaces one usually restricts attention to a linear subspace $X' \subset X^f$, called the normed conjugate or, simply, conjugate space. This linear subspace of X^f is the one made up of all continuous linear functionals. Since every linear transformation defined on a finite-dimensional space is continuous, one has $X' = X^f$ when X is finite dimensional. However, in general X' is a proper linear subspace of X^f.

In Section 4.13 we discussed the transpose of a linear transformation, where the transpose $L^T: Y^f \to X^f$ of a linear transformation $L: X \to Y$ is defined by

$$\langle x, L^T y' \rangle = \langle Lx, y' \rangle$$

for all $x \in X$ and $y' \in Y^f$. The conjugate L' of a continuous linear operator $L: X \to Y$ is the continuous linear transformation $L': Y' \to X'$ defined by

$$\langle x, L'y' \rangle = \langle Lx, y' \rangle$$

for all $x \in X$ and $y' \in Y'$. Note the difference between L' and L^T. L^T is defined for all linear transformations and L' is defined for continuous ones. L^T maps Y^f into X^f and L' maps Y' into X'. If L is continuous, L' is defined and is the restriction of L^T to Y'. Of course, if X and Y are finite dimensional, the $L' = L^T$.

There is a great deal that can be said about normed conjugate spaces and conjugate transformations. Moreover, what can be said is very useful in applications. We shall discuss only the Hilbert space versions of these concepts in Sections 21 and 22.

Part B

Hilbert Spaces

12. INNER PRODUCT SPACES AND HILBERT SPACES

A nice thing about normed linear spaces is that their geometry is much like the familiar two- and three-dimensional Euclidean geometry. Inner product spaces and Hilbert spaces are even nicer because their geometry is even closer to Euclidean geometry. In particular, these latter cases include the concept of orthogonality or perpendicularity. This " nicer " structure of inner product and —especially—Hilbert spaces leads to remarkable simplifications.

We begin with a definition of inner product.

5.12.1 DEFINITION.   Let X be a complex linear space. An *inner product* on X is a mapping that associates to each ordered pair of vectors x, y a scalar, denoted (x,y), that satisfies the following properties:

(IP1) $(x + y, z) = (x,z) + (y,z)$;   (Additivity)
(IP2) $(\alpha x,y) = \alpha(x,y)$;   (Homogeneity)
(IP3)[8] $(x,y) = \overline{(y,x)}$;   (Symmetry)
(IP4) $(x,x) > 0$, when $x \neq 0$.   (Positive Definiteness).

These four properties are to hold for any vectors x, y, z in X and any scalar α. It should be noted that the properties:

(IP5) $(x, y + z) = (x,y) + (x,z)$;
(IP6) $(x,\alpha y) = \bar{\alpha}(x,y)$;

are immediate consequences of (IP1), (IP2), and (IP3). Furthermore, it may appear from (IP4) that we are tacitly assuming (x,x) to be the real. However, (IP3) implies that (x,x) is indeed real. Also note that $(0,0) = (0,x) = (x,0) = 0$ for all x in X.

5.12.2 LEMMA.   *If $(x, y) = 0$ for all $y \in X$, then $x = 0$.*

Proof: $(x,x) = 0$ implies $x = 0$.   ∎

The notion of an inner product on a real linear space is defined similarly. In this case the range of the mapping (x,y) is in the real numbers, that is, the scalar field. Also, (IP3) and (IP6) are simplified. That is, (IP3) becomes $(x,y) = (y,x)$, and (IP6) becomes $(x,\alpha y) = \alpha(x,y)$. It turns out that in the majority of situations one uses complex inner product spaces. *Therefore, unless stated otherwise, all spaces are henceforth complex.* We also caution the reader that a few of the following results in this chapter are not valid when real spaces are used.

---

[8] As usual the bar $\overline{}$ denotes the complex conjugate.

272

The archetype for a real inner product is the dot product of classical vector analysis,

$$(x,y) = x \cdot y = \sum_{i=1}^{3} x_i y_i.$$

Other examples of inner products are presented in the next section.

5.12.3 DEFINITION.   An *inner product space* is defined to be a linear space X together with an inner product defined on X.

Our first task is to show how the inner product generates a norm. Specifically, we shall show that the function

$$\|x\| = (x,x)^{1/2} \tag{5.12.1}$$

defines a norm on X. But before doing this, we need an important inequality.

5.12.4 LEMMA. (SCHWARZ INEQUALITY.)   *Let* (x,y) *be an inner product on a linear space* X. *Then*

$$|(x,y)| \leq \|x\| \, \|y\|, \tag{5.12.2}$$

where $\|x\|$ *and* $\|y\|$ *are defined by* (5.12.1).

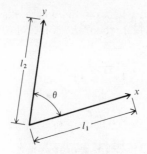

Figure 5.12.1.

In the familiar two-dimensional vector analysis case shown in Figure 5.12.1 we know that for $l_1 = \|x\|$ and $l_2 = \|y\|$ one has $x \cdot y = l_1 l_2 \cos \theta$ and obviously $|x \cdot y| \leq l_1 l_2$. The Schwarz Inequality is simply a generalization of this familiar geometric fact to inner product spaces.

Proof: We will assume that X is a linear space over C, the complex numbers. The same argument applies to linear spaces over R. If either x or y is the origin. the Schwarz Inequality is obviously true. So assume $x \neq 0$ and $y \neq 0$. If α is any complex number, then

$$0 \leq (x - \alpha y, x - \alpha y).$$

In particular, if $\alpha = (x,y)/(y,y)$, then

$$0 \le (x - \alpha y, x - \alpha y) = \|x\|^2 - \frac{|(x,y)|^2}{\|y\|^2}. \quad \blacksquare$$

One might suspect from Figure 5.12.1 that $|(x,y)| = \|x\| \, \|y\|$ if and only if x and y are collinear, that is, the set $\{x,y\}$ is linearly dependent. Inspection of the above proof shows that this is indeed the case. This simple observation has an amazing number of applications, particularly in optimization.

EXAMPLE 1.   Let $X = L_2[0,T]$ and let F be the filter mapping X into itself defined by

$$(Fx)(t) = \int_0^t e^{-(t-\tau)} x(\tau) \, d\tau, \qquad t \in [0,T].$$

Let $(x,y) = \int_0^T x(t) \, \overline{y(t)} \, dt$ denote the inner product on X.

Suppose we want to choose an input $x \in X$ such that $(x,x) = 1$ and $(Fx)(T)$ is maximum. Let $y(t) = e^t$, then

$$(Fx)(T) = e^{-T}(y,x).$$

So from the Schwarz Inequality

$$|(Fx)(T)| \le e^{-T} \|y\| \, \|x\|$$

with the equality being taken when $x = cy$, where c is a constant, that is, when x and y are collinear. In particular, then, the solution of our problem is

$$x(t) = \sqrt{\frac{2}{[e^{2T} - 1]}} \, e^t, \qquad t \in [0,T].$$

Moreover,

$$(Fx)(T) = e^{-T}\sqrt{\tfrac{1}{2}[e^{2T} - 1]} = \sqrt{\tfrac{1}{2}[1 - e^{-2T}]}. \quad \blacksquare$$

We now want to show that $\|x\|$ is a norm on X.

5.12.5 THEOREM.   *Let X be a linear space with inner product (x,y) and define $\|x\|$ by (5.12.1). Then $\|x\|$ is a norm on X.*

Proof: It is very easy to see that $\|x\|$ satisfies properties (N1), (N3), and (N4) for a norm, and all one needs to prove is that $\|x\|$ satisfies the triangle inequality, property (N2). We do this as follows:

$$0 \le \|x + y\|^2 = (x + y, x + y) = \|x\|^2 + 2 \, \mathrm{Re} \, (x,y) + \|y\|^2$$
$$\le \|x\|^2 + 2|(x,y)| + \|y\|^2,$$

and using the Schwarz Inequality we get

$$\|x + y\|^2 \le \|x\|^2 + 2\|x\| \, \|y\| + \|y\|^2 = (\|x\| + \|y\|)^2. \quad \blacksquare$$

Note that $\|x + y\| = \|x\| + \|y\|$ or $\|x + y\| = |\,\|x\| - \|y\|\,|$ if and only if x and y are collinear.

Since every inner product space has a norm, it is a metric space. We therefore adopt the following convention:

> *Whenever one discusses the topological properties of an inner product space, this is in reference to the metric defined by*

$$d(x,y) = \{(x - y, x - y)\}^{1/2}.$$

Given this convention, the first thing to note is that the inner product is a continuous mapping of the product space $(X,d) \times (X,d)$ into the scalar field. We can also now ask if a given inner product space is complete. Some are and some are not, so we make the following definition.

5.12.6 DEFINITION.   A *Hilbert space* is a complete inner product space.

At this point the reader is probably willing to grant that, all else being equal, completeness is a desirable property for an inner product space to have. Actually, the case is much stronger. We shall see that Hilbert spaces have significantly "better" geometric structure than inner product spaces do in general. So carefully note in the rest of this chapter which theorems are true for inner product spaces and which require completeness.

An important fact which we note in passing is that every inner product space has a completion. The concept of the completion of a normed linear space was defined in Section 9. The same concept is applicable in the case of inner product spaces. We shall discuss this further in the exercises in Sections 19 and 21.

One fact of Euclidean geometry is the Parallelogram Law, which we illustrate in Figure 5.12.2. The next theorem shows that the parallelogram law holds for inner product spaces in general.

5.12.7 THEOREM. (PARALLELOGRAM LAW.)   *If X is an inner product space, then*

$$\|x + y\|^2 + \|x - y\|^2 = 2\|x\|^2 + 2\|y\|^2 \tag{5.12.3}$$

for all x and y in X.

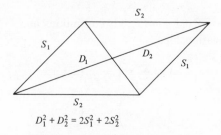

$$D_1^2 + D_2^2 = 2S_1^2 + 2S_2^2$$

Figure 5.12.2. Parallelogram Law.

The proof of this theorem is a simple substitution which we leave to the reader.

It is also a fact that the converse of Theorem 5.12.7 is true. That is, if X is a normed linear space and its norm satisfies (5.12.3), then there is a unique inner product defined on X that generates the norm. In other words, X is an inner product space "in disguise" if and only if its norm satisfies (5.12.3).

5.12.8 THEOREM.   *Let $(X,\|\cdot\|)$ be a complex normed linear space such that*

$$\|x + y\|^2 + \|x - y\|^2 = 2\|x\|^2 + 2\|y\|^2$$

for all x and y in X. Then

$$(x,y) = \tfrac{1}{4}\{\|x + y\|^2 - \|x - y\|^2 + i\|x + iy\|^2 - i\|x - iy\|^2\} \qquad (5.12.4)$$

defines an inner product on X and $\|x\| = (x,x)^{1/2}$ for all x in X. Moreover, the inner product given by (5.12.4) is the only inner product on X that generates the norm.

The proof of this theorem is left as a somewhat tedious exercise.

EXERCISES

1. Show that equality holds in the Schwarz Inequality (5.12.2) if and only if the set $\{x,y\}$ is linearly dependent.

2. Show that the Schwarz Inequality holds even when (IP4) is replaced by: "$(x,x) \geq 0$ for all $x \in X$." When does equality hold in this case?

3. Prove the Parallelogram Law, Theorem 5.12.7.

4. Prove Theorem 5.12.8. Modify Theorem 5.12.8 for the case of a real normed linear space.

5. (a) Use Theorems 5.12.7 and 5.12.8 to show that the complex space C^2 with the norm $\|\cdot\|_p$, $1 \leq p < \infty$ and $p \neq 2$, is not an inner product space.
   (b) Show that the "inequality" in (5.12.3) can go either way.

6. (a) Describe all possible inner products on the complex linear space C.
   (b) Show that the real linear space R^2 can have a (real) inner product that is included not under part (a).

7. Let (x,y) be an inner product on a linear space X.
   (a) Show that for x fixed the mapping $y \to (x,y)$ is a continuous mapping of X into C.
   (b) Show that for y fixed $x \to (x,y)$ is continuous in x.
   (c) Discuss the relationships

$$\left(x, \sum_{n=1}^{\infty} y_n\right) = \sum_{n=1}^{\infty} (x,y_n)$$

$$\left(\sum_{n=1}^{\infty} x_n, y\right) = \sum_{n=1}^{\infty} (x_n, y).$$

[*Hint*: Use Theorem 5.6.2.]

8. Let X be an inner product space. A mapping $q(x,y): X \times X \to C$ is said to be a *sesquilinear functional* if

(i) $q(x_1 + x_2, y) = q(x_1, y) + q(x_2, y)$,
(ii) $q(\alpha x, y) = \alpha q(x, y)$,
(iii) $q(x, y_1 + y_2) = q(x, y_1) + q(x, y_2)$,
(iv) $q(x, \alpha y) = \bar{\alpha} q(x, y)$,

for all x, x_1, x_2, y, y_1, y_2 in X and α in C. A mapping of $Q: X \to C$ is said to be the *quadratic form generated by q* if $Q(x) = q(x,x)$. A sesquilinear functional is said to be *symmetric* if $q(x,y) = \overline{q(y,x)}$. The functional q is *positive* if $q(x,x) \geq 0$ for all x. A sesquilinear functional is bounded if there is a real number k such that $|q(x,y)| \leq k \|x\| \cdot \|y\|$ for all x and y. The quadratic form Q is *bounded* if $|Q(x)| \leq K\|x\|^2$ for all x. Let

$$\|q\| = \inf\{k: |q(x,y)| \leq k\|x\| \cdot \|y\| \text{ for all } x, y\}$$
$$\|Q\| = \inf\{K: |Q(x)| \leq K\|x\|^2 \text{ for all } x\}.$$

(a) Show that

$$q(x,y) = Q(\tfrac{1}{2}(x + y)) - Q(\tfrac{1}{2}(x - y)) + iQ(\tfrac{1}{2}(x + iy)) - iQ(\tfrac{1}{2}(x - iy)).$$

(b) Show that q is symmetric if and only if Q is real-valued.
(c) Show that q is bounded if and only if Q is bounded and that

$$\|q\| \leq \|Q\| \leq 2\|q\|.$$

(d) (Schwarz Inequality.) Let q be positive, then show that

$$|q(x,y)|^2 \leq Q(x)Q(y).$$

(We return to this exercise again in Exercises 2 and 8 of Section 23.)

9. Let x, y be in an inner product space X and assume that $\|\lambda x + (1 - \lambda)y\| = \|x\|$ for all λ, $0 \leq \lambda \leq 1$.
(a) Show that $x = y$.
(b) What happens to part (a) if X has a norm but no inner product? (This shows that spheres in inner product spaces do not have "flat edges.")

10. Let $\{x_1, x_2\}$ be a linearly independent set in an inner product space X. Define $f: C \to R$ by $f(\alpha) = \|x_1 - \alpha x_2\|$.
(a) Where does f take on its minimum value?
(b) Give a geometric interpretation of this.

11. Let $\{x_1, x_2, x_3\}$ be a linearly independent set in an inner product space X and assume that $\{x_1, x_2\}$ satisfy $(x_i, x_j) = \delta_{ij}$, $1 \leq i, j \leq 2$. Define $f: C^2 \to R$ by $f(\alpha_1, \alpha_2) = \|\alpha_1 x_1 + \alpha_2 x_2 - x_3\|$. Now show that f attains its minimum when $\alpha_i = (x_3, x_i)$, $i = 1, 2$. (Compare this with the Gram-Schmidt process in Section 17.)

12. Let $\{x_1, x_2, \ldots, x_n\}$ be vectors from an inner product space X and define

$$G(x_1, x_2, \ldots, x_n) = \det \begin{vmatrix} (x_1, x_1) \ (x_1, x_2) & \cdots & (x_1, x_n) \\ (x_2, x_1) \ (x_2, x_2) & \cdots & (x_2, x_n) \\ \vdots & & \\ (x_n, x_1) \ (x_n, x_2) & \cdots & (x_n, x_n) \end{vmatrix}.$$

Show that $\{x_1, \ldots, x_n\}$ is linearly independent if and only if $G(x_1, \ldots, x_n) \neq 0$.

13. Suppose we consider the Banach space $Blt[H,H]$ made up of all bounded linear transformations of a Hilbert space into itself, where, as before, $\|T\|$ is defined by $\|T\| = \inf\{M : \|Tx\| \leq M\|x\|$ for all $x \in H\}$. Does it follow that

$$\|T + S\|^2 + \|T - S\|^2 = 2\|T\|^2 + 2\|S\|^2$$

for all $T, S \in Blt[H,H]$? (The point of the exercise is to show that $Blt[H,H]$ with the usual norm is not necessarily a Hilbert space.)

14. Show that in a normed linear space X the following inequality holds for all $x, y \in X$:

$$2\|x\|^2 - 4\|x\| \|y\| + 2\|y\|^2 \leq \|x + y\|^2 + \|x - y\|^2$$
$$\leq 2\|x\|^2 + 4\|x\| \|y\| + 2\|y\|^2.$$

Compare this result with the Parallelogram Law.

15. A normed linear space X is said to be *uniformly convex* if whenever $\{x_n\}$ and $\{y_n\}$ are sequences in X with $\|x_n\| = \|y_n\| = 1$ and $\|x_n + y_n\| \to 2$, one also has $\|x_n - y_n\| \to 0$. Show that every inner product space is uniformly convex.

16. Let $A : X \to X$, $B : X \to X$ be linear operators on an inner product space X. Assume that $(x, Ay) = (x, By)$ for all x and y. Show that $A = B$. Is linearity important here?

17. Let y be a fixed point in a Hilbert space H. Describe the operator $L : H \to H$ that satisfies $Lx = (x, y)y$.

13.   EXAMPLES

EXAMPLE 1.   Let $X = C^n$ and define

$$(x, y) = \sum_{i=1}^{n} x_i \bar{y}_i.$$

This is the usual inner product on C^n.   ∎

EXAMPLE 2.   Let $X = C[0,1]$ be the space of continuous complex-valued functions on the interval $[0,1]$ and define

$$(x, y) = \int_0^1 x(t)\overline{y(t)} \, dt.$$

This is an inner product space where the norm is

$$\|x\| = \left(\int_0^1 |x|^2 \, dt \right)^{1/2}.$$

As shown in Appendix D, X is not complete.   ∎

EXAMPLE 3.   The last example was not a Hilbert space, but if we let $X = L_2(I)$ be the space of complex-valued measurable functions x with $\int_I |x|^2 \, dt < \infty$, then X is a Hilbert space when the inner product is given by

$$(x,y) = \int_I x\bar{y} \, dt.$$

The integral here is the Lebesgue integral. (See Appendix D.)   ∎

EXAMPLE 4.   Let $\rho(t)$ be a real-valued, continuous function that satisfies $0 < \rho(t) \le B$ on I. Let X be the complex space $L_2(I)$ and define

$$(x,y) = \int_I x\bar{y}\rho \, dt. \tag{5.13.1}$$

It follows that the product $x\bar{y}$ is in $L_1(I)$. Since $\rho \in L_\infty(I)$, the integrand in (5.13.1) is defined for all x and y in $L_2(I)$. We leave it as an exercise to show that (5.13.1) defines an inner product on X and that X is a Hilbert space.

The function ρ in (5.13.1) is a weighting function. It is not necessary to assume that ρ is continuous, in fact any bounded, measurable, real-valued, positive function in $L_\infty(I)$ would suffice.   ∎

EXAMPLE 5.   Let D be a compact region in the $t_1 t_2$-plane, and let $X = C(D)$ be the space of complex-valued, continuous functions defined on D. Let

$$(x,y) = \iint_D x\bar{y} \, dt_1 \, dt_2.$$

Then X is an inner product space, but X is not complete.   ∎

EXAMPLE 6.   The last example has an analog in higher dimensions. Let D be a compact region in R^n and let $X = C(D)$ be the space of complex-valued, continuous functions defined on D. Let

$$(x,y) = \int_D x\bar{y} \, dt.$$

This space is not complete, but the completion would be $L_2(D)$. (See Appendix D.)   ∎

EXAMPLE 7.   Let D be a compact set in R^3 and let $X = C^2(D)$ be the space of complex-valued functions that have continuous second partial derivatives in D. If $u \in X$ let

$$\nabla u = \left(\frac{\partial u}{\partial x_1}, \frac{\partial u}{\partial x_2}, \frac{\partial u}{\partial x_3} \right).$$

Define

$$(u,v) = \int_D \left[u\bar{v} + \frac{\partial u}{\partial x_1} \frac{\partial \bar{v}}{\partial x_1} + \frac{\partial u}{\partial x_2} \frac{\partial \bar{v}}{\partial x_2} + \frac{\partial u}{\partial x_3} \frac{\partial \bar{v}}{\partial x_3} \right] dx, \qquad (5.13.2)$$

where $x = (x_1,x_2,x_3)$. This is clearly linear in u, also $(u,v) = \overline{(v,u)}$ and $(u,u) \geq 0$. Furthermore, if $(u,u) = 0$, then $\int_D |u|^2 dx = 0$. Since u is continuous, this implies that $u = 0$. Hence (5.13.2) is an inner product on x. The norm is given by

$$\|u\| = \left(\int_D (|u|^2 + |\nabla u|^2) \, dx \right)^{1/2},$$

where $|\nabla u|^2 = \left| \dfrac{\partial u}{\partial x_1} \right|^2 + \left| \dfrac{\partial u}{\partial x_2} \right|^2 + \left| \dfrac{\partial u}{\partial x_3} \right|^2.$ ∎

EXAMPLE 8.   Let $X = C^n$ with (x,y) defined by Example 1. Let I be a compact interval $[a,b]$ and let $U = C(I,X)$ be the space of continuous functions u defined on I with values $u(t)$ in X. If $u = \{u_1,\ldots,u_n\}$ and $v = \{v_1,\ldots,v_n\}$ define

$$(u,v) = \int_I (u(t), v(t)) \, dt \qquad (5.13.3)$$

$$= \int_I \{u_1 \bar{v}_1 + \cdots + u_n \bar{v}_n\} \, dt.$$

Then (u,v) is an inner product on U. The norm of a function $u = \{u_1,\ldots,u_n\}$ is given by

$$\|u\| = \left(\int_I \{|u_1|^2 + \cdots + |u_n|^2\} \, dt \right)^{1/2}.$$

One can replace I with a compact region D in R^m and let $U = C(D,X)$. The integral in (5.13.3) then becomes a volume integral over D.   ∎

EXAMPLE 9.   Let $l_2 = l_2[0,\infty)$ be the space of sequences $x = (x_1,x_2,\ldots)$ of complex numbers such that $\sum_{i=1}^{\infty} |x_i|^2 < \infty$. Define

$$(x,y) = \sum_{i=1}^{\infty} x_i \bar{y}_i.$$

Then this is an inner product on l_2, and l_2 is a Hilbert space. This is referred to as the *usual inner product* on l_2.   ∎

EXAMPLE 10. Let us consider the space $l_2(-\infty,\infty)$ of all bi-sequences $x = (\ldots,x_{-2},x_{-1},x_0,x_1,x_2,\ldots)$ of complex numbers such that $\sum_{-\infty}^{\infty} |x_i|^2 < \infty$. Then

$$(x,y) = \sum_{i=-\infty}^{\infty} x_i \bar{y}_i$$

is an inner product on $l_2(-\infty,\infty)$, and $l_2(-\infty,\infty)$ is a Hilbert space. ∎

EXAMPLE 11. Consider a probability space $(\Omega,\mathscr{F},P)$ and let X denote the collection of all random variables $x(\omega)$ with finite variance $\sigma^2(x)$. That is, if E denotes the expectation, then $\alpha = E(x)$ is finite and

$$\sigma^2(x) = E(|x - \alpha|^2)$$

is finite. An inner product is given on X by

$$(x,y) = E(x\bar{y})$$

and the induced norm is given by $\|x\| = E(|x|^2)^{1/2}$. The space X is then the Hilbert space $L_2(\Omega,\mathscr{F},P)$. (See Exercise 1 and Appendix E.) ∎

EXAMPLE 12. (SOBOLEV SPACES.) This is a continuation of Example 12, Section 3. Let Ω be an open set in R^m and let $u \in C^n(\Omega)$. Define a norm on u by

$$\|u\|_{n,2} = \left[\int_\Omega \sum_{|\alpha| \le n} |D^\alpha u(x)|^2 \, dx \right]^{1/2}. \tag{5.13.4}$$

This number may be $+\infty$. Let $\tilde{C}^n(\Omega)$ denote those functions u in $C^n(\Omega)$ for which $\|u\|_{n,2}$ is finite. It is easy to see that $\tilde{C}^n(\Omega)$ is a normed linear space, and in fact the norm on $\tilde{C}^n(\Omega)$ is generated by the inner product

$$(u,v)_n = \int_\Omega \sum_{|\alpha| \le n} D^\alpha u(x) \overline{D^\alpha v(x)} \, dx.$$

The completion of $\tilde{C}^n(\Omega)$ with respect to this norm is the *Sobolev space* $H^n(\Omega)$.

Another Sobolev space we shall be interested in is $H_0^n(\Omega)$. This is defined as the completion of $C_0^\infty(\Omega)$ (the space of C^∞-functions with compact support in Ω) under the norm (5.13.4).

Both of the spaces $H^n(\Omega)$ and $H_0^n(\Omega)$ are very useful in the study of partial differential operators. We shall see some applications in Sections 7.6–7.8. For other applications we refer the reader to Agmon [1] and Friedman [2].

It should be noted that in general, $H^n(\Omega)$ and $H_0^n(\Omega)$ are different spaces. (See Exercise 4 below.) ∎

EXERCISES

1. In Example 11, show that $x(\omega)$ has finite variance if and only if $x \in L_2(\Omega)$.

2. In Example 11, let H denote the collection of all random variables x in $L_2(\Omega)$ with the property that $E(x) = 0$. Show that H is a closed linear subspace of $L_2(\Omega)$.

3. The following exercises refer to Example 12.
   (a) Let $\{u_k\}$ be a Cauchy sequence in $\tilde{C}^n(\Omega)$. Show that for any α, $|\alpha| \leq n$, $\{D^\alpha u_k\}$ is a Cauchy sequence in $L_2(\Omega)$.
   (b) Show that every element in $H^n(\Omega)$ can be viewed as a function u in $L_2(\Omega)$ with *strong L_2 derivatives of order up to n*, that is, there is a sequence $\{u_k\}$ in $\tilde{C}^n(\Omega)$ such that $\{D^\alpha u_k\}$ is a Cauchy sequence in $L_2(\Omega)$ for $|\alpha| \leq n$, and $u_k \to u$ in $L_2(\Omega)$.
   (c) A function u in $L_2(\Omega)$ is said to have a *weak derivative* u^α in $L_2(\Omega)$ if

$$\int_\Omega \phi(x) u^\alpha(x) \, dx = (-1)^{|\alpha|} \int_\Omega u(x) D^\alpha \phi(x) \, dx$$

   for all ϕ in $C_0^\infty(\Omega)$. A function u has *weak derivatives of order up to n* if it has a weak derivative u^α for every α, $|\alpha| \leq n$. Show that if a function u in $L_2(\Omega)$ has strong derivatives of order up to n, then it has weak derivatives of order up to n.
   (d) Let $W^n(\Omega)$ denote the class of functions in $L_2(\Omega)$ with weak derivatives of order up to n. For u, v in $W^n(\Omega)$ let

$$(u,v)_n = \int_\Omega \sum_{|\alpha| \leq n} D^\alpha u \overline{D^\alpha v} \, dx.$$

   Show that $W^n(\Omega)$ is a Hilbert space with the above inner product.
   (e) Show that $H^n(\Omega) \subset W^n(\Omega)$. (One can show that $H^n = W^n$, see Meyers and Serrin [1] and Friedman [2].)
4. Consider the Sobolev spaces $H^n(\Omega)$ and $H_0{}^n(\Omega)$. Show that $H^0(\Omega) = H_0{}^0(\Omega) = L_2(\Omega)$. (If $n \geq 1$, then H^n is generally different from $H_0{}^n$, see Friedman [2].)
5. Let p satisfy $1 \leq p < \infty$ and let n be an integer. Show that

$$\|u\|_{n, p} = \left[\int_\Omega \sum_{|\alpha| \leq n} |D^\alpha u(x)|^p \, dx \right]^{1/p}$$

   is a norm on an appropriate space. (The completion of these spaces defines the Sobolev spaces $H^{n,p}(\Omega)$ and $H_0{}^{n,p}(\Omega)$, see Friedman [2].)
6. Let D^α denote a partial differential operator with $|\alpha| \leq n$, see Exercise 12, Section 5.3 for notation.
   (a) Show that $D^\alpha : C_0^\infty(\Omega) \to L_2(\Omega)$ is a bounded linear mapping where $L_2(\Omega)$ has the usual norm and $C_0^\infty(\Omega)$ has the norm $\|\cdot\|_{n,2}$.
   (b) Show that D^α has a unique extension to $H_0{}^n(\Omega)$.
   (c) Show that D^α can be similarly defined on $H^n(\Omega)$.

14.   ORTHOGONALITY

As has already been mentioned, the geometry of Hilbert spaces is, roughly speaking, just a generalization of Euclidean geometry. The main reason for this simplicity is that we have an inner product which allows us to introduce the concept of orthogonality.

In order to motivate the definition let us refer again to Figure 5.12.1. The dot product of the two vectors in this figure is given by $x \cdot y = l_1 l_2 \cos \theta$. If $l_1 \neq 0$, $l_2 \neq 0$, then $x \cdot y = 0$ if and only if $\cos \theta = 0$. That is, $x \quad y = 0$ if and only if x and y are perpendicular, or orthogonal, to one another. Based on this observation we make the following definition.

5.14.1 DEFINITION.   Two vectors x and y in an inner product space are said to be *orthogonal* if $(x,y) = 0$. If x and y are orthogonal, this is denoted by $x \perp y$.

We shall say that two subsets A and B of X are *orthogonal* if $x \perp y$ for all x in A and y in B. This will be denoted by $A \perp B$.

An immediate consequence of Definition 5.14.1 is that the familiar Pythagorean Theorem is true in any inner product space.

5.14.2 THEOREM. (PYTHAGOREAN THEOREM.)   *If $x \perp y$ in an inner product space X, then*

$$\|x + y\|^2 = \|x\|^2 + \|y\|^2.$$

(See Figure 5.14.1.)

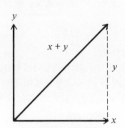

Figure 5.14.1.

Proof: Assume that $x \perp y$, that is, $(x,y) = 0$. Then

$$\|x + y\|^2 = (x + y, x + y) = \|x\|^2 + (y,x) + (x,y) + \|y\|^2$$
$$= \|x\|^2 + \|y\|^2. \quad \blacksquare$$

EXAMPLE 1.   Suppose we make N measurements (N a positive integer) of a quantity $S = S_0 + s$, where S_0 is a known constant, and s is random. Assume further that each measurement is corrupted with additive noise, so that each measurement is of the form

$$m_i = S + n_i, \qquad i = 1, 2, \ldots, N.$$

We can place this situation in a Hilbert space framework by treating $s, n_1, n_2, \ldots,$ $n_N, m_1, m_2, \ldots, m_N$ as vectors in the Hilbert space H made up of all complex-valued random variables x (defined on some underlying probability space) such that

$$E\{|x|^2\} < \infty \quad \text{and} \quad E\{x\} = 0,$$

where E denotes the expectation operation. The inner product on this space is defined by

$$(x,y) = E\{x\bar{y}\},$$

where $\bar{y}$ denotes the complex conjugate of y. In this case, x and y being orthogonal corresponds to x and y being uncorrelated random variables.

We assume here that the random variables $s, n_1, n_2, \ldots, n_N$ are pairwise uncorrelated, that is, orthogonal. Further, we assume that the variances satisfy $(n_1,n_1) = (n_2,n_2) = \cdots = (n_N,n_N) = \sigma^2$ and $(S,S) = \Sigma^2 = |S_0|^2 + \sigma_s^2$ where $\sigma_s^2 = (s,s)$.

Then suppose that we want to form an estimate $\hat{S}$ of S by "averaging" the N measurements

$$\hat{S} = S_0 + \hat{s} = \frac{1}{N}\{m_1 + m_2 + \cdots + m_N\}$$

$$= \frac{1}{N}\{S + S + \cdots + S + n_1 + n_2 + \cdots + n_N\}$$

$$= S_0 + \frac{1}{N}\{s + \cdots + s + n_1 + \cdots + n_N\}.$$

Here the vector $\{s + s + \cdots + s + n_1 + n_2 + \cdots + n_N\}$ has an interesting geometrical interpretation. In particular, $n_1 + n_2 + \cdots + n_N$ is the sum of N pairwise orthogonal vectors of length σ. If σ_N denotes the length of this sum, it follows from the Pythagorean Theorem that $\sigma_N^2 = N\sigma^2$. On the other hand, $s + s + \cdots + s$ is the sum of s with itself N times, and $\|s + \cdots + s\|^2 = N^2\sigma_s^2$. It follows that the "signal-to-noise ratio" in $\hat{S}$ is given by

$$\frac{\|S + S + \cdots + S\|^2}{\|n_1 + \cdots + n_N\|^2} = \frac{N^2\{|S_0|^2 + \sigma_s^2\}}{N\sigma^2} = N\frac{\Sigma^2}{\sigma^2}.$$

In other words, the more measurements (that is, larger N) the more "signal" relative to "noise." If we write

$$\hat{S} = S + D,$$

where D denotes the error, then D is a random variable, $D = (1/N)\{n_1 + \cdots + n_N\}$ and

$$\|D\| = \frac{\sigma}{\sqrt{N}}. \quad \blacksquare$$

One of the most striking differences between Hilbert spaces and Banach spaces arises in approximation theory. Consider the following three part problem: Let M be a proper closed linear subspace in a Banach space B and let $x_0 \in B$ with $x_0 \notin M$. (See Figure 5.14.2.) We first ask, does there exist a point $y_0 \in M$ that is closest to x_0, that is, $\|x_0 - y_0\| \le \|x_0 - y\|$ for all $y \in M$? Second, is y_0 unique? Third, if a y_0 does exist how do we find it?

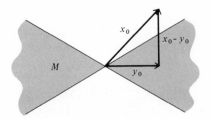

Figure 5.14.2.

Let

$$\delta = \inf\{\|x_0 - y\| : y \in M\}. \tag{5.14.1}$$

Since M is closed in B, it follows (Exercise 18, Section 3.12) that $\delta > 0$. The following result is a direct consequence of the definition of the infimum function and is closely related to the Riesz Theorem (Theorem 5.5.4).

5.14.3 THEOREM.   *Let B be a Banach space and let M be a closed linear subspace of B. Let $x_0 \in B$ and define δ by (5.14.1). Then for each $\eta > 0$ there is a $y \in M$ such that*

$$\delta \le \|x_0 - y\| < \delta + \eta.$$

In other words, there exist approximations to x_0 in M such that $\|x_0 - y\|$ is arbitrarily close to δ. Again, we see this follows from the definition of δ, that is, $\delta = \inf\{\|x_0 - y\| : y \in M\}$. However, Theorem 5.14.3 does *not* say we can actually achieve δ. Since B is complete and M is closed, one might suspect that this would be no problem. Alas, things are not this simple in Banach spaces. The next example illustrates this point.

EXAMPLE 2.   This is a continuation of Example 3, Section 5. Choose x_0 in X but not in M. Let $\alpha_0 = \int_0^1 x_0(t)\, dt$, then $\alpha_0 \ne 0$ since $x_0 \notin M$. If we can find a y_0 in M such that $\|x_0 - y_0\|_\infty \le \|x_0 - y\|_\infty$ for all y in M, then

$$z_0 = \frac{x_0 - y_0}{\|x_0 - y_0\|_\infty}$$

has the property that $\|z_0\|_\infty = 1$, $z_0 \notin M$ and for any y in M one has

$$\|z_0 - y\|_\infty = \left\| \frac{x_0 - y_0}{\|x_0 - y_0\|_\infty} - y \right\|_\infty = \frac{1}{\|x_0 - y_0\|_\infty} \|x_0 - (y_0 + y')\|_\infty ,$$

where $y' = \|x_0 - y_0\|_\infty y$. Since $y_0 + y'$ belongs to M, one then has

$$\|z_0 - y\|_\infty \ge \frac{1}{\|x_0 - y_0\|_\infty} \|x_0 - y_0\|_\infty = 1.$$

But it was shown in Example 3, Section 5 that this is impossible.   ∎

The pathology of Example 2 cannot occur in a Hilbert space.

5.14.4 THEOREM. *Let H be a Hilbert space and let M be a closed linear subspace of H. Let $x_0 \in H$ and define δ by (5.14.1). Then there is one (and only one) $y_0 \in M$ such that*

$$\|x_0 - y_0\| = \delta.$$

Moreover, $x_0 - y_0 \perp M$, that is, $(x_0 - y_0, y) = 0$ for all $y \in M$. Furthermore, y_0 is the only point in M such that $x_0 - y_0 \perp M$.

Again we are faced with a familiar geometric idea. This theorem says (see Figure 5.14.3) that the unique point in M closest to x_0 is found by " dropping a

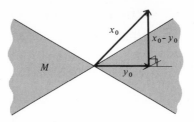

Figure 5.14.3.

perpendicular from x_0 to M." And, it would not be completely incorrect to say that this simple idea is the most widely applied piece of Hilbert space geometry. Moreover, it is important to note that Theorem 5.14.4 is not true for inner product spaces that are not complete.

Proof: First let us show that y_0 exists. Let $\{y_n\}$ be a sequence in M such that

$$\delta \le \|x_0 - y_n\| < \delta + \frac{1}{n}, \qquad (5.14.2)$$

where $\delta = \inf\{\|x_0 - y\| : y \in M\}$. The first thing we do is show that $\{y_n\}$ is a Cauchy sequence. The geometric situation is sketched in Figure 5.14.4. The vectors " below "

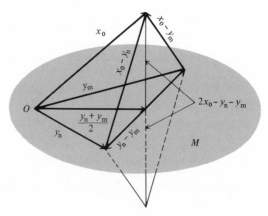

Figure 5.14.4.

the subspace M are shown in order to call attention to the parallelogram with "sides" $(x_0 - y_m)$, $(x_0 - y_n)$ and "diagonals" $(y_n - y_m)$, $(2x_0 - y_m - y_n)$. We know that $\lim_{m \to \infty} \|x_0 - y_m\| = \lim_{n \to \infty} \|x_0 - y_n\| = \delta$, so the length of the sides is approaching δ. On the other hand, M is a convex set; therefore, $(y_n + y_m)/2$ is a point in M. This means that $\|x_0 - (y_n + y_m)/2\| \geq \delta$ or $\|2x_0 - y_n - y_m\| \geq 2\delta$. But then using the Parallelogram Law, we have

$$\|y_n - y_m\|^2 + \|2x_0 - y_n - y_m\|^2 = 2\|x_0 - y_n\|^2 + 2\|x_0 - y_m\|^2,$$

and by (5.14.2) we get

$$\|y_n - y_m\|^2 \leq 2\|x_0 - y_n\|^2 + 2\|x_0 - y_m\|^2 - 4\delta^2$$

$$\leq \frac{2}{n^2} + \frac{2}{m^2} + 4\delta\left(\frac{1}{n} + \frac{1}{m}\right) \to 0,$$

as $m, n \to \infty$. Thus $\{y_n\}$ is a Cauchy sequence in M. Since M is a closed set in a complete space, M itself is complete, so $\{y_n\}$ converges to a point $y_0 \in M$. Moreover,

$$\|x_0 - y_0\| = \lim_{n \to \infty} \|x_0 - y_n\| = \delta,$$

because the norm is a continuous function.

Next let us show that y_0 is unique. Suppose y_0 and y_0' are distinct points in M such that

$$\|x_0 - y_0\| = \|x_0 - y_0'\| = \delta.$$

Then $(y_0 + y_0')/2$ is in M. Again using the Parallelogram Law, we have

$$\delta^2 \leq \left\|x_0 - \frac{y_0 + y_0'}{2}\right\|^2 = 2\left\|\frac{x_0}{2} - \frac{y_0}{2}\right\|^2 + 2\left\|\frac{x_0}{2} - \frac{y_0'}{2}\right\|^2 - \left\|\frac{y_0 - y_0'}{2}\right\|^2$$

$$< \frac{\|x_0 - y_0\|^2}{2} + \frac{\|x_0 - y_0'\|^2}{2} = \delta^2,$$

which is a contradiction.

Finally, let us show that y_0 is the only point in M such that $x_0 - y_0 \perp M$. Let y be any point in M. Then $(y_0 + \alpha y) \in M$ and again from the definition of δ,

$$\delta^2 \leq (x_0 - y_0 - \alpha y, x_0 - y_0 - \alpha y) = \|x_0 - y_0\|^2 - 2 \operatorname{Re}\{\alpha(y, x_0 - y_0)\} + |\alpha|^2(y,y),$$

where α is a scalar. Since $\|x_0 - y_0\| = \delta$, we have

$$0 \leq -2 \operatorname{Re}\{\alpha(y, x_0 - y_0)\} + |\alpha|^2(y,y).$$

Then letting $\alpha = \beta(x_0 - y_0, y)$, where β is real, we have

$$0 \leq -2\beta|(x_0 - y_0, y)|^2 + \beta^2 |(x_0 - y_0, y)|^2(y,y),$$

which holds for all β. But this implies that the coefficient of the linear term in β is zero, that is, $(x_0 - y_0, y) = 0$. Hence $x_0 - y_0 \perp M$. Now suppose that $x_0 - y_0' \perp M$, where $y_0' \in M$. Then $(x_0 - y_0, y) = (x_0 - y_0', y)$ for all $y \in M$. So $(y_0 - y_0', y) = 0$ for $y \in M$. Hence, $y_0 - y_0' \perp M$. But $(y_0 - y_0') \in M$. So $y_0 - y_0' = 0$. This shows that y_0 is the only point in M such that $x_0 - y_0 \perp M$. ∎

Let us illustrate an application of Theorem 5.14.4. It must be said immediately that this example far from exhausts the possible applications of this theorem.

EXAMPLE 3.   Suppose that we want to build a time-invariant linear filter F which is modeled by

$$(Fx)(t) = \int_{-\infty}^{\infty} h(t - \tau)x(\tau)\, d\tau,$$

where the "unit impulse response" h is given by

$$h(t) = \begin{cases} 0, & \text{for} \quad -\infty < t < 0 \\ 1, & \text{for} \quad 0 \le t < 1 \\ 0, & \text{for} \quad 1 \le t < \infty \end{cases}$$

and $x \in L_2(-\infty,\infty)$. (See Figure 5.14.5.) Further suppose that we cannot build this filter itself but must construct an approximation to it. This approximation, call it $\hat{F}$,

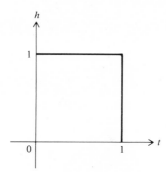

Figure 5.14.5.

will be chosen from those time-invariant linear filters whose unit impulse responses or kernels h are of the form

$$\hat{h}_n(t) = \begin{cases} 0, & t < 0 \\ \alpha_1 e^{-t} + \alpha_2\, te^{-t} + \cdots + \alpha_n\, t^{n-1}e^{-t}, & 0 \le t, \end{cases}$$

where n is a fixed positive integer and the α's are scalars. We assume that $\hat{F}$ is chosen so that the integral

$$\int_0^{\infty} |h(t) - \hat{h}_n(t)|^2\, dt$$

is minimized.

Now in the Hilbert space $L_2[0,\infty]$, the linear subspace spanned by the set $\{e^{-t}, te^{-t}, \ldots, t^{n-1}e^{-t}\}$ is a closed linear subspace M of dimension n. Our approximation problem falls into the framework of Theorem 5.14.4. We want to find the point $\hat{h}_n$ in M that is closest to h. It follows that we want to choose $h - \hat{h}_n$ to be orthogonal to M, so we have the following equations

$$(h - \hat{h}_n, h_i) = 0, \qquad i = 1, 2, \ldots, n,$$

where $h_i(t) = t^{i-1}e^{-t}$. That is,

$$(h, h_i) = \alpha_1(h_1, h_i) + \alpha_2(h_2, h_i) + \cdots + \alpha_n(h_n, h_i), \qquad 1 \le i \le n, \qquad (5.14.3)$$

which is a linear system of n equations and n unknowns $\{\alpha_1, \ldots, \alpha_n\}$. The appropriate α's are the solution of this system of equations. (See Exercise 9.) ∎

In the proof of Theorem 5.14.4 the completeness of H was crucial. The next example illustrates this point.

EXAMPLE 4.   Let us begin this example with a brief synopsis. We start with an incomplete inner product space X that is a dense subspace in a Hilbert space H. Then let z be any point in H that is not in X, and define a subspace M of X as all $y \in X$ such that $(y,z) = 0$. It is then shown that M is a proper closed linear subspace of X. Finally, it is shown that if x_0 is in X but not in M, then there is no $y_0 \in M$ such that

$$\|x_0 - y_0\| = \inf\{\|x_0 - y\| : y \in M\}.$$

Roughly speaking, if there were such a y_0, then $x_0 - y_0$ would be orthogonal to M. This in turn would mean that $x_0 - y_0$ would be a scalar multiple of z, but z is *not* an element of X. Now for the details.

Let X denote the space of all polynomials with complex coefficients. Thus if $x(t) = \sum_{i=0}^{n} a_i t^i$ and $y(t) = \sum_{i=0}^{m} b_i t^i$, define

$$(x,y) = \sum_{i=0}^{\infty} a_i \bar{b}_i.$$

X can be viewed as a dense linear subspace of l_2. (How?) Now define

$$M = \left\{x \in X : \sum_{i=0}^{\infty} (i + 1)^{-2} a_i = 0\right\},$$

that is, z is the sequence $\{(i + 1)^{-2}\} \in l_2$. If we define a mapping $l: X \to C$ by

$$l(x) = l\left(\sum_{i=0}^{n} a_i t^i\right) = \sum_{i=0}^{\infty} (i + 1)^{-2} a_i,$$

it is easy to see that l is a bounded linear mapping. Since M is the null space of l, we see that M is closed. Since l is not the zero mapping, it follows from Theorem 4.12.2 that M is a proper subspace of X, and it has co-dimension one.

Let $x_0 \in X$ with $x_0 \notin M$. If there is a y_0 in M with

$$\|x_0 - y_0\| = \inf\{\|x_0 - y\| : y \in M\} > 0,$$

then $z_0 = x_0 - y_0$ is in X and not in M. Furthermore by using the argument of

Theorem 5.14.4 one can show that $z_0 \perp M$. Since z_0 is in X, we can write it in the form

$$z_0(t) = c_0 + c_1 t + \cdots + c_N t^N$$

for some finite N. For $0 \le i \le N$, let

$$\omega_i(t) = (i + 1)^2 t^i - (N + 2)^2 t^{N+1}.$$

Since $\omega_i \in M$, one has $z_0 \perp \omega_i$, for $0 \le i \le N$. However,

$$z_0 \perp \omega_i \Rightarrow 0 = (z_0, \omega_i) = c_i(i + 1)^2 \Rightarrow c_i = 0.$$

Hence $z_0 = 0$, and this is a contradiction. ∎

The following is an important consequence of Theorem 5.14.4.

5.14.5 COROLLARY.   *Let M and N be closed linear subspaces of a Hilbert space H such that $M \subset N$ and $M \ne N$. Then there is a vector z in N such that $z \ne 0$ and $z \perp M$.*

The proof is simple. In the notation of Theorem 5.14.4 we choose $x_0 \in N$ and $x_0 \notin M$, and note that $z = x_0 - y_0$, where y_0 is the point in M closest to x_0, satisfies the conclusion of the corollary. Needless to say, this corollary, as Theorem 5.14.4, depends on H being complete. This corollary should be compared with the Riesz Theorem (Theorem 5.5.4).

EXERCISES

1. Find the point in the proof of Theorem 5.14.4 where one uses the inner product structure of H. Where does one use the fact that H is complete?

2. Let K be a closed convex set in a Hilbert space H.
   (a) Show that there is one and only one point x_0 in K of minimum norm, that $\|x_0\| \le \|x\|$ for all x in K.
   (b) Show that this fails in an inner product space that is not complete. Recall that a set K is convex if $x_1, x_2 \in K$ implies that $x = \lambda x_1, +(1 - \lambda)x_2 \in K$ for all λ, $0 \le \lambda \le 1$. [*Hint:* Study proof of Theorem 5.14.4.]

3. Consider the Banach space R^2 with norm $\|x\|_1 = |x_1| + |x_2|$.
   (a) Show that every closed convex set in $(R^2, \|\cdot\|_1)$ has a point of minimum norm.
   (b) Show, by example, that this may not be unique.
   (c) What happens with $\|x\|_\infty = \max(|x_1|, |x_2|)$?

4. Let X denote all those functions $x(t)$ in $C[0,1]$ with $x(0) = 0$, and assume that X has the sup-norm. Let K denote the collection of all functions x in X with $\int_0^1 x(t)\, dt = 1$.
   (a) Show that K is convex.
   (b) Show that there is no point in K with minimum norm.

5. Let X denote the linear subspace of $L_2[0,2\pi]$ made up of all trigonometric polynomials of the form

$$x(t) = \sum_{k=-n}^{n} a_k e^{ikt}.$$

Let M be the subspace of X defined by

$$M = \left\{ x \in X: \int_0^{2\pi} tx(t)\, dt = 0 \right\}.$$

Let x_0 be in X with $x_0 \notin M$. Show that there is no point y_0 in M with $\|x_0 - y_0\| = \inf\{\|x_0 - y\|: y \in M\}$.

6. Let L be a bounded linear transformation on a Hilbert space H with $\|L\| \le 1$. Let $x \in H$ and let y_n be the average

$$y_n = \frac{1}{n}[x + Lx + \cdots + L^{n-1}x].$$

The following steps will lead to a proof that there is a y in H with $y = \lim y_n$ in H.

(a) Let K denote the smallest closed convex set in H containing $\{x, Lx, L^2x, \ldots\}$, and let y be the (unique) point in K of minimum norm. (Use Exercise 2.)

(b) Choose $z \in K$ so that $z = \sum_{i=0}^{m-1} \alpha_i L^i x$ where $\alpha_i \ge 0$, $\sum_{i=0}^{m-1} \alpha_i = 1$, $L^0 = I$, and such that $\|z\| \le \|y\| + \varepsilon/2$ where ε is some prescribed positive number. Let $z_n = (1/n)[z + \cdots + L^{n-1}z]$. Show that

$$\|y_n\| \le \|y_n - z_n\| + \|y\| + \varepsilon/2.$$

(c) Show that $\|y_n - z_n\| \le \varepsilon/2$, for n sufficiently large.

(d) Show that $\|y_n\| \le \|y\| + \varepsilon$ for n sufficiently large and hence $\lim \|y_n\| = \|y\|$.

(e) Use the fact that $y_n \in K$ to show that $\lim y_n = y$ in H.

7. (Mean Ergodic Theorem.) Let $(\Omega, \mathscr{B}, \mu)$ be a measure space and let $T: \Omega \to \Omega$ be a one-to-one mapping of Ω into itself and assume that T preserves measure, that is, $\mu(A) = \mu(TA)$ for every measurable set A. For $f \in L_2(\Omega)$ let

$$g_n(x) = \frac{1}{n}[f(x) + f(Tx) + \cdots + f(T^{n-1}x)].$$

Show that there is a function g in $L_2(\Omega)$ such that $g = \lim g_n$ in $L_2(\Omega)$. [Hint: Define $L: L_2 \to L_2$ by $Lf(x) = f(Tx)$. Show that $\|L\| = 1$ and apply Exercise 6.]

8. Let $\phi_a(t) = e^{2\pi i a t}$, where a is real. Let n be a fixed integer. Show that $\phi_a \perp \phi_n$ in $L_2[0,1]$ if and only if $a = m \ne n$, where m is an integer.

9. Show that (5.14.3) always has a solution for $\alpha_1, \alpha_2, \ldots, \alpha_n$. Solve for the α's.

10. Consider the space X of all continuous functions $x(t)$ such that

$$\|x\| = \left\{ \lim_{T \to \infty} \frac{1}{2T} \int_{-T}^{T} |x(t)|^2\, dt \right\}^{1/2} < \infty.$$

Define an inner product on X by

$$(x,y) = \left\{ \lim_{T \to \infty} \frac{1}{2T} \int_{-T}^{T} x(t)\overline{y(t)}\, dt \right\}.$$

Let $\phi_a(t) = e^{iat}$ where a is real. Show that $(\phi_a,\phi_b) = \delta_{ab}$, that is, $\phi_a \perp \phi_b$ whenever $a \neq b$.

11. Let $z = x + y$, where $x \perp y$. Show that (z,x) is real and $(z,x) = \|x\|^2$.

15. ORTHOGONAL COMPLEMENTS AND THE PROJECTION THEOREM

Let X be an inner product space and let M be any subset of X. Define $M^\perp$, the *orthogonal complement* of M, by

$$M^\perp = \{x \in X : (x,y) = 0 \text{ for all } y \in M\}.$$

That is, $M^\perp$ is made up of all points that are orthogonal to every point in M. We shall write $x \perp M$ if $x \in M^\perp$. If $M = \varnothing$, the empty set, then $M^\perp = X$.

EXAMPLE 1.   Let $H = l_2$, and also let M be the set $\{e_1, e_2, e_3, e_4\}$, where $e_1 = \{1,0,0,0,\dots\}$, $e_2 = \{0,1,0,0,\dots\}$, $e_3 = \{0,0,1,0,\dots\}$, and $e_4 = \{0,0,0,1,0,\dots\}$. The orthogonal complement $M^\perp$ is the set of all sequences in l_2 of the form $x = \{0,0,0,0,\xi_5,\xi_6,\xi_7,\dots\}$. ∎

EXAMPLE 2.   Let $(\Omega,\mathscr{F},P)$ be a probability space, and let A be a set in $\mathscr{F}$. Denote the complement of A by A'. Then let M be the linear subspace of $L_2(\Omega,\mathscr{F},P)$ made up of all random variables $x(\omega)$ such that $x(\omega) = 0$ (a.e.) in[9] A'. (See Figure 5.15.1.) That is,

$$M = \left\{ x \in L_2(\Omega,\mathscr{F},P) : \int_{A'} |x(\omega)|^2 \, dP = 0 \right\}.$$

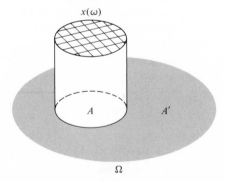

Figure 5.15.1.

---

[9] The abbreviation "a.e." stands for the phrase "almost everywhere" whose technical meaning is discussed in Appendix D.

The orthogonal complement of M is given by

$$M^\perp = \left\{ x \in L_2(\Omega,\mathscr{F},P) : \int_A |x(\omega)|^2 dP = 0 \right\}. \quad \blacksquare$$

EXAMPLE 3.   Suppose that a probability space $(\Omega,\mathscr{F},P)$ can be represented as the square $[0,1] \times [0,1]$ in R^2 with the usual Lebesgue measure structure. Consider the Hilbert space $H = L_2(\Omega,\mathscr{F},P)$, and let y be a point in H and assume that y does not depend on ω_1. (See Figure 5.15.2.) Since the random variable y does not depend on ω_1, the inverse image of any Lebesgue measurable set I in C has a strip form similar to that of the set B shown in Figure 5.15.2. In fact, the collection of all such inverse images generates a sub-σ-algebra, denote it by $\mathscr{B}$, of $\mathscr{F}$. In

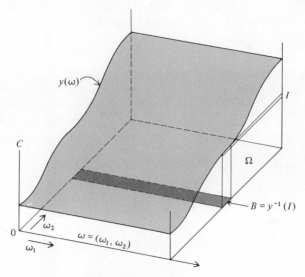

Figure 5.15.2.

this case let us assume that $\mathscr{B}$ is maximal in the following sense: If A is any Lebesgue measurable set in $[0,1]$, then $[0,1] \times A$ is in $\mathscr{B}$. In any event, $(\Omega,\mathscr{B},P_\mathscr{B})$ is a probability space, where $P_\mathscr{B}$ is the restriction of P to $\mathscr{B}$. Let M denote the Hilbert space $L_2(\Omega,\mathscr{B},P_\mathscr{B})$. It should be clear that M is the linear subspace of H made up of all random variables x that do not depend on ω_1.

The orthogonal complement of M is the linear subspace of H made up of all random variables z such that

$$\int_0^1 z(\omega_1,\omega_2)\, d\omega_1 = f(\omega_2) = 0. \quad \text{(a.e.)} \tag{5.15.1}$$

Indeed, z is in $M^\perp$ if and only if $(z,x) = 0$ for all $x \in M$, that is,

$$\int_0^1 \int_0^1 z(\omega_1,\omega_2)\bar{x}(\omega_2)\, d(\omega_1 \times \omega_2) = \int_0^1 f(\omega_2)\bar{x}(\omega_2)\, d\omega_2 = 0,$$

where f is defined by (5.15.1). Since x can be any square integrable function, it follows that $z \in M^\perp$ if and only if $f(\omega_2) = 0$ (a.e.). A sketch of an allowable z is shown in Figure 5.15.3.

The subspace M has an important interpretation in terms of conditional expectations. Suppose that x is any random variable in $L_2(\Omega,\mathscr{F},P)$. The conditional expectation of x with respect to the given random variable y, which we denote by $E^y(x)$, is itself a random variable in $L_2(\Omega,\mathscr{F},P)$. In fact, it is in $L_2(\Omega,\mathscr{B},P_\mathscr{B})$. Indeed, $E^y(x)$ is (compare with Section E.5) the random variable in $L_2(\Omega,\mathscr{B},P_\mathscr{B})$ satisfying the condition

$$\int_B (E^y(x))\, dP_\mathscr{B} = \int_B x\, dP$$

for all sets B in $\mathscr{B}$. It follows that E^y is a mapping of $L_2(\Omega,\mathscr{F},P)$ into itself and the range of E^y is $M = L_2(\Omega,\mathscr{B},P_\mathscr{B})$. We can also characterize the null space of E^y.

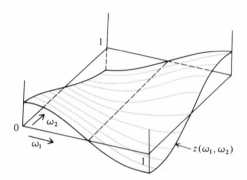

Figure 5.15.3.

Suppose $E^y(x) = 0$, then

$$\int_B 0 \cdot dP_\mathscr{B} = \int_B x(\omega_1,\omega_2)\, d(\omega_1 \times \omega_2)$$

for all $B \in \mathscr{B}$. Since all B are of the form $[0,1] \times A$, we have

$$\int_A \left\{ \int_0^1 x(\omega_1,\omega_2)\, d\omega_1 \right\} d\omega_2 = 0$$

for all A, that is, $\int_0^1 x(\omega_1, \omega_2)\, d\omega_1 = 0$. It follows from (5.15.1) that $E^y(x) = 0$ if and only if $x \in M^\perp$. This example is extended in Example 4, Section 16. ∎

5.15.1 THEOREM.   *Let M be a set in an inner product space X. Then $M^\perp$ is a closed linear subspace of X.*

Proof: It is easy to see that $M^\perp$ is a linear space. In order to show that $M^\perp$ is closed, let $\{x_n\}$ be any convergent sequence in $M^\perp$ with $x_n \to x_0$. If we show that

$x_0 \in M^\perp$, it will follow from the Closed Set Theorem that $M^\perp$ is closed. But if y is any point of M, then

$$(x_0, y) = \lim_{n \to \infty}(x_n, y) = 0$$

owing to the continuity of the inner product and Theorem 3.7.2. (See Exercise 7, Section 12.) ∎

The following corollary is an obvious consequence of the last theorem.

5.15.2 COROLLARY. *If X is complete, then $M^\perp$ is complete.*

Let us consider some elementary properties of orthogonal complements.

5.15.3 THEOREM. *Let M and N be nonempty sets in an inner product space X. The following statements are valid:*

(a) *If $M \subset N$, then $N^\perp \subset M^\perp$.*
(b) *$M \subset M^{\perp\perp}$.*
(c) *If $M \subset N$, then $M^{\perp\perp} \subset N^{\perp\perp}$.*
(d) *$M^\perp = M^{\perp\perp\perp}$.*
(e) *If $x \in M \cap M^\perp$, then $x = 0$.*
(f) *$\{0\}^\perp = X$ and $X^\perp = \{0\}$.*
(g) *If M is a dense subset of X, then $M^\perp = \{0\}$.*

Note that these results do not require completeness.

Proof:
(a) Let $x \in M$ and $y \in N^\perp$. Since $x \in N$, one has $(x, y) = 0$. Since x is an arbitrary point in M, we have $y \in M^\perp$, or $N^\perp \subset M^\perp$.
(b) Let x be any point in M. By definition $x \perp M^\perp$, so clearly $x \in M^{\perp\perp}$.
(c) This follows by applying (a) twice.
(d) Since $M \subset M^{\perp\perp}$, by statement (b), it follows that $M^{\perp\perp\perp} \subset M^\perp$, by statement (a). By applying (b) to $M^\perp$, we get $M^\perp \subset M^{\perp\perp\perp}$. Hence, $M^\perp = M^{\perp\perp\perp}$.
(e) If $x \in M \cap M^\perp$, then $(x,x) = 0$, which implies that $x = 0$.
(f) This is obvious.
(g) If $x \in M^\perp$, then for all $y \in M$ one has

$$\|x - y\|^2 = \|x\|^2 + \|y\|^2 \geq \|x\|^2$$

by the Pythagorean Theorem. Since M is dense we can choose y so that $\|x - y\|$ can be made arbitrarily small. Hence $x = 0$. (Carefully note that this result does not imply that if $M^\perp = \{0\}$, then M is dense in X. Such a statement requires, as we shall see, completeness.) ∎

EXAMPLE 4.   It is shown in Appendix D that the space $C_0^\infty(-\infty,\infty)$, where $C_0^\infty(-\infty,\infty)$ is the subset of $C^\infty(-\infty,\infty)$ made up of functions with compact support, is dense in $L_2(-\infty,\infty)$. Therefore, if M is any linear subspace of $L_2(-\infty,\infty)$ that contains $C_0^\infty(-\infty,\infty)$, then $M^\perp = \{0\}$. A particular example of this is the space

$$M = \left\{ x \in L_2(-\infty,\infty): \int_{-\infty}^{\infty} \omega^2 \, |x(\omega)|^2 \, d\omega < \infty \right\}.$$

Since $C_0^\infty \subset M$, one does have $M^\perp = \{0\}$. This space M arises in the study of the Fourier transform of differential operators. ∎

The orthogonal complement $M^\perp$ is, of course, defined for any set M. However, we shall be primarily interested in the case where M is a linear subspace of X. In fact, M will often be a closed linear subspace of a Hilbert space.

5.15.4 THEOREM.   *Let M be a linear subspace of a Hilbert space H. The following statements are valid:*

(a) $M^{\perp\perp} = \overline{M}$, *where $\overline{M}$ is the closure* of M.
(b) *If M is closed, then $M^{\perp\perp} = M$.*
(c) $M^\perp = \{0\}$ *if and only if M is dense in H.*
(d) *If M is closed and $M^\perp = \{0\}$, then $M = H$.*

Proof:
(a) Let $N = \overline{M}$, where M is a linear subspace of H. Thus, $M \subset N$. By Theorem 5.15.3(b), we have $M \subset M^{\perp\perp}$. Since $M^{\perp\perp}$ is closed (Theorem 5.15.1), one has $\overline{M} = N \subset M^{\perp\perp}$. If $N \neq M^{\perp\perp}$, then by Corollary 5.14.5 there is a nonzero vector z in $M^{\perp\perp}$ such that $z \perp N$. (Here is where we use the completeness of H.) Since $M \subset N$, one has $z \perp M$. Therefore, $z \in M^\perp$. But $z \in M^\perp \cap M^{\perp\perp}$ implies [Theorem 5.15.3(e)] that $z = 0$, a contradiction. Hence, $N = M^{\perp\perp}$.
(b) Follows directly from (a).
(c) The "if" part is simply Theorem 5.15.3(g). Suppose now that $M^\perp = \{0\}$. Then $M^{\perp\perp} = \{0\}^\perp = H$. It follows from (a) that $H = \overline{M}$, so M is dense in H.
(d) Follows directly from (c). ∎

The next example shows that Theorem 5.15.4 does not hold if the space is not complete.

EXAMPLE 5.   In Example 4, Section 14 we showed that if $z_0 \perp M$, then $z_0 = 0$. In other words, $M^\perp = \{0\}$. It follows from Theorem 5.15.3(f) that $M^{\perp\perp} = X$. On the other hand, M is closed but a proper subspace of X. Hence, $M^{\perp\perp} \neq \overline{M} = M$.

Carefully note that this example illustrates the failure of (a), (b), (c), and (d) in Theorem 5.15.4 when X is not complete. ∎

The remainder of this section is devoted to showing that the plausible statement $H = M + M^{\perp}$ is indeed true for Hilbert spaces, see Figure 5.15.4. The preceding example shows that the same statement is not necessarily true if the space is not complete.

To begin with, we have to say a few words about sums of subspaces in inner product spaces. If M and N are linear subspaces of X, the sum $M + N$ is, of course, defined exactly as in Section 4.10. However, now we can investigate the topological properties of $M + N$. Specifically, if M and N are both closed, what can be said about $M + N$? One might suspect that $M + N$ is always closed. Unfortunately, this is not always true as we shall see in the exercises. However, if $M \perp N$, then we can say the following:

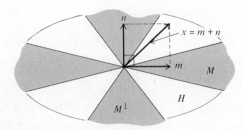

Figure 5.15.4.

5.15.5 THEOREM.   *Let M and N be closed linear subspaces of a Hilbert space H. If $M \perp N$, then $M + N$ is a closed linear subspace of H.*

Proof: Let $\{z_n\}$ be a convergent sequence in $M + N$ with limit $z = \lim z_n$. We wish to show that $z \in M + N$. It follows that $z_n = x_n + y_n$ where $x_n \in M$ and $y_n \in N$. Since $M \perp N$ one has

$$\|z_n - z_m\|^2 = \|x_n - x_m\|^2 + \|y_n - y_m\|^2$$

by the Pythagorean Theorem. Hence $\{x_n\}$ and $\{y_n\}$ are Cauchy sequences in M and N, respectively. Since M and N are complete, the limits $x = \lim x_n$, $y = \lim y_n$ exist and are in M and N, respectively. It follows from the continuity of addition that $z = x + y$. Hence, $M + N$ is closed.   ∎

Two assumptions in the last theorem are critical: (1) H is complete and (2) $M \perp N$. We shall show in the exercises (see Exercise 2, of this section, and Exercise 11, Section 17) that without these assumptions $M + N$ need not be closed.

We are now ready to prove a key theorem of Hilbert space geometry.

5.15.6 THEOREM. (THE PROJECTION THEOREM: FIRST VERSION.)   *Let M be any closed linear subspace of a Hilbert space H. Then $H = M + M^{\perp}$. Moreover, each $x \in H$ can be expressed uniquely $x = m + n$, where $m \in M$ and $n \in M^{\perp}$, and $\|x\|^2 = \|m\|^2 + \|n\|^2$.*

It must be appreciated that this is an extremely important statement. We shall see in the next section that it guarantees the existence of a rich and useful supply of orthogonal projections in any Hilbert space. Moreover, a key difference between inner product spaces and Hilbert spaces is that this theorem is not true for inner product spaces in general. Completeness is needed. We should also mention that a similar statement for Banach spaces is also not true. In particular, suppose that M is a closed linear subspace of a Banach space B. It can happen, see Taylor [2, p. 242], that there is *no* closed linear subspace N disjoint from M such that $B = M + N$. Here again simple geometric intuition works for Hilbert spaces but not elsewhere.

Proof: By Theorem 5.15.5 (which requires the completeness of H) we see that $Y = M + M^{\perp}$ is a closed linear subspace of H. Since $M \subset Y$ and $M^{\perp} \subset Y$, one has, using Theorem 5.15.3(a), $Y^{\perp} \subset M^{\perp}$ and $Y^{\perp} \subset M^{\perp\perp}$, that is, $Y^{\perp} \subset M^{\perp} \cap M^{\perp\perp}$. It follows from Theorem 5.15.3(e) that $Y^{\perp} = \{0\}$. Hence $Y = H$ by Theorem 5.15.4(d).

It follows next from Lemma 4.10.1 that for any $x \in X$ there are unique points $m \in M$ and $n \in M^{\perp}$ such that $x = m + n$. Since $m \perp n$, the Pythagorean Theorem assures us that $\|x\|^2 = \|m\|^2 + \|n\|^2$. ∎

EXAMPLE 6.   Let $H = L_2(-\infty,\infty)$, and let $S_{\tau}: H \to H$ be the shift operation (by τ), that is, $(S_{\tau}x)(t) = x(t - \tau)$. Let x_0 be a fixed point in H, and consider the linear subspace M of H made up of all points x of the form

$$x = \alpha_1 x_0(t - \tau_1) + \cdots + \alpha_n x_0(t - \tau_n) = [\alpha_1 S_{\tau_1} + \cdots + \alpha_n S_{\tau_n}]x_0, \qquad (5.15.2)$$

$n = 1, 2, 3, \ldots$, that is, all (finite) linear combinations of shifted versions of x_0.

It is a fact that $\overline{M}$, the closure of M, has many important applications. For example linear subspaces of this form are used in the study of time-invariant linear systems. Hence, it is of interest to characterize the closed linear subspace $\overline{M}$. We will use the fact that the Fourier transform $\mathscr{F}: L_2(-\infty,\infty) \to L_2(-i\infty,i\infty)$ is an isometric isomorphism, see Example 6, Section 19, and Example 11, Section 22.

Since $\mathscr{F}$ is an isometry characterizing the subspace $\overline{M}$ in $L_2(-\infty,\infty)$ is equivalent to characterizing the subspace $\overline{\mathscr{M}} = \mathscr{F}(\overline{M})$ in $L_2(-i\infty,i\infty)$. Since

$$\mathscr{F}[S_{\tau}x_0](i\omega) = e^{-i\omega\tau}\hat{x}_0(i\omega),$$

where $\hat{x}_0$ is the Fourier transform of x_0, it follows that $\overline{\mathscr{M}}$ is the closure of the subspace $\mathscr{M}$ of $L_2(-i\infty,i\infty)$ made up of all $\hat{x}$ of the form

$$\hat{x}(i\omega) = [\alpha_1 e^{-i\omega\tau_1} + \cdots + \alpha_n e^{-i\omega\tau_n}]\hat{x}_0(i\omega).$$

For each $\hat{x} \in L_2(-i\infty,i\infty)$ we define the support $K(\hat{x})$ to be that subset of $(-i\infty,i\infty)$ on which $\hat{x}(i\omega) \neq 0$. That is, $\hat{x}(i\omega) = 0$ for $(i\omega) \notin K(\hat{x})$. Recall that a function $\hat{x} \in L_2(-i\infty,i\infty)$ is only determined up to a set of measure zero. Therefore, the support set $K(\hat{x})$ is only determined within a set of measure zero.

We now claim that a function $\hat{z}$ lies in $\overline{\mathscr{M}}$ if and only if

$$K(\hat{z}) \subset K(\hat{x}_0) \qquad \text{(a.e)} \qquad\qquad (5.15.3)$$

This means that

$$\{i\omega \in K(\hat{z}): i\omega \notin K(\hat{x}_0)\}$$

is a set of measure zero.

Let us first show that if $\hat{z}$ lies in $\mathcal{M}$, then (5.15.3) holds. Indeed, since

$$\hat{z}(i\omega) = [\alpha_1 e^{-i\omega\tau_1} + \cdots + \alpha_n e^{-i\omega\tau_n}]\hat{x}_0(i\omega)$$

this is obvious. Next, if $\hat{z}$ lies in $\overline{\mathcal{M}}$, then $\hat{z}$ is the limit of a sequence $\{\hat{x}_n\}$ in $\mathcal{M}$. Since $K(\hat{x}_n) \subset K(\hat{x}_0)$, one then has $K(\hat{z}) \subset K(\hat{x}_0)$.

Next let us show that the converse is true, that is, let $\hat{z} \in L_2(-i\infty,i\infty)$ be chosen so that $K(\hat{z}) \subset K(\hat{x}_0)$. Since $\overline{\mathcal{M}}$ is closed, it follows from the Projection Theorem that there is a unique point $\hat{m} \in \overline{\mathcal{M}}$ such that $\hat{z} - \hat{m} \perp \overline{\mathcal{M}}$. In particular, one has

$$f(\tau) = \frac{1}{2\pi} \int_{-\infty}^{\infty} [\hat{z}(i\omega) - \hat{m}(i\omega)]\overline{\hat{x}_0(i\omega)}e^{i\omega\tau}\, d\omega = 0$$

for all $\tau \in (-\infty,\infty)$. However, $f(\tau)$ is merely the inverse Fourier transform of $[\hat{z} - \hat{m}]\overline{\hat{x}}_0$, (compare with Example 6, Section 19). Since $\mathcal{F}^{-1}$ is one-to-one, one has $[\hat{z}(i\omega) - \hat{m}(i\omega)]\overline{\hat{x}_0(i\omega)} = 0$, (a.e.). That is, $\hat{z}(i\omega) = \hat{m}(i\omega)$ (a.e.) on $K(\hat{x}_0)$. Since $K(\hat{z}) \subset K(\hat{x}_0)$, by assumption, and $K(\hat{m}) \subset K(\hat{x}_0)$, by the argument of the last paragraph, we see that $\hat{z} = \hat{m}$. Hence $\hat{z} \in \overline{\mathcal{M}}$.

If $\mathcal{F}^{-1}$ denotes the inverse Fourier transform, we then see that $\overline{M} = \mathcal{F}^{-1}(\overline{\mathcal{M}})$. Note that if $\hat{x}_0(i\omega)$ vanishes only on a set of measure zero, then $\overline{M} = L_2(-\infty,\infty)$. ∎

EXERCISES

1. Let M and N be linear subspaces of a Hilbert space H with $M \perp N$.
   (a) Show that $M^{\perp\perp} \perp N^{\perp\perp}$.
   (b) Is it true that $M^{\perp} \perp N^{\perp}$, or $M^{\perp\perp\perp} \perp N^{\perp\perp\perp}$?

2. Let X be the linear subspace of l_2 generated by the vectors

$$\left\{\left(1,\frac{1}{2},\frac{1}{2^2},\frac{1}{2^3},\ldots\right), e_1,e_2,e_3,\ldots\right\},$$

where $e_i = (\delta_{1i},\delta_{2i},\ldots)$ and $\delta_{ji} = $ Kronecker function. Let M denote those vectors $x = (x_1,x_2,\ldots)$ in X with $x_{2i+1} = 0$ for all i and N denote those vectors with $x_{2i} = 0$ for all i.
   (a) Show that $M \perp N$ and that M and N are closed.
   (b) Show that the vector $(1,1/2,1/2^2,\ldots)$ is not in $M + N$ but it is the limit of vectors z_n in $M + N$. (This shows that Theorem 5.15.5 fails if the underlying space X is not complete.)

3. Let $M_1, \ldots, M_n$ be closed linear subspaces of a Hilbert space H with $M_i \perp M_j$ for $i \neq j$. Show that $M = M_1 + \cdots + M_n$ is closed linear subspace of H.

4. Let $\mathscr{L}$ denote the collection of all closed subspaces of a Hilbert space H. For M, N in $\mathscr{L}$ let $M \wedge N$ denote the usual intersection $M \cap N$ and define

$$M \vee N = (M^\perp \cap N^\perp)^\perp.$$

(a) Show that $M \wedge N$ and $M \vee N$ belong to $\mathscr{L}$ whenever M and N do. (This means that $\mathscr{L}$ is a *lattice*.)
(b) The lattice $\mathscr{L}$ is said to be *modular* if $M \vee (N \wedge K) = (M \vee N) \wedge K$ whenever $M \subseteq K$. Show that $\mathscr{L}$ is modular if and only if H is finite dimensional.
(c) When does one have $M \vee N = M + N$?

16. ORTHOGONAL PROJECTIONS

The algebraic concept of a projection was discussed in Section 4.11. Recall that $P: X \to X$ is a projection if (1) it is linear and (2) $P^2 = P$. It was shown that from three natural points of view this is a reasonable definition of projection. First (Theorem 4.11.2), given a projection P, its range $\mathscr{R}(P)$ and null space $\mathscr{N}(P)$ are disjoint linear subspaces with $X = \mathscr{R}(P) + \mathscr{N}(P)$. Secondly (Theorem 4.11.3), given two disjoint linear subspaces M and N with $X = M + N$, there is a unique projection P such that $\mathscr{R}(P) = M$ and $\mathscr{N}(P) = N$. Thirdly (Theorem 4.11.4), given a linear subspace M, there is a projection P such that $\mathscr{R}(P) = M$; in fact there may be many such projections. Now that X is a normed linear space, one can ask whether or not these projections are continuous. As it happens some are and some are not. Indeed, if the range and null space get arbitrarily "close together" the associated projection is discontinuous.

Orthogonal projections on inner product spaces are a particularly important class of continuous projections with rather obvious geometric antecedents.

5.16.1 DEFINITION. A projection P on an inner product space X is said to be *orthogonal* if its range and null space are orthogonal, that is, $\mathscr{R}(P) \perp \mathscr{N}(P)$.

It follows immediately that if P is an orthogonal projection, then so is $I - P$. As already suggested, orthogonal projections are continuous.

5.16.2 THEOREM. *An orthogonal projection is continuous.*

Proof: Since P is a projection, each x in the inner product space X can be uniquely expressed $x = r + n$, with $r \in \mathscr{R}(P)$ and $n \in \mathscr{N}(P)$. Since P is orthogonal, we have $r \perp n$. It follows from the Pythagorean Theorem that $\|x\|^2 = \|r\|^2 + \|n\|^2$, so $\|Px\|^2 = \|r\|^2 \leq \|x\|^2$ and P is continuous. (We leave it to the reader to show that if $P \neq 0$, then $\|P\| = 1$.) ∎

Notice that the last theorem is valid even if X is not complete.

The next theorem shows that if we start with an orthogonal projection on an inner product space, the geometric situation is without surprise and agrees with intuition.

5.16.3 THEOREM.   *If P is an orthogonal projection on an inner product space X,*
then

(1) $\mathcal{N}(P)$ *and* $\mathcal{R}(P)$ *are closed linear subspaces,*
(2) $\mathcal{N}(P) = \mathcal{R}(P)^{\perp}$ *and* $\mathcal{R}(P) = \mathcal{N}(P)^{\perp}$,
(3) *each* $x \in X$ *can be written uniquely as* $x = r + n$, *where* $r \in \mathcal{R}(P)$ *and* $n \in \mathcal{N}(P)$, *and*
(4) $\|x\|^2 = \|r\|^2 + \|n\|^2$.

Proof: The proof of (3) follows directly from the fact that P is a projection (Theorem 4.11.2). Similarly, (4) follows from the fact that P is orthogonal. Statement (1) follows from the fact that $\mathcal{N}(P)$ and $\mathcal{R}(P)$ are the null spaces of the continuous operators P and $I - P$, respectively.

The proof of (2) is relatively straightforward. Since $\mathcal{N}(P) \perp \mathcal{R}(P)$, it is clear that $\mathcal{N}(P) \subset \mathcal{R}(P)^{\perp}$. Now, we want to show that $\mathcal{N}(P) \supset \mathcal{R}(P)^{\perp}$, and hence, that $\mathcal{N}(P) = \mathcal{R}(P)^{\perp}$. Let x be any point in $\mathcal{R}(P)^{\perp}$. Then there exists a unique $r_0 \in \mathcal{R}(P)$ and $n_0 \in \mathcal{N}(P)$ such that $x = r_0 + n_0$. Since $x \in \mathcal{R}(P)^{\perp}$, one has $(x,r) = 0$ for all $r \in \mathcal{R}(P)$. Then $0 = (r_0 + n_0,r) = (r_0,r) + (n_0,r) = (r_0,r)$ for all $r \in \mathcal{R}(P)$, in particular, for $r = r_0$. Hence, $r_0 = 0$ and $x = n_0 \in \mathcal{N}(P)$, which shows that $\mathcal{N}(P) \supset \mathcal{R}(P)^{\perp}$. A similar argument shows that $\mathcal{R}(P) = \mathcal{N}(P)^{\perp}$.   ∎

Next let M and N be linear subspaces of an inner product space X and assume that $M \perp N$ and $X = M + N$. Since $M \cap N = \{0\}$, it follows from Theorem 4.11.3 that there is a unique projection P with range $\mathcal{R}(P) = M$ and null space $\mathcal{N}(P) = N$. It follows then from Definition 5.16.1 that P is an orthogonal projection.

So far we have not used completeness.

Suppose now we are given a single linear subspace M of an inner product space X. Can we find an orthogonal projection P with the property that the range $\mathcal{R}(P)$ is precisely M? If we can, then M is the null space of the orthogonal projection $I - P$ and, consequently, it is closed. So we obviously have to assume that M is closed. The following result gives an affirmative answer in the case of a Hilbert space.

5.16.4 THEOREM. (THE PROJECTION THEOREM: SECOND VERSION.)   *Let M be any closed linear subspace of a Hilbert space H. Then there is one and only one orthogonal projection P with* $\mathcal{R}(P) = M$.

Proof: By the first version of the Projection Theorem (Theorem 5.15.6) we have

$$H = M + M^{\perp}$$

and every vector x in H can be written uniquely as $x = m + n$, where $m \in M$ and $n \in M^{\perp}$. Now define $P: H \to H$ by

$$P(m + n) = m.$$

It is easy to check that P is an orthogonal projection with $\mathscr{R}(P) = M$. Next let us show that P is unique. Suppose $\hat{P}$ is another orthogonal projection with $\mathscr{R}(\hat{P}) = M$. Since $\hat{P}$ is orthogonal, one has $\mathscr{N}(\hat{P}) = M^{\perp}$. Then

$$Pm = m = \hat{P}m, \qquad m \in M,$$
$$Pn = 0 = \hat{P}n, \qquad n \in M^{\perp}.$$

Hence, $P(m + n) = \hat{P}(m + n)$ for all $x = m + n \in H$, that is, $P = \hat{P}$. ∎

It should be clear that the two versions of the Projection Theorem (Theorems 5.15.5 and 5.16.4) are equivalent. The reason for the name "Projection Theorem" should also be clear after the last result.

Both versions of the Projection Theorem use the fact that H is complete in an essential way. Let us now show that Theorem 5.16.4 fails when the underlying inner product space is not complete.

EXAMPLE 1.   Referring back to Example 4, Section 14, let us assume that P is an orthogonal projection on X with $Px = x$ for all x in M. We shall now show that P is necessarily the identity on X; in other words, $X = \mathscr{R}(P) \neq M$, that is, there is no orthogonal projection of X onto M. First we note that since P is a projection one has $X = \mathscr{R}(P) + \mathscr{N}(P)$, by Theorem 4.11.2. Since $M \subset \mathscr{R}(P)$, one has $\mathscr{R}(P)^{\perp} \subset M^{\perp}$ by Theorem 5.15.3(a). However, we have shown that $M^{\perp} = \{0\}$ in Example 4, Section 14. Hence $\mathscr{N}(P) = \mathscr{R}(P)^{\perp} = \{0\}$, or $\mathscr{R}(P) = \mathscr{N}(P)^{\perp} = X$ by Theorem 5.15.3(f). ∎

In summary, then, geometric intuition "works" completely for orthogonal projections on Hilbert spaces. However, if the space X is not complete, we have to be careful. If we start with an orthogonal projection or if we start with complementary orthogonal subspaces, then there is no difficulty. But, as we have just seen, if we start with a closed linear subspace M in an incomplete space, there may not be an orthogonal projection P with $\mathscr{R}(P) = M$.

Let us return now to the approximation result of Theorem 5.14.4. Another way of formulating this result is to say that there is a mapping of the Hilbert space H into the closed linear subspace M (mapping x_0 onto y_0) with certain properties. We show now that this mapping is, not surprisingly, the orthogonal projection of H onto M.

5.16.5 THEOREM.   *Let M be a closed linear subspace in a Hilbert space H, and let $x_0 \in H$. Further, let P be the orthogonal projection with $\mathscr{R}(P) = M$. Then*

$$\|x_0 - Px_0\| < \|x_0 - y\| \qquad (5.16.1)$$

for all $y \neq y_0 = Px_0$ in M. That is, $\|x_0 - Px_0\| = \inf\{\|x_0 - y\| : y \in M\}$. (See Figure 5.16.1.)

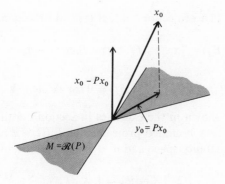

Figure 5.16.1.

Proof: First we note that $x_0 - Px_0 \perp M$. (Why?) Therefore,

$$x_0 - Px_0 \perp Px_0 - y$$

for all y in M. Hence, by the Pythagorean Theorem we get

$$\|x_0 - y\|^2 = \|x_0 - Px_0 + Px_0 - y\|^2 = \|x_0 - Px_0\|^2 + \|Px_0 - y\|^2 \geq \|x_0 - Px_0\|^2.$$

As a matter of fact, the inequality is strict unless $y = y_0 = Px_0$.   ∎

Let us consider some examples of projections.

EXAMPLE 2.   Consider the space $L_2(-a,a)$ where $0 < a \leq \infty$. Let M denote the collection of all even functions, that is, x is in M if and only if $x(-t) = x(t)$, almost everywhere. It is easy to see that M is a linear subspace of $L_2(-a,a)$. Define a mapping $P: L_2 \to L_2$ by $y = Px$, where

$$y(t) = \tfrac{1}{2}[x(t) + x(-t)].$$

It is clear that $M = \mathscr{R}(P)$ and that $Px = x$ for $x \in M$. Hence P is a projection. The null space of P is characterized by $x \in \mathscr{N}(P)$ if and only if $x(t) = -x(-t)$, almost everywhere, that is, x is an odd function. Recall that if a function z is odd and also in $L_1(-a,a)$, then $\int_{-a}^{a} z(t)\,dt = 0$. But if $x \in \mathscr{R}(P)$ and $y \in \mathscr{N}(P)$, then $x\bar{y}$ is odd and in $L_1(-a,a)$, so

$$(x,y) = \int_{-a}^{a} x\bar{y}\,dt = 0,$$

that is, $x \perp y$. We have shown that $\mathscr{R}(P) \perp \mathscr{N}(P)$; hence, P is an orthogonal projection.   ∎

EXAMPLE 3.   Consider the space $L_2(I)$, where I is any interval and let A be a measurable set in I, for example, A may be a subinterval. Define P_A by $y = P_A x$, where

$$y(t) = \begin{cases} x(t), & t \in A, \\ 0, & t \notin A. \end{cases}$$

It is easy to see that P_A is a projection, in fact it is an orthogonal projection with range

$$\mathscr{R}(P_A) = \{x \in L_2(I): x(t) = 0, \text{ for } t \notin A\}$$

and null space

$$\mathscr{N}(P_A) = \{x \in L_2(I): x(t) = 0 \text{ for } t \in A\}. \quad \blacksquare$$

EXAMPLE 4.   It is shown in the exercises in Section E.5 that the conditional expection operator $E^{\mathscr{B}}$ is a bounded projection of $L_2(\Omega,\mathscr{F},P)$ into itself. Recall that if $(\Omega,\mathscr{F},P)$ is a probability space and if $\mathscr{B}$ is a sub-σ-field of $\mathscr{F}$, then

$$\int_B E^{\mathscr{B}}[X] \, dP_{\mathscr{B}} = \int_B X \, dP$$

for all $B \in \mathscr{B}$. Let us show that $E^{\mathscr{B}}$ is an orthogonal projection on $L_2(\Omega,\mathscr{F},P)$. First we recall (Exercise 11, Section E.5) that if $X,\ Y \in L_2(\Omega,\mathscr{F},P)$, then

$$E^{\mathscr{B}}[\overline{E^{\mathscr{B}}[X]}Y] = \overline{E^{\mathscr{B}}[X]}E^{\mathscr{B}}[Y].$$

Thus if $Y \in \mathscr{N}(E^{\mathscr{B}})$, then $Y \perp E^{\mathscr{B}}[X]$ for all X in $L_2(\Omega,\mathscr{F},P)$, that is, $\mathscr{N}(E^{\mathscr{B}}) \perp \mathscr{R}(E^{\mathscr{B}})$. Hence $E^{\mathscr{B}}$ is an orthogonal projection.

We note in passing that one can easily show that

$$\mathscr{R}(E^{\mathscr{B}}) = L_2(\Omega,\mathscr{B},P_{\mathscr{B}}). \quad \blacksquare$$

EXERCISES

1. Contrast the geometric properties of Hilbert spaces with those of incomplete inner product spaces.

2. Let M^n denote the space of $n \times n$ matrices with complex coefficients and assume that M^n has an inner product given by

$$(A,B) = \sum_{i,\,j=1}^{n} a_{ij}\,\overline{b_{ij}} = \text{trace } \overline{B}^{\mathrm{T}} A.$$

Let $A \in M^n$ and let $\{\lambda_1,\dots,\lambda_k\}$ denote the eigenvalues of A, that is, all complex numbers λ with the property that $\det(\lambda I - A) = 0$. Let Γ be any simple closed curve in the complex plane that does not meet any of the eigenvalues $\{\lambda_1, \dots, \lambda_k\}$.
   (a) Show that $P_\Gamma = 1/2\pi i \int_\Gamma (\lambda I - A)^{-1} \, d\lambda$ is in M^n. (P_Γ is defined as the limit of the Riemann sums

$$\frac{1}{2\pi i} \sum_{j=1}^{n} [\lambda_j^* I - A]^{-1}(\lambda_j - \lambda_{j-1}),$$

   where $\{\lambda_0, \lambda_1,\dots,\lambda_n = \lambda_0\}$ is a partition of the curve Γ and λ_i^* is a point on the arc $\lambda_{i-1}\lambda_i$. The limit is taken as the arc lengths $|\lambda_{i-1}\lambda_i|$ tend to zero.)
   (b) Show that P_Γ is a projection on C^n.
   (c) Show that $P_\Gamma = 0$ if Γ does not enclose any of the eigenvalues $\{\lambda_1,\dots,\lambda_k\}$.
   (d) Show that $P_\Gamma = I$ if Γ encloses all of the eigenvalues $\{\lambda_1,\dots,\lambda_k\}$.

3. Let $\{P_1,\ldots,P_m\}$ be a collection of orthogonal projections with $P_iP_j = 0$ for $i \neq j$.
   (a). Show that $Q = P_1 + \cdots + P_m$ is an orthogonal projection.
   (b) What happens if one drops the assumption that $P_iP_j = 0$ for $i \neq j$?

4. Let M and N be closed linear subspaces of a Hilbert space H, and let P and Q denote the orthogonal projections onto M and N, respectively. We say that the ordered pair (M,N) is *compatible* if

$$(M \cap N) + (M \cap N^\perp) = M.$$

   (a) Show that (M,N) is compatible if and only if (N,M) is compatible.
   (b) Show that (M,N) is compatible if and only if P and Q commute.
   (c) Give an example of a pair of incompatible spaces. (This concept is important in quantum mechanics, see Jauch [2, pp. 80–86].)

17. ORTHONORMAL SETS AND BASES: GENERALIZED FOURIER SERIES

Orthonormal Sets

We discussed Hamel bases, a purely algebraic concept, in Section 4.7: Everything said there is, of course, applicable to Banach and Hilbert spaces. In the finite-dimensional case that is almost all that needs to be said. However, with topological structure now present, we have new opportunities open to us. In particular, it is now possible to attach meaning to infinite linear combinations of the form

$$\sum_{i=1}^{\infty} \alpha_i x_i$$

whereas in Chapter 4 we were limited to finite linear combinations. Thus we have the possibility of introducing types of bases which involve topological as well as algebraic structure. Without question the most useful such concept is that of an orthonormal basis in a Hilbert space. We will study it in this section and the next. We start with orthogonal and orthonormal sets.

5.17.1 DEFINITION. A set of points $\{x_\alpha\}$ in an inner product space X is said to be *orthogonal* if $x_\alpha \perp x_\beta$ whenever $\alpha \neq \beta$.

It is possible for a set of orthogonal points to contain the origin 0, since 0 is orthogonal to any point in X.

5.17.2 DEFINITION. A set of points $\{x_\alpha\}$ in an inner product space X is said to be *orthonormal* if

$$(x_\alpha, x_\beta) = \delta_{\alpha\beta}$$

for all α and β, where $\delta_{\alpha\beta}$ is the Kronecker function.

In both these definitions the index α may range over a finite, countably infinite, or uncountably infinite index set. Note that any orthogonal set of nonzero points $\{x_\alpha\}$ can be changed into an orthonormal set by replacing x_α with $x_\alpha/\|x_\alpha\|$.

5.17.3 THEOREM.   *Let $\{x_\alpha\}$ be an orthonormal set of points in an inner product space X. Then $\{x_\alpha\}$ is linearly independent.*

Proof: Let $\{x_1,\ldots,x_n\}$ be any finite set from $\{x_\alpha\}$ and consider

$$\alpha_1 x_1 + \cdots + \alpha_n x_n = 0. \tag{5.17.1}$$

We want to show that the only solution of (5.17.1) is $\alpha_1 = \cdots = \alpha_n = 0$. Let us show that $\alpha_1 = 0$. Indeed,

$$0 = (0,x_1) = (\alpha_1 x_1 + \cdots + \alpha_n x_n, x_1) = \alpha_1(x_1,x_1) + \cdots + \alpha_n(x_n,x_1) = \alpha_1,$$

because $(x_i,x_1) = \delta_{i1}$ for $1 \leq i \leq n$. Similarly, we get $\alpha_2 = \cdots = \alpha_n = 0$.   ∎

5.17.4 DEFINITION.   We shall say that an orthonormal set $B = \{x_\alpha\}$ in an inner product space X is *maximal*[10] if there is no unit vector x_0 in X such that $B \cup \{x_0\}$ is an orthonormal set.

Another way to say the same thing is given in the next lemma.

5.17.5 LEMMA.   *An orthonormal set $B = \{x_\alpha\}$ in an inner product space X is maximal if and only if $x \perp x_\alpha$ for all α implies that $x = 0$.*

The first thing to be said about maximal orthonormal sets is that there are plenty of them available. In fact, one can prove the following fairly strong result.

5.17.6 THEOREM.   *Let $\{x_\alpha\}$ be any orthonormal set in an inner product space X. Then there is a maximal orthonormal set B in X with $\{x_\alpha\} \subset B$.*

The proof of this theorem is a straightforward application of Zorn's lemma (Appendix C) and we shall only outline it here. Let $\{x_\alpha\}$ be an orthonormal set in X, and consider all orthonormal sets in X that contain $\{x_\alpha\}$. Order these sets in the obvious fashion and then verify that Zorn's lemma can be applied.

The main result in this section (Theorem 5.17.8) is that maximal orthonormal sets in Hilbert spaces have a natural geometric interpretation as orthogonal coordinate systems, or generalized Fourier series. Completeness is important here so we give maximal orthonormal sets in complete spaces a special name.

5.17.7 DEFINITION.   A maximal orthonormal set B in a Hilbert space H is referred to as an *orthonormal basis* for H.

---

[10] Many authors use the term "complete orthonormal set." We will not do this here since it seems preferable to reserve the use of the term "complete" for the Cauchy sequence property in metric spaces.

It follows from Theorem 5.17.6 that any orthonormal set in a Hilbert space can be extended to form an orthonormal basis. It can also be shown that any two orthonormal bases of a Hilbert space have the same cardinality, see Exercise 6.

The Fourier Series Theory

The next theorem, then, states the fundamental properties of orthonormal bases. Carefully note how topological as well as algebraic structure comes into play.

5.17.8 THEOREM. (FOURIER SERIES THEOREM.) *Let $\{x_n\}$ be an orthonormal set in a Hilbert space H. Then the following statements are equivalent:*

(a) *$\{x_n\}$ is an orthonormal basis.*
(b) *(Fourier series expansion.) For any x in H one has*

$$x = \sum_n (x,x_n)x_n .$$

(c) *(Parseval Equality.) For any two vectors x and y in H one has*

$$(x,y) = \sum_n (x,x_n)\overline{(y,x_n)}.$$

(d) *For any x in H one has*

$$\|x\|^2 = \sum_n |(x,x_n)|^2.$$

(e) *Let M be any linear subspace of H that contains $\{x_n\}$. Then M is dense in H.*

The coefficients (x,x_n) in the series expansion $x = \sum_n (x,x_n)x_n$ are often called the *Fourier coefficients* of x.

This theorem will be proved momentarily. However before doing this we need three preliminary results. The first of these is the Bessel Inequality. The second is a discussion of the convergence of $\sum_n \alpha_n x_n$ when $\{x_n\}$ is an orthonormal set. The third result gives a formula for computing the values of an orthogonal projection P in terms of an orthonormal basis in the range $\mathcal{R}(P)$.

In order to slightly simplify the following discussion, let us assume for the moment that the orthonormal sets and bases we consider are countable. The uncountable case can be quickly handled afterwards. In fact, we shall show that the theorems that follow are also meaningful for the uncountable case.

5.17.9 LEMMA. (THE BESSEL INEQUALITY.) *Let $\{x_n\}$ be an orthonormal set in an inner product space X. Then for any x in X one has*

$$\sum_n |(x,x_n)|^2 \le \|x\|^2. \tag{5.17.2}$$

Proof: Consider the finite subset $\{x_1,\ldots,x_N\}$ from $\{x_n\}$. Then one has

$$0 \le \left\| x - \sum_{i=1}^{N}(x,x_i)x_i \right\|^2 = \left(x - \sum_{i=1}^{N}(x,x_i)x_i, x - \sum_{j=1}^{N}(x,x_j)x_j \right)$$

$$= (x,x) - \sum_{i=1}^{N}(x,x_i)(x_i,x) - \sum_{j=1}^{N}\overline{(x,x_j)}(x,x_j)$$

$$+ \sum_{i=1}^{N}\sum_{j=1}^{N}(x,x_j)\overline{(x,x_j)}(x_i,x_j)$$

$$= \|x\|^2 - \sum_{i=1}^{N}|(x,x_i)|^2,$$

since $(x_i,x_j) = \delta_{ij}$. Therefore (5.17.2) holds for finite sums. Since the right side of (5.17.2) does not depend on N we see that it also holds for countable sums.   ∎

Next let us take a careful look at series of the form $\sum_n \alpha_n x_n$ where $\{x_n\}$ is an orthonormal set.

5.17.10 LEMMA.   *Let $\{x_n\}$ be a countably infinite orthonormal set in a Hilbert space H. Then the following assertions are valid:*

(a) *The infinite series $\sum_{n=1}^{\infty}\alpha_n x_n$ (where the α_n's are scalars) converges if and only if the series of real numbers $\sum_{n=1}^{\infty}|\alpha_n|^2$ converges.*
(b)[11] *Assume that $\sum_{n=1}^{\infty}\alpha_n x_n$ converges and let*

$$x = \sum_{n=1}^{\infty}\alpha_n x_n = \sum_{n=1}^{\infty}\beta_n x_n.$$

Then $\alpha_n = \beta_n$ for all n and $\|x\|^2 = \sum_{n=1}^{\infty}|\alpha_n|^2$.

Proof:
(a) Suppose that $\sum_{n=1}^{\infty}\alpha_n x_n$ is convergent and let $x = \sum_{n=1}^{\infty}\alpha_n x_n$, that is,

$$\lim_{N \to \infty} \left\| x - \sum_{n=1}^{N}\alpha_n x_n \right\|^2 = 0.$$

Then, since the inner product is continuous, one has

$$(x,x_j) = \left(\sum_{i=1}^{\infty}\alpha_n x_n, x_j \right) = \sum_{i=1}^{\infty}\alpha_n(x_n,x_j) = \alpha_j, \qquad \text{for all } j.$$

Then from Bessel's Inequality one gets

$$\sum_{n=1}^{\infty}|(x,x_n)|^2 = \sum_{n=1}^{\infty}|\alpha_n|^2 \le \|x\|^2,$$

which shows that $\sum_{n=1}^{\infty}|\alpha_n|^2$ converges. [*Note*: We do not need completeness for this part of the proof.]

---

[11] The completeness of H is not crucial for the proof of statement (b).

Next suppose that $\sum_{i=1}^{\infty}$ converges, and let $s_n = \sum_{i=1}^{n} \alpha_i x_i$. It follows that

$$\|s_n - s_m\|^2 = \sum_{i=m+1}^{n} |\alpha_i|^2,$$

so $\{s_n\}$ is a Cauchy sequence. Since H is complete, the sequence of partial sums $\{s_n\}$ is convergent. This completes the proof of statement (a).

(b) Let us first prove that $\|x\|^2 = \sum_{n=1}^{\infty} |\alpha_n|^2$.

$$\|x\|^2 - \sum_{n=1}^{N} |\alpha_n|^2 = \left(x, x - \sum_{n=1}^{N} \alpha_n x_n\right) + \left(x - \sum_{n=1}^{N} \alpha_n x_n, \sum_{n=1}^{N} \alpha_n x_n\right)$$

$$\leq \left\|x - \sum_{n=1}^{N} \alpha_n x_n\right\|^2 \left\{\|x\|^2 + \left\|\sum_{n=1}^{N} \alpha_n x_n\right\|^2\right\}$$

$$\leq 2\|x\|^2 \left\|x - \sum_{n=1}^{N} \alpha_n x_n\right\|^2 \to 0.$$

Hence $\|x\|^2 = \sum_{n=1}^{\infty} |\alpha_n|^2$.

Now if $x = \sum_{n=1}^{\infty} \alpha_n x_n = \sum_{n=1}^{\infty} \beta_n x_n$, then

$$0 = \lim_{N \to \infty} \left[\sum_{n=1}^{N} \alpha_n x_n - \sum_{n=1}^{N} \beta_n x_n\right] = \lim_{N \to \infty} \sum_{n=1}^{N} (\alpha_n - \beta_n) x_n,$$

or $0 = \sum_{n=1}^{\infty} (\alpha_n - \beta_n) x_n$. By the last paragraph we see that $0^2 = \sum_{n=1}^{\infty} |\alpha_n - \beta_n|^2$ or $\alpha_n = \beta_n$ for all n. ∎

Since the use of orthonormal bases involves infinite series, one does have to consider the possibility that the convergence of these series may converge only conditionally. The next corollary shows that this is not an issue.

5.17.11 COROLLARY.   *Let $\{x_i\}$ be a countably infinite orthonormal set in a Hilbert[12] space H. Then the infinite series $\sum_{i=1}^{\infty} \alpha_i x_i$, where the α's are scalars, is convergent if and only if it is unconditionally convergent.*

Proof: If the series is unconditionally convergent, it is certainly convergent. On the other hand, a simple application of Lemma 5.17.10 shows that if $\sum \alpha_i x_i$ is convergent, then any rearrangement of this series is convergent. We leave it as an exercise to show that the limit is independent of rearrangements. ∎

The next thing we wish to look at is the Projection Theorem in terms of an orthonormal basis. More specifically, if $\{x_n\}$ is an orthonormal set in a Hilbert space H, we wish to show that the formula

$$Px = \sum_n (x, x_n) x_n$$

defines an orthogonal projection on H.

Consider first the finite-dimensional case. Let $B = \{x_1, x_2, \ldots, x_n\}$ be a finite orthonormal set in H and let M be the linear subspace spanned by B. Then M is closed because it is finite dimensional. Furthermore, we know from the Projection

---

[12] Completeness is not crucial here.

Theorem (Theorem 5.16.4) that there is a unique orthogonal projection P on H with $\mathscr{R}(P) = M$.

5.17.12 LEMMA. *Let $B = \{x_1, x_2, \ldots, x_n\}$ be a finite orthonormal set in a Hilbert space H, and let M be the finite dimensional linear subspace of H spanned by B. Then the orthogonal projection of H onto M is given by*

$$Px = \sum_{i=1}^{n} (x, x_i) x_i.$$

Proof[13]: It is obvious that P is a linear mapping of H into itself. Now let x be any point in H. Since

$$Px_j = \sum_{i=1}^{n} (x_j, x_i) x_i = \sum_{i=1}^{n} \delta_{ij} x_i = x_j,$$

one has

$$P^2 x = P\left(\sum_{i=1}^{n} (x, x_i) x_i \right) = \sum_{i=1}^{n} (x, x_i) Px_i$$

$$= \sum_{i=1}^{n} (x, x_i) x_i = Px.$$

Hence, P is a projection. Moreover, it is obvious that $\mathscr{R}(P) \subset M$. Conversely, if $x \in M$, then $x = \alpha_1 x_1 + \alpha_2 x_2 + \cdots + \alpha_n x_n$, and it can be seen that $Px = x$, so $\mathscr{R}(P) = M$.

Next we show that P is orthogonal. Let $y \in \mathscr{N}(P)$ and $x \in \mathscr{R}(P)$. We want to show that $(y, x) = 0$. Since $x \in \mathscr{R}(P)$, we have $x = Px$ and

$$(y, x) = (y, Px) = \left(y, \sum_{i=1}^{n} (x, x_i) x_i \right)$$

$$= \sum_{i=1}^{n} (y, x_i) \overline{(x, x_i)} = \sum_{i=1}^{n} (y, x_i)(x_i, x)$$

$$= \left(\sum_{i=1}^{n} (y, x_i) x_i, x \right) = (Py, x) = (0, x) = 0.$$

So P is orthogonal. ∎

5.17.13 COROLLARY. *Let $\{x_1, \ldots, x_n\}$ be an orthonormal set in a Hilbert space H and let $x \in H$. Then for any choice of complex numbers $\{c_1, \ldots, c_n\}$ one has*

$$\left\| x - \sum_{i=1}^{n} (x, x_i) x_i \right\| \le \left\| x - \sum_{i=1}^{n} c_i x_i \right\|. \tag{5.17.3}$$

Proof: This corollary is a direct consequence of Theorem 5.16.5 and the last lemma. ∎

---

[13] The reader should reconsider this proof after the adjoint operator is introduced in Section 22. Also notice that the completeness of H is not really needed here.

Now let us turn to the infinite-dimensional case.

Let $\{x_n\}$ be an orthonormal set in a Hilbert space H. The *closed linear sub-space M of H generated by the* $\{x_n\}$ is defined to be the closure of the linear subspace generated by the $\{x_n\}$. That is, first form the linear subspace generated by the $\{x_n\}$ in the sense of Section 4.6, and then take the closure of this space in the sense of Section 3.12. Recall that the closure of a linear subspace is a closed linear subspace (Theorem 5.5.2).

5.17.14 LEMMA. *Let* $B = \{x_n\}$ *be a countable orthonormal set in a Hilbert*[14] *space H, and let M be the closed linear subspace generated by the set B. Then every vector* $x \in M$ *can be written uniquely as*

$$x = \sum_n (x,x_n)x_n. \tag{5.17.4}$$

Moreover, the mapping P defined by

$$Px = \sum_n (x,x_n)x_n \tag{5.17.5}$$

is the orthogonal projection of H onto M.

Proof: It is clear that any vector of the form (5.17.4) is in M. The problem is to show the converse. So let $x \in M$. Then $x = \lim_{N \to \infty} y_N$, where each y_N is a finite linear combination of vectors in B, that is, $y_N = \sum_{n=1}^{K} \alpha_n x_n$, where the α_n's and K depend on N. By adjoining additional terms if necessary, we can assume that $K \geq N$. It follows from Corollary 5.17.13 that

$$\left\| x - \sum_{n=1}^{K} (x,x_n)x_n \right\| \leq \left\| x - \sum_{n=1}^{K} \alpha_n x_n \right\| = \| x - y_N \|.$$

Since $\| x - y_N \| \to 0$ as $N \to \infty$, this implies that $x = \sum_n (x,x_n)x_n$. The uniqueness of (5.17.4) follows from Lemma 5.17.10 (b).

Let $y = Px$, where P is defined by (5.17.5). By using Lemma 5.17.10 and the Bessel Inequality we get

$$\| Px \|^2 = \left\| \sum_n (x,x_n)x_n \right\|^2 = \sum_n |(x,x_n)|^2 \leq \| x \|^2.$$

It follows that P is continuous. Then the proof that P is an orthogonal projection is essentially the same as the proof of Lemma 5.17.12. One merely calls upon the continuity of P and the continuity of the inner product to justify the interchange with summations.   ∎

Let us now prove the Fourier Series Theorem.

Proof of Fourier Series Theorem:
(a) $\Rightarrow$ (b). Assume that $\{x_n\}$ is a maximal orthonormal set in H and let M be the closed linear subspace of H generated by the set $\{x_n\}$. If $x \in M^{\perp}$, then $x \perp x_n$

---

[14] Completeness is not crucial here.

for all n. But $\{x_n\}$ is maximal, so $x = 0$ and $M^\perp = \{0\}$. It follows from Theorem 5.15.4(c) that $M = H$. (Here we are using the completeness of H!) Therefore, the orthogonal projection onto M is the identity map I and by Lemma 5.17.14 we have

$$Ix = x = \sum_n (x,x_n)x_n$$

for every x in H.

(b) $\Rightarrow$ (c). Let $x = \sum_n (x,x_n)x_n$ and $y = \sum_m (y,x_m)x_m$. Since $(x_n,x_m) = \delta_{nm}$ and the inner product is continuous we get

$$(x,y) = \left(\sum_n (x,x_n)x_n, \sum_m (y,x_m)x_m \right)$$

$$= \sum_n \sum_m (x,x_n)\overline{(y,x_m)}(x_n,x_m)$$

$$= \sum_n (x,x_n)\overline{(y,x_n)}.$$

(c) $\Rightarrow$ (d). This is obvious.

(d) $\Rightarrow$ (a). If $\{x_n\}$ is not maximal, then there is unit vector x_0 such that $x_0 \cup \{x_n\}$ is an orthonormal set. Using (d) and the fact that $(x_0,x_n) = 0$ for all n, we get

$$1 = \|x_0\|^2 = \sum_n |(x_0,x_n)|^2 = 0,$$

a contradiction,

(b) $\Leftrightarrow$ (e). Statement (e) says that the orthogonal projection onto $\overline{M}$ is the identity, and by Lemma 5.17.14 we see that this is equivalent to statement (b). ∎

The following corollary is now a simple exercise. Carefully compare it with Lemma 5.17.14. In particular, the following result requires completeness whereas Lemma 5.17.14 does not.

5.17.15 COROLLARY.   *Let M be any closed linear subspace of a Hilbert space H and let $\{x_n\}$ be an orthonormal basis for M. Then the orthogonal projection P of H onto M is given by $Px = \sum_n (x,x_n)x_n$.*

The Gram-Schmidt Process

Given any countable linearly independent set $\{y_n\}$ in an inner product space, it is always possible, in principle at least, to construct an orthonormal set from it. The construction, which is called the *Gram-Schmidt orthogonalization process*, is rather important, so we shall describe it here. Given the set $\{y_n\}$ we propose to construct an orthonormal set $\{x_n\}$ with the property that x_k is a linear combination of the vectors $y_1, y_2, \ldots, y_k$ for $k = 1, 2, 3, \ldots$. This is done by induction. Let $x_1 = y_1/\|y_1\|$. If $x_1, \ldots, x_k$ have been determined, we define x_{k+1} by

$$x_{k+1} = \alpha \left[y_{k+1} - \sum_{i=1}^{k} (y_{k+1},x_i)x_i \right], \tag{5.17.6}$$

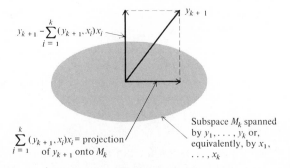

$$y_{k+1} - \sum_{i=1}^{k}(y_{k+1},x_i)x_i$$

$$\sum_{i=1}^{k}(y_{k+1},x_i)x_i = \text{projection}$$
$$\text{of } y_{k+1} \text{ onto } M_k$$

Subspace M_k spanned by $y_1,\ldots,y_k$ or, equivalently, by x_1, $\ldots, x_k$

Figure 5.17.1.

where α is a scalar chosen so that $\|x_{k+1}\| = 1$. This induction step is illustrated in Figure 5.17.1. We leave it to the reader to show that we do indeed generate an orthonormal set in this way.

We mention in passing that the Gram-Schmidt orthogonalization process can sometimes lead to a very sensitive computation. The difficulty arises when the vectors in the original linearly independent set $\{y_n\}$ get very close to being collinear. In that case, the number α can become very large.

EXAMPLE 1.   In C^2 with the usual inner product, let $y_1 = (1,0)$ $y_2 = (1,\varepsilon)$ where $\varepsilon > 0$. Then $x_1 = y_1$ and $x_2 = \alpha[y_2 - (y_2,x_1)x_1] = \alpha(0,\varepsilon)$. Hence $\alpha = 1/\varepsilon$. ∎

EXAMPLE 2.   In l_2 with the usual inner product, let

$$y_n = (y_{1n},y_{2n},\ldots), \qquad n = 1, 2, \ldots,$$

be given with $y_{in} = 1$, for $1 \leq i \leq n$, $y_{n+1,n} = \varepsilon^n$, where $0 < \varepsilon < 1$, and $y_{in} = 0$ for $n + 2 \leq i$. It follows that $x_k = (\delta_{1k},\delta_{2k},\ldots)$, however, the α in (5.17.6) now becomes $\alpha = \varepsilon^{-k}$. This phenomenon can and does cause a problem in computer applications. The reason for this is that the angle between y_n and y_{n+1} becomes very small as n tends to ∞. ∎

So far we have been restricting ourselves to Hilbert spaces with countable orthonormal bases. The next theorem shows that such spaces possess a very useful topological property: They are separable.

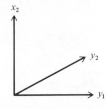

Figure 5.17.2.

5.17.16 THEOREM.   *A Hilbert H has a countable orthonormal basis if and only if it is separable.*

The proof of this is outlined in the exercises.

Because of Theorem 5.17.16, one usually refers to Hilbert spaces with countable orthonormal bases as separable Hilbert spaces.

Uncountable Bases

The next obvious question is "what can we do with nonseparable spaces?" Our basic problem is to say what we mean by the series

$$\sum_\alpha |(x,x_\alpha)|^2 \quad \text{and} \quad \sum_\alpha (x,x_\alpha)x_\alpha \qquad (5.17.7)$$

when the orthonormal set $\{x_\alpha\}$ is not countable. It turns out that (5.17.7) has a very straightforward interpretation. This is shown with the aid of the next result.

5.17.17 LEMMA.   *Let $\{x_\alpha\}$ be an orthonormal set in an inner product space X, and let x be any point in X. Then (x,x_α) is nonzero for at most a countable number of x_α's.*

In other words, if $\{x_\alpha\}$ is uncountable, then "most" of the coefficients (x,x_α) are zero. Needless to say, this result may be somewhat of a shock to geometric intuition.

Proof: Let x be any point in X, and let A denote the set of all x_α such that $|(x,x_\alpha)| > 0$. Then let

$$A_n = \{x_\alpha \in A : |(x,x_\alpha)|^2 > \|x\|^2/n\},$$

where $n = 1, 2, 3, \ldots$. It follows from the version of Bessel's Inequality for countable orthonormal sets already proved, Lemma 5.17.9, that A_n contains at most $(n-1)$ vectors. Since $A = \bigcup_{n=1}^\infty A_n$, it follows that A is at most countably infinite.   ∎

Because of this lemma we see that both series in (5.17.7) contain at most only a countable number of nonzero terms. We now can define when the two series in (5.17.7) converge. Specifically let $\{x_\alpha\}$ be an orthonormal set where α ranges in some index set A. We shall say that[15]

$$x = \sum_{\alpha \in A} (x,x_\alpha)x_\alpha = \sum_\alpha (x,x_\alpha)x_\alpha$$

---

[15] It should be noticed that this definition of convergence is different from that defined in Section 5.4, even in the case where the index set A is the set $1, 2, \ldots$. For example, the new definition of convergence automatically implies unconditional convergence. On the other hand, we only apply this new definition to orthonormal sets so the difference is more apparent than real, see Corollary 5.17.11.

provided for every $\varepsilon > 0$ there is a finite subset $E \subset A$ with the property that

$$\left\| x - \sum_{\alpha \in F} (x, x_\alpha) x_\alpha \right\| \leq \varepsilon$$

for all finite sets $F \subset A$ with $E \subset F$. The definition of the convergence of $\sum_\alpha |(x, x_\alpha)|^2$ is similar.

On the basis of these definitions one can go back through this section and show the following for nonseparable spaces:

(1) Bessel's Inequality (Lemma 5.17.9) is still valid.

(2) Lemma 5.17.10 would be meaningless as stated, but one could say that the series $\sum_\alpha \beta_\alpha x_\alpha$ converges if and only if β_α is nonzero for at most a countable number of α's and $\sum_\alpha |\beta_\alpha|^2 < \infty$.

(3) Lemmas 5.17.12 and 5.17.14 are modified in the same spirit.

(4) Finally the proof of the Fourier Series Theorem carries over in exactly the same way. The same can be said for Corollary 5.17.15.

In short, nonseparable Hilbert spaces offer no problems once Lemma 5.17.17 is available.

Orthonormal Bases versus Hamel Bases

We now have two kinds of bases for Hilbert spaces: Hamel bases and orthonormal bases. Here are a few remarks that one can make about the two:

(1) Hamel basis is a purely algebraic concept and it involves only finite linear combinations. One rarely uses Hamel bases in infinite-dimensional Hilbert spaces.

(2) Orthonormal basis is a combined topological and algebraic concept and it allows countably infinite linear combinations. Orthonormal bases are extremely useful in infinite-dimensional Hilbert spaces.

(3) Any orthonormal basis of a Hilbert space H is a subset of a Hamel basis for H. (Why?)

Series with Nonorthogonal Entries

This entire section up to this point has been devoted to orthonormal bases for Hilbert spaces. By this point the reader should appreciate their "clean" and simple structure. Unfortunately, there are cases where one must abandon this simplicity. Suppose that $\{x_n\}$ is a countably infinite linearly independent set in a Hilbert space H. We do not assume that $\{x_n\}$ is orthonormal or even orthogonal. Let M be the linear subspace of H spanned by $\{x_n\}$ in the sense of Section 4.6, that is, M consists of all finite linear combinations of points in $\{x_n\}$. Next let $\overline{M}$ be the closure of M. All we can say in general about a point $y \in \overline{M}$ is that there exists at least one sequence $\{S_N\}$ in M, where each S_N is a finite linear combination of points in $\{x_n\}$, such that $y = \lim S_N$. Equivalently, there are scalars α_{Ni} such that

$$y = \lim_{N \to \infty} (\alpha_{N1} x_1 + \alpha_{N2} x_2 + \cdots + \alpha_{NN'} x_{N'}).$$

Naturally, if $z \in H$ and

$$z = \beta_1 x_1 + \beta_2 x_2 + \beta_3 x_3 + \cdots,$$

then z is in $\overline{M}$. The very important point to be made here is that the converse is not necessarily true. That is, if y is a point in $\overline{M}$, it does not follow that there exists a convergent infinite series of the form $\alpha_1 x_1 + \alpha_2 x_2 + \cdots$ with limit y. (Compare with Lemma 5.17.14.) This point is illustrated by the next example.

EXAMPLE 3.   Let $\{y, z_1, z_2, z_3, \ldots\}$ be an orthonormal set in a Hilbert space H. Then construct the linearly independent set $\{x_n\}$ by

$$x_n = \left(\cos \frac{1}{n}\right) y + \left(\sin \frac{1}{n}\right) z_n, \qquad n = 1, 2, 3, \ldots.$$

Since $y = \lim_{n \to \infty} x_n$, it follows that $y \in \overline{M}$, where $\overline{M}$ is defined above. Now suppose that y can be written in the form

$$y = \alpha_1 x_1 + \alpha_2 x_2 + \cdots. \tag{5.17.8}$$

Then

$$y = \sum_{n=1}^{\infty} \alpha_n \left\{ \left(\cos \frac{1}{n}\right) y + \left(\sin \frac{1}{n}\right) z_n \right\}$$

and since $\{y, z_1, z_2, \ldots\}$ is an orthonormal set, one has

$$y = \left[\sum_{n=1}^{\infty} \alpha_n \left(\cos \frac{1}{n}\right) \right] y + \sum_{n=1}^{\infty} \alpha_n \left(\sin \frac{1}{n}\right) z_n.$$

Since the two terms on the right-hand side are orthogonal to one another it follows that

$$\sum_{n=1}^{\infty} \alpha_n \left(\cos \frac{1}{n}\right) = 1$$

and

$$\sum_{n=1}^{\infty} \alpha_n \left(\sin \frac{1}{n}\right) z_n = 0.$$

But then Lemma 5.17.10(b) implies that

$$\sum_{n=1}^{\infty} |\alpha_n|^2 \sin^2 \frac{1}{n} = 0^2 = 0,$$

so $0 = \alpha_1 = \alpha_2 = \alpha_3 = \cdots$. This implies that $y = 0$ which is a contradiction, so y cannot be expressed in the form (5.17.8).   ∎

This example illustrates a problem. In particular, we would like to know when every point in $\overline{M}$ can be expressed in the form (5.17.8). Needless to say, Lemma 5.17.14 gives one case. The next theorem gives a more general case.

5.17.18 THEOREM.   *Let $\{x_n\}$ be a countable linearly independent set in a Hilbert space H. Further assume that there are $D > 0$ and $\delta > 0$ such that*

$$\delta^2(|\beta_1|^2 + |\beta_2|^2 + \cdots + |\beta_N|^2) \leq \|\beta_1 x_1 + \cdots + \beta_N x_N\|^2, \qquad (5.17.9)$$

$$\|\beta_1 x_1 + \cdots + \beta_N x_N\|^2 \leq D^2(|\beta_1|^2 + \cdots + |\beta_N|^2) \qquad (5.17.10)$$

for all scalars $\beta_1, \beta_2, \ldots, \beta_N$ and $N = 1, 2, 3, \ldots$. Then each $y \in \overline{M}$, where M is the closed linear subspace of H defined above, can be expressed uniquely in the form

$$y = \alpha_1 x_1 + \alpha_2 x_2 + \cdots,$$

where the α_i's are scalars. Moreover, the coefficients α_n are continuous linear functions of y. In fact, denoting these functions by $\alpha_n = l_n(y)$, there exists a constant $B > 0$ such that $\|l_n\| \leq B$ for all n.

Note that Example 3 contains a violation of (5.17.9) because, for example, $\|x_n - x_{n+1}\|^2 \to 0$ as $n \to \infty$.

Before proving this theorem it is interesting to determine the geometric significance of conditions (5.17.9) and (5.17.10). To begin, we note the $D \geq \|x_N\| \geq \delta$ for $N = 1, 2, \ldots$, so that the x_N's neither get arbitrarily large nor arbitrarily small. It is more interesting to note that (5.17.10) is a kind of orthogonality condition. Let z be any unit vector in H. Then each x_n can be uniquely expressed as $x_n = \gamma_n z + w_n$, where w_n is orthogonal to z and γ_n is a scalar. Then

$$\begin{aligned}
\|\beta_1 x_1 + \cdots + \beta_N x_N\|^2 &= \|(\beta_1\gamma_1 + \cdots + \beta_N\gamma_N)z\|^2 + \|\beta_1 w_1 + \cdots + \beta_N w_N\|^2 \\
&\geq |\beta_1\gamma_1 + \cdots + \beta_N\gamma_N|^2 \|z\|^2 \\
&= |\beta_1\gamma_1 + \cdots + \beta_n\gamma_n|^2
\end{aligned}$$

and from (5.17.10) it follows that

$$D(|\beta_1|^2 + \cdots + |\beta_N|^2)^{1/2} \geq |\beta_1\gamma_1 + \cdots + \beta_N\gamma_N|$$

for all β_n's and N. But this implies (Why?) that $|\gamma_1|^2 + |\gamma_2|^2 + \cdots < \infty$. Hence, $\gamma_n \to 0$ as $n \to \infty$. Since $\gamma_n = (x_n, z)$, we see that x_n's get closer and closer to being orthogonal to z. Moreover, z was any unit vector, so that the x_n's "swing away" from any given direction. Orthonormal sets, of course, do the same thing and more.

Among other things, Inequality (5.17.9) shows that $\|x_N - x_M\|^2 \geq 2\delta^2$ for all N, M. That is, the x_n's cannot get arbitrarily close together. Indeed let $A = \{x_{n_1}, x_{n_2}, \ldots, x_{n_N}\}$ be any finite subset of $\{x_n\}$, and let x_k be a point in $\{x_n\}$ that is not in A. It follows from (5.17.9) that

$$\|x_k - c_1 x_{n_1} - \cdots - c_n x_{n_N}\|^2 \geq \delta^2(1^2 + |c_1|^2 + \cdots + |c_N|^2)^2$$

for all scalars $c_1, c_2, \ldots, c_N$. In other words, x_k is a uniform positive distance away from the finite-dimensional subspace spanned by A.

Proof of Theorem 5.17.18: Let y be any point in $\overline{M}$, and let $y = \lim_{N \to \infty} S_N$, where $S_N = \alpha_{N1} x_1 + \cdots + \alpha_{NN'} x_{N'}$ and N' depends on N. Without any loss of generality we can assume that $N \leq N'$ and that $N' \leq M'$ whenever $N \leq M$. (One need only add extra terms with the α's set equal to 0, if necessary.) Since $\{S_N\}$ is a convergent sequence in H, we have

$$\|S_N - S_M\| = \|\alpha_{N1} x_1 + \cdots + \alpha_{NN'} x_{N'} - \alpha_{M1} x_1 - \cdots - x_{MM'} x_{M'}\| \to 0$$

as $N, M \to \infty$. It follows from (5.17.9) that for $N \geq M$ one has

$$|\alpha_{N1} - \alpha_{M1}|^2 + |\alpha_{N2} - \alpha_{M2}|^2 + \cdots$$
$$+ |\alpha_{NM'} - \alpha_{MM'}|^2 + |\alpha_{N(M'+1)}|^2 + \cdots + |\alpha_{NN'}|^2 \to 0$$

as $N, M \to \infty$. But then the sequence $\{a_N\}$, where

$$a_N = (\alpha_{N1}, \alpha_{N2}, \ldots, \alpha_{NN'}, 0, \ldots)$$

is a Cauchy sequence in l_2, so there is a point $a_0 = (\alpha_{01}, \alpha_{02}, \alpha_{03}, \ldots)$ in l_2 such that $\lim_{N \to \infty} a_N = a_0$. We claim that

$$y = \alpha_{01} x_1 + \alpha_{02} \alpha_2 + \alpha_{03} x_3 + \cdots.$$

Indeed, if

$$z_N = \alpha_{01} x_1 + \alpha_{02} x_2 + \cdots + \alpha_{0N'} x_{N'},$$

then

$$\|y - z_N\| = \|y - S_N + S_N - z_N\| \leq \|y - S_N\| + \|S_N - z_N\|.$$

We know that $\|y - S_N\| \to 0$ as $N \to \infty$. Moreover, from (5.17.10) we have

$$\|S_N - z_N\|^2 = \|(\alpha_{N1} - \alpha_{01}) x_1 + \cdots + (\alpha_{NN'} - \alpha_{0N'}) x_{N'}\|^2$$
$$\leq D^2 \{|\alpha_{N1} - \alpha_{01}|^2 + \cdots + |\alpha_{NN'} - \alpha_{0N'}|^2\}.$$

Since $\lim_{N \to \infty} a_N = a_0$, we conclude that $\|S_N - z_N\| \to 0$ as $N \to \infty$. Hence, $\|y - z_N\| \to 0$ as $N \to \infty$, and $y = \alpha_{01} x_1 + \alpha_{02} x_2 + \cdots$.

Next let us show that this series expression of y is unique. Suppose that

$$y = \alpha_{01} x_1 + \alpha_{02} x_2 + \cdots = \beta_1 x_1 + \beta_2 x_2 + \cdots.$$

Then

$$\|\alpha_{01} x_1 + \cdots + \alpha_{0N} x_N - \beta_1 x_1 - \cdots - \beta_N x_N\| \to 0 \text{ as } N \to \infty.$$

But from (5.17.9) we have

$$\|(\alpha_{01} - \beta_1) x_1 + \cdots + (\alpha_{0N} - \beta_N) x_N\|^2 \geq \delta^2 (|\alpha_{01} - \beta_1|^2 + \cdots + |\alpha_{0N} - \beta_N|^2)$$

for all N. This implies that $\alpha_{0j} = \beta_j$, $j = 1, 2, \ldots$. Hence the series representation is unique.

Finally, we show that the linear functionals l_n are continuous. Let y and z be any two points in $\overline{M}$, with $y = \beta_1 x_1 + \beta_2 x_2 + \cdots$ and $z = \gamma_1 x_1 + \gamma_2 x_2 + \cdots$. Then

$$\|y - z\|^2 = \lim_{N \to \infty} \|(\beta_1 - \gamma_1) x_1 + \cdots + (\beta_N - \gamma_N) x_N\|^2$$

because of the continuity of the norm. It follows from (5.17.9) that

$$\|y - z\|^2 \geq \delta^2 \{|\beta_1 - \gamma_1|^2 + \cdots + |\beta_N - \gamma_N|^2\}$$

so

$$|\beta_n - \gamma_n| \leq \frac{1}{\delta} \|y - z\|, \qquad n = 1, 2, \ldots.$$

Hence, $\|l_n\| \leq 1/\delta = B$ for all n. ∎

EXERCISES

1. (a) Find an example of an infinite series in l_2 that is convergent but not absolutely convergent.
   (b) Show that the collection of absolutely convergent sequences in l_2 forms a dense linear subspace of l_2.

2. How do Theorem 5.4.3 and Corollary 5.17.11 differ?

3. Show that one can drop the assumption of completeness in Corollary 5.17.11.

4. Prove Theorem 5.17.6. [*Hint*: Use Zorn's Lemma.]

5. Show that every orthonormal basis in a finite-dimensional inner product space is also a Hamel basis.

6. Let B_1 and B_2 be two orthonormal bases of a given Hilbert space H. Show that there is a one-to-one mapping ψ of B_1 onto B_2. (This means that B_1 and B_2 have the same cardinal number. This cardinal number is called the *Hilbert dimension* of H.)
   (a) Show that for finite-dimensional spaces H, the Hilbert dimension of H agrees with the (ordinary) dimension.
   (b) What happens in infinite-dimensional spaces?

7. Let $\{e_\alpha : \alpha \in A\}$ be a maximal orthonormal set in a Hilbert space H. Let $\{f_\beta : \beta \in B\}$ be another orthonormal set in H such that for each α in A there is a β in B such that

$$\|e_\alpha - f_\beta\|^2 < \tfrac{1}{2}.$$

   (a) Show that β is uniquely determined.
   (b) Show that there is a one-to-one map of A into B, and that therefore cardinality $(A) \leq$ Cardinality (B). [*Hint*: Determine the mapping by (a). Let $\beta = \alpha$ so that we have $\|e_\alpha - f_\alpha\|^2 < \tfrac{1}{2}$.]
   (c) Show that the mapping in (b) is onto B, that is, $A = B$, as sets.
   (d) Show that $\{f_\beta : \beta \in B\}$ is a maximal orthonormal set for H.

8. In Example 2, show that the angle between y_n and y_{n+1} tends to 0 as n tends to ∞. Use

$$\cos \theta = \frac{(y_n, y_{n+1})}{\|y_n\| \cdot \|y_{n+1}\|}$$

as the definition of the angle.

9. Prove Theorem 5.17.16. [*Hint*: Show that if $\{x_n\}$ is a countable dense set in H, then one can extract a maximal linearly independent subset. Use the Gram-Schmidt process to construct a basis. Going the other way, let $\{y_n\}$ be a countable basis. Show that the collection of all elements of the form

$$x_{nm} = \sum_{\text{finite}} (\alpha_{nm} + i\beta_{nm})y_n,$$

where α_{nm}, β_{nm} are rational, forms a countable dense set in H.]

10. The following is an example of a nonseparable Hilbert space. Let AP denote the collection of all complex-valued functions $x(t)$ defined for $-\infty < t < \infty$ and with the property that

$$\lim_{T \to \infty} \frac{1}{2T} \int_{-T}^{T} |x(t)|^2 \, dt < \infty.$$

(a) Let

$$(x,y) = \lim_{T \to \infty} \frac{1}{2T} \int_{-T}^{T} x(t)\overline{y(t)} \, dt.$$

Show that (x,y) is an inner product on AP.

(b) Let x be a continuous τ-periodic function. Show that $x \in AP$ and that

$$\|x\|^2 = \frac{1}{\tau} \int_{0}^{\tau} |x(t)|^2 \, dt.$$

(c) Show that (x,y) is also given by

$$(x,y) = \lim_{T \to \infty} \frac{1}{T} \int_{0}^{T} x(t)\overline{y(t)} \, dt.$$

(d) Let $\phi_a(t) = e^{iat}$ where a is real. Show that $(\phi_a, \phi_b) = \delta_{ab}$. (Hence AP has an uncountable orthonormal set.)

(e) Let $f(t)$ be *Bohr almost periodic*, that is, f is continuous and for every $\varepsilon > 0$ the set $E(\varepsilon,f) = \{\tau \in R: |f(\tau + t) - f(t)| \le \varepsilon \text{ for all } t\}$ is *relatively dense* in R, which means that there is an $l > 0$ such that every interval of R of length $\ge l$ contains at least one element of $E(\varepsilon,f)$. Show that $f \in AP$. [*Hint*: First note that f is bounded and uniformly continuous on R. Now show that

$$\lim_{T \to \infty} \frac{1}{2T} \int_{-T}^{T} |f(t)|^2 \, dt$$

is finite, see Besicovitch [1].]

(f) Let $f \in AP$. Show that at most a countable number of the Fourier coefficients

$$C_a = \lim_{T \to \infty} \frac{1}{2T} \int_{-T}^{T} f(t)e^{-iat} \, dt$$

are nonzero.

(g) Show that $\sum_a |C_a|^2 \le \|f\|^2$.

For a proof that $\{\phi_a\}$ forms an orthonormal basis for AP see Riesz-Sz. Nagy [1, pp. 256–259].

11. In this exercise you are asked to show that $M + N$ need not be closed in H when M and N are disjoint and nonorthogonal closed linear subspaces of a Hilbert space H. (See Theorem 5.15.5.) Let $H = l_2$ with the usual inner product and let $e_n = (\delta_{1n}, \delta_{2n}, \ldots)$. Then $\{e_1, e_2, \ldots\}$ is an orthonormal basis for l_2. Let M denote the closed linear subspace generated by $\{e_1, e_3, e_5, \ldots\}$, and let N denote the closed linear subspace generated by $\{z_1, z_2, \ldots\}$, where

$$z_n = \alpha_n e_{2n-1} + \frac{1}{n} e_{2n},$$

where $\alpha_n > 0$ and $\alpha_n^2 = 1 - 1/n^2$.
(a) Show that $(z_n, z_m) = \delta_{nm}$ and that $M \cap N = \{0\}$.
(b) Let $y = \sum_{n=1}^{\infty} (1/n)e_{2n}$. Show that $y \notin M + N$ but that $y = \lim y_n$ where $y_n \in M + N$. [*Hint*: Argue by contradiction and show that if $y = x + z$ where $x \in M$ and $z \in N$, then $(z, z_n) = 1$ for all n.]

12. Let $y_n(t) = t^n$ for $n = 0, 1, 2, \ldots$ and $-1 \le t \le 1$.
(a) Show that $\{y_0, y_1, \ldots\}$ is a linearly independent set in $L_2[-1,1]$.
(b) Use the Gram-Schmidt orthogonalization process on $\{y_0, y_1, \ldots\}$ to determine the first four vectors $\{x_0, x_1, x_2, x_3\}$ in the orthonormal set $\{x_0, x_1, \ldots\}$ generated by $\{y_0, y_1, \ldots\}$.

13. Using the notation of Exercise 12, the *Legendre polynomials* are defined by

$$P_n(t) = \left[\frac{2}{2n + 1}\right]^{1/2} x_n(t), \qquad n = 0, 1, \ldots .$$

Show that the following hold:

(a) $\dfrac{d}{dt} P_{n+1} - \dfrac{d}{dt} P_{n-1} = (2n + 1)P_n,$ $\qquad n = 2, 3, \ldots .$

(b) $\dfrac{d}{dt}\left[(1 - t^2)\dfrac{d}{dt} P_n(t)\right] + n(n + 1)P_n(t) = 0,$ $\qquad n = 0, 1, \ldots .$

(c) $P_{n+1}(t) = \dfrac{(2n + 1)tP_n(t) - nP_{n-1}(t)}{n + 1},$ $\qquad n = 1, 2, \ldots .$

(d) $P_n(1) = 1$ and $P_n(-1) = (-1)^n,$ $\qquad n = 0, 1, \ldots .$

(e) $P_n(t) = \dfrac{1}{2^n n!} \dfrac{d^n}{dt^n}(t^2 - 1)^n,$ $\qquad n = 1, 2, \ldots .$

14. Let $\{\phi_n\}$ be an orthonormal set in $L_2[a,b]$. Show that $\{\phi_n\}$ is a basis if and only if

$$\sum_{n=1}^{\infty} \left|\int_a^x \phi_n(t)\, dt\right|^2 = x - a$$

for all x in $[a,b]$. [*Hint*: Show that $\|f\|^2 = \sum_{n=1}^{\infty} |(f, \phi_n)|^2$ for all step functions f.]

15. Let $\{\phi_n\}$ be an orthonormal set in $L_2[a,b]$. Show that $\{\phi_n\}$ is a basis if and only if

$$\sum_{n=1}^{\infty} \left|\int_a^b \int_a^x \phi_n(t)\, dt\, dx\right|^2 = \frac{(b - a)^2}{2}.$$

16. Show that Lemma 5.17.12 holds when M is a finite-dimensional linear subspace of an (incomplete) inner product space. Compare this result with the Projection Theorem (Theorem 5.16.4) and Example 1, Section 16.

17. Let $\{x_n\}$ be an orthonormal set in an inner product space X, and also let $x = \sum_n \alpha_n x_n$. Show that $\alpha_n = (x,x_n)$ and $\|x\|^2 = \sum_n |(x,x_n)|^2$.

18. Let $\{x_n\}$ be an orthonormal set in an inner product space X, and let m be a positive integer. Let x be fixed and show that the set

$$B_m = \{x_n : \|x\|^2 < m\,|(x,x_n)|^2\}$$

contains at most $m - 1$ elements.

19. Let $B = \{x_1,x_2,\ldots\}$ be an orthonormal set in a Hilbert space H. Then let $\{z_n\}$ be a convergent sequence in H, where each z_n is a finite linear combination of points in B, that is, $z_n = \alpha_{n1}x_1 + \alpha_{n2}x_2 + \cdots + \alpha_{nn'}x_{n'}$. Show that for each $j = 1, 2, \ldots$ the sequence $\{\alpha_{nj}\}$ is convergent, that is,

$$\alpha_j = \lim_{n\to\infty} \alpha_{nj}$$

and that

$$\lim_{n\to\infty} z_n = \lim_{n'\to\infty} \sum_{j=1}^{n'} \alpha_{nj} x_j = \sum_{j=1}^{\infty} \alpha_j x_j.$$

20. Is the completeness of H crucial in Corollary 5.17.13?

21. Explain why Corollary 5.17.15 requires completeness whereas Lemma 5.17.14 does not.

22. Use Lemma 5.17.14 and the orthogonal projection P to re-interpret the Bessel Inequality. (See Figure 5.17.3.)

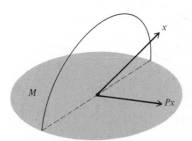

Figure 5.17.3.

18.   EXAMPLES OF ORTHONORMAL BASES

EXAMPLE 1.   Let I be the interval $[0,1]$ and H the complex space $L_2(I)$ with the usual inner product. We claim that the set

$$\phi_n(t) = e^{2\pi int}, \qquad n = 0, \pm1, \pm2, \ldots$$

is an orthonormal set in H. Indeed, if $n \neq m$, then

$$(\phi_n, \phi_m) = \int_0^1 e^{2\pi int} \overline{e^{2\pi imt}} \, dt = \int_0^1 e^{2\pi int} e^{-2\pi imt} \, dt$$

$$= \int_0^1 e^{2\pi i(n-m)t} \, dt = 0.$$

However,

$$\|\phi_n\|^2 = (\phi_n, \phi_n) = \int_0^1 e^{2\pi int} e^{-2\pi int} \, dt = 1.$$

Next we want to show that the set $\{\phi_n : n = 0, \pm 1, \ldots\}$ is a maximal orthonormal set. For this we need the following result.

5.18.1 LEMMA.   *Let $f \in H$ be a continuous real-valued function on $[0,1]$ and assume that*

$$f \perp \phi_n, \qquad n = 0, \pm 1, \ldots.$$

Then $f = 0$.

Proof: Note that if $f \perp \phi_n$ for all n, then $f \perp P$ for any finite linear combination P of the ϕ_n. We will now proceed by contradiction, that is, assume that $f \neq 0$. We shall now construct a finite linear combination P for which $(f, P) > 0$, which gives us the contradiction.

Since $f \neq 0$ one has $f(t_0) \neq 0$ for some $t_0 \in [0,1]$. Say that $f(t_0) > 0$. Since f is continuous, there exists constants $\delta > 0$ and $b > 0$ such that

$$f(t) \geq b, \qquad \text{for } |t - t_0| < \delta. \tag{5.18.1}$$

(We assume that δ is chosen so that $0 \leq t_0 - \delta < t_0 + \delta \leq 1$.) Now let

$$\psi(t) = 1 + \cos 2\pi(t - t_0) - \cos 2\pi\delta$$

and

$$P(t) = [\psi(t)]^N,$$

where N is to be determined. Since

$$2 \cos \theta = (e^{i\theta} + e^{-i\theta}),$$

ψ is a finite linear combination of the ϕ_n. Furthermore, for every $N \geq 0$, P is a finite linear combination of the ϕ_n. Also P is real-valued.

Define k by

$$k = \psi\left(t_0 + \frac{\delta}{2}\right) = 1 + \cos \pi\delta - \cos 2\pi\delta > 1.$$

Since

$$\psi(t) \geq k, \qquad \text{for} \quad |t - t_0| \leq \frac{\delta}{2},$$

one has

$$P(t) > k^N, \qquad \text{for} \quad |t - t_0| \leq \frac{\delta}{2}. \tag{5.18.2}$$

Furthermore, since

$$\psi(t) \geq 1, \qquad \text{for} \quad |t - t_0| \leq \delta,$$

one has

$$P(t) \geq 1, \qquad \text{for} \quad |t - t_0| < \delta. \tag{5.18.3}$$

Similarly, since

$$|\psi(t)| < 1, \qquad \text{for} \quad |t - t_0| > \delta \quad \text{and} \quad t \in I,$$

one has

$$|P(t)| \leq 1, \qquad \text{for} \quad |t - t_0| > \delta \quad \text{and} \quad t \in I. \tag{5.18.4}$$

Since f is a continuous function, it is bounded on I, that is, there is a constant M such that

$$-M \leq f(t) \leq M, \qquad \text{for} \quad t \in I. \tag{5.18.5}$$

By using (5.18.4) and (5.18.5) we get

$$P(t)f(t) \geq -M, \qquad \text{for} \quad |t - t_0| > \delta \quad \text{and} \quad t \in I.$$

And by using (5.18.1) and (5.18.3)

$$P(t)f(t) \geq b \geq -M, \qquad \text{for} \quad \frac{\delta}{2} \leq |t - t_0| < \delta.$$

Hence

$$P(t)f(t) \geq -M, \qquad \text{for} \quad |t - t_0| \geq \frac{\delta}{2} \quad \text{and} \quad t \in I. \tag{5.18.6}$$

Similarly by using (5.18.1) and (5.18.2) we get

$$P(t)f(t) \geq bk^N, \qquad \text{for} \quad |t - t_0| \leq \frac{\delta}{2}. \tag{5.18.7}$$

Since P is a finite linear combination of the ϕ_n one has $f \perp P$. But this gives us

$$0 = (P,f) = \int_0^1 P(t)f(t)\,dt = \left(\int_0^{t_0 - \delta/2} + \int_{t_0 - \delta/2}^{t_0 + \delta/2} + \int_{t_0 + \delta/2}^1 \right) P(t)f(t)\,dt$$

$$\geq -M(1 - \delta) + \delta b k^N.$$

By choosing N sufficiently large, the right-hand side can be made positive and this leads to a contradiction. ∎

5.18.2 LEMMA.   *Let $f \in H$ be a continuous complex-valued function on $[0,1]$ and assume that*

$$f \perp \phi_n, \quad for \quad n = 0, \pm 1, \pm 2, \ldots .$$

Then $f = 0$.

Proof: We apply Lemma 5.18.1 to the two real-valued functions $\operatorname{Re}(f)$ and $\operatorname{Im}(f)$. It is for the reader to show that if $f \perp \phi_n$, then $\operatorname{Re}(f) \perp \phi_n$ and $\operatorname{Im}(f) \perp \phi_n$ for $n = 0, \pm 1, \ldots$. ∎

5.18.3 THEOREM.   *The set $\{\phi_n : n = 0, \pm 1, \pm 2, \ldots\}$ is a maximal orthonormal set in $L_2[0,1]$.*

Proof: We will show that if $f \in L_2[0,1]$ and $f \perp \phi_n$ for all $n = 0, \pm 1, \ldots$, then $f = 0$, almost everywhere. Lemmas 5.18.1 and 5.18.2 establish this when f is continuous. If f is not continuous, let

$$G(t) = -t \int_0^1 f(s)\, ds + \int_0^t f(s)\, ds \quad and \quad F(t) = G(t) - \int_0^1 G(\tau)\, d\tau$$

for $0 \le t \le 1$. It is known (see Appendix D) that F is continuous and differentiable and that $F'(t) = f(t) - \int_0^1 f(s)\, ds$, almost everywhere. By integrating by parts one easily gets

$$\int_0^1 F(t) e^{2\pi i n t}\, dt = 0, \quad for \quad n = 0, \pm 1, \ldots,$$

that is, $F \perp \phi_n$. It follows from the above that $F = 0$, which in turn implies that $f = $ constant. Since $f \perp \phi_0$ one has $f = 0$. ∎

EXAMPLE 2.   Let $I = [a,b]$ be any other bounded interval with $a < b$, and let H be the complex space $L_2(I)$ with the usual inner product. Then

$$\phi_n(t) = (b - a)^{-1/2} \exp\left(2\pi i n \frac{t - a}{b - a}\right), \quad n = 0, \pm 1, \ldots$$

forms a maximal orthonormal set in $L_2(I)$. ∎

EXAMPLE 3.   Let Y denote the linear space made up of all complex-valued functions $f(z)$ defined on the unit circle Γ of the complex plane such that

$$\left| \frac{1}{2\pi i} \oint_\Gamma |f(z)|^2 \frac{dz}{z} \right| < \infty,$$

where the contour of integration is the unit circle, and

$$(f,g) = \frac{1}{2\pi i} \oint_{\Gamma} f(z)\overline{g(z)} \frac{dz}{z},$$

where $\bar{g}$ is the complex conjugate of g. [This is the space of all two-sided z-transforms of sequences in $l_2(-\infty,\infty)$.] We claim that the set $\{\ldots,z^{-2},z^{-1},1,z,z^2,\ldots\}$ is a maximal orthonormal set. In fact, this is merely Example 1 in disguise. ∎

EXAMPLE 4. (MULTIPLE FOURIER SERIES.)   Let I be the rectangle in R^m consisting of all $t = (t_1,\ldots,t_m)$ such that $0 \le t_i \le 1$, $i = 1, \ldots, m$. Let $\mu = (\mu_1,\ldots,\mu_m)$ be an m-vector with integer components and define

$$\mu \cdot t = (\mu,t) = \mu_1 t_1 + \cdots + \mu_m t_m.$$

Let $\|\mu\| = (\mu \cdot \mu)^{1/2}$. Let A denote the class of all such μ. If f is a complex-valued function defined on I we let $\int_I f\, dt$ denote

$$\int_I f\, dt = \int_0^1 \cdots \int_0^1 f(t_1,\ldots,t_m)\, dt_1 \cdots dt_m.$$

The space $L_2(I)$ is the space of all measurable complex-valued functions f for which $\int_I |f|^2\, dt < \infty$. The inner product on $L_2(I)$ is given by

$$(f,g) = \int_I f\bar{g}\, dt.$$

For each μ in A we define

$$\phi_\mu(t) = e^{2\pi i(\mu,\, t)}. \tag{5.18.8}$$

Then $\overline{\phi_\mu(t)} = \phi_{-\mu}(t)$. Let $p = (p_1,\ldots,p_m) = \mu - v$, then

$$(\phi_\mu,\phi_v) = \int_I e^{2\pi i(\mu,\, t)} e^{2\pi i(-v,\, t)}\, dt$$

$$= \int_I e^{2\pi i(\mu - v,\, t)}\, dt$$

$$= \int_I e^{2\pi i(p_1 t_1 + \cdots + p_m t_m)}\, dt$$

$$= \int_0^1 e^{2\pi i p_1 t_1}\, dt_1 \cdot \cdots \cdot \int_0^1 e^{2\pi i p_m t_m}\, dt_m.$$

If $p \neq 0$, then for some coordinate one has $p_j \neq 0$, and

$$\int_0^1 e^{2\pi i p_j t_j}\, dt_j = 0.$$

This implies that $(\phi_\mu,\phi_v) = 0$ if $\mu \neq v$. Similarly one gets

$$(\phi_\mu,\phi_\mu) = \int_I 1\, dt = 1.$$

Hence the family

$$\{\phi_\mu : \mu \in A\}$$

is an orthonormal family. By reasoning similar to that used in Example 1 we see that this family is maximal.

The Fourier series expansion in this case is

$$f = \sum_\mu c_\mu \phi_\mu, \tag{5.18.9}$$

where $c_\mu = (f, \phi_\mu)$. This is sometimes written as

$$f(t_1, \ldots, t_m) = \sum_{(\mu_1, \ldots, \mu_m)} c_{(\mu_1, \ldots, \mu_m)} e^{2\pi i (\mu_1 t_1 + \cdots + \mu_m t_m)}. \tag{5.18.10}$$

For obvious reasons, the notation in (5.18.9) is preferable to that of (5.18.10).   ∎

EXAMPLE 5.   It is easy to see that,

$$e_n = (\delta_{n1}, \delta_{n2}, \ldots), \qquad n = 1, 2, \ldots,$$

defines an orthonormal basis for $l_2[0, \infty)$. Similarly

$$e_n = (\ldots, \delta_{n,-1}, \delta_{n0}, \delta_{n1}, \delta_{n2}, \ldots)$$

defines an orthonormal basis for $l_2(-\infty, \infty)$. (This is probably a good time for the reader to review Example 2 of Section 4.7.)   ∎

EXAMPLE 6.   The *Laguerre functions* form an orthonormal set for $L_2[0, \infty)$ (compare with Exercise 8). These are defined by

$$\phi_n(t) = \frac{1}{n!} e^{-t/2} L_n(t), \qquad n = 0, 1, \ldots,$$

where $L_n(t)$ is the *Laguerre polynomial*

$$L_n(t) = e^t D^n(t^n e^{-t}) = \sum_{k=0}^n (-1)^k \binom{n}{k} n(n-1) \cdots (k+1) t^k$$

and $D^n = d^n/dt^n$.   ∎

EXAMPLE 7.   The *Hermite functions*

$$\phi_n(x) = \frac{e^{-x^2/2}}{[2^n n! \sqrt{\pi}]^{1/2}} H_n(x), \qquad n = 0, 1, \ldots$$

form an orthonormal basis for $L_2(-\infty, \infty)$, where $H_n(x)$ is the *Hermite polynomial*,

$$H_n(x) = (-1)^n e^{x^2} D^n(e^{-x^2}).$$

(See Section 7.14.)   ∎

EXAMPLE 8.   One can construct an orthonormal basis for $L_2(R^m)$ as follows:
Let $n = (n_1,\ldots,n_m)$ be a vector with nonnegative integer entries and also let
$x = (x_1,\ldots,x_m)$ be a point in R^m. Define

$$H_n(x) = H_{n_1}(x_1)H_{n_2}(x_2) \cdots H_{n_m}(x_m),$$

where H_{n_i} is the Hermite polynomial of order n_i. Let

$$n! = (n_1!)(n_2!) \cdots (n_m!), \qquad 2^n = 2^{n_1}2^{n_2} \cdots 2^{n_m},$$

and

$$e^{-|x|^2/2} = e^{-x_1^2/2} \cdot e^{-x_2^2/2} \cdots e^{-x_m^2/2}.$$

Then

$$\phi_n(x) = \frac{e^{-|x|^2/2}}{[2^n n! \sqrt{\pi^n}]^{1/2}} H_n(x)$$

is an orthonormal basis for $L_2(R^m)$. ∎

EXERCISES

1. Prove that the set $\{\phi_n : n = 0, \pm 1, \ldots\}$ defined in Example 2 forms a maximal
   orthonormal set in $L_2[a,b]$. [*Hint*: Define the real-valued function $s = \alpha(t)$ by

$$s = \frac{t - a}{b - a}.$$

   Then $\alpha: [a,b] \rightarrow [0,1]$, and $\alpha^{-1}: [0,1] \rightarrow [a,b]$. Now define an operator K by
   $y = Kx$, where

$$y(t) = (b - a)^{-1/2}x(\alpha(t)).$$

   Show that K is a linear mapping of $L_2[0,1]$ onto $L_2[a,b]$ and $\|Kx\| = \|x\|$.
   Now compare the basis in Example 1 to that of Example 2. What happens if
   $a = b$?]

2. Show that the family

$$\{1\} \cup \{\sqrt{2} \cos 2\pi nt: n = 1, 2, \ldots\} \cup \{\sqrt{2} \sin 2\pi nt: n = 1, 2, \ldots\}$$

   forms an orthonormal basis for $L_2[0,1]$. How is this related to the basis in
   Example 1?

3. Use the methods of Examples 2 and 3 to find a maximal orthonormal set for
   $L_2(I)$ where I is a rectangle in R^m given by

$$I = \{t = (t_1,\ldots,t_m): a_i \leq t_i \leq b_i, i = 1, \ldots, m\},$$

   where $a_i < b_i$.

4. In Exercises 12 and 13 of Section 17 it was shown that the Legendre polynomials

$$P_n(t) = \frac{1}{2^n n!} \frac{d^n}{dt^n} (t^2 - 1)^n, \qquad n = 0, 1, \ldots,$$

form an orthogonal set and that $x_n(t) = [(2n + 1)/2]^{1/2} P_n(t)$ forms an orthonormal set for $L_2[-1,1]$. In this exercise we shall prove that the orthonormal
set $\{x_n : n = 0, 1, \ldots\}$ is maximal. The proof is similar to the argument of
Example 1.
  (a) Show that $f = 0$ is the only continuous, real-valued function f on $[-1,1]$
      for which $\int_{-1}^{1} f(t)t^n \, dt = 0$. In other words, if $y_n(t) = t^n$ we claim that if
      $f \perp y_n$ for all t, then $f = 0$.
  (b) Next show that the only continuous, complex-valued function f on $[-1,1]$
      for which $\int_{-1}^{1} f(t)t^n \, dt = 0$ is $f = 0$, that is, the only continuous, complex-
      valued function for which $f \perp y_n$, $n = 0, 1, \ldots$, is $f = 0$.
  (c) Now show that the orthonormal set $\{x_n : n = 0,1,\ldots\}$ is maximal.

5. Let X denote the collection of all functions $f(z)$ that are analytic for $|z| < 1$ and
such that

$$\iint\limits_{|z| < 1} |f(z)|^2 \, dx \, dy < \infty.$$

  (a) Show that

$$(f,g) = \iint\limits_{|z| < 1} f\bar{g} \, dx \, dy$$

      defines an inner product on X.
  (b) Let $\phi_n(z) = (n/\pi)^{1/2} z^{n-1}$ for $n = 1, 2, \ldots$. Show that $\{\phi_n\}$ forms an ortho-
      normal basis for X.
  (c) Compare the Fourier coefficients of f with the coefficients in the power
      series expansion for f.

6. (Isoperimetric Theorem.) Show that among all simple, closed piecewise smooth
curves of length L in the plane, the circle encloses the maximum area. [*Hint*:
Proceed as follows.
  (a) Let $x = x(s)$, $y = y(s)$, $0 \le s \le L$, be a parametric representation of the
      curve using the arc lengths as a parameter. Let $t \cdot L = s$ and let

$$x(t) = a_0 + \sqrt{2} \sum_{n=1}^{\infty} (a_n \cos 2\pi nt + b_n \sin 2\pi nt)$$

$$y(t) = c_0 + \sqrt{2} \sum_{n=1}^{\infty} (c_n \cos 2\pi nt + d_n \sin 2\pi nt)$$

be the Fourier series expansions for x and y on the interval $0 \le t \le 1$. (See Exercise 2.) Show that

$$\left(\frac{dx}{dt}\right)^2 + \left(\frac{dy}{dt}\right)^2 = \left(\frac{dL}{dt}\right)^2,$$

and

$$L^2 = \int_0^1 \left[\left(\frac{dx}{dt}\right)^2 + \left(\frac{dy}{dt}\right)^2\right] dt = \sum_{n=1}^{\infty} 4\pi^2 n^2 (a_n^2 + b_n^2 + c_n^2 + d_n^2).$$

(b) Show that the area A satisfies

$$A = \int_0^1 x \frac{dy}{dt} dt = \sum_{n=1}^{\infty} 2\pi n(a_n d_n - b_n c_n).$$

(c) Show that $L^2 - 4\pi A \ge 0$ and that equality holds if and only if $a_1 = d_1$, $b_1 = -c_1$, and $a_n = b_n = c_n = d_n = 0$ for $n = 2, 3, \ldots$, which describes the equation of a circle.]

7. Use Example 8 to show that if A is any measurable set in R^n, then $L_2(A)$ has a basis that is at most countable.

8. The object of this exercise is to show that the Laguerre functions form an orthonormal basis for $L_2[(0,\infty)$, see Example 6.
   (a) Show that $\{\phi_n\}$ is an orthonormal set. [Hint: Compute

$$\int_0^{\infty} e^{-t} t^k L_n(t) \, dt = \int_0^{\infty} t^k \frac{d^n}{dt^n} (t^n e^{-t}) \, dt$$

by integrating by parts.]
   (b) Show that $\{\phi_n\}$ is maximal. [Hint: Modify the argument of Example 1, Section 18 by replacing $P(t)$, in Lemma 5.18.1, by a function of the form $e^{t/2} Q(t)$ where $Q(t)$ is an appropriate polynomial in t.]

9. Show that

$$\sum_{n=0}^{\infty} \frac{1}{n!} t^n L_n(x) = (1-t)^{-1} \exp\left[-xt(1-t)^{-1}\right],$$

where $L_n(x)$ is the Laguerre polynomial.

10. In Example 3, Section 14, we used the functions $\{e^{-t}, te^{-t}, t^2 e^{-t}, \ldots\}$.
    (a) How are these related to the Laguerre functions? [Hint: Look at the Gram-Schmidt orthogonalization process.]
    (b) Is it true, in this example, that

$$\lim_{n \to \infty} \int_0^{\infty} |h(t) - \hat{h}_n(t)|^2 \, dt = 0,$$

where h and h_n are defined in Example 3, Section 14?

11. Use the results of this section to re-examine Example 2 of Section 4.7.

12. This is the outline of another proof that the Laguerre functions form a maximal orthonormal set in $L_2[0,\infty)$.

(a) Define $h(x,t)$ and $g(x,t)$ by

$$h(x,t) = (1 - t)^{-1} \exp \frac{-tx}{(1 - t)} = \sum_{n=0}^{\infty} \frac{t^n}{n!} L_n(x)$$

$$g(x,t) = \exp\left(\frac{-x}{2}\right) h(x,t) = (1 - t)^{-1} \exp\left[\tfrac{1}{2}(\{1 + t\}\{1 - t\})x\right]$$

$$= \sum_{n=0}^{\infty} t^n \phi_n(x).$$

Show that

$$\int_0^{\infty} |g(x,t)|^2 \, dx = (1 - t^2)^{-1}$$

and

$$\int_0^{\infty} g(x,t)\phi_n(x) \, dx = t^n.$$

(b) Show that

$$\int_0^{\infty} \left| g(x,t) - \sum_{n=0}^{N} t^n \phi_n(x) \right|^2 dx = \frac{1}{1 - t^2} - \sum_{n=0}^{N} t^{2n}.$$

(c) Show that every function of the form $e^{-\alpha x}$, $(0 < \alpha < \infty)$ can be approximated arbitrarily closely in $L_2[0,\infty)$ by a finite linear combination of Laguerre functions ϕ_n. [*Hint*: Use (b) with $\alpha = \tfrac{1}{2}(\{1 + t\}\{1 - t\})$.]

(d) Show that every function in $L_2[0,\infty)$ can be approximated arbitrarily closely in $L_2[0,\infty)$ by a finite linear combination of functions of the form $e^{-\alpha x}$, $0 < \alpha < \infty$. [*Hint*: Perform a change of variables $y = e^{-x}$ mapping $0 \le x < \infty$ into $[0,1]$ and then use the fact that the polynomials in y are dense in $L_2[0,1]$.]

19. UNITARY OPERATORS AND EQUIVALENT INNER PRODUCT SPACES

We have discussed the question of equivalence at various levels. For example, equivalences between metric spaces (homeomorphisms and isometries), equivalences between linear spaces (isomorphisms), and equivalences between normed linear spaces (topological and isometric isomorphisms). The basic idea of equivalence is the same in each case, that is, there exists a one-to-one mapping of one space onto the other and this mapping preserves the given structure. We now introduce the analog for inner product spaces.

5.19.1 DEFINITION. Let X and Y be two inner product spaces. We say that X and Y are *unitarily equivalent* if there is an isomorphism $\phi: X \to Y$ of X onto Y that preserves inner products, that is

$$(\phi(x_1), \phi(x_2)) = (x_1, x_2)$$

for all $x_1, x_2 \in X$. The mapping ϕ is referred to as a *unitary operator*.

Since unitary operators preserve inner products, one has $\|\phi(x)\| = \|x\|$ for all $x \in X$, so ϕ is also an isometric isomorphism. In fact, the converse is also true.

5.19.2 THEOREM.   *A mapping ϕ is an isometric isomorphism of X onto Y, where X and Y are inner product spaces, if and only if ϕ is a unitary operator.*

Proof: The only issue here is to show that if $\|\phi(x)\| = \|x\|$ for all $x \in X$, then $(\phi(x), \phi(y)) = (x, y)$ for all $x, y \in X$. Since ϕ is linear we have

$$
\begin{aligned}
4(\phi(x), \phi(y)) &= (\phi(x + y), \phi(x + y)) - (\phi(x - y), \phi(x - y)) \\
&\quad + i(\phi(x + iy), \phi(x + iy)) - i(\phi(x - iy), \phi(x - iy)) \\
&= (x + y, x + y) - (x - y, x - y) \\
&\quad + i(x + iy, x + iy) - i(x - iy, x - iy) \\
&= 4(x, y). \quad \blacksquare
\end{aligned}
$$

Needless to say, if two Hilbert spaces are unitarily equivalent, they are essentially the same Hilbert space and differ only in the nature of the points in their underlying sets.

We shall show later (Theorem 5.22.7), after we introduce the concept of the adjoint, that an operator ϕ is unitary if and only if $\phi^{-1} = \phi^*$, where ϕ^* is the adjoint of ϕ.

EXAMPLE 1.   Let X be a finite-dimensional complex inner product space with dim $X = n$. We shall now show that X is unitarily equivalent to C^n with the usual inner product.

Let $\{e_1, e_2, \ldots, e_n\}$ be a Hamel basis for X. By using the Gram-Schmidt orthogonalization process, if necessary, we can assume that $\{e_1, e_2, \ldots, e_n\}$ is an orthonormal basis. It follows from the Fourier Series Theorem that for every $x \in X$ there are complex numbers $x_i = (x, e_i)$, $1 \le i \le n$, such that $x = \sum_{i=1}^{n} x_i e_i$. Now define a mapping $\Phi: X \to C^n$ by

$$
\Phi x = (x_1, \ldots, x_n).
$$

It is clear that Φ is an isomorphism of X onto C^n since $\{e_1, e_2, \ldots, e_n\}$ is a basis. Furthermore, Parseval's Equality (Theorem 5.17.8) assures us that

$$
(x, y) = \sum_n (x, e_i) \overline{(y, e_i)} = \sum_{i=1}^{n} x_i \bar{y}_i = [\Phi x, \Phi y],
$$

where $[\cdot, \cdot]$ denotes the usual inner product on C^n. Hence Φ is a unitary mapping. $\blacksquare$

EXAMPLE 2.   Consider the space $L_2[0,1]$. Let us show that this is unitarily equivalent to $l_2(-\infty, \infty)$. It is shown in Example 1, Section 18, that $\phi_n(t) = e^{2\pi i n t}$ for $n = 0, \pm 1, \ldots$ forms an orthonormal basis for $L_2[0,1]$. By the Fourier Series

Theorem 5.17.8 one can find for every $x \in L_2[0,1]$ a bi-sequence of complex numbers $(\ldots,x_{-1},x_0,x_1,x_2,\ldots)$ where $x_n = (x,\phi_n)$ such that $x = \sum_n x_n \phi_n$. Now define a mapping $F: L_2[0,1] \rightarrow l_2(-\infty,\infty)$ by

$$Fx = (\ldots,x_{-1},x_0,x_1,x_2,\ldots).$$

It is clear that F is linear. It is also one-to-one since $x_n = (x,\phi_n) = 0$ for all n, implies that $x = 0$ [Theorem 5.17.8(e)]. Also the range is all of $l_2(-\infty,\infty)$ by Lemma 5.17.10(a). So we see that F is an isomorphism of $L_2[0,1]$ onto $l_2(-\infty,\infty)$. Furthermore, Parseval's Equality [Theorem 5.17.8(c)] assures us that

$$(x,y) = \sum_n (x,\phi_n)\overline{(y,\phi_n)} = \sum_n x_n \bar{y}_n = [Fx,Fy],$$

where $[\cdot,\cdot]$ denotes the usual inner product on $l_2(-\infty,\infty)$. Hence F is a unitary mapping. ∎

It should be clear that we could replace $L_2[0,1]$ with $L_2(I)$ where I is any interval in R, or more generally, where I is any rectangle or set in R^m. All that is needed in order to get the unitary equivalence is that $L_2(I)$ have a countable orthonormal basis.

EXAMPLE 3.   Let Z be the Hilbert space of z-transforms defined on the unit circle that was introduced in Example 4, Section 4.5. Let U be the mapping of Z into $L_2[0,1]$ defined by

$$(Uf)(t) = f(e^{-2\pi i t}), \qquad t \in [0,1],$$

where f is a point in Z. The basic idea here is to "stretch and wrap" the interval $[0,1]$ around the unit circle. If $u \in L_2[0,1]$, it is easily shown that

$$(U^{-1}u)(z) = u[t(z)],$$

where $t = t(z) = -\arg(z)/2\pi$. The inner product in Z is defined by

$$(f,g)_z = \frac{1}{2\pi i} \oint f(z)\overline{g(z)} \frac{dz}{z}.$$

If u and v are two points in $L_2[0,1]$, then

$$(U^{-1}u, U^{-1}v)_z = \frac{1}{2\pi i} \oint u[t(z)]\overline{v[t(z)]} \frac{dz}{z}$$

and a simple change of variable of integration shows that

$$(U^{-1}u, U^{-1}v)_z = \int_0^1 u(t)\overline{v(t)}\, dt = (u,v)_2$$

for all u, v in $L_2[0,1]$. So U is a unitary operator mapping Z into $L_2[0,1]$. ∎

EXAMPLE 4.   We can combine the operators F and U from the preceding two examples to show that the two-sided z-transform $\mathscr{Z}$ is a unitary operator. We again refer the reader to what has already been said about the two-sided z-transform in

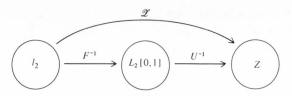

Figure 5.19.1

Example 4, Section 4.5. As cited in that example, $\mathscr{L}$ is a mapping of $l_2(-\infty,\infty)$ into the Hilbert space Z of Example 3 above. So we have the situation illustrated in Figure 5.19.1. Let us show that $\mathscr{L} = U^{-1}F^{-1}$. It follows from the preceding examples that

$$F^{-1}x = \sum_{n=-\infty}^{\infty} x_n e^{2\pi i n t},$$

where $x = \{\ldots,x_{-1},x_0,x_1,x_2,\ldots\}$, and

$$U^{-1}F^{-1}x = \sum_{n=-\infty}^{\infty} x_n z^{-n}.$$

So $\mathscr{L} = U^{-1}F^{-1}$. Hence, $\mathscr{L}$ is unitary. In particular, we have

$$\frac{1}{2\pi i}\oint |f(z)|^2 \frac{dz}{z} = \sum_{n=-\infty}^{\infty} |x_n|^2,$$

where $f = \mathscr{L}x$.   ∎

Some care should be taken in reading the definition of a unitary mapping. A unitary mapping must satisfy four conditions: It must (1) be linear, (2) be one-to-one, (3) be "onto," and (4) preserve inner products.

EXAMPLE 5.   Consider the (left) shift operator S_l on sequences. That is, $y = S_l x$ means $y_n = x_{n+1}$. This is a unitary mapping on $l_2(-\infty,\infty)$. However, S_l is not one-to-one on $l_2[0,\infty)$ since $S_l(1,0,0,\ldots) = 0$.

The (right) shift operator S_r is merely the inverse of S_l on $l_2(-\infty,\infty)$ and it is, of course, a unitary mapping there. On $l_2[0,\infty)$ S_r is defined by $y_n = x_{n-1}$ for $n = 1, 2, \ldots$ and $y_0 = 0$. In this case the range of S_r is not all of $l_2[0,\infty)$ even though this mapping is linear and one-to-one and preserves inner products.   ∎

The Fourier transform is, of course, a famous example of a unitary operator.

EXAMPLE 6. (FOURIER TRANSFORM ON R^1.)   The Fourier transform $\mathscr{F}$ and its inverse $\mathscr{F}^{-1}$ are given by the following equations:

$$\hat{f}(i\omega) = (\mathscr{F}f)(i\omega) = \int_{-\infty}^{\infty} e^{-i\omega x} f(x)\,dx, \tag{5.19.1}$$

$$f(x) = (\mathscr{F}^{-1}\hat{f})(x) = \frac{1}{2\pi}\int_{-\infty}^{\infty} e^{i\omega x}\hat{f}(i\omega)\,d\omega. \tag{5.19.2}$$

We will show in Example 12, Section 22 that $\mathscr{F}$ is a unitary mapping of $L_2(-\infty,\infty)$ onto $L_2(-i\infty,i\infty)$ and that $\mathscr{F}^{-1}$ is given by (5.19.2).

In the literature the transformation

$$g(y) = (Ff)(y) = \frac{1}{\sqrt{2\pi}} \int_{-\infty}^{\infty} e^{-iyx} f(x)\, dx \qquad (5.19.3)$$

is sometimes referred to as the "Fourier transform." It is clear that both F and $\mathscr{F}$ are related and that F^{-1} is given by

$$f(x) = (F^{-1}g)(x) = \frac{1}{\sqrt{2\pi}} \int_{-\infty}^{\infty} e^{iyx} g(y)\, dy. \qquad (5.19.4)$$

Furthermore, since $\mathscr{F}$ is a unitary mapping we see that F is a unitary mapping of $L_2(-\infty,\infty)$ onto itself. ∎

EXAMPLE 7.   In this example we show the use of orthonormal sets and unitary operators in a classic sampling theorem.

This will be one of the few cases where we consider the real instead of the complex Hilbert space $H_t = L_2(-\infty,\infty)$. In particular, the points in H_t are real-valued functions and, of course, the scalars are real too. This restriction to a real space is not necessary, but simplifies things and conforms to the usual applications.

If we take the Fourier transform of points in H_t, we obtain all the complex-valued functions $X(i\omega)$ such that

(a) $\dfrac{1}{2\pi} \displaystyle\int_{-\infty}^{\infty} |X(i\omega)|^2\, d\omega < \infty$,

(b) $\operatorname{Re}[X(i\omega)]$ is an even function about $\omega = 0$, that is,

$$\operatorname{Re}[X(i\omega)] = \operatorname{Re}[X(-i\omega)]$$

for almost all ω, and $\operatorname{Im}[X(i\omega)]$ is an odd function, that is

$$\operatorname{Im}[X(i\omega)] = -\operatorname{Im}[X(-i\omega)]$$

for almost all ω.

Using the natural structure on this set of functions and the real numbers as scalars, we have a real Hilbert space H_ω with the inner product defined by

$$(X,Y) = \frac{1}{2\pi} \int_{-\infty}^{\infty} X(i\omega)\overline{Y(i\omega)}\, d\omega.$$

Note that (X,Y) is real for all X, $Y \in H_\omega$.

It can be shown, using the preceding example, that the Fourier transform $\mathscr{F}$ (in the L_2 sense) is a unitary mapping of H_t onto H_ω; therefore, H_t and H_ω are unitarily equivalent. The usual interpretation is H_t is the time domain and H_ω is the frequency domain associated with real systems.[16]

---

[16] That is, systems that map real-valued input functions into real-valued output functions.

Now let us consider the closed linear subspace M_c of H_t defined by

$$M_c = \{x \in H_t : (\mathscr{F}x)(i\omega) = 0 \text{ for } |\omega| \geq \omega_c\},$$

where $\omega_c > 0$ is a real number. One often says that $x \in M_c$ is "bandlimited to fre-
quencies below $f_c = \omega_c/2\pi$" or "contains no frequency components above f_c."

Since[17] $\hat{x} = \mathscr{F}x$ is in $L_1(-i\infty, i\infty) \cap L_2(-i\infty, i\infty)$, for all $x \in M_c$, it follows
from the theory of the L_1-Fourier transform that $x = \mathscr{F}^{-1}\hat{x}$ is (i) continuous,
(ii) bounded, and (iii) $\lim_{t \to \pm\infty} x(t) = 0$. We hasten to add that there are functions
in H_t satisfying (i), (ii), (iii) that are not in M_c. (See R. Goldberg [1].)

Next let $\mathscr{M}_c$ denote the closed linear subspace of H_ω defined by

$$\mathscr{M}_c = \{\hat{x} \in H_\omega : \hat{x}(i\omega) = 0 \text{ for } |\omega| \geq \omega_c\}.$$

Obviously, $\mathscr{F}(M_c) = \mathscr{M}_c$, so M_c and $\mathscr{M}_c$ are themselves unitarily equivalent
Hilbert spaces.

We state that the set $\{\phi_n\}$ in M_c, where

$$\phi_n(t) = \frac{1}{\sqrt{\pi}} \frac{\sin(\omega_c t - n\pi)}{(\omega_c t - n\pi)}, \qquad n = \ldots, -1, 0, 1, 2, \ldots,$$

is an orthonormal set. We also state that $\hat{\phi}_n = \mathscr{F}\phi_n$ is given by

$$\hat{\phi}_n(i\omega) = \begin{cases} \sqrt{\dfrac{\pi}{\omega_c}}\, e^{-inT\omega}, & |\omega| \leq |\omega_c| \\[2mm] & \qquad\qquad\qquad\qquad n = \ldots, -1, 0, 1, 2, \ldots, \\[2mm] 0, & |\omega| > |\omega_c|, \end{cases}$$

where $T = \pi/\omega_c$. The set $\{\hat{\phi}_n\}$ is, of course, an orthonormal set in $\mathscr{M}_c$. Moreover,
it follows from Example 2, Section 18 that $\{\hat{\phi}_n\}$ is maximal in $\mathscr{M}_c$. Thus, $\{\phi_n\}$ is
maximal in M_c. It follows, then, that if x is any point in M_c, it can be expressed
in the form

$$x = \sum_{k=-\infty}^{\infty} (x, \phi_k)\phi_k.$$

Since $\phi_k(nT) = (1/\sqrt{\pi})\delta_{kn}$, where δ_{kn} is the Kronecker function, one would
suspect that for continuous functions x, one has

$$x(kT) = \frac{1}{\sqrt{\pi}} (x, \phi_k),$$

and this is in fact the case. In order to show this one would start with

$$\lim_{N \to \infty} \left\| x - \sum_{k=-N}^{N} (x, \phi_k)\phi_k \right\| = 0$$

[17] It is in L_2 and has compact support; therefore, it is also in L_1.

and use the continuity of x together with the inequality

$$\left| \sum_{\substack{k=-N \\ k \neq n}}^{N} (x,\phi_k)\phi_k \right| \leq K |\omega_c t - n\pi|$$

for $-\pi < \omega_c t - n\pi < \pi$, where $K > 0$. We leave the details as an exercise.

In summary, we have that any continuous $x \in M_c$ can be written

$$x(t) = \sum_{k=-\infty}^{\infty} x(kT) \frac{\sin(\omega_c t - k\pi)}{(\omega_c t - k\pi)}. \tag{5.19.5}$$

The usual interpretation of (5.19.5) is that a bandlimited signal can be completely recovered from samples of its value [that is, $x(kT)$, $k = \ldots, -1, 0, 1, 2, \ldots$] as long as the samples are taken frequently enough. Finally, we remark that this is just one of a large family of results concerning "sampling," see Beutler [1]. ∎

EXAMPLE 8.   Let $H = L_2(-\infty,\infty)$ and consider the mapping T of H into itself defined by

$$(Tx)(t) = \begin{cases} x(t), & \text{for } t \geq 0 \\ -x(t), & \text{for } t < 0. \end{cases}$$

T is obviously linear. Moreover,

$$(Tx,Ty) = \int_{-\infty}^{0} \overline{y(t)}x(t)\, dt + \int_{0}^{\infty} \overline{y(t)}x(t)\, dt = (x,y).$$

Furthermore, $T^2 = I$, so T is invertible. It follows that T is a unitary operator.

Let P_+ denote the orthogonal projection defined by

$$(P_+ x)(t) = \begin{cases} x(t), & \text{for } t \geq 0 \\ 0, & \text{for } t < 0 \end{cases}$$

and let P_- denote the orthogonal projection defined by

$$(P_- x)(t) = \begin{cases} 0, & \text{for } t \geq 0 \\ x(t), & \text{for } t < 0. \end{cases}$$

Note that $P_+ + P_- = I$ and $P_+ P_- = P_- P_+ = 0$. In addition,

$$T = \lambda_1 P_+ + \lambda_2 P_-,$$

where $\lambda_1 = 1$ and $\lambda_2 = -1$. If we let M_+ denote those functions $x(t)$ in $L_2(-\infty,\infty)$ that vanish for $t < 0$, and M_- denote those functions that vanish for $t > 0$, then $Tx = x$ for all $x \in M_+$ and $Tx = -x$ for all $x \in M_-$. Furthermore, one has $M_+^\perp = M_-$, $M_-^\perp = M_+$ and $H = M_+ + M_-$.

If we consider continuous linear mappings of $L_2(-\infty,\infty)$ into itself that can be represented in the form

$$z(t) = \int_{-\infty}^{\infty} k(t - \tau)x(\tau)\, d\tau, \qquad t \in (-\infty,\infty),$$

where $k \in L_1(-\infty,\infty) \cap L_2(-\infty,\infty)$, then the k's that correspond to causal systems are exactly those that satisfy the equation $Tk = k$. By using the Fourier transform $\mathscr{F}$ the equation $Tk = k$ becomes

$$T\mathscr{F}^{-1}\mathscr{F}k = \mathscr{F}^{-1}\mathscr{F}k, \quad \text{or} \quad (\mathscr{F}T\mathscr{F}^{-1})\hat{k} = \hat{k},$$

where $\hat{k} = \mathscr{F}k$ is the Fourier transform of the "unit impulse response" k.

Let $\mathscr{T} = \mathscr{F}T\mathscr{F}^{-1}$, and let us see how $\mathscr{T}$ transforms a real-valued function r in $L_2(-i\infty,i\infty)$. First the inverse Fourier transform of r has an even real part and an odd imaginary part. The $\mathscr{T}$ operating on $\mathscr{F}^{-1}r$ converts the even real part to an odd function and the odd imaginary part to an even function. But the Fourier transform of a complex-valued function with odd real part and even imaginary part is a function with zero real part. So $\mathscr{T}$ maps real-valued functions into imaginary-valued functions.

If we let $\hat{k} = \rho + i\sigma$, where ρ and σ are, respectively, the real and imaginary parts of $\hat{k}$, then it follows that $\mathscr{T}\hat{k} = \hat{k}$ becomes

$$\mathscr{T}\rho + i\mathscr{T}\sigma = \rho + i\sigma.$$

But $\mathscr{T}\rho$ is imaginary-valued and $i\mathscr{T}\sigma$ is real-valued, so the above equation becomes

$$\rho = i\mathscr{T}\sigma$$
$$\sigma = -i\mathscr{T}\rho.$$

In other words, as is well-known, the real and imaginary parts of $\hat{k}$ are interdependent for causal systems. ∎

Finally let us briefly return to the topic of equivalent operators, which we began to study in Section 4.9.

Clearly, everything said in that purely algebraic context is still applicable in normed linear spaces, Banach spaces, and here in inner product spaces. It is also obvious that now with topological structure present one not only wants linear transformations to be equivalent in an algebraic sense but also in a topological sense. Whereas isomorphisms were used in Section 4.9, in this new context one would usually use topological and isometric isomorphisms. The classic case of this is two operators being unitarily equivalent.

5.19.3 DEFINITION.   Let $T: X_1 \to X_1$ and $S: X_2 \to X_2$ be linear operators where X_1 and X_2 are inner product spaces. The operators T and S are said to be *unitarily equivalent* if there exists a unitary mapping U of X_1 onto X_2 such that

$$T = U^{-1}SU \quad \text{and} \quad S = UTU^{-1}.$$

This situation is illustrated in Figure 5.19.2.

Undoubtedly two of the best known examples of pairs of operators being unitarily equivalent arises in the study of time-invariant linear operators on $L_2(-\infty,\infty)$ and $l_2(-\infty,\infty)$, as we now see.

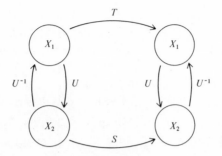

Figure 5.19.2

EXAMPLE 9. Let $X_1 = L_2(-\infty,\infty)$, and let T be a bounded linear time-invariant transformation of X_1 into itself. We also know from Example 6 the Fourier transform $\mathscr{F}$ is a unitary mapping of $L_2(-\infty,\infty)$ onto $L_2(-i\infty,i\infty)$. The operator $S\colon L_2(-i\infty,i\infty) \to L_2(-i\infty,i\infty)$ defined by

$$S = \mathscr{F}T\mathscr{F}^{-1}$$

is obviously unitarily equivalent to T. The fundamental fact about S is that S has a simple form. In particular,

$$(S\hat{x})(i\omega) = T(i\omega)\hat{x}(i\omega),$$

where $T(i\omega)$ is the transfer function associated with the operator T. In other words, S is an operation of multiplication by a function, see Bochner [1].

It should be remarked that if causality is left aside there is a one-to-one correspondence between bounded linear time-invariant mappings T of $L_2(-\infty,\infty)$ into itself and measurable essentially bounded complex-valued functions $T(i\omega)$ defined on the imaginary axis. ∎

EXAMPLE 10. Let $X = l_2(-\infty,\infty)$, and let T be a bounded linear time-invariant transformation of X into itself. Let Z be the inner product space of complex-valued functions defined on the unit circle of the complex plane used above in Example 3. We showed in Example 4 that the two-sided z-transform $\mathscr{Z}\colon X \to Z$ is a unitary operator. It follows that the operator $S\colon Z \to Z$ defined by

$$S = \mathscr{Z}T\mathscr{Z}^{-1}$$

is unitarily equivalent to T. As with Example 9, the fundamental result here is that

$$(Sf)(z) = T(z)f(z)$$

for all z such that $|z| = 1$, where $T(z)$ is the transfer function associated with the operator T. ∎

EXERCISES

1. Show that $l_2(-\infty,\infty)$ and $l_2(0,\infty)$ are unitarily equivalent.

2. Show that every separable Hilbert space is unitarily equivalent with l_2 or C^n for some $n \geq 0$.

3. Let U be a unitary mapping on an inner product space X. Show that $\|U\| = 1$.

4. Let $I = [a,b]$ be a bounded interval and define $F: L_2(I) \to L_2(I)$ by $y = Fx$ or $y(t) = f(t)x(t)$, where $f \in L_\infty(I)$. Show that F is a unitary mapping if and only if $|f(t)| = 1$, almost everywhere.

5. Let X be an inner product space and let B be a Banach space that is the completion of X. (See Exercise 6, Section 9.) Show that B is a Hilbert space.

6. Let U be a unitary operator on a Hilbert space H and define
$$A_n = (1/n)[I + U + \cdots + U^{n-1}] \qquad \text{for} \quad n = 1, 2, \ldots .$$
Let $M = \{x - Ux : x \in H\}$.
(a) Show that if $y \in M$, then $\|A_n y\| \to 0$ as $n \to \infty$.
(b) Show that if $y \in \overline{M}$, then $\|A_n y\| \to 0$ as $n \to \infty$.
(c) Show that if $x \in M^\perp$, then $Ux = x$.
(d) Show that $P = {}_s\lim A_n$, where P is the orthogonal projection onto
$$M^\perp = \mathcal{N}(I - U).$$
(e) Show that $\|A_n - P\| \to 0$.

7. Let H_1 and H_2 be two Hilbert spaces with the same Hilbert dimension. (Compare with Exercise 6, Section 17.) Show that H_1 and H_2 are unitarily equivalent.

8. Let P be an orthogonal projection on a Hilbert space H. Also let $U_t = e^{itP}$, $-\infty < t < \infty$.
(a) Show that for each t, U_t is a unitary operator.
(b) When is P a unitary operator?

20.   SUMS AND DIRECT SUMS OF HILBERT SPACES

We introduced the concepts of sum and direct sum in Section 4.10. Again, since Hilbert spaces are linear spaces, everything said about sums and direct sums in Chapter 4 applies here. The purpose of this section is to see what else can be said now that we have the topological structure generated by an inner product. More specifically we shall be interested in the question of sums of mutually orthogonal closed linear subspaces of a Hilbert space.

In Section 15 we discussed the properties of sums of finitely many mutually orthogonal closed linear subspaces (see Theorem 5.15.5 and Exercise 3, Section 15). So in this section we shall focus our attention on infinite sums of these spaces. For simplicity we shall consider only countable sums. The discussion in Section 17 on uncountable sums also applies here and the results we describe below readily extend to uncountable sums.

5.20.1 DEFINITION.   A collection $\{M_n\}$ of sets in a Hilbert space H is said to be mutually orthogonal if $M_n \perp M_m$ whenever $n \neq m$. If $\{M_n\}$ is a collection of mutually orthogonal closed linear subspaces of a Hilbert space H we shall define the (topological) sum

$$M = \sum_n M_n = M_1 + M_2 + \cdots$$

to be the closure of the linear subspace of H generated by $\{M_n\}$.

It should be noted here that this definition does *not* agree with the algebraic concept introduced in Section 4.10. The algebraic sum of these spaces would consist of all finite sums of vectors from $\{M_n\}$, whereas the topological sum also includes limits of sequences of these finite sums. Because of Theorem 5.15.5 and Exercise 3, Section 15 we see that the topological sum and the algebraic sum agree when the collection $\{M_n\}$ is finite. It is only when the collection of spaces $\{M_n\}$ is infinite that these concepts differ.

The next theorem shows that each point in $M_1 + M_2 + M_3 + \cdots$ can be expressed as a series.

5.20.2 THEOREM. (ORTHOGONAL STRUCTURE THEOREM.) *Let $\{M_n\}$ be a countable collection of mutually orthogonal, closed linear subspaces of a Hilbert space H, and let $M = M_1 + M_2 + \cdots$ be the sum as defined above. Then each $x \in M$ can be expressed uniquely as*

$$x = \sum_n x_n$$

where

$$x_n \in M_n, \qquad \text{for all } n,$$

and

$$\|x\|^2 = \sum_n \|x_n\|^2.$$

Furthermore, if $x_n \in M_n$ and $\sum_n \|x_n\|^2 < \infty$, then there exists an $x \in M$ such that $x = \sum_n x_n$.

Carefully note that this is a statement about the orthogonal structure of *Hilbert* spaces, not inner product spaces in general. Indeed, it is not difficult to find a collection $\{M_n\}$ of mutually orthogonal closed linear subspaces of an incomplete inner product space X such that Theorem 5.20.2 is not satisfied.[18]

Proof: This proof is very much like the proof of Lemma 5.17.14. Since $x \in M$, there is a sequence $\{y_N\}$ converging to x, where each y_N can be expressed as

$$y_N = x_{N1} + x_{N2} + \cdots + x_{NK}, \qquad N = 1, 2, \ldots$$

where $x_{Nn} \in M_n$, and K depends on N. For simplicity we set $x_{Nn} = 0$ for $n > K$. Since M_n is closed, there is (Theorem 5.14.4) a unique point $x_n \in M_n$ such that $\|x - x_n\| \le \|x - m\|$ for all $m \in M_n$ and $x - x_n \perp M_n$. In particular, then $\|x - x_n\| \le \|x - x_{Nn}\|$ for all N and n. Since $M_n + M_m$ is closed (Theorem 5.15.5), there is a unique point $z \in M_n + M_m$ such that $\|x - z\| \le \|x - m\|$ for all $m \in M_n + M_m$ and $x - z \perp M_n + M_m$. Let us show that $z = x_n + x_m$. Since $z \in M_n + M_m$, there is a unique decomposition $z = z_n + z_m$, where $z_n \in M_n$ and $z_m \in M_m$. Because $x - z \perp M_n + M_m$, we have $(x - z, m) = 0$ for all $m \in M_n + M_m$.

---

[18] The topological sum $\Sigma_n M_n$ is defined just as in a Hilbert space.

Hence, $(x - z_n - z_m, m_n) = 0$ for all $m_n \in M_n$. Therefore, $(x - z_n, m_n) = 0$ for all $m_n \in M_n$. It follows that $z_n = x_n$. Similarly, one has $z_m = x_m$, or $z = x_n + x_m$.

In general, then,

$$\|x - x_1 - x_2 - \cdots - x_K\| \leq \|x - y_N\|.$$

Since $x = \lim_{N \to \infty} y_N$, it follows that

$$x = \sum_n x_n.$$

The uniqueness of the expansion and the equality $\|x\|^2 = \sum_n \|x_n\|^2$ follows from Lemma 5.17.10(b), and the last part of the theorem is a consequence of Lemma 5.17.10(a). ∎

EXAMPLE 1. The importance of the Orthogonal Structure Theorem will become evident in the next chapter. However, we can give a preview here. Let H be a Hilbert space and let $L: H \to H$ be a bounded linear operator. Assume that there exists a collection $\{M_n\}$ of mutually orthogonal nontrivial closed linear subspaces of H with $H = \sum_n M_n$. Assume further that there is a sequence $\{\lambda_n\}$ of scalars with the property that $Lx_n = \lambda_n x_n$ for all $x_n \in M_n$. Finally, assume that the λ_n's satisfy $|\lambda_n| \leq |\lambda_1|$ for all n. Now if $x \in H$ and if $x = \sum_n x_n$, where $x_n \in M_n$, then $Lx = L(\sum_n x_n) = \sum_n Lx_n = \sum_n \lambda_n x_n$. The series $\sum_n \lambda_n x_n$ converges since

$$\sum_n \|\lambda_n x_n\|^2 \leq \sum_n |\lambda_1|^2 \|x_n\|^2 = |\lambda_1|^2 \|x\|^2.$$

The last inequality also shows that $\|Lx\| \leq |\lambda_1| \|x\|$, in other words, $\|L\| \leq |\lambda_1|$. One can actually show that $\|L\| = |\lambda_1|$. Indeed, if $x \in M_1$ is chosen so that $\|x\| = 1$, one then has

$$|\lambda_1| = |\lambda_1| \|x\| = \|\lambda_1 x\| = \|Lx\| \leq \|L\| \cdot \|x\| = \|L\|. \quad ∎$$

Next let us turn to the question of direct sums of inner product spaces. Let X_1 and X_2 be two inner product spaces. The direct sum $X_1 \oplus X_2$, as a linear space, was introduced in Section 4.10. We will use, perhaps somewhat confusingly, exactly the same notation to denote the *direct sum* of the inner product spaces X_1 and X_2. If $(\cdot, \cdot)_1$ and $(\cdot, \cdot)_2$ are the inner products, X_1 and X_2, respectively, then the inner product on $X_1 \oplus X_2$ is defined by

$$(x, y) = (x_1, y_1)_1 + (x_2, y_2)_2,$$

where x and y are the ordered pairs $\{x_1, x_2\}$ and $\{y_1, y_2\}$, respectively, with $x_1, y_1 \in X_1$ and $x_2, y_2 \in X_2$. We leave it to the reader to show that this does define an inner product. We also remark that if X_1 and X_2 are complete, then so is $X_1 \oplus X_2$.

If $\{X_1, \ldots, X_n\}$ is a finite collection of inner product spaces, we define $X_1 \oplus X_2 \oplus \cdots \oplus X_n$ in the obvious way. However, if $\{X_i\}$ is a countably infinite collection, we have to say a few things.

Let $\{X_i\}$ be a countable collection of inner product spaces, with $(\,\cdot\,,\,\cdot\,)_i$ denoting the inner product on X_i. Now define the set

$$X = X_1 \times X_2 \times X_3 \times \cdots$$

as follows. Each point $x \in X$ is a sequence, $x = \{x_i\}$, where $x_i \in X_i$, $i = 1, 2, \ldots,$ and such that

$$\sum_i \|x_i\|_i^2 < \infty.$$

Using the natural structure on X, it is a linear space. We claim that

$$(x,y) = \sum_i (x_i, y_i)_i, \tag{5.20.1}$$

where $x = \{x_1, x_2, \ldots\}$ and $y = \{y_1, y_2, \ldots\}$ are in X, defines an inner product on X. Let us prove that the series (5.20.1) is absolutely convergent. By using the Schwarz Inequality twice, we get

$$\sum_i |(x_i, y_i)_i| \leq \sum_i \|x_i\|_i \|y_i\|_i$$

$$\leq \left(\sum_i \|x_i\|_i^2 \right)^{1/2} \left(\sum_i \|y_i\|_i^2 \right)^{1/2}$$

showing that the series in (5.20.1) is absolutely convergent. The rest of the proof that (5.20.1) defines an inner product is straightforward.

We denote X equipped with the inner product (5.20.1) by

$$X = X_1 \oplus X_2 \oplus X_3 \oplus \cdots$$

and refer to it as the *direct sum* of $\{X_i\}$.

Let us now prove that if all the X_i's are complete, then X is complete. Let $\{x^n\}$ be a Cauchy sequence in X. Since

$$\|x_i - y_i\|_i \leq \|x - y\|,$$

each $\{x_i^n\}$ is a Cauchy sequence in X_i. Since X_i is complete, each x_i^n converges, say that $x_i^n \to y_i$ as $n \to \infty$. Now let $y = \{y_1, y_2, \ldots\}$. We want to show that $y \in X$. Since $\{x^n\}$ is a Cauchy sequence, for every $\varepsilon > 0$ one can find an integer N such that

$$\sum_i \|x_i^n - x_i^m\|_i^2 \leq \varepsilon^2,$$

whenever $n, m \geq N$. If we let $m \to \infty$, the last inequality becomes (compare with Exercise 9, Section 2)

$$\sum_i \|x_i^n - y_i\|_i^2 \leq \varepsilon^2, \tag{5.20.2}$$

for $n \geq N$. In other words, $(y - x^n)$ is in X for $n \geq N$. Since $y = (y - x^n) + x^n$, it follows that $y \in X$. Furthermore, it follows from (5.20.2) that $x^n \to y$ in X.

The last thing to do is show the connection between sums and direct sums. Suppose that $M_1, M_2, \ldots$ are mutually orthogonal, closed linear subspaces of a Hilbert space H. Just as in Section 4.10, we want to compare $M_S = M_1 + M_2 + \cdots$ and $M_{DS} = M_1 \oplus M_2 \oplus \cdots$. It goes without saying that these are different spaces. On the other hand, we shall see that they are unitarily equivalent.

Let ψ be the mapping of M_S into M_{DS} defined as follows: We know from the Orthogonal Structure Theorem that each $x \in M_S$ can be uniquely expressed as $x = x_1 + x_2 + \cdots$, where $x_i \in M_i$, $i = 1, 2, \ldots$. We define $\psi(x) = \{x_1, x_2, x_3, \ldots\}$.

It follows that ψ is linear and maps M_S into M_{DS}. Moreover, ψ preserves norms, and the range is all of M_{DS}. Hence, ψ is a unitary mapping and we have just proved the following result.

5.20.3 THEOREM. *If X is complete, the mapping $\psi: M_S \to M_{DS}$ discussed above is a unitary mapping of M_S onto M_{DS}, and M_S and M_{DS} are unitarily equivalent.*

Because of this theorem, there is a widespread practice of referring to a *sum* of closed, mutually orthogonal linear subspaces of a Hilbert space as a *direct sum*. This is an innocent abuse of terminology, but the correct use should be understood.

EXERCISES

1. Extend the results of this section to uncountable sums.
2. Show by example that Theorem 5.20.2 does not hold (in general) for inner product spaces that are not complete. Where is the completeness of H used in the proof of the Orthogonal Structure Theorem?

21. CONTINUOUS LINEAR FUNCTIONALS

The concept of linear functionals on a linear space was discussed in Section 4.12. Continuous linear functionals on normed linear spaces were briefly introduced in Section 11 of this chapter. Now we want to discuss linear functionals on inner product spaces. The main result of this section will be to show that every continuous linear functional on a Hilbert space H (completeness is important) has a particularly simple representation. In other words, it is easy to "get our hands on" all the continuous linear functionals.

Let X be an inner product space and let y be a fixed element in X. Define a mapping l by

$$l(x) = (x, y).$$

We claim that l is a bounded (that is, continuous) linear functional and $\|l\| = \|y\|$, where $\|l\|$ is the operator norm defined in Section 8. The linearity follows directly from the definition of inner product. Also, by the Schwarz Inequality we get

$$|l(x)| = |(x, y)| \leq M \|x\|, \qquad \text{for all } x \in X,$$

where $M = \|y\|$. Hence $\|l\| \leq \|y\|$. However, $|l(y)| = \|y\| \|y\|$ so $\|l\| \geq \|y\|$. Thus, $\|l\| = \|y\|$.

We have shown, then, that a continuous linear functional l is naturally associated with each vector y in an inner product space. The extremely important fact about Hilbert spaces—as opposed to inner product spaces—is that this is true the other way around. That is, given any bounded linear functional l there exists a vector y such that

$$l(x) = (x,y), \qquad \text{for all } x \in H.$$

5.21.1 THEOREM. (RIESZ REPRESENTATION THEOREM.) *Let H be a Hilbert space and let l be a bounded linear functional on H. Then there is one and only one vector $y \in H$ such that*

$$l(x) = (x,y), \qquad \text{for all } x \in H.$$

The vector y is sometimes called the *representation* of l. However, l and y are different objects, l a linear functional on H, and y a point in H.

Proof: If $l = 0$, then choose $y = 0$. Now assume that $l \neq 0$ and let $M = \mathcal{N}(l)$, the null space of l. Since l is linear, M is a linear subspace of H. Since l is continuous, M is closed. Furthermore, $M \neq H$, because $l \neq 0$. By Corollary 5.14.5, there is a nonzero vector $z \in H$ such that $z \perp M$. We can assume $\|z\| = 1$. We will now show that the desired vector y is given by $y = \alpha z$ for some nonzero scalar α.

Since $z \notin M$, one has $l(z) \neq 0$. We shall now show that

$$l(x) = (x, \overline{l(z)}z)$$

for all $x \in H$. By the Projection Theorem (which requires completeness) one has $H = M + M^\perp$, so every vector x in H can be uniquely written as $x = m + n$ where $m \in M$ and $n \in M^\perp$. We know from Theorem 4.12.2 that dim $(M^\perp) = 1$. So we can write $n = \beta z$, that is, $x = m + \beta z$, for some scalar β. See Figure 5.21.1.

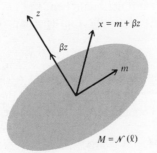

$$x = m + \beta z$$
$$\beta z$$
$$m$$
$$M = \mathcal{N}(\ell)$$

Figure 5.21.1.

Since $\|z\| = 1$, we have

$$l(x) = l(m) + \beta l(z) = 0 + \beta l(z)\|z\|$$
$$= (m, \overline{l(z)}z) + (\beta z, \overline{l(z)}z)$$
$$= (m + \beta z, \overline{l(z)}z) = (x,y),$$

where $y = \overline{l(z)}z$.

The uniqueness of y follows from the fact that if $(x,y) = (x,y')$ for all x, then for $x_0 = y - y'$ one has

$$0 = (x_0,y - y') = (y - y',y - y'),$$

hence $y = y'$. ∎

There is a variation of the Riesz Representation Theorem which is very useful in the study of differential operators, as shown in Section 7.8. We state it here and leave the proof as an exercise (see Exercise 10).

5.21.2 THEOREM. (LAX-MILGRAM THEOREM.) *Let $B[u,v]$ be a sesquilinear functional on a Hilbert space H and assume that there are positive constants a and b such that*

$$|B[u,v]| \leq a\|u\| \cdot \|v\|$$
$$b\|u\|^2 \leq |B[u,u]|$$

for all u, v in H. Let l be any bounded linear functional on H. Then there exist unique points u_0 and v_0 in H such that

$$l(x) = B[x,v_0] = \overline{B[u_0,x]}$$

for all x in H.

EXAMPLE 1.   We have already discussed the basic Hilbert space approximation concept in Theorems 5.14.4, 5.16.5, and Lemma 5.17.14. In all cases, we are interested in some closed linear subspace M of a Hilbert space H. The variations on this basic theme arise from the way we characterize the subspace M. Here we show how continuous linear functionals can be used. In particular, if l_1 is a continuous linear functional on H, then $\mathcal{N}(l_1)$, the null space of l_1, is a closed linear subspace of H. If l_2 is another continuous linear functional, then $\mathcal{N}(l_1) \cap \mathcal{N}(l_2)$ is also a closed linear subspace. In general, if $\{l_\alpha\}$ is a collection (perhaps uncountable) of continuous linear functionals, then

$$M = \bigcap_\alpha \mathcal{N}(l_\alpha)$$

is a closed linear subspace.

Now if x_0 is a point in H, and if we want to find the unique point $y_0 \in M$ closest to x_0, we know from Theorem 5.14.4 that $x_0 - y_0$ is orthogonal to M. (See Figure 5.21.2.)

Let us assume that $M = \mathcal{N}(l_1) \cap \mathcal{N}(l_2) \cap \cdots \cap \mathcal{N}(l_n)$, that is, we are considering a finite collection of functionals. We know from the Riesz Representation Theorem that for each l_i there is a $y_i \in H$ such that $l_i(x) = (x,y_i)$ for all $x \in H$, $i = 1, 2, \ldots, n$. Hence,

$$M = \{x \in H: (x,y_1) = (x,y_2) = \cdots = (x,y_n) = 0\}.$$

In other words, $M = \{y_1,y_2,\ldots,y_n\}^\perp$. Moreover, $M^\perp$ is the linear subspace spanned by the set $\{y_1,y_2,\ldots,y_n\}$. (Why?) Since $(x_0 - y_0) \in M^\perp$, it follows that $(x_0 - y_0)$ can be expressed as

$$x_0 - y_0 = \alpha_1 y_1 + \cdots + \alpha_n y_n,$$

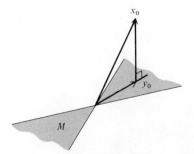

Figure 5.21.2

where the α's are scalars. Then since $y_0 \in M$, we have

$$(x_0 - y_0, y_1) = (x_0, y_1) = \alpha_1(y_1, y_1) + \cdots + \alpha_n(y_n, y_1)$$
$$\vdots \qquad\qquad \vdots \qquad\qquad \vdots \qquad\qquad \vdots$$
$$(x_0 - y_0, y_n) = (x_0, y_n) = \alpha_1(y_1, y_n) + \cdots + \alpha_n(y_n, y_n).$$

Everything in the above system of equations is known except the α's, so one merely solves for the α's and, then,

$$y_0 = x_0 - \alpha_1 y_1 - \cdots - \alpha_n y_n.$$

Let us note that there may be more than one solution for the α's. This would occur if the collection $\{y_1, \ldots, y_n\}$ is not linearly independent. ∎

EXAMPLE 2.   The reasoning of Example 1 can be used if M is a hyperplane defined by functionals. For example, assume that a second-order time-varying linear system is modeled by the equations

$$\begin{bmatrix} x_1(t) \\ x_2(t) \end{bmatrix} = \begin{bmatrix} \phi_{11}(t,t_0) & \phi_{12}(t,t_0) \\ \phi_{21}(t,t_0) & \phi_{22}(t,t_0) \end{bmatrix} \begin{bmatrix} x_1(t_0) \\ x_2(t_0) \end{bmatrix} + \int_{t_0}^{t} \begin{bmatrix} \phi_{11}(t,\tau) & \phi_{12}(t,\tau) \\ \phi_{21}(t,\tau) & \phi_{22}(t,\tau) \end{bmatrix} \begin{bmatrix} b_1 \\ b_2 \end{bmatrix} u(\tau)\, d\tau.$$

Suppose that starting at $(x_1(t_0), x_2(t_0))$ we want to choose the input u so that we arrive at prescribed point $(x_1(T), x_2(T))$ at time T. Moreover, suppose we want to choose the u with the smallest L_2-norm that does this. It follows that u must satisfy the equations

$$\int_{t_0}^{T} h_1(\tau)u(\tau)\, d\tau = A_1 \tag{5.21.1}$$

and

$$\int_{t_0}^{T} h_2(\tau)u(\tau)\, d\tau = A_2, \tag{5.21.2}$$

where

$$h_i(\tau) = \phi_{i1}(T,\tau)b_1 + \phi_{i2}(T,\tau)b_2, \qquad\qquad i = 1, 2$$

and

$$A_i = x_i(T) - \phi_{i1}(T,t_0)x_1(t_0) - \phi_{i2}(T,t_0)x_2(t_0), \qquad i = 1, 2.$$

Figure 5.21.3

Equations (5.21.1) and (5.21.2) define a hyperplane M in $L_2[t_0,T]$. (See Figure 5.21.3.) It follows that the input u of minimum norm can be written

$$u = \alpha_1 h_1 + \alpha_2 h_2$$

and

$$(\alpha_1 h_1 + \alpha_2 h_2, h_1) = A_1,$$
$$(\alpha_1 h_1 + \alpha_2 h_2, h_2) = A_2,$$

or

$$\begin{bmatrix} \alpha_1 \\ \alpha_2 \end{bmatrix} = \frac{1}{(h_1,h_1)(h_2,h_2) - (h_1,h_2)(h_2,h_1)} \begin{bmatrix} (h_2,h_2) & -(h_2,h_1) \\ -(h_1,h_2) & (h_1,h_1) \end{bmatrix} \begin{bmatrix} A_1 \\ A_2 \end{bmatrix}.$$ ∎

EXERCISES

1. Let $C[0,1]$ be considered as a linear subspace of $L_2[0,1]$, where $L_2[0,1]$ has the usual inner product. Define $l: C[0,1] \to C$ by $l(x) = x(\frac{1}{2})$.
   (a) Show that l is an unbounded linear functional.
   (b) Show that l is a bounded linear functional when $C[0,1]$ has the sup-norm.

2. For each $f \in L_2[0,1]$ let $\phi(t)$ be the solution of $y' + ay = f$ that satisfies $\phi(0) = 0$, where a is a constant. Define $l: L_2[0,1] \to C$ by

$$l(f) = \int_0^1 \phi(t)\, dt.$$

   (a) Show that l is a bounded linear functional.
   (b) Find the representation of l as $l(f) = (f,g)$.

3. For each $f \in L_2[0,1]$ let $\phi(t)$ be the solution of $y'' + ay' + by = f$ that satisfies $\phi(0) = \phi'(0) = 0$ where a and b are constants. Define $l: L_2[0,1] \to C$ by

$$l(f) = \int_0^1 \phi(t)\, dt.$$

   (a) Show that l is a bounded linear functional.
   (b) Find the representation of l as $l(f) = (f,g)$.
   [*Hint:* Distinguish between the two cases where the roots of $r^2 + ar + b = 0$ are the same or where they differ.]

4. Consider the real inner product space R^n with the inner product given by

$$(x,y) = \sum_{i,j} x_i a_{ij} y_j,$$

where $A = (a_{ij})$ is a real symmetric $n \times n$ matrix that satisfies $\sum_{i,j} x_i a_{ij} x_j > 0$ when $x = (x_1,\ldots,x_n) \neq 0$. Find a representation for $l: R^n \to R$ when l is given by
(a) $l(x) = x_1$,
(b) $l(x) = x_1 + x_2$,
(c) $l(x) = \sum_{i=1}^n x_i b_i$,   ($b_i$ real).

5. The time-varying network in Figure 5.21.4 satisfies the differential equation

$$R \frac{dx}{dt} + \frac{1}{c(t)} x = u(t),$$

where R is the resistance, $c(t)$ the capacitance, $u(t)$ the input voltage, $x(t)$ the charge density, and $v(t) = x(t)/c(t)$. We assume that $c(t)$ is positive and continuous for $0 \leq t \leq 1$ and that $x(0) = 0$.
(a) Show that the set of inputs $u(t)$, with the property that $u \in L_2[0,1]$ and $v(1) = 1$ satisfy $l(u) = \alpha$ for some functional l and some constant α. What is the representation of l?
(b) Find the input u in $L_2[0,1]$ with the property that $v(1) = 1$ and $\int_0^1 |u|^2 \, dt =$ minimum.

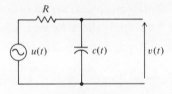

Figure 5.21.4

6. (Generalization of Exercise 5.) Consider the differential system

$$\frac{dx}{dt} = Ax + Bu, \tag{5.21.3}$$

where x is a real n-vector, u a real m-vector, A a real $n \times n$ matrix and B a real $n \times m$ matrix. Assume that there is at least one controller $u(t)$ in $L_2[0,1]$ such that the corresponding solution of (5.21.3) satisfies

$$x(0) = x^0 \quad \text{and} \quad x_1(1) = \alpha, \tag{5.21.4}$$

where $x = (x_1, x_2, \ldots, x_n)$.
(a) Show that the set of all $u(t)$ in $L_2[0,1]$ that satisfy (5.21.4) can be represented by $l(u) = \beta$ for some continuous linear functional l and some constant β.
(b) Find the input $u(t)$ in $L_2[0,1]$ such that (5.12.4) holds and $\int_0^1 \|u\|^2 \, dt =$ minimum, where $\|u\|^2 = |u_1|^2 + \cdots + |u_m|^2$.

7. The Riesz Representation Theorem (Theorem 5.21.1) defines a mapping L of H' (the space of all bounded linear functionals on H) onto H.
   (a) Show that this mapping is one-to-one.
   (b) Show that L satisfies

$$L(f+g) = L(f) + L(g), \qquad (f,g \in H')$$
$$L(\alpha f) = \bar{\alpha} L(f), \qquad (f \in H', \alpha \in C).$$

   (c) Using the definition of norm from Section 8 on H', show that for all f in H' one has $\|L(f)\| = \|f\|$.
   (d) Is L an isometric isomorphism?

8. (Completion of an inner product space.) Let X be an inner product space. Show that the completion of X as a Banach space (compare with Exercise 7, Section 9) is a Hilbert space. [*Hint*: Let $\{x_n\}$ and $\{y_n\}$ be two Cauchy sequences in X. Show that

$$(\{x_n\},\{y_n\}) = \lim (x_n,y_n)$$

defines an inner product on the space of all Cauchy sequences.]

9. Locate the first point in the proof of the Riesz Representation Theorem where the completeness of the Hilbert space H is used. Are there any other places in this proof where completeness is used?

10. Prove the Lax-Milgram Theorem (Theorem 5.21.2). [*Hint*: For v fixed show that there is a y in H with $B[x,v] = (x,y)$ for all $x \in H$. Define A by $y = Av$, and show that A is a topological isomorphism mapping H onto itself. If $l(x) = (x,y)$, then $l(x) = B[x, A^{-1}y]$. Now apply the same reasoning to $\overline{B[u,x]}$.]

11. (a) Show that the Riesz Representation Theorem is valid for real Hilbert spaces.
    (b) What happens to Exercise 7 in the case of real Hilbert spaces?

12. Let H be a Hilbert space of complex-valued functions defined on a set S. One says that H is a *proper functional Hilbert space* if for every $s \in S$ the mapping $\phi_s: H \to C$ given by

$$\phi_s(x) = x(s)$$

is a bounded linear functional on H. H is said to have a *reproducing kernel* $K(t,s)$ if there is a complex-valued function $K(t,s)$ defined on $S \times S$ with the properties:
    (i) For each t, the function of s, $K(s,t)$ lies in H; and
    (ii) For each $x \in H$ and each $t \in S$ one has $x(t) = (x, K(\cdot, t))$.
    (a) Show that H is a proper functional Hilbert space if and only if H has a reproducing kernel.
    (b) Assume that H has a reproducing kernel $K(s,t)$ and let $\{x_n\}$ be an orthonormal basis for H. Show that

$$K(s,t) = \sum_n x_n(s)\overline{x_n(t)}.$$

13. Let H denote all functions $f(z)$ that are analytic for $|z| < 1$ and such that

$$\iint_D |f(z)|^2 \, dx \, dy < \infty,$$

where D is the unit disk $\{z: |z| < 1\}$ and $z = x + iy$.

(a) Show that H is a Hilbert space when the inner product is given by

$$(f,g) = \iint_D f(z)\overline{g(z)} \, dx \, dy.$$

(b) Show that H is a proper functional Hilbert space. (The reproducing kernel for this Hilbert space is called the Bergman kernel. See Goffman and Pedrick [1] and Nehari [1].)

14. Let A be a subset of a Hilbert space H with the property that $\overline{V(A)}$, the closure of the span of A, is H. Let l be a bounded linear functional on H. Show that l is uniquely determined if one knows $l(a)$ for all $a \in A$.

Part C
Special Operators

22. THE ADJOINT OPERATOR

It is appropriate here to state what our major objective will be in the next chapter. Suppose that K is a bounded linear operator on a Hilbert space H. For example, K might be given by

$$(Ky)(t) = \int_I k(t,\tau)x(\tau)\, d\tau$$

on $L_2(I)$. We wish to get a better geometric picture of the behavior of K. In many important cases it turns out that the restrictions of K to certain linear subspaces are particularly simple operators. (For example, the restriction of K to its null space is the zero operator.) By piecing together these subspaces and simple operators in a suitable manner, we get a global picture of the operator K. However, before we can start any serious analysis of linear operators on Hilbert spaces it is absolutely necessary that we have the concept of the adjoint operator.

Let $K: H \to H$ be a bounded, linear operator on a Hilbert space H, and let y be a fixed element in H. Now consider the form

$$l(x) = (Kx,y).$$

The mapping l is obviously a linear functional on H. Furthermore, from the Schwarz Inequality we have $|l(x)| = |(Kx,y)| \le \|Ky\|\,\|x\|$, and by the definition of the operator norm this becomes

$$|l(x)| \le \|K\|\,\|y\|\,\|x\|,$$

that is, l is bounded. Therefore, by the Riesz Representation Theorem there is a unique y^* in H such that

$$(Kx,y) = (x,y^*) \tag{5.22.1}$$

for all $x \in H$. Thus, given a $y \in H$ there is a unique y^* associated with it. In other words, we have a mapping of H into itself.

5.22.1 Definition.   Let $K^*: H \to H$ be the mapping defined by (5.22.1) so that $y^* = K^*y$. K^* is referred to as the *adjoint* of K.

It follows from this definition that

$$(Kx,y) = (x,K^*y) \tag{5.22.2}$$

for all $x, y \in H$. Moreover, K^* is the only mapping satisfying (5.22.2). Indeed, if K_1 and K_2 are two transformations satisfying (5.22.2),

$$(x,K_1 y) = (x,K_2 y)$$

for all $x, y \in H$. It follows that $K_1 = K_2$. (Why?)

We leave it to the reader to show the following:

$$I^* = I,$$
$$0^* = 0,$$
$$(S + T)^* = S^* + T^*,$$
$$(\alpha T)^* = \bar{\alpha}T^*,$$
$$(ST)^* = T^*S^*.$$

In addition, we have the following facts about the adjoint.

5.22.2 THEOREM. *Let $K: H \to H$ be a bounded linear operator on a Hilbert space H. Then the adjoint operator K^* is a bounded, linear operator and $\|K\| = \|K^*\|$. Moreover, $(K^*)^* = K$.*

First Part of Proof: First we note that

$$\begin{aligned}(x, K^*(\alpha_1 y_1 + \alpha_2 y_2)) &= (Kx, \alpha_1 y_1 + \alpha_2 y_2) \\ &= \bar{\alpha}_1 (Kx, y_1) + \bar{\alpha}_2 (Kx, y_2) \\ &= \bar{\alpha}_1 (x, K^* y_1) + \bar{\alpha}_2 (x, K^* y_2) \\ &= (x, \alpha_1 K^* y_1 + \alpha_2 K^* y_2),\end{aligned}$$

which shows that K^* is linear.

We will now show that $\|K^*\| \le \|K\|$. Note that

$$\begin{aligned}\|K^* y\|^2 &= (K^* y, K^* y) = (KK^* y, y) \\ &\le \|KK^* y\| \cdot \|y\| \le \|K\| \cdot \|K^* y\| \cdot \|y\|.\end{aligned}$$

Hence $\|K^* y\| \le \|K\| \cdot \|y\|$, which implies that

$$\|K^*\| \le \|K\|.$$

If we replace K by K^* in Equation (5.22.2) this allows us to define the second adjoint K^{**}. That is,

$$(K^* x, y) = (x, K^{**} y).$$

Since

$$\overline{(x, Ky)} = (Ky, x) = (y, K^* x) = \overline{(K^* x, y)}$$

for all x and y, it follows that

$$(x, Ky) = (x, K^{**} y)$$

for all x and y. But this implies that $K^{**} = K$.

We can now complete the proof of Theorem 5.22.2.

Second Part of Proof: We have shown that $\|K^*\| \le \|K\|$. Now if we replace K by K^* we get $\|K\| = \|K^{**}\| \le \|K^*\|$. Hence $\|K\| = \|K^*\|$. ∎

5.22.3 COROLLARY. $\|K^*K\| = \|KK^*\| = \|K\|^2 = \|K^*\|^2$

Proof: It follows from Theorem 5.8.4 and the foregoing theorem that

$$\|K^*K\| \le \|K^*\| \, \|K\| = \|K\| \, \|K\| = \|K\|^2 = \|K^*\|^2. \qquad (5.22.3)$$

Furthermore,

$$\|Kx\|^2 = (Kx,Kx) = (K^*Kx,x) \le \|K^*Kx\| \, \|x\| \le \|K^*K\| \, \|x\|^2,$$

so we have

$$\|K\|^2 \le \|K^*K\|. \qquad (5.22.4)$$

Combining (5.22.3) and (5.22.4) yields the equality $\|K^*K\| = \|K\|^2$. The remainder of the theorem is proved in an obvious way. ∎

EXAMPLE 1. Let $H = C^n$ with the inner product given by

$$(x,y) = \sum_{i=1}^{n} x_i \bar{y}_i.$$

Let $K: C^n \to C^n$ be a linear operator that is represented by an $n \times n$ matrix of complex coefficients (k_{ij}). That is, if $y = (y_1,\ldots,y_n)$ and $x = (x_1,\ldots,x_n)$, then

$$y = Kx \Leftrightarrow y_i = \sum_{j=1}^{n} k_{ij} x_j, \qquad i = 1, \ldots, n.$$

What is the adjoint K^*? Assume that K^* is represented by the $n \times n$ matrix with coefficients (l_{ij}). The equation $(Kx,y) = (x,K^*y)$ then becomes

$$\sum_{i=1}^{n} \left(\sum_{j=1}^{n} k_{ij} x_j \right) \bar{y}_i = \sum_{i=1}^{n} x_i \overline{\left(\sum_{j=1}^{n} l_{ij} y_j \right)}.$$

Since

$$\sum_i \sum_j k_{ij} x_j \bar{y}_i = \sum_j \sum_i k_{ij} x_j \bar{y}_i = \sum_j x_j \sum_i k_{ij} \bar{y}_i$$

$$= \sum_j x_j \overline{\left(\sum_i \bar{k}_{ij} y_i \right)} = \sum_i x_i \overline{\left(\sum_j \bar{k}_{ji} y_j \right)}$$

(where in the last step we merely change the summation indices), it follows that $l_{ij} = \bar{k}_{ji}$. That is, the matrix for K^* is found by (a) taking the transpose of (k_{ij}) and (b) taking the complex conjugate of each entry of (k_{ij}). ∎

EXAMPLE 2. Let $H = R^n$ with the inner product given by $(x,y) = \sum_{i=1}^{n} x_i y_i$. The adjoint of a matrix operator is the transpose. ∎

EXAMPLE 3. Let I be an interval and let $k: I \times I \to C$ be such that

$$\int_I \int_I |k(s,t)|^2 \, ds \, dt < \infty.$$

Define $K: L_2(I) \to L_2(I)$ by $z = Kx$ where

$$z(s) = \int_I k(s,t)x(t)\, dt,$$

and assume that $L_2(I)$ has the usual inner product

$$(x,y) = \int_I x\bar{y}\, dt.$$

We will show that K^* is also an integral operator. In this case we get

$$(Kx,y) = \int_I \left[\int_I k(s,t)x(t)\, dt \right] \overline{y(s)}\, ds = \int_I \int_I k(s,t)x(t)\overline{y(s)}\, ds\, dt$$

$$= \int_I x(t) \left[\int_I \overline{k(s,t)}y(s)\, ds \right] dt = (x, K^*y).$$

Hence, after interchanging the s and t variables, we get

$$(K^*y)(s) = \int_I \overline{k(t,s)}y(t)\, dt = \int_I k^*(s,t)y(t)\, dt,$$

that is, $k^*(s,t) = \overline{k(t,s)}$. ∎

EXAMPLE 4.   The foregoing example is particularly interesting when K represents a Volterra integral operator

$$y(t) = \int_0^t k(t,\tau)x(\tau)\, d\tau. \qquad (5.22.5)$$

Here we take $I = [0,T]$. If we set $k(t,\tau) = 0$ for $t < \tau$, then (5.22.5) becomes

$$y(t) = \int_I k(t,\tau)x(\tau)\, d\tau.$$

According to the last example the adjoint is given by

$$(K^*y)(t) = \int_I \overline{k(\tau,t)}y(\tau)\, d\tau = \int_t^T \overline{k(\tau,t)}y(\tau)\, d\tau. \qquad (5.22.6)$$

In other words, the adjoint of a Volterra integral operator is also a Volterra integral operator. However, if K depends on the "past," then K^* depends on the "future."

The Volterra integral operator in (5.22.5) is a particular example of a causal operator, and (5.22.6) is an example of an anticausal operator (see Example 7). ∎

EXAMPLE 5.   On $L_2(-\infty,\infty)$ consider the multiplication operator

$$F: x(t) \to f(t)x(t),$$

where $|f(t)| \leq B < \infty$ for almost all t. F is a bounded linear operator and it is easy to show that

$$\|F\| = \operatorname*{ess.\ sup.}_{t \in R} |f(t)| = \|f\|_\infty.$$

Furthermore, the adjoint mapping is given by

$$F^*: y(t) \to \overline{f(t)}y(t).$$

Indeed

$$(Fx,y) = \int_{-\infty}^{\infty} f(t)x(t)\overline{y(t)}\, dt = \int_{-\infty}^{\infty} x(t)\overline{\overline{f(t)}y(t)}\, dt$$

$$= (x,F^*y). \quad \blacksquare$$

If $T: X \to X$ is a linear transformation and M is a linear subspace of X such that $T(M) \subset M$, we say that M is *invariant under* T. In this case the restriction of T to M is a mapping of M into itself.

EXAMPLE 6. Consider the space $L_2(-\infty,\infty)$ and let I be a interval in R. Let

$$M = \{x \in L_2(-\infty,\infty): x(t) = 0 \text{ for } t \notin I\}.$$

If F is a multiplication operator

$$F: x(t) \to f(t)x(t),$$

where $\|f\|_\infty < \infty$, then $F(M) \subset M$. $\quad \blacksquare$

EXAMPLE 7. In this example we show that if L is causal, then L^* is anticausal. Let $H = L_2(-\infty,\infty)$ and let L be a bounded linear mapping of H into itself. We know from Example 4, Section 4.3, that L is causal if and only if each linear subspace H_T is invariant under L, where H_T is the linear subspace of H made up of all functions x such that $x(t) = 0$ (almost everywhere) for $t < T$. Let $P_T: H \to H$ be defined by

$$(P_T x)(t) = \begin{cases} x(t), & \text{for } T \le t < \infty \\ 0, & \text{for } -\infty < t < T. \end{cases}$$

Obviously P_T is an orthogonal projection with $\mathscr{R}(P_T) = H_T$. Furthermore we see that $P_T^* = P_T$.

We claim that H_T is invariant under L if and only if $LP_T = P_T LP_T$. To prove this first assume that $L(H_T) \subset H_T$. Then $LP_T x$ is a point in H_T for any $x \in H$. But the restriction of a projection to its range is the identity operator on the range, so $LP_T x = P_T LP_T x$ for all $x \in H$.

Next we assume that $LP_T = P_T LP_T$. Since $\mathscr{R}(P_T) = H_T$, the linear subspace H_T is exactly the set of all points of the form $P_T x$ with $x \in H$. Thus the set of all points of the form $LP_T x$ with $x \in H$ is $L(H_T)$. Since $LP_T = P_T LP_T$, it follows that $y = P_T y$ for all $y \in L(H_T)$. In other words, $L(H_T) \subset \mathscr{R}(P_T) = H_T$.

If we let $Q_T = I - P_T$, then it can be shown in a manner similar to the above argument that L^* is anticausal if and only if $L^* Q_T = Q_T L^* Q_T$ for each T.

Assuming, then, that $LP_T = P_T LP_T$ for each T, we have

$$(y,LP_T x) = (y,P_T LP_T x), \qquad \text{for all } x, y \in H.$$

So

$$(P_T L^* y,x) = (P_T L^* P_T y,x).$$

Then using $P_T = I - Q_T$, we eventually obtain

$$0 = (Q_T L^* Q_T y, x) - (L^* Q_T y, x)$$

for all $x, y \in H$. Hence, $Q_T L^* Q_T = L^* Q_T$ showing that L^* is anticausal. A simple reversal of the above steps shows that if L^* is anticausal then L is causal. ∎

Suppose T is linear transformation of a Hilbert space H into itself. Sometimes it happens that a closed linear space M and its orthogonal complement $M^\perp$ are both invariant under T, that is, $T(M) \subset M$ and $T(M^\perp) \subset M^\perp$. When this happens we say that M reduces T. We say that M "reduces" T because then T is completely characterized by its restrictions to M and $M^\perp$. It often happens that these restrictions of T are simpler than T itself.

Going further, we sometimes can find a family of closed mutually orthogonal subspaces $\{M_n\}$ such that $H = M_1 + M_2 + M_3 + \cdots$, in the sense of Section 20, such that each M_n reduces T, and such that the restriction of T to each M_n is a simple operator. This idea is developed further in the next chapter.

5.22.4 THEOREM.   *Let T be a continuous linear transformation of a Hilbert space H into itself. A closed linear subspace M of H is invariant under T if and only if $M^\perp$ is invariant under T^*.*

Proof: Suppose M is invariant under T, that is, $T(M) \subset M$. Then $(y, Tx) = 0$ for all $y \in M^\perp$ and $x \in M$. Thus $(T^* y, x) = 0$ for all $y \in M^\perp$ and $x \in M$. Hence $T^* y \in M^\perp$ for all $y \in M^\perp$, which shows that $M^\perp$ is invariant under T^*. Also if $M^\perp$ is invariant under T^*, essentially the same argument shows that M is invariant under T. ∎

5.22.5 COROLLARY.   *A closed linear subspace M of H reduces T if and only if M is invariant under both T and T^*.*

Proof: One merely recalls [Theorem 5.15.4(b)] that $M^{\perp\perp} = M$ and $T^{**} = T$. The proof is then trivial. ∎

The preceding two results (Theorem 5.22.4 and Corollary 5.22.5) are our first use of the adjoint. They are by no means our last. The next theorem shows that T^* can be used to characterize the range and null space of T and vice versa.

5.22.6 THEOREM.   *Let T be a bounded linear transformation of a Hilbert space H into itself. Then*

$$\overline{\mathscr{R}(T)} = \{\mathscr{N}(T^*)\}^\perp \tag{5.22.7}$$

and

$$\{\mathscr{N}(T)\}^\perp = \overline{\mathscr{R}(T^*)},$$

where $\mathscr{N}(T)$ and $\mathscr{N}(T^*)$ are the null spaces of T and T^*, respectively; and $\overline{\mathscr{R}(T)}$ and $\overline{\mathscr{R}(T^*)}$ are the closures of the ranges of T and T^*, respectively.

Proof: Since $T^{**} = T$ it will suffice to prove (5.22.7). A point $z \in H$ is in $\mathscr{R}(T)^{\perp}$ if and only if $(z,Tx) = 0$ for all $x \in H$. But by definition of the adjoint, it follows

$$(T^*z,x) = 0$$

for all $x \in H$. So $z \in \{\mathscr{R}(T)\}^{\perp}$ if and only if $T^*z = 0$. Hence, we have shown that $\{\mathscr{R}(T)\}^{\perp} = \mathscr{N}(T^*)$.

However, now $\mathscr{R}(T)$ may not be closed, so it is not necessarily the case that $\{\mathscr{R}(T)\}^{\perp\perp} = \mathscr{R}(T)$; but, we do always have that $\{\mathscr{R}(T)\}^{\perp\perp} = \overline{\mathscr{R}(T)}$, which proves (5.22.7). ∎

The concept of the adjoint mapping offers an elegant way of characterizing unitary operators.

5.22.7 THEOREM.   *Let* $U: H \to H$ *be a bounded linear operator on a Hilbert space H. Then U is a unitary operator if and only if* $UU^* = U^*U = I$, *that is, if and only if* $U^* = U^{-1}$.

Proof: If U is a unitary mapping, then

$$(U^*Ux,y) = (Ux,Uy) = (x,y),$$

for all x and y in H. Hence $U^*U = I$. Similarly, we see that $UU^* = I$.

Now assume that $U^*U = I$. Then

$$(x,y) = (U^*Ux,y) = (Ux,Uy),$$

for all x and y in H, so U is unitary.   ∎

EXAMPLE 8.   Let $H = L_2(-i\infty,i\infty)$, and let T denote the linear transformation of H into itself defined by $Y = TX$ where

$$Y(i\omega) = \frac{1}{1 + i\omega} X(i\omega), \qquad i\omega \in (-i\infty,i\infty).$$

Then the adjoint of T can be represented by the equation

$$Z(i\omega) = \frac{1}{1 - i\omega} W(i\omega), \qquad i\omega \in (-i\infty,i\infty).$$

Since the (transfer) function is nonzero for all $i\omega$, the null space of T^*, $\mathscr{N}(T^*)$, is the trivial subspace $\{0\}$. Hence, $\{\mathscr{N}(T^*)\}^{\perp} = H$, and from Theorem 5.22.6 it follows that $\overline{\mathscr{R}(T)} = H$. Of course, in this simple case we can see directly that this is so. Indeed,

$$\mathscr{R}(T) = \left\{ Y \in H: Y(i\omega) = \frac{1}{1 + i\omega} X(i\omega) \text{ with } X \in H \right\}$$

which is a subspace that is dense in H. (Why?) On the other hand, one does have $\mathscr{R}(T) \neq H$. ∎

EXAMPLE 9.   Theorem 5.22.6 can be used in approximation. Recall that Theorem 5.14.4 is the key result concerning approximation in Hilbert spaces. Now it often happens that the closed linear subspace M containing the approximation y_0 is specified as the null space or range of a bounded linear transformation. Let us briefly consider these two cases of this approximation problem.

Case I: $M = \mathcal{N}(T)$.

Suppose that T is a bounded linear transformation of a Hilbert space H into itself. Then $M = \mathcal{N}(T)$ is a closed linear subspace of H. If $x_0 \in H$, then (by Theorem 5.14.4) there is a unique $y_0 \in M$ such that $\|x_0 - y_0\| \le \|x_0 - y\|$ for all $y \in M$ and $x_0 - y_0 \in M^{\perp}$. It follows from Theorem 5.22.6 that $(x_0 - y_0)$ is in $\overline{\mathcal{R}(T^*)}$. Of course, if $\mathcal{R}(T^*)$ is closed, then there exists a $z_0 \in H$ such that $x_0 - y_0 = T^*z_0$. However, we cannot assume in general that $\mathcal{R}(T^*)$ is closed. Hence, all we can say for sure is that there exists a sequence $\{z_n\}$ in H such that

$$x_0 - y_0 = \lim_{n \to \infty} T^*z_n.$$

In order to simplify the discussion, let us assume that there is a $z_0 \in H$ such that $x_0 - y_0 = T^*z_0$. The point y_0 is in the subspace $M = \mathcal{N}(T)$ if and only if

$$Tx_0 = TT^*z_0. \tag{5.22.8}$$

So our problem reduces to finding a z_0 that satisfies (5.22.8). Then given such a z_0 we can use $y_0 = x_0 - T^*z_0$ to get y_0.

Case II: $M = \overline{\mathcal{R}(T)}$.

Here again it may be that $\mathcal{R}(T)$ is not closed. If it is not, we can say that there exists a sequence $\{z_n\}$ in H such that

$$\lim_{n \to \infty} \|x_0 - Tz_n\| = \text{dist}\,(x_0, M).$$

Again to simplify matters let us assume that there is a $z_0 \in H$ such that

$$\|x_0 - Tz_0\| \le \|x_0 - Tz\|$$

for all $z \in H$. Then $(x_0 - Tz_0) \perp M$, so

$$0 = (x_0 - Tz_0, Tz) = (T^*x_0 - T^*Tz_0, z),$$

for all $z \in H$. It follows that

$$T^*x_0 = T^*Tz_0. \tag{5.22.9}$$

Given, then, a z_0 that satisfies (5.22.9) the approximation y_0 to x_0 is given by

$$y_0 = Tz_0. \quad \blacksquare$$

EXAMPLE 10.   Let $H = L_2(-\infty, \infty)$, and consider the delay or shift $S_\tau : H \to H$, where $(S_\tau x)(t) = x(t - \tau)$ for $t \in (-\infty, \infty)$. It is clear that S_τ has a continuous inverse; indeed, $S_\tau^{-1} = S_{-\tau}$. In fact S_τ is a unitary operator, for

$$(S_\tau x, S_\tau y) = \int_{-\infty}^{\infty} \overline{y(t - \tau)} x(t - \tau)\, dt = \int_{-\infty}^{\infty} \overline{y(t)} x(t)\, dt = (x, y),$$

for all $x, y \in H$. Moreover,

$$(S_\tau x, y) = \int_{-\infty}^{\infty} \overline{y(t)}x(t - \tau)\, dt = \int_{-\infty}^{\infty} \overline{y(t + \tau)}x(t)\, dt,$$

for all $x, y \in H$, so the adjoint of S_τ is defined by

$$(S_\tau{}^* y)(t) = y(t + \tau), \qquad \text{for } t \in (-\infty, \infty).$$

Hence,

$$S_\tau{}^* = S_{-\tau} = S_\tau{}^{-1}. \quad \blacksquare \tag{5.22.10}$$

EXAMPLE 11. In this example we show that the adjoint of a linear time-invariant operator is also time-invariant. Let L be a time-invariant continuous linear transformation of $L_2(-\infty, \infty)$ into itself. Recall that L being time-invariant means

$$S_\tau L = L S_\tau, \qquad \text{for all } -\infty < \tau < \infty,$$

where S_τ denotes the delay or shift $(S_\tau x)(t) = x(t - \tau)$. Then, using (5.22.10), we get

$$(x, L^* S_\tau y) = (Lx, S_\tau y) = (S_{-\tau} Lx, y)$$

and

$$(x, S_\tau L^* y) = (S_{-\tau} x, L^* y) = (L S_{-\tau} x, y) = (S_{-\tau} Lx, y)$$

for all $x, y \in L_2(-\infty, \infty)$. Note that the last interchange is a consequence of L being time-invariant. In any event, we have $(x, L^* S_\tau y) = (x, S_\tau L^* y)$ for all x, y in $L_2(-\infty, \infty)$, so $L^* S_\tau = S_\tau L^*$; therefore, L^* is time-invariant. $\quad \blacksquare$

EXAMPLE 12. (FOURIER TRANSFORM ON R^1.) In this example we shall prove that the Fourier transform F, as defined by (compare with Example 6, Section 19)

$$\hat{f}(y) = (Ff)(y) = \frac{1}{\sqrt{2\pi}} \int_{-\infty}^{\infty} e^{-iyx} f(x)\, dx, \tag{5.22.11}$$

is a unitary mapping of $L_2(-\infty, \infty)$ onto itself and that the inverse is given by

$$f(x) = (F^{-1}\hat{f})(x) = \frac{1}{\sqrt{2\pi}} \int_{-\infty}^{\infty} e^{iyx} \hat{f}(y)\, dy. \tag{5.22.12}$$

Actually the representation for F and F^{-1} given by (5.22.11) and (5.22.12) is valid for functions f and $\hat{f}$ in $L_1(-\infty, \infty) \cap L_2(-\infty, \infty)$. In order to discuss arbitrary functions in $L_2(-\infty, \infty)$ we need a different representation, namely,

$$\hat{f}(y) = (Ff)(y) = \frac{1}{\sqrt{2\pi}} \frac{d}{dy} \int_{-\infty}^{\infty} \frac{e^{-ixy} - 1}{-ix} f(x)\, dx \tag{5.22.13}$$

and

$$f(x) = (F^{-1}\hat{f})(x) = \frac{1}{\sqrt{2\pi}} \frac{d}{dx} \int_{-\infty}^{\infty} \frac{e^{ixy} - 1}{iy} \hat{f}(y)\, dy. \tag{5.22.14}$$

If f and $\hat{f}$ belong to $L_1(-\infty,\infty)$ then we can bring the differentiation inside the integral and then (5.22.13) and (5.22.14) reduce to (5.22.11) and (5.22.12). The transformation F represented by (5.22.13) is oftentimes referred to as the Fourier-Plancherel transform.

Let us now show that F and F^{-1} are unitary operators on $L_2(-\infty,\infty)$. For this purpose we shall denote the operator defined by (5.22.14) as G. We will then show that F and G are unitary and that $F^* = G$, which implies that $F^{-1} = G$.

Now define

$$H(y,x) = \frac{1}{\sqrt{2\pi}} \frac{e^{-ixy} - 1}{-ix}, \qquad K(x,y) = \frac{1}{\sqrt{2\pi}} \frac{e^{ixy} - 1}{iy}$$

and let

$$\phi_r(x) = \begin{cases} +1, & 0 \le x \le r, \quad 0 \le r, \\ -1, & r \le x \le 0, \quad r \le 0, \\ 0, & \text{otherwise.} \end{cases}$$

Now for $r \ge 0$ one has

$$(F\phi_r)(y) = \frac{1}{\sqrt{2\pi}} \frac{d}{dy} \int_0^r \frac{e^{-ixy} - 1}{-ix} \, dx = \frac{1}{\sqrt{2\pi}} \int_0^r e^{-ixy} \, dx$$

$$= H(r,y).$$

Similarly for $r \le 0$ one has $(F\phi_r)(y) = H(r,y)$. Likewise we get $(G\phi_r)(x) = K(r,x)$.

Since $\phi(y) = \text{Im } H(r,y)\overline{H(s,y)}$ is an odd function in $L_1(-\infty,\infty)$, one has $\int_{-\infty}^{\infty} \phi(y) \, dy = 0$. Hence

$$(F\phi_r, F\phi_s) = \int_{-\infty}^{\infty} H(r,y)\overline{H(s,y)} \, dy = \frac{1}{2\pi} \int_{-\infty}^{\infty} \frac{\cos(s-r)y - \cos sy - \cos ry + 1}{y^2} \, dy.$$

By using the trigonometric identity $\cos\theta = 1 - 2\sin^2\theta/2$ and by changing variables we get

$$(F\phi_r, F\phi_s) = \frac{1}{2\pi} \{|r| + |s| - |r - s|\} \int_{-\infty}^{\infty} \frac{\sin^2 u}{u^2} \, du.$$

Since $\int_{-\infty}^{\infty} \frac{\sin^2 u}{u^2} \, du = \pi$, we get

$$(F\phi_r, F\phi_s) = \begin{cases} \min\{|r|, |s|\}, & \text{if } rs \ge 0 \\ 0, & \text{if } rs \le 0 \end{cases} = (\phi_r, \phi_s).$$

Similarly one has

$$(G\phi_r, G\phi_s) = (\phi_r, \phi_s).$$

Furthermore by a simple change of variables we get

$$(F\phi_r, \phi_s) = (\phi_r, G\phi_s).$$

If f and g are now finite linear combinations of the functions ϕ_r, that is, if they are step-functions, then one has

$$(Ff,Fg) = (f,g),$$
$$(Gf,Gg) = (f,g), \tag{5.22.15}$$
$$(Ff,g) = (f,Gg).$$

However the step functions are dense in $L_2(-\infty,\infty)$, therefore (5.22.15) is valid for all f and g in $L_2(-\infty,\infty)$. This shows that F and G are unitary and that $F^* = G$.

Next, one can prove that the Fourier transform F is given by (5.22.11) for all $f \in L_2(-\infty,\infty)$ if one interprets the integral in (5.22.11) in the following sense:

$$\frac{1}{\sqrt{2\pi}} \int_{-\infty}^{\infty} e^{-iyx} f(x)\, dx = \lim_{N\to\infty} \frac{1}{\sqrt{2\pi}} \int_{-N}^{N} e^{-iyx} f(x)\, dx,$$

where *lim* means "limit in the mean," that is,

$$\int \left| \hat{f}(y) - \frac{1}{\sqrt{2\pi}} \int_{-N}^{N} e^{-iyx} f(x)\, dx \right|^2 dy \to 0$$

as $N \to \infty$. You are asked to prove this in Exercise 13 below.

The reader is probably aware of the importance of the Fourier transform. It is a fundamental tool in the operational calculus of differential operators. The following theorem is the cornerstone of this theory.

5.22.8 THEOREM.   *Let P and Q be the linear operators defined by*

$$P: u(x) \to i\,\frac{du}{dx}$$

$$Q: u(x) \to xu(x),$$

where the domains are

$\mathscr{D}_P = \{u \in L_2(-\infty,\infty): u$ *is absolutely continuous and* $u' \in L_2(-\infty,\infty)\}$
$\mathscr{D}_Q = \{u \in L_2(-\infty,\infty\}: xu(x) \in L_2(-\infty,\infty)\}.$

Then the Fourier transform F sets up a one-to-one correspondence between $\mathscr{D}_P$ *and* $\mathscr{D}_Q$ *in such a way that*

$$P = FQF^{-1} \quad \text{and} \quad Q = F^{-1}PF.$$

Proof: The first step is to show that if $u \in \mathscr{D}_Q$, then $Fu \in \mathscr{D}_P$ and $PFu = FQu$. Let $u \in \mathscr{D}_Q$, then $\int_{-\infty}^{\infty} |u(x)|\, dx < \infty$, by Exercise 14 below. Thus $v = Fu$ is given by

$$v(y) = (Fu)(y) = \frac{1}{\sqrt{2\pi}} \int_{-\infty}^{\infty} e^{-iyx} u(x)\, dx.$$

However, $FQu \in L_2(-\infty, \infty)$ and

$$(FQu)(y) = \frac{1}{\sqrt{2\pi}} \frac{d}{dy} \int_{-\infty}^{\infty} \frac{e^{-iyx} - 1}{-ix} \, xu(x) \, dx$$

$$= i \frac{d}{dy} \frac{1}{\sqrt{2\pi}} \int_{-\infty}^{\infty} (e^{-iyx} - 1)u(x) \, dx$$

$$= i \frac{d}{dy} \frac{1}{\sqrt{2\pi}} \int_{-\infty}^{\infty} e^{-iyx}u(x) \, dx$$

$$= (PFu)(y).$$

Hence $Fu \in \mathcal{D}_P$ and $FQu = PFu$.

The second step is to show that if $v \in \mathcal{D}_P$, then $F^{-1}v \in \mathcal{D}_Q$ and $F^{-1}Pv = QF^{-1}v$. Let $v \in \mathcal{D}_P$, then

$$\lim_{x \to \pm\infty} v(x) = 0,$$

by Exercise 15 below. Furthermore $F^{-1}Pv$ is in $L_2(-\infty, \infty)$ and if we integrate by parts we get

$$(F^{-1}Pv)(x) = \frac{d}{dx} \frac{1}{\sqrt{2\pi}} \int_{-\infty}^{\infty} \frac{e^{ixy} - 1}{iy} \, i \frac{dv(y)}{dy} \, dy$$

$$= \frac{d}{dx} \frac{-1}{\sqrt{2\pi}} \int_{-\infty}^{\infty} \frac{e^{ixy}(ixy) - (e^{ixy} - 1)}{y^2} \, v(y) \, dy$$

$$= \frac{d}{dx} \frac{x}{\sqrt{2\pi}} \int_{-\infty}^{\infty} \frac{e^{ixy} - 1}{iy} \, v(y) \, dy + \frac{d}{dx} \frac{1}{\sqrt{2\pi}} \int_{-\infty}^{\infty} \frac{e^{ixy} - 1 - ixy}{y^2} \, v(y) \, dy.$$

Since

$$\frac{1}{\sqrt{2\pi}} \int_{-\infty}^{\infty} \frac{e^{ixy} - 1}{iy} \, v(y) \, dy = -\frac{d}{dx} \frac{1}{\sqrt{2\pi}} \int_{-\infty}^{\infty} \frac{e^{ixy} - 1 - ixy}{y^2} \, v(y) \, dy,$$

we get

$$(F^{-1}Pv)(x) = x \frac{d}{dx} \frac{1}{\sqrt{2\pi}} \int_{-\infty}^{\infty} \frac{e^{ixy} - 1}{iy} \, v(y) \, dy = x(F^{-1}v)(x).$$

Hence $F^{-1}v \in \mathcal{D}_Q$ and $F^{-1}Pv = QF^{-1}v$. ∎

EXAMPLE 13. (FOURIER TRANSFORM ON R^n.)   On R^n the Fourier transform takes on the form (compare with Example 6, Section 19)

$$\hat{f}(y) = (\mathscr{F}f)(y) = \int_{R^n} e^{-ix \cdot y} f(x) \, dx \tag{5.22.16}$$

$$f(x) = (\mathscr{F}^{-1}f)(x) = \frac{1}{(2\pi)^n} \int_{R^n} e^{ix \cdot y} \hat{f}(y) \, dy, \tag{5.22.17}$$

where $x = (x_1, \ldots, x_n)$ and $y = (y_1, \ldots, y_n)$ are points in R^n and

$$x \cdot y = x_1 y_1 + \cdots + x_n y_n.$$

Equations (5.22.16) and (5.22.17) are valid for f and $\hat{f}$ in $L_1(R^n) \cap L_2(R^n)$, and they are also valid for all f and $\hat{f}$ in $L_2(R^n)$ provided that we compute the integral as a limit in the mean of integrals over bounded regions.

One can prove that

$$(\mathscr{F}f, \mathscr{F}g) = (2\pi)^n (f, g)$$

for all f, g in $L_2(R^n)$. This is done by applying the Fourier transform on R^1 to each of the variables $x_1, x_2, \ldots, x_n$ successively. Similarly one gets

$$(\mathscr{F}^{-1}f, \mathscr{F}^{-1}g) = (2\pi)^{-n} (f, g).$$

By the same method one can prove the following theorem.

5.22.9 THEOREM.   *Let P_k and Q_k be the linear operators defined by*

$$P_k: u(x_1, \ldots, x_n) \to i \frac{\partial u}{\partial x_k}$$

$$Q_k: u(x_1, \ldots, x_n) \to x_k u(x_1, \ldots, x_n),$$

where the domains are

$$\mathscr{D}_{P_k} = \{u \in L_2(R^n) : P_k u \in L_2(R^n)\}$$

$$\mathscr{D}_{Q_k} = \{u \in L_2(R^n) : Q_k u \in L_2(R^n)\}.$$

Then the Fourier transform $\mathscr{F}$ sets up a one-to-one correspondence between $\mathscr{D}_P$ and $\mathscr{D}_Q$ in such a way that

$$P_k = \mathscr{F} Q_k \mathscr{F}^{-1} \quad and \quad Q_k = \mathscr{F}^{-1} P_k \mathscr{F}.$$

As a corollary to this we can prove that if $u \in L_2(R^n)$ with

$$D^\alpha u = \frac{\partial^\alpha u}{\partial x_1^{\alpha_1} \cdots \partial x_n^{\alpha_n}} \in L_2(R^n),$$

then

$$(\mathscr{F} D^\alpha u)(y) = (iy)^\alpha \hat{u}(y), \tag{5.22.18}$$

where $\hat{u} = \mathscr{F}u$ and $y^\alpha = y_1^{\alpha_1} \cdots y_n^{\alpha_n}$.   ∎

EXERCISES

1. For each $x \in L_2[0,1]$ let $y = Tx$ be the solution of $y' + ay = x$ that satisfies $y(0) = 0$, where a is a constant. Determine the adjoint T^*.

2. For each x in $L_2[0,1]$ let $y = Tx$ be the solution of $y'' + ay' + by = x$ that satisfies $y(0) = y(1) = 0$, where a and b are constants. Determine T^*. Is it ever true that $T = T^*$?

3. Let L be a bounded linear operator on a Hilbert space H. Verify the following relationships:

$$\mathcal{N}(L^*) = \mathcal{N}(LL^*); \qquad \overline{\mathcal{R}(L)} = \overline{\mathcal{R}(LL^*)}.$$

4. Use Theorem 5.22.6 to show that $\mathcal{R}(L) = H$ if and only if L^* has a continuous inverse. [*Hint*: Use also the Closed Graph Theorem, Exercise 13, Section 8.]

5. Let $L: H \to H$ be a topological isomorphism. Show that $(L^{-1})^* = (L^*)^{-1}$.

6. Let H be a Hilbert space and consider * as a mapping of $Blt[H,H]$ into itself where $*: L \to L^*$. Show that * is one-to-one and its range is all of $Blt[H,H]$. Is * linear? Does * preserve norms? (Compare this with Exercise 7, Section 21.)

7. The adjoint was defined for operators on a Hilbert space. Can this be extended to operators on an (incomplete) inner product space? If not, why not?

8. Let $T: l_2 \to l_2$ be given by $T(x_1, x_2, \ldots) = (x_1, \tfrac{1}{2}x_2, \ldots, (1/n)x_n, \ldots)$. Determine T^*.

9. Let $L: H \to H$ be a bounded linear mapping of H onto H. Show that L is an isometry if and only if L^* is an isometry.

10. Define a mapping $L: l_2 \to l_2$ by $(y_1, y_2, \ldots) = L(x_1, x_2, \ldots)$ where

$$y_n = \frac{x_1 + \cdots + x_n}{n}.$$

Show that L is a bounded linear operator with $\|L\| = (\sum_{n=1}^{\infty} 1/n^2)^{1/2}$. What is L^*?

11. Define a mapping $A: l_2 \to l_2$ by $(y_1, y_2, \ldots) = A(x_1, x_2, \ldots)$ where

$$y_n = \sum_{j=1}^{n} a_{nj} x_j.$$

Assume that A is a time-invariant operator and $\sum_{n=1}^{\infty} |a_{n1}|^2 < \infty$. Show that A is a bounded linear operator with $\|A\| = (\sum_{n=1}^{\infty} |a_{n1}|^2)^{1/2}$. What is A^*?

12. Show that if a sequence $\{L_n\}$ converges uniformly to L, where the L_n's and L are bounded linear transformations on a Hilbert space H, then the sequence $\{L_n^*\}$ converges uniformly to L^*.

13. Let $\hat{f}$ be the Fourier transform of f given by (5.19.3). Show that $\|\hat{f} - g_N\| \to 0$ where

$$g_N(y) = \frac{1}{\sqrt{2\pi}} \int_{-N}^{N} e^{-iyx} f(x) \, dx.$$

14. Show that if $u \in \mathcal{D}_Q$, then $\int_{-\infty}^{\infty} |u(x)| \, dx < \infty$. [*Hint*: Let $v(x) = xu(x)$, then $v \in L_2(-\infty, \infty)$ and by Schwarz Inequality

$$u(x) = \frac{1}{x} v(x) \in L_1(-\infty, -1) \cap L_1(1, \infty).$$

Since $u \in L_2(-1, 1)$ one has $u \in L_1(-\infty, \infty)$.]

15. Show that if $v \in \mathcal{D}_P$, then

$$\lim_{x \to \pm\infty} v(x) = 0.$$

[*Hint*: Note that

$$|v(x)|^2 - |v(0)|^2 = \int_0^x \frac{d}{dy} |v(y)|^2 \, dy = \int_0^x v(y)\overline{v'(y)} \, dy + \int_0^x v'(y)\overline{v(y)} \, dy.$$

Now use the fact that v and v' are in $L_2(-\infty,\infty)$ to show that the limits as $x \to \pm\infty$ exist.]

16. Prové Theorem 5.22.9.

17. Prove Equation (5.22.18).

18. Let $g = Sf$ be defined by

$$g(x) = \sqrt{\frac{2}{\pi}} \int_0^\infty \sin xy \, f(y) \, dy.$$

Show that S is a unitary mapping of $L_2[0,\infty)$ onto itself and that $S^{-1} = S$.

19. Let $g = Cf$ be defined by

$$g(x) = \sqrt{\frac{2}{\pi}} \int_0^\infty \cos xy \, f(y) \, dy.$$

Show that C is a unitary mapping of $L_2[0,\infty)$ onto itself and that $C^{-1} = C$.

20. (Hankel Transform.) Let $g = Hf$ be defined by

$$g(x) = \int_0^\infty (xy)^{1/2} J_\nu(xy) f(y) \, dy,$$

where J_ν is the Bessel function of the first kind of order ν, and where $\nu \geq -\frac{1}{2}$. Show that H is a unitary mapping on $L_2[0,\infty)$ with $H^{-1} = H$. (See Stone [1, p. 110] for more details.)

21. (Watson Transform.) Let $w(r)$ be a function with the property that $r^{-1}w(r)$ is in $L_2(a,b)$ and

$$\int_a^b \frac{w(sx)\overline{w(tx)}}{x^2} \, dx = \begin{cases} \min\{|s|, |t|\}, & \text{if } st \geq 0 \\ 0, & \text{if } st \leq 0. \end{cases}$$

Define the Watson transform $g = Wf$ by

$$g(x) = \frac{d}{dx} \int_a^b \frac{w(xy)}{y} f(y) \, dy.$$

(a) Show that W is a unitary mapping of $L_2(a,b)$ onto itself and that W^{-1} is given by

$$f(y) = \frac{d}{dy} \int_a^b \frac{\overline{w(xy)}}{x} g(x) \, dx.$$

[*Hint*: Study Example 12 carefully.]

(b) Show that the Watson transform is a generalization of the Fourier-Plancherel transform.

22. Let $L: C^3 \to C^3$ be represented by the matrix

$$
\begin{bmatrix}
a & \dfrac{i}{\sqrt{4}} & \dfrac{-1+2i}{\sqrt{12}} \\[2ex]
b & \dfrac{1+i}{\sqrt{4}} & \dfrac{1-i}{\sqrt{12}} \\[2ex]
c & \dfrac{-1}{\sqrt{4}} & \dfrac{2-i}{\sqrt{12}}
\end{bmatrix}.
$$

Determine a, b, and c so that L is unitary. (Assume that C^3 has the usual inner product.)

23. Let H_1 and H_2 be Hilbert spaces with inner products $(\cdot\,, \cdot)_1$ and $(\cdot\,, \cdot)_2$ respectively. Let $L: H_1 \to H_2$ be a bounded linear operator. Define the *adjoint* $L^*: H_2 \to H_1$ so that

$$(y,Lx)_2 = (L^*y,x)_1$$

holds for all x in H_1 and all y in H_2.

(a) Show that L^* is a bounded linear operator.

(b) Show that $L = L^{**}$.

(c) Show that $\|L\| = \|L^*\|$.

(d) Show that a bounded linear operator $U: H_1 \to H_2$ is unitary if and only if $U^*U = I_1$ and $UU^* = I_2$, where I_1 and I_2 are the identity operators on H_1 and H_2, respectively.

23.   NORMAL AND SELF-ADJOINT OPERATORS

This section is devoted to the elementary properties of normal and self-adjoint operators.

5.23.1 DEFINITION.   Let H be a Hilbert space, and let $T: H \to H$ be a bounded linear transformation. T is said to be *normal* if $TT^* = T^*T$, that is, if T commutes with its adjoint.

5.23.2 DEFINITION.   Let H be a Hilbert space, and let $T: H \to H$ be a bounded linear transformation, T is said to be *self-adjoint* if $T = T^*$.

Obviously another characteristic of self-adjointness is given by the relationship $(Tx,y) = (x,Ty)$ for all x, y in H.

5.23.3 LEMMA.   *If T is self-adjoint, then it is normal.*

Proof: The proof of this lemma is trivial. ∎

We have discussed a number of classes of operators so far. Their relation to normal and self-adjoint operators is illustrated in Figure 5.23.1.

Normal and self-adjoint operators are important for at least two reasons. First, many physical systems—but not all—can be modeled mathematically using these operators. Secondly, normal and self-adjoint operators have an especially simple structure. This latter point is not self-evident. Indeed, its explanation is the subject of the next chapter.

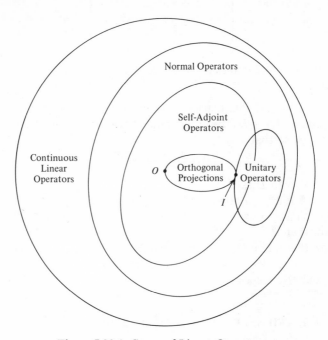

Figure 5.23.1. Space of Linear Operators.

Self-Adjoint Operators

Let us begin with some examples of self-adjoint operators.

EXAMPLE 1. Let $H = C^n$ be the space of ordered n-tuples of complex numbers with the usual inner product. Let T be the linear transformation represented by the matrix $[T]$. T^* is the linear transformation represented by $\overline{T}^T$, that is, the complex conjugate of the transpose. Therefore, T is self-adjoint if and only if $[T] = [\overline{T}]^T$, that is, if and only if $[T]$ is a Hermitian matrix. ∎

EXAMPLE 2. The operator T on H defined by $Tx = \alpha x$, where α is a scalar, is self-adjoint if and only if α is real. ∎

EXAMPLE 3.   Consider the integral operator K given by

$$y(t) = \int_I k(t,s)x(s)\,ds$$

or $L_2(I)$ where $\int_I \int_I |k(t,s)|^2\,ds\,dt < \infty$. It follows from Example 3, Section 22 that K is self-adjoint if and only if $k(t,s) = \overline{k(s,t)}$.   ∎

If A and B are self-adjoint operators on H, then so is $A + B$. Similarly, αA, where α is real, is self-adjoint whenever A is. So the set of all self-adjoint operators on H forms a *real* normed linear space. The norm is, of course, the operator norm. Note that this *real* linear space is not a linear subspace of the *complex* space $Blt[H,H]$, the space of bounded linear operators on H. However, we can say the following.

5.23.4 THEOREM.   *The set of all self-adjoint operators on H is a closed set in $Blt[H,H]$.*

Proof: Let $\{L_n\}$ be a sequence of self-adjoint operators with $\|L_n - L\| \to 0$, where L is a bounded linear operator on H. We want to show that

$$(Lx,y) = (x,Ly)$$

for all x, y in H. Since L_n is self-adjoint we get

$$
\begin{aligned}
|(Lx,y) - (x,Ly)| &= |(Lx,y) - (L_n x,y) + (x,L_n y) - (x,Ly)| \\
&= |((L - L_n)x,y) + (x, (L_n - L)y)| \\
&\le 2\|L - L_n\|\,\|x\| \cdot \|y\| \to 0, \text{ as } n \to \infty.
\end{aligned}
$$

Hence L is self-adjoint.   ∎

It is not true that if A and B are self-adjoint, then AB is self-adjoint. However, one can say the following:

5.23.5 THEOREM.   *If A and B are self-adjoint operators on a Hilbert space H, then AB is self-adjoint if and only if $AB = BA$.*

We leave the proof of this theorem as an exercise.

If T is self-adjoint, then

$$(x,Tx) = (Tx,x) = \overline{(x,Tx)}$$

for all $x \in H$. In other words, (x,Tx) is real-valued on H. Actually this fact is one characterization of self-adjoint operators.

5.23.6 THEOREM.   *Let $T\colon H \to H$ be a bounded linear operator on a Hilbert space H. T is self-adjoint if and only if (x,Tx) is real-valued for all x in H.*

Proof: In light of the comments above we need only show that (x,Tx) real implies that T is self-adjoint. Suppose

$$(x,Tx) = \overline{(x,Tx)}$$

for all $x \in H$. It follows that

$$(x,Tx) = (Tx,x)$$

for all $x \in H$. Since

$$
\begin{aligned}
4(x,Ty) &= (x + y, T(x + y)) - (x - y, T(x - y)) \\
&\quad + i(x + iy, T(x + iy)) - i(x - iy, T(x - iy)) \\
&= (T(x + y), x + y) - (T(x - y), x - y) \\
&\quad + i(T(x + iy), x + iy) - i(T(x - iy), x - iy) \\
&= 4(Tx,y),
\end{aligned}
$$

it follows that $T = T^*$. ∎

EXAMPLE 4. Let $H = L_2(-i\infty,i\infty)$ and consider a linear time-invariant system whose transfer function is $T(i\omega)$. The equation

$$Y(i\omega) = T(i\omega)X(i\omega), \qquad i\omega \in (-i\infty,i\infty),$$

models the system in the frequency domain. The operator is self-adjoint if and only if $T(i\omega)$ is real for (almost) all $i\omega$. ∎

The last theorem leads to a method of ordering self-adjoint operators.

5.23.7 DEFINITION. A bounded linear self-adjoint operator T on a Hilbert space H is said to be *positive* if $(x,Tx) \geq 0$ for all x in H. We denote this by $T \geq 0$ or $0 \leq T$. It is *strictly positive* if $(x,Tx) > 0$ for all $x \neq 0$. We shall denote this by $T > 0$ or $0 < T$.

We shall, then, write $A \leq B$ if $0 \leq B - A$. We leave it to the reader to show that " $\leq$ " does indeed define a partial ordering on the set of all self-adjoint operator on H. Similarly, we write $A < B$ if $0 < B - A$.

Recall that the norm of T is given by any one of the following:

$$\|T\| = \sup\{\|Tx\| : \|x\| = 1\},$$

$$\|T\| = \sup\{\|Tx\| : \|x\| \leq 1\},$$

$$\|T\| = \sup\left\{\frac{\|Tx\|}{\|x\|} : x \neq 0\right\},$$

$$\|T\| = \inf\{B : \|Tx\| \leq B\|x\| \text{ for all } x\}.$$

The next theorem gives two new formulas for computing $\|T\|$ when T is self-adjoint.

5.23.8 THEOREM.   *Let $T: H \to H$ be a bounded linear self-adjoint operator on a Hilbert space H. Then $\|T\|$ is given by*

$$\|T\| = \sup\{|(Tx,x)|: \|x\| = 1\},$$

or

$$\|T\| = \sup\{|(Tx,y)|: \|x\| = \|y\| = 1\}.$$

Proof: We shall prove the first statement and leave the second as an exercise. Let

$$\alpha = \sup\{|(Tx,x)|: \|x\| = 1\}.$$

Since

$$|(Tx,x)| \le \|T\| \|x\|^2,$$

for all $x \in H$, it follows that $\alpha \le \|T\|$.

Let us now show that $\|T\| \le \alpha$. For this we let $\beta > 0$, then using the fact that $(Tx,Tx) = (T^2x,x)$ we get

$$4\|Tx\|^2 = (T(\beta x + \beta^{-1}Tx), \beta x + \beta^{-1}Tx) \\ - (T(\beta x - \beta^{-1}Tx), \beta x - \beta^{-1}Tx).$$

From the definition of α we get

$$4\|Tx\|^2 \le \alpha\|\beta x + \beta^{-1}Tx\|^2 + \alpha\|\beta x - \beta^{-1}Tx\|^2 \qquad (5.23.1)$$
$$\le 2\alpha(\beta^2\|x\|^2 + \beta^{-2}\|Tx\|^2),$$

where the last step is an application of the Parallelogram Law. If $\|Tx\| \ne 0$, we set $\beta^{-2} = \|x\|/\|Tx\|$ and (5.23.1) becomes

$$\|Tx\|^2 \le \alpha\|Tx\| \|x\|,$$

that is,

$$\|Tx\| \le \alpha\|x\|. \qquad (5.23.2)$$

If $\|Tx\| = 0$, then (5.23.2) is obviously true; therefore, $\|T\| \le \alpha$.   ∎

As the next theorem shows, a projection is orthogonal if and only if it is self-adjoint.

5.23.9 THEOREM.   *A continuous projection P on a Hilbert space H is orthogonal if and only if it is self-adjoint.*

Proof: First suppose that P is orthogonal. It follows that if x is any point in H, there exists a unique $r \in \mathscr{R}(P)$ and $n \in \mathscr{N}(P)$ such that $x = r + n$ and $r \perp n$. Then $(x,Px) = (r + n, P(r + n)) = (r + n,r) = (r,r)$ which is real for all x. Hence, from Theorem 5.23.6, P is self-adjoint.

Now suppose that P is a self-adjoint projection on H. Again for any $x \in H$ we can uniquely write $x = r + n$. We want to show that $r \perp n$. But $(r,n) = (Pr,n) = (r,P^*n) = (r,Pn) = (r,0) = 0$. Hence $\mathscr{R}(P) \perp \mathscr{N}(P)$, so P is an orthogonal projection. ∎

EXAMPLE 5. Let M be a closed linear subspace in a Hilbert space H, and let P_1 and P_2 be the orthogonal projection of H onto M and $M^\perp$, respectively. Then

$$T = \lambda_1 P_1 + \lambda_2 P_2 \tag{5.23.3}$$

is self-adjoint if and only if λ_1 and λ_2 are real. This is an especially important example, for it is a major fact of linear analysis that (5.23.3) is the basic model for *all* self-adjoint operators. This will be made amply clear in the next chapter. ∎

EXAMPLE 6. We say that a mapping Y of $L_2(-\infty,\infty)$ into itself is *passive* if

$$\mathrm{Re}\left\{\int_{-\infty}^{t} \bar{x}(\tau)(Yx)(\tau)\, d\tau\right\} \geq 0,$$

for all $x \in L_2(-\infty,\infty)$ and all $t \in R$, see Youla, Castriota and Carlin [1]. The motivation for this definition comes from network theory. Now, we will suppose we have a network as shown in Figure 5.23.2. Here v denotes the voltage across the

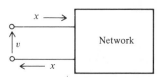

Figure 5.23.2.

terminals, and x denotes the current flowing through them. We assume that the relation between v and x can be modeled by an equation $x = Yv$, where Y is a linear mapping of $L_2(-\infty,\infty)$ into itself. This network is passive if the net energy supplied is positive at all times, that is, if

$$\mathrm{Re}\left\{\int_{-\infty}^{t} \bar{v}(\tau)x(\tau)\, d\tau\right\} = \mathrm{Re}\left\{\int_{-\infty}^{t} \bar{v}(\tau)(Yv)(\tau)\, d\tau\right\} \geq 0,$$

for all applied voltages v, resulting currents x, and times t.

It is an interesting fact that if Y is passive, it is causal. We will demonstrate this fact here. For each time T let X_T denote the linear subspace of $L_2(-\infty,\infty)$ made up of all functions x that vanish to the left of T. Since Y is linear, it follows (Example 4, Section 4.3) that Y is causal if and only if each subspace X_T is invariant under Y.

Let P_T denote the orthogonal projection of $L_2(-\infty,\infty)$ onto $X_T^\perp$. Then we can restate the passivity of Y by

$$\mathrm{Re}(P_T v, Yv) \geq 0,$$

for all $v \in L_2(-\infty,\infty)$, and all $T \in (-\infty,\infty)$.

Let T be fixed, and let v_1 be any point in X_T and v_2 be any point in $L_2(-\infty,\infty)$. If $v = \alpha v_1 + v_2$, where α is a scalar, then $P_T v = P_T v_2$ and $Yv = \alpha Yv_1 + Yv_2$, so

$$\text{Re}(P_T v, Yv) = \text{Re}(P_T v_2, \alpha Yv_1) + \text{Re}(P_T v_2, Yv_2) \geq 0. \qquad (5.23.4)$$

Since α is arbitrary we see that the only way (5.23.4) can hold is if

$$(P_T v_2, Yv_1) = 0.$$

Since P_T is self-adjoint, we have

$$(v_2, P_T Yv_1) = 0.$$

But v_2 is arbitrary, so

$$P_T Yv_1 = 0.$$

In other words, $Yv_1 \in X_T$, or X_T is invariant under Y. Since T was arbitrary, it follows that Y is causal. ∎

Normal Operators

Again, normal operators are those operators that commute with their adjoint, and the self-adjoint operators are a subset of the normal ones. Let us begin by considering an example of a normal operator.

EXAMPLE 7. Let $T = \lambda_1 P_1 + \lambda_2 P_2$, where P_1 and P_2 are the orthogonal projections onto M and $M^\perp$, as in Example 5.

Since $T^* = \bar{\lambda}_1 P^* + \bar{\lambda}_2 P^* = \bar{\lambda}_1 P_1 + \bar{\lambda}_2 P_2$, we have

$$T^*T = (\bar{\lambda}_1 P_1 + \bar{\lambda}_2 P_2)(\lambda_1 P_1 + \lambda_2 P_2) = |\lambda_1|^2 P_1 + |\lambda_2|^2 P_2$$

and

$$TT^* = (\lambda_1 P_1 + \lambda_2 P_2)(\bar{\lambda}_1 P_1 + \bar{\lambda}_2 P_2) = |\lambda_1|^2 P_1 + |\lambda_2|^2 P_2,$$

so T is normal for any complex numbers λ_1 and λ_2. This is an especially important example, for it is another major fact of linear analysis that T represents the basic model for all normal operators. This also will be made amply clear in the next chapter. ∎

EXAMPLE 8. In this example we show that continuous time-invariant linear operators are normal. Let T be a continuous time-invariant linear mapping of $H = L_2(-\infty,\infty)$ into itself. We know (Example 10, Section 5.22) that T^* is time-invariant. But we recall (Example 5, Section 4.10) that continuous time-invariant linear operators on $L_2(-\infty,\infty)$ commute; therefore, $TT^* = T^*T$ and T is normal. ∎

The following theorem is a convenient characterization of normal operators.

5.23.10 THEOREM. *A bounded linear operator L on a Hilbert space H is normal if and only if $\|L^*x\| = \|Lx\|$ for every $x \in H$.*

Proof: First assume that L is normal. Then one has $(LL^*x,x) = (L^*Lx,x)$, for all x in H, which implies that $(L^*x,L^*x) = (Lx,Lx)$ for all x in H. In other words, $\|L^*x\| = \|Lx\|$.

Now assume that $\|L^*x\| = \|Lx\|$, for all x in H. By reversing the reasoning above, one gets

$$((LL^* - L^*L)x,x) = 0$$

for all x in H. We need the next lemma to conclude that $LL^* - L^*L = 0$.

5.23.11 LEMMA. *Let M be a linear operator on a complex inner product space X. If $(Mx,x) = 0$, for all x in X, then $M = 0$.*

Proof: Since $(Mx, x) = 0$ for all x in X, we note that for every x and y in X and any scalars α and β one has

$$0 = (M(\alpha x + \beta y), (\alpha x + \beta y)) - |\alpha|^2(Mx,x) - |\beta|^2(My,y)$$
$$= \alpha\bar{\beta}(Mx,y) + \bar{\alpha}\beta(My,x).$$

If we choose $\alpha = \beta = 1$, this fact becomes

$$(Mx,y) + (My,x) = 0,$$

and for $\alpha = i$, $\beta = 1$ one gets

$$i(Mx,y) - i(My,x) = 0.$$

It follows that $(Mx,y) = 0$, for all x and y, which implies that $M = 0$. This completes the proof of the lemma and Theorem 5.23.10. ∎

In Theorem 5.23.4 we indicated how the class of self-adjoint operators fits into the linear space of bounded linear operators $Blt[H,H]$. We can say the following in the case of the class of normal operators.

5.23.12 THEOREM. *The class of all normal operators on a Hilbert space H is a closed subset of $Blt[H,H]$; moreover, it is closed under scalar multiplication.*

Proof: Let $\{L_n\}$ be a sequence of normal operators with $\|L_n - L\| \to 0$, where L is a bounded linear operator on H. We want to show that L is normal. But

$$\|LL^* - L^*L\| \leq \|LL^* - L_nL_n^*\| + \|L_nL_n^* - L_n^*L_n\| + \|L_n^*L_n - L^*L\|$$
$$\leq \|LL^* - L_nL_n^*\| + \|L_n^*L_n - L^*L\|$$
$$\leq \|(L - L_n)(L^* - L_n^*) + (L - L_n)L_n^* + L_n(L^* - L_n^*)\|$$
$$\quad + \|(L^* - L_n^*)(L - L_n) + L_n^*(L - L_n) + (L^* - L_n^*)L_n\|$$
$$\leq \|L - L_n\| \|L^* - L_n^*\| + \|L - L_n\| \|L_n^*\| + \|L_n\| \|L^* - L_n^*\|$$
$$\quad + \|L^* - L_n^*\| \|L - L_n\| + \|L_n^*\| \|L - L_n\| + \|L^* - L_n^*\| \|L_n\|.$$

It follows from Exercise 12, Section 5.22 that the right-hand side of the inequality converges to zero as $n \to \infty$, so $\|LL^* - L^*L\| = 0$ showing that L is normal. The fact that L is normal implies that αL is normal for all scalars is trivial. ∎

If A and B are normal, contrary to the self-adjoint case, it does not follow that $A + B$ is normal, nor, of course, does it follow that AB is normal. Not too surprisingly, commutativity is important.

5.23.13 THEOREM. *If A and B are normal operators on a Hilbert space H such that one commutes with the adjoint of the other, then $A + B$, AB, and BA are normal.*

We see that the proof of this theorem is straightforward as soon as one notes that $AB^* = B^*A \Leftrightarrow BA^* = A^*B$.

The next theorem is a sometimes more useful one whose proof is not trivial.

5.23.14 THEOREM. *If A and B are normal operators on a Hilbert space H such that $AB = BA$, then $A + B$, AB, and BA are normal.*

The proof of this theorem is available in Fuglede [1].

In general, if T is a bounded linear transformation, we have $\|T^2\| \le \|T\|^2$. In the case of normal operators, we always get the equality.

5.23.15 THEOREM. *If L is a normal operator on a Hilbert space H, then*

$$\|L^2\| = \|L\|^2.$$

Proof: It follows from Theorem 5.23.10 that $\|L^2x\| = \|L^*Lx\|$, for all $x \in H$. So $\|L^2\| = \|L^*L\|$. But from Corollary 5.22.3 we know that $\|L^*L\| = \|L\|^2$. ∎

EXERCISES

1. By referring to the examples in Section 22, derive some examples of
   (a) self-adjoint operators,
   (b) normal operators.

2. In the notation of Exercise 8, Section 12, show that $\|q\| = \|Q\|$, when q is a bounded, symmetric sesquilinear functional on a complex inner product space X. [*Hint:* Study the proof of Theorem 5.23.8.]

3. (Cayley Transform.) Let A be a self-adjoint operator on a Hilbert space H and define U by

$$U = \frac{A - iI}{A + iI}$$

under the assumption that $A + iI$ is invertible.
   (a) Show that U is a unitary operator.
   (b) Show that

$$A = i\frac{I + U}{I - U}.$$

4. Let $H = L_2[0,1]$. Define $K: H \to H$ as follows: For each x in $L_2[0,1]$ let $y = Kx$ be the solution $y(t)$ of

$$y'' + a_1 y' + a_2 y = x$$

that satisfies $y(0) = 0$, $y'(0) = 0$. Show that K can be represented by an integral operator. Find K^*. Find conditions on the coefficients a_1 and a_2 in order that K be self-adjoint.

5. Let I be any interval and let f be a complex-valued bounded function on I. Let H be the complex space $L_2(I)$ with the usual inner product. Define $F: H \to H$ by

$$Fx(t) = f(t)x(t).$$

Show that F is a bounded linear normal operator on H.

6. Let L be a bounded linear operator that is unitarily equivalent to multiplication by a bounded transfer function. Show that L is normal.

7. Define $A: C^2 \to C^2$ by $y = Ax$ or

$$\begin{pmatrix} y_1 \\ y_2 \end{pmatrix} = \begin{pmatrix} a & b \\ c & d \end{pmatrix} \begin{pmatrix} x_1 \\ x_2 \end{pmatrix},$$

where a, b, c, d are complex coefficients.
   (a) Find necessary and sufficient conditions on the coefficients in order that A be normal, self-adjoint, or unitary.
   (b) Assume that A is normal. Find a polar decomposition of A. (See Exercise 16.)

8. Let $q(x,y)$ be a bounded sesquilinear functional on a Hilbert space H.
   (a) Show that $q(x,y) = (Lx,y)$ for some bounded linear operator L. [Hint: Use the Riesz Representation Theorem to get $q(x,y) = (x,y^*) = (x,L^*y)$ and show that L^* is a bounded linear operator.]
   (b) Show that q is symmetric if and only if L is self-adjoint.

9. Where are the invertible operators located in Figure 5.23.1? The operators with bounded inverse? Where are the positive and strictly positive operators located in Figure 5.23.1?

10. Prove Theorem 5.23.5.

11. Let $A = (a_{ij})$ be a 2×2 complex matrix operator. Give conditions on the entries a_{ij} in order that A be normal.

12. Show that Lemma 5.23.11 fails in a real inner product space. [Hint: Consider an antisymmetric matrix on R^n.]

13. If $T = A + iB$, where A and B are self-adjoint operators on a Hilbert space H, then this is said to be the *Cartesian decomposition* of T.
   (a) Show that every bounded linear operator on H has a Cartesian decomposition. [Hint: $2A = T + T^*$, $2iB = T - T^*$.]
   (b) Show that the Cartesian decomposition is unique.
   (c) Compute T^* in terms of A and B.
   (d) Show that T is normal if and only if A and B commute.

14. Let L be a normal operator on a Hilbert space H and let $L = A + iB$ be the Cartesian decomposition of L. Show that

$$\max\{\|A\|^2, \|B\|^2\} \le \|L\|^2 \le \|A\|^2 + \|B\|^2.$$

15. Let L be a positive self-adjoint operator on a Hilbert space H. This exercise will lead to a proof of the fact that L has a *positive square root*; that is, there is a positive self-adjoint operator R that satisfies $R^2 = L$. (Compare this with Exercise 7, Section 3.6)

   (a) First assume that $L \le I$ and set $M = I - L$, and $R = I - S$. Then $R^2 = L$ becomes $S = \frac{1}{2}(M + S^2)$. Let $S_0 = 0$, $S_1 = 2^{-1}M$,

   $$S_{n+1} = \tfrac{1}{2}(M + S_n^2), \qquad n = 1, 2, \dots.$$

   Show that S_n and $S_n - S_{n-1}$ are polynomials in M with nonnegative, real coefficients.

   (b) Show that $S_n \ge 0$ and $S_n - S_{n-1} \ge 0$ for all n.
   (c) Show that $\|S_n\| \le 1$ for all n.
   (d) Show that for each $x \in H$ the sequence $\{S_n x\}$ converges. Let $Sx = \lim S_n x$.
   (e) Show that this operator S satisfies $S = \frac{1}{2}(M + S^2)$.
   (f) Drop the restriction that $L \le I$.

16. Let T be a normal operator and assume that $T = RU = UR$, where R is a positive self-adjoint operator and U is a unitary operator. Then RU, or UR, is said to be the *polar decomposition* of T. This exercise will lead to a proof that every normal operator T has a polar decomposition.

   (a) Let R be the positive square root of $TT^* = T^*T$.
   (b) If $y = Rx$, let $Uy = Tx$. This defines U on the range of R, say $U: \mathscr{R}(R) \to H$. Show that $\|Uy\| = \|y\|$.
   (c) Show that U can be extended to all of H so that the extension is a unitary mapping.
   (d) Show that $T = RU$.
   (e) Show that R and U commute.
   (f) Show that the polar decomposition is unique.

17. (Continuation of Exercise 16.) Let T be any bounded linear operator on a Hilbert space H.

   (a) Show that there are positive self-adjoint operators R_1 and R_2 and unitary operators U_1 and U_2 so that $T = R_1 U_1 = U_2 R_2$.
   (b) Show that R_1, U_1 and R_2, U_2 are unique. [*Note:* There is an analog between operators on a Hilbert space H and complex numbers that is often useful. The basic idea is to view the operation of taking the adjoint as analogous to that of taking the complex conjugate. Then the self-adjoint operators are analogous to the real numbers, for $A = A^*$. Positive operators are analogous to nonnegative real numbers and unitary transformations are analogous to numbers of unit magnitude, for $U^*U = UU^* = I$. Just as any complex number, any operator T has a Cartesian and polar decomposition.]

18. Show that if a unitary operator $U: H \to H$ is positive, then $U = I$, the identity transformation.

19. If $T: H \to H$ is any bounded linear operator on the Hilbert space H, show that T^*T and TT^* are self-adjoint.

20. Show that if T is normal, then $\lambda I - T$ is normal for all complex numbers λ.

21. Let $L = {}_s \lim L_n$ and L_n be bounded linear operators on a Hilbert space H.
    (a) Show that if L_n is self-adjoint for all n, then L is self-adjoint. [*Hint*: Study Theorem 5.23.4.]
    (b) Show that if L_n is normal for all n, then L is normal. [*Hint*: Study Theorem 5.23.12.]

22. Do the relationships $A \leq B$ or $A < B$ define a partial ordering (refer to Appendix C) on the collection of all self-adjointed operators on a Hilbert space H? Is it ever a total ordering?

23. Let ξ be a random variable with range in a Hilbert space H and such that $E(\|\xi\|^2) < \infty$. Define the *covariance operator* A by

$$E((x,\xi)\overline{(y,\xi)}) = (Ax,y),$$

where $x, y \in H$. Show that A is a bounded, positive, self-adjoint linear operator.

24. Calculate $\|A\|$, where

    (a) $A = \begin{bmatrix} a & c \\ c & b \end{bmatrix}$,

    (b) $A = \begin{bmatrix} a & d & f \\ d & b & e \\ f & e & c \end{bmatrix}$,

    when the entries are all real.

25. Let A be a self-adjoint operator on a Hilbert space H, and let

$$U = e^{iA} = \sum_{n=0}^{\infty} \frac{(iA)^n}{n!}.$$

    (a) Show that U is a unitary operator.
    (b) Show that $U^n = e^{inA}$ for every integer n.

26.[19] Let A be a bounded self-adjoint operator on a Hilbert space H and let

$$U_t = e^{itA} = \sum_{n=0}^{\infty} \frac{(itA)^n}{n!}.$$

    (a) Show that for each real number t, U_t is a unitary operator.
    (b) Show that $U_t U_s = U_{t+s}$.
    (c) Show that the mapping $t \to U_t$ is continuous, where the space of operators has the usual operator norm.
    (d) Discuss the meaning of the equality

$$\left.\frac{dU_t}{dt}\right|_{t=0} = \lim_{t \to 0} \frac{U_t - I}{t} = iA$$

    in terms of topologies on the space of operators. (See Section 8.)

---

[19] In this exercise, one constructs a continuous group U_t of unitary operators in terms of a given self-adjoint operator A. It is possible to turn this around, that is, given the continuous group U_t of unitary operators one can construct the "infinitesimal generator" A by means of $dU_t/dt = iA$, and show that $U_t = e^{itA}$, see Dunford and Schwartz [1].

27. (Scattering operators.) Let A and B be two bounded linear self-adjoint operators on a separable Hilbert space H. Define U_t and V_t by $U_t = e^{-itA}$ and
$V_t = e^{-itB}$. Assume that the limits

$$\lim_{t \to +\infty} V_t^* U_t x = x_- = \Omega_- x$$

and

$$\lim_{t \to -\infty} V_t^* U_t x = x_+ = \Omega_+ x$$

exist for each $x \in H$. Let $R_\pm$ denote the range of $\Omega_\pm$ and assume that $R_+ = R_-$.
The *scattering operators* are defined by

$$S = \Omega_-^* \Omega_+$$

and

$$T = \Omega_+ \Omega_-^*.$$

(Compare with Jauch [1].) The object here is to show that S is a unitary
operator.
(a) Show that $\|\Omega_\pm x\| = \|x\|$ and that $\|\Omega_\pm^* x\| = \|x\|$ for all $x \in H$. (Explain
why this fact alone does *not* show that S and T are unitary.)
(b) Show that $\Omega_+^* \Omega_+ = I$ and $\Omega_-^* \Omega_- = I$.
(c) Show that $\Omega_+ \Omega_+^* = \Omega_- \Omega_-^* = P$, where P is the orthogonal projection
onto $R_+ = R_-$.
(d) Show that $SS^* = S^*S = I$.
(e) Show that $TT^* = T^*T = P$.

28. Let $y = Kx$ be a positive self-adjoint operator on $L_2[a,b]$ that is given by
$y(t) = \int_a^b k(t,\tau)x(\tau)\,d\tau$, where $k(t,\tau)$ is real-valued and continuous.
(a) Show that $k(t,t) \geq 0$ for $a \leq t \leq b$.
(b) Show that the converse need not be true. That is, construct a kernel
$k(t,\tau)$ that satisfies $k(t,t) \geq 0$ for $a \leq t \leq b$ such that the corresponding
operator K is self-adjoint but not positive.

24.   COMPACT OPERATORS

The compact operators form another important class of linear operators. As
we shall see below they are operators with finite- or, in a meaningful sense, almost
finite-dimensional ranges. They are neither included in nor include the class of
normal operators or, for that matter, the class of self-adjoint operators. The situation (for infinite-dimensional spaces) is illustrated in Figure 5.24.1. As we shall see
in the next chapter, operators that are both normal and compact yield about the
closest thing to a finite-dimensional structure that one can have on an infinite-
dimensional space.

Since the elementary properties of compact operators are not dependent on the
presence of an inner product, we shall abandon Hilbert space structure for this
section and return to Banach spaces.[20]

[20] Compact operators can be defined on normed linear spaces, but many results require completeness, so we just assume it at the outset.

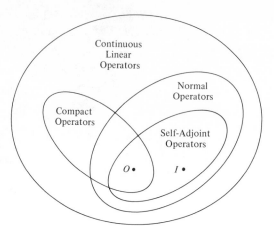

Figure 5.24.1.

5.24.1 DEFINITION.   Let X and Y be two Banach spaces and let $L: X \to Y$ be a linear transformation. L is said to be *compact*[21] if $L(D)$ lies in a compact subset of Y, where $D = \{x \in X: \|x\| \leq 1\}$.

The following theorem states that a compact operator is continuous; however, there are plenty of continuous operators that are not compact. For example, consider the identity mapping I on any infinite-dimensional space.

5.24.2 THEOREM.   *Let* $L: X \to Y$ *be a compact linear transformation of a Banach space* X *into a Banach space* Y. *Then* L *is continuous.*

Proof: The validity of this theorem follows almost immediately from Definition 5.24.1. Since $L(D)$ lies in a compact set, it is totally bounded, hence bounded (Lemma 3.16.3). That is, there exists an $M < \infty$ such that $\sup\{\|x\|: x \in L(D)\} \leq M$. It follows that $\|Lx\| \leq M\|x\|$ for all $x \in X$; therefore, L is bounded (that is, continuous).   ∎

An important example of a compact operator is described in the next theorem.

5.24.3 THEOREM.   *Let* $L: X \to Y$ *be a linear operator where the range* $\mathcal{R}(L)$ *is finite-dimensional. Then* L *is compact.*

Proof: By Theorem 5.10.7 we see that the unit ball in $\mathcal{R}(L)$ is compact. Therefore a ball of any radius is compact, from which it is easily seen that L is compact (Compare with Example 3, Section 10.)   ∎

As we said, any compact operator comes close to having a finite-dimensional range.

---

[21] Some authors use the phrase "completely continuous" instead.

5.24.4 THEOREM. *Let* $L: X \to Y$ *be a compact linear transformation, where* X *and* Y *are Banach spaces. Then given any* $\varepsilon > 0$, *there exists a finite-dimensional subspace* M *of* $\mathcal{R}(L)$ *such that*

$$\inf\{\|Lx - m\| : m \in M\} \leq \varepsilon \|x\|.$$

In other words, the finite-dimensional subspace M comes within ε (in the above sense) of being the range of L. Presumably, the smaller ε is, the larger the dimension of M must be.

Proof: Let $\varepsilon > 0$ be given. Since $L(D)$ is contained in a compact set, where D is the closed unit ball in X, there is an ε-net in $\mathcal{R}(L) \cap L(D)$. Let M be the linear subspace of Y generated by this ε-net. It follows that M is finite dimensional. Moreover, $\text{dist}(Lz, M) \leq \varepsilon$ for all $z \in D$. Then if x is any point in X it follows that

$$\inf \left\{ \left\| L \frac{x}{\|x\|} - m \right\| : m \in M \right\} \leq \varepsilon$$

so

$$\inf\{\|Lx - m'\| : m' \in M\} \leq \varepsilon \|x\|,$$

where $m' = \|x\| m$. ∎

The following theorem presents a number of equivalent formulations for compactness of an operator.

5.24.5 THEOREM. *Let* $L: X \to Y$ *be a linear operator, where* X *and* Y *are Banach spaces. Then the following statements are equivalent:*
  (a) *L is compact.*
  (b) *If B is any bounded set in X, then $L(B)$ lies in a compact subset of Y.*
  (c) *If B is any bounded set in X, then $L(B)$ lies in a sequentially compact subset of Y.*
  (d) *If $\{x_n\}$ is any bounded sequence in X, then $\{Lx_n\}$ contains a convergent subsequence in Y.*
  (e) *If B is any bounded set in X, then $L(B)$ is a totally bounded set in Y.*

Proof: Since the equivalence of (b), (c), (d), and (e) follows from the characterization of compactness in Section 3.17, we shall prove only that (a) $\Leftrightarrow$ (b).
  It is obvious that (b) $\Rightarrow$ (a). Let us show that (a) $\Rightarrow$ (b). Let B be any bounded set in X. Then there is a real number $k > 0$ such that

$$\|x - 0\| = \|x\| \leq k, \qquad \text{for all } x \in B.$$

Let $D = \{x \in X : \|x\| \leq 1\}$. Then $B \subset kD$, where

$$kD = \{kx \in X : \|x\| \leq 1\} = \{x \in X : \|x\| \leq k\}.$$

Since $L(B) \subset L(kD) = kL(D)$ and since $\overline{L(kD)} = k\overline{L(D)}$, it follows that $L(kD)$ lies in a compact set in Y. Hence, $L(B)$ lies in a compact set in Y. ∎

Let us now consider some examples of compact and noncompact operators.

EXAMPLE 1.   Let $\phi_1, \ldots, \phi_n, \psi_1, \ldots, \psi_n$ be elements of $L_2(I)$ and let

$$k(t,s) = \sum_{i=1}^{n} \alpha_i \phi_i(t)\psi_i(s),$$

where the α_i's are scalars. Define $y = Kx$ by

$$y(t) = \int_I k(t,s)x(s)\, ds.$$

Since every point y in $\mathcal{R}(K)$ is given by

$$y(t) = \sum_{i=1}^{n} \beta_i \phi_i(t),$$

where $\beta_i = \alpha_i \int_I \psi_i(s)x(s)\, ds$, we see that $\mathcal{R}(K)$ has dimension less than or equal to n. Hence K is compact.   ∎

EXAMPLE 2.   Every linear operator defined on a finite-dimensional normed linear space is compact.   ∎

EXAMPLE 3.   Consider the multiplication operator

$$F: x(t) \rightarrow f(t)x(t)$$

on $L_2(I)$, where f is a bounded measurable function. We have seen elsewhere (Example 2, Section 7) that F is a bounded linear operator and $\|F\| \leq \|f\|_\infty$. We will now show that F is compact if and only if $f(t) = 0$ almost everywhere, that is, $\|f\|_\infty = \|F\| = 0$.

It is clear that $\|f\|_\infty = 0$ implies that F is the zero operator and, therefore, compact. Going the other way now, assume on the contrary that $\|f\|_\infty \neq 0$.

If f is continuous, then there are positive numbers α, β such that

$$|f(t)| \geq \alpha, \qquad t \in J, \tag{5.24.1}$$

where J is an interval of length β.

If $x \in L_2(I)$ and $x(t) = 0$ for $t \notin J$ one has

$$\|Fx\|_2^2 = \int_I |f(t)|^2 |x(t)|^2\, dt \geq \int_J |f(t)|^2 |x(t)|^2\, dt$$

$$\geq \alpha^2 \|x\|_2^2.$$

Now choose an orthonormal sequence $\{x_n\}$ in $L_2(I)$ such that $x_n(t) = 0$ for $t \notin J$. (Why can we do this?) One then has $\|x_n - x_m\| = \sqrt{2}$ for $n \neq m$ and

$$\|Fx_n - Fx_m\| \geq \sqrt{2}\,\alpha, \qquad n \neq m$$

by the above. Hence $\{Fx_n\}$ cannot contain a convergent subsequence. Therefore, F is not compact.

If f is not continuous, then for some integer n the set

$$A_n = \left\{ t: |f(t)| \geq \frac{1}{n} \right\}$$

has positive measure. For this n, let $\alpha = 1/n$ and let J be a subset of A_n with measure $\beta > 0$. One can then repeat the above argument and show that F is not compact. ∎

EXAMPLE 4.   Let $H = l_2$, and let K denote the linear transformation of H into itself defined by

$$y_n = \alpha_n x_n, \qquad n = 1, 2, 3, \ldots,$$

where $y = Kx$, $x = \{x_1, x_2, x_3, \ldots\}$, $y = \{y_1, y_2, y_3, \ldots\}$, and the α_n's are scalars.
We claim that K is compact if and only if the α_n's satisfy the condition

$$\lim_{n \to \infty} |\alpha_n| = 0. \tag{5.24.2}$$

First assume that K is compact and that $|\alpha_n| \geq \varepsilon > 0$ for all n. Then, let

$$e_m = \{\delta_{1m}, \delta_{2m}, \delta_{3m}, \ldots\},$$

where δ_{ij} is the Kronecker function. Then

$$Ke_m = (\alpha_1 \delta_{1m}, \alpha_2 \delta_{2m}, \ldots) = (0, 0, \ldots, \alpha_m, 0, \ldots),$$

and for $m \neq n$

$$\|Ke_m - Ke_n\|^2 = |\alpha_m|^2 + |\alpha_n|^2 \geq 2\varepsilon^2.$$

Hence, $\{Ke_m\}$ does not contain any subsequence that is convergent, and we contradict the fact that K is compact.

If K is compact and (5.24.2) fails, then there is an $\varepsilon > 0$ and a subsequence $\{\alpha_{n_k}\}$ with $|\alpha_{n_k}| \geq \varepsilon$. By using the sequence $\{e_{n_k}\}$ and the above argument we arrive at a similar contradiction. Hence, K compact implies that (5.24.2) holds.

On the other hand, assume that (5.24.2) holds, and then let $A = K(D)$, where $D = \{x: \|x\| \leq 1\}$. We shall now show that A has compact closure by applying Exercise 1, Section 3.17. Since $\|Kx\| \leq \{\max |\alpha_n|\} \|x\|$, we see that A is bounded. If $y \in A$, then

$$\sum_{n=N}^{\infty} \|y_n\|^2 = \sum_{n=N}^{\infty} \|\alpha_n x_n\|^2$$

$$\leq \max_{N \leq n} |\alpha_n|^2 \left\{ \sum_{n=1}^{\infty} \|x_n\|^2 \right\}$$

$$\leq \max_{N \leq n} |\alpha_n|^2 \to 0, \text{ as } N \to \infty.$$

Hence, $\sum_{n=N}^{\infty} \|y_n\|^2 \to 0$ uniformly as $N \to \infty$, so K is compact. ∎

EXAMPLE 5.   Let $\{\phi_n\}$ and $\{\psi_n\}$ be orthonormal sets in the Hilbert space $H = L_2(I)$, and consider the linear mapping K of H into itself defined by

$$Kx = \sum_{n=1}^{\infty} \left\{ \alpha_n \phi_n(t) \int_a^b \overline{\psi}_n(\tau) x(\tau)\, d\tau \right\} = \int_a^b \left(\sum_{n=1}^{\infty} \alpha_n \phi_n(t) \overline{\psi_n(t)} \right) x(\tau)\, d\tau$$

$$= \int_a^b k(t,\tau) x(\tau)\, d\tau,$$

where $\overline{\psi}_n$ is the complex conjugate of ψ_n and the α_n's are scalars.

We claim that K is compact if and only if

$$\lim_{n \to \infty} |\alpha_n| = 0. \tag{5.24.3}$$

However, the proof of this assertion is a simple variation of the argument of the preceding example. ∎

The proofs of the next two results are elementary and are outlined in the exercises.

5.24.6 THEOREM.   *Let $A: X \to Y$ and $B: X \to Y$ be compact linear operators, where X and Y are Banach spaces. Then $A + B$ is compact.*

5.24.7 THEOREM.   *Let $A: X \to X$ be a compact linear operator, where X is a Banach space. Let $B: X \to X$ be a bounded linear operator. Then AB and BA are compact.*

It is obvious that for every scalar α, the operator αA is compact whenever A is compact.

The next theorem shows that the class of all compact operators is a closed subset of $Blt[X,Y]$.

5.24.8 THEOREM.   *Let X and Y be Banach spaces and also let $L_n: X \to Y$, $n = 1, 2, \ldots$, be a sequence of compact linear operators converging to a bounded linear operator $L: X \to Y$, that is, $\|L_n - L\| \to 0$ as $n \to \infty$. Then L is a compact linear operator. Hence the space of compact operators forms a closed linear subspace of $Blt[X,Y]$.*

Proof: Let $\{L_n\}$ be a sequence of compact linear operators such that $\|L_n - L\| \to 0$ as $n \to \infty$. We know that L is a bounded linear transformation. The issue is to show that it is compact. We shall do this by showing that $L(D)$ is totally bounded in Y [see Theorem 5.24.5(e)]. Let $\varepsilon > 0$ be given and choose N so that $\|L - L_N\| < \varepsilon$. This means that

$$\|Lx - L_N x\| \le \|L - L_N\|\, \|x\| \le \varepsilon \|x\|$$

for all $x \in X$. Since L_N is compact, the set $L_N(D)$ is totally bounded, so it contains an ε-net $\{y_1, y_2, \ldots, y_n\}$.

It is easy to check that the set $\{y_1, y_2, \ldots, y_n\}$ forms a 2ε-net for $L(D)$. ∎

The next example is very important and should be studied carefully.

EXAMPLE 6. Consider the space $L_2(I)$ where I is the finite interval $[a,b]$. Let $y = Kx$ be the integral operator given by

$$y(t) = \int_I k(t,s)x(s) \, ds,$$

where the kernel $k(t,s)$ satisfies

$$\int_I \int_I |k(t,s)|^2 \, dt \, ds < \infty. \qquad (5.24.4)$$

In other words, k is an element of $L_2(I \times I)$. [Recall that (5.24.4) is satisfied if $k(t,s)$ is continuous and I is compact.] It was shown in Examples 2 and 4 of Section 18 that

$$\phi_n(t) = (b - a)^{-1/2} \exp\left(2\pi i n \frac{t - a}{n - a}\right), \qquad n = 0, \pm 1, \pm 2, \ldots$$

forms an orthonormal basis for $L_2(I)$ and that

$$\psi_n(t,s) = \phi_n(t)\overline{\phi_n(s)}$$

forms an orthonormal basis for $L_2(I \times I)$. The Fourier Series Theorem for $L_2(I \times I)$ then tells us that

$$\|k_N - k\| \to 0, \qquad \text{as } N \to \infty,$$

where

$$k_N = \sum_{n=-N}^{N} (k,\psi_n)\psi_n.$$

If we let K_N denote the integral operator with kernel k_N, then we have

$$\|K - K_N\| \le \|k - k_N\| \to 0, \qquad \text{as } N \to \infty.$$

Since the dimension of $\mathcal{R}(K_N)$ is $2N + 1$, it follows that K_N is compact and, as a consequence of Theorem 5.24.8, we see that K is compact.

Integral operators of this form arise naturally in many applications. ∎

EXAMPLE 7. The conclusion of the last example is valid even when the interval I is nonfinite. The argument is exactly the same. The only difference is in terms of the representation for the orthonormal basis for $L_2(I \times I)$. Thus, if $I = [0,\infty)$ we could take

$$\phi_n(t) = \frac{1}{n!} e^{-t/2} L_n(t),$$

where $L_n(t)$ is the Laguerre polynomial (see Exercise 10, Section 18) and

$$\psi_n(t,s) = \phi_n(t)\overline{\phi_n(s)}.$$

This is discussed further in the exercises. ∎

EXERCISES

1. In Theorem 5.24.5, we showed that (e) ⇒ (a). Show that this implication may fail if Y is not complete.

2. In Example 1 it was shown that dim $\mathscr{R}(K) \leq n$. Is it possible to have dim $\mathscr{R}(K) < n$? In Example 6 we state that dim $\mathscr{R}(K_N) = 2N + 1$. How do these examples differ?

3. Let $k(s,t)$ be a measurable function on $I \times I$ such that

$$\int_I |k(s,t)| \, dt \leq M,$$

for all s in I. Define the integral transformation K by $y = Kx$, where

$$y(s) = \int_I k(s,t)x(t) \, dt.$$

Show that K is a bounded linear operator on $L_\infty(I)$. Now assume that I is a compact set and $\int_I 1 \, dt < \infty$, and that the mapping $s \to k(s, \cdot)$ is a continuous mapping of I into $L_1(I)$. Then show that K is compact. [*Hint*: For the last part, show that $\mathscr{R}(K) \subset C$, the space of continuous functions, and then apply the Arzela-Ascoli Theorem.]

4. Let H be a Hilbert space with a countable orthonormal basis $\{e_1, e_2, \ldots\}$, that is, $(e_n, e_m) = \delta_{nm}$. Let $L: X \to X$ be a compact linear operator and define $L_n: X \to X$ by

$$L_n x = \sum_{i,j=1}^{n} (x, e_i)(Le_i, e_j)e_j.$$

Show that L_n is compact and that $\|L_n - L\| \to 0$ as $n \to \infty$. (This shows that every compact linear operator on a separable Hilbert space is the limit of operators with finite-dimensional range. This is an unresolved question for compact operators on a Banach space.) How does this result compare with Theorem 5.24.4?

5. Prove Theorem 5.24.6.

6. Prove Theorem 5.24.7. [*Hint*: Recall that a continuous function preserves convergent sequences and that a bounded function preserves bounded sequences.]

7. Consider the Hilbert space AP defined in Exercise 10, Section 17, with inner product

$$(x,y) = \lim_{T \to \infty} \frac{1}{2T} \int_{-T}^{T} x(t)\overline{y(t)} \, dt.$$

Let $x_0 \in AP$ and define $y = Lx$ by

$$y(t) = \lim_{T \to \infty} \frac{1}{2T} \int_{-T}^{T} x_0(t - \tau)x(\tau) \, d\tau.$$

(a) Show that L is a bounded linear mapping of AP into AP.

(b) Show that if $x_0(t) = e^{iat}$ for some real number a, then L is self-adjoint and compact. (Compactness is not easily verified.)

8. Let $y = Ax$ be an operator on $l_2[0,\infty)$ defined by

$$y_n = \sum_{m=0}^{\infty} a_{nm} x_m,$$

where $\sum_{m,n=0}^{\infty} |a_{mn}|^2 < \infty$. Show that $A: l_2 \to l_2$ and that A is compact.

9. Let $y = Ax$ be an operator on $l_2(-\infty,\infty)$ defined by

$$y_n = \sum_{m=-\infty}^{\infty} \alpha_m x_{n-m}.$$

Assume that the sequence $\{\alpha_m\}$ is chosen so that A is a bounded linear operator with range in $l_2(-\infty,\infty)$. Assume also that α_m is real and nonnegative for all m.

(a) When is the range of A finite dimensional?

(b) When is the operator A compact?

10. Let $L: X \to X$ be an isometric isomorphism, where X is a normed linear space.

(a) Show that L is compact if and only if X is finite dimensional.

(b) Is the same conclusion valid if L is only a topological isomorphism?

11. Let $L: X \to X$ be a bounded linear operator that satisfies $\|Lx\| \geq \alpha \|x\|$ for all x, where $\alpha > 0$. Show that L is compact if and only if X is finite dimensional.

12. Let $L: X \to Y$ be a compact operator and let M be a linear subspace of X. Show that the restrictions of L to M, that is, $L: M \to Y$, is compact.

13. (Hilbert-Schmidt operators.) Let $\{x_n\}$ be an orthonormal basis for a Hilbert space H. A bounded linear operator $T: H \to H$ is said to be a *Hilbert-Schmidt operator* if $\sum_n \|Tx_n\|^2 < \infty$. The number

$$\||T\|| = \left(\sum_n \|Tx_n\|^2 \right)^{1/2}$$

is called the *Hilbert-Schmidt norm* of T.

(a) Show that $\||T\||$ does not depend on the choice of basis.

(b) Show that $\|T\| \leq \||T\||$, where $\|T\|$ denotes the usual norm of T.

(c) Show that the integral operators in Example 4 are Hilbert-Schmidt operators. (For a converse statement, see Dunford and Schwartz [1, Part 2 p. 1083].)

(d) Show that every Hilbert-Schmidt operator is compact and is the limit (in the Hilbert-Schmidt norm) of a sequence of operators with finite-dimensional range.

(e) See Dunford and Schwartz [1, Part 2, p. 1020] for a representation of the Hilbert-Schmidt norm in the case of a one-to-one matrix operator.

14. Where would the unitary operators fit in Figure 5.24.1? (Distinguish between the finite-dimensional and the infinite-dimensional cases.)

15. Under what conditions will the operators K discussed in Examples 4 and 5 be Hilbert-Schmidt operators?

16. Show that an orthogonal projection P on a Hilbert space H is compact if and only if the range of P is finite dimensional.

17. Show that if T is a continuous time-invariant mapping of $L_2(-\infty,\infty)$ into itself, then T is compact if and only if $T = 0$. (Assume here that $T = \mathscr{F}^{-1}T(i\omega)\mathscr{F}$, where $\mathscr{F}$ is the Fourier transform.)

18. Verify the assertion that the operator K in Example 5 is compact if and only if (5.24.3) holds.

25.   FOUNDATIONS OF QUANTUM MECHANICS

One of the major triumphs of the theory of Hilbert spaces is that it affords a framework for discussing and analyzing the mathematical theory of quantum mechanics. In the 1920s W. Heisenberg and E. Schrödinger developed seemingly different theories to explain the quantum effect of atomic physics. The Heisenberg theory is based on (infinite-dimensional) matrix methods whereas the Schrödinger theory is based on the properties of the differential operators appearing in the wave equation. It soon became apparent, however, that the two theories were actually equivalent[22] and that this equivalence is a consequence of the theory of linear operators on a Hilbert space.

We do not propose to discuss these two theories here. That is adequately treated in a myriad of books which have appeared in the last half century. Instead we would like to present a concise mathematical description[23] of the foundations of quantum mechanics.

Let us begin with the concept of the energy of a physical system. This is an example of an "observable" in physics. In classical mechanics, the energy of a system is a real-valued function of the phase coordinates of the system. In quantum mechanics the energy of a system is identified with an appropriate self-adjoint operator.[24] In general, the observables (such as position, velocity, momentum, and so on) of physics are identified with appropriate self-adjoint operators on a Hilbert space, specifically an infinite-dimensional separable Hilbert space.

There are at least two ways to explain this identification. The first explanation and undoubtedly the simplest, is to observe that the theory of self-adjoint operators adequately describes the pertinent physical phenomena. This then reduces the identification question to one of mathematical modeling.

The second explanation occurs when one examines the "propositional calculus of quantum mechanics." Picture the collection of all yes-no experiments, or propo-

[22] In fact Schrödinger, himself, was one of the first people to make this observation.

[23] The description we present here is by no means the only possible point of view. Moreover, while it is possible to base this description on a physically reasonable and mathematically rigorous foundation, we simply do not have the space to do that here. For more details we refer the reader to the excellent book by J. M. Jauch [2].

[24] For this example the appropriate self-adjoint operator happens to be unbounded. Therefore we shall postpone a more detailed discussion of the energy operator until Section 7.12.

sitions, which are used to describe a physical system. For example, one might ask whether the energy of the system is positive. In classical mechanics this collection of propositions can be identified with the collection of all[25] subsets of the phase space. In this case, an answer would be " yes " if the phase coordinates of the system were within a prescribed subset.

One of the fundamental facts in the quantum theory is that the outcome of two yes-no experiments A and B may depend on the order in which A and B are measured. This is not true in classical mechanics. Therefore one cannot expect to identify the yes-no experiments of quantum mechanics in the same manner. However, one can identify these experiments with the orthogonal projections on a Hilbert space. Let us denote this identification by $a \leftrightarrow P_a$ where a represents a yes-no experiment and P_a is an orthogonal projection with range $\mathscr{R}(P_a)$ and null space $\mathscr{N}(P_a)$. This identification is further explained in Table 1.

TABLE 1
IDENTIFICATION OF YES-NO EXPERIMENTS WITH ORTHOGONAL PROJECTIONS

PROPOSITIONAL NOTATION	HEURISTIC MEANING	HILBERT SPACE NOTATION
a	proposition "a"	P_a = associated orthogonal projection
b	proposition "b"	P_b = associated orthogonal projection
$a \subset b$	a implies b	$P_a \subset P_b$, that is P_b is an extension of P_a
a'	not a	$P_{a'} = I - P_a$
$a \cap b$	a and b	$P_a \cap P_b$ is the orthogonal projection onto $P_a(\mathscr{R}(P_b))$.
$a \cup b$	a or b	$P_a \cup P_b$ is the orthogonal projection onto $[P_{a'} \cdot (\mathscr{N}(P_b))]^\perp$

The next concept is that of the "state" of a physical system. In classical deterministic mechanics the state is a point in the phase space, or a delta function. In classical statistical mechanics the state is a probability function defined on the subsets of the phase space, or equivalently on the collection of yes-no experiments. Therefore, in quantum mechanics a *state* is defined as a probability function defined on the collection of yes-no experiments.

By using the fact that the yes-no experiments have been identified with the orthogonal projections on a Hilbert space H, it is possible to find a representation for the states of a quantum-mechanical system, but first we need a few definitions.

If L is a self-adjoint operator on a Hilbert space H, then the *trace* of L is given by

$$\operatorname{tr} L = \sum_n (Lx_n, x_n), \tag{5.25.1}$$

where $\{x_n\}$ is an orthonormal basis for H. The sum in (5.25.1) may be unbounded, in which case the trace is said to be $+\infty$. In the exercises the reader is asked to show that this definition of $\operatorname{tr} L$ does not depend on the choice of basis $\{x_n\}$.

---

[25] Strictly speaking, all "Borel" subsets.

Now let $W: H \to H$ be a linear operator. We shall say that W, is a *density operator* if W is self-adjoint, tr $W = 1$, and

$$0 \le (W^2x,x) \le (Wx,x), \qquad x \in H. \tag{5.25.2}$$

Inequality (5.25.2) is sometimes noted as $0 \le W^2 \le W$, by Definition 5.23.7.

It was shown by Gleason [1] that the states of a quantum-mechanical system can be put into one-to-one correspondence with the density operators. Moreover, if a state p is identified with a density operator W then the value of p on a yes-no experiment a is given by

$$p(a) = \text{tr } WP_a \, (= p(P_a)),$$

where P_a is the orthogonal projection associated with a.

A state p is said to be a *pure state* if the associated density operator W satisfies $W = W^2$.

The proof of the following theorem is an easy exercise.

5.25.1 THEOREM. *Let p be a state and let W be the associated density operator. Then the following statements are equivalent:*

(a) *p is a pure state, that is, $W^2 = W$.*

(b) *W is an orthogonal projection with a one-dimensional range.*

(c) *There is a unit vector e in H such that $p(P) = (Pe,e)$ for every orthogonal projection P.*

(d) *There is an orthogonal projection P that satisfies $p(P) = 1$.*

As a consequence of this result we see that every pure state p can be identified with a unit vector e in H so that the formula $p(P) = (Pe,e)$ is valid for every orthogonal projection P.

As already noted the observables in quantum mechanics are identified with self-adjoint operators. It should be mentioned that the most interesting observables are identified with unbounded self-adjoint operators so we have to postpone further discussion of this until Section 7.12. These identifications are summarized in Table 2.

TABLE 2
IDENTIFICATION OF QUANTUM-MECHANICAL OBJECTS WITH HILBERT SPACE OBJECTS

PHYSICAL OBJECT	CORRESPONDING HILBERT SPACE OBJECT
Yes-no experiment	Orthogonal projection
State	Density operator
Pure state	Unit vector
Observable	Self-adjoint operator

A final concept we shall note here is that of "expected value." Let A be a self-adjoint operator (observable) and let W be a density operator (state). Then the *expected value* of the observable A with respect to the state W, is given by

$$E(A) = \text{tr } WA. \tag{5.25.3}$$

If W represents a pure state and e is a unit vector with $We = e$, then one has

$$E(A) = (Ae, e). \tag{5.25.4}$$

This notion of expected value is related to the probabilistic concept of mathematical expectations. We must, however, postpone a more detailed discussion of it until we examine the spectral theorem. (See Section 7.12.)

Dynamical Equations

We shall use the Schrödinger version of the equations of motion for a quantum-mechanical system. A discussion of other versions can be found in Jauch [2, Chap. 10].

The dynamics of the system are described in terms of a one-parameter group of unitary operators

$$U_t = e^{-iAt},$$

where A is self-adjoint. We note that U_t then satisfies

$$U_t U_s = U_{t+s} \quad \text{and} \quad U_t^* = U_{-t}.$$

If p is a pure state and is represented by a unit vector ϕ, in the sense of Theorem 5.25.1, then the time evolution of ϕ is given by

$$\phi_t = U_t \phi = e^{-iAt}\phi.$$

One then has

$$\frac{d}{dt}\phi_t = \lim_{h \to 0} \frac{1}{h}(\phi_{t+h} - \phi_t)$$

$$= \lim_{h \to 0} \frac{1}{h}(e^{-iAh} - I)e^{-iAt}\phi = -iA\phi_t,$$

where the limiting behavior above is discussed in Exercise 7 below. In other words, the equations of motion become

$$\frac{d}{dt}\phi_t = -iA\phi_t, \tag{5.25.5}$$

where A is a self-adjoint operator. We shall return to this in Section 7.12.

EXERCISES

1. Show that the formula for trace does not depend on the basis.
2. Show that the expected value E, with respect to a pure state p, satisfies

$$E(A + B) = E(A) + E(B)$$
$$E(\lambda A) = \bar{\lambda}E(A).$$

3. Let $\{x_n\}$ be an orthonormal set and let $\{\lambda_n\}$ be a sequence of real numbers that satisfy $\lambda_n > 0$ and $\sum_n \lambda_n = 1$. Define $p(P)$ by

$$p(P) = \sum_n \lambda_n(x_n, Px_n),$$

where P is an arbitrary orthogonal projection.
(a) Show that p is a state by determining the associated density operator.
(b) Show that $p = \sum_n \lambda_n p_n$, where $\{p_n\}$ is a sequence of pure states.

4. A state p is said to be a *mixture* if there are two distinct states p_1 and p_2 and positive numbers λ_1, λ_2 such that $\lambda_1 + \lambda_2 = 1$ and $p = \lambda_1 p_1 + \lambda_2 p_2$. Show that every state is either a mixture or a pure state.

5. Let p be a state and define the *dispersion function* $\sigma(a) = p(a) - p^2(a)$ for every proposition a. Let $\sigma = \sup\{\sigma(a): a$ is a proposition$\}$ be the *overall dispersion*. A state is *dispersion-free* if $\sigma = 0$. Show that every dispersion-free state is pure.

6. Let $A = (a_{ij})$ be a self-adjoint matrix operator on the finite-dimensional Hilbert space C^n, where C^n has the usual inner product. Show that

$$\text{tr } A = \sum_{i=1}^{n} a_{ii}.$$

7. Let A be a bounded self-adjoint operator on a Hilbert space H. Show that the limit

$$\lim_{h \to 0} \frac{1}{h} (e^{-iAh} - I) = -iA$$

exists in the norm topology on the space of bounded linear operators. [This proves Equation (5.25.5) for bounded operators A.]

8. Let P be an orthogonal projection in a Hilbert space H with dim $\mathcal{R}(P) = k$. Show that tr $P = k$.

9. Let A be a bounded linear operator on a Hilbert space H. Show that

$$\text{tr } A^*A = \sum_n \|Ax_n\|^2$$

where $\{x_n\}$ is an orthonormal basis in H.

10. Let A be a self-adjoint operator with $0 \leq \text{tr } A < \infty$. Show that $\|A\| \leq \text{tr } A$.

11. Let K be a self-adjoint integral operator on $L_2(I)$ given by

$$y(t) = \int_I k(t,s)x(s)\, ds,$$

where $\int_I \int_I |k(t,s)|^2\, dt\, ds < \infty$. Show that tr $K = \int_I k(t,t)\, dt$.

12. Let A be a bounded self-adjoint operator on l_2. Assume that there is a unitary operator U such that $UAU^{-1} = \Delta$ where $\Delta = \text{diag}\,(\lambda_1, \lambda_2, \ldots)$ is a diagonal matrix. Show that tr $A = \sum_n \lambda_n$.

SUGGESTED REFERENCES

Akhiezer and Glazman [1]
Banach [1]
Day [1]
Dunford and Schwartz [1]
Edwards [1]
Goffman and Pedrick [1]
Halmos [3]
Hewitt and Stromberg [1]
Indritz [1]

Kolmogorov and Fomin [1]
Krasnosel'skii and Rutickii [1]
Naimark [1]
von Neumann [1]
Porter [1]
Simmons [1]
Taylor [2]
Wilansky [1]
Zaanen [1]

6

Analysis of Linear Operators (Compact Case)

1. INTRODUCTION

In this chapter we will be concerned with the (spectral) analysis of continuous linear operators on a complex Hilbert space. More precisely, we will be primarily concerned with a special class of continuous linear operators, namely, compact operators.

The first part of this chapter is devoted to an illustrative example where we describe this analysis. The reader will see that this analysis, the spectral analysis, is basically a geometric study of the behavior of linear operators. Our purpose in discussing this example early in the chapter is twofold:

(1) to show the genesis of the eigenvalue problem for linear operators, and
(2) to give a spectral analysis of certain finite-dimensional operators.

Both of these aspects will play a central role in our study of operators on infinite-dimensional spaces.

The remainder of the chapter is devoted to finding an appropriate generalization of the finite-dimensional methods to (Hilbert) spaces of arbitrary dimension. In order to do this it will be necessary to find an appropriate generalization of the concept of an eigenvalue. This leads to the notion of the spectrum of an operator, which is defined in Section 5.

There are several reasons for concentrating on compact operators at this point. As might be expected, one of the big divisions in spectral theory is the distinction between the finite- and the infinite-dimensional cases. However, within the infinite-dimensional case itself there are also different levels of complexity. Unbounded linear operators call for a spectral theory of great subtlety. The spectral theory for continuous linear operators is less subtle but certainly not child's play. However, the spectral theory for compact operators on infinite-dimensional spaces is relatively simple. It is a more or less enriched version of the finite-dimensional theory, and it offers the beginner very few unpleasant shocks.

One of the reasons, then, that we start with compact linear operators is that they are easy. Another reason is that understanding of the spectral theory for compact operators is a wonderful stepping stone to the understanding of the spectral theory for general bounded and unbounded linear operators. Finally, and most important of all, a great number of linear operators that occur in practice are compact operators defined on Hilbert spaces.

Actually the most powerful results we obtain will not be applicable to compact operators in general but to compact normal operators. Although many useful compact operators are also normal, it is also true that many are not. This is just a fact of life that has to be lived with. At the end of the chapter we will show one method of treating compact operators that are not normal. In any event, by the end of this chapter the reader should begin to understand why one usually prefers, when given a choice, to work with normal operators.

Part A

An Illustrative Example

2.   GEOMETRIC ANALYSIS OF OPERATORS

Let L be a linear operator defined on a complex Hilbert space H. The idea of a geometric analysis of L is to break up H into a number of parts (perhaps infinitely many) in such a way that the operation of L on each part is particularly simple. A simple example of this was presented in the last chapter when we discussed orthogonal projections. That is, if $P:H \to H$ is an orthogonal projection on a Hilbert space H, then

(i) P is the identity operator on $\mathscr{R}(P)$ and
(ii) P is the zero operator on $\mathscr{N}(P)$.

Knowing (i) and (ii) and knowing that $H = \mathscr{R}(P) + \mathscr{N}(P)$, we know what P does to an arbitrary element of H.

The decomposition of the operator P into the two parts given in (i) and (ii) illustrates in a minuscule manner how we propose to analyze an arbitrary linear operator L.

Let us illustrate this further with a slightly more complicated operator, which we define in Equation (6.2.2). But first we need the following concept.

6.2.1 DEFINITION.[1]   We shall say that a family of continuous projections $\{P_1, \ldots, P_m\}$ is a *resolution of the identity* if (i) the projections are orthogonal, (ii) $P_i P_j = 0$ if $i \neq j$, and (iii) $I = P_1 + \cdots + P_m$.

6.2.2 LEMMA.   *Let $\{P_1, \ldots, P_m\}$ be a resolution of the identity in a Hilbert space H. Then*

$$H = \mathscr{R}(P_1) + \cdots + \mathscr{R}(P_m). \tag{6.2.1}$$

The proof uses mathematical induction on m. For $m = 1$, it is obvious, and for $m = 2$ it follows from the Projection Theorem (Theorem 5.16.4). We leave the details as an exercise.

Let $\{P_1, \ldots, P_m\}$ be a resolution of the identity on H and let $\{\lambda_1, \ldots, \lambda_m\}$ be a family of complex numbers. Assume that $\lambda_i \neq \lambda_j$ for $i \neq j$, and let

$$L = \lambda_1 P_1 + \cdots + \lambda_m P_m. \tag{6.2.2}$$

Without any loss of generality we can assume that $P_i \neq 0$ for $1 \leq i \leq m$.

[1] Later we shall give another definition which will include this one as a special case. (See Section 8.)

Before giving a geometric analysis of the operator L, we note that it is always normal.

6.2.3 LEMMA. *Let $\{P_1,\dots,P_m\}$ be a resolution of the identity. Then the linear operator L given by (6.2.2) is continuous and normal. Moreover, L is self-adjoint if and only if all the λ_i's are real.*

The following theorem gives a geometric analysis of L.

6.2.4 THEOREM. *Let $\{P_1,\dots,P_m\}$ be a resolution of the identity, where $P_i \neq 0$ for $1 \le i \le m$. Then the space H can be decomposed as*

$$H = \mathscr{R}(P_1) + \cdots + \mathscr{R}(P_m),$$

where $\mathscr{R}(P_i) \perp \mathscr{R}(P_j)$ for $i \neq j$, and the operator $L = \lambda_1 P_1 + \cdots + \lambda_m P_m$ agrees with $\lambda_i I$ on $\mathscr{R}(P_i)$, and any vector $x \in H$ can be expressed (uniquely) as $x = x_1 + \cdots + x_m$ where $x_i \in \mathscr{R}(P_i)$ and

$$Lx = \lambda_1 x_1 + \cdots + \lambda_m x_m. \tag{6.2.3}$$

Proof: Lemma 6.2.2 assures us that the space H can be decomposed as indicated. Since $P_j P_i = 0$ for $i \neq j$, it follows that $\mathscr{R}(P_i) \subset \mathscr{N}(P_j) = \mathscr{R}(P_j)^\perp$ for $i \neq j$. Therefore, if $x \in \mathscr{R}(P_i)$, then $x \in \mathscr{N}(P_j)$ for $i \neq j$ and

$$Lx = (\lambda_1 P_1 + \cdots + \lambda_m P_m)(x) = \lambda_i P_i x = \lambda_i Ix.$$

Hence L agrees with $\lambda_i I$ on $\mathscr{R}(P_i)$. Equation (6.2.3) follows from the linearity of L. ∎

EXERCISES

1. Let $\{P_1,\dots,P_m\}$ be a resolution of the identity on a Hilbert space H. Show that

$$\mathscr{R}(P_1) = \bigcap_{j=2}^{m} \mathscr{N}(P_j)$$

and

$$\mathscr{R}(P_i) = \bigcap_{j \neq i} \mathscr{N}(P_j).$$

2. Let $\{Q_1,\dots,Q_m\}$ be a family of nonzero orthogonal projections on a Hilbert space H with the property that

$$H = V(\mathscr{R}(Q_1) \cup \cdots \cup \mathscr{R}(Q_m)).$$

   (a) Show that there is a resolution of the identity $\{P_1,\dots,P_n\}$ such that $P_1 = Q_1$ and $\mathscr{R}(Q_i) \subset \mathscr{R}(P_i)$, $i = 1, 2, \dots, n \le m$.
   (b) Is it possible to have $m = n$?

3. Prove Lemma 6.2.2.

4. Prove Lemma 6.2.3.

5. Let $\{Q_1,\ldots,Q_m\}$ be a family of orthogonal projections on a Hilbert space H and set

$$L = \lambda_1 Q_1 + \cdots + \lambda_m Q_m.$$

(a) Show that L is linear and continuous.
(b) Is is possible for L to be nonnormal?

6. Let $\{P_1,\ldots,P_m\}$ be a resolution of the identity where $m \geq 2$. Show that $\{Q_2,\ldots,Q_m\}$ is a resolution of the identity where $Q_2 = P_1 + P_2$ and $Q_i = P_i$, $3 \leq i \leq m$.

7. Let $\{e^{i2\pi nt} : n = 0, \pm 1,\ldots\}$ be an orthonormal basis for $L_2[0,1]$. Let P_n be the orthogonal projection onto $V(e^{i2\pi nt})$ and Q_n the orthogonal projection onto $V(\{e^{i2\pi mt} : n < |m|\})$. Show that $\{P_n : |n| \leq N\} \cup \{Q_N\}$ is a resolution of the identity. (For example, the operator $L = \sum_{n=-N}^{N} P_n$ represents a low-pass filter with gain $= 1$ and phase shift $= 0$ in the passband.)

8. Let $\{Q_1,\ldots,Q_m\}$ be a resolution of the identity on a Hilbert space H. Let $\{A_1,\ldots,A_n\}$ be a partition of the set of integers $\{1,2,\ldots,m\}$. Assume that $A_i \neq \varnothing$ for $i = 1, 2, \ldots, n$. Define P_i by

$$P_i = \sum_{j \in A_i} Q_j.$$

Show that $\{P_1,\ldots,P_n\}$ is a resolution of the identity. [*Hint*: Use mathematical induction.]

3. GEOMETRIC ANALYSIS. THE EIGENVALUE-EIGENVECTOR PROBLEM

Theorem 6.2.4 does give a geometric interpretation of an operator L that can be expressed as a finite linear combination of orthogonal projections. Unfortunately, linear operators are generally not given in this convenient form. Therefore, two questions arise.

(i) Under what conditions can a linear operator L be represented as a finite linear combination of orthogonal projections?
(ii) If one can do this, how can one find the scalars λ_i and the corresponding projections P_i?

The first question will require a little work, and we shall give a partial answer in the next section. The answer to the second question is, however, rather easy.

6.3.1 THEOREM. *Let L be a linear operator on a Hilbert space H and assume that there is a resolution of the identity $\{P_1,\ldots,P_m\}$, with $P_i \neq 0$ for $i = 1, \ldots, m$, and that there is an m-tuple of distinct scalars $\{\lambda_1,\ldots,\lambda_m\}$ such that*

$$L = \lambda_1 P_1 + \cdots + \lambda_m P_m.$$

Then the only scalars λ for which the equation

$$\lambda x - Lx = 0 \tag{6.3.1}$$

has a nontrivial solution x are $\lambda_1, \lambda_2, \ldots, \lambda_m$. Moreover, if $\lambda = \lambda_i$, then the corresponding solution x must lie in $\mathscr{R}(P_i)$ and vice versa, that is $\mathscr{R}(P_i) = \mathscr{N}(\lambda_i I - L)$.

The λ_i's are referred to as the *eigenvalues* of L and $\mathscr{N}(\lambda_i I - L)$ is called the *eigenmanifold associated with λ_i*. Solving Equation (6.3.1), or equivalently $Lx = \lambda x$, is sometimes referred to as the *eigenvalue-eigenvector problem* for the operator L. We shall return to these concepts in Section 5.

Proof: We use Theorem 6.2.4. Let $H = \mathscr{R}(P_1) + \cdots + \mathscr{R}(P_m)$ be the decomposition of H. For each x in H we let $x = x_1 + \cdots + x_m$, where $x_i \in \mathscr{R}(P_i)$. Then

$$Lx = \lambda_1 x_1 + \cdots + \lambda_m x_m.$$

If $Lx = \lambda x$, then, by using the fact that $(\lambda_j x_j, x_i) = 0$, if $i \neq j$ we have

$$\lambda_i(x_i, x_i) = \lambda(x_i, x_i),$$

indeed,

$$\lambda_i(x_i, x_i) = (\lambda_1 x_1 + \cdots + \lambda_m x_m, x_i) = (Lx, x_i)$$
$$= (\lambda x, x_i) = \lambda(x_1 + \cdots + x_m, x_i) = \lambda(x_i, x_i),$$

for $i = 1, \ldots, m$. Since the $\{\lambda_1, \ldots, \lambda_m\}$ are distinct, the above equality will hold only if $\lambda = \lambda_i$ or $x_i = 0$.

Now if $\lambda \neq \lambda_i$, then it must be that $x_i = 0$ for all i, so then it also must be that $x = x_1 + \cdots + x_m = 0$. Hence the only solution of (6.3.1) in this case is the trivial solution.

On the other hand, if $\lambda = \lambda_i$, then any nonzero vector x in $\mathscr{R}(P_i)$ is a solution of (6.3.1).

Let us now show that if $\lambda = \lambda_i$, then the corresponding solution x of (6.3.1) *must* lie in $\mathscr{R}(P_i)$. [This will prove that $\mathscr{R}(P_i) = \mathscr{N}(\lambda_i I - L)$.] Since the λ_i's are distinct and

$$0 = (\lambda_i I - L)(x_i + \cdots + x_m) = (\lambda_i - \lambda_1)x_1 + \cdots + 0 \cdot x_i + \cdots + (\lambda_i - \lambda_m)x_m,$$

we see that $x_j = 0$ for $j \neq i$ and $x = x_i$, that is, $x \in \mathscr{R}(P_i)$. ∎

In the next section we shall consider a finite-dimensional problem. This will illustrate the techniques we wish to develop later. But before we do this, let us note that the geometric analysis described in Theorem 6.2.4 leads naturally to the eigenvalue-eigenvector problem, Equation (6.3.1). In the next section we shall show how the solution of the eigenvalue-eigenvector problem (for finite-dimensional self-adjoint operators) leads back to the geometric analysis.

Finally, the conclusion of Theorem 6.2.4 can be reformulated another way in terms of eigenvalues and eigenvectors. In practice, this version seems to be more useful, so we ask the reader to take particular note of it.

6.3.2 THEOREM.   *Let* $\{P_1,\ldots,P_m\}$ *be a resolution of the identity on a complex Hilbert space* H *and let* $L = \sum_{i=1}^{m} \lambda_i P_i$, *where* $\{\lambda_1,\ldots,\lambda_m\}$ *are distinct scalars. Then there exists an orthonormal basis* $\{e_n\}$ *of eigenvectors of* L, *that is,* $Le_n = \mu_n e_n$, *where* μ_n *is one of the numbers* $\{\lambda_1,\ldots,\lambda_m\}$. *Moreover, every vector* $x \in H$ *can be written in the form* $x = \sum_n (x,e_n)e_n$ *and* $Lx = \sum_n \mu_n(x,e_n)e_n$.

Proof: Let $H = \mathscr{R}(P_1) + \cdots + \mathscr{R}(P_m)$ be the decomposition of H generated by the resolution of the identity. Let A_i be an orthogonal basis for $\mathscr{R}(P_i)$, $1 \le i \le m$. It follows from Theorem 6.3.1 that each vector in A_i is an eigenvector for L. Thus, let $A = A_1 \cup \cdots \cup A_m$. We claim A is a basis for H. However, this follows from Lemma 6.2.2 and the Orthogonal Structure Theorem. The rest of the theorem now follows from the Fourier Series Theorem (Theorem 5.17.8) and the fact that L is linear and continuous. ∎

EXERCISES

1. What happens to Theorem 6.3.1 if the scalars $\{\lambda_1,\ldots,\lambda_m\}$ are not distinct? What happens if $P_i = 0$ for some i?

2. Let $\{e_n\}$ be an orthonormal set in a Hilbert space H and let $\{\mu_1,\ldots,\mu_m\}$ be a collection of scalars. Define $L\colon H \to H$ by

$$Lx = \sum_{n=1}^{m} \mu_n(x,e_n)e_n.$$

   (a) Show that L is a bounded linear operator.
   (b) Show that L is normal.
   (c) Show that

$$\|L\| = \max\{|\mu_n| : n = 1,\ldots,m\}.$$

3. Let $\{e_n\}$ be an orthonormal set in a Hilbert space H and let $\{\mu_n\}$ be a collection of scalars from a bounded set D. Define $L\colon H \to H$ by

$$Lx = \sum_n \mu_n(x,e_n)e_n.$$

   (a) Show that L is a bounded linear operator.
   (b) What is L^*?
   (c) Show that L is normal.
   (d) Show that

$$\|L\| = \sup_n |\mu_n|.$$

4. Let L be a self-adjoint operator on a Hilbert space H. Show that if L is an orthogonal projection, then the only eigenvalues of L are 0 and/or 1.

4.   A FINITE-DIMENSIONAL PROBLEM

Let $L\colon H \to H$ be a bounded linear operator on a complex Hilbert space H. Assume that L is self-adjoint and that the range of L is finite dimensional. We will now show that there is a resolution of the identity $\{P_0,\ldots,P_m\}$ on H and a family of distinct scalars $\{\lambda_0,\ldots,\lambda_m\}$ such that $L = \sum_{i=0}^{m} \lambda_i P_i$. By then using Theorem 6.2.4 or Theorem 6.3.2 one arrives at a geometric analysis of L.

First we note that the range $\mathcal{R}(L)$ is invariant under L. Since L is self-adjoint, one has $\mathcal{N}(L) = \mathcal{R}(L)^{\perp}$. Indeed if $x \in \mathcal{N}(L)$ and $y \in H$, then we see that $x \perp Ly$ since $(x,Ly) = (Lx,y) = (0,y) = 0$. Let P_0 be the orthogonal projection onto $\mathcal{N}(L)$ and let $\lambda_0 = 0$. We note that L is a one-to-one mapping of $\mathcal{R}(L)$ onto $\mathcal{R}(L)$. Let $\{\lambda_1,\ldots,\lambda_m\}$ denote the nonzero eigenvalues of L and assume that the λ_i's are distinct. Let R_i denote the set

$$R_i = \{x \in \mathcal{R}(L): Lx = \lambda_i x, \ 1 \le i \le m.\}$$

Each R_i is then a finite-dimensional linear subspace of $\mathcal{R}(L)$ and, by Theorem 5.10.3, it is a closed linear subspace of $\mathcal{R}(L)$ and hence of H. We let P_i, $1 \le i \le m$, be the orthogonal projection onto R_i. We claim that $\{P_0,P_1,\ldots,P_m\}$ is a resolution of the identity on H and that $L = \sum_{i=0}^{m} \lambda_i P_i$.

First we observe that each P_i is an orthogonal projection, by construction. Next we claim that $P_i P_j = 0$ if $i \ne j$. In order to prove this we need the following result.

6.4.1 LEMMA. *Let $L: H \to H$ be a self-adjoint operator. Then the eigenvalues of L are real.*

Proof: Let λ be an eigenvalue of L and let x be a nonzero vector that satisfies $Lx = \lambda x$. One then has

$$0 = (Lx,x) - (x,Lx) = (\lambda x,x) - (x,\lambda x) = (\lambda - \bar{\lambda})(x,x),$$

which implies that $\lambda = \bar{\lambda}$. ∎

We now note that $P_i P_j = 0$ for $i \ne j$ is equivalent to saying that $R_i \perp R_j$ for $i \ne j$. To show this, we let λ_i and λ_j be distinct eigenvalues and choose nonzero eigenvectors x and y in R_i and R_j, respectively. One then has

$$0 = (Lx,y) - (x,Ly) = (\lambda_i x,y) - (x,\lambda_j y) = (\lambda_i - \lambda_j)(x,y),$$

which implies that $(x,y) = 0$.

Let $Q = P_0 + \cdots + P_m$. It is easy to see that Q is an orthogonal projection (compare with Exercise 3, Section 5.16). To show that $\{P_0,\ldots,P_m\}$ is a resolution of the identity, we want to show that $Q = I$, or equivalently that $M = H$, where $M = \mathcal{R}(Q)$. Since $L(M) \subset M$, it follows (Theorem 5.22.4) that $L: M^{\perp} \to M^{\perp}$. Since $M^{\perp} \subset \mathcal{R}(L)$, it follows that L is a one-to-one mapping of $M^{\perp}$ onto itself. We want to show that $M^{\perp}$ is trivial, that is, $M^{\perp} = \{0\}$. This is, however, an immediate consequence of the following result.

6.4.2 THEOREM. *Let Y be a finite-dimensional complex Hilbert space, with $\dim Y = n \ge 1$. Let $L: Y \to Y$ be a linear mapping of Y into Y. Then L has at least one eigenvalue.*

(In applying this theorem to our problem, we merely observe that the mapping $L: M^{\perp} \to M^{\perp}$ has *no* eigenvalues. It follows then that dim $M^{\perp} = 0$.)

Proof: Let $\{e_1, \ldots, e_n\}$ be a basis for Y. Any vector x in Y can be written as $x = \sum_{i=1}^{n} x_i e_i$. If we identify x with the n-tuple of complex numbers $(x_1, \ldots, x_n)$ and y with $(y_1, \ldots, y_n)$, one can then express the relationship $y = Lx$ by

$$y_i = \sum_{j=1}^{n} l_{ij} x_j, \tag{6.4.1}$$

where $Le_j = \sum_{i=1}^{n} l_{ij} e_i$. In matrix notation, (6.4.1) can be written as

$$[y] = [L][x],$$

where $[y]$ and $[x]$ are column vectors with entries $(y_1, \ldots, y_n)$ and $(x_1, \ldots, x_n)$ respectively and $[L]$ is the $n \times n$ matrix with entries (l_{ij}). The eigenvalue problem $Lx - \lambda x = 0$ then becomes

$$\begin{pmatrix} l_{11} - \lambda & l_{12} & \cdots & l_{1n} \\ l_{21} & l_{22} - \lambda & \cdots & l_{2n} \\ \vdots & & & \\ l_{n1} & l_{n2} & \cdots & l_{nn} - \lambda \end{pmatrix} \begin{pmatrix} x_1 \\ x_2 \\ \vdots \\ x_n \end{pmatrix} = \begin{pmatrix} 0 \\ 0 \\ \vdots \\ 0 \end{pmatrix} \tag{6.4.2}$$

or

$$\sum_{j=1}^{n} (l_{ij} - \lambda \delta_{ij}) x_j = 0, \quad i = 1, \ldots, n.$$

By Cramer's Rule, Equation (6.4.2) has a nontrivial solution $(x_1, \ldots, x_n)$ if and only if the determinant

$$p(\lambda) = \det(l_{ij} - \lambda \delta_{ij}) \tag{6.4.3}$$

vanishes at λ. Since $p(\lambda)$ is a polynomial of degree $n \geq 1$ it has at least one zero $\hat{\lambda}$. This zero $\hat{\lambda}$ is then an eigenvalue of L. ∎

We have thus shown that $\{P_0, \ldots, P_m\}$ forms a resolution of the identity. This means that $H = R_0 + R_1 + \cdots + R_m = R_0 \oplus R_1 \oplus \cdots \oplus R_m$, where $R_0 = \mathcal{N}(L)$. That is, any vector $x \in H$ can be written uniquely as $x = x_0 + \cdots + x_m$, where $x_i \in R_i$.

Since

$$Lx = Lx_0 + \cdots + Lx_m = \lambda_0 x_0 + \cdots + \lambda_m x_m$$
$$= (\lambda_0 P_0 + \cdots + \lambda_m P_m)x,$$

it follows that $L = \sum_{i=0}^{m} \lambda_i P_i$.

Let us summarize what we have just proven.

6.4.3 THEOREM.   *Let $L: H \to H$ be a bounded, linear self-adjoint operator on a Hilbert space H. Assume that the range of L is finite-dimensional. Then there is a resolution of the identity $\{P_0, P_1, \ldots, P_m\}$ and corresponding real numbers $\{\lambda_0, \lambda_1, \ldots, \lambda_m\}$ such that $L = \sum_{i=0}^{m} \lambda_i P_i$.*

We see then that Theorem 6.2.4 or Theorem 6.3.2 can be applied to the operator L given above. These two theorems (Theorems 6.2.4 and 6.3.2) are prototypes of the Spectral Theorem, which is discussed in Section 11.

Before we turn to the study of general operators let us note here that there is a third way of formulating the Spectral Theorem. We shall do this for linear operators on a finite-dimensional space.

6.4.4 THEOREM.   *Let H be a finite-dimensional complex Hilbert space and let $L: H \to H$ be a self-adjoint linear operator. Then there is a basis in H such that the matrix that represents L in this basis is a diagonal matrix.*

Proof: By the analysis above, we see that the operator L satisfies the hypotheses of Theorem 6.3.2. Let $\{e_1,\ldots,e_n\}$ be an orthonormal basis of eigenvectors of L so that $Le_n = \mu_n e_n$ for some eigenvalue μ_n. It follows then that L is represented by the diagonal matrix

$$\begin{bmatrix} \mu_1 & & 0 \\ & \ddots & \\ 0 & & \mu_n \end{bmatrix}$$

in this basis.   ∎

There is yet another way of interpreting the last result. This may be called the "transfer function representation" for L. It follows from the Parseval Equality (see the Fourier Series Theorem) that the mapping

$$U: x \to (x_1,\ldots,x_n),$$

where $x_i = (x,e_i)$ is a unitary mapping of H onto C^n. This means that one can write

$$L = U^{-1} \Lambda U, \quad \text{or} \quad \Lambda = ULU^{-1},$$

where Λ is the diagonal matrix $\Lambda = \text{diag}(\mu_1,\ldots,\mu_n)$. The representation Λ is sometimes called the "transfer function" of L.

EXAMPLE 1.   Let X denote the collection of all random variables ξ defined on a probability space $(\Omega,\mathscr{F},P)$ with range in a finite-dimensional complex Hilbert space H and satisfying the following conditions:

(i) $E[\xi] = 0$,
(ii) $E[|\xi|^2] < \infty$,
(iii) $E[|(x,\xi)|^2] > 0$ for all $x \in H$ with $x \neq 0$.

Here we use E to denote the mathematical expectation. Furthermore we note that if $x \in H$ and $\xi \in X$, then

$$E[(x,\xi)] = (x,E[\xi]) = (x,0) = 0. \tag{6.4.4}$$

Now let ξ^1, ξ^2, and ξ^3 be three random variables in X with the same distribution functions, that is,

$$P[\xi^1(\omega) \in A] = P[\xi^2(\omega) \in A] = P[\xi^3(\omega) \in A], \qquad (6.4.5)$$

for all measurable sets $A \subset H$. Assume further that

$$\xi^1 \text{ and } \xi^2 \text{ are stochastically independent} \qquad (6.4.6)$$

and that there is a self-adjoint operator L on H that satisfies

$$(Lx,x) > 0 \text{ for all } x \in H \text{ with } x \neq 0, \text{ and such that} \qquad (6.4.7)$$

$\xi^1 + \xi^2$ and $\sqrt{2L}\xi^3$ have the same distribution functions, that is,

$$P[(\xi^1(\omega) + \xi^2(\omega)) \in A] = P[\sqrt{2L}\xi^3(\omega) \in A] \qquad (6.4.8)$$

for all measurable sets $A \subset H$. We claim that[2] $L = I$, the identity operator.

Let us now prove that $L = I$. For this purpose we shall use Theorem 6.4.3 which assures us that $L = \sum_{i=0}^{m} \lambda_i P_i$, where $\{P_0, P_1, \ldots, P_m\}$ is a resolution of the identity on H. Let $x \in \mathcal{R}(P_i)$ with $x \neq 0$. Now by (6.4.7) we see that

$$0 < (Lx,x) = (\lambda_i x,x) = \lambda_i(x,x).$$

Hence $\lambda_i > 0$ for $i = 0, 1, \ldots, m$.

Since ξ^1, ξ^2, and ξ^3 have the same distribution functions we see that

$$E[|(x,\xi^1)|^2] = E[|(x,\xi^2)|^2] = E[|(x,\xi^3)|^2].$$

Furthermore, since ξ^1 and ξ^2 are independent, the real-valued random variables $\text{Re}(x,\xi^1)$ and $\text{Re}(x,\xi^2)$ are independent for each $x \in H$. Also $\text{Im}(x,\xi^1)$ and $\text{Im}(x,\xi^2)$ are independent. Hence

$$E[(x,\xi^1)\overline{(x,\xi^2)} + \overline{(x,\xi^1)}(x,\xi^2)] = 0, \qquad (6.4.9)$$

since

$$
\begin{aligned}
E[(x,&\xi^1)\overline{(x,\xi^2)} + \overline{(x,\xi^1)}(x,\xi^2)] \\
&= 2E[\text{Re}(x,\xi^1) \cdot \text{Re}(x,\xi^2) + \text{Im}(x,\xi^1)\text{Im}(x,\xi^2)] \\
&= 2\{E[\text{Re}(x,\xi^1)]E[\text{Re}(x,\xi^2)] + E[\text{Im}(x,\xi^1)]E[\text{Im}(x,\xi^2)]\} \\
&= 2\{\text{Re } E[(x,\xi^1)]\text{Re } E[(x,\xi^2)] + \text{Im } E[(x,\xi^1)]\text{Im } E[(x,\xi^2)]\} \\
&= 0
\end{aligned}
$$

by (6.4.4).

Let ξ denote either ξ^1, ξ^2, or ξ^3. The relationship

$$E[(x,\xi)\overline{(y,\xi)}] = q(x,y)$$

defines a sesquilinear functional on H (see Exercise 8, Section 5.12). By using (ii)

---

[2] There is nothing mysterious about the factor $\sqrt{2}$ used above. If this did not appear in Equation (6.4.8), then one would show instead that $L = \sqrt{2}I$.

and the Schwarz Inequality it is easy to see that q is bounded. It follows from Exercise 8, Section 5.23, that there is a bounded linear operator S on H with

$$q(x,y) = (Sx,y)$$

which then gives

$$E[|(x,\xi)|^2] = (Sx,x). \qquad (6.4.10)$$

It follows from (iii) that

$$(Sx,x) > 0 \text{ for all } x \in H, \ x \neq 0. \qquad (6.4.11)$$

By applying the self-adjointness of L together with (6.4.8) and (6.4.9) we get

$$\begin{aligned} E[|(\sqrt{2}Lx,\xi^3)|^2] &= E[|(x, \sqrt{2}L\xi^3)|^2] \\ &= E[|(x,\xi^1 + \xi^2)|^2] = E[|(x,\xi^1) + (x,\xi^2)|^2] \\ &= E[|(x,\xi^1)|^2] + E[|(x,\xi^2)|^2]. \end{aligned}$$

By using (6.4.10), the last equation can be written as

$$(S\sqrt{2}Lx, \sqrt{2}Lx) = (Sx,x) + (Sx,x)$$

or

$$(Sx,x) = (SLx,Lx).$$

If we replace x by Lx, above, we get

$$(Sx,x) = (SL^2x,L^2x).$$

By repeating this step we get

$$(Sx,x) = (SL^n x,L^n x), \qquad n = 1, 2, \ldots. \qquad (6.4.12)$$

Let us now show that the eigenvalues $\lambda_0, \lambda_1, \ldots, \lambda_m$ of L must satisfy $1 \leq \lambda_i$ for all i. Indeed, if this were not true, say that $\lambda_i < 1$, then by choosing $x \in \mathscr{R}(P_i)$ and $x \neq 0$ we get $Lx = \lambda_i x$, $L^2 x = \lambda_i^2 x$, and $L^n x = \lambda_i^n x$. Hence

$$|(SL^n x, L^n x)| \leq \|S\| \|L^n x\|^2 = \|S\| \lambda_i^{2n} \|x\| \to 0$$

as $n \to \infty$. Thus (6.4.12) implies that $(Sx,x) = 0$, which contradicts (6.4.11).

The final step is to show that the eigenvalues λ_i satisfy $\lambda_i \leq 1$. We do this by studying the inverse L^{-1}.

Since the eigenvalues λ_i are positive, the inverse L^{-1} exists and must be continuous. (Here we use Theorem 5.10.4 and the fact that L^{-1} exists if and only if $\lambda = 0$ is *not* an eigenvalue.) Furthermore, L^{-1} must be self-adjoint, as can easily be checked. Finally we note that if one applies Theorem 6.4.3 to L^{-1}, one gets

$$L^{-1} = \sum_{i=0}^{m} \lambda_i^{-1} P_i,$$

where the λ_i's and P_i's are the same quantities used for the operator L.

One can replace x with $L^{-n}x$ in (6.4.12) and thereby get

$$(Sx,x) = (SL^{-n}x, L^{-n}x).$$

If we repeat the argument following Equation (6.4.12) we conclude that $1 \le \lambda_i^{-1}$ for all i, that is, $\lambda_i \le 1$ for all i. Hence $\lambda_i = 1$ for all i and $L = I$.

The results of this example can be extended, see Prokhorov and Fish [1]. ∎

EXERCISES

1. Let H be a complex Hilbert space with finite dimension n and let $L: H \to H$ be a self-adjoint operator. Let $\{e_1,\ldots,e_n\}$ be any orthonormal basis in H and let (l_{ij}) denote the matrix representation of L with respect to this basis. Let $p(\lambda) = \det(l_{ij} - \lambda\delta_{ij})$. Show that the eigenvalues of L are precisely zeros of $p(\lambda)$. Factor $p(\lambda)$ as

$$p(\lambda) = (-1)^n(\lambda - \lambda_1)^{m_1} \cdots (\lambda - \lambda_k)^{m_k},$$

where the roots $\{\lambda_1,\ldots,\lambda_k\}$ are distinct, $m_i \ge 1$, and $m_1 + \cdots + m_k = n$. For $1 \le i \le k$, let $R_i = \{x \in H: Lx = \lambda_i x\}$.
(a) Show that dim $R_i = m_i$.
(b) Show that $L = \lambda_1 P_1 + \cdots + \lambda_k P_k$, where P_i is the orthogonal projection onto R_i.

2. Let T be the matrix operator

$$T = \begin{pmatrix} 3 & 1 & -2 \\ 1 & -5 & 4 \\ -2 & 4 & 1 \end{pmatrix}$$

on C^3. Find the eigenvalues and eigenvectors for T. Determine the corresponding resolution of the identity for T and express the corresponding projections as matrix operators.

3. Let $\phi_1, \ldots, \phi_n, \psi_1, \ldots, \psi_n \in L_2(I)$ and let $k(s,t) = \sum_{i=1}^k \phi_i(s)\overline{\psi_i(t)}$. Assume that $k(s,t) = \overline{k(t,s)}$.
(a) Show that the integral operator $y = Kx$, where

$$y(s) = \int_I k(s,t)x(t)\, dt,$$

is self-adjoint and has a finite-dimensional range.
(b) Determine the eigenvalues and eigenvectors of K.
(c) Assume that the eigenvalues are nonnegative. Do there exist functions $\{\xi_1,\ldots,\xi_k\}$ in $L_2(I)$ such that $k(s,t) = \sum_{i=1}^k \xi_i(s)\overline{\xi_i(t)}$? What happens if the eigenvalues are negative?

4. Let $H = l_2$ be the space of sequences $x = (x_1, x_2, \ldots)$ of complex numbers with $\sum_{i=1}^{\infty} |x_i|^2 < \infty$ and with the usual inner product. Let K be any $n \times n$ matrix and define the infinite matrix L by

$$L = \begin{pmatrix} K & 0 \\ 0 & 0 \end{pmatrix} = \begin{pmatrix} k_{11} & \cdots & k_{1n} 0 & \cdots \\ k_{21} & \cdots & k_{2n} 0 & \cdots \\ \vdots & & & \\ k_{n1} & \cdots & k_{nn} 0 & \cdots \\ 0 & \cdots & 0 & \cdots \\ \vdots & & \vdots & \end{pmatrix}.$$

What is L^*? Show that $L = L^*$ if and only if $K = K^*$. Determine the eigenvalues and eigenvectors for L.

5. Let L be the matrix operator

$$\begin{pmatrix} 4 + 2i & -2 + 2i & -2 + 2i \\ -2 + 2i & 4 + 5i & -2 - i \\ -2 + 2i & -2 - i & 4 + 5i \end{pmatrix}$$

on C^3. Show that L is normal. Show that there are eigenvectors $\{e_1, e_2, e_3\}$ and corresponding eigenvalues $\{\lambda_1, \lambda_2, \lambda_3\}$ for L so $\{e_1, e_2, e_3\}$ forms an orthonormal basis. [*Answer:* $\lambda_1 = 6 + 6i$, $e_1 = (0, 1/\sqrt{2}, -1/\sqrt{2})$; $\lambda_2 = 6i$, $e_2 = (1/\sqrt{3}, 1/\sqrt{3}, 1/\sqrt{3})$, $\lambda_3 = 6$, $e_3 = (2/\sqrt{6}, -1/\sqrt{6}, -1/\sqrt{6})$.]

6. What happens in Example 1 if one drops assumption (6.4.7)?

7. Let L and M be two self-adjoint matrix operators on a finite-dimensional Hilbert space H and assume that $LM = ML$. Show that there is a single unitary mapping $U: H \to H$ such that

$$ULU^{-1} \quad \text{and} \quad UMU^{-1}$$

are diagonal matrices. (That is, if L and M commute they have a common diagonalization.) What happens if L and M do not commute?

8. Let $\{L_1, \ldots, L_k\}$ be a family of self-adjoint matrix operators on a finite-dimensional Hilbert space H and assume that $L_i L_j = L_j L_i$ for $i, j = 1, 2, \ldots, k$. Show that there is a single unitary mapping $U: H \to H$ such that

$$UL_1 U^{-1}, UL_2 U^{-1}, \ldots, UL_k U^{-1}$$

are diagonal matrices.

9. Find the eigenvalues and a corresponding orthonormal set of eigenvectors for the following matrix operators acting on C^n.

(a) $\begin{bmatrix} 1 & -i \\ i & 1 \end{bmatrix}$

(b) $\begin{bmatrix} 1 & 1 - i \\ 1 + i & 3 \end{bmatrix}$

(c) $\begin{bmatrix} 0.5 & 0.3 \\ 0.3 & 0.5 \end{bmatrix}$

(d)[3] $\begin{bmatrix} 0 & -1 & -2 \\ 1 & 0 & -3 \\ 2 & 3 & 0 \end{bmatrix}$

[3] The matrix in (d) is normal but not self-adjoint. However, it still has an orthonormal basis of eigenvectors. Does this suggest a theorem?

(e) $\begin{bmatrix} 1 & i & 0 \\ -i & 1 & i \\ 0 & -i & 1 \end{bmatrix}$

(f) $\dfrac{1}{45} \begin{bmatrix} 89 & 2 & 20 & 0 \\ 2 & 86 & -40 & 0 \\ 20 & -40 & 95 & 0 \\ 0 & 0 & 0 & 4 \end{bmatrix}$

(Eigenvalues are $\lambda = 1,2,3,4$.)

(g) $\dfrac{1}{24\sqrt{6}} \begin{bmatrix} 24\sqrt{6} & 0 & 0 & 0 & 0 \\ 0 & 6\sqrt{6} & 72 & -18 & -18 \\ 0 & 72 & 8\sqrt{6} & 2\sqrt{6} & 2\sqrt{6} \\ 0 & -18 & 2\sqrt{6} & 17\sqrt{6} & -7\sqrt{6} \\ 0 & -18 & 2\sqrt{6} & -7\sqrt{6} & 17\sqrt{6} \end{bmatrix}$

(Eigenvalues are $\lambda = 1,-1,2$.)

10. In each of the following exercises you are asked to show that the operator K, given by

$$y(t) = \int_{-\pi}^{\pi} k(t,\tau)x(\tau)\, d\tau,$$

is a self-adjoint operator on $L_2[-\pi,\pi]$ with finite-dimensional range. You are also asked to determine all the nonzero eigenvalues $\{\lambda_1,\lambda_2,\ldots,\lambda_m\}$ and corresponding projections $\{P_1,\ldots,P_m\}$ so that $K = \lambda_1 P_1 + \cdots + \lambda_m P_m$.
   (a) $k(t,\tau) = 4\cos(t-\tau)$.
   (b) $k(t,\tau) = 1 + \cos(t-\tau)$.
   (c) $k(t,\tau) = 1 + \cos(t-\tau) + \sin 2(t+\tau)$.
   (d) $k(t,\tau) = \sin 3(t-\tau)$.
   (e) $k(t,\tau) = \sum_{n=1}^{N} \{a_n \cos n(t-\tau) + b_n \sin n(t-\tau)\}$.

11. Let P be an orthogonal projection on a finite-dimensional Hilbert space H.
   (a) Use the theorems of this section to give a spectral analysis of P.
   (b) Express P as a diagonal matrix operator.

Part B

The Spectrum

5. THE SPECTRUM OF A LINEAR TRANSFORMATION

In the last three sections we have considered several aspects of the eigenvalue-eigenvector problem for linear operators. It will be helpful to reformulate some of these concepts in a slightly more general context.

6.5.1 DEFINITION. Let T be a linear transformation with its domain $\mathscr{D}(T)$ and range $\mathscr{R}(T)$ contained in a linear space X. A scalar λ such that there does exists an $x \in \mathscr{D}(T)$, $x \neq 0$, satisfying the equation $Tx = \lambda x$, is said to be an *eigenvalue* of T. If λ is an eigenvalue of T, any nonzero $x \in \mathscr{D}(T)$, satisfying the equation $Tx = \lambda x$, is said to be an *eigenvector* of T corresponding to the eigenvalue λ. If λ is an eigenvalue of T, the null space of the transformation $\lambda I - T$, $\mathscr{N}(\lambda I - T)$, is said to be the *eigenmanifold* (*eigenspace*) *corresponding to* the eigenvalue λ. The dimension of the eigenmanifold is called the *multiplicity* of the eigenvalue λ.

Note that this definition applies to all linear operators, continuous or not. It is important though to emphasize that the solution x of $Tx = \lambda x$ must be nonzero and must lie in $\mathscr{D}(T)$.

The reader may suspect that a knowledge of the eigenvalues and eigenvectors of a linear operator may be enough for a geometric analysis. Unfortunately this is not the case.

EXAMPLE 1. Let $H = L_2(-\infty,\infty)$ and define $T: H \to H$ by $y = Tx$, where

$$y(t) = \int_{-\infty}^{t} e^{-(t-\tau)}x(\tau)\, d\tau. \qquad (6.5.1)$$

One can show that $\mathscr{D}(T) = H$ and $\mathscr{R}(T) \subset H$. Since

$$\int_{-\infty}^{t} e^{-(t-s)}e^{i\omega s}\, ds = \frac{1}{1 + i\omega}\, e^{i\omega t}$$

one might be tempted to say that $x(t) = e^{i\omega t}$ is an eigenvector of T with corresponding eigenvalue $\lambda = (1 + i\omega)^{-1}$. However, $e^{i\omega t}$ is *not* in $L_2(-\infty,\infty)$, so this is not correct. In fact, T does not have any eigenvectors *in* $L_2(-\infty,\infty)$. Nevertheless, one can give a geometric analysis of T by using the Fourier transform. Such an analysis requires knowledge of the spectrum of T. ∎

In order to motivate the definition of the spectrum, let us return to the concept of an eigenvalue. First we observe that a scalar λ is an eigenvalue for T if and only if

411

the linear transformation $\lambda I - T$ is not one-to-one, that is, if and only if the null space $\mathcal{N}(\lambda I - T)$ is nontrivial.

Of course, if $(\lambda I - T)$ is not one-to-one, it is not invertible. It does not even have an inverse defined on its range. Now it can happen that $(\lambda I - T)$ is one-to-one, that is, λ is *not* an eigenvalue, and $(\lambda I - T)$ still does not have an inverse defined on X. Or, bringing in topological considerations, the inverse may exist but not be continuous. Indeed, there are a number of possibilities for $(\lambda I - T)$. It turns out that a key to the analysis of a linear operator T is the study of the subset of λ's for which $\lambda I - T$ fails to have a continuous inverse. This subset of λ's is called the spectrum of T. Let us now be more precise.

6.5.2 DEFINITION.   Let T be a linear transformation whose domain $\mathcal{D}(T)$ and range $\mathcal{R}(T)$ are contained in a complex Banach space X. The set of all λ such that the range of the transformation $(\lambda I - T)$ is dense in X and such that $(\lambda I - T)$ has a continuous inverse defined on its range is said to be the *resolvent set* of T and denoted by $\rho(T)$. The set of all complex numbers that are not in the resolvent set is said to be the *spectrum* of T and denoted by $\sigma(T)$.

Needless to say, there are several ways that a complex number λ can fail to be in the resolvent set $\rho(T)$. This fact leads us to a subdivision of the spectrum.

(a) The *point spectrum* of a linear transformation T is the subset of all λ's, denoted by $P\sigma(T)$, for which the transformation $(\lambda I - T)$ is *not* one-to-one. That is, the point spectrum is exactly the set of all eigenvalues.

(b) The *continuous spectrum* of a linear transformation T is the subset of all λ's, denoted by $C\sigma(T)$, for which the transformation $(\lambda I - T)$ has its range dense in X, is one-to-one, and for which the inverse defined on the range is *not* continuous.

(c) The *residual spectrum* of a linear transformation T is the subset of all λ's, denoted by $R\sigma(T)$, for which the transformation $(\lambda I - T)$ is one-to-one but does not have its range dense in X.

The first things to note are that $P\sigma(T)$, $C\sigma(T)$, $R\sigma(T)$ are pairwise disjoint and that

$$\sigma(T) = P\sigma(T) \cup C\sigma(T) \cup R\sigma(T).$$

Figure 6.5.1 may aid the reader in remembering the contents of the above definitions.
    This may seem like a big jump from the set of all eigenvalues of an operator, but as the story of spectral analysis unfolds the need for each aspect of this definition reveals itself. Again, the definition has been stated in a more general form than needed here. In particular, it has been stated for Banach spaces even though we are only interested in Hilbert spaces here. Also, $\mathcal{D}(T)$ is not required to be all of X. This latter situation arises, for example, when one considers, as we do in Chapter 7, unbounded operators such as differential operators.

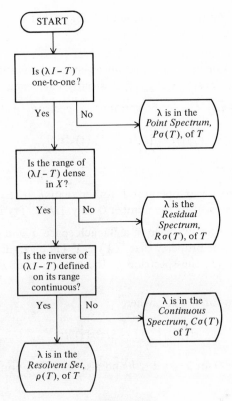

Figure 6.5.1.

EXERCISES

1. Let $T: X \to X$ be a bounded linear operator on a complex Banach space X. We say that a complex number λ belongs to the *approximate point spectrum* of T, denoted $(AP\sigma(T)$, if there is a sequence of vectors $\{x_n\}$ in X with $1 \le \|x_n\|$ for all n and

$$\|(\lambda I - T)x_n\| \to 0$$

as $n \to \infty$. Prove the following:
   (a) $\lambda \in AP\sigma(T)$ if and only if $(\lambda I - T)$ is not bounded below.
   (b) $AP\sigma(T) \subset \sigma(T)$.
   (c) $C\sigma(T) \cup P\sigma(T) \subset AP\sigma(T)$.
   (d) If the residual spectrum $R\sigma(T)$ is empty, then $AP\sigma(T) = \sigma(T)$.

2. Examples of spectra will be discussed in Section 6. Here we consider the Volterra integral operator (of the first kind), $y = Kx$, where

$$y(t) = \int_0^t k(t,s)x(s)\, ds.$$

Assume that the kernel $k(t,s)$ is continuous for $0 \le s \le t \le 1$. Show that

$K: L_2[0,1] \to L_2[0,1]$ and that K is bounded. Show that $\sigma(K) = \{0\}$ and either 0 is an eigenvalue or $0 \in C\sigma(K)$.

3. Let $L: X \to X$ be a linear operator on X and let $\{\lambda_1, \lambda_2, \ldots\}$ be a set of distinct eigenvalues for L. Further, let x_n be an eigenvector associated with λ_n, $n = 1, 2, \ldots$. Show that the set $\{x_1, x_2, \ldots\}$ is linearly independent.

4. Show that $\lambda = (1 + i\omega)^{-1}$ is in the continuous spectrum of the operator

$$y(t) = \int_{-\infty}^{t} e^{-(t-\tau)} x(\tau)\, d\tau$$

given in Example 1.

5. Construct a nonzero linear operator L (other than K in Exercise 2) with the property that $\sigma(L)$ consists of the number 0 only. [*Hint*: Try it for $L: C^2 \to C^2$.]

6. Let X be a closed linear subspace in a Banach space Y and let $L: Y \to Y$ be a linear operator with the property that $L(X) \subset X$. Let $\sigma_Y(L)$ denote the spectrum of $L: Y \to Y$ and $\sigma_X(L)$ the spectrum of the restriction of L to X. Show that $\sigma_X(L) \subset \sigma_Y(L)$.

7. Find the eigenvalues for $y = Kx$, where

$$y(t) = \int_{-1}^{1} (1 - 3t\tau) x(\tau)\, d\tau.$$

8. Assume that $k(t,\tau) \geq 0$ for $a \leq t, \tau \leq b$ and also let λ be a nonzero eigenvalue of $y = Kx$, where

$$y(t) = \int_{a}^{b} k(t,\tau) x(\tau)\, d\tau.$$

Assume that the corresponding eigenfunction $\phi(t)$ is positive for $a \leq t \leq b$. Show that $\mathcal{N}(\lambda I - K)$ is one-dimensional.

9. Let L be a self-adjoint operator on a Hilbert space H and assume that $\sigma(L)$ is one of the following three sets: $\{0\}$, $\{1\}$, $\{0,1\}$. Show that L is an orthogonal projection.

6.   EXAMPLES OF SPECTRA

In this section we determine the spectra for a number of widely applied linear operators. Although our main interest in this and the next chapter is in compact operators, we will not limit our attention to such operators here. This will, among other things, allow the reader to develop an understanding of the place of compact operators within the class of linear operators in general.

EXAMPLE 1. (FINITE-DIMENSIONAL CASE.)   Let X be a finite-dimensional Banach space and let T be a linear transformation of X into itself. Since (Theorem 4.7.7)

$$\dim[\mathcal{N}(\lambda I - T)] + \dim[\mathcal{R}(\lambda I - T)] = \dim X,$$

it follows that $(\lambda I - T)$ is one-to-one if and only if $\mathscr{R}(\lambda I - T) = X$. Therefore the residual spectrum of T is empty. Further, if $(\lambda I - T)$ is one-to-one, it has an inverse defined on X. Since any linear transformation defined on a finite-dimensional space is continuous (Theorem 5.10.4) $(\lambda I - T)^{-1}$ is continuous and the continuous spectrum of T is empty. Hence, we have shown that for the finite-dimensional case, T has a pure point spectrum, that is, $\sigma(T) = P\sigma(T)$. If $[T]$ is the matrix representing T relative to a basis or coordinate system $\{x_1,\ldots,x_n\}$ of X, then we have

$$P\sigma(T) = \{\lambda \in C : \det(\lambda I - [T]) = 0\}$$

which we know is nonempty and contains at most n complex numbers λ, see Section 4.

This example demonstrates why the reader, who is familiar with finite-dimensional linear operator theory only, may never have been confronted with the continuous and residual spectra before. ∎

Shift operators on sequence spaces have many applications. For example, z-transform techniques are based on shift operators. The next four examples develop the spectrum for four important cases.

EXAMPLE 2. (RIGHT SHIFT ON $l_2(-\infty,\infty)$.) Let H be the Hilbert space $l_2(-\infty,\infty)$ with the usual inner product.

We define the right shift operator $S_r : H \to H$ by $y = S_r x$, where

$$y_k = x_{k-1}, \qquad k = \ldots, -1, 0, 1, 2, \ldots, \tag{6.6.1}$$

for $x = \{\ldots,x_{-1},x_0,x_1,\ldots\}$ and $y = \{\ldots,y_{-1},y_0,y_1,\ldots\}$. That is, S_r shifts the sequence x to the right by one position. Needless to say, this operator is a building block with which difference equations are formed. For example, the difference operator Δ defined by

$$\Delta x = x_k + 2x_{k-1} + 5x_{k-2}, \qquad k = \ldots, -1, 0, 1, 2, \ldots,$$

can be written

$$\Delta = S_r^0 + 2S_r + 5S_r^2,$$

where we use the convention $S_r^0 = I$.

First let us see where $\lambda I - S_r$ is one-to-one. Let x be a point in H such that

$$(\lambda I - S_r)x = 0,$$

that is,

$$\lambda x_k - x_{k-1} = 0, \qquad k = \ldots, -1, 0, 1, 2, \ldots.$$

If $\lambda = 0$, it is obvious that S_r is one-to-one. If $\lambda \neq 0$, x is of the form

$$x = \{\ldots,c\lambda^2,c\lambda,c,c\lambda^{-1},c\lambda^{-2},\ldots\},$$

where c is a constant. But for arbitrary nonzero λ, this sequence is in H only if $c = 0$.

Therefore, $\lambda I - S_r$ is one-to-one for all λ. It follows that the point spectrum of S_r is empty.

Next let us consider the range of $(\lambda I - S_r)$. We consider three cases: $|\lambda| > 1$, $|\lambda| < 1$, and $|\lambda| = 1$.

$|\lambda| > 1$ Case

In this case the range of $(\lambda I - S_r)$ is all H. Indeed, let $y = \{y_k\}$ be any point in H. We can show that $x = \{x_k\}$, where

$$x_k = \frac{y_k}{\lambda} + \frac{y_{k-1}}{\lambda^2} + \frac{y_{k-2}}{\lambda^3} + \cdots, \quad k = \cdots, -1, 0, 1, 2, \ldots, \tag{6.6.2}$$

is a preimage of y. First let us show that x is in H. Since $\{y_k\}$ is in H and $|\lambda| > 1$, the infinite series (6.6.2) converges for all k. Then

$$\sum_{k=-N}^{N} |x_k|^2 = \sum_{k=-N}^{N} \left\{ \sum_{j=0}^{\infty} \frac{y_{k-j}}{\lambda^{j+1}} \right\} \left\{ \sum_{l=0}^{\infty} \frac{\bar{y}_{k-l}}{\bar{\lambda}^{l+1}} \right\},$$

where the bar denotes complex conjugate. Changing the order of summation, a step that can be justified, yields

$$\sum_{k=-N}^{N} |x_k|^2 \le \sum_{j=0}^{\infty} \sum_{l=0}^{\infty} \frac{1}{|\lambda^{j+1}| \, |\bar{\lambda}^{l+1}|} \left| \sum_{k=-N}^{N} y_{k-j} \bar{y}_{k-l} \right|.$$

But using the Schwarz Inequality one has

$$\left| \sum_{k=-N}^{N} y_{k-j} \bar{y}_{k-l} \right| \le \sqrt{\sum_{k=-\infty}^{\infty} |y_k|^2} \sqrt{\sum_{k=-\infty}^{\infty} |\bar{y}_k|^2}$$

for all N. So

$$\sum_{k=-\infty}^{\infty} |x_k|^2 \le \left\{ \sum_{j=0}^{\infty} \frac{1}{|\lambda^{j+1}|} \right\}^2 \sum_{k=-\infty}^{\infty} |y_k|^2. \tag{6.6.3}$$

That is, $\{x_k\}$ is in H for arbitrary $\{y_k\}$ in H. Noting that

$$\lambda x_k - x_{k-1} = \lambda \left\{ \frac{y_k}{\lambda} + \frac{y_{k-1}}{\lambda^2} + \cdots \right\} - \left\{ \frac{y_{k-1}}{\lambda} + \frac{y_{k-2}}{\lambda^2} + \cdots \right\} = y_k$$

we see that $x = \{x_k\}$ is a preimage of $y = \{y_k\}$. We have shown, then, that $\lambda I - S_r$ maps H onto itself for all $|\lambda| > 1$. We have also exhibited the inverse of $\lambda I - S_r$ in (6.6.2). Moreover, (6.6.3) shows that this inverse is continuous. Hence, any λ with $|\lambda| > 1$ is in the resolvent set of S_r.

$|\lambda| < 1$ Case

In this case again the range of $\lambda I - S_r$ is all of H. In fact, if $y = \{y_k\}$ is an arbitrary point in X, then its preimage $x = \{x_k\}$ is given by

$$x_k = -y_{k+1} - \lambda y_{k+2} - \lambda^2 y_{k+3} - \cdots, \quad k = \ldots, -1, 0, 1, 2, \ldots.$$

This can be shown using an argument that is analogous to the argument used above in the $|\lambda| > 1$ case. Similarly, $(\lambda I - S_r)$ has a continuous inverse defined on X. Hence, any λ with $|\lambda| < 1$ is in the resolvent set of S_r.

(It is interesting to note that in the $|\lambda| > 1$ case, x_k is independent of y_n for $n > k$, whereas in the $|\lambda| < 1$ case x_k is independent of y_n for $n \leq k$. In linear systems theory one would say that in one case the inverse $(\lambda I - S_r)^{-1}$ is causal and in the other it is anticausal.)

$|\lambda| = 1$ Case

This case is slightly more complicated. First let us consider the $\lambda = 1$ case. To begin with, the range of $I - S_r$ is not all of H. Indeed, if $x = \{x_k\}$ is any point in H and if y_k is defined by

$$y_k = x_k - x_{k-1},$$

then it follows that

$$\sum_{k=-N}^{M} y_k = (x_{-N} - x_{-N-1}) + (x_{-N+1} - x_{-N}) + \cdots + (x_{M-1} - x_{M-2}) + (x_M - x_{M-1})$$

$$= x_M - x_{-N-1}.$$

Since $x \in H$ one has $x_M \to 0$ and $x_{-M-1} \to 0$ as $M \to \infty$. Hence,

$$\sum_{k=-\infty}^{\infty} y_k = 0.$$

Let M denote the subspace of H defined by

$$M = \left\{ y = \{y_k\} \in H : \sum_{k=-\infty}^{\infty} y_k = 0 \right\}.$$

This is a proper subset of H and the range of $I - S_r$ is contained in it; therefore, the range of $I - S_r$ is not all of H. Nor, as a matter of fact, is the range of $I - S_r$ all of M. For example, let $\{y_k\}$ be the sequence

$$y_k = 0 \quad \text{for } k < 0$$

$$y_0 = 1$$

$$y_1 = \frac{1}{\sqrt{2}} - 1$$

$$\cdots$$

$$y_k = \frac{1}{\sqrt{k+1}} - \frac{1}{\sqrt{k}}$$

$$\cdots.$$

This sequence is M, but a corresponding sequence $\{x_k\}$ satisfying

$$y_k = x_k - x_{k-1}$$

must be of the form

$$x_k = c \quad \text{for } k < 0$$

$$x_0 = 1 + c$$

$$x_1 = \frac{1}{\sqrt{2}} + c$$

$$\cdots$$

$$x_k = \frac{1}{\sqrt{k+1}} + c$$

$$\cdots,$$

where c is a constant. But for no constant c is the sequence in H, therefore, $I - S_r$ does not map H onto M.

It turns out that the range of $I - S_r$ is the subspace **R** of M made up of all sequences $\{y_k\}$ such that

$$\sum_{k=-\infty}^{\infty} |y_k + y_{k-1} + y_{k-2} + \cdots|^2 < \infty$$

or, equivalently,

$$\sum_{k=-\infty}^{\infty} |y_{k+1} + y_{k+2} + y_{k+3} + \cdots|^2 < \infty.$$

The fact that each term in these series exists and that the two series have the same limit follow from the relationship

$$\sum_{k=-\infty}^{\infty} y_k = 0.$$

Let us now show that $\mathbf{R} \supset \mathcal{R}(I - S_r)$. Let $\{x_k\}$ be any point in H, and define y_k by

$$x_k = y_k + x_{k-1}, \qquad k = \ldots, -1, 0, 1, 2, \ldots.$$

One then has

$$x_k = x_{k-N-1} + y_k + y_{k-1} + \cdots + y_{k-N}.$$

Since

$$\lim_{N \to \infty} x_{k-N-1} = 0,$$

one gets

$$x_k = \sum_{i=0}^{\infty} y_{k-i} = y_k + y_{k-1} + \cdots, \qquad k = \ldots, -1, 0, 1, \ldots.$$

Therefore, $\{y_k\}$ is in $\mathbf{R}$. Next let us show that $\mathbf{R} \subset \mathscr{R}(I - S_r)$. Let $\{y_k\}$ be any point in $\mathbf{R}$ and consider the sequence

$$x_k = +y_k + y_{k-1} + \cdots = -y_{k+1} - y_{k+2} - \cdots.$$

Since $\{y_k\}$ is in $\mathbf{R}$, $\{x_k\}$ is in H. Moreover, $\{x_k\}$ is obviously a preimage of $\{y_k\}$. Hence, the range of $(I - S_r)$ is $\mathbf{R}$.

It can be shown that $\mathbf{R}$ is dense in H. Let $\{z_k\}$ be an arbitrary point in H. Then given an $\varepsilon > 0$, there exists an integer N such that

$$\sum_{k=-\infty}^{-N-1} |z_k|^2 < \varepsilon \qquad \text{and} \qquad \sum_{k=N+1}^{\infty} |z_k|^2 < \varepsilon.$$

Let $\{y_k\}$ be a sequence in H such that $y_k = z_k$ for $-N \leq k \leq N$. Further let

$$c = \sum_{k=-N}^{N} z_k.$$

Let K be an integer such that

$$\frac{|c|^2}{K} \leq \varepsilon.$$

Then let

$$y_{N+1} = -\frac{c}{K}, \, y_{N+2} = -\frac{c}{K}, \, \ldots, \, y_{N+K} = -\frac{c}{K}.$$

It follows that

$$\sum_{k=N+1}^{N+K} |y_k|^2 = K \frac{|c|^2}{K^2} \leq \varepsilon.$$

Let all other entries in the sequence $\{y_k\}$ be zero. It is easily seen that $\{y_k\}$ is in $\mathbf{R}$ and that

$$\sum_{k=-\infty}^{\infty} |z_k - y_k|^2 \leq 3\varepsilon.$$

Hence, $\mathbf{R}$ is dense in H.

Thus we know what the range of $I - S_r$ is, and we know how to represent the inverse of $I - S_r$ defined on its range. Let us now show that the inverse is not continuous. Let y^N be the following sequences with $4N^2$ nonzero entries:

$$y^N = \left\{ \ldots, 0, 0, \frac{+1}{2N}, \frac{+1}{2N}, \ldots, \frac{+1}{2N}, 0, \frac{-1}{2N}, \frac{-1}{2N}, \ldots, \frac{-1}{2N}, 0, 0, \ldots \right\}.$$

It can be seen that y^N is in the range of $I - S_r$ for each $N = 1, 2, \ldots$ and $\|y^N\| = 1$. On the other hand,

$$(I - S_r)^{-1} y^N = x^N,$$

where

$$x_k^N = 0 \quad \text{for} \quad k < -2N^2,$$

$$x_{-2N^2}^N = \frac{1}{2N}, \ x_{-2N^2+1}^N = \frac{2}{2N}, \ \ldots, \ x_{-1}^N = \frac{2N^2}{2N},$$

$$x_0^N = \frac{2N^2}{2N}, \ x_1^N = \frac{2N^2-1}{2N}, \ \ldots, \ x_{2N^2-1}^N = \frac{1}{2N},$$

$$x_k^N = 0. \quad \text{for} \quad k \geq 2N^2.$$

Since

$$\sum_{k=-\infty}^{\infty} |x_k^N|^2 = \frac{1}{2N^2} \{1 + 2^2 + 3^2 + \cdots + (2N^2)^2\}$$

$$= \frac{1}{2N^2} \frac{2N^2(2N^2 + 1)(4N^2 + 1)}{6},$$

by Exercise 1.5.5, it follows that $\|(I - S_r)^{-1}y^N\| \to \infty$ as $N \to \infty$. Thus, $(I - S_r)^{-1}$ is not continuous, and $\lambda = 1$ is in the continuous spectrum of S_r.

Finally, let us show that each λ with $|\lambda| = 1$ is in the continuous spectrum of S_r. Again we have

$$y_k = \lambda x_k - x_{k-1}, \quad k = \ldots, -1, 0, 1, 2, \ldots,$$

where $|\lambda| = 1$. Let $x_k = \lambda^{-k}z_k$ and $y_k = \lambda^{-k+1}w_k$, then

$$\lambda^{-k+1}w_k = \lambda^{-k+1}z_k - \lambda^{-k+1}z_{k-1}$$

or

$$w_k = z_k - z_{k-1}.$$

In other words, we have the $\lambda = 1$ case again. Thus $|\lambda| = 1$ is in the continuous spectrum of S_r. The situation is sketched in Figure 6.6.1. The point and residual spectra are empty. ∎

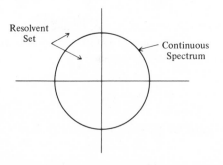

Figure 6.6.1. Spectrum of S_r or S_l on $l_2(-\infty, \infty)$.

EXAMPLE 3. (RIGHT SHIFT ON $l_2[0,\infty]$.) We consider the right shift again, but this time S_r is defined on $H = l_2[0,\infty)$. In particular, if $x = \{x_0, x_1, x_2, \ldots\}$, then

$$y = S_r x = \{0, x_0, x_1, \ldots\}.$$

Surprisingly enough this slight change does cause a change in the spectrum.

First let us show that $\lambda I - S_r$ is one-to-one for all λ. Let $(\lambda I - S_r)x = 0$, that is,

$$\lambda x_0 = 0$$

$$\lambda x_1 - x_0 = 0$$

$$\cdots$$

$$\lambda x_n - x_{n-1} = 0$$

$$\cdots.$$

It is easily seen that the only solution to this equation is $x = 0$. Therefore, $\lambda I - S_r$ is one-to-one, that is, the point spectrum is empty.

For $|\lambda| > 1$ essentially the same argument as that used in the preceding example shows that λ is in the resolvent set of S_r. Similarly, a simple variation of argument shows that $|\lambda| = 1$ is in the continuous spectrum of S_r. On the other hand, we have a new situation for the $|\lambda| < 1$ case. First note that if $y_k = \lambda x_k - x_{k-1}$, then

$$\sum_{k=0}^{N} \lambda^k y_k = (\lambda x_0) + \lambda(\lambda x_1 - x_0) + \lambda^2(\lambda x_2 - x_1) + \cdots + \lambda^N(\lambda x_N - x_{N-1})$$

$$= \lambda^{N+1} x_N \to 0 \quad \text{as} \quad N \to \infty.$$

Hence

$$\sum_{k=0}^{\infty} \lambda^k y_k = 0.$$

Thus when $|\lambda| < 1$, the range of $(\lambda I - S_r)$ is orthogonal to $\Lambda = \{1, \lambda, \lambda^2, \ldots\}$. We can show that the range of $\lambda I - S_r$ is exactly the subspace $M = \{y \in H : y \perp \Lambda\}$. Indeed we will let $\{y_k\}$ be any point in M, and consider the sequence $\{x_k\}$ given by

$$x_0 = -y_1 - \lambda y_2 - \lambda^2 y_3 - \cdots$$

$$x_1 = -y_2 - \lambda y_3 - \lambda^2 y_4 - \cdots$$

$$\cdots$$

$$x_k = -y_{k+1} - \lambda y_{k+2} - \lambda^2 y_{k+3} - \cdots$$

$$\cdots$$

By using the argument preceding Inequality (6.6.3) one can show that the sequence $\{x_k\}$ is in H whenever $\{y_k\}$ is in M. Since

$$\sum_{k=0}^{\infty} \lambda^k y_k = 0,$$

it follows that

$$x_0 = +\frac{1}{\lambda} y_0$$

$$x_1 = +\frac{1}{\lambda} y_1 + \frac{1}{\lambda^2} y_0$$

$$\cdots$$

$$x_k = +\frac{1}{\lambda} y_k + \frac{1}{\lambda^2} y_{k-1} + \cdots + \frac{1}{\lambda^{k+1}} y_0$$

$$\cdots,$$

and this sequence is a preimage for y. Thus the range of $\lambda I - S_r$ is M, which is not dense in H, so $|\lambda| < 1$ is in the residual spectrum of S_r. The situation is shown in Figure 6.6.2. ∎

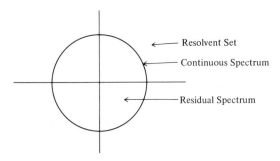

Figure 6.6.2. Spectrum of S_r on $l_2[0, \infty)$.

EXAMPLE 4. (LEFT SHIFT ON $l_2(-\infty, \infty)$.) In this example we will consider the left shift operator S_l on $H = l_2(-\infty, \infty)$. The operator S_l shifts a sequence $x = \{...,x_{-1},x_0,x_2,...\}$ to the left by one position. That is, if $y = S_l x$, where $y = \{...,y_{-1},y_0,y_1,y_2,...\}$, then

$$y_k = x_{k+1}, \qquad k = \dots, -1, 0, 1, 2, \dots.$$

(We note in passing that S_l is the adjoint as well as the inverse of S_r.) It is a simple matter to show that the spectrum of S_l is the same as that of S_r in Example 2. The argument is almost the same. ∎

EXAMPLE 5. (LEFT SHIFT on $l_2[0,\infty)$.) Here we will consider S_l defined $H = l_2[0,\infty)$. That is,

$$S_l\{x_0,x_1,x_2,\dots\} = \{x_1,x_2,x_3,\dots\}.$$

In this case, as in Example 4, S_l is the adjoint of S_r. But now it is only the left inverse of S_r. First let us see where $(\lambda I - S_l)$ is one-to-one. Let $x \in H$ be such that $(\lambda I - S_l)x = 0$, that is,

$$\lambda x_k - x_{k+1} = 0, \qquad k = 0, 1, 2, \dots.$$

It follows that any sequence $\{x_k\}$ satisfying this difference equation is of the form

$$\{c,\lambda c,\lambda^2 c,\ldots\},$$

where c is a constant. A nontrivial sequence of this form is in H if and only if $|\lambda| < 1$. Thus, for any λ satisfying $|\lambda| < 1$, $\lambda I - S_l$ is not one-to-one, and λ is in the point spectrum of S_l. Again using arguments similar to those used in the preceding examples, one can show that $|\lambda| = 1$ is in the continuous spectrum of S_l and $|\lambda| > 1$ is in the resolvent set. The situation is sketched in Figure 6.6.3.   ▌

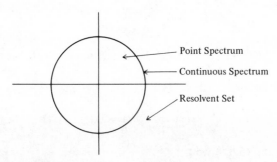

Figure 6.6.3. Spectrum of S_l on l_2 $(0, \infty)$.

So much for examples involving the shift operator. Another operator which plays an important role in linear analysis is the derivative operator. The next three examples treat three cases of differentiation defined in a Hilbert space.

EXAMPLE 6.   Let X be the linear subspace of $L_2(-\infty,\infty)$ made up of all absolutely continuous functions x such that dx/dt is in $L_2(-\infty,\infty)$. We note that it can be shown that X is dense in $L_2(-\infty,\infty)$. We consider the differential operator $y = Dx$, or

$$y = \frac{dx}{dt},$$

where $\mathscr{D}(D) = X$ and $\mathscr{R}(D) \subset L_2(-\infty,\infty)$. Let us see first where $\lambda I - D$ is one-to-one. Let $x \in X$ be such that $(\lambda I - D)x = 0$, that is,

$$\frac{dx}{dt} = \lambda x.$$

We know from the theory of differential equations that all (absolutely continuous) solutions of this equation are of the form

$$x(t) = ce^{\lambda t}, \qquad t \in (-\infty,\infty),$$

where c is a constant. Clearly x is in $L_2(-\infty,\infty)$ if and only if $c = 0$. Therefore $(\lambda I - D)$ is one-to-one for all λ.

Next let us consider the range of $(\lambda I - D)$, that is, all y such that

$$y = \lambda x - \frac{dx}{dt} \qquad \text{(a.e.)}$$

where, recall a.e. is the abbreviation for "almost everywhere."

First consider the case Re $\lambda \neq 0$. Let y be an arbitrary point in $L_2(-\infty,\infty)$. We claim that a preimage of y is given by

$$x(t) = -\int_{-\infty}^{t} e^{\lambda(t-\tau)}y(\tau)\, d\tau, \qquad \text{for Re } \lambda < 0 \qquad (6.6.4)$$

and by

$$x(t) = +\int_{t}^{\infty} e^{\lambda(t-\tau)}y(\tau)\, d\tau, \qquad \text{for Re } \lambda > 0. \qquad (6.6.5)$$

Indeed, by differentiating (6.6.4) we get

$$\frac{dx}{dt} = -\lambda \int_{-\infty}^{t} e^{\lambda(t-\tau)}y(\tau)\, d\tau - y(t) = \lambda x(t) - y(t) \qquad \text{(a.e.)}$$

and by differentiating (6.6.5) we get

$$\frac{dx}{dt} = +\lambda \int_{t}^{\infty} e^{\lambda(t-\tau)}y(\tau)\, d\tau - y(t) = \lambda x(t) - y(t) \qquad \text{(a.e.)}$$

So we see that $(\lambda I - D)$ is a mapping of X onto $L_2(-\infty,\infty)$ for all λ with Re $\lambda \neq 0$. Moreover, (6.6.4) and (6.6.5) are representations of $(\lambda I - D)^{-1}$ and it can be shown that both correspond to continuous transformations. Thus, any λ with Re $\lambda \neq 0$ is in the resolvent set of D.

Now consider the case Re $\lambda = 0$. First, we let $\lambda = 0$. We then want to solve the equation

$$y(t) = -\frac{dx}{dt} \qquad \text{(a.e.)}$$

Next, we note that

$$\int_{-a}^{b} y(t)\, dt = -\int_{-a}^{b} \frac{dx}{dt}\, dt = x(-a) - x(b).$$

However, we can show that $\lim_{b\to\infty} x(b) = 0$ and $\lim_{a\to\infty} x(-a) = 0$ (see Exercise 15, Section 5.22). Thus, every point in the range of D satisfies the condition

$$\lim_{N\to\infty} \int_{-N}^{N} y(t)\, dt = 0.$$

Let M be the linear subspace defined by

$$M = \left\{ y \in L_2(-\infty,\infty): \lim_{N\to\infty} \int_{-N}^{N} y(t)\, dt = 0 \right\}.$$

Since the range $\mathscr{R}(D)$ is contained in M and M is a proper subspace of $L_2(-\infty,\infty)$ D is not a mapping of X onto $L_2(-\infty,\infty)$. Further, it can be shown that $\mathscr{R}(D)$ is not all of M. In fact, let

$$y(t) = \begin{cases} 0 & \text{for } t < 0 \\ -2 & \text{for } 0 \le t < 1 \\ \dfrac{1}{t^{3/2}} & \text{for } 1 \le t < \infty. \end{cases}$$

The function y is in M. Now if y has a preimage x in X

$$-\frac{dx}{dt} = y \qquad \text{(a.e.)}$$

which implies that

$$\begin{aligned} x(t) &= c_0 = \text{constant}, & t < 0 \\ x(t) &= c_0 - 2t, & 0 \le t < 1 \\ x(t) &= c_0 - 2t^{-1/2}, & 1 \le t. \end{aligned}$$

Obviously such an $x(t)$ is not in $L_2(-\infty,\infty)$ for any choice of constants. So y does not have a preimage in X. It can be shown that $\mathscr{R}(D)$ is the subspace of M made up of all y such that

$$\int_{-\infty}^{t} y(\tau)\, d\tau$$

is in $L_2(-\infty,\infty)$. Moreover, it can be shown that this subspace is dense in $L_2(-\infty,\infty)$. Finally, it can be shown that the inverse of D defined on its range can either be represented by

$$x(t) = -\int_{-\infty}^{t} y(\tau)\, d\tau$$

or

$$x(t) = +\int_{t}^{\infty} y(\tau)\, d\tau.$$

Since the mapping is not continuous, it follows that $\lambda = 0$ is in the continuous spectrum of D.

If $\operatorname{Re} \lambda = 0$, but $\lambda \ne 0$, we can transform the problem to the $\lambda = 0$ case just considered. Let $\lambda = i\omega$; then the equation

$$y = i\omega x - \frac{dx}{dt}$$

can be changed to

$$u(t) = -\frac{dv}{dt}$$

by setting $y(t) = e^{i\omega t}u(t)$ and $x(t) = e^{i\omega t}v(t)$.

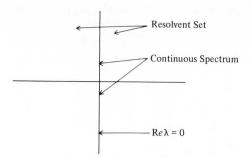

Figure 6.6.4. Spectrum of D on L_2 $(-\infty, \infty)$.

In summary, then, Re $\lambda \neq 0$ is in the resolvent set and Re $\lambda = 0$ is in the continuous spectrum of D (see Figure 6.6.4). ▮

EXAMPLE 7. Here we let X be the linear subspace of $L_2[0,\infty)$ made up of all functions x that are absolutely continuous on $[0,\infty)$ and such that dx/dt is in $L_2[0,\infty)$. Moreover, we require that $x(0) = 0$. Since we will be interested in the operator $D = d/dt$, this latter requirement simply limits the solutions we consider of the differential equation

$$y = \lambda x - \frac{dx}{dt}$$

to those with zero initial conditions.

Since $x(0) = 0$, the only solution to the homogeneous equation

$$\lambda x - \frac{dx}{dt} = 0$$

in X is the trivial solution. It follows that $(\lambda I - D)$ is one-to-one for all λ. A simple change in the argument of the preceding example shows that all λ's with Re $\lambda = 0$ are in the continuous spectrum of D. It is also easily shown on the basis of the preceding example that all λ's with Re $\lambda < 0$ are in the resolvent set of D. In fact, in this case the inverse of $(\lambda I - D)$ is represented by

$$x(t) = -\int_0^t e^{\lambda(t-\tau)}y(\tau)\,d\tau,$$

and this is a continuous transformation. The situation is different now for Re $\lambda > 0$. One might be tempted to say that

$$x(t) = +\int_t^\infty e^{\lambda(t-\tau)}y(\tau)\,d\tau, \tag{6.6.6}$$

which is a bounded linear transformation defined on $L_2[0,\infty)$, is the inverse of $(\lambda I - D)$ for Re $\lambda > 0$. However, it is not necessarily the case that one has

$$x(0) = +\int_0^\infty e^{-\lambda\tau}y(\tau)\,d\tau = 0. \tag{6.6.7}$$

As a matter of fact, the range of $(\lambda I - D)$ is not even dense in $L_2[0,\infty)$. We claim that the range is the subspace **R** of $L_2[0,\infty)$ that is orthogonal to $e^{-\lambda t}$. Indeed, if

$$y = \lambda x - \frac{dx}{dt},$$

then $x(0) = 0$ implies that

$$\int_0^\infty e^{-\lambda t}\left[\lambda x - \frac{dx}{dt}\right] dt = 0.$$

So $\mathcal{R}(\lambda I - D) \subset \mathbf{R}$. Going the other way, if $y \in \mathbf{R}$, it follows immediately that (6.6.6) yields a preimage $x \in X$. So $\mathbf{R} \subset \mathcal{R}(\lambda I - D)$. Thus $\mathbf{R} = \mathcal{R}(\lambda I - D)$. Since **R** is not dense in $L_2[0,\infty)$, it follows that λ's with Re $\lambda > 0$ are in the residual spectrum of D. The situation is sketched in Figure 6.6.5.   ∎

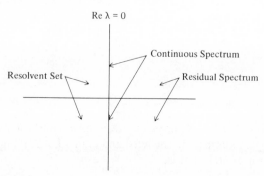

Figure 6.6.5. Spectrum of D on L_2 $[0,\infty)$, with Boundary Condition $x(0) = 0$.

EXAMPLE 8.   This example is the same as the preceding one except for the fact that $x(0)$ is no longer required to be zero. It is easily shown that each λ with Re $\lambda = 0$ is in the continuous spectrum of D. However, everywhere else things are different. For Re $\lambda < 0$, $e^{\lambda t}$ is a nontrivial solution of

$$(\lambda I - D)x = 0.$$

So each λ with Re $\lambda < 0$ is in the point spectrum of D. However, for Re $\lambda > 0$, $\lambda I - D$ has a continuous inverse defined on $L_2[0,\infty)$ that can be represented by

$$x(t) = +\int_t^\infty e^{\lambda(t-\tau)}y(\tau)\,d\tau.$$

So each λ with Re $\lambda > 0$ is in the resolvent set of D. The situation is shown in Figure 6.6.6.   ∎

We invite the reader to note the interesting parallels between the shift operators and the derivative operators.

The next example shows the connection between the spectrum of a linear time-invariant system and its frequency response.

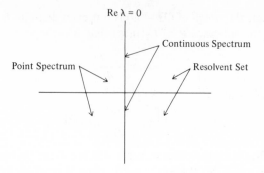

Figure 6.6.6. Spectrum of D on $L_2[0, \infty)$, with No Boundary Condition.

EXAMPLE 9.   Let $H = L_2(-\infty, \infty)$ and let T be the bounded linear mapping of H into itself defined by

$$(Tx)(t) = \int_{-\infty}^{t} h(t - \tau)x(\tau)\, d\tau, \qquad (6.6.8)$$

where $h(t) = 0$ for $t < 0$ and

$$h(t) = A_1 e^{a_1 t} + \cdots + A_n e^{a_n t} \qquad \text{for } t > 0$$

and the coefficients $A_1, \ldots, A_n$ are complex numbers and the exponents $a_1, \ldots, a_n$ are distinct complex numbers in the left-hand plane. Let $H(s)$ denote the one-sided Laplace transform of $h(t)$, that is,

$$H(s) = \int_{0}^{\infty} h(t)e^{-st}\, dt.$$

Then we assert that the continuous spectrum of T is given by

$$C\sigma(T) = \{\lambda: \lambda = H(i\omega) \text{ for some extended real number } \omega\}.$$

We say extended real number because $H(\infty) = 0$ is also a point in the continuous spectrum. Further, we assert that the rest of the complex plane is in the resolvent set. A typical situation is shown in Figure 6.6.7. We shall merely sketch the proof of

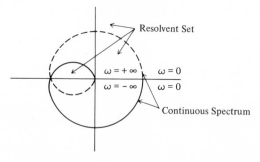

Figure 6.6.7.

these assertions here because this proof is very similar in spirit to that used in Example 6. We note that $H(s)$ is a rational function, that is, $H(s) = P(s)/Q(s)$, where $P(s)$ and $Q(s)$ are polynomials. It follows that

$$\frac{1}{\lambda - H(s)} = \frac{Q(s)}{\lambda Q(s) - P(s)}.$$

Using the partial function expansion, we have

$$\frac{1}{\lambda - H(s)} = \frac{Q(s)}{\lambda Q(s) - P(s)} = B_0(\lambda) + \frac{B_1(\lambda)}{s + \mu_1(\lambda)} + \cdots + \frac{B_n(\lambda)}{s + \mu_n(\lambda)}. \quad (6.6.8)$$

(Implicitly we have assumed that we do not have repeated roots for the above value of λ. In general, this is not the case, and the partial fraction expansion becomes slightly more involved. However, we ignore this detail here.)

Assume in (6.6.8) that $-\mu_1(\lambda), \ldots, -\mu_k(\lambda)$ are in the left-hand plane and $-\mu_{k+1}(\lambda), \ldots, -\mu_n(\lambda)$ are in the right-hand plane. Then $(\lambda I - T)$ has a continuous inverse defined on all of $L_2(-\infty, \infty)$. That is, λ is in the resolvent set. This inverse is represented by

$$(\lambda I - T)^{-1} y = B_0(\lambda) y(t) + B_1(\lambda) \int_{-\infty}^{t} e^{-\mu_1(\lambda)(t-\tau)} y(\tau)\, d\tau + \cdots$$

$$+ B_k(\lambda) \int_{-\infty}^{t} e^{-\mu_k(\lambda)(t-\tau)} y(\tau)\, d\tau + \cdots$$

$$- B_{k+1}(\lambda) \int_{t}^{\infty} e^{-\mu_{k+1}(\lambda)(t-\tau)} y(\tau)\, d\tau + \cdots$$

$$- B_n(\lambda) \int_{t}^{\infty} e^{-\mu_n(\lambda)(t-\tau)} y(\tau)\, d\tau.$$

However, if one of the roots is on the imaginary axis, that is, if $\lambda = H(i\omega)$ for some real number ω, then $(\lambda I - T)^{-1}$ is not continuous. Similarly, T^{-1} (that is, $\lambda = 0$) is not continuous. ∎

EXERCISES

1. We will let H_r be the Hilbert space made up of all doubly infinite sequences $x = \{\ldots, x_{-1} x_0 x_2, \ldots\}$ such that

$$\sum_{k=-\infty}^{\infty} |x_k|^2 r^{-2k} < \infty,$$

where r is a real number and the inner product is given by

$$(x,y) = \sum_{k=-\infty}^{\infty} x_k \bar{y}_k r^{-2k}.$$

Discuss the spectra of the left and right shifts on this space. How do these spectra behave as a function of r?

2. Let $l_{2,r}$ denote the Hilbert space made up of all sequences $x = \{x_0, x_1, x_2, \ldots\}$ such that

$$\sum_{k=0}^{\infty} |x_k|^2 r^{-2k} < \infty,$$

where r is a real number and the inner product is given by

$$(x, y) = \sum_{k=0}^{\infty} x_k \bar{y}_k r^{-2k}.$$

Discuss the spectra of the left and right shifts on this space. How do these spectra behave as a function of r?

3. Let $L_2^{\sigma}(-\infty, \infty)$ denote the Hilbert space made up of complex-valued functions x such that

$$\int_{-\infty}^{\infty} |x(t)| e^{-2\sigma t}\, dt < \infty,$$

where σ is a real number and the inner product is given by

$$(x, y) = \int_{-\infty}^{\infty} x(t) \bar{y}(t) e^{-2\sigma t}\, dt.$$

Discuss the spectrum of the derivative operator $D = d/dt$, where the domain of D is a proper subset of $L_2^{\sigma}(-\infty, \infty)$ defined as in Example 6.

4. Similarly to Exercise 3, consider operator D defined in the Hilbert space $L_2^{\sigma}[0, \infty)$. Assume that $x \in \mathscr{D}(D)$ implies $x(0) = 0$. (See Example 7.)

5. Repeat Exercise 4 with the boundary condition $x(0) = 0$ removed (see Example 8).

6. Discuss what happens if, in Example 9, $L_2(-\infty, \infty)$ is replaced by $L_2[0, \infty)$. [*Hint*: Review Examples 6, 7, and 8.]

7. Let $X = L_2(I)$, where I is an interval. Let $f(t)$ be a continuous complex-valued function defined on I. Define $F: \mathscr{D}(F) \to X$ by $y = Fx$, where $y(t) = f(t)x(t)$ and $\mathscr{D}(F) = \{x \in X : fx \in X\}$. Show that if $f(t)$ is bounded, then F is bounded. Now, show that the spectrum of F consists of continuous spectrum only and that $C\sigma(F) = \overline{f(I)}$.

8. Show that if a linear operator has a finite-dimensional range, then its entire spectrum is pure point spectrum.

9. Discuss the spectrum of the following operators on $l_2(-\infty, \infty)$:
   (a) $S_r + 2I$.
   (b) $S_r + \lambda I$.
   (c) $\beta S_r + \lambda I$, where $\beta \neq 0$.
   (d) S_r^2.
   (e) $\alpha S_r^2 + \beta S_r + 2I$, where $\alpha \neq 0$.

10. Repeat Exercise 9 but now on $l_2[0, \infty)$.

11. Repeat Exercises 9 and 10 for the operator S_l. Do your answers suggest a general theorem?

12. Consider the operator $T = S_r + S_l$ on $l_2[0,\infty)$. Show that $\sigma(T)$ is the real interval $[-2, 2]$ and that $\sigma(T) = C\sigma(T)$.

13. Consider the operator Φ on $l_2(0,\infty)$ given by

$$\Phi(x_1, x_2, \ldots) = (\phi(1)x_1, \phi(2)x_2, \ldots),$$

where $\sup_n |\phi(n)| < \infty$.

(a) Show that $\phi(n) \in P\sigma(\Phi)$ for all n and that $\sigma(\Phi)$ is the closure of $P\sigma(\Phi)$.
(b) Is Φ self-adjoint? It not, when is Φ self-adjoint?
(c) Is Φ normal? If not, when is Φ normal?

14. Consider the operator $L = T + \Phi$ on $l_2(0,\infty)$, where T and Φ are given in Exercises 12 and 13. Assume that $\phi(n)$ is real-valued.
(a) Show that if $\lambda \in \sigma(L)$, then $|\lambda| \leq 2 + \|\Phi\|$, where $\|\Phi\| = \sup_n |\phi(n)|$.

(b) Show that L is self-adjoint.
(c) Show that L is not compact.
(d) Assume that $\phi(n)$ is real-valued and that $\phi(n) \to \lambda_0$ as $n \to \infty$. Show that $C\sigma(L)$ is the interval $[-2 - \lambda_0, 2 - \lambda_0]$. What is $\sigma(L)$?
(e) Assume that $\phi(n) \to 0$ and that $|\phi(n)| > \sqrt{2}$ for some n. Show that L has at least one eigenvalue.

7. PROPERTIES OF THE SPECTRUM

Let T be a continuous linear transformation mapping a complex Banach space into itself. In this section we show that the spectrum of T is a compact subset of the complex plane lying in the closed ball $\{z: |z| \leq \|T\|\}$. It follows, then, that if $|\lambda| > \|T\|$, then λ is in the resolvent set $\rho(T)$. We can see from the examples of the preceding section that this result does not hold for discontinuous operators.

The following theorem is useful in reaching the desired result.

6.7.1 THEOREM. *Let T be a continuous linear transformation of a Banach space X into itself such that $\|T\| < 1$. Then $(I - T)^{-1}$ exists and is continuous. Moreover,*

$$(I - T)^{-1} = I + T + T^2 + T^3 + \cdots = \sum_{n=0}^{\infty} T^n, \qquad (6.7.1)$$

where the convergence is in terms of the uniform topology and

$$\|(I - T)^{-1}\| \leq (1 - \|T\|)^{-1}. \qquad (6.7.2)$$

Proof: Recall that $Blt[X,X]$ denotes the normed linear space made up of all continuous linear mappings of X into itself, and the norm on $Blt[X,X]$ is the operator norm. Since X is complete, $Blt[X,X]$ is complete.

Since $\|T^n\| \leq \|T\|^n$ and since $\|T\| < 1$, it follows that the series $\sum_{n=0}^{\infty} T^n$ is absolutely convergent, that is, $\sum_{n=0}^{\infty} \|T^n\|$ is convergent. It is then a consequence of Theorem 5.4.2 that the series $\sum_{n=0}^{\infty} T^n$ is convergent.

Let $S_N = \sum_{n=0}^{N} T^n$, and $S_\infty = \sum_{n=0}^{\infty} T^n$. We want to show that S_∞ is the inverse of $(I - T)$, that is,

$$(I - T)S_\infty = S_\infty(I - T) = I.$$

However,

$$(I - T)S_N = I - T^{N+1} \to I,$$

since $\|T^{N+1}\| \to 0$ as $N \to \infty$. On the other hand,

$$(I - T)S_N \to (I - T)S_\infty,$$

since

$$\|(I - T)S_N - (I - T)S_\infty\| \leq \|I - T\| \cdot \|S_N - S_\infty\| \to 0.$$

Since limits are unique, one has $(I - T)S_\infty = I$. Similarly one shows that $S_\infty(I - T) = I$. The proof of the inequality (6.7.2) is left as an easy exercise. ∎

Please note that this theorem does *not* say that $\|T\| < 1$ is a necessary condition for the existence and continuity of $(I - T)^{-1}$. It merely says that it is a sufficient condition.

6.7.2 THEOREM. (NEUMAN EXPANSION.) *Let T be a continuous linear mapping of a Banach space X into itself. If $|\lambda| > \|T\|$, then λ is in the resolvent set of T. Moreover, $(\lambda I - T)^{-1}$ is given by*

$$(\lambda I - T)^{-1} = \sum_{n=0}^{\infty} \lambda^{-n-1} T^n \tag{6.7.3}$$

and

$$\|(\lambda I - T)^{-1}\| \leq (|\lambda| - \|T\|)^{-1}. \tag{6.7.4}$$

Proof: We want to show that $(\lambda I - T)$ has a continuous inverse. But

$$(\lambda I - T) = \frac{1}{\lambda}\left(I - \frac{1}{\lambda}T\right).$$

Since $\|(1/\lambda)T\| < 1$, it follows from the previous theorem that $(I - (1/\lambda)T)$ has a continuous inverse. Hence,

$$(\lambda I - T)^{-1} = +\lambda^{-1}\left(I - \frac{1}{\lambda}T\right)^{-1} = \sum_{n=0}^{\infty} \lambda^{-n-1} T^n,$$

which is continuous. The proof of (6.7.4) is left as an exercise. ∎

We see, then, that the resolvent set of a continuous operator is not empty and the spectrum is bounded.

6.7.3 THEOREM.   *Let T be a continuous linear mapping of a Banach space X into itself. If $\lambda \in \rho(T)$ and $|\mu| < \|(\lambda I - T)^{-1}\|^{-1}$, then $\lambda + \mu \in \rho(T)$. In particular, the resolvent set $\rho(T)$ is an open set in the complex plane.*

Proof: Let λ be any point in the resolvent set $\rho(T)$. Since

$$(\lambda + \mu)I - T = (\lambda I - T)[I + \mu(\lambda I - T)^{-1}],$$

it follows from Theorem 6.7.1 that $[I + \mu(\lambda I - T)^{-1}]$ has a continuous inverse provided

$$\|\mu(\lambda I - T)^{-1}\| = |\mu| \, \|(\lambda I - T)^{-1}\| < 1.$$

Since $(\lambda I - T)^{-1}$ is continuous, the theorem now follows from Theorem 6.7.2.   ∎

In summary, then, we can say the following:

6.7.4 THEOREM.   *Let T be a continuous linear transformation of a complex Banach space X into itself. The spectrum of T is a compact subset[4] of the complex plane lying in the closed ball $\{z: |z| \leq \|T\|\}$.*

There remains one technical point to be mentioned; namely, the spectrum of a bounded linear operator on a (nontrivial) Banach space is never empty. The reader interested in the proof of this general fact is referred to Taylor [2, p. 261]. We shall show that this is the case for compact normal operators (Theorem 6.10.16).

EXERCISES

1. One can actually show that the spectrum of a bounded linear operator T lies in the circle $\{\lambda: |\lambda| \leq \lim \sup \|T^n\|^{1/n}\}$, which is contained in $\{\lambda: |\lambda| \leq \|T\|\}$. The following steps lead to a proof of this fact: First show that whenever the series in (6.7.3) converges, it converges to $(\lambda I - T)^{-1}$. Use standard real analysis to show that the series $\sum_{n=0}^{\infty} |\lambda|^{-n-1} \|T^n\|$ converges for $|\lambda| > \lim \sup \|T^n\|^{1/n}$. This implies that $(\lambda I - T)^{-1}$ exists and is continuous for $|\lambda| > \lim \sup \|T^n\|^{1/n}$.

2. We will let $p(r)$ be a polynomial in r with complex coefficients, that is, let $p(r) = a_0 r^n + \cdots + a_n$, where $a_0 \neq 0, n \geq 0$. Let $L: X \to X$ be a bounded linear operator on a complex Banach space X and define $p(L)$ by

$$p(L) = a_0 L^n + \cdots + a_n I.$$

Show that the spectra of L and $p(L)$ are related by

$$\sigma(p(L)) = p(\sigma(L)) = \{p(\lambda): \lambda \in \sigma(L)\}.$$

3. Let $L: X \to X$ be a bounded linear mapping of X onto X and assume that L^{-1} exists and is continuous. Show that

$$\sigma(L^{-1}) = \sigma(L)^{-1} = \left\{\frac{1}{\lambda}: \lambda \in \sigma(L)\right\}.$$

---

[4] Recall that a subset of the complex plane is compact if and only if it is closed and bounded.

4. Let $L: X \to X$ be a linear operator, where X is a Banach space. The *spectral radius* of L is defined by

$$r_\sigma(L) = \sup\{|\lambda|: \lambda \in \sigma(L)\}.$$

(a) Show that $r_\sigma(L) \leq \|L\|$.

(b) Show that if L is the matrix operator $\begin{pmatrix} 1 & 1 \\ 0 & 1 \end{pmatrix}$, then $r_\sigma(L) < \|L\|$.

(In Exercise 3, Section 6.11 we will show that if L is a compact normal operator on a Hilbert space, then $r_\sigma(L) = \|L\|$.)

5. Let L be a bounded linear operator on a Hilbert space H. We define the *numerical range* of L to be the set of complex numbers

$$W(L) = \{(Lx, x): \|x\| = 1\}.$$

(a) Show that the spectrum $\sigma(L)$ lies in $\overline{W(L)}$.

(b) Show that if $d(\lambda, \overline{W(L)}) = \rho > 0$, then $\|(\lambda I - L)^{-1}\| \leq 1/\rho$.

(c) Show that $W(L)$ is convex.

(d) Show that

$$W(\alpha L + \beta I) = \alpha W(L) + \beta$$

when $\alpha \neq 0$.

(e) Show that $W(ULU^{-1}) = W(L)$ when U is a unitary operator.

6. Let L be a bounded linear operator on a Hilbert space H. Let x and y be fixed elements in H.

(a) Show that the function

$$f(\lambda) = ((\lambda I - L)^{-1}x, y)$$

is an analytic function for $\lambda \in \rho(L)$. Let $R_\lambda = (\lambda I - L)^{-1}$.

(b) Show that for λ, μ in $\rho(L)$ one has

$$R_\lambda - R_\mu = (\mu - \lambda)R_\lambda R_\mu$$
$$R_\lambda R_\mu = R_\mu R_\lambda.$$

(c) Show that if $\mu \in \rho(L)$ and $|\mu - \lambda| \, \|R_\mu\| < 1$, then $\lambda \in \rho(L)$ and

$$R_\lambda = \sum_{n=0}^{\infty} (\mu - \lambda)^n R_\mu^{n+1}.$$

(d) Show that

$$\frac{d^n}{d\lambda^n} R_\lambda = (-1)^n n! \, R_\lambda^{n+1}.$$

[That is, R_λ is analytic for $\lambda \in \rho(L)$.]

7. Two sequences $\{x_n\}$ and $\{y_n\}$ are said to form a *biorthonormal sequence* if $(x_n, y_m) = 0$ for $n \neq m$ and $(x_n, y_n) = 1$. A biorthonormal sequence is said to be *maximal* if finite linear combinations of the x_n's, as well as those of the y_n's, are dense in the basic Hilbert space H.

Let $\{x_n\}$ and $\{y_n\}$ be a maximal biorthonormal sequence in a Hilbert space H.

(a) Show that if $(x,y_n) = 0$ for all n, then $x = 0$.
(b) Show that if $(x_n,y) = 0$ for all n, then $y = 0$.
(c) Show that if $\sum_n (x,y_n)x_n$ or $\sum_n (x,x_n)y_n$ converges, then $x = \sum_n (x,y_n)x_n$ or $x = \sum_n (x,x_n)y_n$, respectively.

8. Let $\{e_n\}$ be an orthonormal basis in a Hilbert space H and let $\{x_n\}$ be a sequence in H with the property that there is a θ, $0 \le \theta < 1$, such that

$$\left\| \sum_n a_n(e_n - x_n) \right\|^2 \le \theta^2 \sum_n |a_n|^2 \qquad (6.7.5)$$

for every sequence $\{a_n\}$ of complex numbers.
(a) Show that

$$Kx = \sum_n (x,x_n)(e_n - x_n)$$

defines a bounded linear operator on H with $\|K\| \le \theta$.
(b) Let $T = I - K$ and show that

$$(1 - \theta) \|x\| \le \|Tx\| \le (1 + \theta) \|x\|, \qquad x \in H.$$

(c) Show that $x_n = Te_n$.
(d) Let $y_n = (T^{-1})^*e_n$. Show that $\{x_n\}$ and $\{y_n\}$ forms a maximal biorthonormal sequence.
(e) Show that for each $x \in H$ one has

$$x = \sum_n (x,x_n)y_n \quad \text{and} \quad x = \sum_n (x,y_n)x_n.$$

9. Let $x_n(t) = (2\pi)^{-1/2} \exp(i\lambda_n t)$, $n = 0, \pm 1, \pm 2, \ldots$ for $-\pi \le t \le \pi$. [Recall that $e_n(t) = (2\pi)^{-1/2} \exp(int)$, $n = 0, \pm 1, \ldots$ forms an orthonormal basis for $L_2(-\pi, \pi)$.] Assume that

$$M = \sup_n |\lambda_n - n| < \frac{\log 2}{\pi}.$$

Show that $\{x_n\}$ and $\{e_n\}$ satisfy (6.7.5) with $\theta = e^{M\pi} - 1 < 1$.

10. We will let L, S, and T be bounded linear operators on a Hilbert space H where $L = T + S$. Assume S^{-1} exists and is compact, and that $\|T\| \|S^{-1}\| < 1$. Show that L^{-1} exists and is compact.

11. Let W be a density operator on a Hilbert space H. (That is, W is a self-adjoint with $0 \le W^2 \le W$ and tr $W = 1$. See Section 5.25.)
(a) Show that the spectrum $\sigma(W)$ lies in the interval $[0,1]$.
(b) Show that if $1 \in \sigma(W)$, then $\sigma(W) = \{0,1\}$ and W represents a pure state.
(c) Show that if λ is an eigenvalue with $\frac{1}{2} < \lambda \le 1$, then dim $\mathcal{N}(W - \lambda I) = 1$.
(d) Show that if λ is an eigenvalue with $(n + 1)^{-1} < \lambda \le n^{-1}$, then dim $\mathcal{N}(W - \lambda I) \le n$.

12. Prove Inequalities (6.7.2) and (6.7.4). In Theorem 6.7.3 estimate

$$\|[(\lambda + \mu)I - T]^{-1}\|$$

in terms of $|\mu|$ and $\|(\lambda I - T)^{-1}\|$.

13. Consider the Volterra integral operator K given by

$$y(t) = \int_0^t k(t,\tau)x(\tau)\, d\tau,$$

on $(C[0,T], \|\cdot\|_\infty)$, where $\|\cdot\|_\infty$ denotes the sup-norm. Assume that $k(t,\tau)$ is continuous for $0 \le \tau \le t \le 1$ and satisfies $|k(t,\tau)| \le M$.
(a) Show that if $x \in C[0,T]$, then $|Kx(t)| \le Mt\,\|x\|_\infty$.
(b) Show that

$$|K^n x(t)| \le (Mt^n/n!)\,\|x\|_\infty.$$

(c) Show that

$$\left\|\left(\frac{1}{\lambda}K\right)^n\right\| \le M\left(\frac{T}{\lambda}\right)^n \frac{1}{n!}.$$

(d) Show that $(\lambda I - K)^{-1} = \sum_{n=0}^\infty \lambda^{-n-1}K^n$, for all $\lambda \ne 0$. That is, the Neuman series converges for all $\lambda \ne 0$.
(e) Show that $\sigma(K) = \{0\}$.

14. Consider the Volterra integral operator K in Exercise 13 but now on the space $L_2[0,T]$. Show that $\sigma(K) = \{0\}$.

15. (Perturbation Theory). Suppose we wish to solve the equation

$$(I - K)x = y$$

for a given y in a Banach space X. Assume that $(I - K)^{-1}$ exists and is continuous. Assume further that there is another operator K_0 near K and that the equation

$$(I - K_0)x_0 = y$$

is easier to solve. In this exercise you are asked to show that the "approximate" solution x_0 is close to x in the following precise sense:
(a) Assume that $\|K - K_0\| \le \delta$ and $\delta\,\|(I - K)^{-1}\| < 1$. Show that $(I - K_0)$ has a continuous inverse and that

$$\|x - x_0\| \le \delta\|(I - K_0)^{-1}\|\,\|y\|.$$

[Hint: Show that $x - x_0 = (I - K_0)^{-1}(K - K_0)y$.]
(b) Assume that $\delta\,\|(I - K_0)^{-1}\| \le r < 1$. Show that

$$\|x - x_0\| \le \frac{r}{1 - r}\|x_0\|.$$

16. (a) Use Exercise 15 to find an approximate solution to the problem

$$x(t) = \int_0^{0.5} k(t, \tau)x(\tau) \, d\tau + f(t),$$

where $k(t,\tau) = \sin(t\tau)$ and $f(t) = t^{-1}(\cos t/2 - 1) + 1$. Use the integral equation

$$x_0(t) = \int_0^{0.5} k_0(t,\tau)x_0(\tau) \, d\tau + f(t),$$

where $k_0(t,\tau) = t\tau$. [*Hint*: Show that

$$x_0(t) = Ct + f(t)$$

for an appropriate choice of *C*.]

(b) Show that $\|x - x_0\| \le 0.002$, where the norm is the sup-norm.
(c) Use the above information to guess the solution $x(t)$.

Part C

Spectral Analysis

8. RESOLUTIONS OF THE IDENTITY

In the first part of this chapter we studied bounded linear operators that could be expressed as a finite linear combination of projections. The projections formed a "resolution of the identity." The study of more general operators will require an expanded version of this concept. That is, we need a concept of a resolution of the identity which will allow us to treat infinite collections of orthogonal projections instead of only finite collections. This, however, raises a very important convergence problem, which we have seen once before, see Section 5.8. Let us illustrate this problem by means of a simple example.

EXAMPLE 1.   Consider the Hilbert space $l_2 = l_2(0,\infty)$ and let $\{e_n\}$ be the complete orthonormal basis for l_2 defined by $e_n = (\delta_{1n}, \delta_{2n}, \dots)$, that is, $e_1 = (1, 0, \dots)$, and so on.

Define $P_n: l_2 \to l_2$ by

$$P_n x = (x, e_n)e_n, \qquad n = 1, 2, 3, \dots.$$

Each P_n is, of course, an orthogonal projection on l_2. Moreover, we have that

$$P_i P_j = 0 \qquad \text{for} \quad i \neq j.$$

And if x is any *fixed* element in l_2, then by the Fourier Series Theorem one has

$$Ix = x = \sum_{n=1}^{\infty} (x, e_n)e_n = \sum_{n=1}^{\infty} P_n x$$

$$= P_1 x + P_2 x + \cdots,$$

where the convergence is in terms of the norm on l_2. It may appear, then, that one can write the identity operator as

$$I = P_1 + P_2 + \cdots = \sum_{n=1}^{\infty} P_n. \tag{6.8.1}$$

However, this is an example of a series that converges strongly but not uniformly (see Section 5.8). That is, if we set

$$S_N = \sum_{n=1}^{N} P_n, \qquad N = 1, 2, \dots,$$

then $I = \underset{N \to \infty}{s\,\lim}\, S_N$, but $\|S_N - I\| = 1$. In order to prove that $\|S_N - I\| = 1$, we use the

Fourier Series Theorem to note that

$$\|(I - S_N)x\|^2 = \left\| \sum_{n=N}^{\infty} (x, e_n)e_n \right\| = \sum_{n=N}^{\infty} |(x, e_n)|^2 \leq \sum_{n=1}^{\infty} |(x, e_n)|^2 = \|x\|^2.$$

Hence, we see that $\|I - S_N\| \leq 1$. However,

$$\|(I - S_N)e_{N+1}\| = \|(0,\ldots,0,1,0,\ldots)\| = 1.$$

Hence $\|I - S_N\| = 1$ for *every* N. Therefore, it is *not* true that $\{S_N\}$ converges to I in terms of the uniform topology.

We see then that (6.8.1) is not correct but that one does have

$$I = {}_s\sum_{n=1}^{\infty} P_n. \quad \blacksquare \tag{6.8.2}$$

We are now ready to define a resolution of the identity.

6.8.1 DEFINITION.   A sequence of operators $\{P_n\}$ on a Hilbert space H is said to be a *resolution of the identity* if it is true that (i) each P_n is an orthogonal projection, (ii) $P_n P_m = 0$ if $n \neq m$, and (iii)

$$I = {}_s\sum_n P_n.$$

We do not rule out the possibility that $\{P_n\}$ may be finite. Therefore, Definition 6.8.1 includes Definition 6.2.1 as a special case. The above definition is not the most general definition of a resolution of the identity. In particular, when one considers noncompact linear operators, resolutions of the identity involving uncountable sets of projections are needed. However, for present purposes this definition is general enough.

Let us now show the relation between a resolution of the identity $\{P_n\}$ on a Hilbert space H and the corresponding subspaces $\{\mathcal{R}(P_n)\}$ of H.

6.8.2 THEOREM.   (a) *Let $\{P_n\}$ be a resolution of the identity on a Hilbert space H and $\mathcal{R}(P_n)$ be the range of P_n. Then $\mathcal{R}(P_n) \perp \mathcal{R}(P_m)$ for $n \neq m$ and*

$$H = \sum_n \mathcal{R}(P_n).$$

(b) *Conversely, let $\{R_n\}$ be a collection of closed linear subspaces of a Hilbert space H with $R_n \perp R_m$ for $m \neq n$ and such that $H = \sum_n R_n$. Let P_n be the orthogonal projection of H onto R_n. Then $\{P_n\}$ is a resolution of the identity on H.*

Proof: (a) Let $x \in \mathcal{R}(P_n)$ and $y \in \mathcal{R}(P_m)$, where $n \neq m$. Then since P_m is self-adjoint and $P_m P_n = 0$ one has

$$(x,y) = (P_n x, P_m y) = (P_m P_n x, y) = (0, y) = 0.$$

Hence $\mathscr{R}(P_n) \perp \mathscr{R}(P_m)$ for $n \neq m$. Next let M be the closed linear subspace given by

$$M = \mathscr{R}(P_1) + \mathscr{R}(P_2) + \cdots = \left\{ x \in H : x = \sum_{n=1}^{\infty} x_n, \, x_n \in \mathscr{R}(P_n) \right\},$$

where we used the Orthogonal Structure Theorem 5.20.2. If $x \in M^{\perp}$, then

$$x = Ix = P_1 x + P_2 x + \cdots = 0,$$

hence $M^{\perp} = \{0\}$. It follows from Theorem 5.15.4(d) that $M = H$.

(b) First, it is clear that $P_m P_n = 0$ if $n \neq m$. Secondly, it follows from Orthogonal Structure Theorem 5.20.2 that every $x \in H$ can be written uniquely as $x = x_1 + x_2 + \cdots$, where $x_n \in R_n$. Thus, since $P_n x = x_n$, it also follows that $Ix = P_1 x + P_2 x + \cdots$, that is, $\{P_n\}$ is a resolution of the identity. ∎

EXERCISES

1. Let P_T denote the orthogonal projection of $L_2(-\infty, \infty)$ onto $L_2(-\infty, T]$. Let T_n be a strictly monotone sequence (that is, $T_n > T_{n-1}$) with $T_n \to \infty$ and define Q_n by $Q_0 = P_{T_0}$, $Q_n = P_{T_n} - P_{T_{n-1}}$. Show that $\{Q_0, Q_1, \ldots\}$ is a resolution of the identity on $L_2(-\infty, \infty)$.

2. Let P_n denote the projection onto the nth-coordinate in the Hilbert space $l_2(-\infty, \infty)$, and let

$$Q_n = {}_s \sum_{k=-\infty}^{n} P_k.$$

Let T be a bounded linear operator on $l_2(-\infty, \infty)$. Show that the following statements are equivalent:
(a) T is causal.
(b) For all n one has:

$$Q_n x = Q_n y \Rightarrow Q_n T x = Q_n T y.$$

(c) For all n one has:

$$Q_n x = 0 \Rightarrow Q_n T x = 0.$$

(d) For all n one has $T(\mathscr{N}(Q_n)) \subseteq \mathscr{N}(Q_n)$.
Where is the linearity of T required in the above list?

3. Let $\{e_n\}$ be an orthonormal set in a Hilbert space H and define $P_n : H \to H$ by $P_n x = (x, e_n) e_n$.
(a) Show that P_n is an orthogonal projection and that $P_n P_m = 0$ whenever $n \neq m$.
(b) Show that $\{P_n\}$ is a resolution of the identity if and only if $\{e_n\}$ is a maximal orthonormal set.
(c) Use the Fourier Series Theorem to give other characterizations of the statement that $\{P_n\}$ is a resolution of the identity.

9.   WEIGHTED SUMS OF PROJECTIONS

In the last section we considered resolutions of the identity $\{P_n\}$. In this section we consider a special class of linear operators that can be constructed using resolutions of the identity. In particular, we consider weighted sums of projections.

The reason for the importance of weighted sums of projections is that many linear operators with important applications turn out to be weighted sums of projections in disguise.

6.9.1 DEFINITION.   Let H be a Hilbert space and let $\{P_n\}$ be a resolution of the identity defined on H. Further, let $\{\lambda_n\}$ be a sequence of scalars. A transformation of the form

$$Tx = \sum_{n=1}^{\infty} \lambda_n P_n x, \qquad x \in \mathscr{D}(T), \tag{6.9.1}$$

where

$$\mathscr{D}(T) = \left\{ x \in H \colon \lim_{N \to \infty} \sum_{n=1}^{N} \lambda_n P_n x \quad \text{exists} \right\}$$

is said to be a *weighted sum of projections*.

Note that we do have to be careful here about the domain of T. The obvious reason is that we have placed no constraints on the set $\{\lambda_n\}$. If we let T_N denote the continuous linear operator

$$T_N = \sum_{n=1}^{N} \lambda_n P_n, \qquad N = 1, 2, \ldots,$$

it follows from the definition that T is the strong limit of the sequence $\{T_N\}$. We do not rule out the possibility that T may be unbounded.

6.9.2 LEMMA.   *A weighted sum of projections is linear.*

Proof: Let x_1 and x_2 be any two points in $\mathscr{D}(T)$, and let

$$y_1 = \lim_{N \to \infty} \sum_{n=1}^{N} \lambda_n P_n x_1, \qquad y_2 = \lim_{N \to \infty} \sum_{n=1}^{N} \lambda_n P_n x_2.$$

We want to show that for any scalars α_1 and α_2, the limit

$$\lim_{N \to \infty} \sum_{n=1}^{N} \lambda_n P_n(\alpha_1 x_1 + \alpha_2 x_2)$$

exists and is equal to $\alpha_1 y_1 + \alpha_2 y_2$. But

$$\left\| \alpha_1 y_1 + \alpha_2 y_2 - \sum_{n=1}^{N} \lambda_n P_n(\alpha_1 x_1 + \alpha_2 x_2) \right\| \leq |\alpha_1| \left\| y_1 - \sum_{n=1}^{N} \lambda_n P_n x_1 \right\|$$

$$+ |\alpha_2| \left\| y_2 - \sum_{n=1}^{N} \lambda_n P_n x_2 \right\|.$$

Completion of the proof is now left to the reader.   ∎

According to our definition of resolution of the identity, it is possible that $P_n = 0$ for certain n. In order to avoid having to handle this special, yet trivial case, we assume here that only nontrivial projections occur in our resolutions of the identity. Also let us assume that H is infinite dimensional. Further, let us assume that $\{P_n\}$ is infinite. Otherwise the problems we consider next are, by and large, trivial.

The next lemma provides a simple characterization of the situation where the domain of T is all of H.

6.9.3 LEMMA. $\mathscr{D}(T) = H$ *if and only if the set* $\{|\lambda_1|, |\lambda_2|, \ldots\}$ *is bounded.*

Proof: Suppose first that the set $\{|\lambda_1|, |\lambda_2|, \ldots\}$ is bounded; that is, there exists a real number $M \geq 0$ such that $|\lambda_n| \leq M$ for all n. Let us show that

$$y_N = \sum_{n=1}^{N} \lambda_n P_n x$$

is a Cauchy sequence for arbitrary $x \in H$. Since $\{P_n\}$ is a resolution of the identity one has $x = \sum_n P_n x$ and $\|x\|^2 = \sum_n \|P_n x\|^2$ by the Orthogonal Structure Theorem. Hence the sequence of partial sums $z_N = \sum_{n=1}^{N} \|P_n x\|^2$ is a Cauchy sequence. Since

$$\|y_N - y_M\|^2 = \left\| \sum_{M+1}^{N} \lambda_n P_n x \right\|^2 = \sum_{M+1}^{N} |\lambda_n|^2 \|P_n x\|^2 \leq M^2 \|z_N - z_M\|,$$

it follows that $\{y_N\}$ is a Cauchy sequence and therefore convergent, since H is complete. It follows then that $\mathscr{D}(T) = H$.

Now assume that $\mathscr{D}(T) = H$. We must show that the set $\{|\lambda_1|, |\lambda_2|, \ldots\}$ is bounded. We argue by contradiction, that is, we show that if $\{|\lambda_1|, |\lambda_2|, \ldots\}$ is not bounded, then $\mathscr{D}(T) \neq H$.

If $\{|\lambda_n|\}$ is not bounded, there exists a subsequence $\{|\lambda_{n_1}|, |\lambda_{n_2}|, |\lambda_{n_3}|, \ldots\}$ such that $|\lambda_{n_k}| \geq k$. Corresponding to each λ_{n_k} is the projection P_{n_k} with nontrivial range $\mathscr{R}(P_{n_k})$. Let $x_{n_k} \in \mathscr{R}(P_{n_k})$ and $\|x_{n_k}\| = 1$. It is easily shown that x given by

$$x = \sum_k \frac{1}{\lambda_{n_k}} x_{n_k}$$

is a point in H. On the other hand, the series

$$\sum_n \lambda_n P_n x = \sum_k x_{n_k}$$

obviously does not converge since it is the sum of mutually orthogonal unit vectors. Therefore, x is not in $\mathscr{D}(T)$ and $\mathscr{D}(T) \neq H$.   ∎

We can show that $\mathscr{D}(T)$ is dense in H no matter what the set $\{\lambda_n\}$ is.

6.9.4 LEMMA. $\overline{\mathscr{D}(T)} = H$.

Proof: Let x be an arbitrary point in H and let ε be any positive number. We must show that there exists an $x_\varepsilon \in \mathcal{D}(T)$ such that $\|x - x_\varepsilon\| < \varepsilon$. Thus, since $\mathcal{R}(P_1) + \mathcal{R}(P_2) + \cdots = H$ and these ranges are mutually orthogonal, we can write x uniquely as

$$x = x_1 + x_2 + \cdots,$$

where $x_i \in \mathcal{R}(P_i)$, $i = 1, 2, \ldots$. Furthermore, there exists an integer N such that

$$\|x - x_1 - x_2 - \cdots - x_N\| < \varepsilon.$$

But $(x_1 + x_2 + \cdots + x_N) \in \mathcal{D}(T)$. So $x_\varepsilon = x_1 + \cdots + x_N$, will do. ∎

We can also easily state necessary and sufficient conditions for the continuity of T.

6.9.5 LEMMA. *T is continuous if and only if the set $\{|\lambda_1|, |\lambda_2|, \ldots\}$ is bounded. Moreover, $\|T\| = \sup\{|\lambda_1|, |\lambda_2|, \ldots\}$.*

Proof: Suppose $\{|\lambda_1|, |\lambda_2|, \ldots\}$ is bounded and $M \geq 0$ is a bound. Let x be any point in H. Then by the Orthogonal Structure Theorem one has

$$\|Tx\|^2 = \left\| \lim_{N \to \infty} \sum_{n=1}^{N} \lambda_n P_n x \right\|^2 \tag{6.9.2}$$

$$= \lim_{N \to \infty} \sum_{n=1}^{N} |\lambda_n|^2 \|P_n x\|^2$$

$$\leq M^2 \lim_{N \to \infty} \sum_{n=1}^{N} \|P_n x\|^2 = M^2 \|x\|^2.$$

So T is continuous and $\|T\| \leq M$.

Next suppose that T is continuous. Then

$$\|Tx\| \leq \|T\| \, \|x\|$$

for all $x \in H$. Let x_n be a point in $\mathcal{R}(P_n)$. It follows that $Tx_n = \lambda_n x_n$ and $|\lambda_n| \leq \|T\|$ for all n. Thus $\{|\lambda_n|\}$ is bounded by $\|T\|$. Finally let $\alpha = \sup_n \{|\lambda_n|\}$. We see that $\alpha \leq \|T\|$. But Inequality (6.9.2) also holds for $M = \alpha$, so $\|T\| \leq \alpha$. ∎

Putting the preceding lemmas together we have the following theorem.

6.9.6 THEOREM. *A weighted sum of projections T is continuous if and only if $\mathcal{D}(T) = H$.*

The spectrum of a weighted sum of projections has some particularly interesting properties. It is easily seen that the operator $(\lambda I - T)$ is defined by

$$(\lambda I - T)x = \sum_{n=1}^{\infty} (\lambda - \lambda_n)P_n x \qquad x \in \mathcal{D}(T).$$

Thus $\lambda I - T$ is itself a weighted sum of projections.

6.9.7 LEMMA. $\lambda I - T$ is one-to-one if and only if $\lambda \neq \lambda_n$ for all n.

Proof: Let $x \in \mathscr{D}(T)$ be a point such that $(\lambda I - T)x = 0$. Let $x = x_1 + x_2 + \cdots$, where $x_n \in \mathscr{R}(P_n)$. Then

$$(\lambda I - T)x = \sum_n (\lambda - \lambda_n)x_n = 0.$$

If $\lambda \neq \lambda_n$, for all n, then $x_1 = x_2 = \cdots = 0$. So $(\lambda I - T)$ is one-to-one. On the other hand, if $\lambda = \lambda_n$, for some n, then $(\lambda_n I - T)x_n = 0$ for $\|x_n\| \neq 0$, so $(\lambda_n I - T)$ is not one-to-one.   ∎

It follows immediately, of course, that the point spectrum of T is the set $\{\lambda_1, \lambda_2, \ldots\}$.

Next let us investigate the range of $\lambda I - T$.

6.9.8 LEMMA.   *The range of $(\lambda I - T)$ is dense in H if and only if $\lambda \neq \lambda_n$, $n = 1, 2, \ldots$.*

Proof: If $\lambda = \lambda_n$ for some n, it is clear that the range of $\lambda I - T$ is orthogonal to $\mathscr{R}(P_n)$. Since $P_n \neq 0$, $\mathscr{R}(P_n)$ is a nontrivial closed subspace and $\mathscr{R}(\lambda I - T)$ is not dense in H.

Next suppose that $\lambda \neq \lambda_n$, for all n, and let y be any point in H. Given any $\varepsilon > 0$, we must find an x_ε in $\mathscr{D}(\lambda I - T)$ such that $\|y - y_\varepsilon\| < \varepsilon$, where $y_\varepsilon = (\lambda I - T)x_\varepsilon$. Let

$$y = y_1 + y_2 + \cdots,$$

where $y_n \in \mathscr{R}(P_n)$, $n = 1, 2, \ldots$. Since $y \in H$, there is an integer N such that

$$\|y - y_1 - y_2 - \cdots - y_N\| < \varepsilon.$$

We let $x_\varepsilon = \sum_{n=1}^{N} (\lambda - \lambda_n)^{-1} y_n$. Then x_ε is in $\mathscr{D}(\lambda I - T)$ and

$$y_\varepsilon = y_1 + \cdots + y_N = (\lambda I - T)x_\varepsilon.$$ ∎

Lemmas 6.9.7 and 6.9.8 show us that if λ is not in the point spectrum, it is either in the continuous spectrum or the resolvent set of T, that is, the residual spectrum is always empty.

Next let us find out when the range of $(\lambda I - T)$ is all of H.

6.9.9 LEMMA. $\mathscr{R}(\lambda I - T) = H$ if and only if the set $\{|\lambda - \lambda_1|, |\lambda - \lambda_2|, \ldots\}$ is *bounded away from zero, that is, there exists a $\delta > 0$ such that $|\lambda - \lambda_n| \geq \delta > 0$ for all n.*

Proof: Suppose that $\{|\lambda - \lambda_n|\}$ is bounded away from zero and y is any point in H. Let $y = y_1 + y_2 + \cdots$, where $y_n \in \mathscr{R}(P_n)$. We assert that

$$x = \sum_n (\lambda - \lambda_n)^{-1} y_n$$

is a preimage of y. Moreover, the inverse of $(\lambda I - T)$ in this case can be represented by

$$(\lambda I - T)^{-1} y = \sum_n (\lambda - \lambda_n)^{-1} P_n y,$$

that is, $(\lambda I - T)^{-1}$ is a weighted sum of projections. We leave these details as an exercise.

Now let us show that $\mathcal{R}(\lambda I - T) = H$ implies that $\{|\lambda - \lambda_n|\}$ is bounded away from 0. Let x be any point in $\mathcal{D}(\lambda I - T)$ and $y = (\lambda I - T)x$. By definition

$$y = \lim_{N \to \infty} \sum_{n=1}^{N} (\lambda - \lambda_n) P_n x.$$

Since λ is not (Lemma 6.9.7) an eigenvalue, $(\lambda - \lambda_n)^{-1} P_n$ is continuous for each n. Hence

$$(\lambda - \lambda_k)^{-1} P_k y = \lim_{N \to \infty} (\lambda - \lambda_k)^{-1} P_k \sum_{n=1}^{N} (\lambda - \lambda_n) P_n x.$$

From the definition of resolution of the identity the right-hand side reduces to $P_k x$. Thus

$$(\lambda - \lambda_k)^{-1} P_k y = P_k x.$$

It follows that $y \in \mathcal{R}(\lambda I - T)$ implies that infinite sum

$$\sum_k (\lambda - \lambda_k)^{-1} P_k y = \sum_k P_k x \qquad (6.9.3)$$

exists.

Now if $\{|\lambda - \lambda_n|\}$ is not bounded away from zero, it is easy to pick a y in H for which the sum in (6.9.3) does not exist. That is, $\{|\lambda - \lambda_n|\}$ not bounded away from zero implies that $\mathcal{R}(\lambda I - T) = H$, or equivalently, $\mathcal{R}(\lambda I - T) = H$ implies $\{|\lambda - \lambda_n|\}$ is bounded away from zero. ∎

In any event, if $\lambda \neq \lambda_n$, for all n, the operator $(\lambda I - T)$ has an inverse defined on its range.

6.9.10 LEMMA. *If $\lambda \neq \lambda_n$ for all n, $(\lambda I - T)$ has an inverse defined on its range and*

$$(\lambda I - T)^{-1} y = \sum_n (\lambda - \lambda_n)^{-1} P_n y \qquad (6.9.4)$$

for $y \in \mathcal{R}(\lambda I - T)$. Moreover, $(\lambda I - T)^{-1}$ is continuous if and only if $\{|\lambda - \lambda_n|\}$ is bounded away from zero.

The proof of this lemma is an obvious combination of the preceding lemmas.

We see, then, that λ is in the resolvent set of T if and only if $\{|\lambda - \lambda_n|\}$ is bounded away from zero. Further, λ is in the continuous spectrum of T if and only if (i) $\lambda \neq \lambda_n$ for all n and (ii) $\{|\lambda - \lambda_n|\}$ is not bounded away from zero. In other words, λ is a point of accumulation of the set $\{\lambda_n\}$ but λ is not in $\{\lambda_n\}$.

So far we have considered both bounded and unbounded weighted sums of projections. For the remainder of this section we restrict our attention to bounded weighted sums of projections.

6.9.11 THEOREM.   *If*

$$T = {}_s\sum_n \lambda_n P_n,$$

where $\{P_n\}$ is a resolution of the identity, and if T is bounded, then the adjoint of T is given by

$$T^* = {}_s\sum_n \bar{\lambda}_n P_n, \tag{6.9.5}$$

where $\bar{\lambda}_n$ denotes the complex conjugate of λ_n.

Proof: Let x and y be any points in H, and consider

$$(y, Tx) = \left(y, \lim_{N \to \infty} \sum_{n=1}^{N} \lambda_n P_n x\right).$$

Since the inner product is continuous, we have

$$(y, Tx) = \lim_{N \to \infty} \{(y, \lambda_1 P_1 x) + \cdots + (y, \lambda_N P_N x)\}.$$

But P_n is self-adjoint, so $(y, \lambda_n P_n x) = (\bar{\lambda}_n P_n y, x)$ for all n; therefore,

$$(y, Tx) = \lim_{N \to \infty} \{(\bar{\lambda}_1 P_1 y, x) + \cdots + (\bar{\lambda}_N P_N y, x)\}.$$

Again using the continuity of the inner product, we have

$$\left(\lim_{N \to \infty} \sum_{n=1}^{N} \bar{\lambda}_n P_n y, x\right) = \lim_{N \to \infty} \{(\bar{\lambda}_1 P_1 y_1, x) + \cdots + (\bar{\lambda}_N P_N y, x)\}.$$

Finally, then

$$(y, Tx) = (T^* y, x)$$

for all $x, y \in H$, where T^* is given by (6.9.5).   ∎

The following result is an obvious consequence of the preceding one.

6.9.12 COROLLARY.   *A bounded linear operator T that is the weighted sum of projections,*

$$T = \sum_{n=1}^{\infty} \lambda_n P_n,$$

is self-adjoint if and only if all the λ's are real.

If some or all of the λ's are nonreal, T is still normal. Recall that a normal operator is one that commutes with its adjoint, that is, $TT^* = T^*T$.

6.9.13 THEOREM.   *A bounded linear operator T that is the weighted sum of projections is normal.*

We leave the proof of this theorem to the reader.

This is an important result for it shows that the only operators that we can hope to express as weighted sums of projections are normal ones.

The bulk of the remainder of this chapter will be concerned with compact operators. Let us see when weighted sums of projections are compact.

6.9.14 THEOREM.   *A weighted sum of projections is compact if* (i) *for every nonzero λ_n the range of P_n, $\mathscr{R}(P_n)$, is finite dimensional and* (ii) *for every real number $\alpha > 0$ the number of λ_n's with $|\lambda_n| \geq \alpha$ is finite.*

Proof: Suppose that T is a weighted sum of projections satisfying (i) and (ii). We must show that T is compact. Recall that a compact transformation maps bounded sets into compact sets. Let T_N be

$$T_N = \sum_{n=1}^{N} \lambda_n P_n,$$

where $\{\lambda_1,\dots,\lambda_N\}$ are all the λ's such that $|\lambda_n| \geq \varepsilon$, where $\varepsilon > 0$. By our hypotheses N is finite and the range of T_N is finite dimensional. Moreover, since

$$\|(T - T_N)x\|^2 = \left\| \sum_{n=N+1}^{\infty} \lambda_n P_n x \right\|^2 \leq \sup_{n \geq N+1} |\lambda_n|^2 \sum_n \|P_n x\|^2 \leq \varepsilon^2 \|x\|^2,$$

one has $\|T - T_N\| \leq \varepsilon$. It follows now from Theorems 5.24.3 and 5.24.8 that T is compact.   ∎

The converse of this theorem is also true. That is, if a weighted sum of projections is compact, then (i) and (ii) follow. We will not prove this here because we plan to consider a more general case in a later section.

The rather significant result that we will obtain shortly is that *every* compact normal operator is a weighted sum of projections.

EXERCISES

1. Let $T = \sum_n \lambda_n P_n$ be a continuous weighted sum of projections on a Hilbert space H. Show that there exists an orthonormal basis of eigenvectors $\{x_1, x_2, \dots\}$ for H. Let $\{\mu_1, \mu_2, \dots\}$ be the corresponding eigenvalues.
   (a) How are the μ_i's related to the λ_n's?
   (b) Show that

   $$Tx = \sum_n \mu_n(x, x_n)x_n$$

   for all $x \in H$.
   (c) What happens if T is not continuous?

2. Consider the operator

$$\Phi: (x_1, x_2, \ldots) \to (\phi(1)x_1, \phi(2)x_2, \ldots)$$

on $l_2(0, \infty)$.

(a) Show that Φ is a weighted sum of projections.

(b) What is the spectrum of Φ?

(c) Assume that $\phi(n) \neq 0$ for all n. Show that Φ^{-1} exists and that Φ^{-1} is a weighted sum of projections. What is the spectrum of Φ^{-1}?

(d) Assume further that $|\phi(n)| \to \infty$ as $n \to \infty$. Show that Φ^{-1} is compact.

(e) Assume that $\phi(n) \to \lambda_0$ as $n \to \infty$, where λ_0 is finite. Show that $(\Phi - \lambda_0 I)$ is compact.

3. (Continuation of Exercise 2.) Let $L = T + \Phi = S_r + S_l + \Phi$. (See Exercises 13 and 14 of Section 6.) Assume that $|\phi(n)| \to +\infty$ as $n \to \infty$. Show that L^{-1} exists and is compact. [*Hint*: Use Exercise 10, Section 7 with $S = \Phi + \lambda I$, for an appropriate choice of λ.]

10. SPECTRAL PROPERTIES OF COMPACT, NORMAL, AND SELF-ADJOINT OPERATORS

In this section we first investigate the spectral properties of compact operators. Then we investigate the spectral properties of self-adjoint and normal operators. In the next section we will combine the results of this and the previous section to get the Spectral Theorem.

A. Compact Operators

The following theorems state the spectral properties of compact linear operators which we will need later.

6.10.1 THEOREM. *Let T be a compact linear transformation of a Hilbert space H into itself and let $\lambda \neq 0$. Then the null space $\mathcal{N}(\lambda I - T)$ is finite dimensional.*

Proof: The compact operator T maps $\mathcal{N}(\lambda I - T)$ into $\mathcal{N}(\lambda I - T)$. Moreover, the restriction of T to $\mathcal{N}(\lambda I - T)$ is λI. The restriction of a compact operator is a compact operator; therefore, λI is compact. It follows from Theorem 5.10.7 (or Exercise 11, Section 5.24) that $\mathcal{N}(\lambda I - T)$ is finite dimensional. ∎

6.10.2 THEOREM. *Let T be a compact linear transformation of a Hilbert space H into itself and let $\lambda \neq 0$. Then λ is either an eigenvalue of T or λ is in the resolvent set $\rho(T)$. [That is $\lambda \neq 0$ is never in the continuous spectrum $C\sigma(T)$ or the residual spectrum $R\sigma(T)$.]*

Proof[5]: Choose λ with $\lambda \neq 0$. Suppose $\lambda \in \sigma(T)$. First let us show that λ cannot be in the continuous spectrum. We shall do this by assuming that $(\lambda I - T)$ is

---

[5] This proof is long and technical and the reader may wish to skip it on his first reading.

one-to-one and then show that $\lambda I - T$ is bounded below, that is, there exists a constant $m > 0$ such that $\|(\lambda I - T)x\| \geq m \|x\|$ for all x. [This shows that any time $(\lambda I - T)$ has an inverse, this inverse is continuous, and the scalar λ cannot be in the continuous spectrum.]

We argue by contradiction. Suppose there is a sequence of unit vectors $\{x_n\}$ such that $\|\lambda x_n - T x_n\| \to 0$ as $n \to \infty$. Since T is compact, $\{T x_n\}$ contains a convergent subsequence, which we shall also denote by $\{T x_n\}$. Let $z = \lim_{n \to \infty} T x_n$.

Since

$$z - \lambda x_n = (z - T x_n) + (T x_n - \lambda x_n)$$

we have

$$\|z - \lambda x_n\| \leq \|z - T x_n\| + \|T x_n - \lambda x_n\|.$$

But both sequences on the right converge to zero. Hence

$$z = \lim_{n \to \infty} \lambda x_n,$$

or, using the fact that $\lambda \neq 0$, one has

$$\frac{1}{\lambda} z = \lim_{n \to \infty} x_n.$$

Since, $\|x_n\| = 1$ we have $\|z\| = |\lambda|$, thus $z \neq 0$. Since T is continuous one has

$$T\left(\frac{1}{\lambda} z\right) = \lim_{n \to \infty} T(x_n) = z.$$

In other words, z is an eigenvector of T. But this is a contradiction, for we have assumed $(\lambda I - T)$ is one-to-one. Hence we have shown that there does exist a $m > 0$ such that $\|(\lambda I - T)x\| \geq m\|x\|$ for all x, and $(\lambda I - T)^{-1}$ must be continuous. [*Note*: $\lambda \neq 0$ was important.]

Next let us show that λ is not in the residual spectrum of T. Recall that λ is in the residual spectrum of T if $(\lambda I - T)$ is one-to-one and the range of $(\lambda I - T)$ is not dense in H. We will again argue by contradiction. We will suppose that $(\lambda I - T)$ is one-to-one and $\overline{\mathscr{R}(\lambda I - T)} \neq H$. Let $X_0 = H$, $X_1 = (\lambda I - T)X_0$, $X_2 = (\lambda I - T)X_1$, and $X_{n+1} = (\lambda I - T)X_n$. It can be seen that $X_0 \supset X_1 \supset X_2 \supset X_3 \supset \cdots$. The rest of this proof depends on the fact that $X_1 \neq X_0$ implies that X_{n+1} is a proper closed linear subspace of X_n for all n. For the moment let us assume that this has been shown. It follows, then, that there is an $x \in X_0$ such that $\|x_0\| = 1$ and $x_0 \perp X_1$, by Corollary 5.14.5. Furthermore, there is an $x_1 \in X_1$ such that $\|x_1\| = 1$ and $x_1 \perp x_2$. In fact, there is an $x_n \in X_n$ such that $\|x_n\| = 1$ and $x_n \perp X_{n+1}$ for all n. It can be seen that $\{x_n\}$ is an orthonormal sequence. Let $n > m$, then

$$\frac{1}{\lambda}(Tx_m - Tx_n) = x_m + \left\{-x_n - \left[\frac{(\lambda I - T)x_m - (\lambda I - T)x_n}{\lambda}\right]\right\}.$$

But the term

$$\left\{ -x_n - \left[\frac{(\lambda I - T)x_m - (\lambda I - T)x_n}{\lambda} \right] \right\}$$

is a point in X_{m+1}, call it $-x$; therefore,

$$\frac{1}{\lambda}(Tx_m - Tx_n) = x_m - x.$$

Since $\|x_m\| = 1$ and $x_m \perp X_{m+1}$, one has

$$\|Tx_m - Tx_n\| \geq |\lambda|,$$

which shows that the sequence $\{Tx_n\}$ cannot contain a convergent subsequence. This contradicts the assumption that T is compact. Hence, $\overline{\mathscr{R}(\lambda I - T)} = H$ and λ is not in the residual spectrum of T.

We are not finished yet with the proof. We still have to show $X_1 \neq X_0$ implies that X_{n+1} is a proper closed linear subspace of X_n for all n. First, let us show that $\mathscr{R}(\lambda I - T)$ is closed for all $\lambda \neq 0$.

6.10.3 LEMMA.   *The range of* $(\lambda I - T)$ *is a closed linear subspace of* H *for all* $\lambda \neq 0$.

Proof: Let $\{y_n\}$ be any convergent sequence in $\mathscr{R}(\lambda I - T)$, and let $y_0 = \lim_{n \to \infty} y_n$. We want to show that $y_0 \in \mathscr{R}(\lambda I - T)$. Since $\{y_n\} \in \mathscr{R}(\lambda I - T)$, there is at least one sequence $\{x_n\}$ in H such that $(\lambda I - T)x_n = y_n$ for all n. Let us show that the sequence $\{x_n\}$ is bounded. Since $\lambda I - T$ is continuous, its null space $\mathscr{N}(\lambda I - T)$ is closed. Then $H = \mathscr{N}(\lambda I - T) + \mathscr{N}(\lambda I - T)^{\perp}$. With no loss in generality we can assume that $\{x_n\}$ is in $\mathscr{N}(\lambda I - T)^{\perp}$. (Why?) Now $(\lambda I - T)$ restricted to the closed subspace $\mathscr{N}(\lambda I - T)^{\perp}$ is one-to-one. So, repeating the argument used to prove the first part of Theorem 6.10.2, we know that there exists a constant $m > 0$ such that $\|(\lambda I - T)x\| \geq m\|x\|$ for all $x \in \mathscr{N}(\lambda I - T)^{\perp}$. Since $\{y_n\}$ is convergent, there is a bound $M > 0$ such that $\|y_n\| \leq M$ for all n. Then $M \geq \|(\lambda I - T)x_n\| \geq m\|x_n\|$ or $\|x_n\| \leq M/m$ for all n, showing that $\{x_n\}$ is bounded. Since T is compact, $\{x_n\}$ contains a subsequence, which we denote by $\{x_n\}$, such that $\{Tx_n\}$ is convergent. Then

$$\lambda x_n = y_n + Tx_n. \tag{6.10.1}$$

Since $\{y_n\}$ converges to y_0, both sequences on the right of (6.10.1) are convergent and $\lambda \neq 0$, so $\{x_n\}$ is convergent. Let $x_0 = \lim_{n \to \infty} x_n$. Since $(\lambda I - T)$ is continuous one has

$$(\lambda I - T)\left(\lim_{n \to \infty} x_n \right) = \lim_{n \to \infty}(\lambda I - T)x_n$$

or

$$(\lambda I - T)x_0 = y_0.$$

Thus $y_0 \in \mathscr{R}(\lambda I - T)$ and $\mathscr{R}(\lambda I - T)$ is closed.   ∎

The above lemma shows that the space X_1 constructed in Theorem 6.10.2 is closed. A slight variation on it shows that X_n is closed for all n. Thus we do not have to distinguish between $\overline{X}_n$ and X_n.

We are now ready to finish the proof of Theorem 6.10.2. We want to show that X_{n+1} is a proper closed linear subspace of X_n for all n. We argue by induction. By our hypotheses, X_1 is a proper closed linear subspace of X_0. Assume that X_k is a proper closed linear subspace of X_{k-1} for $1 \leq k \leq n$ and we will now show that this implies that X_{n+1} is a proper closed linear subspace of X_n. In any event, we have that $X_{n+1} \subset X_n$. So if X_{n+1} is not a proper closed linear subspace of X_n, we have $X_{n+1} = X_n$. That is, $(\lambda I - T)X_n = X_n$. Since $(\lambda I - T)$ is one-to-one, we have $X_n = (\lambda I - T)^{-1}X_n = X_{n-1}$ which is a contradiction. Therefore, if $X_1 \neq X_0$, then X_{n+1} is a proper linear subspace of X_n. This completes the proof of Theorem 6.10.2. ∎

The next theorem shows that $\lambda = 0$ is the only possible point of accumulation for the spectrum of a compact operator.

6.10.4 THEOREM.   *Let T be a compact linear transformation of a Hilbert space into itself, and let $\alpha > 0$. Then the number of eigenvalues λ with $|\lambda| \geq \alpha$ is finite.*

Proof: We argue by contradiction. Suppose that there is an $\alpha_0 > 0$ such that the number of eigenvalues λ with $|\lambda| \geq \alpha_0$ is infinite. It follows (Why?) that the spectrum of T must contain at least one nonzero point of accumulation, call it λ_0. So there must be a sequence $\{\lambda_n\}$ of eigenvalues such that $\lim_{n \to \infty} \lambda_n = \lambda_0$. Let x_n be an eigenvector associated with λ_n, $n = 1, 2, \ldots$. The set $\{x_1, x_2, \ldots\}$ is linearly independent (see Exercise 3, Section 5). Let X_n be the finite dimensional and, therefore, closed linear subspace spanned by $\{x_1, x_2, \ldots, x_n\}$. We know from the Riesz Theorem (Theorem 5.5.4) that there is a sequence $\{y_n\}$ with $y_n \in X_n$, $\|y_n\| = 1$, and $\text{dist}(y_n, X_{n-1}) \geq \frac{1}{2}$ $(n = 2, 3, \ldots)$. If $n > m$, then

$$\frac{1}{\lambda_n} T y_n - \frac{1}{\lambda_m} T y_m = y_n + \left(-y_m - \frac{\lambda_n y_n - T y_n}{\lambda_n} + \frac{\lambda_m y_m - T y_m}{\lambda_m} \right)$$

$$= y_n - z,$$

where $z \in X_{n-1}$. (Why?) Therefore,

$$\left\| \frac{1}{\lambda_n} T y_n - \frac{1}{\lambda_m} T y_m \right\| \geq \text{dist}(y_n, X_{n-1}) \geq \frac{1}{2}.$$

But we can use the above inequality together with $\lim_{n \to \infty} \lambda_n = \lambda_0 \neq 0$ to show that the sequence $\{Ty_n\}$ does not contain a convergent subsequence. This contradicts the fact that T is compact. Hence, the assumption about α_0 leads to a contradiction. ∎

The next corollary should be obvious.

6.10.5 COROLLARY.   *Let T be a compact operator on a Hilbert space H. Then the spectrum of T is (at most) countably infinite and $\lambda = 0$ is the only possible point of accumulation.*

As far as the point $\lambda = 0$ is concerned, we cannot say too much. If T is compact, $\lambda = 0$ can be in the resolvent set or any part of the spectrum. However, if $\lambda = 0$ is in the resolvent set, H must be finite dimensional.

B.   Normal and Self-Adjoint Operators

Now let us consider operators that are normal but not necessarily compact. Recall that every self-adjoint operator is normal; therefore, anything that is said about normal operators applies also to self-adjoint operators.

6.10.6 THEOREM.   *Let T be a normal transformation of a Hilbert space H into itself. If $x \in H$ is an eigenvector of T associated with an eigenvalue λ, then x is an eigenvector of T^*, the adjoint of T, associated with an eigenvalue $\bar{\lambda}$. Furthermore,*

$$\mathcal{N}(\lambda I - T) = \mathcal{N}(\bar{\lambda}I - T^*).$$

Proof: From Theorem 5.23.10 we know that T is normal if and only if $\|Tx\| = \|T^*x\|$ for all x. Moreover, if T is normal, then $\lambda I - T$ is normal. Hence, $\|(\lambda I - T)x\| = 0$ if and only if $\|(\bar{\lambda}I - T^*)x\| = 0$.   ∎

6.10.7 THEOREM.   *Let T be a normal operator mapping a Hilbert space H into itself. Then the null spaces $\mathcal{N}(\lambda I - T)$ and $\mathcal{N}(\mu I - T)$ are orthogonal to one another whenever $\lambda \neq \mu$.*

Proof: Let $x \in \mathcal{N}(\lambda I - T)$ and $y \in \mathcal{N}(\mu I - T)$. We want to show that $(x,y) = 0$. By using the last theorem and the fact that $(Tx,y) = (x,T^*y)$ we get $(\lambda x,y) = (x,\bar{\mu}y)$ or $(\lambda - \mu)(x,y) = 0$. Hence $(x,y) = 0$.   ∎

Recall (Corollary 5.22.5) that a closed linear subspace M reduces a bounded linear operator T if and only if M is invariant under T and T^*. We can say more when T is normal.

6.10.8 THEOREM.   *Let T be a normal transformation of a Hilbert space H into itself. Then for each complex number λ the closed linear subspace $\mathcal{N}(\lambda I - T)$ reduces T.*

Proof: Let $M = \mathcal{N}(\lambda I - T)$. Since $(\lambda I - T)$ is continuous, there is no question about M being closed. We have to show that $T(M) \subset M$ and $T(M^\perp) \subset M^\perp$. If λ is not an eigenvalue of T, then $M = \{0\}$ and $M^\perp = H$. In this case, then, the theorem is clearly true. Assume that λ is an eigenvalue of T. Since M is the eigen-manifold associated with λ, we immediately have that $T(M) \subset M$. Let $x \in M$ and $y \in M^\perp$, then $(x, Ty) = (T^*x, y)$. Theorem 6.10.6 assures us that $T^*(M) \subseteq M$, hence we get $(x,Ty) = 0$, for all $x \in M$ and $y \in M^\perp$. Continuing further, this shows that $T(M^\perp) \subset M^\perp$.   ∎

6.10.9 COROLLARY.   *If $\{M_n\}$ is a family of eigenmanifolds of a normal operator T, then $M = M_1 + M_2 + M_3 + \cdots$ reduces T.*

Proof: From Theorem 6.10.7 we know that the M_n's are pairwise orthogonal The rest of the proof should be obvious.   ∎

6.10.10 THEOREM.   *The residual spectrum of a normal operator is empty.*

Proof: Let T be a normal operator mapping a Hilbert space H into itself. We have to show that if $(\lambda I - T)$ is one-to-one, then the range $\mathscr{R}(\lambda I - T)$ is dense in H. Let y be a point in H that is orthogonal to $\mathscr{R}(\lambda I - T)$. That is,

$$(\lambda x - Tx, y) = 0 \qquad \text{for all } x \text{ in } H.$$

Since $(x, \bar{\lambda} y - T^* y) = 0$ for all x in H, it follows that $(\bar{\lambda} I - T^*) y = 0$, that is $y \in \mathscr{N}(\bar{\lambda} I - T^*)$. It now follows from Theorem 6.10.6 that $y = 0$. Therefore, since $R(\lambda I - T)^{\perp} = \{0\}$ we note that $R(\lambda I - T)$ is dense in H, see Theorem 5.15.4(c).   ∎

Needless to say, it also follows that the residual spectrum of a self-adjoint operator is empty.

6.10.11 COROLLARY.   *A complex number λ is in the spectrum of a normal operator T if and only if there exists a sequence $\{x_n\}$, $\|x_n\| = 1$ for all n, such that $\|(\lambda I - T) x_n\| \to 0$ as $n \to \infty$. In other words, the operator $(\lambda I - T)$ is not bounded below.*

The proof of this corollary is left to the reader as an easy but not completely trivial exercise. (Also, see Exercise 1, Section 6.5.)

As anyone familiar with the theory of Hermitian matrices would suspect the spectrum of a self-adjoint operator is confined to the real line.

6.10.12 THEOREM.   *The spectrum of a self-adjoint operator T is a subset of the real interval $[-\|T\|, \|T\|]$.*

Proof: We can use Corollary 6.10.11. Let us show that if λ is not real, then there exists a constant $m > 0$ such that $\|(\lambda I - T) x\| \geq m \|x\|$ for all x. It will follow from Corollary 6.10.11 that λ is in the resolvent set of T.

Assume that $\lambda = \rho + i\sigma$, where $\sigma \neq 0$. Then a simple calculation gives

$$\begin{aligned}
\|(\lambda I - T) x\|^2 &= (\lambda x - Tx, \lambda x - Tx) \\
&= (\rho x - Tx, \rho x - Tx) + (i\sigma x, i\sigma x) \\
&\geq |\sigma|^2 \|x\|^2.
\end{aligned}$$

Hence $\lambda I - T$ is bounded below and λ is in the resolvent set $\rho(T)$. Therefore the spectrum of T is real. It follows now from Theorem 6.7.4 that $\sigma(T)$ lies in the interval $[-\|T\|, \|T\|]$.   ∎

C.   Compact Self-Adjoint Operators

We turn now to a statement about the existence of eigenvalues for compact self-adjoint operators. Before giving this, though, let us recall that the norm of a self-adjoint operator T is given by

$$\|T\| = \sup\{|(Tx, x)| : \|x\| = 1\}. \tag{6.10.2}$$

(See Theorem 5.23.8.)

6.10.13 THEOREM. *Let T be a compact, self-adjoint operator on a nontrivial Hilbert space H. Then T has an eigenvalue λ with $|\lambda| = \|T\|$.*

Proof: It follows from (6.10.2) that there is a sequence $\{x_n\}$ in H with $\|x_n\| = 1$ and $|(Tx_n,x_n)| \to \|T\|$. Since T is compact we can find a subsequence of $\{Tx_n\}$ that converges in H; furthermore, since the sequence of complex numbers $\{(Tx_n,x_n)\}$ lies in a closed bounded set, we can find a subsequence of $\{(Tx_n,x_n)\}$ that converges in the complex plane. By calling this subsequence $\{x_n\}$, one then has

$$(Tx_n,x_n) \to \lambda \qquad \text{and} \qquad Tx_n \to x,$$

where $|\lambda| = \|T\|$ and $x \in H$.

If $\|T\| = 0$, the conclusion of the theorem is trivial. Assume now that $T \neq 0$, which implies that $\lambda \neq 0$. One then has

$$
\begin{aligned}
0 \leq \|Tx_n - \lambda x_n\|^2 &= \|Tx_n\|^2 + \|\lambda x_n\|^2 - \lambda(\overline{Tx_n,x_n}) - \bar{\lambda}(Tx_n,x_n) \\
&\leq (\|T\|^2 + |\lambda|^2)\|x_n\|^2 - \lambda(\overline{Tx_n,x_n}) - \bar{\lambda}(Tx_n,x_n) \\
&= 2|\lambda|^2 - \lambda(\overline{Tx_n,x_n}) - \bar{\lambda}(Tx_n,x_n).
\end{aligned}
$$

Since the right side tends to 0 as $n \to \infty$, we see that

$$Tx_n - \lambda x_n \to 0.$$

Hence $\lambda x_n \to x$, or $x_n \to (1/\lambda)x$. Hence $\|x\| = |\lambda| \neq 0$. Also

$$T\left(\frac{1}{\lambda}x\right) = T(\lim x_n) = \lim Tx_n = x,$$

or $Tx = \lambda x$. ∎

6.10.14 COROLLARY. *Let T be a compact, self-adjoint operator on a Hilbert space H. If T has no eigenvalues, then $H = \{0\}$.*

D.   Compact Normal Operators

We have just seen that a compact self-adjoint operator on a nontrivial Hilbert space has at least one eigenvalue. Our object here is to show that the same conclusion is valid for compact normal operators.

Let T be a normal operator on a Hilbert space H. We know then (by Exercise 13, Section 5.23) that there are commuting self-adjoint operators A and B such that

$$T = A + iB \quad \text{and} \quad T^* = A - iB. \tag{6.10.3}$$

Furthermore, one has (Exercise 14, Section 5.23)

$$\max(\|A\|, \|B\|) \leq \|T\| = \|T^*\| \quad \text{and} \quad \|T\|^2 \leq \|A\|^2 + \|B\|^2.$$

We can use the Cartesian decomposition of T in (6.10.3) to determine whether T is compact.

6.10.15 LEMMA. *Let T be a normal operator on a Hilbert space H and let $T = A + iB$ be the Cartesian decomposition of T. Then T is compact if and only if both A and B are compact. Furthermore, T is compact if and only if T^* is compact.*

Proof: First we note that

$$\|Tx\|^2 = \|Ax\|^2 + \|Bx\|^2$$

for all x in H. Indeed, since $AB = BA$ one has

$$
\begin{aligned}
\|Tx\|^2 &= ((A + iB)x, (A + iB)x) \\
&= (Ax,Ax) + (Ax,iBx) + (iBx,Ax) + (iBx,iBx) \\
&= \|Ax\|^2 - i(Ax,Bx) + i(Bx,Ax) + \|Bx\|^2 \\
&= \|Ax\|^2 - i(BAx,x) + i(ABx,x) + \|Bx\|^2 \\
&= \|Ax\|^2 + \|Bx\|^2.
\end{aligned}
$$

It follows, then, that a sequence $\{Tx_n\}$ is a Cauchy sequence if and only if both the sequences $\{Ax_n\}$ and $\{Bx_n\}$ are Cauchy sequences. (Why?) Hence T is compact if and only if both A and B are compact.

Since $T^* = A - iB$, it follows from the above that T is compact if and only if T^* is compact. ∎

Let us now study the relationships between the eigenvalues of A and B.

Let $\lambda = \alpha + i\beta$ be an eigenvalue for T. We recall (Theorem 6.10.6) that $\bar\lambda$ is then an eigenvalue of T^*. In addition, one can show that α and β are eigenvalues of A and B, respectively. Indeed, if x satisfies $Tx = \lambda x$, then $T^*x = \bar\lambda x$ and

$$Ax = \tfrac{1}{2}(T + T^*)x = \tfrac{1}{2}(\lambda + \bar\lambda)x = \alpha x,$$

$$Bx = \frac{1}{2i}(T - T^*)x = \frac{1}{2i}(\lambda - \bar\lambda)x = \beta x.$$

This also shows that

$$\mathcal{N}(\lambda I - T) = \mathcal{N}(\bar\lambda I - T^*) \subseteq \mathcal{N}(\alpha I - A),$$

and

$$\mathcal{N}(\lambda I - T) = \mathcal{N}(\bar\lambda I - T^*) \subseteq \mathcal{N}(\beta I - B).$$

In order to get further information concerning the eigenvalues of T we have to study the relationship between the eigenspaces $\mathcal{N}(\alpha I - A)$ and $\mathcal{N}(\beta I - B)$. For this purpose let us now assume that T is compact and normal and that α is a nonzero eigenvalue of A. Let $x \in \mathcal{N}(\alpha I - A)$. Then

$$(\alpha I - A)Bx = B(\alpha I - A)x = 0,$$

which shows that B maps $\mathcal{N}(\alpha I - A)$ into itself. That is,

$$B: \mathcal{N}(\alpha I - A) \to \mathcal{N}(\alpha I - A)$$

and B is a compact self-adjoint operator on this subspace. Furthermore, it follows from Theorem 6.10.1 that $\mathcal{N}(\alpha I - A)$ is finite dimensional. Therefore, we can find an orthonormal basis of eigenvectors $\{e_1, e_2, \ldots, e_n\}$ of B in $\mathcal{N}(\alpha I - A)$ such that the mapping B can be represented by a diagonal matrix in terms of this basis

(Theorem 6.4.4). Let $\{\beta_1, \beta_2, \ldots, \beta_n\}$ be the entries in this diagonal matrix. Let us now show that the complex numbers

$$\alpha + i\beta_1, \alpha + i\beta_2, \ldots, \alpha + i\beta_n$$

are eigenvalues for T. Indeed, let e_j be a nonzero vector in $\mathcal{N}(\alpha I - A)$ that satisfies $Be_j = \beta_j e_j$. Hence

$$
\begin{aligned}
Te_j &= (A + iB)e_j = Ae_j + iBe_j \\
&= \alpha e_j + i\beta_j e_j = (\alpha + i\beta_j)e_j,
\end{aligned}
$$

for $j = 1, 2, \ldots, n$.

If we had started instead with a nonzero eigenvalue β of B, then one can show that A is a compact self-adjoint operator that maps $\mathcal{N}(\beta I - B)$ into itself. One can then construct an orthonormal basis for $\mathcal{N}(\beta I - B)$ so that the restriction of A to this subspace can be expressed as a diagonal matrix

$$\operatorname{diag}(\alpha_1, \alpha_2, \ldots, \alpha_m).$$

By the same reasoning used above one can then show that the complex numbers

$$\alpha_1 + i\beta, \alpha_2 + i\beta, \ldots, \alpha_m + i\beta$$

are eigenvalues for T.

It is easy, then, to see that if α is a nonzero eigenvalue of A, then for some β

$$\mathcal{N}(\lambda I - T) \text{ is nonempty,} \qquad \text{where } \lambda = \alpha + i\beta. \tag{6.10.4}$$

Moreover, one has

$$\mathcal{N}(\lambda I - T) = \mathcal{N}(\alpha I - A) \cap \mathcal{N}(\beta I - B). \tag{6.10.5}$$

Similarly if we start with a nonzero eigenvalue β for B, then for some α (6.10.4) and (6.10.5) are valid. Moreover, (6.10.4) and (6.10.5) are valid for every eigenvalue of T.

6.10.16 THEOREM.   *Let T be a compact normal operator on a nontrivial Hilbert space H. Then T has an eigenvalue λ with*

$$\max(\|A\|, \|B\|) \leq |\lambda|,$$

where $T = A + iB$ is the Cartesian decomposition of T, see Figure 6.10.1.

Proof: If $T = 0$, then $A = B = 0$ and $\lambda = 0$ is an eigenvalue satisfying the conclusion of the theorem.

Now assume that $T \neq 0$ and say that

$$\|A\| = \max(\|A\|, \|B\|) > 0.$$

Theorem 6.10.13 assures us that there is an eigenvalue α for A with the property that $|\alpha| = \|A\|$. The above discussion leads to the conclusion that there is a β such that $\lambda = \alpha + i\beta$ is an eigenvalue of T. Finally, we note that $|\lambda| \geq |\alpha| = \max(\|A\|, \|B\|)$. ∎

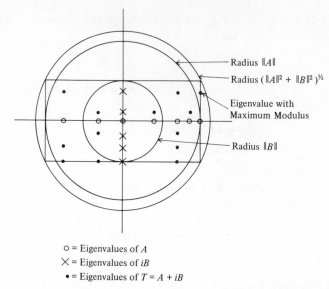

○ = Eigenvalues of A
✕ = Eigenvalues of iB
● = Eigenvalues of $T = A + iB$

Figure 6.10.1. Eigenvalues of $T = A + iB$, T is Compact and Normal.

6.10.17 COROLLARY. *Let T be a compact normal operator on a Hilbert space H. If T has no eigenvalues then $H = \{0\}$.*

Theorem 6.10.16, then, assures us of the existence of at least one eigenvalue for T. If λ_0 is the eigenvalue of T with maximum modulus, that is,

$$|\lambda_0| = \max\{|\lambda| : \lambda \text{ is an eigenvalue for } T\},$$

then one can show that $|\lambda_0| = \|T\|$ (see Exercise 3, Section 11.)

And now for the Spectral Theorem.

EXERCISES

1. Prove Corollary 6.10.11.

2. Show that the spectrum of a unitary operator U lies on the unit circle $\{z : |z| = 1\}$.

3. (a) Construct a compact normal operator $T = A + iB$ with the property that there is at least one eigenvalue λ that satisfies $|\lambda|^2 = \|A\|^2 + \|B\|^2$.
   (b) Construct a compact normal operator $T = A + iB$ with the property that every eigenvalue λ satisfies $|\lambda|^2 < \|A\|^2 + \|B\|^2$. For your example, compute the spectral radius $r_\sigma(T)$ and the norm T.
   (c) Construct a compact normal (nonself-adjoint) operator $T = A + iB$ with the property that every eigenvalue λ satisfies $|\lambda| \leq \max(\|A\|, \|B\|)$.

4. Let A be an observable (that is, a bounded self-adjoint operator) and let p be a state with associated density operator W. Assume that the spectrum of A lies in the interval $[a,b]$. Show that $E(A)$, the expected value of A with respect to the state p, satisfies

$$a \leq E(A) \leq b.$$

11.   THE SPECTRAL THEOREM

The purpose of this section is to prove the following result:

6.11.1 THEOREM. (SPECTRAL THEOREM. FIRST VERSION.)   *Let T be a compact normal operator on a Hilbert space H. Then there is a resolution of the identity $\{P_n\}$ and a sequence of complex numbers $\{\lambda_n\}$ such that*

$$T = \sum_n \lambda_n P_n, \tag{6.11.1}$$

where the convergence in (6.11.1) *is in terms of the uniform operator norm topology.*

The expression (6.11.1) is sometimes called the *spectral decomposition* of T.

Proof: Let $\{\lambda_1, \lambda_2, \ldots\}$ denote the collection of *all* eigenvalues of T. This collection is at most countable by Corollary 6.10.5. Let P_n be the orthogonal projection onto $M_n = \mathcal{N}(\lambda_n I - T)$. Since $M_n \perp M_m$ for $n \neq m$, it follows that $P_n P_m = 0$ for $n \neq m$. Let

$$Q =_s \sum_n P_n.$$

Then Q is the orthogonal projection onto

$$M = M_1 + M_2 + \cdots.$$

We want to show that $Q = I$, or equivalently that $M^\perp = 0$. It follows from Corollary 6.10.9 that $T(M^\perp) \subset M^\perp$. Let S denote the restriction of T to $M^\perp$, that is, $S: M^\perp \to M^\perp$. Then S is compact and normal, and any eigenvalue of S is an eigenvalue of T. However, S has *no* eigenvalues. Therefore, it follows from Corollary 6.10.17 that $M^\perp = \{0\}$.

We have shown that $\{P_n\}$ is a resolution of the identity. Let us now show that $T = \sum_n \lambda_n P_n$. For this it will be convenient to order the eigenvalues so that $|\lambda_1| \geq |\lambda_2| \geq \cdots$. Let

$$S_N = \sum_{n=1}^{N} \lambda_n P_n.$$

Since $\{P_n\}$ is a resolution of the identity, one has

$$H = M_1 + M_2 + \cdots.$$

As a consequence of the Orthogonal Structure Theorem, every vector $x \in H$ can be written uniquely as

$$x = x_1 + x_2 + \cdots = \sum_n x_n,$$

where $x_n \in M_n$, and $\|x\|^2 = \sum_n \|x_n\|^2$. It follows that

$$Tx = \lambda_1 x_1 + \lambda_2 x_2 + \cdots = \sum_n \lambda_n x_n,$$

and

$$(T - S_N)x = \sum_{n=N+1}^{\infty} \lambda_n x_n.$$

Hence

$$\|(T - S_N)x\|^2 = \sum_{n=N+1}^{\infty} |\lambda_n|^2 \|x_n\|^2$$

$$\leq |\lambda_{N+1}|^2 \sum_{n=N+1}^{\infty} \|x_n\|^2 \leq |\lambda_{N+1}|^2 \|x\|^2.$$

Therefore $\|T - S_N\| \leq |\lambda_{N+1}| \to 0$ by Theorem 6.10.4. ∎

Actually some other versions of the Spectral Theorem are more practical in applications. Probably the most useful is the eigenvalue-eigenvector representation.

6.11.2 THEOREM. (SPECTRAL THEOREM. SECOND VERSION.)   *Let T be a compact normal operator on a Hilbert space H. Then there exists a (orthonormal) basis of eigenvectors $\{e_n\}$ and corresponding eigenvalues $\{\mu_n\}$ such that if $x = \sum_n (x,e_n)e_n$ is the Fourier expansion for x, then*

$$Tx = \sum_n \mu_n(x,e_n)e_n. \qquad (6.11.2)$$

Proof: We use the notation of the last theorem. With $M_n = \mathcal{N}(\lambda_n I - T)$, let $\{f_k^{(n)}\}$ be an orthonormal basis[6] for M_n. Then

$$Tf_k^{(n)} = \lambda_n f_k^{(n)}. \qquad (6.11.3)$$

Let $\{f\}$ denote the union of all these $\{f_k^{(n)}\}$. Renumber the collection $\{f\}$ to get the family $\{e_1,e_2,\ldots\}$ and let $\{\mu_1,\mu_2,\ldots\}$ be the corresponding eigenvalues given by (6.11.3). The only thing we have to prove is that the family $\{f\}$, or $\{e_1,e_2,\ldots\}$ is a basis, that is, a maximal orthonormal set, in H.

First this family is orthonormal. That is, if n is fixed, then $f_i^{(n)} \perp f_j^{(n)}$ for $i \neq j$, by construction. Also, if $n \neq m$, then $f_i^{(n)} \perp f_j^{(m)}$ for any i and j, since $M_n \perp M_m$. Since each vector $f_k^{(n)}$ is a unit vector we see that $\{e_1,e_2,\ldots\}$ is an orthonormal set.

Next we claim that this family is maximal. Indeed if $x \perp e_n$ for all n, then $x \perp f_k^{(n)}$ for all n and k. That is, $x \perp M_n$ for all n, or $x \perp H$. Hence $x = 0$.

The proof of (6.11.2) is a simple adaptation of the argument of the last theorem. ∎

The last theorem admits another interpretation which can be viewed as the third version of the Spectral Theorem. For this we shall assume that the Hilbert space H is separable.[7] This means that the mapping $U: H \to l_2$ given by

$$U: x \to ((x,e_1),(x,e_2),\ldots)$$

---

[6] The subspace M_n is, of course, finite dimensional when $\lambda_n \neq 0$. If $\lambda = 0$ is an eigenvalue, then the corresponding null space $\mathcal{N}(T)$ may be infinite dimensional. In fact, if H is not separable, then $\lambda = 0$ is necessarily an eigenvalue and $\mathcal{N}(T)$ must have an uncountable orthonormal basis.

[7] Separability is not really necessary. It just makes things simpler.

is a unitary mapping. The operator T is then transformed into an operator Λ on l_2 by the equation

$$T = U^{-1}\Lambda U \quad \text{or} \quad \Lambda = UTU^{-1}, \tag{6.11.4}$$

see Figure 6.11.1. Also Λ is the diagonal matrix $\Lambda = \mathrm{diag}(\mu_1, \mu_2, \ldots)$. This representation Λ is sometimes called the "transfer function" of T.

In summary, then, every compact normal operator is a compact weighted sum of projections in disguise. Moreover, this fact can be used to view or represent compact normal operators in (at least) three ways: weighted sums of projections, eigenvalue-eigenvector representation, unitary equivalence to operation with a diagonal matrix or multiplication by a transfer function.

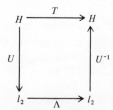

Figure 6.11.1.

Now one point must be made. The Spectral Theorem presented here is not the most general one possible. This should not be surprising at all, for even the weighted sums of projections discussed in Section 9 can be used to represent some noncompact normal operators. In fact, if we generalized from weighted sums of projections to "weighted integrals of projections," we would be able to represent *all* normal operators. Likewise, the "transfer function" representation can be very successfully generalized (for example, Fourier transform and z-transform methods). However, a few mathematical difficulties arise here and there. On the other hand, the Eigenvalue-Eigenvector representation really cannot be developed much further. All this, however, is another story, beyond the scope of this book. The only generalization we will present (Section 14) concerns nonnormal compact operators.

EXAMPLE 1. (THE RAYLEIGH-RITZ METHOD.) The Rayleigh-Ritz Method, which we now describe, is a technique for finding the eigenvalues of a compact normal operator T. In this example we will assume that T is actually self-adjoint and positive, in addition to being compact. The extension of the method to arbitrary compact self-adjoint operators, or compact normal operators, is discussed in the exercises.

So then let $T: H \to H$ be a compact, self-adjoint, positive operator on a Hilbert space H. Recall that the positivity means that

$$(Tx, x) \geq 0, \qquad \text{for all } x \in H. \tag{6.11.5}$$

We see then that if μ is an eigenvalue of T, then $|\mu| \leq \|T\|$ and Equation (6.11.5) implies that $\mu \geq 0$. Now Theorem 6.10.13 tells us that

$$\mu_1 = \|T\|$$

is an eigenvalue of T. Let e_1 be an eigenvector of T associated with μ_1 and also let $M_1 = V(e_1)$ be the one-dimensional linear space spanned by e_1. Then M_1 reduces T, therefore T maps $M_1^\perp$ into $M_1^\perp$. Let T_2 denote the restriction of T to $M_1^\perp$. T_2 is, of course, compact and self-adjoint. So if we apply Theorem 6.10.13 again we see that

$$\mu_2 = \|T_2\|$$

is an eigenvalue of both T_2 and T. This process now continues. Let e_2 be an eigenvector associated with μ_2 and let $M_2 = V(e_1,e_2)$. Then T maps $M_2^\perp$ into $M_2^\perp$. Therefore, if we let T_3 denote the restriction of T to $M_2^\perp$, then

$$\mu_3 = \|T_3\|$$

is another eigenvalue of T.

It can easily be seen that if we continue in this way we can then find all the eigenvalues of T. With these preliminaries behind us, we are now prepared to give the Rayleigh-Ritz Formula for the eigenvalues, which is merely successive applications of Theorem 5.23.8, or Equation (6.10.2).

First we note that

$$\mu_1 = \sup_{(x,x)=1} (Tx,x). \tag{6.11.6}$$

Next we have

$$\mu_2 = \sup_{\substack{(x,x)=1 \\ (x,e_1)=0}} (Tx,x). \tag{6.11.7}$$

Indeed, the condition $(x,e_1) = 0$ in Equation (6.11.7) is precisely the condition that restricts T to the closed linear subspace $M_1^\perp$. Hence Equation (6.11.7) also can be written as

$$\mu_2 = \sup_{(x,x)=1} \{(Tx,x) : x \in M_1^\perp\} = \|T_2\|.$$

In general, if the eigenvalues $\mu_1, \mu_2, \ldots, \mu_n$ are known with corresponding eigenvectors $e_1, e_2, \ldots, e_n$, then μ_{n+1} is given by

$$\mu_{n+1} = \sup_{\substack{(x,x)=1 \\ (x,e_1)=\cdots=(x,e_n)=0}} (Tx,x). \tag{6.11.8}$$

This formula can easily be proved by a direct application of mathematical induction.

Before the reader becomes too enamoured with this method a somewhat subtle limitation should be noted. Equation (6.11.8) does require that we know the eigenvectors $e_1, \ldots, e_n$, but the Rayleigh Ritz Method does not give any clue for determining these eigenvectors.

It is possible to circumvent this deficiency by using certain approximation techniques. We refer the reader to the work of Aronszajn [1] for more details. ∎

EXAMPLE 2. (FREDHOLM ALTERNATIVES.)   Let T be a compact normal operator on a Hilbert space H. Let y be given in H and we now seek a solution of the equation

$$x = Tx + y. \tag{6.11.9}$$

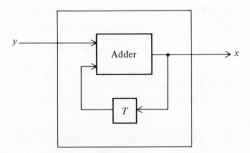

Figure 6.11.2.

This can be viewed as a black-box problem, as shown in Figure 6.11.2.

The Fredholm alternatives tell us precisely when it is possible to solve this problem.

(a) *If 1 is not an eigenvalue of T, then there is precisely one solution x for every y in H.* The solution is of course given by

$$x = (I - T)^{-1}y.$$

(See Exercise 25 for more details.)

(b) *If 1 is an eigenvalue of T, then there is a solution of (6.11.9) if and only if $y \perp \mathcal{N}(I - T)$. In this case, if x^* is any solution of (6.11.9), then every other solution is of the form*

$$x = x^* + c_1 e_1 + \cdots + c_n e_n, \tag{6.11.10}$$

where $\{e_1, \ldots, e_n\}$ is an orthonormal basis for $\mathcal{N}(I - T)$.

The first alternative (a) follows from the fact that if 1 is not an eigenvalue of T, then 1 is in the resolvent set of T.

The second alternative (b) follows from the fact that Equation (6.11.9) has a solution if and only if y is in the range of $I - T$. Since $\mathcal{R}(I - T) = \mathcal{N}(I - T)^{\perp}$ we see that Equation (6.11.9) has a solution if and only if $y \perp \mathcal{N}(I - T)$.

The proof of Equation (6.11.10) is left as an exercise. ∎

EXERCISES

1. Let $K: L_2(I) \to L_2(I)$ be an integral operator $y = Kx$, where

$$y(t) = \int_I k(t,s)x(s)ds.$$

Assume that I is compact and $k(t,s)$ is continuous. Show that K is compact. Show that the eigenfunctions corresponding to nonzero eigenvalues can be chosen to be continuous. What happens to eigenfunctions corresponding to the eigenvalue $\lambda = 0$?

2. Use Mathematical Induction to prove Equation (6.11.8).

3. Let T be a compact normal operator on a Hilbert space H.
   (a) Show that there is an eigenvalue λ_0 that satisfies

$$|\lambda_0| = \max\{|\lambda|: \lambda \text{ is an eigenvalue of } T\}.$$

   (b) Show that $|\lambda_0| = \|T\|$.
   (c) Show that (6.10.2) holds for this T.
   (d) Let H be a two-dimensional complex Hilbert space. Show that λ_0 satisfies one of the following:

$$|\operatorname{Re} \lambda_0| = \|A\| \quad \text{or} \quad |\operatorname{Im} \lambda_0| = \|B\|.$$

   What happens if H has dimension ≥ 3?

4. Let $T = A + iB$ be a compact normal operator. When is it true that

$$\|T\|^2 = \|A\|^2 + \|B\|^2?$$

5. Let T be a compact self-adjoint operator on a Hilbert space H and also let $T = \sum_n \lambda_n P_n$ be the spectral decomposition of T. Then the nonzero eigenvalues $\{\lambda_n\}$ can be partitioned into two sets Λ_+ and Λ_-, the positive and the negative eigenvalues. The operators

$$T_+ = \sum_{\lambda_n \in \Lambda_+} \lambda_n P_n, \qquad T_- = - \sum_{\lambda_n \in \Lambda_-} \lambda_n P_n$$

   are called the *positive and negative parts of* T.
   (a) Show that $T = T_+ - T_-$.
   (b) Show that $(T_+ x, x) \geq 0$ and $(T_- x, x) \geq 0$ for all x in H.
   (c) Show that $T_+ T_- = T_- T_+ = 0$.
   (d) Let $|T| = T_+ + T_-$. Show that $T \leq |T|$ and $-T \leq |T|$.

6. Let T be a compact self-adjoint operator on a Hilbert space H.
   (a) Show that the positive eigenvalues of T can be found by Equations (6.11.6), (6.11.7), and (6.11.8).
   (b) Show that the negative eigenvalues of T can be found by replacing "sup" by "inf" in these three equations.

7. Use the results of Exercise 6 and Section 6.10D to discuss a method for finding the eigenvalues of a compact normal operator.

8. Let L be a compact normal operator on a Hilbert space H and let

$$L = \sum_n \lambda_n P_n$$

   be the decomposition of L as a weighted sum of projections. Assume that $\lambda_n \neq 0$ for all n. Show that the polar decomposition of L is given by $L = RU$, where

$$R = \sum_n |\lambda_n| P_n \qquad \text{and} \qquad U = \sum_n \lambda_n |\lambda_n|^{-1} P_n.$$

   What happens if $\lambda = 0$ is an eigenvalue of L? Show that the Cartesian decomposition of L is given by $L = A + iB$ when

$$A = \sum_n \operatorname{Re}(\lambda_n) P_n \qquad \text{and} \qquad B = \sum_n \operatorname{Im}(\lambda_n) P_n.$$

9. Complete the proof of the Fredholm alternative (b) by verifying Equation (6.11.10).

10. Extend the Fredholm alternatives to compact nonnormal operators by proving the following:
    (a) If 1 is not an eigenvalue of T, then there is precisely one solution x of $x - Tx = y$ for every y in H.
    (b) If 1 is an eigenvalue of T, then $x - Tx = y$ has a solution if and only if $y \perp \mathcal{N}(I - T^*)$.

11. Let $k(t,s) \in L_2(I \times I)$ and define $y = Kx$ by
$$y(t) = \int_I k(t,s)x(s)\, ds.$$
    Assume that $k(t,s) = \overline{k(s,t)}$.
    (a) Show that K is a compact self-adjoint operator on $L_2(I)$ and then show that $\|K\| \le \|k\|_2$.
    (b) Let $\{e_n(t)\}$ be an orthonormal basis of eigenvectors for K with associated eigenvalues $\{\mu_n\}$. Assume that $|\mu_1| \ge |\mu_2| \ge \cdots$. Show that
$$k(t,s) = \sum_n \mu_n e_n(t)\overline{e_n(s)}, \qquad (6.11.11)$$
    where the convergence above is in $L_2(I \times I)$.
    (c) Show that
$$\|k\|_2 = \left(\int_I \int_I |k(t,s)|^2\, dt\, ds \right)^{1/2} = \sum_n |\mu_n|^2$$
    (d) Show that $\|K\| = |\mu_1|$.
    (e) Characterize those operators K for which one has $\|K\| = \|k\|_2$.

12. (Continuation of Exercise 11.) Assume that I is closed and bounded and that $k(t,s)$ is continuous in t and s. In this exercise we will show that the series in Equation (6.11.11) converges to $k(t,s)$ uniformly in t and s provided the operator K is positive and $k(t,s)$ is real-valued.
    (a) Show that if μ_n is a nonzero eigenvalue, then the associated eigenfunction $e_n(t)$ can be chosen to be continuous and real-valued.
    (b) Let $k_N(t,s) = \sum_{n=1}^N \mu_n e_n(t)e_n(s)$ and $h_N(t,s) = k(t,s) - k_N(t,s)$. Show that $h_N(t,t) \ge 0$ for all t.
    (c) Show that there is a M such that
$$k_N(t,t) \le k(t,t) \le M$$
    for all t and all N.
    (d) Show that
$$\left| \sum_{i=m}^n \mu_i e_i(t)e_i(s) \right|^2 \le M \sum_{i=m}^n \mu_i e_i(t)^2 \to 0$$
    as $m, n \to \infty$, uniformly in s for each fixed t. [That is, the convergence in Equation (6.11.11) is uniform in each variable separately.]
    (e) Show that the convergence in Equation (6.11.11) is pointwise.
    (f) Show that the convergence in Equation (6.11.11) is uniform in both t and s. [*Hint*: Use Dini's Theorem, from Section D.4.]

13. Let $T: H \to H$ be a compact self-adjoint operator on a Hilbert space H and let $T = \sum_n \lambda_n P_n$ be the spectral decomposition of T. For $\lambda \in (-\infty, \infty)$ let

$$Q_\lambda = {}_s\sum_{\lambda_n \leq \lambda} P_n,$$

that is, $Q_\lambda x = \sum_{\lambda_n \leq \lambda} P_n x$ for all $x \in H$.
(a) Show that for each λ, Q_λ is an orthogonal projection.
(b) Show that $Q_\lambda \leq Q_\mu$ if $\lambda \leq \mu$.
(c) Show that

$$0 = {}_s\lim_{\lambda \to -\infty} Q_\lambda, \qquad I = {}_s\lim_{\lambda \to +\infty} Q_\lambda.$$

(Q_λ is sometimes referred to as a *spectral family*.)

14. Let L be a self-adjoint operator on a Hilbert space H. Let $\{\phi_n\}$ be an orthonormal collection of eigenvectors of L and let M denote the closed linear subspace of H generated by $\{\phi_n\}$. Assume that every eigenvector of L lies in M.
(a) Show that if $M = H$ (that is, $\{\phi_n\}$ is an orthonormal basis for H), then L is a weighted sum of projections. Show that $\sigma(L)$ is the closure of $P\sigma(L)$.
(b) Show that if the continuous spectrum of L contains a nontrivial interval, then $M \neq H$, that is, $\{\phi_n\}$ is not a basis for H.

15. Consider $L = S_r + S_l + \Phi$ on $l_2(0, \infty)$, where S_r and S_l are the right and left shift operators, Φ is the *Coulomb perturbation*

$$\Phi(x_1, x_2, \ldots, x_n, \ldots) = 2b\left(x_1, \frac{1}{2}x_2, \ldots, \frac{1}{n}x_n, \ldots\right),$$

where $b > 0$.
(a) Show that the eigenvalues of L are

$$\lambda_k = 2\left[1 + \left(\frac{b}{k}\right)^2\right]^{1/2}, \qquad k = 1, 2, \ldots.$$

(b) Let $\{\phi_k\}$ be the associated eigenvector with $\|\phi_k\|_2 = 1$. Show that $\{\phi_k\}$ is *not* a basis for $l_2(0, \infty)$. [*Hint:* Use Exercise 14 and Exercise 14 of Section 6.]

16. Let W be a density operator, that is, W is self-adjoint with $0 \leq W^2 \leq W$ and $\operatorname{tr} W = 1$.
(a) Show that W is compact.
(b) Show that one can write $W = \sum_n \lambda_n W_n$, where W_n are density operators representing pure states and $\sum_n \lambda_n = 1$.
(c) Let e_n be a unit vector in $\mathcal{R}(W_n)$, and let A be an observable, that is, self-adjoint operator. Show that the expected value of A is $E(A) = \sum_n \lambda_n (Ae_n, e_n)$.

17. Let L and M be two compact normal operators on a Hilbert space H that commute, that is, $LM = ML$. Show that there is a resolution of the identity $\{P_n\}$ such that

$$L = \sum_n \lambda_n P_n, \qquad M = \sum_n \mu_n P_n$$

for appropriate choice of $\{\lambda_n\}$ and $\{\mu_n\}$.

18. Let $\{L_1,\ldots,L_k\}$ be a collection of compact normal operators on a Hilbert space H that satisfy $L_iL_j = L_jL_i$ for all i, j. Show that there is a resolution of the identity $\{P_n\}$ such that

$$L_i = \sum_n \lambda_n^{(i)}P_n, \qquad i = 1, \ldots, k,$$

where $\{\lambda_n^{(i)}\}$ depends on L_i.

19. Let $\{U\}$ be a family of unitary operators on a Hilbert space of finite dimension n. Assume that $\{U\}$ is a commutative family, that is, if U and V belong to $\{U\}$, then $UV = VU$. Show that there is an orthonormal basis of common eigenvectors. [*Hint*: Use Mathematical Induction on the dimension n.]

20. (Continuation of Exercise 19.) Let U_t, $-\infty < t < \infty$, be a commutative family of unitary operators on a finite-dimensional Hilbert space H that satisfy:

$$U_0 = I, \qquad U_s U_t = U_{s+t}, \quad \text{and} \quad U_s x \to U_t x \qquad \text{as} \quad s \to t$$

for every $x \in H$. Let $\{\phi_1,\ldots,\phi_n\}$ be an orthonormal basis of eigenvectors and let $\rho_k(t)$ satisfy

$$U_t \phi_k = \rho_k(t)\phi_k, \qquad k = 1, \ldots, n.$$

(a) Show that $\rho_k(t) = \exp(iw_k t)$ for appropriate choice of w_k.
(b) Show that in terms of this basis U_t is the matrix operator $U_t = e^{itA}$, where $A = \mathrm{diag}\{w_1,\ldots,w_n\}$.

21. Show that the conclusions of Example 1, Section 4 can be extended to an infinite-dimensional Hilbert space H provided one assumed that the operator L is a compact self-adjoint strictly positive operator. What happens if one only assumes L to be compact self-adjoint and positive?

22. What conclusion could one draw in Example 1, Section 4 if one assumes L to be compact and normal?

23. Let A be a compact self-adjoint positive operator on a Hilbert space H and let $\{\mu_1,\mu_2,\ldots\}$ be an enumeration of the eigenvalues of A, including multiplicity. Show that $\mathrm{tr}\, A = \sum_n \mu_n$.

24. Find the eigenvectors and eigenvalues for $y(t) = \int_{-\pi}^{\pi} k(t,\tau)x(\tau)\, d\tau$, where

$$k(t,\tau) = \sum_{n=0}^{\infty} [a_n \cos nt + b_n \sin nt],$$

where $\sum_{n=0}^{\infty} (|a_n|^2 + |b_n|^2) < \infty$.

25. Consider the equation $(\lambda I - T)x = y$, where T is a compact normal operator. Let $\{e_n\}$ be an orthonormal basis of eigenvectors for T with corresponding eigenvalues $\{\lambda_n\}$. Assume that $\lambda \neq 0$.
(a) Show that if λ is not an eigenvalue of T, then for every y in the Hilbert space H there is a solution x of $(\lambda I - T)x = y$ and it is given by

$$x = \sum_{n=1}^{\infty} \frac{(y,e_n)}{\lambda - \mu_n} e_n.$$

(b) Show that if λ is an eigenvalue of T, then $(\lambda I - T)x = y$ has a solution if and only if $y \perp \mathcal{N}(\lambda I - T)$. Show that if $y \perp \mathcal{N}(\lambda I - T)$, then a solution is given by

$$x^* = \sum_{n=1}^{\infty} \frac{(y,e_n)}{\lambda - \mu_n} e_n,$$

where the terms involving eigenvectors in $\mathcal{N}(\lambda I - T)$ drop out since $(y,e_n) = 0$ for these. What is the general solution of $(\lambda I - T)x = y$ when λ is an eigenvalue of T?

12.   FUNCTIONS OF OPERATORS (OPERATIONAL CALCULUS)

Let T be a compact normal operator on a Hilbert space H and express T as a weighted sum of projections

$$T = \sum_n \lambda_n P_n$$

as indicated in the Spectral Theorem. The operator T^2 is also a compact operator and, furthermore, one has

$$T^2 = \sum_n \lambda_n^2 P_n.$$

To see this we note that

$$T^2 x = T(Tx) = T\left(\sum_n \lambda_n P_n x\right) = \sum_m \lambda_m P_m \left(\sum_n \lambda_n P_n x\right)$$

$$= \sum_{m,n} \lambda_m \lambda_n P_m P_n x$$

$$= \sum_n \lambda_n^2 P_n x$$

since $P_m P_n = 0$ when $m \neq n$ and $P_n P_n = P_n$.

Similarly one has

$$T^N = \sum_n \lambda_n^N P_n,$$

where N is any positive integer. In fact if

$$p(z) = \sum_{i=0}^{N} \alpha_i z^i$$

is any polynomial in z, then

$$p(T) = \sum_n p(\lambda_n) P_n,$$

where

$$p(T) = \sum_{i=0}^{N} \alpha_i T^i \quad \text{and} \quad T^0 = I.$$

We also know from Lemma 6.9.11 that

$$T^* = \sum_n \bar{\lambda}_n P_n.$$

It follows, then, that if

$$p(z,\bar{z}) = \sum_{i,\,j=1}^{n} \alpha_{ij} z^i \bar{z}^j$$

is a polynomial in the variables z and $\bar{z}$, then

$$p(T,T^*) = \sum_n p(\lambda_n, \bar{\lambda}_n) P_n,$$

where

$$p(T,T^*) = \sum_{i,\,j=1}^{n} \alpha_{ij} T^i T^{*j}.$$

We also know from Lemma 6.9.7 that the operator T is one-to-one if and only if $\lambda_n \neq 0$ for all n. In this case T^{-1} is defined on the range $\mathcal{R}(T)$ and by Lemma 6.9.10 one has

$$T^{-1} = \sum_n \lambda_n^{-1} P_n \qquad (x \in \mathcal{R}(T)). \tag{6.12.1}$$

Furthermore one has

$$T^{-N} = \sum_n \lambda_n^{-N} P_n \qquad (x \in \mathcal{R}(T^N)),$$

where N is a positive integer. In general, if $p(z)$ is a polynomial in z with no zeros on the spectrum of T, then one has

$$p(T)^{-1} = \sum_n p(\lambda_n)^{-1} P_n.$$

As a consequence of these observations one can easily prove the following theorem.

6.12.1 THEOREM. *Let T be a compact normal operator on a Hilbert space H and let*

$$T = \sum_n \lambda_n P_n$$

be the decomposition of T as a weighted sum of projections.

(a) *If $p(z)$ and $q(z)$ are two polynomials in Z, where $q(z)$ has no zeros on the spectrum $\sigma(T)$, and $r(z) = p(z) \cdot q(z)^{-1}$, then*

$$p(T)q(T)^{-1} = \sum_n r(\lambda_n) P_n.$$

(b) *If $p(z,\bar{z})$ and $q(z,\bar{z})$ are two polynomials in z and $\bar{z}$, where $q(z,\bar{z})$ has no zeros on $\sigma(T) \times \sigma(T^*)$, and $r = pq^{-1}$, then*

$$r(T,T^*) = \sum_n r(\lambda_n, \bar{\lambda}_n) P_n.$$

This operational calculus can be extended to discuss continuous (even discontinuous) functions of z. That is, if $f(z)$ is a continuous function defined on the spectrum $\sigma(T)$, then

$$f(T) = \sum_n f(\lambda_n) P_n.$$

The main problem here is defining $f(T)$. The reader who is interested in pursuing this further is referred to Dunford and Schwartz [1; Section 7.3], Simmons [1], and Taylor [1].

There is one more point we would like to bring up here and that is the question of the square root of a positive compact self-adjoint operator T. In this case the eigenvalues λ_n are all real and nonnegative, and therefore, the positive square root $\sqrt{\lambda_n}$ is well-defined. It should be clear that in this case one has

$$T^{1/2} = \sum_n \sqrt{\lambda_n} P_n. \tag{6.12.2}$$

EXERCISES

1. Let A be any bounded linear operator on a Hilbert space H.
   (a) Show that the series

   $$e^A = I + A + \frac{A^2}{2!} + \cdots = \sum_{n=0}^{\infty} \frac{A^n}{n!}$$

   converges absolutely and represents a bounded linear operator.
   (b) Show that e^A commutes with A.

2. Let $A = \sum_n \lambda_n P_n$ be the spectral decomposition of a compact normal operator. Show that $e^A = \sum_n e^{\lambda_n} P_n$. What is $\sin A$, $\cos A$? Is it true that

   $$e^{iA} = \cos A + i \sin A?$$

3. Prove Equation (6.12.2).

4. Let A be a bounded linear operator on a Hilbert space H and assume that $\sigma(A)$ lies in the left half of the complex plane.
   (a) Show that $\|\exp At\| \to 0$ as $t \to +\infty$.
   (b) Use this to show that if u is a solution of

   $$\frac{du}{dt} = Au, \tag{6.12.3}$$

   then $\|u(t)\| \to 0$ as $t \to \infty$. (Show that $u(t) = (\exp At)u_0$ is a solution of (6.12.3) that satisfies $u(0) = u_0$.)

13.   APPLICATIONS OF THE SPECTRAL THEOREM

In this section we shall present a number of applications of the Spectral Theorem.

EXAMPLE 1. (MATCHED FILTER.)   Suppose that we wish to select a linear filter L so that a certain signal-to-noise ratio is maximized. In particular, let us assume that L is to be selected from among those linear filters that can be modeled mathematically in the form $y = Lx$, or

$$y(t) = \int_0^t g(t)x(t - \tau)\, d\tau, \qquad t \in [0,T],$$

where x is the input, y is the output, and the weighting function g is in $L_2[0,T]$. We assume that we are given an input signal $S(t)$ and a noise random process $N(\omega,t)$ (see Figure 6.13.1). At the final time $t = T$, the output is the sum of

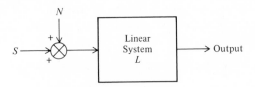

Figure 6.13.1.

$$s = \int_0^T g(\tau)S(T - \tau)\, d\tau$$

and the random variable

$$n(\omega) = \int_0^T g(\tau)N(\omega,T - \tau)\, d\tau.$$

Our problem is to pick g so as to maximize the signal-to-noise ratio

$$\frac{|s|^2}{E\{|n|^2\}},$$

where E denotes the mathematical expectation. Therefore, we assume that $S \in L_2[0,T]$. Then s can be viewed as the inner product between the points $g(\tau)$ and $S_R(\tau) = \overline{S(T - \tau)}$ in the Hilbert space $L_2[0,T]$. Furthermore, let us assume that the noise $N(\omega, t)$ satisfies

$$E\left\{\int_0^T \int_0^T g(\tau_1)\overline{g(\tau_2)}\overline{N(\omega,T - \tau_2)}N(\omega,T - \tau_1)\, d\tau_1\, d\tau_2\right\}$$

$$= \int_0^T \int_0^T g(\tau_1)\overline{g(\tau_2)}E\{\overline{N(\omega,T - \tau_2)}N(\omega,T - \tau_1)\}\, d\tau_1\, d\tau_2,$$

and that the function

$$W(\tau_1,\tau_2) = E\{\overline{N(\omega,T - \tau_2)}N(\omega,T - \tau_1)\}$$

satisfies the condition

$$\int_0^T \int_0^T |W(\tau_1,\tau_2)|^2\, d\tau_1\, d\tau_2 < \infty.$$

We then obtain

$$E\{|n|^2\} = (Wg, g),$$

where W is the linear transformation of $L_2[0,T]$ into itself defined by

$$h(\tau_2) = \int_0^T W(\tau_1,\tau_2)g(\tau_1)\,d\tau_1.$$

The transformation W is, of course, by assumption, known.

We then have

$$\frac{|s|^2}{E\{|n|^2\}} = \frac{|(g,S_R)|^2}{(Wg,g)} \tag{6.13.1}$$

and we now have the problem stated entirely in terms of the Hilbert space $L_2[0,T]$. Moreover, W is a compact, positive self-adjoint transformation. Therefore it has a unique positive self-adjoint square root, say that $W = A^2$. We then have

$$(Wg,g) = (A^2g,g) = (Ag,Ag) = (\phi,\phi),$$

where $\phi = Ag$. Let us now assume that there is a function Φ in $L_2[0,T]$ with the property that $S_R = A\Phi$, that is, S_R lies in the range of A. One then has

$$\frac{|s|^2}{E\{|n|^2\}} = \frac{|(g,A\Phi)|^2}{(\phi,\phi)} = \frac{|(\phi,\Phi)|^2}{(\phi,\phi)}.$$

Then using Schwarz's Inequality we have

$$\frac{|s|^2}{E\{|n|^2\}} = \frac{|(\phi,\Phi)|^2}{(\phi,\phi)} \le (\Phi, \Phi).$$

Moreover, the equality will be taken on if and only if $\phi = k\Phi$, where k is a nonzero scalar. Thus a function g that maximizes the signal-to-noise ratio exists and is given by $k\Phi = Ag$ or equivalently $kS_R = Wg$. We can make this more explicit by using the Spectral Theorem.

Since W is a compact positive self-adjoint transformation, there exists an orthonormal system $\{w_1,w_2,w_3,\ldots\}$ and a sequence of nonnegative real numbers $\{\lambda_1^2,\lambda_2^2,\ldots\}$ with $\lambda_n^2 \to 0$ as $n \to \infty$ such that

$$Wx = \sum_{n=1}^{\infty} \lambda_n^2(x,w_n)w_n.$$

Furthermore, the square root A is given by

$$Ax = \sum_{n=1}^{\infty} \lambda_n(x,w_n)w_n.$$

The solution of the problem $kS_R = Wg$ is then given by

$$k\sum_{n=1}^{\infty} (S_R,w_n)w_n = \sum_{n=1}^{\infty} \lambda_n^2(g,w_n)w_n,$$

or equivalently

$$g = \sum_{n=1}^{\infty} (g,w_n)w_n = k\sum_{n=1}^{\infty} \lambda_n^{-2}(S_R,w_n)w_n.$$

If any of the eigenvalues $\{\lambda_n\}$ are zero, then the last equation still is valid provided the corresponding coefficient (S_R, w_n) vanishes, or equivalently, provided $S_R \perp \mathcal{N}(W)$. However, since we have assumed S_R to belong to the range of A (and ipso facto to be the range of W) we see that $S_R \perp \mathcal{N}(W)$. (Why?) ∎

EXAMPLE 2. (KARHUNEN-LOEVE EXPANSION.)   Let $[a,b]$ be a finite interval. For $t \in [a,b]$ let $X(t)$ denote a random process with

$$E\{X(t)\} = 0, \qquad E\{|X(t)|^2\} < \infty, \tag{6.13.2}$$

and where the covariance function

$$r(t,s) = E\{X(t)\overline{X(s)}\} \tag{6.13.3}$$

is continuous (see Example 1, Section E.6).

Let f be a complex-valued function defined on $[a,b]$. We shall define the random variable $I = \int_a^b f(t)X(t)\,dt$ as follows: Let $P : a = t_0 < t_1 < \cdots < t_n = b$ be a partition of $[a,b]$ (see Section D.2) and let $|P| = \max |t_i - t_{i-1}|$. Let $I(P)$ be the random variable given by

$$I(P) = \sum_{i=1}^{n} f(t_i)X(t_i)(t_i - t_{i-1}).$$

If it happens that $E\{|I(P) - I|^2\} \to 0$ as $|P| \to 0$, then we shall define I as

$$I = \int_a^b f(t)X(t)\,dt.$$

In the exercises the reader is asked to show that if $f(t)$ is continuous and if the covariance function $r(t,s)$ is continuous, then the integral $\int_a^b f(t)X(t)dt$ exists and that

$$E\left\{\int_a^b f(t)X(t)\,dt\right\} = 0. \tag{6.13.4}$$

Furthermore, if g is also continuous, then one can show that

$$E\left\{\int_a^b f(t)X(t)\,dt \int_a^b \overline{g(s)X(s)}\,ds\right\} = \int_a^b \int_a^b f(t)\overline{g(s)}E\{X(t)\overline{X(s)}\}\,dt\,ds$$

$$= \int_a^b \int_a^b f(t)\overline{g(s)}r(t,s)\,dt\,ds \tag{6.13.5}$$

and

$$E\left\{\int_a^b f(t)X(t)\,dt \cdot \overline{X(s)}\right\} = \int_a^b f(t)r(t,s)\,dt. \tag{6.13.6}$$

6.13.1 THEOREM.   *Let $X(t)$ be a random process defined on a finite interval $[a,b]$ satisfying (6.13.2). Assume that the covariance function $r(t, s)$ given by (6.13.3) is continuous. Then one can write*

$$X(t) = \sum_{n=1}^{\infty} Y_n \phi_n(t), \qquad a \le t \le b, \tag{6.13.7}$$

where $\{\phi_n\}$ is an orthonormal family of eigenfunctions of the integral operator R given by

$$y(t) = \int_a^b r(t,s)x(s)\,ds \qquad (6.13.8)$$

and moreover $\{\phi_n\}$ forms a basis for $\mathcal{N}(R)^\perp$. The random variables Y_n in (6.13.7) are given by $Y_n = \int_a^b \overline{\phi_n(t)}X(t)\,dt$ and satisfy $E\{Y_n\} = 0$ and $E\{Y_n \overline{Y_m}\} = \delta_{nm}\lambda_m$, where λ_m is the eigenvalue associated with ϕ_m. Finally the series in (6.13.7) converges in the mean square sense to $X(t)$, that is,

$$E\left\{\left| X(t) - \sum_{n=1}^N Y_n\,\phi_n(t) \right|^2\right\} \to 0$$

as $N \to \infty$ for all t in $[a,b]$.

Proof: We note that the integral operator R given by (6.13.8) is compact since $r(t,s)$ is continuous, see Example 6, Section 5.24. Also $r(t,s) = \overline{r(s,t)}$, so R is self-adjoint. Let $\{\phi_n\}$ be an orthonormal collection of eigenfunctions of R associated with the nonzero eigenvalues $\{\lambda_n\}$. Then $\phi_n(t)$ is continuous and real-valued (see Exercise 12, Section 11) and the random variable $Y_n = \int_a^b \overline{\phi_n(t)}X(t)\,dt$ exists and by (6.13.4) one has $E\{Y_n\} = 0$. Furthermore, (6.13.5) implies that

$$E\{Y_n \overline{Y_m}\} = \int_a^b \int_a^b \overline{\phi_n(t)}\phi_m(s)r(t,s)\,ds\,dt$$

$$= \int_a^b \overline{\phi_n(t)}\lambda_m\,\phi_m(t)\,dt = \lambda_m\,\delta_{nm}.$$

Next let $S_N(t) = \sum_{n=1}^N Y_n\,\phi_n(t)$. Then by a straightforward application of (6.13.5) and (6.13.6), together with the fact that the eigenvalues of R are real, we get

$$E\{|X(t) - S_N(t)|^2\} = E\{(X(t) - S_N(t))(\overline{X(t)} - \overline{S_N(t)})\}$$

$$= r(t,t) - \sum_{n=1}^N \lambda_n\,\phi_n(t)\overline{\phi_n(t)}.$$

It is shown in Exercise 12, Section 11 that

$$r(t,s) = \lim_{N\to\infty} \sum_{n=1}^N \lambda_n\,\phi_n(t)\overline{\phi_n(s)},$$

therefore, we conclude that

$$E\{|X(t) - S_N(t)|^2\} \to 0$$

as $N \to 0$. ∎

EXAMPLE 3. (THE KARHUNEN-LOEVE EXPANSION FOR DISCRETE RANDOM PRO-CESSES.)   The expansion described in the last example is also valid when the interval $[a,b]$ is replaced by a discrete countable set say $t = 1, 2, \ldots$ In this case, a somewhat different notation is customarily employed.

Let $\{X_n : n = 1, 2, \ldots\}$ be a discrete random process with

$$E\{X_n\} = 0 \quad \text{and} \quad E\{|X_n|^2\} < \infty. \tag{6.13.9}$$

Define the covariance matrix $\Gamma = (\gamma_{nm})$ by

$$\gamma_{nm} = E\{X_n \bar{X}_m\}, \qquad n, m = 1, 2, \ldots,$$

and assume that

$$\sum_{n,\, m} |\gamma_{nm}|^2 < \infty. \tag{6.13.10}$$

6.13.2 THEOREM.   *Let $\{X_n : n = 1, 2, \ldots\}$ be a discrete random process satisfying (6.13.9) and assume that the covariance matrix Γ satisfies (6.13.10). Then one can write*

$$X = \{X_1, X_2, \ldots\} = \sum_{k=1}^{\infty} Y_k \phi_k, \tag{6.13.11}$$

where $\phi_k = \{\phi_k^{(1)}, \phi_k^{(2)}, \ldots\}$ is an element of l_2 and the collection of $\{\phi_k\}$ is an orthonormal family of eigenvectors for the matrix operator Γ given by $y = \Gamma x$, and, moreover, $\{\phi_k\}$ forms a basis for $\mathcal{N}(\Gamma)^{\perp}$. Furthermore the random variables Y_k in (6.13.11) are given by $Y_k = \sum_{n=1}^{\infty} \phi_k^{(n)} X_n$ and satisfy $E\{Y_k\} = 0$ and $E\{Y_k \bar{Y}_l\} = \delta_{kl} \lambda_k$, where λ_k is the eigenvalue associated with ϕ_k. Finally, the series in (6.13.11) converges to $X = \{X_1, X_2, \ldots\}$ in the mean-square sense, that is,

$$E\left\{ \left| X_n - \sum_{k=1}^{K} Y_k \phi_k^{(n)} \right|^2 \right\} \to 0$$

as $K \to \infty$, for all $n = 1, 2, \ldots$.

The proof of this theorem, which we shall leave as an exercise, follows the argument used in Theorem 6.13.1. The only noteworthy difference is to show that the series

$$Y_k = \sum_{n=1}^{\infty} \phi_k^{(n)} X_n$$

converges to a random variable Y_k. ∎

EXERCISES

1. This exercise will lead to a proof that the integral $\int_a^b f(t) X(t)\, dt$ is defined when f and X are continuous. We use the notation of Example 2.
   (a) Let $P = a = t_0 < t_1 < \cdots < t_n = b$ and $P' = a = t_0' < t_1' < \cdots < t_m' = b$ be two partitions of $[a,b]$. Show that

$$E(I(P)\overline{I(P')}) = \sum_{i,\, j} f(t_i) f(t_j') R(t_i, t_j')(t_i - t_{i-1})(t_j' - t_{j-1}')$$

$$\to \int_a^b \int_a^b f(t) f(t') R(t, t')\, dt\, dt'$$

as $|P|, |P'| \to 0$.

(b) Show that $E(|I(P) - I(P')|^2) \to 0$ as $|P|, |P'| \to 0$.

(c) Use the completeness of $L_2(\Omega, \mathcal{F}, P)$, where $(\Omega, \mathcal{F}, P)$ is the underlying probability space, to conclude that $I(P)$ has a limit in L_2 as $|P| \to 0$.

2. Using the notation of Example 2, show that if $E(X(t)) = 0$ for all t, then $E(\int_a^b f(t) X(t)\, dt) = 0$, when f and X are continuous.

3. Using the notation of Example 2, show that if f, g, and X are continuous, then

$$E\left(\int_a^b f(t) X(t)\, dt \int_a^b \overline{g(s) X(s)}\, ds \right) = \int_a^b \int_a^b f(t) \overline{g(s)} r(t,s)\, dt\, ds$$

$$E\left(\int_a^b f(t) X(t)\, dt \cdot \overline{X(s)} \right) = \int_a^b f(t) r(t,s)\, dt.$$

4. Prove Theorem 6.13.2.

5. Let $Y(t)$ be a random process defined on a finite interval $[a,b]$ with $E\{|Y(t)|^2\} < \infty$, where $E(Y(t))$ and $E(Y(t) \overline{Y(s)})$ are continuous functions. Show that

$$Y(t) = E(Y(t)) + \sum_{n=1}^{\infty} Y_n\, \phi_n(t),$$

where Y_n and ϕ_n have structure similar to that defined in Theorem 6.13.1.

6. Let $x(\omega,t)$ be a complex-valued function defined for $\omega \in [0,W]$ and $t \in [a,b]$ and satisfying:

$$\int_0^W x(\omega,t)\, d\omega = 0, \qquad \int_0^W |x(\omega,t)|^2\, d\omega < \infty,$$

for all $t \in [a,b]$. Also assume that

$$k(t,s) = \int_0^W x(\omega,t)\, \overline{x(\omega,s)}\, d\omega$$

is continuous. Show that one can express x in the form

$$x(\omega,t) = \sum_{n=1}^{\infty} Y_n(\omega)\, \phi_n(t),$$

where

$$Y_n(\omega) = \int_a^b x(\omega,t)\, \overline{\phi_n(t)}\, dt.$$

7. Use Exercises 5 and 6 to study the function $x(\omega,t) = \exp(i\omega t)$.

14.   NONNORMAL OPERATORS

So far we have been concentrating on compact normal operators. But suppose we have a compact operator that is not normal. What can we do? Clearly we cannot expect to express it as a weighted sum of projections, for all weighted sums of (orthogonal) projections are normal. Equivalently, we cannot expect the eigenvectors to form an orthonormal set. As a matter of fact, all linear operators on finite-dimensional spaces are compact and it is well known that even there, the ones that are not normal can lead to difficulties. (The reader may be familiar with the Jordan canonical form.) Not too surprisingly, things can be more difficult in the case of infinite-dimensional spaces. For example, one may be able to show that a nonnormal compact operator is similar to the operation of multiplication by a (transfer) function, but it is impossible for it to be unitarily equivalent to such an operator, for then it would be normal. In any event, we shall avoid all of these difficulties by taking a slightly different approach. The two main advantages of this approach are that (1) it is applicable to all compact operators, normal or not, and (2) it involves only orthonormal sets of vectors. In fact, we shall show in this section the every compact operator T can be represented in the form

$$Tx = \sum_{n=1}^{\infty} \mu_n(x,x_n)y_n,$$

where the μ_n's are nonnegative real numbers and $\{x_n\}$ and $\{y_n\}$ are orthonormal sets.

6.14.1. THEOREM.   *Let T be compact transformation of a Hilbert space H into itself. Then there exist two orthonormal systems $\{x_n\}$ and $\{y_n\}$ and a sequence of nonnegative real numbers $\{\mu_1,\mu_2,\mu_3,\ldots\}$ such that*

$$Tx = \sum_n \mu_n(x,x_n)y_n, \tag{6.14.1}$$

where convergence is in terms of the uniform topology, that is, $\|T - S_N\| \to 0$ as $N \to \infty$, where

$$S_N x = \sum_{n=1}^{N} \mu_n(x,x_n)y_n. \tag{6.14.2}$$

Proof: Whether T is normal or not, the operator T^*T is compact (Theorem 5.24.7) and self-adjoint. Moreover, T^*T is nonnegative, that is

$$(x,T^*Tx) = (Tx,Tx) \geq 0$$

for all $x \in H$. Therefore, it follows that the eigenvalues of T^*T are real and nonnegative. Let $\{\mu_1^2,\mu_2^2,\ldots\}$ denote these eigenvalues, where $\mu_n \geq 0$ for all n. For convenience we assume that $\mu_1 \geq \mu_2 \geq \mu_3 \geq \cdots$. Then, using the eigenvalue-eigenvector representation for T^*T, we have

$$T^*Tx = \sum_{n=1}^{\infty} \mu_n^2(x,x_n)x_n,$$

where a given eigenvalue is repeated according to its multiplicity and $\{x_n\}$ is an orthonormal basis of eigenvectors. This operator T^*T has a unique nonnegative square root R given by

$$Rx = \sum_{n=1}^{\infty} \mu_n(x,x_n)x_n.$$

For $\mu_n \neq 0$, let

$$y_n = \frac{1}{\mu_n} Tx_n.$$

Then

$$(y_n, y_m) = \frac{1}{\mu_n \mu_m}(Tx_n, Tx_m) = \frac{(x_n, T^*Tx_m)}{\mu_n \mu_m} = \frac{\mu_m}{\mu_n}(x_n, x_m)$$

so $\{y_n\}$ is an orthonormal system. We now extend the class $\{y_n\}$ to an orthonormal basis in H. One then has

$$Tx_n = \mu_n y_n$$

for all μ_n's even for $\mu_n = 0$. Define S_N by (6.14.2) and let us show that $\|T - S_N\| \to 0$ as $N \to \infty$.

We know that each $x \in H$ can be expressed uniquely in the form

$$x = (x,x_1)x_1 + \cdots + (x,x_N)x_N + \sum_{n=N+1}^{\infty}(x,x_n)x_n.$$

Then

$$Tx - S_N x = \sum_{n=N+1}^{\infty}(x,x_n)Tx_n = \sum_{n=N+1}^{\infty}\mu_n(x,x_n)y_n.$$

Since the $\{y_n\}$ form an orthonormal system, one has

$$\|Tx - S_N x\|^2 = \sum_{n=N+1}^{\infty}\mu_n^2|(x,x_n)|^2 \le \mu_{N+1}\|x\|^2.$$

Therefore $\|T - S_N\| \le \mu_{N+1} \to 0$, as asserted. ∎

We leave it to the reader to show that the μ_n^2's are also the eigenvalues of TT^* and that the y_n's are eigenvectors of TT^*.

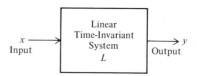

Figure 6.14.1.

EXAMPLE 1.   This example and the next are concerned with a classical systems identification problem that arises in control engineering or systems engineering, see Truxal [1, pp. 437–438]. In simple terms, one is given a black box and the ability to record inputs and the corresponding outputs (see Figure 6.14.1). With this information one seeks a method for characterizing a mathematical model for the black box. Usually one can make some assumptions about the format of the mathematical model, and for our purposes we assume that this is a convolution operator on $L_2[0,T]$, that is a linear time-invariant operator of the form

$$y(t) = \int_0^t h(\tau)x(t - \tau)\, d\tau, \qquad t \in [0,T], \tag{6.14.4}$$

where[8] $h \in L_2[0,T]$ and $0 < T < \infty$. However, we do not know h. Our problem is to determine h by running appropriate experiments with an input x and the corresponding output y. We will then try to use the ordered pair $\{x,y\}$ to determine h, or perhaps, to estimate h.

To begin with let us change our view of (6.14.4). Instead of it representing a mapping of x's into y's, we view it as a mapping X of h's into y's. Thus each input x yields a mapping $X: L_2[0,T] \to L_2[0,T]$, given by $y = Xh$. Then if we have chosen our experiment input x so that X is invertible, determination of h is, in principle at least, simple; indeed, $h = X^{-1}y$.

However, an interesting problem arises. First note that for any x, the operator X is compact. In fact,

$$\int_0^T \int_0^T |x(t-\tau)|^2 \, d\tau \, dt < \infty,$$

where $x(t-\tau) = 0$ for $\tau > t$. So not only is X compact, it is also a Hilbert-Schmidt operator. In any event, using the decomposition of Theorem 6.14.1, we have that

$$Xh = \sum_{n=1}^{\infty} \mu_n(h,x_n)y_n,$$

where $\mu_n \to 0$ as $n \to \infty$. (Why?) If X is one-to-one, then $\mu_n \neq 0$ for all n and (see Exercise 2)

$$X^{-1}y = \sum_{n=1}^{\infty} \frac{1}{\mu_n}(y,y_n)x_n \qquad (6.14.5)$$

for all y in the range of X. We see then that if X is one-to-one (which is an *a priori* condition on the input x), then h is given by (6.14.5), where y is the corresponding output.

Since in this problem one has $\mu_n \to 0$, it immediately follows from (6.14.5) that X^{-1} is not continuous no matter what experiment input x we use. This can be important. If for some reason (and there always is one) an error is made in measuring y and one has $y + z$ instead, then

$$\|X^{-1}(y+z) - X^{-1}(y)\|$$

can be very large even when $\|z\|$ is small. ∎

EXAMPLE 2. (CONTINUATION OF EXAMPLE 1.) One often proceeds with the identification of L by first performing a correlation type of operation between the input x and the output y; in particular, let

$$\phi_{yx}(\tau) = \int_\tau^T x(t-\tau)y(t) \, dt, \qquad \tau \in [0,T]. \qquad (6.14.6)$$

But (6.14.6) represents the adjoint X^* of X, so we have $X^*y = X^*Xh$. If the input x could be chosen so that $X^*X = I$, we would have

$$h = X^*y. \qquad (6.14.7)$$

---

[8] Since $L_2[0,T] \subseteq L_1[0,T]$, we also have $h \in L_1[0,T]$. So we are assured that L is a bounded operator.

However, this is impossible because X is compact. Indeed

$$X^*Xh = \sum_{n=1}^{\infty} \mu_n \bar{\mu}_n (h,x_n)x_n$$

and $\mu_n \bar{\mu}_n$ cannot be unity for all n. On the other hand, it is often possible to choose the input x so that (6.14.7) is a suitable approximation; in particular, x can be chosen so that the restriction of X^*X to an appropriate finite-dimensional subspace containing h is approximately I. See Truxal [1] for further details. ∎

EXAMPLE 3. (ε-CAPACITY OF A LINEAR CHANNEL.)[9]   Consider the linear communication channel L operating as shown in Figure 6.14.2. We assume that the

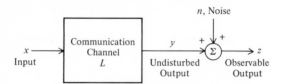

Figure 6.14.2.

input x, the undisturbed output y, the noise n, and the observable output z are all in the Hilbert space $L_2(I)$, where $I = [-T/2, T/2]$, $T > 0$, Further, L is a linear transformation of $L_2(I)$ into itself. We will view $\|\cdot\|^2$ as the energy of the various signals. The only thing we know about the noise n is that $\|n\| < \varepsilon$, where $\varepsilon > 0$. We assume that the inputs x are restricted in maximum energy. That is, the allowable set of inputs S_I is given by

$$S_I = \{x \in L_2(I): \|x\| < E\},$$

where $0 < E < \infty$.

Suppose that we want to send signals through this system in such a way that on the basis of an observation z one can say exactly which input x was used. If, for example, one of two inputs is sent, then the two possibilities are shown in Figure 6.14.3. In (a), if a z is observed that comes from the shaded region (the intersection

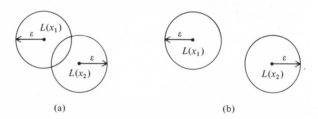

(a)   (b)

Figure 6.14.3.

[9] This example was suggested by the work of Prosser and Root [1].

of the two ε-balls), then one cannot say whether x_1 or x_2 was sent. Whereas, in (b) one can say unambiguously on the basis of z which input was used. In general, given a maximum input energy E^2 and a noise level ε, one would like to choose a set of inputs (that is, messages) $M \subseteq S_I$ such that $\|L(x_\alpha) - L(x_\beta)\| \geq 2\varepsilon$, $x_\alpha, x_\beta \in M$, and $x_\alpha \neq x_\beta$. Since it is desirable to have as many inputs (messages) available as possible, one is interested in how large M can be. Since $L_2(I)$ is separable, we can immediately say that M is at most countably infinite. (Why?) However, for most practical communication channels L is compact, so we can say that M is usually finite. (Why?) In any event, if L is compact and $N(\varepsilon,E)$ is the maximum number of messages that can be unambiguous in the above sense, then

$$C(\varepsilon,E) = \log_2 N(\varepsilon,E)$$

is referred to as the ε-capacity of the channel.

Suppose the communication channel L can be modeled mathematically by the integral operator

$$(Lx)(t) = \int_I h(t - \tau)x(\tau)\, d\tau, \qquad t \in I, \tag{6.14.8}$$

where the weighting or unit impulse response function h is defined on $(-\infty,\infty)$ and

$$\int_{-\infty}^{\infty} |h(t)|^2\, dt < \infty, \tag{6.14.9}$$

that is, $h \in L_2(-\infty,\infty)$. It follows that (6.14.8) represents a continuous linear transformation of $L_2(I)$ into itself. Moreover, since we are considering a finite time interval, we have

$$\int_I \int_I |h(t - \tau)|^2\, d\tau\, dt < \infty, \tag{6.14.10}$$

that is, L is compact.

Since L is compact, we have

$$Lx = \sum_{n=1}^{\infty} \mu_n(x,x_n)y_n,$$

where we assume that $\mu_1 \geq \mu_2 \geq \mu_3 \geq \cdots$, and the image of the closed ball of radius E is a compact ellipsoid $\mathscr{E}$ whose semi-axes are $\mu_n E y_n$, $n = 1, 2, \ldots$. $N(\varepsilon,E)$ is, then, the maximum number of pairwise disjoint open ε-balls that can be fitted into $\mathscr{E}$. Prosser and Root [1] have shown that

$$\prod_{i=1}^{n(2\varepsilon)} \left(\frac{\mu_i E}{2\varepsilon}\right) < N(\varepsilon,E) < \prod_{i=1}^{n(\varepsilon/\sqrt{2})} \left(\frac{2\sqrt{2}\mu_i E}{\varepsilon}\right),$$

where $n(c)$ denotes the number of μ_n's such that $\mu_n > c$. They have also investigated the behavior of the ε-capacity of (6.14.8) as $T \to \infty$ and $\varepsilon \to 0$ and given estimates of this ε-capacity in terms of the Fourier transform of $h(t)$. ∎

EXERCISES

1. How is the decomposition of T, as given in this section, related to the decomposition in the Spectral Theorem when T is compact and normal?

2. If T is compact and one-to-one show that the inverse of T defined on its range can be represented by

$$T^{-1}y = \sum_{n=1}^{\infty} \frac{1}{\mu_n}(y,y_n)x_n,$$

where the domain of T^{-1} is all y for which the above series converges.

3. Let T be a compact (not necessarily normal) operator. Show that the transformation $(\lambda I - T)$ can be represented by

$$(\lambda I - T)x = \sum_{n=1}^{\infty} \sigma_n(x,x_n)y_n, \qquad \text{for all } x \in H,$$

where $\{x_1,x_2,...\}$ and $\{y_1,y_2,...\}$ are orthonormal systems and the σ_n's are nonnegative real numbers. [*Hint*: Show that

$$(\bar{\lambda}I - T^*)(\lambda I - T)x = \sum_{n=1}^{\infty} (|\lambda|^2 - \mu_n)(x,x_n)x_n,$$

for all $x \in H$, where $|\lambda|^2 - \mu_n \geq 0$. Now use the argument of Theorem 6.14.1.]

4. Discuss Theorem 6.14.1 for the case of the Volterra integral operator $y = Kx$, where

$$y(t) = \int_a^t k(t,\tau)x(\tau)\, d\tau$$

on $L_2(a,b)$ with

$$\int_a^b \int_a^b |k(t,\tau)|^2 \, dt \, d\tau < \infty.$$

5. Define $k(t,\tau)$ by

$$k(t,\tau) = \begin{cases} 2t - \tau - 1, & 0 \leq \tau \leq t \leq 1, \\ \tau - 1, & 0 \leq t \leq \tau \leq 1. \end{cases}$$

Use Theorem 6.14.1 to analyze the operator

$$y(t) = \int_0^1 k(t,\tau)x(\tau)\, d\tau.$$

[See Exercise 1(f), Section 7.2.]

SUGGESTED REFERENCES

Aronszajn [1]

Ash [1]

Bachman and Narici [1]

Bochner [1]

Edwards [2]

Gikhman and Skorokhod [1]

Hilbert [1]

Lorch [1]

Paley and Wiener [1]

Prosser and Root [1]

Riesz and Sz. Nagy [1]

Stone [1]

Truxal [1]

Zygmund [1]

Also see references at the end of Chapter 5.

7

Analysis of Unbounded Operators

1. INTRODUCTION

In the last chapter we discussed the concept of weighted sums of projections. We saw that every compact normal operator can be expressed as a weighted sum of projections. Compact operators are, of course, bounded operators.

In this chapter we turn our attention to unbounded operators. We are interested in knowing when an unbounded linear operator can be represented as a weighted sum of projections. You will recall (Exercise 1, Section 6.9) that if L is a bounded linear operator that can be represented as a weighted sum of projections, then there exists an orthonormal basis of eigenvectors $\{x_1 x_2, ...\}$ for L, with corresponding eigenvalues $\{\mu_1 \mu_2, ...\}$, such that

$$x = \sum_n (x, x_n) x_n$$

and

$$Lx = \sum_n \mu_n (x, x_n) x_n.$$

The same phenomenon applies to certain unbounded linear operators (see Exercise 1 in this section).

To repeat, we wish to know when an unbounded operator can be represented as a weighted sum of projections. A sufficient condition, which will handle all the situations we shall be interested in, is given in Theorem 7.1.1 below. However, before formulating this theorem, it is necessary to introduce a technical condition concerning unbounded linear operators.

We shall say that a linear operator L is *an operator on a Hilbert space H* if both the domain $\mathscr{D}_L$ and the range $\mathscr{R}_L$ lie in H and $\mathscr{D}_L$ is dense in H. One sometimes says that L is *densely defined*. In the remainder of this chapter we shall consider only operators that are densely defined.

As we shall see, this restriction to densely defined operators is not serious. The following test will suffice in many cases. Assume that H is the Hilbert space $L_2(R^n)$. It is shown in Appendix D that the space of infinitely differentiable functions with compact support $C_0^\infty(R^n)$ is dense in $L_2(R^n)$. It follows then that if L is an operator with $L: \mathscr{D}_L \to L_2(R^n)$ and $C_0^\infty(R^n) \subset \mathscr{D}_L$, then L is densely defined. More generally, if Ω is an open set in R^n and $H = L_2(\Omega)$, then the space $C_0^\infty(\Omega)$ is dense in $L_2(\Omega)$ (Exercise 4, Section D.12). Thus any linear operator $L: \mathscr{D}_L \to L_2(\Omega)$ with $C_0^\infty(\Omega) \subset \mathscr{D}_L$ is densely defined.

7.1.1 THEOREM. *Let L be a linear operator on a Hilbert space H. If there is a complex number λ_0 in the resolvent set of L for which $(\lambda_0 I - L)^{-1}$ is compact and normal,*[1] *then L can be expressed as a weighted sum of projections.*

---

[1] This is a slight abuse of terminology since the range of $(\lambda_0 I - L)$ may be only dense in H and not all of H. Therefore $(\lambda_0 I - L)^{-1}$ would be defined only on a dense subset of H. But if $(\lambda_0 I - L)^{-1}$ is compact it is also continuous. Therefore, it has a unique extension to H which we also denote by $(\lambda_0 I - L)^{-1}$. The hypotheses then ask that this extension be normal. (It is automatically compact.)

486

Proof: Since $(\lambda_0 I - L)^{-1}$ is compact and normal, there is a weighted sum of projections $\sum_n \lambda_n P_n$ with

$$(\lambda_0 I - L)^{-1} = \sum_n \lambda_n P_n,$$

by the first version of the Spectral Theorem. By taking the inverse, we get

$$\lambda_0 I - L = \sum_n \lambda_n^{-1} P_n,$$

or

$$L = \sum_n (\lambda_0 - \lambda_n^{-1}) P_n,$$

where the last equality holds on $\mathcal{D}_L$. ∎

The hypothesis of the last theorem is so important that it warrants a special definition.

7.1.2 DEFINITION.   Let L be a linear operator on a Hilbert space H. We shall say that L has a *compact-normal resolvent* if there is a λ_0 in the resolvent set of L for which $(\lambda_0 I - L)^{-1}$ is compact and normal.

While the last theorem gives a sufficient[2] condition that an operator can be represented as a weighted sum of projections, it is only the starting point for our analysis. The problem we wish to solve is the following:

Given the operator L, how can we determine whether it has a compact-normal resolvent? More specifically, what conditions on L itself will guarantee that L has a compact-normal resolvent?

Our problem, then, is twofold. Given a complex number λ_0 in the resolvent set of L,

(1) when is $(\lambda_0 I - L)^{-1}$ normal, or self-adjoint, and
(2) when is $(\lambda_0 I - L)^{-1}$ compact?

The rest of the chapter is concerned with the study of these questions.

EXERCISES

1. Let L be a linear operator on a Hilbert space H with a compact-normal resolvent. Show that there exists an orthonormal basis $\{x_1, x_2, \ldots\}$ in H and a sequence of complex numbers $\{\mu_1, \mu_2, \ldots\}$ such that
   (a) $Lx_n = \mu_n x_n$.
   (b) $Lx = \sum_n \mu_n (x, x_n) x_n$ for every vector x in the domain of L.
   (c) Show that the domain of L is contained in the linear subspace

   $$\{x \in H: \sum_n |\mu_n (x, x_n)|^2 < \infty\}.$$

   (d) Show that L is bounded if and only if the sequence $\{\mu_1, \mu_2, \ldots\}$ is bounded.

---

[2] It is not a necessary condition for the simple reason that many *noncompact* operators can be expressed as weighted sums of projections.

2. Let L be a linear operator on a Hilbert space H with a compact-normal resolvent.
   (a) Show that for every λ in the resolvent set of L the operator $(\lambda I - L)^{-1}$ is compact and normal.
   (b) Show that L has at most a countable number of points in its spectrum.

2. GREEN'S FUNCTIONS

For a certain class of operators, especially certain differential operators, the inverse operator can be represented as an integral operator. For these operators we have available a test for a compact-normal resolvent. To be specific assume that H is the Hilbert space $L_2(I)$ and

$$((\lambda_0 I - L)^{-1} u)(x) = \int_I g(x,y) u(y) \, dy$$

for all u in H. Then the kernel $g(x,y)$ is said to be the *Green's function* for $\lambda_0 I - L$, or $(\lambda_0 I - L)^{-1}$.

Recall (Example 6, Section 5.24) that if

$$\int_I \int_I |g(x,y)|^2 \, dx \, dy < \infty, \tag{7.2.1}$$

then $(\lambda_0 I - L)^{-1}$ is compact. In particular, if I is a compact set and if g is continuous or bounded, then we see that $(\lambda_0 I - L)^{-1}$ is compact. Also recall that if

$$g(x,y) = \overline{g(y,x)}, \tag{7.2.2}$$

then $(\lambda_0 I - L)^{-1}$ is self-adjoint.[3] Thus, if (7.2.1) and (7.2.2) are both satisfied, then L has a compact-normal resolvent. Let us now look at some examples.[4]

EXAMPLE 1. Consider the Hilbert space $L_2[0,1]$ and let $Lu = -u''$, where the domain $\mathcal{D}_L$ consists of all functions u in $L_2[0,1]$ such that u' is absolutely continuous, $u'' \in L_2[0,1]$, and $u(0) = u(1) = 0$. Let us show that L^{-1} is an integral operator given by

$$(L^{-1} v)(t) = \int_0^1 g(t,\tau) v(\tau) \, d\tau, \tag{7.2.3}$$

where

$$g(t,\tau) = \begin{cases} (1 - \tau)t, & 0 \leq t \leq \tau \leq 1, \\ (1 - t)\tau, & 0 \leq \tau < t \leq 1. \end{cases}$$

If we can verify that (7.2.3) holds for this $g(t,\tau)$, then it follows that L^{-1} is compact, since $g(t,\tau)$ is bounded. Also, L^{-1} is self-adjoint since $g(t,s)$ satisfies (7.2.2).

---

[3] We could replace (7.2.2) with the corresponding property for normality. We will not do this here since all of our applications deal with the self-adjoint case.

[4] In our examples we will not explain how one can *derive* the given Green's functions. If one is interested in this question, an excellent discussion can be found in Courant and Hilbert [1, Vol. 1, pp. 351–388]. As we shall ultimately see (see Section 5) the actual derivation of the Green's function is not important from our point of view. Most of the information we seek concerning linear differential operators can be derived without the use of the Green's function.

Now let $v \in L_2[0,1]$ and define u by

$$u(t) = \int_0^1 g(t,\tau)v(\tau)\,d\tau = \int_0^t \tau v(\tau)\,d\tau + t\int_t^1 v(\tau)\,d\tau - t\int_0^1 \tau v(\tau)\,d\tau.$$

It is easy to see that $u(0) = u(1) = 0$. Since

$$u'(t) = \int_t^1 v(\tau)\,d\tau - \int_0^1 \tau v(\tau)\,d\tau \qquad \text{and} \qquad u''(t) = -v(t),$$

we see that $u \in \mathscr{D}_L$ and $Lu = v$. Since L is one-to-one we see that L^{-1} is given by (7.2.3). ∎

The last example is an example of a Sturm-Liouville operator. These operators will be studied in greater generality in Section 5.

In the next example we shall consider a second-order differential operator on an infinite interval.

EXAMPLE 2. Consider the operator

$$Lu = u'' - (1 + x^2)u$$

on $L_2(-\infty,\infty)$. We shall let the domain $\mathscr{D}_L$ be the collection of all C^2-functions $u(x)$ with $Lu \in L_2(-\infty,\infty)$ and $u(x) \to 0$ as $x \to \pm\infty$. It is shown in Exercise 7 that L is one-to-one.

We claim that the inverse is given by

$$(L^{-1}v)(x) = \int_{-\infty}^{\infty} g(x,y)v(y)\,dy, \tag{7.2.4}[5]$$

where

$$g(x,y) = \begin{cases} \pi^{-1/2}\exp\left[\dfrac{(x^2+y^2)}{2}\right]\displaystyle\int_{-\infty}^x e^{-t^2}\,dt\int_y^{\infty} e^{-t^2}\,dt, & x \le y, \\[3mm] \pi^{-1/2}\exp\left[\dfrac{(x^2+y^2)}{2}\right]\displaystyle\int_{-\infty}^y e^{-t^2}\,dt\int_x^{\infty} e^{-t^2}\,dt, & y < x. \end{cases}$$

Since $g(x,y) = \overline{g(y,x)}$ we see that the integral operator given by (7.2.4) is self-adjoint. It is possible to show directly with appropriate estimates that

$$\int_{-\infty}^{\infty}\int_{-\infty}^{\infty} |g(x,y)|^2\,dx\,dy < \infty. \tag{7.2.5}$$

We will not do this here. However, in Exercise 7, Section 14, we shall actually compute the integral in (7.2.5) by means of the Spectral Theorem.

[5] The representation in (7.2.4) is strictly speaking an extension of L^{-1} to all of $L_2(-\infty,\infty)$.

Through a straightforward but somewhat laborious computation one can show that if v is continuous and u is determined by

$$u(x) = \int_{-\infty}^{\infty} g(x,y)v(y) \, dy,$$

then $Lu = v$. Furthermore, if $v \in L_2(-\infty,\infty)$, then $u(x) \to 0$ as $x \to \pm\infty$. Thus L^{-1} is given by (7.2.4). ∎

EXAMPLE 3.   Consider the Laplacian operator

$$\Delta u = \frac{\partial^2 u}{\partial x^2} + \frac{\partial^2 u}{\partial y^2}$$

on $L_2(D)$, where D is the unit disk in R^2,

$$D = \{(x,y): x^2 + y^2 \le 1\}.$$

Assume that the domain $\mathscr{D}_\Delta$ of Δ consists of all C^2-functions $u(x,y)$ with the property that $\Delta u \in L_2(D)$ and $u = 0$ on ∂D, the boundary of D. We claim that the inverse Δ^{-1} is given by

$$(\Delta^{-1}v)(x,y) = \int_D g(x,y; \xi,\eta)v(\xi,\eta) \, d\xi \, d\eta, \qquad (7.2.6)$$

where

$$g(x,y; \xi,\eta) = -\frac{1}{2\pi} \log \frac{\sigma_2}{\sigma_1 \sigma_3}, \qquad (7.2.7)$$

σ_i is the distance between the points P_{i-1} and P_i, where $P_0 = (0,0)$, $P_1 = (\xi,\eta)$, $P_2 = (x,y)$, and

$$P_3 = \left(\frac{\xi}{\xi^2 + \eta^2}, \frac{\eta}{\xi^2 + \eta^2} \right).$$

Refer to Figure 7.2.1.

We shall not prove here that the integral operator given by (7.2.6) does indeed represent Δ^{-1}. We shall leave that as an exercise. One point should be made though.

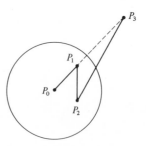

Figure 7.2.1. Unit Disk in R^2.

The kernel g is not continuous in D; with (x,y) held fixed, g has singularities at $(\xi,\eta) = (x,y)$ and $(\xi,\eta) = (0,0)$. However, when v is continuous in the interior of D, the integral in (7.2.6) is well defined (see Exercise 3). ∎

EXERCISES

1. In the following problems you are given a second-order differential operator L, an interval I, and boundary conditions for restricting the domain of L as was done in Examples 1 and 2. You are also given a function $g(t,\tau)$ and you are asked to show that it is a Green's function for L.

(a) $Lu = u''$, $I = [0,1]$,

$u(0) = u'(1) = 0$,

$$g(t,\tau) = \begin{cases} t, & 0 \le t \le \tau \le 1 \\ \tau, & 0 \le \tau \le t \le 1. \end{cases}$$

(b) $Lu = u''$, $I = [-1,1]$,

$u(-1) = u(1) = 0$,

$g(t,\tau) = -\tfrac{1}{2}(|t - \tau| + \tau t - 1)$.

(c) $Lu = u''$, $I = [0,1]$,

$u(0) = u(1),\ u'(0) = u'(1)$,

$$(1 - e)g(t,\tau) = \begin{cases} e^{t-\tau} + e^{1} \cdot e^{\tau-t}, & 0 \le \tau \le t \le 1, \\ e^{1} \cdot e^{t-\tau} + e^{\tau-t}, & 0 \le t \le \tau \le 1. \end{cases}$$

(d) $(Lu)(t) = tu''(t) + u'(t)$, $I = [0, 1]$,

$u(1) = 0,\ u(0)$ finite,

$$g(t,\tau) = \begin{cases} -\log \tau, & 0 \le t \le \tau \le 1 \\ -\log t, & 0 \le \tau \le t \le 1. \end{cases}$$

(e) $(Lu)(x) = tu''(t) + u'(t) - (1/2 + t/4)u(t)$,

$I = (0,+\infty)$, $u(0)$ is finite,

$u(t) \to 0$ as $t \to +\infty$,

$$g(t,\tau) = \begin{cases} \exp\left(\dfrac{\tau + t}{2}\right) \displaystyle\int_{\tau}^{\infty} \dfrac{e^{-r}}{r}\, dr, & 0 < t \le \tau, \\[3mm] \exp\left(\dfrac{\tau + t}{2}\right) \displaystyle\int_{t}^{\infty} \dfrac{e^{-r}}{r}\, dr, & 0 < \tau \le t. \end{cases}$$

(f) $Lu = \tfrac{1}{2}u''$, $I = [0,1]$,

$u(0) + u(1) = 0,\ u'(0) = 0$,

$$g(x,y) = \begin{cases} 2x - y - 1, & 0 \le y \le x \le 1, \\ y - 1, & 0 \le x \le y \le 1. \end{cases}$$

[Note: In this case L^{-1} is not self-adjoint. Is L^{-1} normal?]

2. Verify Equation (7.2.5).

3. (a) Let $r = (\xi^2 + \eta^2)^{1/2}$ and D be the unit disk in R^2. Show that the following integrals are finite:

$$\int_D \log r \, d\xi \, d\eta, \qquad \int_D [\log r]^2 \, d\xi \, d\eta.$$

[Hint: Change to polar coordinates.]

(b) Use the last result to show that if v is continuous inside D, then the integral in (7.2.6) is well defined.

(c) Show that

$$\int_D \int_D |g(x,y; \xi,\eta)|^2 d\xi \, d\eta \, dx \, dy < \infty,$$

where g is given by (7.2.7).

(d) Show that $g(x,y; \xi,\eta) = g(\xi,\eta; x,y)$, where g is given by (7.2.7).

4. Show that if v is continuous in the interior of D and $v \in L_2(D)$ and if u is defined by

$$u(x,y) = \int_D g(x,y; \xi,\eta)v(\xi,\eta) \, d\xi \, d\eta,$$

where g is given by (7.2.7), then $\Delta u = v$ in D, and $u = 0$ on ∂D.

5. Consider the Laplacian operator

$$\Delta_3 u = \frac{\partial^2 u}{\partial x_1^2} + \frac{\partial^2 u}{\partial x_2^2} + \frac{\partial^2 u}{\partial x_3^2}$$

on $L_2(D)$, where D is the unit ball in R^3 and $x = (x_1,x_2,x_3) \in R^3$. Show that the Green's function for Δ_3 is given by

$$g(x,\xi) = \frac{1}{4\pi} \left[\frac{1}{\sigma_2} - \frac{1}{\sigma_1 \sigma_3} \right],$$

where σ_i is the distance between the points P_{i-1} and P_i and $P_0 = (0,0,0)$, $P_1 = (\xi_1,\xi_2,\xi_3)$, $P_2 = (x_1,x_2,x_3)$, and $P_3 = (\xi_1/|\xi|^2,\xi_2/|\xi|^2,\xi_3/|\xi|^2)$, where $|\xi|^2 = \xi_1^2 + \xi_2^2 + \xi_3^2$.

6. Show that the last exercise can be extended to the Laplacian operator

$$\Delta_n u = \sum_{i=1}^n \frac{\partial^2 u}{\partial x_i^2}$$

on $L_2(D)$, where D is the unit ball in R^n, $n \geq 3$. In this case, the Green's function becomes

$$g(x,\xi) = \frac{1}{(n-2)\tau_{n-1}} \left[\frac{1}{\sigma_2^{n-2}} - \frac{1}{\sigma_1^{n-2}\sigma_3^{n-2}} \right],$$

where σ_i is defined as in Exercise 5 and τ_{n-1} is the surface area of the $(n-1)$-dimensional unit sphere in R^n.

7. Show that the operator $Lu = u'' - (1 + x^2)u$ given in Example 2 is one-to-one. [*Hint*: Let $u(x) = \sum_{n=0}^{\infty} a_n x^n$ be a solution of $Lu = 0$. Show that no solution of this form lies in $L_2(-\infty,\infty)$ with $u(x) \to 0$ as $x \to \pm\infty$.]

3. SYMMETRIC OPERATORS

In this section we turn our attention to the first problem posed in the introductory section of this chapter. This is, let L be a linear operator on a Hilbert space H and let λ_0 be a complex number from the resolvent set of L. We now seek conditions on L itself in order that $(\lambda_0 I - L)^{-1}$ be self-adjoint.[6]

It seems natural to ask that the operator L satisfy the relationship

$$(Lx,y) = (x,Ly) \tag{7.3.1}$$

for all x and y in the domain $\mathscr{D}_L$. The last equation is very important, and we shall say that a linear operator L is *symmetric* if (7.3.1) holds for all x and y in $\mathscr{D}_L$. Some authors say that a linear operator L is *formally self-adjoint* if (7.3.1) holds. We prefer to use the term "symmetric" here and reserve the use of "self-adjoint" for a more specialized concept which will be defined in Section 10.

One point should be emphasized before we proceed and that is that the concept of symmetry for an operator L depends on the domain of definition $\mathscr{D}_L$ for L. This is an important observation because with many of the differential operators we shall study there is wide latitude in the choice of the domain of definition. For these operators, the domain of definition is oftentimes prescribed by means of certain boundary conditions and, consequently, these boundary conditions play an important role in determining whether the operator is symmetric or not.

We will consider some examples of symmetric linear operators shortly. However, before doing this it is useful to make note of the following results concerning the eigenvectors of a symmetric operator.

7.3.1 LEMMA. *Let λ be an eigenvalue for a symmetric linear operator L defined on a Hilbert space H. Then λ is real.*

Proof: Let $x \neq 0$ be an eigenvector associated with λ. By using (7.3.1) one then has

$$(\lambda - \bar{\lambda})(x,x) = \lambda(x,x) - \bar{\lambda}(x,x) = (\lambda x,x) - (x,\lambda x)$$
$$= (Lx,x) - (x,Lx) = 0.$$

Since $(x,x) \neq 0$, one has $\lambda = \bar{\lambda}$, that is, λ is real. ∎

7.3.2 LEMMA. *Let L be a symmetric linear operator defined on a Hilbert space H and let x_1 and x_2 be eigenvectors of L with associated eigenvalues λ_1 and λ_2. If $\lambda_1 \neq \lambda_2$, then $x_1 \perp x_2$.*

---

[6] Please note that we are not seeking conditions on L in order that $(\lambda_0 I - L)^{-1}$ be normal. We restrict out attention to self-adjoint operators because the examples of interest fall into this category. A discussion of normality for general (unbounded) linear operators can be found in Riesz and Sz. Nagy [1; p. 349] and Stone [1; pp. 311–331].

Proof: Since $Lx_i = \lambda_i x_i$, $i = 1, 2$, it follows from (7.3.1) that

$$(\lambda_1 - \lambda_2)(x_1, x_2) = \lambda_1(x_1, x_2) - \lambda_2(x_1, x_2) = (\lambda_1 x_1, x_2) - (x_1, \lambda_2 x_2)$$
$$= (Lx_1, x_2) - (x_1, Lx_2) = 0.$$

Since $\lambda_1 \neq \lambda_2$, one has $(x_1, x_2) = 0$, or $x_1 \perp x_2$. ∎

Thus the geometry of the eigenvectors for unbounded symmetric operators is similar to that for bounded self-adjoint operators. A very interesting, although not surprising, situation arises when a collection of eigenvectors for a symmetric operator forms an orthonormal basis for the Hilbert space.

7.3.3 THEOREM. *Let L be a symmetric linear operator on a Hilbert space H. Let $\{x_n\}$ be an orthonormal collection of eigenvectors for L with associated eigenvalues $\{\mu_n\}$. If $\{x_n\}$ is an orthonormal basis for H, then for every $x \in \mathscr{D}_L$ one has*

$$Lx = L(\textstyle\sum_n (x, x_n)x_n) = \sum_n L(x, x_n)x_n$$
$$= \textstyle\sum_n (x, x_n)Lx_n = \sum_n \mu_n(x, x_n)x_n. \qquad (7.3.2)^7$$

Proof: The proof of this theorem is very simple. It is based on the Fourier Series Theorem. If $x \in H$, then $x = \sum_n (x, x_n)x_n$, since $\{x_n\}$ is an orthonormal basis for H. Similarly, if $x \in \mathscr{D}_L$, then

$$Lx = \textstyle\sum_n (Lx, x_n)x_n = \sum_n (x, Lx_n)x_n$$
$$= \textstyle\sum_n (x, \mu_n x_n)x_n = \sum_n \mu_n(x, x_n)x_n$$
$$= \textstyle\sum_n (x, x_n)Lx_n. \qquad ∎$$

Even though the last theorem is a relatively simple result, it is nevertheless a very practical result. The reason for this is that for certain linear operators (especially for differential operators arising in boundary value problems) it is relatively easy to determine the collection of all eigenvectors. The only hypothesis of Theorem 7.3.3 that causes some concern in applications is showing that a given orthonormal collection of eigenvectors forms an orthonormal basis for the Hilbert space. In order to solve this problem one can either use the techniques of Section 5.18 or one can use some of the results presented in Section 5.

EXERCISES

1. Show that the operators defined in Exercises 1(a), (b), (c), (d), and (e), Section 2 are symmetric.

2. Show that the operator defined in Exercise 1 (f), Section 2 is not symmetric.

---

[7] Equation (7.3.2) holds for all bounded operators L. What is important here is that L may be unbounded.

3. Consider the operator $\tilde{L}u = u''$ on $L_2[0,1]$ with boundary condition $u'(0) = 0$.
   (a) Show that every real number λ is an eigenvalue for $\tilde{L}$.
   (b) Does $\tilde{L}$ have other eigenvalues?
   (c) Show that $\tilde{L}$ is an extension of the operator $2L$, where L is defined in Exercise 1 (f), Section 2.
   (d) Is $\tilde{L}$ symmetric?

4. Find the eigenvalues and eigenvectors for the operators defined in Exercise 1, Section 2.

5. Find the eigenvalues and eigenvectors for the operator defined in Example 1, Section 2.

6. Find the eigenvalues and eigenvectors for the Laplacian operator defined in Example 3, Section 2. [*Hint*: Express the Laplacian operator in polar coordinates (r,θ) and separate the variables.]

4. EXAMPLES OF SYMMETRIC OPERATORS

The following operators are important in quantum mechanics.

EXAMPLE 1. (THE POSITION OPERATOR Q.) Consider the Hilbert space $L_2(-\infty,\infty)$. Let

$$\mathcal{D}_Q = \left\{u(x) \in L_2(-\infty,\infty): \int_{-\infty}^{\infty} x^2|u(x)|^2\, dx < \infty\right\}$$

and define $Q: \mathcal{D}_Q \to L_2(-\infty,\infty)$ by

$$(Qu)(x) = xu(x). \tag{7.4.1}$$

Since $C_0^\infty(-\infty,\infty) \subset \mathcal{D}_Q$, we see that Q is densely defined. Furthermore, if $u, v, \in \mathcal{D}_Q$, then

$$(Qu,v) = \int_{-\infty}^{\infty} xu(x)\overline{v(x)}\, dx = \int_{-\infty}^{\infty} u(x)\overline{xv(x)}\, dx = (u,Qv).$$

Thus Q is symmetric.

We will show in the exercises that the spectrum of the operator Q is only a continuous spectrum and that it is precisely the real line R. ∎

EXAMPLE 2. (THE POSITION OPERATOR Q_k.) Consider now the Hilbert space $L_2(R^n)$, where $x = (x_1,\ldots,x_n) \in R^n$. For $k = 1, 2, \ldots, n$ let

$$\mathcal{D}_{Q_k} = \left\{u(x) \in L_2(R^n): \int_{R^n} x_k^2|u(x)|^2\, dx < \infty\right\},$$

where $dx = dx_1 \ldots dx_n$, and define Q_k by

$$Q_k: u(x_1,\ldots,x_n) \to x_k u(x_1,\ldots,x_n), \tag{7.4.2}$$

where $u(x) \in \mathcal{D}_{Q_k}$. It is easy to see that Q_k is a symmetric operator. Also, it is easy to show that the spectrum of Q_k, which is a continuous spectrum only, consists of the real line R. ∎

EXAMPLE 3. (POTENTIAL OPERATOR $V(Q)$.)   For $k = 1, \ldots, n$, let Q_k be given by the last example. Let $V(x_1, \ldots, x_n) = V(x)$ be a real-valued, continuous function defined for $x = (x_1, \ldots, x_n) \in R^n$ and define $V(Q) = V(Q_1, \ldots, Q_n)$ by

$$V(Q): u(x) \rightarrow V(x)u(x). \qquad (7.4.3)$$

If we define the domain of $V(Q)$ by

$$\mathscr{D}_{V(Q)} = \{u(x) \in L_2(R^n): V(x)u(x) \in L_2(R^n)\},$$

then it is easy to see that $V(Q)$ is symmetric.

It is not necessary to assume that $V(x)$ is continuous. The function may even have poles in R^n. An example, which arises in some applications in mechanics, occurs when $n = 3l$,

$$r_{0k} = [x_{3k-2}{}^2 + x_{3k-1}{}^2 + x_{3k}{}^2]^{1/2}, \qquad 1 \leq k \leq l,$$
$$r_{jk} = [(x_{3j-2} - x_{3k-2})^2 + (x_{3j-1} - x_{3k-1})^2 + (x_{3j} - x_{3k})^2]^{1/2},$$

then

$$V(x) = \sum_{j \neq k} \frac{C_{jk}}{r_{jk}},$$

where C_{jk} are real numbers. The corresponding operator $V(Q)$, in this case, is called a *Coulomb potential*. This type of potential arises in quantum mechanics when one is studying the interaction between l electrons and a nucleus. ∎

EXAMPLE 4. (THE MOMENTUM OPERATOR P.)   Consider the Hilbert space $L_2(-\infty, \infty)$. Define $P: \mathscr{D}_P \rightarrow L_2(-\infty, \infty)$ by

$$P: u(x) \rightarrow -iu'(x), \qquad (7.4.4)$$

where the domain $\mathscr{D}_P$ is the space $C_0{}^1(-\infty, \infty)$, the space of all C^1-functions with compact support. Since $C_0{}^1(-\infty, \infty)$ is dense in $L_2(-\infty, \infty)$ we see that P is densely defined. Furthermore, if $u, v \in \mathscr{D}_P$, then by integrating by parts we get

$$(Pu, v) = -\int_{-\infty}^{\infty} iu'(x)\overline{v(x)} \, dx$$

$$= -i \lim_{R \rightarrow +\infty} (u(R)\overline{v(R)} - u(-R)\overline{v(-R)}) - \int_{-\infty}^{\infty} u(x)\overline{iv'(x)} \, dx$$

$$= (u, Pv).$$

Hence P is symmetric. The spectrum of this operator is analyzed in Exercise 5. ∎

EXAMPLE 5. (THE MOMENTUM OPERATOR P_k.)   Consider now the Hilbert space $L_2(R^n)$. For $k = 1, 2, \ldots, n$ define P_k by

$$P_k: u(x_1, \ldots, x_n) \rightarrow -i \frac{\partial}{\partial x_k} u(x_1, \ldots, x_n), \qquad (7.4.5)$$

where the domain of P_k is $C_0^\infty(R^n)$. By integrating by parts one can easily show that P_k is symmetric.

The operator P_k^2 is also of interest. This is, of course, the operator

$$P_k^2: u(x_1,\dots,x_n) \to \frac{\partial^2}{\partial x_k^2}\, u(x_1,\dots,x_n), \tag{7.4.6}$$

which is also symmetric on $C_0^\infty(R^n)$. ∎

EXAMPLE 6. (THE LAPLACIAN OPERATOR Δ_n.)   The Laplacian operator Δ_n can be represented in terms of the momenta operators P_k, $k = 1, 2, \dots, n$,

$$\Delta_n = \sum_{k=1}^{n} P_k^2,$$

which means that

$$\Delta_n u = \frac{\partial^2 u}{\partial x_1^2} + \cdots + \frac{\partial^2 u}{\partial x_n^2},$$

for $u \in C_0^\infty(R^n)$.

In order to show that Δ_n is symmetric we use *Green's formula* (see Taylor [1, pp. 458–459])

$$\int_\Omega (u\overline{\Delta_n v} - \bar{v}\Delta_n u)\, dx = \int_{\partial\Omega} \left(u\,\frac{\overline{\partial v}}{\partial \nu} - \frac{\partial u}{\partial \nu}\,\bar{v} \right) ds,$$

where Ω is a bounded region in R^n, $\partial/\partial\nu$ denotes the outward-normal-derivative and $\partial\Omega$ is the boundary of Ω. Now let $u, v \in C_0^\infty(R^n)$ and choose Ω large enough to contain the support of both u and v. One then has $u = v = 0$ on $\partial\Omega$ and it follows from Green's formula that

$$(\Delta_n u, v) - (u, \Delta_n v) = \int_{R^n} [(\Delta_n u)\bar{v} - u(\overline{\Delta_n v})]\, dx$$

$$= \int_\Omega [(\Delta_n u)\bar{v} - u(\overline{\Delta_n v})]\, dx = 0.$$

Hence Δ_n is symmetric. ∎

EXERCISES

1. (a) Show that the operator Q in (7.4.1) has no point spectrum.
   (b) Show that every $\lambda \in R$ is in the continuous spectrum of Q.
   (c) Show that every nonreal complex number is in the resolvent set of Q. (Compare this with Exercise 7, Section 6.6.)

2. Let $f: R \to C$ be a continuous function and define $f(Q)$ by

$$f(Q): u(x) \to f(x)u(x),$$

where the domain of $f(Q)$ is

$$\mathscr{D} = \{u(x) \in L_2(-\infty,\infty): f(x)u(x) \in L_2(-\infty,\infty)\}.$$

(a) Show that $f(Q)$ is symmetric when $f(x)$ is real valued.

(b) Show that the spectrum of $f(Q)$ is $\overline{f(R)}$.

(c) Show that λ is an eigenvalue of $f(Q)$ if and only if the set $f^{-1}(\{\lambda\})$ has positive measure.

(d) Let λ_0 be an eigenvalue of $f(Q)$. Show that the null space $\mathcal{N}(f(Q) - \lambda_0 I)$ is infinite dimensional.

3. (a) Show that the operator Q_k given by (7.4.2) is symmetric.

(b) Show that the spectrum of Q is the real line R and that this is a continuous spectrum.

4. (a) Show that the operator $V(Q)$ given by (7.4.3) is symmetric when $V(x)$ is a real-valued continuous function.

(b) Show that a Coulomb potential $V(Q)$ is a symmetric operator on an appropriate domain of definition.

5. Analyze the spectrum of
   (a) The momentum operator P.
   (b) The momentum operator P_k.
   [*Hint*: Use Example 6, Section 6.6.]

5.   STURM-LIOUVILLE OPERATORS

The operators we shall study in this section arise in the context of second-order ordinary linear differential equations of the form

$$-(pu')' + qu = kv,$$

or

$$(pu')' + (\lambda k - q)u = 0,$$

subject to appropriate boundary conditions. Operators of this type are rather common in applications. They arise in the study of vibrating strings, vibrating membrane, resonance in a cavity, transmission lines, wave guides as well as optimal control problems. We begin with the definition of a (nonsingular) Sturm-Liouville operator.

7.5.1 DEFINITION.   Let $p(x)$, $p'(x)$, $q(x)$, and $\omega(x)$ be continuous real-valued functions on the finite interval $a \le x \le b$, where $a < b$, and assume that $p(x) > 0$ and $\omega(x) > 0$ for $a \le x \le b$. Let H be the complex Hilbert space

$$\left\{ u(x) : \int_a^b |u(x)|^2 \omega(x) \, dx < \infty \right\},$$

where the inner product on H is given by

$$(u,v) = \int_a^b u(x)\overline{v(x)}\omega(x) \, dx. \tag{7.5.1}$$

[The term $\omega(x)$ in (7.5.1) is called a weight function.] Now let $\mathscr{D}$ denote the collection of all C^2-functions $u(x)$ in H satisfying the boundary conditions:

$$B_1 u = B_2 u = 0, \tag{7.5.2}$$

where

$$\begin{aligned} B_1 u &= \beta_1 u(a) + \gamma_1 u'(a), \\ B_2 u &= \beta_2 u(b) + \gamma_2 u'(b), \end{aligned} \tag{7.5.3}$$

$\beta_1, \beta_2, \gamma_1, \gamma_2$, are real constants with

$$|\beta_1| + |\gamma_1| > 0, \qquad |\beta_2| + |\gamma_2| > 0. \tag{7.5.4}$$

Under these conditions the operator

$$Lu = \frac{1}{\omega}\left[-(pu')' + qu\right] \tag{7.5.5}$$

is said to be a *Sturm-Liouville operator* if $\mathscr{D} = \mathscr{D}_L$. Since $C_0^\infty[a,b] \subset \mathscr{D}$, we see that L is densely defined.

7.5.2 LEMMA.   *Let L be a Sturm-Liouville operator. Then for every real number λ, the operator $L - \lambda I$ is a Sturm-Liouville operator with $\mathscr{D}_{(L-\lambda I)} = \mathscr{D}_L$.*

Proof: Since

$$(L - \lambda I)u = \frac{1}{\omega}\left[-(pu')' + (q - \lambda\omega)u\right].$$

the conclusion is obvious.   ∎

7.5.3. THEOREM.   *Let L be a Sturm-Liouville operator. Then L is symmetric.*

Proof: Let L be given by (7.5.5) and let u and v be in $\mathscr{D}$. Then

$$(Lu,v) - (u, Lv) = \int_a^b \{-(pu')'\bar{v} + qu\bar{v} + (p\bar{v}')'u - qu\bar{v}\}\, dx$$

$$= \int_a^b \{-(pu')'\bar{v} + (p\bar{v}')'u\}\, dx.$$

By integrating by parts we get

$$(Lu,v) - (v,Lu) = p(b)[u(b)\overline{v'(b)} - u'(b)\overline{v(b)}] - p(a)[u(a)\overline{v'(a)} - u'(a)\overline{v(a)}]. \tag{7.5.6}$$

Since β_i and γ_i in (7.5.3) are real, we see that $\bar{u}$ and $\bar{v}$ also lie in $\mathscr{D}$. If we apply B_1 to u and $\bar{v}$, we get

$$\begin{aligned} \beta_1 u(a) + \gamma_1 u'(a) &= 0, \\ \beta_1 \bar{v}(a) + \gamma_1 \bar{v}'(a) &= 0. \end{aligned} \tag{7.5.7}$$

It follows from (7.5.4) that at least one of the terms β_1, γ_1 is nonzero. Hence the determinant of the system (7.5.7) must vanish, that is,

$$\begin{vmatrix} u(a) & u'(a) \\ \bar{v}(a) & \bar{v}'(a) \end{vmatrix}^2 = 0.$$

If we would apply B_2 to u and $\bar{v}$, we would get a system of equations similar to (7.5.7). From this we would conclude, in an analogous manner, that

$$\begin{vmatrix} u(b) & u'(b) \\ \bar{v}(b) & \bar{v}'(b) \end{vmatrix}^2 = 0.$$

If the last two equalities are inserted in (7.5.6) we see that $(Lu,v) = (u,Lv)$. Hence L is symmetric. ∎

Carefully note that the symmetry of L in the last theorem is determined by the form of (7.5.5) *along with* the boundary conditions (7.5.2).

Let us now study the eigenvalues of the Sturm-Liouville operator L given by (7.5.5) and (7.5.2). The equation $Lu = \lambda u$ reduces to

$$u'' + \frac{p'}{p} u' + \frac{\lambda \omega - q}{p} u = 0. \tag{7.5.8}$$

Let $u_1(x,\lambda)$ and $u_2(x,\lambda)$ be a fundamental system for (7.5.8), that is, $u_1(x,\lambda)$ and $u_2(x,\lambda)$ are two linearly independent solutions of (7.5.8). We claim that λ is an eigenvalue for L if and only if

$$\Delta(\lambda) = \begin{vmatrix} B_1 u_1 & B_1 u_2 \\ B_2 u_1 & B_2 u_2 \end{vmatrix} = 0. \tag{7.5.9}$$

Indeed, every solution of (7.5.8) can be written as $u = c_1 u_1 + c_2 u_2$, where c_1 and c_2 are complex numbers. The boundary conditions (7.5.2) then become

$$\begin{aligned} B_1 u = c_1 B_1 u_1 + c_2 B_1 u_2 = 0, \\ B_2 u = c_1 B_2 u_1 + c_2 B_2 u_2 = 0. \end{aligned} \tag{7.5.10}$$

If λ is an eigenvalue, then there is a nontrivial solution (c_1,c_2) of (7.5.10); hence the determinant $\Delta(\lambda)$ must vanish. Conversely, if the determinant $\Delta(\lambda)$ does vanish, then we can find a nontrivial solution (c_1,c_2) of (7.5.10). This then defines an eigenfunction $u = c_1 u_1 + c_2 u_2$ for the eigenvalue λ.

If 0 is not an eigenvalue of L, then L is one-to-one and the inverse L^{-1} is defined on the range $\mathscr{R}_L$. Let us now show that L^{-1} can be represented as an integral operator where the Green's function is continuous. In the proof of the next theorem we shall give a method for constructing the Green's function of any Sturm-Liouville operator that does not have 0 as an eigenvalue.

7.5.4 THEOREM. *Let L be a Sturm-Liouville operator given by (7.5.5) and (7.5.2) and assume that 0 is not an eigenvalue of L. Then there is a continuous function $g(x,y)$ defined for $a \leq x, y \leq b$ such that*

$$(L^{-1}v)(x) = \int_a^b g(x,y)v(y)\omega(y)\,dy$$

for all v in the range $\mathcal{R}_L$. Furthermore, $g(x,y)$ can be chosen so that it is real-valued and $g(x,y) = g(y,x)$. Consequently, L^{-1} is a compact, self-adjoint[8] *operator.*

Proof: For $j = 1, 2$ let $u_j(x)$ be a nontrivial real solution of the boundary-value problem:

$$pu'' + p'u' - qu = 0, \qquad B_j u_j = 0.$$

Since 0 is not an eigenvalue of L, it follows from the consideration above that $u_1(x)$ and $u_2(x)$ are linearly independent, and $B_i u_j \neq 0$ for $i \neq j$. Let

$$W(x) = \begin{vmatrix} u_1(x) & u_2(x) \\ u_1'(x) & u_2'(x) \end{vmatrix}$$

be the Wronskian for these two solutions. Then the general solution of $Lu = v$ or

$$-(pu')' + qu = \omega v$$

becomes

$$u(x) = c_1 u_1(x) + c_2 u_2(x) + z(x),$$

where

$$z(x) = \int_a^x \frac{u_1(x)u_2(y) - u_2(x)u_1(y)}{p(a)W(a)} v(y)\omega(y)\, dy.$$

We want to choose the coefficients so that when v is continuous then u is in $\mathcal{D}$, that is, u satisfies the boundary conditions (7.5.2). If we set $c_2 = 0$, and

$$c_1 = -\int_a^b \frac{u_2(y)}{p(a)W(a)} v(y)\omega(y)\, dy,$$

then the solution $u(x)$ becomes

$$u(x) = -\int_x^b \frac{u_1(x)u_2(y)}{p(a)W(a)} v(y)\omega(y)\, dy - \int_a^x \frac{u_2(x)u_1(y)}{p(a)W(a)} v(y)\omega(y)\, dy$$

$$= \int_a^b g(x,y)v(y)\omega(y)\, dy,$$

where $g(x,y)$ satisfies

$$p(a)W(a)g(x,y) = \begin{cases} -u_1(x)u_2(y), & a \leq x \leq y \leq b, \\ -u_2(x)u_1(y), & a \leq y \leq x \leq b. \end{cases}$$

It is easy to verify that this function u satisfies the boundary conditions (7.5.2). Furthermore the kernel $g(x,y)$ is evidently continuous.

Since u_1 and u_2 are real-valued functions we see that $g(x,y)$ is real-valued and $g(x,y) = g(y,x)$. Thus, as noted in Section 2, the operator L^{-1} is compact and self-adjoint. ∎

[8] See footnote concerning Theorem 7.1.1.

What happens if 0 is an eigenvalue of L? Then, of course, L does not have an inverse. However, if λ is a real number and is not an eigenvalue of L, then $(\lambda I - L)^{-1}$ exists. If we now combine Lemma 7.5.2 and the last theorem we get the next result.

7.5.5 COROLLARY. *Let L be a Sturm-Liouville operator given by (7.5.5) and assume that a real number λ is not an eigenvalue of L. Then there is a Green's function $g(x,y,\lambda)$, defined for $a \leq x, y \leq b$, continuous in x and y and such that*

$$((\lambda I - L)^{-1}v)(x) = \int_a^b g(x,y,\lambda)v(y)\omega(y)\,dy$$

for all v in the range for $\lambda I - L$. Moreover the operator $(\lambda I - L)^{-1}$ is compact and self-adjoint.

Let L be a given Sturm-Liouville operator. With the above corollary in mind we pose the question: "Does there exist a real number λ with the property that λ is not an eigenvalue of L?" The answer is definitely yes.

7.5.6 THEOREM. *Let L be a Sturm-Liouville operator. Then L has at most a countable number of eigenvalues, all of which are real.*

Proof: It follows from Lemma 7.3.2 that the eigenvectors of L, associated with distinct eigenvalues, are mutually orthogonal. Since the Hilbert space H is separable (Example 12, Section 3.12) there are at most a countable number of nonzero mutually orthogonal vectors. Hence L has at most a countable number of eigenvalues. ∎

In the exercises you will be asked to show that the eigenvalues of L are actually bounded below.

We can now state our main result.

7.5.7 THEOREM. *Let L be a Sturm-Liouville operator. Then there exists a sequence of real numbers $\{\mu_1, \mu_2, \ldots\}$ and an orthonormal basis $\{\phi_1, \phi_2, \ldots\}$ of H such that each ϕ_n is a C^2-function and $L\phi_n = \mu_n \phi_n$.*

Moreover, each real number μ appears at most twice in the sequence $\{\mu_1, \mu_2, \ldots\}$. Furthermore, each function u in H can be written in terms of the Fourier series

$$u = \sum_{n=1}^{\infty} (u,\phi_n)\phi_n.$$

If $u \in \mathcal{D}_L$, then

$$\sum_{n=1}^{\infty} |(u,\phi_n)|^2 \mu_n^2 < \infty \qquad (7.5.11)$$

and

$$Lu = \sum_{n=1}^{\infty} \mu_n(u,\phi_n)\phi_n.$$

Proof: Choose a real number λ so that λ is not an eigenvalue of L. The operator $(\lambda I - L)^{-1}$ is then compact and self-adjoint by Corollary 7.5.5. Now apply the Spectral Theorem to $(\lambda I - L)^{-1}$ and let $\{\tilde{\mu}_1, \tilde{\mu}_2, \ldots\}$ be the eigenvalues of $(\lambda I - L)^{-1}$ with corresponding eigenfunctions $\{\phi_1, \phi_2, \ldots\}$. It follows from Theorem 7.1.1, that if

$$\mu_n = \lambda - \tilde{\mu}_n^{-1},$$

then $\{\mu_1, \mu_2, \ldots\}$ are the eigenvalues of L with corresponding eigenfunctions $\{\phi_1, \phi_2, \ldots\}$.

Each ϕ_n is a C^2-function since it is the solution of a second-order differential equation (7.5.8) with continuous coefficients. Also, each eigenvalue appears at most twice since Equation (7.5.8) has at most two linearly independent solutions for each λ.

The characterization of the domain $\mathscr{D}_L$ in terms of (7.5.11) we shall leave as an exercise. ∎

A word on methodology seems warranted. In this section we constructed a Green's function for the inverse of a Sturm-Liouville operator, and you will be asked to work out some specific examples in the exercises. In practice one seldom seeks the eigenvalues and eigenfunctions in this fashion. It is almost always easier to work directly with the differential equation and boundary conditions.

EXERCISES

1. In the following problems you are given a second-order differential operator L, an interval I, and boundary conditions so that L becomes a Sturm-Liouville operator. You are asked to find the collection of eigenfunctions and associate eigenvalues for L.

   (a) $Lu = u''$, $I = [0,1]$,

   $u(0) = u'(1) = 0$.

   (b) $Lu = u''$, $I = [0,1]$,

   $u(0) = u(1) = 0$.

   (c) $Lu = \tau u''$, $I = [0,l]$,

   $u(0) = u(l) = 0$ $(\tau > 0)$.

   (d) $Lu = u''$, $I = [-1,1]$,

   $u(-1) = u(1) = 0$.

   (e) $Lu = u''$, $I = [0,1]$,

   $u'(0) = u'(1) = 0$.

2. Extend the theory of Sturm-Liouville operators to operators of the form (7.5.5) with the more general boundary conditions, namely,

$$B_1 u = B_2 u = 0,$$

where

$$B_1 u = \beta_{11} u(a) + \beta_{12} u'(a) + \beta_{13} u(b) + \beta_{14} u'(b)$$

and

$$B_2 u = \beta_{21} u(a) + \beta_{22} u'(a) + \beta_{23} u(b) + \beta_{24} u'(b),$$

where the β_{ij}'s are real,

$$\text{rank} \begin{pmatrix} \beta_{11} & \beta_{12} & \beta_{13} & \beta_{14} \\ \beta_{21} & \beta_{22} & \beta_{23} & \beta_{24} \end{pmatrix} = 2,$$

and

$$p(a)[\beta_{13}\beta_{24} - \beta_{14}\beta_{23}] = p(b)[\beta_{11}\beta_{22} - \beta_{12}\beta_{21}].$$

(See Hellwig [1, pp. 39–47] and Coddington and Levinson [1, Chaps. 7, 11].)

3. This exercise follows the format of Exercise 1, but uses the conclusion of Exercise 2.

   (a) $Lu = u''$, $I = [0,1]$,

   $u(0) = u(1)$, $u'(0) = u'(1)$.

   (b) $Lu = u'' - u$, $I = [0,1]$,

   $u(0) - u'(0) + u(1) = 0$,

   $u(0) + u'(0) + 2u'(1) = 0$.

4. If the term $p(x)$ in (7.5.5) vanishes at one of the endpoints or if the interval $[a,b]$ becomes infinite, then the Sturm-Liouville operator is said to be *singular*. For some of these singular operators the theory is similar to that of nonsingular Sturm-Liouville operators. Many of the classical functions arise in this way. We follow the format of Exercise 1. For these problems we consider the Hilbert space $L_2(I)$ with the inner product given by $(u,v) = \int_I u\bar{v}\omega \, dx$, where $\omega > 0$ is a weight function.

   (a)   (Legendre Polynomials): $Lu = [(x^2 - 1)u']'$, $I = [-1,1]$, $\omega(x) = 1$, $u(1)$, and $u(-1)$ are finite.

   (b)   (Tchebychev Polynomials): $Lu = (1 - x^2)u'' - xu'$, $I = [-1,1]$, $\omega(x) = (1 - x^2)^{-1/2}$, $u(1)$ and $u(-1)$ are finite.

   (c)   (Laguerre Polynomials): $Lu = xu'' + (1 - x)u'$, $I = [0,\infty)$, $\omega(x) = e^{-x}$, $u(1)$ are finite.

5. The *violin operator* is defined to be the Sturm-Liouville operator

$$Lu = -\frac{\tau}{\rho} u''$$

on $0 \le x \le l$ with $u(0) = u(l) = 0$. For this operator τ denotes the tension in the violin string, ρ the density of the string, and l the length of the string. Let $\lambda_1 < \lambda_2 < \cdots$ denote the eigenvalues of L.

(a) Show that $\lambda_n = n^2\alpha$, for $n = 1, 2, \ldots$ and an appropriate choice of α.

(b) The first eigenvalue λ_1 is defined to be the *pitch* of the violin operator. Show that one can increase the pitch by either increasing the tension, decreasing the length, or decreasing the density.

6. Other problems can sometimes be reduced to Sturm-Liouville problems, as we now illustrate. Consider the Brownian motion $X(t)$. That is, $X(t)$ is a real stochastic process defined for $0 \le t \le 1$ and satisfying:

$$X(0) = 0;$$

$$E(X(t)) = 0, \qquad \text{for } 0 \le t \le 1;$$

$$E(|X(t)|^2) = t^2, \qquad \text{for } 0 \le t \le 1;$$

$$k(t,s) = E(X(t)X(s)) = \min(t,s).$$

(a) Show that the eigenvalues of the integral operator K with kernel $k(t,s)$ are solutions of

$$\lambda\phi(t) = \int_0^t s\phi(s)\, ds + \int_t^1 t\phi(s)\, ds,$$

where $\phi(0) = 0$.

(b) Show by taking two derivatives this reduces to the Sturm-Liouville problem

$$\phi'' + \frac{1}{\lambda}\phi = 0,$$

$$\phi(0) = 0, \qquad \phi'(1) = 0.$$

(c) Find the eigenvalues and corresponding eigenvectors.

(d) For an interpretation see the Karhunen-Loeve expansion (Theorem 6.13.1).

6. GÅRDING'S INEQUALITY

There is another—almost direct—way of showing that a linear operator L has a compact resolvent and that is by showing that L satisfies a Gårding inequality. In order to explain this we need some additional concepts.

7.6.1 DEFINITION. Let X_1 be a linear subspace in a Banach space $(X_2, \|\cdot\|_2)$ and assume that $\|\cdot\|_1$ is a norm on X_1. We shall say that the space $(X_1, \|\cdot\|_1)$ is *compact in* $(X_2, \|\cdot\|_2)$ if the unit ball in X_1

$$B_1 = \{x \in X_1 : \|x\|_1 \le 1\}$$

lies in a compact subset of $(X_2, \|\cdot\|_2)$. The following characterization of this concept is a simple consequence of the Compactness Theorem (Theorem 3.17.13).

7.6.2 THEOREM.   *Let X_1 be a linear subspace of a Banach space $(X_2, \|\cdot\|_2)$ and let $\|\cdot\|_1$ be a norm on X_1. Then the following statements are equivalent:*

(a) *$(X_1, \|\cdot\|_1)$ is compact in $(X_2, \|\cdot\|_2)$.*

(b) *Every subset of X_1 that is bounded in the $\|\cdot\|_1$-norm is relatively compact in the $\|\cdot\|_2$-norm.*

(c) *Let $\{x_n\}$ be any sequence in X_1 with $\|x_n\|_1 \leq b < \infty$ for all n. Then there is a subsequence of $\{x_n\}$ that converges in $(X_2, \|\cdot\|_2)$.*

(d) *Every subset of X_1 that is bounded in the $\|\cdot\|_1$-norm is totally bounded in the $\|\cdot\|_2$-norm.*

EXAMPLE 1.[9]   The Sobolev spaces $H_0{}^n(\Omega)$, for $n = 0, 1, \ldots$, are defined in Example 12, Section 5.13. Recall that $H_0{}^n(\Omega)$ is the completion of $C_0{}^\infty(\Omega)$ with respect to the norm $\|\cdot\|_n$ generated by the inner product

$$(u,v)_n = \int_\Omega \sum_{|\alpha| \leq n} D^\alpha u(x) \overline{D^\alpha v(x)} \, dx.$$

It is easy to see that the spaces are decreasing with n, that is,

$$\cdots \subset H_0{}^{n+1} \subset H_0{}^n \subset \cdots \subset H_0{}^1 \subset H_0{}^0 = L_2.$$

Furthermore one can prove the following result.

7.6.3 THEOREM. (RELLICH'S THEOREM.)   *Let k and l be nonnegative integers with $l > k$, and let Ω be an open bounded domain in R^m. Then $(H_0{}^l(\Omega), \|\cdot\|_l)$ is compact in $(H_0{}^k(\Omega), \|\cdot\|_k)$.*

Proof: We shall prove this for the case $l = 1$ and $k = 0$. The other cases are treated in the exercises.

Before proceeding with the proof it will be helpful if the reader reviewed the results of Exercise 29, Section 3.17, and Exercise 16, Section 5.6, as well as Example 4, Section 3.17. Let us make note of some of the properties of the mollifier operator defined in Exercise 16, Section 5.6. First recall that

$$J_\varepsilon : L_2(\Omega) \to L_\infty(\Omega)$$

is a bounded linear operator. This means that there is a constant K such that

$$\sup_{x \in \Omega} |J_\varepsilon u(x)| \leq K \left(\int_\Omega |u(x)|^2 \, dx \right)^{1/2} \tag{7.6.1}$$

for all $u \in L_2(\Omega)$. Furthermore, if $u \in C_0{}^\infty(\Omega)$, then

$$D^\alpha(J_\varepsilon u) = (-1)^{|\alpha|} J_\varepsilon(D^\alpha u).$$

---

[9] This is a rather long and technical example which may be skipped on the first reading. However, it is important for later examples.

Hence there is a constant $K_{|\alpha|}$, depending on $|\alpha|$, such that

$$\sup_{x \in \Omega} |D^\alpha (J_\varepsilon u(x))| \le K_{|\alpha|} \left(\int_\Omega |D^\alpha u(x)|^2 \, dx \right)^{1/2} \tag{7.6.2}$$

If $v \in C^\infty(\Omega)$, then the gradient of v is the vector

$$\nabla v = \left(\frac{\partial v}{\partial x_1}, \ldots, \frac{\partial v}{\partial x_m} \right)$$

and

$$|\nabla v|^2 = \sum_{i=1}^m \left| \frac{\partial v}{\partial x_i} \right|^2 = \sum_{|\alpha|=1} |D^\alpha v|^2.$$

Now by applying (7.6.2) we get

$$\sup_{x \in \Omega} |\nabla J_\varepsilon v(x)|^2 \le \sum_{|\alpha|=1} \sup_{x \in \Omega} |D^\alpha (J_\varepsilon v(x))|^2 \le K_1^2 \sum_{|\alpha|=1} \int_\Omega |D^\alpha v(x)|^2 \, dx,$$

or in terms of the norm $\| \cdot \|_1$ we get

$$\sup_{x \in \Omega} |\nabla J_\varepsilon v(x)| \le K_1 \|v\|_1. \tag{7.6.3}$$

Now let $A = \{u\}$ be a bounded set in $(H_0^1(\Omega), \| \cdot \|_1)$. Since $C_0^\infty(\Omega)$ is dense in $(H_0^1(\Omega), \| \cdot \|_1)$, for every $\eta > 0$ we can find a set $A_\eta = \{v\}$ in $C_0^\infty(\Omega)$ such that for each u in A there is a v in A_η with $\|u - v\|_1 \le \eta$. It is easy to see that for each η, A_η is a bounded set in $(H_0^1(\Omega), \| \cdot \|_1)$. If we can show that for every $\eta > 0$, the set A_η is totally bounded in $(H_0^0(\Omega), \| \cdot \|_0)$, it will follow that A is totally bounded in $(H_0^0(\Omega), \| \cdot \|_0)$, by Exercise 29, Section 3.17.

Now fix $\eta > 0$ and consider A_η, which can be described as a collection $\{v\}$ in $C_0^\infty(\Omega)$ with the property that there is a B with $\|v\|_1 \le B$ for all $v \in A_\eta$. We will now show that for every $\varepsilon > 0$ the set $J_\varepsilon(A_\eta)$ is totally bounded in $L_2 = H_0^0$. We do this by showing that the set of functions $J_\varepsilon(A_\eta)$ is pointwise compact and equi-continuous, see Example 4, Section 3.17.

If $v \in A_\eta$, then it follows from (7.6.1) that

$$|J_\varepsilon v(x)| \le K \|v\|_0 \le K \|v\|_1 \le KB.$$

Thus $J_\varepsilon(A_\eta)$ is pointwise compact. Similarly, by the Mean Value Theorem[10] one has

$$|J_\varepsilon v(x) - J_\varepsilon v(y)| = |\nabla (J_\varepsilon v) \cdot (x - y)|,$$

where the $\cdot$ is the usual dot product and the gradient $\nabla(J_\varepsilon v)$ is evaluated at some point on the line segment connecting x and y. By applying the Schwarz Inequality to this dot product and by using (7.6.3) we get

$$|J_\varepsilon v(x) - J_\varepsilon v(y)| \le \sup_{x \in \Omega} |\nabla J_\varepsilon v| \cdot |x - y| \le K_1 \|v\|_1 |x - y|,$$

---

[10] See Taylor [1, p. 224].

where $|x - y|^2 = \sum_{i=1}^m |x_i - y_i|^2$. Since $\|v\|_1 \le B$ we see that $J_\varepsilon(A_\eta)$ is equicontinuous. Finally we note that (see Exercise 16, Section 5.6)

$$|J_\varepsilon v(x) - v(x)| = \left| \int_{|\xi| \le 1} \rho(\xi)[v(x - \varepsilon\xi) - v(x)] \, d\xi \right|$$

$$= \left| \int_{|\xi| \le 1} \rho(\xi)\nabla v \cdot \varepsilon\xi \, d\xi \right|$$

$$\le \varepsilon \int_{|\xi| \le 1} \rho(\xi) |\nabla v| \, |\xi| \, d\xi$$

$$\le \sup_{x \in \Omega} |\nabla v| \cdot \varepsilon \le K \|v\|_1 \varepsilon \le KB\varepsilon.$$

Thus

$$\left\{ \int_\Omega |J_\varepsilon v(x) - v(x)|^2 \, dx \right\}^{1/2} \le KB |\Omega|^{1/2} \varepsilon,$$

where $|\Omega|$ denotes the Lebesgue measure of Ω. The result for the case $l = 1$ and $k = 0$ now follows from Exercise 16, Section 5.6. ∎

Now let $(X_1, \| \cdot \|_1)$ be compact in $(X_2, \| \cdot \|_2)$ and let T be a linear operator on X_2. Assume that the domain $\mathcal{D}(T)$ lies in X_1. We shall say that T satisfies a *Gårding Inequality* if there is a constant $a > 0$ such that

$$a\|x\|_1 \le \|Tx\|_2, \qquad x \in \mathcal{D}(T). \tag{7.6.4}$$

If we consider T as a mapping from $(\mathcal{D}(T), \| \cdot \|_1)$ to $(\mathcal{R}(T), \| \cdot \|_2)$, then Equation (7.6.4) says that T is bounded below. Therefore the inverse

$$T^{-1} : (\mathcal{R}(T), \| \cdot \|_2) \to (\mathcal{D}(T), \| \cdot \|_1)$$

exists and is bounded, in particular,

$$\|T^{-1}y\|_1 \le a^{-1}\|y\|_2, \qquad y \in \mathcal{R}(T).$$

However, because of the compactness property of $(X_1, \| \cdot \|_1)$ and $(X_2, \| \cdot \|_2)$, T^{-1} is also a compact operator since it maps the unit ball

$$B_2 \cap \mathcal{R}(T) = \{ y \in \mathcal{R}(T) : \|y\|_2 \le 1 \}$$

into a bounded set in $(X_1, \| \cdot \|_1)$. This in turn implies (Theorem 5.24.2) that if we consider the norm $\| \cdot \|_2$ on $\mathcal{D}(T)$, then

$$T^{-1} : (\mathcal{R}(T), \| \cdot \|_2) \to (\mathcal{D}(T), \| \cdot \|_2)$$

is compact. Let us summarize this.

7.6.4 THEOREM.   *Let $(X_1, \| \cdot \|_1)$ be compact in $(X_2, \| \cdot \|_2)$, where $(X_2, \| \cdot \|_2)$ is a Banach space and let T be a linear operator on X_2. Assume that the domain $\mathcal{D}(T)$*

lies in X_1. If T satisfies a Gårding Inequality (7.6.4), then T^{-1} exists and is a compact operator. Furthermore, there are constants A, B such that

$$\|T^{-1}y\|_1 \leq A\|y\|_2, \qquad \|T^{-1}y\|_2 \leq B\|y\|_2$$

for all $y \in \mathscr{R}(T)$. If, in addition $\mathscr{R}(T)$ is dense in X_2, then T^{-1} has a unique continuous extension to all of X_2.

The fact that T^{-1} can be extended to all of X_2, when $\mathscr{R}(T)$ is dense in X_2 is a standard result for linear operators; see Exercise 3, Section 5.6.

There is a particular form of the Gårding Inequality which we shall encounter in our study of symmetric differential operators. In order to explain this, let L be a symmetric operator on a Hilbert space H, with inner product (,). Let H_0 be a linear subspace of H, and let (,)$_0$ and $\|\cdot\|_0$ be another inner product and associated norm on H_0. Assume that the domain $\mathscr{D}(L)$ lies in H_0 and that $(H_0, \|\cdot\|_0)$ is compact in $(H, \|\cdot\|)$.

Now define $B[u,v]$ by

$$(u, Lv) = B[u,v].$$

We note that $B[u,v]$ is a sesquilinear functional defined for $u \in H$, $v \in \mathscr{D}(L)$. Also, since L is symmetric, then $B[u,u]$ is real-valued. (Why ?)

Now assume that there are real constants c and k with $c > 0$ and such that

$$B[u,u] \geq c\|u\|_0 - k\|u\|, \qquad u \in \mathscr{D}(L). \tag{7.6.5}$$

Equation (7.6.5) can be rewritten as

$$(u, Lu + ku) \geq c\|u\|_0{}^2.$$

By using the Schwarz Inequality for $(\cdot, \cdot)$ we get

$$c\|u\|_0{}^2 \leq \|(L + kI)u\| \cdot \|u\|.$$

However, since $(H_0, \|\cdot\|_0)$ is compact in $(H, \|\cdot\|)$ there is (by Exercise 3 in this section) a constant $b > 0$ such that

$$\|u\| \leq b\|u\|_0.$$

By applying the last inequality to (7.6.6) we get

$$cb^{-1}\|u\|_0 \leq \|(L + kI)u\|,$$

that is, $(L + kI)$ satisfies a Gårding Inequality. For this reason Equation (7.6.5) is also referred to as a *Gårding Inequality*.

Let us summarize these observations.

7.6.5 COROLLARY. *Let L be a symmetric operator on a Hilbert space H with inner product (,). Let (,)$_0$ be an inner product on $\mathscr{D}(L)$ and assume that $(\mathscr{D}(L), \|\cdot\|_0)$ be compact in $(H, \|\cdot\|)$. If there are real constants c and k with $c > 0$ and such that*

$$B[u,u] \geq c\|u\|_0{}^2 - k\|u\|^2, \tag{7.6.7}$$

then $(L + kI)$ satisfies a Gårding Inequality and $(L + kI)^{-1}$ is a compact operator.

EXERCISES

1. Prove Theorem 7.6.2. [*Hint*: Recall the equivalent versions of compactness in Section 3.17.]

2. Let $X_2 = C[0,1]$ and $X_1 = C^1[0,1]$ be given, where
$$\|x\|_2 = \sup\{|x(t)|: 0 \le t \le 1\},$$
and
$$\|x\|_1 = \|x\|_2 + \|x'\|_2.$$
Show that $(X_1, \|\cdot\|_1)$ is compact in $(X_2, \|\cdot\|_2)$. [*Hint*: Apply the Arzela-Ascoli Theorem.]

3. Let $(X_1, \|\cdot\|_1)$ be compact in $(X_2, \|\cdot\|_2)$. Show that there is a $b \ge 0$ such that
$$\|x\|_1 \le b\|x\|_2, \qquad x \in X_1.$$
[*Hint*: Show that the identity mapping $I: X_1 \to X_2$ is compact and therefore continuous.]

4. (a) Prove Theorem 7.6.3 for the case $l = k + 1$, $k \ge 1$. [*Hint*: Use mathematical induction.]
   (b) Prove Theorem 7.6.3 for the general case, $l > k$.

5. Show that Theorem 7.6.3 can be extended to the Sobolev spaces $H^l(\Omega)$.

6. The Hölder space $C^\alpha[0,1]$ with norm $\|\cdot\|_\alpha$ are defined in Example 10, Section 5.3. Show that if $0 < \alpha < \beta \le 1$, then $(C^\beta[0,1], \|\cdot\|_\beta)$ is compact in $(C^\alpha[0,1], \|\cdot\|_\alpha)$.

7. Let X_1 be a linear subspace of a Banach space $(X_2, \|\cdot\|_2)$, and consider the given norm $\|\cdot\|_2$ on X_1. Show that $(X_1, \|\cdot\|_2)$ is compact in $(X_2, \|\cdot\|_2)$ if and only if X_1 is finite dimensional.

8. Show that Corollary 7.6.5 can be generalized to a nonsymmetric operator L when (7.6.7) is replaced by
$$\text{Re } B[u,u] \ge c\|u\|_0^2 - k\|u\|^2.$$

7.   ELLIPTIC PARTIAL DIFFERENTIAL OPERATORS

In the next section we shall study the Dirichlet problem, which is a boundary-value problem for certain elliptic partial differential operators. Our objective in this section is to show that, under appropriate conditions, these operators satisfy a Gårding Inequality. We shall use the notation of *Example 12, Section 5.3*.

Consider the differential operator of order $2m$,

$$Lu = \sum_{0 \le |\alpha|, |\beta| \le m} (-1)^{|\alpha|} D^\alpha(a^{\alpha\beta}(x)D^\beta u), \qquad (7.7.1)$$

where the coefficients $a^{\alpha\beta}$ are assumed to be sufficiently differentiable in a region Ω in R^n, say that $a^{\alpha\beta}$ belongs to C^m. If u and v belong to $C_0^{2m}(\Omega)$, then by integrating by parts we get

$$(v,Lu) = \int_\Omega v\overline{Lu}\,dx = \sum_{0 \le |\alpha|,|\beta| \le m} (-1)^{|\alpha|} \int_\Omega v\overline{D^\alpha(a^{\alpha\beta}D^\beta u)}\,dx$$

$$= \sum_{0 \le |\alpha|,|\beta| \le m} (-1)^{|\alpha|}(-1)^{|\alpha|+|\beta|} \int_\Omega D^\beta(\overline{a^{\alpha\beta}D^\alpha v})\bar{u}\,dx$$

$$= (L^*v,u),$$

where[11]

$$L^*v = \sum_{0 \le |\alpha|,|\beta| \le m} (-1)^{|\beta|}D^\beta(\overline{a^{\alpha\beta}}(x)D^\alpha v).$$

If $L = L^*$, then L is symmetric on the domain $C_0^{2m}(\Omega) \subset L_2(\Omega)$. For example, if L consists only of higher-order terms

$$Lu = \sum_{|\alpha|=|\beta|=m} (-1)^m D^\alpha(a^{\alpha\beta}D^\beta u),$$

then L is symmetric whenever $a^{\alpha\beta} = \overline{a^{\beta\alpha}}$. (Compare this with Example 1, Section 5.23.) In particular, the Laplacian operator

$$\Delta = \sum_{i=1}^n \frac{\partial^2}{\partial x_i^2}$$

and the fourth-order biharmonic operator

$$L = \sum_{i=1}^n \frac{\partial^4}{\partial x_i^4}$$

are symmetric operators on $C_0^2(\Omega)$ and $C_0^4(\Omega)$, respectively.

For any differential operator L given by Equation (7.7.1), we define the sesquilinear functional $B[u,v]$ by

$$B[u,v] = (u,Lv) = (L^*u,v), \tag{7.7.2}$$

where $u, v \in C_0^{2m}(\Omega)$. Let us note here that if we integrate by parts $B[u,v]$ takes on the form

$$B[u,v] = \sum_{0 \le |\alpha|,|\beta| \le m} (D^\alpha u, a^{\alpha\beta}D^\beta v), \tag{7.7.3}$$

where $(\cdot,\cdot)$ denotes the usual inner product on $L_2(\Omega)$ and $u, v \in C_0^{2m}(\Omega)$. Because of Exercise 3, Section 5.13, $B[u,v]$ is well defined for all $u, v \in H_0^m(\Omega)$.

Our objective now is to show that B satisfies a Gårding Inequality of the form (7.6.5). In order to simplify our discussion, we shall assume henceforth that the coefficients $a^{\alpha\beta}$ are constant. A more general discussion, which includes the case of bounded continuous coefficients, can be found in Friedman [2, pp. 32–37].

---

[11] For now we use L^* only as a notational convenience. In Section 10 we shall define the adjoint for certain unbounded operators. At that time L^* will take on an added significance.

Our main hypothesis concern the highest-order coefficients $a^{\alpha\beta}$ with $|\alpha| = |\beta| = m$.

7.7.1 DEFINITION.   Let L be a differential operator given by (7.7.1), where the coefficients $a^{\alpha\beta}$ are constant. We shall say that L is *strongly elliptic* if there is a positive constant c_0 such that

$$(-1)^m \operatorname{Re}\left(\sum_{|\alpha|=|\beta|=m} a^{\alpha\beta}\xi^\alpha\xi^\beta\right) \geq c_0 |\xi|^{2m}, \qquad \xi \in R^n, \qquad (7.7.4)$$

where $\xi^\alpha = \xi_1^{\alpha_1}\cdots\xi_n^{\alpha_n}$ and $|\xi|^2 = \xi_1^2 + \cdots + \xi_n^2$. In this notation the expression $|\xi|^{2m}$ can be written as

$$|\xi|^{2m} = \sum_{|\alpha|=|\beta|=m} \delta^{\alpha\beta}\xi^\alpha\xi^\beta, \qquad (7.7.5)$$

where $\delta^{\alpha\beta}$ is the Kronecker delta function.

The Laplacian operator Δ furnishes examples of strongly elliptic operators. One sees easily that $(-1)^m\Delta^m$ is strongly elliptic where $m = 1, 2, \ldots$ and

$$\Delta^m u = \Delta(\Delta^{m-1}u)$$

is defined by induction.

Before stating the main result, let us recall the notation for the norm $\|u\|_m$, where

$$\|u\|_m = \left(\sum_{|\alpha|\leq m} \int_\Omega |D^\alpha u|^2 \, dx\right)^{1/2}.$$

7.7.2 THEOREM.   *Let L be given by (7.7.1), where the coefficients $a^{\alpha\beta}$ are constant, and assume that L is strongly elliptic. Let B be given by (7.7.2). Then there are real numbers c_1 and k_0 with $c_1 > 0$ and such that*

$$\operatorname{Re} B[u,u] \geq c_1\|u\|_m^2 - k_0\|u\|_0^2 \qquad (7.7.6)$$

for all $u \in C_0^\infty(\Omega)$.

Before proving Inequality (7.7.6) let us note that if L is symmetric on $C_0^{2m}(\Omega)$, then

$$B[u,u] = (u,Lu)$$

is real-valued for all u in $C_0^{2m}(\Omega)$. (Why?) So in this case, Inequality (7.7.6) becomes the Gårding Inequality (7.6.5).

The proof of this theorem relies heavily on the Fourier transform (see Example 13, Section 5.22). Recall that if $U = \mathscr{F}u$ is the Fourier transform of u given by

$$U(\xi) = \int_{R^n} e^{-ix\cdot\xi}u(x)\,dx$$

and $V = \mathscr{F}v$, where $u, v \in C_0^\infty(\Omega) \subset C_0^\infty(R^n)$, then

$$\mathscr{F}(D^\alpha u) = (i\xi)^\alpha U(\xi) \qquad (7.7.7)$$

and

$$\int_{R^n} U(\xi)\overline{V(\xi)}\, d\xi = (2\pi)^n \int_{R^n} u(x)\overline{v(x)}\, dx. \tag{7.7.8}$$

We will also need the following two lemmas, which will be proved in the exercises.

7.7.3. LEMMA. *Let Ω be a bounded domain in R^n. Then there exists an $\varepsilon_0 > 0$ such that for any ε with $0 < \varepsilon < \varepsilon_0$ there is a constant c_ε such that*

$$\|u\|_{m-1}^2 \le \varepsilon \|u\|_m^2 + c_\varepsilon \|u\|_0^2 \tag{7.7.9}$$

for all $u \in C_0^m(\Omega)$.

The important thing to note in that last lemma is that ε_0 and c_ε do not depend on u.

7.7.4 LEMMA. *Let Ω be a bounded domain in R^n and assume that α and β satisfy $|\alpha| \le m$, $|\beta| \le m - 1$. Then there is a constant K such that*

$$\left| \int_\Omega D^\alpha u \overline{D^\beta u}\, dx \right| = \left| \int_\Omega D^\beta u \overline{D^\alpha u}\, dx \right| \le K \|u\|_m \|u\|_{m-1}$$

for all u in $C_0^\infty(\Omega)$.

Again it is important to note that the constant K is independent of u.

Proof of Theorem 7.7.2: Let $B = \tilde{B} + R$, where B is given by (7.7.3) and

$$\tilde{B}[u,v] = \sum_{|\alpha|=|\beta|=m} (D^\alpha u, a^{\alpha\beta} D^\beta v),$$

$$R[u,v] = \sum_{\substack{|\alpha| \le m, |\beta| \le m \\ |\alpha|+|\beta| < 2m}} (D^\alpha u, a^{\alpha\beta} D^\beta v).$$

By repeated applications of Lemma 7.7.4 we get

$$|R[u,u]| \le K_1 \|u\|_m \|u\|_{m-1},$$

where K_1 depends on m and n, but not on u. By using the Fourier transform together with (7.7.7) and (7.7.8), we get

$$\tilde{B}[u,u] = \sum_{|\alpha|=|\beta|=m} (D^\alpha u, a^{\alpha\beta} D^\beta u)$$

$$= (2\pi)^{-n} \sum_{|\alpha|=|\beta|=m} ((i\xi)^\alpha U, a^{\alpha\beta}(i\xi)^\beta U)$$

$$= (2\pi)^{-n} \int_{R^n} |U(\xi)|^2 \left[\sum_{|\alpha|=|\beta|=m} (-1)^m a^{\alpha\beta} \xi^\alpha \xi^\beta \right] d\xi.$$

Thus by (7.7.4) we get

$$\text{Re } \tilde{B}[u,u] \ge (2\pi)^{-n} c_0 \int_{R^n} |U(\xi)|^2 |\xi|^{2m}\, d\xi.$$

Now by applying the inverse Fourier transform together with (7.7.5) we get

$$\text{Re } \tilde{B}[u,u] \geq c_0 \sum_{|\alpha|=m} (D^\alpha u, D^\alpha u).$$

Combining these two inequalities we get

$$\text{Re } B[u,u] \geq \text{Re } \tilde{B}[u,u] - |R[u,u]|$$

$$\geq c_0 \sum_{|\alpha|=m} (D^\alpha u, D^\alpha u) - K_1 \|u\|_m \|u\|_{m-1}. \tag{7.7.10}$$

Since

$$0 \leq \left(\frac{K_1}{2\sigma} \|u\|_{m-1} - \sigma \|u\|_m \right)^2$$

for all $\sigma > 0$, we get

$$K_1 \|u\|_m \|u\|_{m-1} \leq \sigma^2 \|u\|_m^2 + \frac{K_1^2}{4\sigma^2} \|u\|_{m-1}^2,$$

and by applying Lemma 7.7.3 we then get

$$K_1 \|u\|_m \|u\|_{m-1} \leq \sigma^2 \|u\|_m^2 + \frac{K_1^2}{4\sigma^2} (\varepsilon \|u\|_m^2 + c_\varepsilon \|u\|_0^2).$$

In other words,

$$K_1 \|u\|_m \|u\|_{m-1} \leq \eta \|u\|_m^2 + K_2 \|u\|_0^2, \tag{7.7.11}$$

where η can be made arbitrarily small (at the expense of making K_2 large).

Similarly one has

$$\|u\|_m^2 = \|u\|_{m-1}^2 + \sum_{|\alpha|=m} (D^\alpha u, D^\alpha u)$$

$$\leq \sum_{|\alpha|=m} (D^\alpha u, D^\alpha u) + \varepsilon \|u\|_m^2 + c_\varepsilon \|u\|_0^2. \tag{7.7.12}$$

Now by applying (7.7.11) with $\eta \leq c_0/3$ and (7.7.12) with $\varepsilon = \frac{1}{3}$, to (7.7.10) we get

$$\text{Re } B[u,u] \geq \tfrac{1}{3} c_0 \|u\|_m^2 - k_0 \|u\|_0^2,$$

for an appropriate choice of k_0. ∎

Inequality (7.7.6) can actually be extended to hold for all $u \in H_0^m(\Omega)$.

7.7.5 COROLLARY.   *Under the hypotheses and notation of Theorem 7.7.2, one has*

$$\text{Re } B[u,u] \geq c_1 \|u\|_m^2 - k_0 \|u\|_0^2$$

for all $u \in H_0^m(\Omega)$.

Proof: Let $u \in H_0^m(\Omega)$ and choose a sequence $\{u_j\}$ in $C_0^\infty(\Omega)$ with the property that

$$\|u - u_j\|_m \to 0 \qquad as \qquad j \to \infty.$$

Since

$$\big| \|u_j\|_0 - \|u\|_0 \big| \le \|u_j - u\|_0 \le \|u_j - u\|_m$$

and

$$\big| \|u_j\|_m - \|u\|_m \big| \le \|u_j - u\|_m,$$

we see that $\|u_j\|_m \to \|u\|_m$ and $\|u_j\|_0 \to \|u\|_0$ as $j \to \infty$. Furthermore, if we apply the Schwarz Inequality to (7.7.3) we get

$$|B[u,v]| \le \sum_{0 \le |\alpha|, |\beta| \le m} \|D^\alpha u\|_0 \, \|a^{\alpha\beta} D^\beta v\|_0$$

$$\le K \|u\|_m \|v\|_m$$

for some constant K. Thus

$$|B[u_j, u_j] - B[u,u]| = |B[u_j, u_j - u] + B[u_j - u, u]|$$

$$\le K(\|u_j\|_m + \|u\|_m)\|u_j - u\|_m$$

$$\le \tilde{K}\|u_j - u\|_m$$

since $\|u_j\|_m$ is bounded. Hence $B[u_j, u_j] \to B[u,u]$ as $j \to \infty$. Since Inequality (7.7.6) holds for each u_j, it also holds in the limit. ∎

EXERCISES

The first five exercises will lead to a proof of Lemma 7.7.3.

1. In addition to the hypotheses of Lemma 7.7.3 assume that for $i < j$ one has

$$\sum_{|\alpha| = i} \int_\Omega |D^\alpha u|^2 \, dx \le \varepsilon \sum_{|\beta| = j} \int_\Omega |D^\beta u|^2 \, dx + \frac{c}{\varepsilon^{i/(j-i)}} \int_\Omega |u|^2 \, dx \qquad (7.7.13)$$

for all $u \in C_0^j(\Omega)$, where c depends on j and Ω. Show that Inequality (7.7.9) holds.

Figure 7.7.1.

2. In this exercise we shall prove Inequality (7.7.13) for $i = 1$, $j = 2$, and $n = 1$, that is, Ω is an interval with length $|\Omega|$. First divide Ω into intervals each of length $\le \sqrt{\varepsilon}$ and $\ge \sqrt{\varepsilon/2}$. Let (a,b) denote a typical subinterval and set $\alpha = (b - a)/4$. Note that $\alpha^2 \le \varepsilon/16$ and $\alpha^{-2} \le 64/\varepsilon$. Let x_1 and x_2 be arbitrary points in $(a, a + \alpha)$ and $(a + 3\alpha, b)$, respectively.

(a) Show that for any $x \in (a,b)$ one has

$$|Du(x)| \leq \frac{|u(x_1)| + |u(x_2)|}{2\alpha} + \int_a^b |D^2 u(\xi)| \, d\xi. \qquad (7.7.14)$$

(b) Show that

$$\alpha^2 |Du(x)| \leq \tfrac{1}{2} \int_\alpha^b |u(\xi)| \, d\xi + \alpha^2 \int_a^b |D^2 u(\xi)| \, d\xi. \qquad (7.7.15)$$

[*Hint*: Integrate Inequality (7.7.14) with respect to x_1 and x_2 over the intervals $(a, a + \alpha)$ and $(a + 3\alpha, b)$.]

(c) Show that

$$\int_a^b |Du|^2 \, dx \leq \frac{c}{\alpha^2} \int_a^b |u|^2 \, dx + c\alpha^2 \int_a^b |D^2 u|^2 \, dx. \qquad (7.7.16)$$

[*Hint*: Apply the Schwartz Inequality to (7.7.15) and then integrate with respect to x.]

(d) Prove (7.7.13) for the case $i = 1, j = 2, n = 1$.

3. Use Exercise 2 to prove (7.7.13) for the case $i = 1, j = 2, n \geq 2$, and Ω is a cube with sides parallel to the coordinate planes.
   [*Hint*: Note that (7.7.16) holds when $Du = \partial u / \partial x_i$ and the integration is with respect to the x_i-variable. Now integrate (7.7.16) with respect to the other variables.]

4. Use Exercise 3 to prove (7.7.13) for an arbitrary bounded domain Ω.

5. Prove (7.7.13) in general. [*Hint*: Note that (7.7.13) is trivially true for $i = 0$. Now apply mathematical induction on j.]

6. Prove Lemma 7.7.4. [*Hint*: Apply the Schwarz Inequality.]

8.   THE DIRICHLET PROBLEM

The Dirichlet problem is one of the fundamental problems of mathematical physics. Before giving a mathematical formulation of this problem it is helpful to recall some of the underlying physical problems.

EXAMPLE 1. (STATIONARY MEMBRANE.)   Consider a membrane stretched over a bounded region Ω in the plane R^2. Let $u(x,y)$ denote the displacement of the membrane when some external force $f(x,y)$ is acting on the membrane. The relationship between u and f is then given by

$$\frac{\partial^2 u}{\partial x^2} + \frac{\partial^2 u}{\partial y^2} = f \qquad \text{in } \Omega,$$

$$u = 0 \qquad \text{on } \partial\Omega,$$

where $\partial\Omega$ denotes the boundary of Ω.   ∎

EXAMPLE 2. (ELECTROSTATIC FIELD.)   Let $f(x,y,z)$ denote the electric charge density of a region Ω in R^3. If $u(x,y,z)$ denotes the potential of the electrostatic field generated by f, then u and f are related by

$$\Delta u = \frac{\partial^2 u}{\partial x^2} + \frac{\partial^2 u}{\partial y^2} + \frac{\partial^2 u}{\partial z^2} = f. \tag{7.8.1}$$

Let us assume that there is a function g, defined on the boundary $\partial\Omega$, and we seek a solution u with the property that

$$u = g \qquad \text{on } \partial\Omega. \tag{7.8.2}$$

This would arise if the boundary were insulated or contained a fixed charge.

The boundary-value problem (7.8.1), (7.8.2) can be reduced to (7.8.1), (7.8.3), where

$$u = 0 \qquad \text{on } \partial\Omega, \tag{7.8.3}$$

provided the boundary $\partial\Omega$ and the function g are sufficiently smooth. This is done as follows: If $\partial\Omega$ is of class C^2 and g is a C^2-function on $\partial\Omega$, then there is a C^2-function v defined on $\Omega \cup \partial\Omega$ with the property that $v = g$ on $\partial\Omega$, see Friedman [2, p. 38]. If we set $w = u - v$, then the boundary-value problem (7.8.1), (7.8.2) is equivalent to

$$\Delta w = f - \Delta v \qquad \text{in } \Omega$$
$$w = 0 \qquad \text{on } \partial\Omega. \quad \blacksquare$$

EXAMPLE 3. (DEFORMED PLATE.)   The physical properties of the deformed plate are similar to those of the stationary membrane. However, in this case the relationship between the displacement $u(x,y)$ and the external force $f(x,y)$ is given by the biharmonic equation

$$\frac{\partial^4 u}{\partial x^4} + 2\frac{\partial^4 u}{\partial x^2 \, \partial y^2} + \frac{\partial^4 u}{\partial y^2} = f \qquad \text{in } \Omega.$$

If the plate is fastened at the boundary, $\partial\Omega$, the boundary conditions take on the form

$$u = 0, \quad \frac{\partial u}{\partial n} = 0 \qquad \text{on } \partial\Omega,$$

where $\partial u/\partial n$ denotes the outward normal derivative of u. $\blacksquare$

Let us now formulate the Dirichlet problem. We shall give two formulations. In the first, we refer to a classical solution whereas in the second we introduce the concept of a weak solution. The reader will see that every classical solution of the Dirichlet problem is also a weak solution. We shall not attempt to discuss the converse question (namely, when is a weak solution also a classical solution) here. This would take us too far afield. Instead we refer the reader to excellent treatments in Agmon [1], Bers, John, and Schechter [1], Friedman [2], and Garabedian [1].

7.8.1 DEFINITION.   Let Ω be a bounded region in R^n and let L be an elliptic operator of order $2m$ on Ω given by

$$Lu = \sum_{0 \le |\alpha|, |\beta| \le m} (-1)^{|\alpha|} D^\alpha (a^{\alpha\beta} D^\beta u). \tag{7.8.4}$$

The *classical Dirichlet problem* is the following: Let f be a continuous function on Ω. Find a function u in $C^{2m}(\overline{\Omega})$ with the property that

$$Lu = f \quad \text{in } \Omega \tag{7.8.5}$$

$$\frac{\partial^j u}{\partial n^j} = 0 \quad \text{on } \partial\Omega, \quad 0 \le j \le m - 1, \tag{7.8.6}$$

where $\partial^j u / \partial n^j$ denote the outward normal derivatives of u. Such a function u would then be called a *classical solution* of the Dirichlet problem.[12]

If u is a classical solution of the Dirichlet problem and $\phi \in C_0^\infty(\Omega)$, then $(\phi, f) = (\phi, Lu)$. If we integrate the last equality by parts we get

$$(\phi, f) = B[\phi, u],$$

where

$$B[\phi, u] = \sum_{0 \le |\alpha|, |\beta| \le m} (D^\alpha \phi, a^{\alpha\beta} D^\beta u).$$

Furthermore, one can show that if the boundary $\partial\Omega$ is sufficiently smooth and if u is a classical solution of the Dirichlet problem, then $u \in H_0^m(\Omega)$. (See Friedman [2, pp. 39–40].) We will not do this here, but instead we shall use this fact to motivate the following definition.

7.8.2 DEFINITION.   Let Ω be a bounded region in R^n and let L be an elliptic operator of order $2m$ on Ω given by (7.8.4). The *generalized Dirichlet problem for L* is the following: Let f be a function in $L_2(\Omega)$. Find a function u with the property that

$$(\phi, f) = B[\phi, u], \quad \text{for all } \phi \in C_0^\infty(\Omega), \tag{7.8.7}$$

and

$$u \in H_0^m(\Omega). \tag{7.8.8}$$

In this case, u is called a *(weak) solution* of the Dirichlet problem, or sometimes a solution of the generalized Dirichlet problem.

For any complex number λ the generalized Dirichlet problem for $L + \lambda I$ is defined similarly. In this case the sesquilinear functional $B[\phi, u]$ is replaced by

$$B_\lambda[\phi, u] = (\phi, (L + \lambda I)u).$$

Let us now prove the following existence theorem, which gives a solution of the generalized Dirichlet problem.

---

[12] The boundary conditions (7.8.6) are called *homogeneous* boundary conditions. *Inhomogeneous* boundary conditions would arise if we set $\partial^j u / \partial n^j = g_j$. However, the inhomogeneous boundary conditions can be reduced to homogeneous boundary conditions by techniques similar to those discussed in Example 2. (See Friedman [2, p. 38].)

7.8.3 Theorem.   *Let L be given by (7.8.4) where the coefficients $a^{\alpha\beta}$ are constant, and assume that L is strongly elliptic. Then there is a real number k_0 such that for any $k \geq k_0$ the generalized Dirichlet problem for $L + kI$ has a unique solution. Moreover, $(L + kI)^{-1}$ exists and is a compact operator.*

Proof: The proof of this result relies on the Lax-Milgram Theorem (Theorem 5.21.2) as well as Corollary 7.7.5.

Let $B[u,v]$ be given by Equation (7.7.3). As noted in the proof of Corollary 7.7.5, one has

$$|B[u,v]| \leq K\|u\|_m\|v\|_m$$

for all $u, v \in H_0^m(\Omega)$. Furthermore, if k is any real number, then

$$|B_k[u,v]| \leq |B[u,v]| + |(u,kv)|$$
$$\leq K\|u\|_m\|v\|_m + |k|\,\|u\|_0\|v\|_0$$
$$\leq (K + k)\|u\|_m\|v\|_m.$$

Now let $c_1 > 0$ and k_0 be given by Gårdings Inequality, that is,

$$\operatorname{Re} B[u,u] \geq c_1\|u\|_m^2 - k_0\|u\|_0^2$$

for all $u \in H_0^m(\Omega)$. If $k \geq k_0$, then

$$|B_k[u,u]| \geq \operatorname{Re} B_k[u,u] = \operatorname{Re} B[u,u] + (u,ku)$$
$$\geq c_1\|u\|_m^2 + (k - k_0)\|u\|_0^2 \geq c_1\|u\|_m^2.$$

Hence the sesquilinear function $B_k[u,v]$ and the Hilbert space $H_0^m(\Omega)$ satisfies the hypotheses of the Lax-Milgram Theorem (Theorem 5.21.2).

Now let $l: H_0^m(\Omega) \to C$ be given by

$$l(\phi) = (\phi,f),$$

where $f \in L_2(\Omega)$. Since

$$|l(\phi)| \leq \|\phi\|_0\|f\|_0 \leq \|f\|_0\|\phi\|_m,$$

we see that l is a bounded linear functional. Thus by the Lax-Milgram Theorem there is a unique $u \in H_0^m(\Omega)$ with

$$l(\phi) = (\phi,f) = B_k[\phi,u]$$

for all ϕ in $H_0^m(\Omega)$. This establishes the existence of a weak solution u.

Finally, if L is symmetric the compactness of $(L + kI)^{-1}$ follows from Corollary 7.6.5. The general case (where L is not symmetric) follows from Exercise 8, Section 6.   ∎

One can say more about the generalized Dirichlet problem if the operator L contains only highest-order terms,

$$Lu = \sum_{|\alpha| = |\beta| = m} (-1)^m D^\alpha(a^{\alpha\beta} D^\beta u). \qquad (7.8.9)$$

7.8.4 COROLLARY.   *In addition to the hypotheses of Theorem* 7.8.3 *assume that L is given by Equation* (7.8.9). *Then the generalized Dirichlet problem for L has a unique solution. Moreover* L^{-1} *exists and is a compact operator.*

Proof: If one examines the proof of Theorem 7.7.2 carefully one sees that the constant k_0, which appears in Equation (7.7.6), can be chosen to be zero when L is given by (7.8.9). This corollary now follows directly from the last theorem. ∎

It should be noted that the only property of L that was critical in the proof of the last theorem (and corollary) was that L satisfied the Gårding Inequality (7.7.6). We really did not use the fact that L had constant coefficients, other than to prove (7.7.6). This suggests that the methods described here may be extended to differential operators with variable coefficients, and this is indeed the case. We refer the reader to Agmon [1] and Friedman [2] for these details.

We see, then, that if L is an elliptic partial differential operator that satisfies the Gårding Inequality (7.7.6), then L has a compact resolvent. If L is also symmetric,[13] then it has a compact, self-adjoint resolvent, and one can use the eigenvalue-eigenfunction representation of L as described in Section 1. The problem of *finding* the eigenvalues and eigenfunctions requires further analysis of the individual operator. However, there is a technique of finding the eigenvalues and eigenfunctions which is very useful and deserves special mention. This is the method of separation[14] of variables, which we describe in the following example.

EXAMPLE 4.   Let Ω be the unit square

$$\{(x,y): 0 \le x \le 1, 0 \le y \le 1\}$$

in R^2 and consider the Laplacian operator Δ on Ω, with boundary conditions $u = 0$ on $\partial\Omega$. We seek solutions of the eigenvalue-eigenvector problem:

$$\begin{aligned} \Delta u &= \lambda u, & \text{on } \Omega, \\ u &= 0, & \text{on } \partial\Omega. \end{aligned} \qquad (7.8.10)$$

Let us assume now that the solution of (7.8.10) is of the form

$$u(x,y) = X(x)\,Y(y).$$

Equation (7.8.10) then becomes

$$\frac{d^2 X}{dx^2}\, Y + X\, \frac{d^2 Y}{dy^2} = \lambda XY,$$

or by dividing by XY we get

$$\frac{1}{X}\frac{d^2 X}{dx^2} = \lambda - \frac{1}{Y}\frac{d^2 Y}{dy^2}. \qquad (7.8.11)$$

---

[13] If L is not symmetric, then one can use the theory of nonnormal compact operators presented in Section 6.14.

[14] Let not the ox and ass plow together. (Origin unknown.)

Since the left side of (7.8.11) depends on x and the right side depends on y we see that each side must be constant, that is,

$$\frac{1}{X}\frac{d^2X}{dx^2} = \mu = \lambda - \frac{1}{Y}\frac{d^2Y}{dy^2}$$

or

$$\frac{d^2X}{dx^2} = \mu X, \qquad \frac{d^2Y}{dy^2} = (\lambda - \mu)Y. \tag{7.8.12}$$

The boundary conditions $u = 0$ on $\partial\Omega$ now become

$$X(0) = X(1) = 0, \qquad Y(0) = Y(1) = 0. \tag{7.8.13}$$

In other words, we have reduced the planar problem (7.8.10) to solving two Sturm-Liouville problems (7.8.12)–(7.8.13). We see then that $\mu = n^2\pi^2$ for $n = 0, 1, 2, \ldots$ and $(\lambda - \mu) = m^2\pi^2$ for $m = 0, 1, 2, \ldots$. The solution of (7.8.12)–(7.8.13) is

$$\begin{aligned} X(x) &= \cos n\pi x, & \mu &= n^2\pi^2, & n &= 0, 1, \ldots, \\ Y(y) &= \cos m\pi y, & \lambda - \mu &= m^2\pi^2, & m &= 0, 1, \ldots. \end{aligned}$$

Consequently the solution of (7.8.10) is then

$$u(x,y) = \cos n\pi x \cos m\pi y, \qquad \lambda = (n^2 + m^2)\pi^2,$$

$n = 0, 1, \ldots, m = 0, 1, \ldots$. ∎

EXERCISES

1. Show that the eigenvalues for the problem

$$\begin{aligned} \Delta u &= \lambda u & \text{on } \Omega, \\ u &= 0 & \text{on } \partial\Omega, \end{aligned}$$

where

$$\Omega = \{(x_1, x_2, x_3) : 0 \le x_i \le 1, i = 1, 2, 3\}$$

are of the form

$$\lambda = (l^2 + m^2 + n^2)\pi^2,$$

where $l = 0, 1, \ldots, m = 0, 1, \ldots, n = 0, 1, \ldots$. Find the corresponding eigenfunctions. [*Hint*: Let $u(x_1, x_2, x_3) = V_1(x_1)W(x_2, x_3)$ and reduce this to Example 4.]

2. Find the eigenvalues and eigenfunctions for the Laplacian operator Δu on

$$\Omega = \{(x_1, \ldots, x_n) : 0 \le x_i \le 1, 1 \le i \le n\},$$

with $u = 0$ on $\partial\Omega$.

3. Let Ω be the unit square as given in Example 4. In each of the following exercises you are given a function $f(x,y)$ and are asked to express the solution $u(x,y)$ of

$$\frac{\partial^2 u}{\partial x^2} + \frac{\partial^2 u}{\partial y^2} = f(x,y) \qquad \text{on } \Omega,$$

$$u = 0 \qquad \text{on } \partial\Omega,$$

as a Fourier expansion in terms of the eigenfunctions of Δ.
(a) $f(x,y) = x$.
(b) $f(x,y) = \sin \pi(x + y)$.
(c) $f(x,y) = x(1 - y)$.
(d) $f(x,y) = x + y$. [*Hint*: Use (a).]

4. Find the eigenvalues and eigenfunctions for the Laplacian operator Δu on

$$\Omega = \{(x,y): x^2 + y^2 \leq a^2\}$$

with $u = 0$ on $\partial\Omega$. [*Hint*: Use polar coordinates and separate variables.]

5. Find the eigenvalues and eigenfunctions for the Laplacian operator Δu on the sphere

$$\Omega = \{(x_1,x_2,x_3): x_1^2 + x_2^2 + x_3^2 \leq a^2\}$$

with $u = 0$ on $\partial\Omega$. [*Hint*: Use spherical coordinates and separate variables.]

6. Find the eigenvalues and eigenfunctions for the Laplacian operator Δu on the cylinder

$$\Omega = \{(x_1,x_2,x_3): x_1^2 + x_2^2 \leq a^2, 0 \leq x_3 \leq h\}$$

with $u = 0$ on $\partial\Omega$. [*Hint*: Use cylindrical coordinates and separate variables.]

7. Find the eigenvalues and eigenfunctions for the operator

$$Lu = \frac{\partial^2 u}{\partial x^2} + \frac{\partial^2 u}{\partial y^2} + 2\frac{\partial u}{\partial x}$$

on

$$\Omega = \{(x,y): 0 \leq x \leq 1, 0 \leq y \leq 1\}$$

with $u = 0$ on $\partial\Omega$. Is the operator L symmetric?

8. Consider the biharmonic operator

$$\Delta^2 u = \frac{\partial^4 u}{\partial x^4} + 2\frac{\partial^4 u}{\partial x^2 \partial y^2} + \frac{\partial^4 u}{\partial y^4}$$

on the disk $\Omega = \{(x,y): x^2 + y^2 \leq a^2\}$.
(a) Show that if $\Delta^2 u = \lambda u$, where $u = 0$ on $\partial\Omega$ and $u \neq 0$, then $\lambda > 0$. [*Hint*: Use the Gårding Inequality.]
(b) Find the eigenvalues and eigenfunctions of $\Delta^2 u$ on Ω with $u = 0$ on $\partial\Omega$. [*Hint*: Let $\lambda = k^4$ and write $\Delta^2 u - \lambda u = 0$ as $(\Delta - k^2)(\Delta + k^2)u = 0$. Now use polar coordinates to separate variables.]

9.   THE HEAT EQUATION AND WAVE EQUATION

Let L be an elliptic differential operator of order $2m$ defined on some region $\Omega \subset R^n$. In this section we shall discuss two partial differential equations

$$\frac{\partial u}{\partial t} = Lu,$$

and

$$\frac{\partial^2 u}{\partial t^2} = Lu.$$

The first equation is called the *heat equation* and the second is called the *wave equation*. The problem is to find a solution $u(x,t)$ with $x \in \Omega$, $t \geq 0$ subject to appropriate boundary conditions, which we shall formulate momentarily. However, before doing this, there is another assumption which warrants some discussion.

We shall assume later that the boundary conditions that determine L are homogeneous and that L is a symmetric operator with a compact self-adjoint resolvent. The conditions under which this assumption is satisfied are discussed in the last two sections. For such an operator, recall that there exists a set of eigenvalues $\{\mu_1,\mu_2,\ldots\}$ and corresponding eigenvectors (or eigenfunctions) $\{\phi_1,\phi_2,\ldots\}$ such that the eigenvectors form an orthonormal basis for $L_2(\Omega)$. Furthermore, if $u = \sum_n (u,\phi_n)\phi_n$ lies in the domain of L, then

$$Lu = \sum_n \mu_n(u,\phi_n)\phi_n.$$

Heat Equation

Let $u(x,t)$ denote the temperature at a point $x \in \Omega \subseteq R^n$ at some time t. Let us assume that the initial temperature distribution

$$u(x,0) = f(x), \qquad x \in \Omega \tag{7.9.1}$$

is known and that the temperature is normalized so that on the boundary $\partial\Omega$ one has

$$u(x,t) = 0, \qquad x \in \partial\Omega, \qquad t \geq 0. \tag{7.9.2}$$

We now seek to solve the heat equation

$$\frac{\partial u}{\partial t} = Lu \tag{7.9.3}$$

subject to the boundary conditions (7.9.1) and (7.9.2).

Let us assume that

$$u(x,t) = U(x)V(t).$$

Then Equation (7.9.3) becomes

$$\frac{1}{V}\frac{dV}{dt} = \frac{1}{U}LU. \tag{7.9.4}$$

Since the left side of Equation (7.9.4) depends only on t and the right side depends only on x, we see that they must be constant. That is, Equation (7.9.4) becomes

$$LU = \lambda U, \tag{7.9.5}$$

$$\frac{dV}{dt} = \lambda V. \tag{7.9.6}$$

The boundary conditions (7.9.2) can now be written as

$$U = 0 \quad \text{on } \partial\Omega. \tag{7.9.7}$$

Now assume that the operator L with the homogeneous boundary conditions (7.9.7) is a symmetric operator with a compact self-adjoint resolvent and let $\{U_1, U_2, \ldots\}$ be an orthonormal basis of eigenvectors with corresponding eigenvalues $\{\mu_1, \mu_2, \ldots\}$. The eigenvalues are, of course real, and

$$V(t) = e^{\mu_n t}$$

is the solution of (7.9.6). Hence $u(x,t)$ takes on the form

$$u(x,t) = e^{\mu_n t} U_n(x), \qquad n = 1, 2, \ldots.$$

The general solution of (7.9.2), (7.9.3) would then be

$$u(x,t) = \sum_{n=1}^{\infty} d_n e^{\mu_n t} U_n(x),$$

where the coefficients d_n are to be determined by (7.9.1). That is,

$$f(x) = u(x,0) = \sum_{n=1}^{\infty} d_n U_n(x).$$

If we assume that $f \in L_2(\Omega)$ then $d_n = (f, U_n)$ by the Fourier Series Theorem. Hence the general solution of (7.9.1), (7.9.2), (7.9.3) is

$$u(x,t) = \sum_{n=1}^{\infty} (f, U_n) e^{\mu_n t} U_n(x). \quad \blacksquare$$

Wave Equation

Let $u(x,t)$ denote the displacement coordinate of some wave phenomenon at a point $x \in \Omega \subset R^n$ at some time t. For example, this may be a vibrating string, a vibrating membrane, a resonating cavity or one of a multitude of other phenomena. Let us assume that the initial distribution of u and u_t is given by

$$u(x,0) = f(x), \qquad x \in \Omega \tag{7.9.8}$$

and

$$\frac{\partial u}{\partial t}(x,0) = g(x), \qquad x \in \Omega. \tag{7.9.9}$$

Also assume that on the boundary $\partial\Omega$ one has

$$u(x,t) = 0, \qquad x \in \partial\Omega, \qquad t \geq 0. \tag{7.9.10}$$

We now seek to solve the wave equation

$$\frac{\partial^2 u}{\partial t^2} = Lu \tag{7.9.11}$$

subject to the boundary conditions (7.9.8), (7.9.9), and (7.9.10).

To do this we can use the method of separation of variables as was done for the heat equation. The analysis will differ in that the equation for $V(t)$ becomes

$$\frac{d^2 V}{dt^2} = \lambda V.$$

Now assume that the operator L satisfies the conditions for the heat equation analysis, and in addition, that the eigenvalues $\{\mu_1, \mu_2, \ldots\}$ of L are positive. Thus if $\lambda = \mu_n$, $V(t)$ becomes

$$V(t) = a_n e^{i\sigma_n t} + b_n e^{-i\sigma_n t},$$

where $\sigma_n = \sqrt{\mu_n}$ and a_n and b_n are arbitrary. The general solution of (7.9.10), (7.9.11) then becomes

$$u(x,t) = \sum_{n=1}^{\infty} (a_n e^{i\sigma_n t} + b_n e^{-i\sigma_n t}) U_n(x), \tag{7.9.12}$$

where a_n and b_n are to be determined by (7.9.8) and (7.9.9). That is,

$$u(x,0) = f(x) = \sum_{n=1}^{\infty} (a_n + b_n) U_n(x),$$

$$u_t(x,0) = g(x) = \sum_{n=1}^{\infty} i(\sigma_n a_n - \sigma_n b_n) U_n(x).$$

If f and g belong to $L_2(\Omega)$, then the Fourier Series Theorem implies that

$$a_n + b_n = (f, U_n),$$
$$i\sigma_n a_n - i\sigma_n b_n = (g, U_n),$$

or equivalently,

$$a_n = \tfrac{1}{2}\left(f - \frac{i}{\sigma_n} g, U_n\right),$$

$$b_n = \tfrac{1}{2}\left(f + \frac{i}{\sigma_n} g, U_n\right). \tag{7.9.13}$$

If we incorporate (7.9.13) into (7.9.12) we get the general solution of (7.9.8)–(7.9.11). ∎

EXERCISES

1. Analyze the solution of the wave equation in the case where some of the eigenvalues of L are negative.

2. Solve the heat equation

$$\frac{\partial u}{\partial t} = k \frac{\partial^2 u}{\partial x^2}, \qquad 0 < x < l, \qquad 0 < t,$$

$$u(x,0) = f(x), \qquad u(0,t) = u(l,t) = 0$$

on $L_2[0,l]$. [*Hint*: Expand $f(x)$ in terms of a suitable orthonormal basis in $L_2[0,l]$.]

3. Solve the heat equation

$$\frac{\partial u}{\partial t} = (x + 1)^2 \frac{\partial^2 u}{\partial x^2}, \qquad 0 < x < 1, \qquad 0 < t,$$

$$u(x,0) = f(x), \qquad u(0,t) = u(1,t) = 0$$

on $L_2[0,1]$.

4. Solve the wave equation

$$\frac{\partial^2 u}{\partial t^2} = k^2 \frac{\partial^2 u}{\partial x^2}, \qquad 0 < x < l, \qquad 0 < t,$$

$$u(x,0) = f(x), \qquad u_t(x,0) = g(x),$$

$$u(0,t) = u(1,t) = 0$$

on $L_2[0,1]$.

5. Solve the wave equation

$$\frac{\partial^2 u}{\partial t^2} = (x + 1)^2 \frac{\partial^2 u}{\partial x^2}, \qquad 0 < x < 1, \qquad 0 < t,$$

$$u(x,0) = f(x), \qquad u_t(x,0) = 0,$$

$$u(0,t) = u(1,t) = 0,$$

on $L_2[0,1]$.

6. Solve the heat equation

$$\frac{\partial u}{\partial t} = \Delta u, \qquad x = (x_1, x_2, x_3) \in \Omega, \qquad 0 < t,$$

$$u(x,0) = \|x\| = (x_1^2 + x_2^2 + x_3^2)^{1/2},$$

$$u(0,t) = 0,$$

$$u(x,t) = 0, \qquad \text{for } x \in \partial\Omega,$$

where Ω is the sphere of radius 1 centered at the origin and Δ is the Laplacian operator.

7. Solve the wave equation

$$\frac{\partial^2 u}{\partial t^2} = k^2 \Delta u, \qquad x = (x_1, x_2, x_3) \in \Omega, \qquad 0 < t,$$

$$u(x,0) = f(x), \qquad u_t(x,0) = g(x),$$

$$u(0,t) = 0,$$

$$u(x,t) = 0, \qquad \text{for } x \in \partial\Omega,$$

where Ω is as given in Exercise 6.

10.   SELF-ADJOINT OPERATORS

In many applications, especially in quantum mechanics, one is interested in studying a given linear operator L by means of its spectrum. More specifically, if L is a symmetric operator one is interested in knowing whether L can be written as a weighted sum of projections.[15] The answer, as we shall see in Example 1, requires that we look carefully at the domain $\mathcal{D}_L$ of the operator L. We begin by first defining the concept of the adjoint of an unbounded linear operator.

Let L be a linear operator on a Hilbert space H. Recall that this means that L is linear and its domain $\mathcal{D}_L$ is dense in H. The adjoint operator L^* will be defined so that the equation

$$(Lx,y) = (x,L^*y) \tag{7.10.1}$$

is "valid." However, we have to be more precise. First let us define $\mathcal{D}^*$ as follows:

$$\mathcal{D}^* = \{y \in H : (Lx,y) = (x,z) \text{ for some } z \in H \text{ and all } x \in \mathcal{D}_L\}. \tag{7.10.2}$$

Let $y \in \mathcal{D}^*$ and assume that w and z satisfy

$$(Lx,y) = (x,w) = (x,z)$$

for all x in $\mathcal{D}_L$. One then has $(x, w - z) = 0$ for all x in $\mathcal{D}_L$, or $w - z \perp \mathcal{D}_L$, which implies that $w - z \perp \overline{\mathcal{D}}_L = H$. Hence $w - z = 0$, or $w = z$. This means that the mapping $y \to z$ given in the definition of $\mathcal{D}^*$ maps y onto precisely one point z. We call this mapping L^* and let $\mathcal{D}^* = \mathcal{D}_{L^*}$ be the domain of L^*. Furthermore, it is easy to show that L^* is linear and we ask the reader to do this.

7.10.1 THEOREM.   *Let L be a densely defined operator on a Hilbert space H and let L^* and $\mathcal{D}^*$ be given by (7.10.1) and (7.10.2). Then L^* is a linear operator on $\mathcal{D}^*$.*

---

[15] We are purposely making a simplification here. Many of the operators of mathematical physics have a nontrivial continuous spectrum and therefore these operators require a Spectral Theorem which uses weighted "integrals" of projections instead of weighted sums of projections. In any case, the concept of self-adjointness which we introduce in this section is the same regardless of the form of the Spectral Theorem.

EXAMPLE 1.   Let L be a weighted sum of projections on a separable Hilbert space H. That is,

$$Lx = \sum_{n=1}^{\infty} \lambda_n P_n x,$$

where the domain of L is given by

$$\mathcal{D}(L) = \left\{ x \in H : \lim_{N \to \infty} \sum_{n=1}^{N} \lambda_n P_n x \text{ exists} \right\}.$$

The object of this example is to show that the adjoint L^* is given by

$$L^* x = \sum_{n=1}^{\infty} \bar{\lambda}_n P_n x, \qquad (7.10.3)$$

and (more importantly) that the domain $\mathcal{D}(L^*)$ of L^* satisfies

$$\mathcal{D}(L^*) = \mathcal{D}(L).$$

Since $\{P_n\}$ is a resolution of the identity, it follows that the closed linear spaces $\mathcal{R}(P_n)$ are mutually orthogonal and that

$$H = \mathcal{R}(P_1) + \mathcal{R}(P_2) + \cdots .$$

Without any loss in generality we can assume that $\mathcal{R}(P_n)$ is one-dimensional. Therefore, if e_n is a unit vector in $\mathcal{R}(P_n)$, then $\{e_n\}$ is an orthonormal basis for H. The Fourier Series Theorem then assures us that any vector $x \in H$ can be expressed uniquely as

$$x = \sum_{n=1}^{\infty} (x, e_n) e_n$$

and that

$$(x, y) = \sum_{n=1}^{\infty} (x, e_n)\overline{(y, e_n)}.$$

Furthermore, we note that $x \in \mathcal{D}(L)$ if and only if

$$\sum_{n=1}^{\infty} |\lambda_n (x, e_n)|^2 < \infty.$$

Since $P_n x = (x, e_n) e_n$ we see that

$$Lx = \sum_{n=1}^{\infty} \lambda_n (x, e_n) e_n = \sum_{n=1}^{\infty} \lambda_n P_n x.$$

Thus if $x \in \mathcal{D}(L)$ and $y \in H$, then

$$(Lx, y) = \sum_{n=1}^{\infty} \lambda_n (x, e_n)\overline{(y, e_n)}.$$

Furthermore, if there is a $z \in H$ such that

$$(Lx, y) = (x, z) = \sum_{n=1}^{\infty} (x, e_n)\overline{(z, e_n)}$$

for all $x \in \mathscr{D}(L)$, then

$$(z,e_n) = \bar{\lambda}_n(y,e_n).$$

That is, if $z = L^*y$, then

$$L^*y = \sum_{n=1}^{\infty} \bar{\lambda}_n(y,e_n)e_n = \sum_{n=1}^{\infty} \bar{\lambda}_n P_n y.$$

We see, then, that L^* does satisfy (7.10.3).

Let us now show that $\mathscr{D}(L) = \mathscr{D}(L^*)$. First we note that the above argument shows that if $y \in \mathscr{D}(L)$, then $y \in \mathscr{D}(L^*)$. Now assume that $y \in \mathscr{D}(L^*)$. Then there is a $z \in H$ such that for all x in $\mathscr{D}(L)$ one has

$$0 = (Lx,y) - (x,z) = \sum_{n=1}(x,e_n)[\overline{\bar{\lambda}_n(y,e_n) - (z,e_n)}].$$

Since $\mathscr{D}(L)$ is dense in H, this implies that $\bar{\lambda}_n(y,e_n) = (z,e_n)$ for all n, so

$$z = \sum_{n=1}^{\infty} \bar{\lambda}_n(y,e_n)e_n = L^*y.$$

Now by the Parseval Equality we get

$$\sum_{n=1}^{\infty} |\bar{\lambda}_n(y,e_n)|^2 < \infty.$$

Hence $y \in \mathscr{D}(L)$. ∎

We would like to show that $\mathscr{D}^*$ is in general dense in H, that is, L^* is densely defined; however, this requires the concept of a "closed" operator. This concept has been discussed in Exercises 9–14 of Section 5.6. and Exercises 13–14 of Section 5.8. We shall give the definition again; however, it would be helpful if the reader would quickly review these earlier exercises.

7.10.2 DEFINITION. Let L be a densely defined operator on a Hilbert space H, with the domain $\mathscr{D}_L$. We shall say that L is a *closed* operator if for every sequence $\{x_n\}$ in $\mathscr{D}_L$, with the property that both limits $x = \lim x_n$ and $y = \lim Lx_n$ exist, one has $x \in \mathscr{D}_L$ and $y = Lx$.

It is important to note that the concept of a closed operator does depend on the domain of the operator. It may happen that an operator L on $\mathscr{D}_L$ is not closed, but that there is an extension of L to $\hat{L}$ on $\mathscr{D}_{\hat{L}}$, where $\hat{L}$ is closed. In this case we shall say that the operator L is *closable*. The concept of closable operators is discussed further in the exercises. The reader may be assured that just about all linear operators of interest (bounded operators, differential operators, multiplicative operators, integral operators) are either closed or closable.

7.10.3 THEOREM. *Let L be a densely defined linear operator on a Hilbert space H with domain $\mathscr{D}_L$ and let L^* denote the adjoint with domain $\mathscr{D}^*$. If L is a closable operator, then L^* is a closed densely defined operator on H.*

Proof: We will prove that L^* is a closed operator. The fact that $\mathcal{D}^*$ is dense in H will be shown in Exercise 3, in this section.

In order to show that L^* is closed we let $\{y_n\}$ be a sequence in $\mathcal{D}^*$ with the property that both limits $y = \lim y_n$ and $z = \lim L^* y_n$ exist. We then must show that $y \in \mathcal{D}^*$ and $z = L^* y$. However, the continuity of the inner product gives us

$$(Lx,y) = \lim(Lx,y_n) = \lim(x,L^*y_n) = (x,z)$$

for all $x \in \mathcal{D}_L$. Hence, $y \in \mathcal{D}^*$ and $z = L^* y$.   ∎

Since the adjoint operator L^* is closed and densely defined one can define the second adjoint L^{**} by the equation

$$(L^*y,z) = (y,L^{**}z).$$

By combining this with (7.10.1) one suspects that $L = L^{**}$. If L is closed, then—as we shall see in the exercises—one does have $L = L^{**}$. The proof of the last equation is not a complete triviality for unbounded operators since one must prove two things: (1) the domains $\mathcal{D}_L$ and $\mathcal{D}_{L^{**}}$ are the same; (2) the operators L and L^{**} agree.

We are now prepared to define self-adjointness. However, before doing this let us observe that the following theorem is a characterization of symmetric operators.

7.10.4 THEOREM.   *A linear operator $L: \mathcal{D}_L \to H$ is symmetric if and only if $\mathcal{D}_L \subseteq \mathcal{D}_{L^*}$ and $Lx = L^*x$ for all $x \in \mathcal{D}_L$.*

7.10.5 DEFINITION.   A densely defined operator $L: \mathcal{D}_L \to H$ is said to be *self-adjoint* if $\mathcal{D}_L = \mathcal{D}_{L^*}$ and $Lx = L^*x$ for all $x \in \mathcal{D}_L$.

It follows from Example 1 that a weighted sum of projections is symmetric if and only if it is self-adjoint and this occurs if and only if the λ_n's are real.

We will conclude this section with a few examples. However, before turning to these let us observe that it is possible to define the concept of "normality" for unbounded operators. We will not do that here but instead we refer the reader to Stone [1, p. 311 ff].

EXAMPLE 2. (THE POSITION OPERATOR Q.)   This operator is defined in Example 1, Section 4 and was shown to be symmetric. In order to show that it is self-adjoint we let $v \in \mathcal{D}_{Q^*}$ and set $v^* = Q^*v$. We wish to show that

$$v^*(x) = xv(x). \tag{7.10.4}$$

Since $(Qu,v) = (u,v^*)$ for all u in $\mathcal{D}_Q$, we have

$$\int_{-\infty}^{\infty} u(x)\overline{[xv(x) - v^*(x)]}\, dx = 0, \qquad u \in \mathcal{D}_Q.$$

That is $xv(x) - v^*(x) \perp u(x)$ for all $u \in \mathscr{D}_Q$. Therefore, since $\mathscr{D}_Q$ is dense, one has $xv(x) - v^*(x) = 0$, which is (7.10.4). ∎

EXAMPLE 3. The Position Operator Q_k defined in Example 2, Section 4 is self-adjoint. This can be proved by repeating the reasoning of the last example. ∎

EXAMPLE 4. The Potential Operator $V(Q)$ defined in Example 3, Section 4 is self-adjoint. ∎

EXAMPLE 5. The Momentum Operator P defined in Example 4, Section 4 is symmetric but not self-adjoint. The reason for this is that the domain $C_0^1(-\infty,\infty)$ is not large enough. However, if we enlarge the domain, which means that we extend P to the space $\mathscr{D}_P$ of all absolutely continuous functions u in $L_2(-\infty,\infty)$ with $u' \in L_2(-\infty,\infty)$, then we can show that P is self-adjoint.

First we recall that if $u \in \mathscr{D}_P$, then

$$\lim_{t \to \pm\infty} u(t) = 0, \tag{7.10.5}$$

see Exercise 15, Section 5.22.

Now if $u, v \in \mathscr{D}_P$, then by integrating by parts we get

$$(Pu,v) - (u,Pv) = \int_{-\infty}^{\infty} [-iu'(\xi)\overline{v(\xi)} - u(\xi)(\overline{-iv'(\xi)})] \, d\xi$$

$$= \lim_{T \to \infty} \int_{-T}^{T} [-iu'(\xi)\overline{v(\xi)} - iu(\xi)\overline{v'(\xi)}] \, d\xi$$

$$= \lim_{T \to \infty} -i[u(T)\overline{v(T)} - u(-T)\overline{v(-T)}] = 0,$$

by (7.10.5). Hence, P is symmetric on this larger domain.

The fact that P is self-adjoint now follows easily from Theorem 5.22.8 and Example 1. ∎

EXAMPLE 6. The Momentum Operator P_k defined in Example 5, Section 4 is self-adjoint on the domain $\mathscr{D}_{P_k}$ defined in Theorem 5.22.9. This can be seen by an obvious adaptation of the argument used in the last example. ∎

EXERCISES

1. Show that the adjoint operator L^* is linear. (Does your argument use the fact that L is linear?)

2. Let $L: \mathscr{D}_L \to H$ be a linear operator and let G_L denote the graph of L, that is,

$$G_L = \{\{x,Lx\} \in H \oplus H: x \in \mathscr{D}_L\}.$$

   (a) Show that G_L is a linear subspace of $H \oplus H$.
   (b) Show that L is a closed operator if and only if G_L is a closed linear subspace of $H \oplus H$.

(c) Define $V: H \oplus H \to H \oplus H$ by $V\{x,y\} = \{y,-x\}$. Show that (7.10.1) can be rewritten as

$$(V\{x,Lx\}, \{y,L^*y\}) = 0, \tag{7.10.6}$$

where $x \in \mathcal{D}_L$ and $y \in \mathcal{D}_{L^*}$.

(d) Use (7.10.6) to show that $G_{L^*} = V(G_L)^{\perp}$.

(e) Show that L is closed if and only if $G_L = V(G_{L^*})^{\perp}$.

3. The following argument, which uses Exercise 2, will lead to a proof that $\mathcal{D}_{L^*}$ is dense in H when L is closed (Theorem 7.10.3). Let $z \perp \mathcal{D}_{L^*}$.
   (a) Show that $\{0,z\} \perp V(G_{L^*})$ in $H \oplus H$.
   (b) Show that $\{0,z\} \in G_L$ when L is closed.
   (c) Show that $z = 0$.

4. Extend Exercise 3 to show that $\mathcal{D}_{L^*}$ is dense in H when L is closable.

5. Let L be a closed, densely defined operator on a Hilbert space H. Use Exercise 2 and Theorem 7.10.3 to show that $L = L^{**}$.

6. Let L be a closable, densely defined operator on a Hilbert space H with domain $\mathcal{D}_L$. We shall say that an operator $\bar{L}$ with domain $\mathcal{D}_{\bar{L}}$, is the *closure* of L if
   (a) $\bar{L}$ is a closed operator,
   (b) $\bar{L}$ is an extension of L, and
   (c) every other closed linear operator of L that is an extension of L is also an extension of $\bar{L}$.
   Show that L has a closure and that the closure is uniquely defined. Show that L^{**} is the closure of L.

7. (a) Show that every symmetric linear operator is closable.
   (b) Show that a linear operator L is symmetric if and only if L^* is an extension of L^{**}.

8. Let L be a densely defined operator on a Hilbert space H and let G_L denote the graph of L. Let $\bar{G}_L$ denote the closure of G_L in $H \oplus H$. Assume that $\bar{G}_L$ has the property that (x,y) and (x,y') are in $\bar{G}_L$, then $y = y'$. In other words, the x-coordinate determines the y-coordinate, which we write as $y = \bar{L}x$. Show that L is closable and that $\bar{L}$ is the closure of L.

9. Let $\{e_n : n = 0,1,\ldots\}$ be an orthonormal basis for an infinite-dimensional separable Hilbert space H and let $\mathcal{D}$ denote the collection of all vectors x in H for which $\sum_{n=0}^{\infty} n|(x,e_n)|^2 < \infty$. The *creation and annihilation operators*, A and $\tilde{A}$, are defined on $\mathcal{D}$ by

$$Ax = \sum_{n=0}^{\infty} \sqrt{n+1}(x,e_{n+1})e_n$$

and

$$\tilde{A}x = \sum_{n=1}^{\infty} \sqrt{n}(x,e_{n-1})e_n.$$

(a) Show that A and $\tilde{A}$ are densely defined.

(b) Show that for $x, y \in \mathcal{D}$ one has $(Ax,y) = (x,\tilde{A}y)$. (The operators A and $\tilde{A}$ are related to P and Q as we shall see in Section 14.)

10. Let $f: R \to R$ be a continuous real-valued function on R and define $f(Q)$ by

$$f(Q): u(x) \to f(x)u(x),$$

where the domain of $f(Q)$ is

$$\{u(x) \in L_2(-\infty,\infty): f(x)u(x) \in L_2(-\infty,\infty)\}.$$

(a) Show that $f(Q)$ is self-adjoint.
(b) Does $f(Q)$ remain self-adjoint if we drop the assumption that $f(x)$ be continuous?

11. Consider $L = S_r + S_l + \Phi$ on $l_2(0,\infty)$, where S_r and S_l are the right and left shift operators, Φ is given by

$$\Phi(x_1,x_2,\ldots) = (\phi(1)x_1,\phi(2)x_2,\ldots),$$

and $\phi(n)$ is real-valued with $|\phi(n)| \to \infty$ as $n \to \infty$.
(a) Define the domain of L and show that L is self-adjoint.
(b) Show that L has a compact self-adjoint resolvent. (See Exercise 3, Section 6.9.)
(c) Show that the spectrum of L is only point spectrum.

12. (Continuation of Exercise 11.) Assume that $\phi(n) \to \lambda_0$ as $n \to \infty$.
(a) Show that L does not have a compact self-adjoint resolvent.
(b) Let $\{\lambda_n\}$ be the eigenvalues of L. Show that the sequence $\{\lambda_n\}$ has no limit, finite or infinite.

13. Show that a weighted sum of projections is closable.

14. Extend the results of Example 1, to nonseparable Hilbert spaces.

15. Let P be a self-adjoint projection on a Hilbert space H. Show that P is continuous.

16. Show that a projection P on a Hilbert space H is orthogonal if and only if it is self-adjoint. (Compare with Theorem 5.23.9.)

11. THE CAYLEY TRANSFORM

We see, then, that if L is a symmetric operator that is not self-adjoint, then $\mathscr{D}_L \subseteq \mathscr{D}_{L^*}$ but $\mathscr{D}_L \neq \mathscr{D}_{L^*}$. This suggests that it may be possible to extend L, that is, enlarge $\mathscr{D}_L$ in such a way that the extension is self-adjoint. It is the purpose of this section to develop a theory for determining when such a symmetric operator has a self-adjoint extension. We will show, for example, that every symmetric differential operator with real coefficients has a self-adjoint extension (Corollary 7.11.7).

The main tool in this section will be the *Cayley transform*

$$M = (L - iI)(L + iI)^{-1}, \tag{7.11.1}$$

where L is a symmetric operator. Before proceeding, we note here, for reference, that the *inverse Cayley transform* is given by

$$L = i(I + M)(I - M)^{-1}. \tag{7.11.2}$$

In Exercise 1 we will show that if L is a symmetric operator on a Hilbert space H, then the Cayley transform M is well defined. Indeed, if $\mathscr{D}_M$ is the closure of the linear subspace

$$\{y = (L + iI)x \colon x \in \mathscr{D}_L\},$$

then for $y = (L + iI)x \in \mathscr{D}_M$, My is given by

$$My = (L - iI)x. \tag{7.11.3}$$

Let $\mathscr{R}_M = M(\mathscr{D}_M)$ denote the range of M. We will also show that $\|My\| = \|y\|$ for all $y \in \mathscr{D}_M$. That is, the Cayley transform M is an isometry on $\mathscr{D}_M$ and, therefore, $\mathscr{R}_M$ is a closed linear subspace of H. (This does not mean that M is a unitary operator since it may happen that $\mathscr{D}_M \neq H$.)

This ends our discussion of the general case where L is symmetric. We now ask, what happens to M when L is a self-adjoint operator? The answer is given in the next theorem.

7.11.1 THEOREM.   *Let* $L \colon \mathscr{D}_L \to H$ *be a symmetric operator and let* M *be the Cayley transform of* L. *Then the following statements are equivalent*:

(a) L *is self-adjoint.*

(b) M *is a unitary operator.*

(c) $\mathscr{D}_M = H$ *and* $\mathscr{R}_M = H$.

(d) $\mathscr{D}_M^{\perp} = \{0\} = \mathscr{R}_M^{\perp}$.

The proof of this theorem is presented in Exercise 2 in this section.

Some further properties of the inverse Cayley transform are discussed in the exercises.

In order to see how the Cayley transform can be used in the study of symmetric operators, let us assume that we are given an operator L that is symmetric but not necessarily self-adjoint. This means that

$$L \subseteq L^{**} \subseteq L^*, \tag{7.11.4}$$

where we use the notation $A \subseteq B$ to depict the fact that B is an extension of A. The proof of (7.11.4) is not difficult. The relationship $L \subseteq L^*$ follows from the fact that L is symmetric. The relationship $L \subseteq L^{**}$ and $L^{**} \subseteq L^*$ follows from the fact that L^{**} is the closure of L, by Exercises 6–7 of Section 10.

Now if L is not self-adjoint, then its domain $\mathscr{D}_L$ is strictly smaller than $\mathscr{D}_{L^*}$. This suggests that we may try to extend L in order to "make it" self-adjoint. Furthermore, if $\tilde{L}$ is some extension of L, that is, $L \subseteq \tilde{L}$, then it follows from the definition of the adjoint that $\tilde{L}^* \subseteq L^*$. Thus an extension of L will not only enlarge $\mathscr{D}_L$ but it will also shrink $\mathscr{D}_{L^*}$. Finally, if $\tilde{L}$ is a symmetric extension of L, then we must have

$$L \subseteq \tilde{L} \subseteq \tilde{L}^* \subseteq L^*. \tag{7.11.5}$$

If we can choose the extension $\tilde{L}$ in such a way that $\mathscr{D}\tilde{L} = \mathscr{D}\tilde{L}^*$, then we see that $\tilde{L} = \tilde{L}^*$, that is, $\tilde{L}$ is self-adjoint.

The problem we now wish to study is under which conditions does a symmetric operator have a self-adjoint extension. We will see shortly that just about all of the symmetric operators discussed in this chapter have self-adjoint extensions.

The proof of the following lemma is left as an exercise.

7.11.2 LEMMA. *Let L be a symmetric linear operator on a Hilbert space H and define the Cayley transform M by* (7.11.1). *Then the following statements are valid.*

(a) *If $\tilde{L}$ is a symmetric extension of L and $\tilde{M}$ is the Cayley transform of $\tilde{L}$, then $\tilde{M}$ is an extension of M.*

(b) *If $\tilde{M}$ is an isometric extension of M and if $\tilde{L}$ is the inverse Cayley transform of $\tilde{M}$, then $\tilde{L}$ is a symmetric extension of L.*

Let L be a symmetric operator on a Hilbert space H and let M be the Cayley transform of L. The subspaces $\mathscr{D}_M{}^\perp$ and $\mathscr{R}_M{}^\perp$ are called the *deficiency subspaces* of L. Let $m = \dim \mathscr{D}_M{}^\perp$ and $n = \dim \mathscr{R}_M{}^\perp$. Then (m,n) are called the *deficiency indicies* of L. It follows from Theorem 7.11.1 (d) that L is self-adjoint if and only if the deficiency indicies are $(0,0)$.

We can now determine which symmetric operators have self-adjoint extensions.

7.11.3 THEOREM. *Let (m,n) denote the deficiency indicies of a symmetric operator L. Then L has a self-adjoint extension if and only if $m = n$.*

Proof: First assume that L has a self-adjoint extension and let $\tilde{L}$ denote this extension. Let M and $\tilde{M}$ be the Cayley transforms of L and $\tilde{L}$, respectively. Since $\tilde{M}$ is an isometric extension of M it is evident that the deficiency indicies for $\tilde{L}$ are of the form $(m - p, n - p)$. However, $\tilde{L}$ is self-adjoint and therefore

$$m - p = n - p = 0,$$

or $m = n = p$.

Now assume that $m = n$, and let $M: \mathscr{D}_M \to \mathscr{R}_M$ be the Cayley transform of L. Since $\dim \mathscr{D}_M{}^\perp = \dim \mathscr{R}_M{}^\perp$ there is an isometric mapping N of $\mathscr{D}_M{}^\perp$ onto $\mathscr{R}_M{}^\perp$, by Exercise 12, Section 5.19. Since

$$H = \mathscr{D}_M + \mathscr{D}_M{}^\perp = \mathscr{R}_M + \mathscr{R}_M{}^\perp,$$

we define $\tilde{M}: H \to H$ by

$$\tilde{M}(x + y) = Mx + Ny,$$

where $x \in \mathscr{D}_M$ and $y \in \mathscr{D}_M{}^\perp$. It is easy to see that $\tilde{M}$ is an isometric extension of M and that $\tilde{M}$ is a unitary mapping. It follows from Theorem 7.11.1 and Lemma 7.11.2 that the inverse Cayley transform $\tilde{L}$ is a self-adjoint extension of L. ∎

We shall give here one criterion that a symmetric operator L have equal deficiency indicies. Other criteria are presented in the exercises below.

7.11.4 DEFINITION.   A mapping $J: H \to H$ is said to be a *conjugation* if $J^2 = I$ and $(Jx,Jy) = (y,x)$ for all x and y in H.

For example, if H is a complex L_2-space and $Jf = \bar{f}$, where the bar denotes complex conjugation, then J is a conjugation in the sense of Definition 7.11.4.

The following lemma is an easy exercise for the reader.

7.11.5 LEMMA.   *Let J be a conjugation on a Hilbert space H. Then J is a one-to-one mapping of H onto itself with $J^{-1} = J$. Furthermore one has*

$$J(x + y) = Jx + Jy, \qquad x, y \in H,$$
$$J(\alpha x) = \bar{\alpha} J x, \qquad \alpha \in C, x \in H.$$

7.11.6 THEOREM.   *Let L be a symmetric operator on a Hilbert space H and assume that there is a conjugation J on H with the property that $JLx = LJx$ for all $x \in \mathcal{D}_L$. Then L has a self-adjoint extension.*

Proof: First we note that $JL = LJ$ on $\mathcal{D}_L$ implies that J is a one-to-one mapping of $\mathcal{D}_L$ onto itself. Furthermore since

$$J(L + iI)x = (L - iI)Jx$$

for all $x \in \mathcal{D}_L$, we see that J is a one-to-one mapping of

$$\mathcal{R}_+ = \{y = (L + iI)x : x \in \mathcal{D}_L\} \text{ onto } \mathcal{R}_- = \{y = (L - iI)x : x \in \mathcal{D}_L\}.$$

Hence J is a one-to-one mapping of $\mathcal{D}_M = \bar{\mathcal{R}}_+$ onto $\mathcal{R}_M = \bar{\mathcal{R}}_-$, where M denotes the Cayley transform of L.

Now if $z \in \mathcal{D}_M^\perp$, then for all $x \in \mathcal{D}_L$ one has $((L + iI)x,z) = 0$. Furthermore,

$$(Jz, (L - iI)x) = (Jz, J^2(L - iI)x)$$
$$= (J(L - iI)x,z) = ((L + iI)Jx,z) = 0.$$

Hence J is a one-to-one mapping of $\mathcal{D}_M^\perp$ onto $\mathcal{R}_M^\perp$. Finally, this implies that $\dim \mathcal{D}_M^\perp = \dim \mathcal{R}_M^\perp$, so L has equal deficiency indicies and therefore has a self-adjoint extension.   ∎

The following corollary is now immediate.

7.11.7 COROLLARY.   *Let L be a differential or integral operator on the complex space $L_2(\Omega)$. Assume that L has real coefficients. If L is symmetric, then L has a self-adjoint extension.*

Proof: We simply note that $LJu = JLu$ for all $u \in \mathcal{D}_L$, where $Ju = \bar{u}$.   ∎

EXERCISES

1. Let $L: \mathcal{D}_L \to H$ be a symmetric operator and let $M = (L - iI)(L + iI)^{-1}$ be the Cayley transform of L.

   (a) Show that $\|(L + iI)x\| = \|(L - iI)x\| \geq \|x\|$ for all $x \in \mathcal{D}_L$.

   (b) Let $\mathcal{D}_M$ denote the closure of $\{y = (L + iI)x: x \in \mathcal{D}_L\}$. Show that for $y = (L + iI)x \in \mathcal{D}_M$ one has $My = (L - iI)x$.

   (c) Show that $\|My\| = \|y\|$ for all $y \in \mathcal{D}_M$.

2. This will lead to a proof of Theorem 7.11.1.

   (a) Use the results of Section 5.15 to show that 7.11.1 (c) and 7.11.1 (d) are equivalent.

   (b) Use Theorem 5.19.2 to show that 7.11.1 (b) and 7.11.1 (c) are equivalent.

   (c) Define $\mathcal{R}_+$ and $\mathcal{R}_-$ by

   $$\mathcal{R}_+ = \{y = (L + iI)x: x \in \mathcal{D}_L\}, \qquad \mathcal{R}_- = \{y = (L - iI)x: x \in \mathcal{D}_L\}.$$

   Assume that 7.11.1 (a) holds. Let $z \in \mathcal{R}_+^{\perp}$ and show that z is in the domain of $(L - iI)$ and that $(L - iI)z \perp \mathcal{D}_L$. Hence $\mathcal{R}_+^{\perp} = \{0\}$ and, similarly, $\mathcal{R}_-^{\perp} = \{0\}$. Then show that 7.11.1 (c) holds.

   (d) Assume that both 7.11.1 (b) and 7.11.1 (c) hold and let $x \in \mathcal{D}_{L^*}$ and $x^* = Lx$. Show that

   $$x = (I - M)\frac{x - ix^*}{2}, \qquad x^* = i(I + M)\frac{x - ix^*}{2}.$$

   Next show that $L = i(I + M)(I - M)^{-1}$, that is, if $z = (I - M)w$ for some $w \in H$, then $z \in \mathcal{D}_L$ and $Lz = i(I + M)w$. Finally show that $x \in \mathcal{D}_L$ and $x^* = Lx$.

3. (a) Prove Lemma 7.11.2.

   (b) Prove Lemma 7.11.5.

4. (a) Let M be the Cayley transform of a symmetric operator L on a Hilbert space H. Show that the range of $M - I$ is dense in H.

   (b) Let $M: \mathcal{D}_M \to \mathcal{R}_M$ be an isometric operator, where $\mathcal{D}_M$ and $\mathcal{R}_M$ are closed linear subspaces of a Hilbert space H. Assume that the range of $M - I$ is dense in H, and define L by Equation (7.11.2). Show that L is a closed densely defined symmetric operator.

5. (Semibounded Symmetric Transformations.) A symmetric operator S on a Hilbert space H is said to be semibounded if there is a real number α such that

   $$(Sx,x) \geq \alpha(x,x) \qquad \text{for all } x \text{ in } \mathcal{D}_S. \tag{7.11.6}$$

   (a) Let S satisfy for (7.11.6) and let $T = S - \alpha I$. Show that T is a symmetric operator with $(Tx,x) \geq (x,x)$ for all x in $\mathcal{D}_T$. Show that S has a self-adjoint extension if and only if T has a self-adjoint extension.

   (b) Let $\langle x,y \rangle = (Tx,y)$. Show that $\langle x,y \rangle$ defines a new inner product on $\mathcal{D}_T$. Let $\|\|x\|\| = (\langle x,x \rangle)^{1/2}$ be the norm and show that $\|\|x\|\| \geq \|x\|$.

(c) Let H_0 denote the completion of the inner product space $(\mathcal{D}_T, \langle \cdot, \cdot \rangle)$. Show that H_0 can be identified with a linear subspace of H with the property that $\mathcal{D}_T \subseteq H_0 \subseteq H$. Show that for all x in H_0 one has $|||x||| \geq \|x\|$.

(d) Let $y \in H$ be fixed. Show that the linear functional $l_y(x) = (x, y)$ satisfies

$$|l_y(x)| = |(x, y)| \leq \|x\| \|y\| \leq |||x||| \cdot \|y\|$$

for all $x \in H_0$. Hence, there is a $z \in H_0$ such that

$$l_y(x) = \langle x, z \rangle.$$

Let $B \colon H \to H_0$ be the mapping $z = By$. Show that B is a bounded linear operator on H. Show that B is self-adjoint and that $A = B^{-1}$ exists.

(e) Show that A is self-adjoint and that A is an extension of T.

6. Show that a symmetric operator may have several self-adjoint extensions. For example, consider

$$Lu = u'',$$

where $\mathcal{D}_L = \{u \in L_2(0,1) \colon u(0) = u'(0) = u(1) = u'(1) = 0\}$.

7. A symmetric operator L, with the property that L^{**} is self-adjoint, is said to be *essentially self-adjoint*. (Recall that L^{**} is the closure of L, see Exercise 6, Section 10.)

(a) Show that an essentially self-adjoint operator has a unique self-adjoint extension.

(b) Show that a symmetric operator on a Hilbert space H is essentially self-adjoint if and only if $\mathcal{R}_+$ and $\mathcal{R}_-$ are dense in H.

8. Let L be a densely defined bounded symmetric operator on a Hilbert space H. Show that L is essentially self-adjoint.

9. Let M be the Cayley transform of a self-adjoint operator L, and assume that L^{-1} exists and is densely defined.

(a) Show that L^{-1} is symmetric.

(b) Show that the Cayley transform of L^{-1} is $-M^{-1}$.

(c) Show that L^{-1} is self-adjoint.

10. Let L be essentially self-adjoint on a Hilbert space H and, $\mathcal{D}_L$ the domain of L. Show that for all complex numbers λ with $\operatorname{Im} \lambda \neq 0$, the space $(L - \lambda I)(\mathcal{D}_L)$ is dense in H.

11. (Converse of Exercise 10.) Let L be a symmetric operator on a Hilbert space H with domain $\mathcal{D}_L$. Assume that for some complex number λ with $\operatorname{Im} \lambda \neq 0$ the spaces $(L - \lambda I)(\mathcal{D}_L)$ and $(L - \bar{\lambda} I)(\mathcal{D}_L)$ are dense in H. Show that L is essentially self-adjoint. [*Hint:* First note that if $\mathcal{D}$ is dense in H and if T is a bounded linear operator with $\|T\| < 1$, then $(I + T)(\mathcal{D})$ is dense in H. Next show that

$$(L + iI) = \frac{1}{c + 1} [cT + I](L - \bar{\lambda} I)$$

for an appropriate choice of c, $0 < c < 1$, where

$$T = (L - \lambda I)(L - \bar{\lambda} I).]$$

12. Let S be a symmetric operator and let T be an essentially self-adjoint operator on a common domain $\mathscr{D}$. Assume that there is an ε, $0 < \varepsilon < 1$, and a constant K such that

$$\|Su\| \leq \varepsilon\|Tu\| + K\|u\|$$

for all $u \in \mathscr{D}$. Show that $S + T$ is essentially self-adjoint. [*Hint*: Choose r so that $\varepsilon + K|r|^{-1} < 1$ and show that

$$\|S(T \pm irI)^{-1}\| < 1.$$

Then apply Exercise 10 while noting that

$$(S + T \pm irI)\mathscr{D} = \{S(T \pm irI)^{-1} + I\}\{T \pm irI\}\mathscr{D} \quad \text{is a dense in } H.]$$

13. Extend Exercise 12 to show that if T is self-adjoint on $\mathscr{D}$, then $S + T$ is self-adjoint on $\mathscr{D}$.

14. Let S be a symmetric operator and let T be an essentially self-adjoint operator on a common domain $\mathscr{D}$. Assume that there are nonnegative constants α and β such that

$$\|Su\|^2 \leq \alpha(u,Tu) + \beta\|u\|^2.$$

Show that $S + T$ is essentially self-adjoint.

15. Let L be a symmetric operator on a Hilbert space H and assume that there exists an orthonormal basis consisting entirely of eigenvectors of L. Show that L is essentially self-adjoint. (Use this fact to compare Theorem 7.3.3 with the Spectral Theorem.)

16. Show that the linear operator

$$Lu = i\frac{du}{dt}$$

with domain

$$\mathscr{D}_L = \{u \in L_2(0,\infty): u(0) = 0 \text{ and } u' \in L_2(0,\infty)\}$$

is symmetric on $L_2(0,\infty)$ but that it has no self-adjoint extensions.

17. Let $L = \sum_n \lambda_n P_n$ be a weighted sum of projections, where the λ_n's are real. Show that the Cayley transform is given by

$$M = \sum_n \left(\frac{\lambda_n - i}{\lambda_n + i}\right)P_n.$$

12. QUANTUM MECHANICS, REVISITED

Let us now illustrate how the concepts of self-adjoint operators and self-adjoint extensions of symmetric operators are used in the study of quantum mechanics. In this section we will show how the energy function in a classical-mechanical system becomes a self-adjoint operator in the corresponding quantum-mechanical system. In the next section we shall look at the Heisenberg Uncertainty

Relation and, finally, in Section 14 we shall analyze the quantum-mechanical harmonic oscillator.

Recall that in Section 5.25 we gave a brief introduction into the foundations of quantum mechanics. We noted that the observables of a quantum-mechanical system can be identified with the self-adjoint operators on an infinite-dimensional separable Hilbert space H. Of course, all such Hilbert spaces are unitarily equivalent. Therefore, when one is trying to study a particular quantum-mechanical system, one has a choice for the mathematical model.

It is customary to make a " standard " choice for certain quantum-mechanical systems, in particular for a system consisting of l particles interacting with one another. This type of system arises in atomic physics.

In the classical-mechanical system the differential equations of motion can be represented in the Hamiltonian form

$$\frac{dq}{dt} = \frac{\partial H}{\partial p}, \qquad \frac{dp}{dt} = -\frac{\partial H}{\partial q},$$

where $q = (q_1,\ldots,q_n)$ represent the position coordinates, $p = (p_1,\ldots,p_n)$ represent the " generalized " momenta coordinates, and $H = H(q,p)$ is the Hamiltonian function. For the quantum-mechanical model one then makes the identification $(x_1,\ldots,x_n) = (q_1,\ldots,q_n)$,

$$q_k \leftrightarrow Q_k,$$
$$p_k \leftrightarrow hP_k,$$

for $1 \leq k \leq n$, where Q_k and P_k are the operators on $L_2(R^n)$ given in Examples 2 and 5 of Section 4. As shown in Section 10, the operators P_k and Q_k are self-adjoint. The Schrödinger operator for this quantum-mechanical system then becomes

$$S = H(Q,hP) = H(Q_1,\ldots,Q_n,hP_1,\ldots,hP_n), \qquad (7.12.1)$$

where h is a universal constant. For example, in many cases the Hamiltonian is of the form

$$H(q,p) = (p_1{}^2 + \cdots + p_n{}^2) + V(q_1,\ldots,q_n),$$

where V is a real-valued potential function. Then S is uniquely determined by

$$S = -h^2 \Delta_n + V(Q),$$

where Δ_n is the n-dimensional Laplacian operator given in Example 6, Section 4 and $V(Q)$ is the potential operator described in Example 3, Section 4. Since S is a symmetric differential operator on $C_0{}^\infty(R^n)$ with real coefficients we see that S has a self-adjoint extension.

The dynamical equations of motion [Equation (5.25.5)] for this system then become

$$\frac{\partial \phi}{\partial t} = -iS\phi,$$

where ϕ is a function of $(x_1,\ldots,x_n)$ and t.

EXAMPLE 1. (HYDROGEN ATOM.) For the hydrogen atom one has a single electron interacting with a nucleus. Thus $n = 3$, $\Delta_n = \Delta_3$, and the potential V becomes the Coulomb potential

$$V(q_1, q_2, q_3) = -k[q_1{}^2 + q_2{}^2 + q_3{}^2]^{-1/2},$$

where k is a positive constant. ∎

EXAMPLE 2. (HELIUM ATOM.) For the helium atom one has two electrons interacting with a nucleus. Thus $n = 6$, $\Delta_n = \Delta_6$, and the potential V becomes

$$V(q_1, q_2, q_3, q_4, q_5, q_6) = k[|u - v|^{-1} - 2|u|^{-1} - 2|v|^{-1}],$$

where k is a constant, $u = (q_1, q_2, q_3)$, $v = (q_4, q_5, q_6)$, and $|\ |$ denotes the Euclidean norm, that is, $|u| = [q_1{}^2 + q_2{}^2 + q_3{}^2]^{1/2}$.

For a detailed study of hydrogen and helium operators we refer the reader to Hellwig [1] and Kato [1]. ∎

We conclude our story by considering two applications in quantum mechanics. The first is a discussion of the Heisenberg Uncertainty Theorem and the second is a discussion of the quantum-mechanical harmonic oscillator.

13. HEISENBERG UNCERTAINTY THEOREM

Let L and M be two observables (that is, self-adjoint operators) for a quantum-mechanical system and let N be defined by

$$iN = LM - ML.$$

Assume that L and M are so defined that the commutator N given above is densely defined. [*Note*: N is symmetric.]

Consider now a pure state that is represented (in the sense of Theorem 5.25.1) by a unit vector e in H. This means that the expected value of L and M in state e is given[16] by

$$\alpha = E(L) = (Le, e),$$
$$\beta = E(M) = (Me, e).$$

The *deviation* of the observable L in state e is given by

$$(\Delta\alpha)^2 = E((L - \alpha I)^2) = ((L - \alpha I)^2 e, e)$$
$$= \|(L - \alpha I)e\|^2,$$

since L is self-adjoint. Similarly, one has

$$(\Delta\beta)^2 = \|(M - \beta I)e\|^2.$$

7.13.1 THEOREM. *In the above notation one has*

$$(\Delta\alpha)(\Delta\beta) \geq \tfrac{1}{2}|E(N)| = \tfrac{1}{2}|(Ne, e)|. \tag{7.13.1}$$

---

[16] We assume that the vector e lies in the common domain of L, M, and N.

Proof: Let $T = L - \alpha I + i\rho(M - \beta I)$, where ρ is real. Then T is densely defined and

$$T^* = (L - \alpha I) - i\rho(M - \beta I).$$

Furthermore,

$$TT^* = (L - \alpha I)^2 + \rho N + \rho^2 (M - \beta I)^2.$$

Hence,

$$0 \le (T^*e, T^*e) = (TT^*e, e)$$
$$= ((L - \alpha I)^2 e, e) + \rho(Ne, e) + \rho^2 ((M - \beta I)^2 e, e).$$

That is,

$$0 \le (\Delta\alpha)^2 + \rho E(N) + \rho^2 (\Delta\beta)^2. \tag{7.13.2}$$

Since (7.13.2) is valid for all real ρ, the discriminant[17]

$$[E(N)]^2 - 4(\Delta\alpha)^2 (\Delta\beta)^2$$

is nonpositive, that is,

$$|E(N)| \le 2(\Delta\alpha)(\Delta\beta). \quad \blacksquare$$

EXAMPLE 1. (HEISENBERG UNCERTAINTY PRINCIPLE.) Equation (7.13.1) reduces to the Heisenberg Uncertainty Principle when L is the position operator Q_k and M is the momentum operator P_k on $L_2(R^n)$. In this case, it is easy to see that

$$(Q_k P_k - P_k Q_k)u = iu$$

for all functions u in $C_0{}^2(R^n)$. In other words, the operator N becomes the identity and (7.13.1) becomes

$$(\Delta\alpha)(\Delta\beta) \ge \tfrac{1}{2}.$$

If instead we replace M with hP_k, where h is a constant, then (7.13.1) becomes

$$(\Delta\alpha)(\Delta\beta) \ge \frac{h}{2},$$

which is the original result of Heisenberg.

EXERCISES

1. Theorem 7.13.1 can be extended to quantum-mechanical states given by a density operator W. In this case, one has $\alpha = E(L) = \operatorname{tr} WL$, $\beta = E(M) = \operatorname{tr} WM$, $(\Delta\alpha)^2 = \operatorname{tr}[W(L - I)^2]$, and $(\Delta\beta)^2 = \operatorname{tr}[W(M - \beta I)^2]$. Also Equation (7.13.1) becomes

$$(\Delta\alpha)(\Delta\beta) \ge \tfrac{1}{2}|\operatorname{tr} WN|. \tag{7.13.3}$$

---

[17] Compare this argument with the proof of the Schwarz Inequality.

We now outline the proof:

(a) Show that

$$\operatorname{tr} WTT^* = \operatorname{tr} T^*WT > 0.$$

(b) Show that

$$\operatorname{tr} WTT^* = \operatorname{tr} W(L - \alpha I)^2 + \rho \operatorname{tr} WN + \rho^2 \operatorname{tr} W(M - \beta I)^2.$$

(c) Prove (7.13.3).

2. The relationship

$$LMu - MLu = iu$$

is called the *Heisenberg commutation property*. We saw that Q_k and P_k satisfy this property. (It can be shown that if L and M satisfy this property, then they are unbounded, see von Neumann [1].)

14.   THE HARMONIC OSCILLATOR

The Hamiltonian differential equation for a harmonic oscillator in classical mechanics is

$$\frac{dx}{dt} = \frac{\partial H}{\partial y}, \qquad \frac{dy}{dt} = -\frac{\partial H}{\partial x},$$

where $H = y^2 + \omega^2 x^2$, x is position, y is momentum, and ω is a physical constant.

In the quantum-mechanical system we replace x by the position operator Q and y by the momentum operator P; compare with Section 12. The Hamiltonian function H then becomes the Schrödinger operator

$$S = P^2 + \omega^2 Q^2,$$

or

$$Su = -\frac{d^2 u}{dx^2} + \omega^2 x^2 u$$

for $-\infty < x < \infty$. The operator S is, then, a singular Sturm-Liouville operator, since the interval is now infinite. We consider S in the Hilbert space $L_2(-\infty, \infty)$.

Let us first show that S is symmetric on $C_0^2(-\infty, \infty)$, where C_0^2 denotes the C^2 functions with compact support. If u, v are in C_0^2, then by integrating by parts we get

$$(Su, v) - (u, Sv) = \int_{-\infty}^{\infty} [-u''(x) + \omega^2 x^2 u(x)]\bar{v}(x) - u(x)[-\bar{v}''(x) + \omega^2 x^2 v(x)]\, dx$$

$$= \int_{-\infty}^{\infty} [-u''(x)\bar{v}(x) + u(x)\bar{v}''(x)]\, dx$$

$$= \int_{-\infty}^{\infty} (+u'(x)\bar{v}'(x) - u'(x)\bar{v}'(x))\, dx = 0.$$

Hence, S is symmetric on $C_0^2(-\infty,\infty)$. Furthermore, since S is real it has a self-adjoint extension, which we shall denote by S. Its domain, which includes C_0^2, we shall denote by $\mathscr{D}_s$.

Our objective now is to construct an orthonormal basis of eigenfunctions for S. We shall see that this basis has the form

$$\phi_n(x) = c_n H_n(\sqrt{\omega}x)\exp\left(\frac{-\omega x^2}{2}\right), \qquad n = 0, 1, \ldots,$$

where H_n is the Hermite polynomial of degree n and c_n is a normalization factor.

Thus we seek solutions of

$$Su = \lambda u, \tag{7.14.1}$$

where $u \neq 0$ and $u \in L_2(-\infty,\infty)$. If we make a change of variable, replacing $\sqrt{\omega}x$ by x and λ by $\omega\lambda$, Equation (7.14.1) becomes

$$-\frac{d^2u}{dx^2} + x^2u = \lambda u. \tag{7.14.2}$$

Let us now make another change of variable with $u = v\exp(-x^2/2)$ and we see that (7.14.2) becomes

$$v'' - 2xv' + \lambda v = 0. \tag{7.14.3}$$

Let $H_n(x)$ denote the function

$$H_n(x) = (-1)^n\exp(x^2)D^n\exp(-x^2), \tag{7.14.4}$$

where $D^n v = d^n v/dx^n$. It is easy to verify that $H_n(x)$ is a polynomial of degree n. We will now show [see Equation (7.14.8) below] that $H_n(x)$ is a solution of (7.14.3) with $\lambda = 2n$.

7.14.1 THEOREM.   *The Hermite polynomials satisfy*:

$$H_n'(x) = 2xH_n(x) - H_{n+1}(x); \tag{7.14.5}$$

$$H_{n+1}(x) - 2xH_n(x) + 2nH_{n-1}(x) = 0; \tag{7.14.6}$$

$$H_n'(x) = 2nH_{n-1}(x); \tag{7.14.7}$$

$$H_n''(x) - 2xH_n'(x) + 2nH_n(x) = 0; \tag{7.14.8}$$

$$\int_{-\infty}^{\infty} H_n(x)H_m(x)\exp(-x^2)\,dx = \delta_{mn}\sqrt{\pi}\,2^n n!; \tag{7.14.9}$$

for $n = 1, 2, \ldots$.

Proof: If we differentiate (7.14.4), we get (7.14.5). Now let $a(x) = \exp(-x^2)$. Then by repeated differentiation we get

$$D^{n+1}a(x) + 2xD^n a(x) + 2nD^{n-1}a(x) = 0, \tag{7.14.10}$$

$n = 1, 2, \ldots$. If we now multiply (7.14.10) through by $(-1)^{n+1}\exp(x^2)$, we get (7.14.6). By combining (7.14.5) and (7.14.6) we get (7.14.7). If we now differentiate (7.14.5) and replace $H_{n+1}'(x)$ by using (7.14.7), we get (7.14.8).

In order to prove (7.14.9) we first note that for any polynomial $P(x)$ one has

$$P(x) \exp(-x^2) \to 0 \qquad \text{as } x \to \pm\infty.$$

Now let $m \leq n$. Then, by repeated integration by parts we get

$$\int_{-\infty}^{\infty} H_m(x)H_n(x) \exp(-x^2)\, dx = \int_{-\infty}^{\infty} H_m(x)(-1)^n D^n \exp(-x^2)\, dx$$

$$= \int_{-\infty}^{\infty} \exp(-x^2) D^n H_m(x)\, dx.$$

Thus, if $m < n$, then $D^n H_m(x) = 0$, and the integral above vanishes. If $m = n$, the last integral becomes

$$n!\, d_n \int_{-\infty}^{\infty} \exp(-x^2)\, dx = n!\, d_n \sqrt{\pi},$$

where d_n is the coefficient of x^n in $H_n(x)$. In the exercises the reader is asked to show that $d_n = 2^n$. ∎

If we now retrace our steps and use (7.14.9) we see that the *Hermite functions*

$$\phi_n(x) = (\sqrt{\pi}\, 2^n n!)^{-1/2} H_n(x) \exp\left(\frac{-x^2}{2}\right), \qquad n = 0, 1, \ldots$$

form an orthonormal collection of eigenfunctions for the operator $[-d^2u/dx^2 + x^2u]$ given in Equation (7.14.2). The corresponding eigenvalues are $\lambda_n = 2n$. Also, $\{\phi_n(\sqrt{\omega}x)\}$ forms an orthonormal collection of eigenfunctions for the operator S and the corresponding eigenvalues are $\lambda_n = 2n\omega^{-1}$.

It remains to show that the collection $\{\phi_n(x)\}$ forms a basis for $L_2(-\infty,\infty)$. This is discussed in the exercises.

EXERCISES

1. Show that the coefficient of x^n in $H_n(x)$ is 2^n. [*Hint*: Use (7.14.5) and mathematical induction.]

2. Show that $H_n(-x) = (-1)^n H_n(x)$.

3. Let $G(x,t) = \exp(2tx - t^2)$. Show that

$$G(x,t) = \sum_{n=0}^{\infty} \frac{1}{n!} H_n(x)t^n.$$

[$G(x,t)$ is called the *generating function* for the Hermite polynomials.]

4. Find the first five Hermite polynomials. [*Note*: $H_0(x) = 1$.]

5. Show that the Hermite functions form an orthonormal basis for $L_2(-\infty,\infty)$. [*Hint*: Note that if $f \in L_2(-\infty,\infty)$, then $f(x) = f_e(x) - f_o(x)$ where f_e and f_o are even and odd functions. Now use the change of independent variables $y^2 = x$ and the fact that the Laguerre functions form an orthonormal basis for $L_2[(0,\infty)$. See Exercises 8 and 12 in Section 5.18.]

6. Let $\{\phi_n : n = 0, 1, \ldots\}$ be the Hermite functions on $L_2(-\infty, \infty)$ and let

$$\mathscr{D} = \left\{ u : \sum_{n=0}^{\infty} n \, |(u, \phi_n)|^2 < \infty \right\}.$$

Define the creation and annihilation operators by $\tilde{A}$ and A by

$$A\phi_n = \sqrt{n}\,\phi_{n-1}, \qquad n = 1, 2, \ldots$$
$$A\phi_0 = 0,$$
$$\tilde{A}\phi_n = \sqrt{n+1}\,\phi_{n+1}, \qquad n = 0, 1, \ldots.$$

(a) Show that $A^* = \tilde{A}$.

(b) Show that $A = \dfrac{1}{\sqrt{2}}(Q + iP)$, and $\tilde{A} = \dfrac{1}{\sqrt{2}}(Q - iP)$ on $\mathscr{D}$.

7. Using the notation of Example 2, Section 2, show that

$$\int_{-\infty}^{\infty} \int_{-\infty}^{\infty} |g(x,y)|^2 \, dx \, dy = \sum_{n=0}^{\infty} \frac{1}{(2n-1)^2}.$$

SUGGESTED REFERENCES

Agmon [1]

Bers [1]

Coddington and Levinson [1]

Courant and Hilbert [1]

Friedman [1], [2]

Garabedian [1]

S. Goldberg [1]

Hellwig [1]

Hille and Phillips [1]

Jauch [1], [2]

John and Schechter [1]

Kato [1], [2]

Lanczos [1]

Meyers and Serrin [1]

Mikhlin [1]

Schwartz [1]

Sobolev [1]

Also see references at end of Chapters 5 and 6.

Appendices

Appendix A

The Hölder, Schwarz, and Minkowski Inequalities

The purpose of this appendix, as the name suggests, is to prove the Hölder, Schwarz, and Minkowski Inequalities. We will consider these inequalities in three settings: (1) finite sums; (2) infinite sums; and (3) integrals. Let us now state the inequalities for each of these settings.

Schwarz and Hölder Inequality $1 < p < \infty$ and $p^{-1} + q^{-1} = 1$.

1. FINITE SUMS:

$$\sum_{i=1}^{n} |x_i y_i| \leq \left(\sum_{i=1}^{n} |x_i|^p \right)^{1/p} \left(\sum_{i=1}^{n} |y_i|^q \right)^{1/q}$$

2. INFINITE SUMS: Given that $\sum_{i=1}^{\infty} |x_i|^p < \infty$ and $\sum_{i=1}^{\infty} |y_i|^q < \infty$, then

$$\sum_{i=1}^{\infty} |x_i y_i| \leq \left(\sum_{i=1}^{\infty} |x_i|^p \right)^{1/p} \left(\sum_{i=1}^{\infty} |y_i|^q \right)^{1/q}.$$

3. INTEGRALS: Given that $\int_{\Omega} |x|^p \, dt < \infty$ and $\int_{\Omega} |y|^q \, dt < \infty$, then

$$\int_{\Omega} |xy| \, dt \leq \left(\int_{\Omega} |x|^p \, dt \right)^{1/p} \left(\int_{\Omega} |y|^q \, dt \right)^{1/q}.$$

The special case $p = q = 2$ is often referred as the *Schwarz Inequality*.

Minkowski Inequality $1 \leq p < \infty$.

1. FINITE SUMS:

$$\left(\sum_{i=1}^{n} |x_i \pm y_i|^p \right)^{1/p} \leq \left(\sum_{i=1}^{n} |x_i|^p \right)^{1/p} + \left(\sum_{i=1}^{n} |y_i|^p \right)^{1/p}.$$

2. INFINITE SUMS: Given that $\sum_{i=1}^{\infty} |x_i|^p < \infty$ and $\sum_{i=1}^{\infty} |y_i|^p < \infty$, then

$$\left(\sum_{i=1}^{\infty} |x_i \pm y_i|^p \right)^{1/p} \leq \left(\sum_{i=1}^{\infty} |x_i|^p \right)^{1/p} + \left(\sum_{i=1}^{\infty} |y_i|^p \right)^{1/p}.$$

3. INTEGRALS: Given that $\int_{\Omega} |x|^p \, dt < \infty$ and $\int_{\Omega} |y|^p \, dt < \infty$, then

$$\left(\int_{\Omega} |x \pm y|^p \, dt \right)^{1/p} \leq \left(\int_{\Omega} |x|^p \, dt \right)^{1/p} + \left(\int_{\Omega} |y|^p \, dt \right)^{1/p}.$$

The above inequalities can be rewritten in a succinct form if we introduce the following norm notation:

1. FINITE SUMS: $\|x\|_p = \left(\sum_{i=1}^n |x_i|^p\right)^{1/p}$, where $x = (x_1, \ldots, x_n)$.

2. INFINITE SUMS: $\|x\|_p = \left(\sum_{i=1}^\infty |x_i|^p\right)^{1/p}$, where $x = (x_1, x_2, \ldots)$.

3. INTEGRALS: $\|x\|_p = \left(\int_\Omega |x|^p \, dt\right)^{1/p}$.

The Minkowski Inequality then becomes

$$\|x \pm y\|_p \le \|x\|_p + \|y\|_p,$$

where $x \pm y$ denotes either $(x_1 \pm y_1, \ldots, x_n \pm y_n)$, or $(x_1 \pm y_1, x_2 \pm y_2, \ldots)$, or $x(t) \pm y(t)$ as the case may be.

In order to prove these inequalities we will use the following lemma.

A.1 LEMMA.   *Let a and b be nonnegative real numbers. Then*

$$ab \le \frac{a^p}{p} + \frac{b^q}{q}, \tag{A.1}$$

where $1 < p < \infty$ and $p^{-1} + q^{-1} = 1$.

Proof: In the (ξ, η)-plane consider the curve $\eta = \xi^{p-1}$, or equivalent, $\xi = \eta^{q-1}$. Let

$$A_1 = \int_0^a \xi^{p-1} \, d\xi = \frac{a^p}{p} \quad \text{and} \quad A_2 = \int_0^b \eta^{q-1} \, d\eta = \frac{b^q}{q}.$$

If we interpret A_1 and A_2 as areas, as shown in Figure A.1, then it is clear that $ab \le A_1 + A_2$. ∎

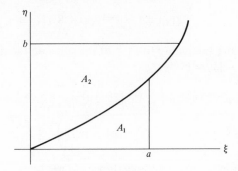

Figure A.1.

The proof of the Hölder Inequality for finite sums is now easy. Since[1]

$$\frac{|x_i|}{\|x\|_p} \frac{|y_i|}{\|y\|_q} \leq \frac{|x_i|^p}{p\|x\|_p^p} + \frac{|y_i|^q}{q\|y\|_q^q}, \tag{A.2}$$

$$\sum_{i=1}^{n} \frac{|x_i|}{\|x\|_p} \frac{|y_i|}{\|y\|_q} \leq \frac{1}{p\|x\|_p^p} \sum_{i=1}^{n} |x_i|^p + \frac{1}{q\|y\|_q^q} \sum_{i=1}^{n} |y_i|^q = \frac{1}{p} + \frac{1}{q} = 1.$$

Hence,

$$\sum_{i=1}^{n} |x_i y_i| = \sum_{i=1}^{n} |x_i| |y_i| \leq \|x\|_p \|y\|_q.$$

The proof of the Hölder Inequality for infinite sums is a straightforward extension of the result for finite sums. Indeed,

$$\sum_{i=1}^{N} |x_i y_i| \leq \left(\sum_{i=1}^{N} |x_i|^p \right)^{1/p} \left(\sum_{i=1}^{N} |y_i|^q \right)^{1/q}$$

$$\leq \left(\sum_{i=1}^{\infty} |x_i|^p \right)^{1/p} \left(\sum_{i=1}^{\infty} |y_i|^q \right)^{1/q}.$$

Now let $N \to \infty$ on the left side and one gets the Hölder Inequality for infinite sums.

The proof of the Hölder Inequality for integrals is similar. Here we replace (A.2) with

$$\frac{|x(t)|}{\|x\|_p} \frac{|y(t)|}{\|y\|_q} \leq \frac{|x(t)|^p}{p\|x\|_p^p} + \frac{|y(t)|^q}{q\|y\|_q^q}$$

and then integrate to get the desired result.

Minkowski Inequality follows from the Hölder Inequality. We will present here the argument for finite sums and ask the reader to verify that the same reasoning applies to infinite sums and integrals.

First note that

$$(|a| + |b|)^p = (|a| + |b|)^{p-1}|a| + (|a| + |b|)^{p-1}|b|.$$

Now set $a = x_i$ and $b = y_i$ and sum over i. Then

$$\sum_{i=1}^{n} |x_i \pm y_i|^p \leq \sum_{i=1}^{n} (|x_i| + |y_i|)^p$$

$$= \sum_{i=1}^{n} (|x_i| + |y_i|)^{p-1}|x_i| + \sum_{i=1}^{n} (|x_i| + |y_i|)^{p-1}|y_i|.$$

Now apply the Hölder Inequality to each of the sums on the right side of the above equation. Since $(p-1)q = p$, one gets

$$\sum_{i=1}^{n} (|x_i| + |y_i|)^p \leq \left(\sum_{i=1}^{n} (|x_i| + |y_i|)^p \right)^{1/q} \left(\sum_{i=1}^{n} |x_i|^p \right)^{1/p}$$

$$+ \left(\sum_{i=1}^{n} (|x_i| + |y_i|)^p \right)^{1/q} \left(\sum_{i=1}^{n} |y_i|^p \right)^{1/p}. \tag{A.3}$$

[1] We assume here that $\|x\|_p \neq 0$ and $\|y\|_q \neq 0$. If one of these happened to be zero, the Hölder Inequality is trivially true.

If $\left(\sum_{i=1}^{n} (|x_i| + |y_i|)^p\right)^{1/q} \neq 0$, we can divide both sides of (A.3) by it, thereby getting

$$\left(\sum_{i=1}^{n} |x_i \pm y_i|^p\right)^{1/p} \leq \left(\sum_{i=1}^{n} (|x_i| + |y_i|)^p\right)^{1/p}$$

$$\leq \left(\sum_{i=1}^{n} |x_i|^p\right)^{1/p} + \left(\sum_{i=1}^{n} |y_i|^p\right)^{1/p},$$

which is the Minkowski Inequality. If $\left(\sum_{i=1}^{n} (|x_i| + |y_i|)^p\right)^{1/q} = 0$, then the Minkowski Inequality is trivially true.

There is also a Hölder Inequality for the case $p = 1$ and $q = \infty$ as well as a Minkowski Inequality for $p = \infty$. In both cases the proofs are elementary.

Hölder Inequality $p = 1, q = \infty$.

1. FINITE OR INFINITE SUMS: Given $\sum_i |x_i| < \infty$ and $\sup_i |y_i| < \infty$, then

$$\sum_i |x_i y_i| \leq \left(\sum_i |x_i|\right) \left(\sup_i |y_i|\right).$$

2. INTEGRALS: Given $\int_\Omega |x|\, dt < \infty$ and $\text{ess.}_t \sup |y(t)| < \infty$, then

$$\int_\Omega |xy|\, dt \leq \left(\int_\Omega |x|\, dt\right) (\text{ess.} \sup_t |y(t)|),$$

where ess. sup is defined in Appendix D.

Minkowski Inequality $p = \infty$.

1. FINITE OR INFINITE SEQUENCES: Given $\sup_i |x_i| < \infty$ and $\sup_i |y_i| < \infty$, then

$$\sup_i |x_i + y_i| \leq \sup_i |x_i| + \sup_i |y_i|.$$

2. FUNCTIONS: Given $\text{ess.} \sup_t |x(t)| < \infty$ and $\text{ess.} \sup_t |y(t)| < \infty$, then

$$\text{ess.} \sup_t |x(t) + y(t)| \leq \left(\text{ess.} \sup_t |x(t)|\right) + \left(\text{ess.} \sup_t |y(t)|\right).$$

Appendix B

Cardinality

Let X be a set. Then card (X), the cardinal number of X, is merely the number of elements in X. For finite sets, this is, of course, a rather elementary idea. For infinite sets, the concept of cardinal number is a bit more complicated.

We will not define cardinal number here.[1] Instead we shall define when one has

$$\text{card } (X) = \text{card } (Y),$$
$$\text{card } (X) \leq \text{card } (Y),$$
$$\text{card } (X) < \text{card } (Y).$$

These definitions will be adequate for our purposes.

B.1 DEFINITION. Let X and Y be two sets. We shall say that

$$\text{card } (X) = \text{card } (Y)$$

if there is a one-to-one mapping of X onto Y. We say that

$$\text{card } (X) \leq \text{card } (Y)$$

if there is a one-to-one mapping of X into Y. If we have card $(X) \leq$ card (Y), then we say that

$$\text{card } (X) < \text{card } (Y)$$

if every one-to-one mapping ϕ of X into Y is not onto, that is, the range $\phi(X)$ is a proper subset of Y.

Before we look at some examples, there is one very important theorem which we should prove.

B.2 THEOREM. (BERNSTEIN.) *Let X and Y be sets. If card $(X) \leq$ card (Y) and card $(Y) \leq$ card (X), then card $(X) =$ card (Y).*

Proof: At first glance it may appear that there is nothing to prove. However, the theorem is not that trivial. We are given one-to-one mappings (see Figure B.1)

$$f: X \to Y \qquad \text{and} \qquad g: Y \to X.$$

(Remember that neither mapping may be onto.) The problem is then to construct a one-to-one mapping h of X *onto* Y.

---

[1] A definition can be found in Wilder [1, p. 99].

552

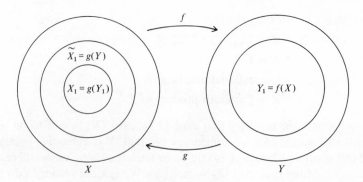

Figure B.1.

Let $Y_1 = f(X)$, $\tilde{X}_1 = g(Y)$, and $X_1 = g(Y_1)$. If one had $Y_1 = Y$, or equivalently, $X_1 = \tilde{X}_1$, then f would be a one-to-one mapping of X onto Y. On the other hand, if we could construct a one-to-one mapping k of X onto $\tilde{X}_1$, then

$$h(x) = g^{-1}(k(x))$$

would be a one-to-one mapping of X onto Y. We now proceed to construct such a mapping k. But first recall that the composition $g(f(x))$ is a one-to-one mapping.

Now define sets $\tilde{X}_n$ and X_n for $n = 1, 2, \ldots$, as follows:

$$\tilde{X}_1 = g(Y), \qquad X_1 = g(f(X))$$
$$\tilde{X}_2 = g(f(\tilde{X}_1)), \qquad X_2 = g(f(X_1))$$
$$\vdots \qquad\qquad \vdots$$
$$\tilde{X}_{n+1} = g(f(\tilde{X}_n)), \qquad X_{n+1} = g(f(X_n)).$$

Since $X_1 \subset \tilde{X}_1$, we see that $X_n \subset \tilde{X}_n$ for all n. Furthermore, since $\tilde{X}_1 \subset X$ we see that $\tilde{X}_2 \subset X_1$. Therefore, $\tilde{X}_{n+1} \subset X_n$ for all n.

Let $X_0 = X$ and now define a mapping $k : X \to X$ as follows:

$$k(x) = g(f(x)), \quad \text{if } x \in X_n - \tilde{X}_{n+1}, \quad n = 0, 1, \ldots,$$
$$k(x) = x, \quad \text{if } x \in \tilde{X}_n - X_n, \quad n = 1, 2, \ldots,$$
$$k(x) = x, \quad \text{if } x \in \bigcap_{n=1}^{\infty} X_n.$$

It is easy to see that from the way the sets X_n and $\tilde{X}_n$ were constructed, k maps $(X_n - \tilde{X}_{n+1})$ onto $(X_{n+1} - \tilde{X}_{n+2})$ for $n = 0, 1, \ldots$. Since the composition $g(f(x))$ is one-to-one over each of these sets, we see that k is a one-to-one mapping on X. Furthermore, it is a simple exercise to see that the range of k is

$$\bigcup_{n=1}^{\infty}(X_n - \tilde{X}_{n+1}) \cup \bigcup_{n=1}^{\infty}(\tilde{X}_n - X_n) \cup \left(\bigcap_{n=1}^{\infty} X_n\right) = \tilde{X}_1.$$

Hence $h(x) = g^{-1}(k(x))$ is a one-to-one mapping of X onto Y. ∎

EXAMPLE 1.   Consider the sets

$$Z = \{\ldots, -1, 0, 1, 2, \ldots\},$$
$$N = \{1, 2, \ldots\},$$
$$Q = \text{rational numbers},$$
$$Z \times Z = \text{Cartesian product of } Z.$$

Any set X with the property that card $(X) \leqslant$ card (N) is said to be *countable*. When card $(X) =$ card (N), we sometimes say that X is *countably infinite*. If one has card $(N) <$ card (X), then X is said to be *uncountable* or *uncountably infinite*.

It is easy to show that card $(Z) =$ card (N). We ask the reader to do this.

Let us show that card $(Z) =$ card $(Z \times Z)$. The mapping $n \to (n, 1)$ defines a one-to-one mapping of Z into $Z \times Z$ so we see that card $(Z) \leq$ card $(Z \times Z)$. A one-to-one mapping of $Z \times Z$ into Z is suggested by Figure B.2. This shows that

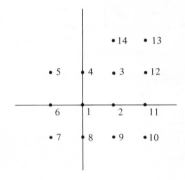

Figure B.2.

card $(Z \times Z) \leq$ card (Z). Hence card $(Z) =$ card $(Z \times Z)$.

Let p/q denote a rational number, where p and q have no common divisors. Then the mapping $p/q \to (p, q)$ defines a one-to-one mapping of Q into $Z \times Z$ so

$$\text{card } (Q) \leq \text{card } (Z \times Z) = \text{card } (Z).$$

Since card $(Z) \leq$ card (Q) (Why?) we see that card $(Z) =$ card (Q). ∎

EXAMPLE 2.   Let X by any nonempty set and let $P(X)$ denote the collection of all subsets of X. One can show that

$$\text{card } (X) < \text{card } (P(X)),$$

see Wilder [1, pp. 102–103]. For finite sets the last inequality becomes $n < 2^n$. ∎

EXAMPLE 3.   Let I be the real interval $[0, 1)$. Let us show that

$$\text{card } (N) < \text{card } (I).$$

In order to do this let ϕ be a one-to-one mapping of N into I. We want to show that there is a number $r \in I$ but $r \notin \phi(N)$.

For each $n \in N$, $\phi(n)$ is then a real number and we express it in terms of its decimal expression.[2]

Then

$$\phi(1) = 0.\, d_1{}^1 \, d_2{}^1 \, d_3{}^1 \, d_4{}^1 \cdots$$
$$\phi(2) = 0.\, d_1{}^2 \, d_2{}^2 \, d_3{}^2 \, d_4{}^2 \cdots$$
$$\vdots$$
$$\phi(n) = 0.\, d_1{}^n \, d_2{}^n \, d_3{}^n \, d_4{}^n \ldots,$$

where $d_i{}^n$ represents one of the integers $0, 1, \ldots, 9$. Let

$$r = 0.\, r_1 \, r_2 \, r_3 \, r_4 \ldots$$

be a real number chosen so that

$$r_1 \neq d_1{}^1, r_2 \neq d_2{}^2, r_3 \neq d_3{}^3, \ldots.$$

Then $r \neq \phi(1)$, $r \neq \phi(2)$, ... In other words, $r \in I$, but $r \notin \phi(N)$. ∎

EXERCISES

1. The mapping $n \to n$ defines a one-to-one mapping of $N_e = \{2,4,6,\ldots\}$ into $N = \{1,2,3,\ldots\}$. Explain why this does not prove that card $(N_e) <$ card (N).

2. Show that card $(R) =$ card (R^2), where R denotes the real line

3. One can define a finite set as follows: A set X is finite if every one-to-one mapping ϕ of X into itself is necessarily onto, that is, $\phi(X) = X$. Using this definition, show that the union of two finite sets is finite. Also, show that $N = \{1,2,\ldots\}$ is not finite.

4. Show that the countable union of countable sets is countable. That is, if X_n is a countable set for $n \in N$, then $X = \bigcup_{n=1}^{\infty} X_n$ is countable.

5. Show that card $(X) =$ card $(X \times X)$ for any infinite set X.

6. Assume that Y is a nonfinite set and also that card $(X) \leqslant$ card (Y). Show that card $(X \times Y) =$ card (Y). (This fact motivates the equality

$$\text{card } (X) + \text{card } (Y) = \text{card } (Y),$$

which occurs in cardinal arithmetic.)

---

[2] We agree not to use a decimal expansion that terminates in an infinite sequence of 9's.

Appendix C

Zorn's Lemma

The phrase "Zorn's lemma" is a misnomer. It is really an axiom that is used oftentimes in analysis. It is actually equivalent to at least eight other similar statements, including the Axiom of Choice, see Kelley [1, pp. 31–36].

In this appendix we shall formulate Zorn's lemma, as well as the Axiom of Choice. We shall also give one illustration of how Zorn's lemma can be used as a logical tool.

C.1 DEFINITION. A *relation* R on a set X is a subset R of the Cartesian product $X \times X$. We say that x is *R-related* to y, or xRy, if (x,y) belongs to R.

In this appendix we shall be concerned with "orderings." These are relations with additional properties which we prescribe below. For this reason we shall replace xRy with $x \leq y$.

C.2 DEFINITION. A relation $\leq$ on X is said to be *partial ordering* if (i) $x \leq x$ for $x \in X$; (ii) if $x \leq y$ and $y \leq x$, then $x = y$; (iii) $x \leq z$ whenever $x \leq y$ and $y \leq z$. A partial ordering is said to be a *total ordering* on X if, in addition, (iv) for any two points $x, y \in X$ one has either $x \leq y$ or $y \leq x$. A set X is said to be *partially ordered* if it has a partial ordering defined on it. A subset Z of a partially ordered set X is said to be a *chain* if for any two points $x, y \in Z$ one has either $x \leq y$ or $y \leq z$, that is, the restriction of the partial ordering relation to Z yields a total ordering.

C.3 DEFINITION. Let $\leq$ be a partial ordering on a set X and let $A \subset X$. We shall say that a point $x \in X$ is an *upper bound* for A if one has $a \leq x$ for all $a \in A$. If the set X itself has an upper bound $\tilde{x}$, we shall call $\tilde{x}$ a *maximal element* for X.

We can now state Zorn's lemma.

ZORN'S LEMMA. *If every chain in a partially ordered set X has an upper bound, then X has a maximal element.*

Although we shall not offer a "proof," it is interesting to note that Zorn's lemma is equivalent to the following:

AXIOM OF CHOICE. *Let A be an index set. For every $a \in A$, let X_a denote some nonempty set. Then there is a function f defined on A such that $f(a) \in X_a$ for all $a \in A$.*

556

Let us now illustrate how Zorn's lemma can be used.

Let A be a linearly independent set in a linear space Y. Let us show that there is a Hamel basis $H \subset Y$ that contains A. Let X denote the collection of all linearly independent sets B in Y with the property that $A \subset B$. We say that $B_1 \le B_2$ if $B_1 \subset B_2$. It is easy to see that $\le$ is a partial ordering on X. Let $Z = \{B_i : i \in I\}$ denote a chain in X. (Here I denotes some index set.) Let $B = \bigcup_{i \in I} B_i$. If we can show $B \in X$, it is clear that B is an upper bound for Z. In order to show that $B \in X$ we must show that $A \subset B$ (this is, of course, obvious) and that B is a linearly independent set. Let $\{y_1, \ldots, y_n\}$ be any finite collection from B. Then for some indices $\{i_1, \ldots, i_n\}$ one has $y_k \in B_{i_k}$. Since Z is a chain the sets $\{B_{i_1}, \ldots, B_{i_n}\}$ are comparable in the sense that either $B_{i_k} \subseteq B_{i_m}$ or $B_{i_m} \subseteq B_{i_k}$. Say that the indices are chosen so that

$$B_{i_1} \subseteq B_{i_2} \subseteq \cdots \subseteq B_{i_n}.$$

Then $\{y_1, \ldots, y_n\}$ belongs to B_{i_n}. Since B_{i_n} is linearly independent, we see that the only solution of

$$\alpha_1 y_1 + \cdots + \alpha_n y_n = 0$$

is $\alpha_1 = \alpha_2 = \cdots = \alpha_n = 0$. Hence, B is linearly independent.

We see then that Z has an upper bound, so by Zorn's lemma X has a maximal element H. It is now easily checked that H is a Hamel basis for Y.

EXERCISES

1. Show that a total ordering on a set is an equivalence relation (by Chapter 2). What about the converse?

2. Give an example of a partial ordering that is not a total ordering.

3. Give an example of a relation that is not a partial ordering.

Appendix D
Integration
and Measure Theory

1. INTRODUCTION

One of the most important concepts in the world of mathematics is that of the integral. In its most primitive form the integral was used by the early Greeks in their development of Euclidean geometry. For example, the problem of determining the area of a region as simple as the inside of a circle was solved by means of an integration process, that is, by summing the areas of disjoint rectangles contained in the circle. However, it was not until after Descartes' work in 1637 on analytic geometry that mathematicians could begin to view the integral as an object of analysis.

Descartes' work paved the way for the discovery of the calculus by Leibniz and Newton around 1665. At that time a big argument ensued as to who discovered the calculus first, and the mathematicians of Germany and England split off into warring camps each with their respective champion. It is now believed that Newton's work slightly preceded the work of Leibniz. However, it is the notation and the viewpoint of Leibniz that was adopted by the mathematical world, and his symbols "$\int$" and "d" are still used today.

Leibniz developed the calculus using the concept of the "infinitesimal." At first there was much confusion over the "nature of infinitesimals" and the question of "adding infinitesimals." However, today this is well understood.

It was Cauchy and Riemann who first gave a systematic definition of an integral in the first half of the nineteenth century. This integral is now named after Riemann. Around the turn of the century, Lebesgue, in his doctoral dissertation, gave a more general treatment of integration which has resulted in a second revolution in the field of analysis. One of the objectives of this appendix is to develop the theory of Lebesgue integration. In this development we shall be particularly interested in the relationship between the Lebesgue and Riemann integrals.

The difficulty with the Riemann integral is that the space of Riemann integrable functions is not complete, when considered as a metric space. One can view the space of Lebesgue integrable functions as the completion of the space of Riemann integrable functions and the Lebesgue integral as an extension of the Riemann integral. In fact, it is possible to define the Lebesgue integral in exactly this fashion, but we will not do that here. We will use instead the Daniell approach in developing the Lebesgue integral. Our starting point will be based on knowledge of certain elementary properties of the Riemann integral for continuous functions. We will define the Lebesgue integral in terms of the Riemann integral. It will be evident from this approach that the Lebesgue integral and the Riemann integral of a continuous function are the same.

558

There are other ways of defining the Lebesgue integral which also use the Daniell approach. They differ from what we will do here in that they begin from a different starting point. These other methods are important and we shall discuss them in Section 10.

It must be remembered that the Lebesgue integral is developed primarily to satisfy certain theoretical questions. Its importance lies in the structure as described by the limit theorems of Section 8. We do not pretend that it is always easy to compute with the Lebesgue integral. This is not our purpose.

2.   THE RIEMANN INTEGRAL

Assume that we are given a bounded function $f(t)$ defined on a finite closed interval $a \le t \le b$; that is, $m \le f(t) \le M$ for all t in the interval, see Figure D.2.1. We can assume that $M = \sup\{f(t) : a \le t \le b\}$ and $m = \inf\{f(t) : a \le t \le b\}$. A partition P of the interval $[a,b]$ is a *finite* collection of points $\{t_0, \ldots, t_n\}$ such

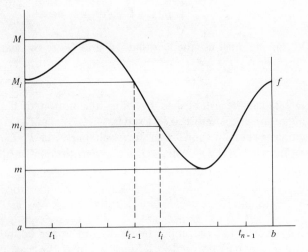

Figure D.2.1.

that $a = t_0 < t_1 < \cdots < t_n = b$. Since f is bounded on $[a,b]$ it is also bounded on the ith subinterval $[t_{i-1}, t_i]$. Consequently the numbers

$$M_i = \sup\{f(t) : t_{i-1} \le t \le t_i\}$$
$$m_i = \inf\{f(t) : t_{i-1} \le t \le t_i\}$$

do exist, and $m \le m_i \le M_i \le M$ for all $i = 1, 2, \ldots, n$. Let $\Delta t_i = t_i - t_{i-1}$.

For each partition P we can form the upper and lower sums:

$$U(P,f) = \sum_{i=1}^{n} M_i \, \Delta t_i,$$

$$L(P,f) = \sum_{i=1}^{n} m_i \, \Delta t_i.$$

The functions U, L satisfy the following inequalities:

$$m(b - a) \leq L(P,f) \leq U(P,f) \leq M(b - a).$$

Thus for f fixed, $U(P,f)$ forms a set of real numbers which is bounded above and below. Consequently, this set has an infimum (and supremum). So we define the *upper integral* of f by

$$\overline{\int_a^b} f \, dt = \inf U(P,f),$$

where the inf is taken over all possible partitions of the interval $[a,b]$. Similarly, we define the lower integral by

$$\underline{\int_a^b} f \, dt = \sup L(P,f).$$

One can show that for all bounded functions f one has

$$\underline{\int_a^b} f \, dt \leq \overline{\int_a^b} f \, dt. \tag{D.2.1}$$

To prove (D.2.1), one must use the fact that if P and Q are two partitions of $[a,b]$, then

$$L(P,f) \leq U(Q,f). \tag{D.2.2}$$

However, (D.2.2) can be established by using the notion of a refinement of a partition. This argument is outlined in Exercises 1–2.

The inequality (D.2.2) shows that for each partition P, $L(P,f)$ is a lower bound for the family $\{U(Q,f)\}$, where Q is any partition of $[a,b]$. Consequently, it is smaller than the greatest lower bound, or

$$L(P,f) \leq \overline{\int_a^b} f \, dt. \tag{D.2.3}$$

Finally, (D.2.3) shows that the upper integral is an upper bound for the family $L(P,f)$, where P is any partition of $[a,b]$. Since $\underline{\int_a^b} f \, dt$ is the least upper bound for the family it follows that (D.2.1) is true.

If it happens that the equality holds in (D.2.1), then we say that f is *Riemann integrable* (*R-integrable*) and the integral of f is defined to be the common value

$$\int_a^b f \, dt = \overline{\int_a^b} f \, dt = \underline{\int_a^b} f \, dt.$$

We shall let $\mathscr{R} = \mathscr{R}([a,b])$ denote the class of R-integrable functions on the interval $[a,b]$.

Summarizing, then, we see that *every* bounded function has an upper and lower integral. A certain subclass $\mathscr{R}$ (where the two integrals are the same) is the class of Riemann-integrable functions.

The first question that naturally arises is to determine which functions lie in $\mathscr{R}$. A partial answer to this question is given in the following existence theorems, which are usually proven in an elementary calculus course. (See Exercise 5.) A complete answer is stated in Exercise 2, Section D.9.

D.2.1 THEOREM.   *Let $f(t)$ be continuous for $a \le t \le b$. Then $f \in \mathcal{R}$.*

Let us summarize some of the elementary properties of the Riemann integral.

D.2.2 THEOREM.

(a) *If $f \in \mathcal{R}$ ([a,b]) and $a \le c \le b$, then $f \in \mathcal{R}$ ([a,c]), $f \in \mathcal{R}$ ([c,b]), and*

$$\int_a^b f\, dt = \int_a^c f\, dt + \int_c^b f\, dt.$$

(b) *If $f_1, f_2 \in \mathcal{R}$ and $\alpha_1, \alpha_2 \in R$, then $f = \alpha_1 f_1 + \alpha_2 f_2 \in \mathcal{R}$ and*

$$\int_a^b f\, dt = \alpha_1 \int_a^b f_1\, dt + \alpha_2 \int_a^b f_2\, dt.$$

(c) *If $f \in \mathcal{R}$ and $f(t) \ge 0$ on $[a,b]$, then $\int_a^b f\, dt \ge 0$.*

(d) *If $f \in \mathcal{R}$ and $m \le f(t) \le M$ on $[a,b]$, then*

$$m(b - a) \le \int_a^b f\, dt \le M(b - a).$$

(e) *If f is continuous on $[a,b]$, then there is a $\xi \in [a,b]$ such that*

$$\int_a^b f\, dt = f(\xi)[b - a].$$

(f) *If $f \in \mathcal{R}$, then $|f| \in \mathcal{R}$, where $|f|(t) = |f(t)|$, and*

$$\left| \int_a^b f\, dt \right| \le \int_a^b |f|\, dt.$$

(g) *If $f, g \in \mathcal{R}$ and $f(t) \le g(t)$, then $\int_a^b f\, dt \le \int_a^b g\, dt$.*

Since these facts are proven in most books on elementary calculus, we shall not discuss them here.

Before proceeding further, we should note that there are (bounded) functions which are not Riemann integrable. Indeed,

$$f(t) = \begin{cases} 1, & \text{if } t \text{ is rational,} \\ 0, & \text{if } t \text{ is irrational,} \end{cases} \tag{D.2.5}$$

is one such function. For if P is any partition of the interval $[a,b]$, then $M_i = 1$ and $m_i = 0$ for all i. Consequently, $U(P,f) = 1(b - a)$ and $L(P,f) = 0(b - a)$. Thus $\overline{\int}_a^b f\, dt = (b - a)$ and $\underline{\int}_a^b f\, dt = 0$. We do show later that this particular function is Lebesgue integrable.

While the above definition of the Riemann integral is satisfactory from a theoretical point of view, it is not practical for computation. We need some additional information for this purpose.

Let $P: a = t_0 < t_1 < \cdots < t_n = b$ be a partition of $[a,b]$. Define the *norm of P* by

$$|P| = \max_{1 \le i \le n} |\Delta t_i| = \max_{1 \le i \le n} |t_i - t_{i-1}|.$$

D.2.3 THEOREM. *Let f be an R-integrable function on $[a, b]$. Let $\{P_m\}$ be a sequence of partitions such that $|P_m| \to 0$ as $m \to \infty$. Then*

$$L(P_m, f) \to \int_a^b f\, dt$$

and

$$U(P_m, f) \to \int_a^b f\, dt$$

as $m \to \infty$.

In applying this theorem, one quite often chooses P_{m+1} to be a refinement of P_m, that is, the set P_{m+1} contains the set P_m. In this case, the sequences $\{L(P_m, f)\}$ and $\{U(P_m, f)\}$ are monotone. In fact, if P_m is the partition

$$P_m: a = t_0 < t_1 < \cdots < t_n = b$$

and the functions U_m and L_m are defined by

$$U_m(t) = \begin{cases} f(a), & t = a, \\ M_i, & t_{i-1} < t \le t_i, i = 1, 2, \ldots, n, \end{cases}$$

$$L_m(t) = \begin{cases} f(a), & t = a, \\ m_i, & t_{i-1} < t \le t_i, i = 1, 2, \ldots, n, \end{cases}$$

then one can easily show that

$$L_1(t) \le L_2(t) \le \cdots \le f(t) \le \cdots \le U_2(t) \le U_1(t).$$

Therefore,

$$L(P_1, f) \le L(P_2, f) \le \cdots \le \int_a^b f\, dt \le \cdots \le U(P_2, f) \le U(P_1, f).$$

A special case of this occurs when the partition points are equally spaced.

D.2.4 COROLLARY. *Let f be a Riemann-integrable function on $[a,b]$. Further, let P_m be the partition formed by decomposing $[a,b]$ into m equal parts, that is, $P_m = \{t_0, t_1, \ldots, t_m\}$, where $t_i = a + i(b - a)/m$. Then*

$$L(P_m, f) \to \int_a^b f\, dt,$$

$$U(P_m, f) \to \int_a^b f\, dt.$$

EXERCISES

1. Let P and Q be partitions of the interval $a \le t \le b$. We say that $P \subseteq Q$ is every point of P is a point of Q. Show that for every bounded function f one has $L(P,f) \le L(Q,f) \le U(Q,f) \le U(P,f)$ when $P \subset Q$.

2. Use the result of Exercise 1 to prove (D.2.2). [*Hint*: First show that if P and Q are two partitions, then there is a partition $\hat{P}$ that satisfies $P \subseteq \hat{P}$, $Q \subseteq \hat{P}$.]

3. Use the Corollary D.2.4 and mathematical induction to show that

$$\int_0^b t \, dt = \tfrac{1}{2} b^2.$$

[That is, if P_m is defined as in Corollary D.2.4 and $f(t) = t$; show that

$$U(P_m, f) = \frac{b^2}{m^2} \sum_{i=1}^m (i) = \frac{b^2}{2} \frac{m+1}{m} \to \frac{b^2}{2}$$

as $m \to \infty$.]

4. One can also show that any bounded monotone function has a Riemann integral. Let f be a bounded increasing function on $[a,b]$. If $P = \{t_0, t_1, \ldots, t_n\}$ is any partition and M_i and m_i are defined as above, then $M_i = f(t_i)$ and $m_i = f(t_{i-1})$. [If the points of P are equally spaced, then $U(P,f) - L(P,f) = [f(b) - f(a)] \Delta t$, where $\Delta t = t_i - t_{i-1} = (b-a)/n$.] Show that there is a sequence of partitions $\{P_m\}$ such that $U(P_m, f) - L(P_m, f) \to 0$. Use this to show that f is R-integrable.

5. The following steps will lead to a proof of Theorem D.2.1: For $a \le t \le b$ define

$$G(t) = \int_{\underline{a}}^t f \, dt, \qquad H(t) = \overline{\int}_a^t f \, dt,$$

where f is continuous.

(a) Show that for $a \le t \le t + h \le b$ one has

$$G(t+h) - G(t) = \int_{\underline{t}}^{t+h} f \, dt = f(t^*) \cdot h$$

for some t^* with $t \le t^* \le t + h$, and

$$H(t+h) - H(t) = \overline{\int}_t^{t+h} f \, dt = f(t^{**}) \cdot h$$

for some t^{**} with $t \le t^{**} \le t + h$.

(b) Show that $\dfrac{dG}{dt} = \dfrac{dH}{dt} = f$.

(c) Show that $G(t) = H(t)$ for all t, $a \le t \le b$. That is,

$$\left(\int_{\underline{a}}^b f \, dt = \overline{\int}_a^b f \, dt. \right)$$

3. A PROBLEM WITH THE RIEMANN INTEGRAL

The Riemann integral does have at least one serious shortcoming. That is, it is possible for a sequence of functions, $\{f_n\} \subset \mathscr{R}$ to converge to a function f, which may even be bounded, but which is not R-integrable. For example, let

$$f_n(t) = \lim_{m \to \infty} [\cos (n! \, \pi t)]^{2m} = \begin{cases} 1, & \text{if } t = \dfrac{k}{n!}, \, k \text{ an integer} \\ 0, & \text{otherwise} \end{cases}$$

for $n = 1, 2, \ldots$. Then $f_n(t) \to f(t)$, where

$$f(t) = \begin{cases} 1, & \text{if } t \text{ is rational,} \\ 0, & \text{if } t \text{ is irrational,} \end{cases}$$

which is the function given in (D.2.5). As shown above, $f(t)$ is not in $\mathscr{R}$. However, $f_n \in \mathscr{R}$ and $\int_b^a f_n(t) \, dt = 0$ for all n.

Although $f \notin \mathscr{R}$, it would seem natural to define $\int_b^a f(t) \, dt$ by the relationship:

$$\lim_{n \to \infty} \int_a^b f_n(t) \, dt = \int_a^b \left\{ \lim_{n \to \infty} f_n(t) \right\} dt = \int_a^b f(t) \, dt. \tag{D.3.1}$$

In other words, it seems natural to enlarge the class of integrable functions by defining the integral of f, the limit of a sequence f_n, by means of (D.3.1). It is shown below that this leads to a consistent notion for the integral, and the new class of integrable functions are the Lebesgue integrable functions.

Actually we shall not use sequences as the vehicle for defining the Lebesgue integral, but instead we shall use infinite series. Since sequences and series are correlative concepts, it really does not matter which concept we use to extend the integral. We choose to use series only because certain technical conditions are easier to formulate in this context.

4. THE SPACE C_0

We begin our construction of the Lebesgue integral by isolating the properties of the Riemann integral that are essential for our theory. First we let $C_0 = C_0(R,R)$ denote the class of all continuous real-valued functions defined on R with *compact support*. This means that $\phi \in C_0$ if and only if $\phi: R \to R$ is continuous and there is a bounded interval $[a,b]$, which may depend on ϕ, such that $\phi(t) = 0$ for $t \notin [a,b]$. The graph of a typical function in C_0 appears in Figure D.4.1. It follows from Theorem D.2.1 that for any continuous function ϕ, the integral $\int_a^b \phi \, dt$ is defined and therefore we define $\int \phi \, dt$ by

$$\int \phi \, dt = \int_{-\infty}^{\infty} \phi \, dt = \int_a^b \phi \, dt, \tag{D.4.1}$$

where ϕ vanishes outside $[a,b]$. The integral $\int \phi \, dt$ is then defined for every function in C_0.

The space C_0 has an important property.

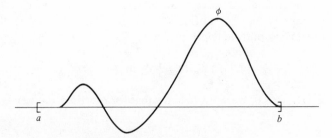

Figure D.4.1. A Function in C_0.

(I.1) *If ϕ and ψ are two functions in C_0, and α and β are two real numbers, then $\phi \vee \psi$, $\phi \wedge \psi$, and $\alpha\phi + \beta\psi$ are in C_0, where*

$$(\phi \vee \psi)(t) = \max \, [\phi(t),\psi(t)],$$
$$(\phi \wedge \psi)(t) = \min \, [\phi(t),\psi(t)],$$
$$(\alpha\phi + \beta\psi)(t) = \alpha\phi(t) + \beta\psi(t).$$

In addition to property (I.1), the integral $\int \phi \, dt$ satisfies the following properties:

(I.2) *If ϕ and ψ are two functions in C_0 and α and β are two real numbers, then*

$$\int (\alpha\phi + \beta\psi) \, dt = \alpha \int \phi \, dt + \beta \int \psi \, dt.$$

(I.3) *If ϕ is a function in C_0 with $\phi \geq 0$, then $\int \phi \, dt \geq 0$.*

(I.4) (Dini's Theorem) *If $\{\phi_n\}$ is a decreasing sequence of nonnegative functions in C_0 (that is, $0 \leq \phi_{n+1} \leq \phi_n$) and $\lim_{n \to \infty} \phi_n(t) = 0$ for all t, then*

$$\lim_{n \to \infty} \int \phi_n \, dt = 0.$$

We shall give a proof of Dini's Theorem shortly. Before we do this though, let us examine some of the consequences of Properties (I.1), (I.2), and (I.3) which will be used below. Let $\phi \in C_0$. It follows that

$$\phi^+ = \phi \vee 0, \qquad \phi^- = (-\phi) \vee 0 = -(\phi \wedge 0), \qquad |\phi| = \phi^+ + \phi^-$$

are also in C_0. Furthermore,

$$\left| \int \phi \, dt \right| = \left| \int \phi^+ \, dt - \int \phi^- \, dt \right| \leq \int \phi^+ \, dt + \int \phi^- \, dt = \int |\phi| \, dt.$$

Also, if ϕ, ψ are in C_0 and $\phi \leq \psi$, then $0 \leq \psi - \phi$ so that

$$0 \leq \int (\psi - \phi) \, dt = \int \psi \, dt - \int \phi \, dt, \qquad \text{that is, } \int \phi \, dt \leq \int \psi \, dt.$$

Let us now prove Dini's Theorem. Since $\phi_1 \in C_0$, there is a bounded interval $[a,b]$ with the property that ϕ_1 vanishes outside $[a,b]$. Since $0 \leq \phi_n \leq \phi_1$ for all n we see that ϕ_n vanishes outside $[a,b]$. Hence

$$\int \phi_n \, dt = \int_a^b \phi_n \, dt$$

for all n. Since $\int \phi_n \, dt \geq 0$, we want to show that for every $\varepsilon > 0$, there is an N such that

$$\int \phi_n \, dt \leq \varepsilon(b-a)$$

for all $n \geq N$. This will be accomplished once we show that for every $\varepsilon > 0$, there is an N (independent of t) such that

$$\phi_n(t) \leq \varepsilon, \qquad a \leq t \leq b, \qquad n \geq N. \tag{D.4.2}$$

If (D.4.2) were not true, then this would mean that we can find an $\varepsilon_0 > 0$ such that for every N one can find an $n \geq N$ and a t_n in $[a,b]$ such that

$$\varepsilon_0 < \phi_n(t_n). \tag{D.4.3}$$

Since the interval $[a,b]$ is (sequentially) compact (Theorem 3.17.14) we can find a convergent subsequence of $\{t_n\}$, call it $\{t_{n'}\}$ and let $t_0 = \lim t_{n'}$. Since $\phi_{n'}(t_0) \to 0$ as $n' \to \infty$ we can find an M such that $\phi_M(t_0) < \varepsilon_0/4$. It follows from the continuity of ϕ_M that there is a $\delta > 0$ such that $\phi_M(t) \leq \varepsilon_0/2$ for all t satisfying $|t - t_0| < \delta$. Since the sequence $\{\phi_n\}$ is decreasing, this implies that

$$\phi_{n'}(t) \leq \frac{\varepsilon_0}{2}, \qquad |t - t_0| < \delta, n' \geq M. \tag{D.4.4}$$

By combining (D.4.3) and (D.4.4) we see that $|t_{n'} - t_0| \geq \delta$ which contradicts the fact that $t_0 = \lim t_{n'}$; therefore, (D.4.2) is true. ∎

We ask the reader to note that the theory of the Lebesgue integral, which we now turn to, depends only on properties (I.1), (I.2), (I.3), and (I.4). This is important because if one replaces C_0 with another class of functions, and if one replaces the Riemann integral $\int \phi \, dt$ with another integral, and if one does it in such a way that properties (I.1), (I.2), (I.3), and (I.4) still hold, then one can mimic the theory we now present and extend the integral to a larger class of functions. This approach is what is called the Daniell approach. We will return to some variations on this theme in Section 10.

5. NULL SETS

The first step in the construction of the Lebesgue integral is to introduce a weaker form of convergence, namely "convergence almost everywhere." This will be defined precisely in the next section, but it means roughly that a sequence of functions converges everywhere except for a negligible set, or null set. In this section we wish to study the concept of a "null set."

D.5.1 DEFINITION. A set $E \subset R$ is said to be a *null set* if there is an increasing sequence $\{\phi_n\}$ in C_0 such that $\{\phi_n(t)\}$ diverges for each t in E while

$$\lim \int \phi_n \, dt < +\infty.$$

The following theorem gives a characterization of null sets in terms of infinite series.

D.5.2 THEOREM. *The following statements are equivalent:*

(a) *E is a null set.*

(b) *There is a sequence $\{\psi_n\}$ in C_0 such that $\sum_n \psi_n(t)$ diverges for each t in E while $\sum_n \int |\psi_n| dt < \infty$.*

(c) *There is a sequence $\{\psi_n\}$ in C_0 such that $\sum_n |\psi_n(t)|$ diverges for each t in E while $\sum_n \int |\psi_n| dt < \infty$.*

Proof: (a) $\Rightarrow$ (c). Assume that E is a null set and let $\{\phi_n\}$ be a sequence satisfying the conditions of Definition D.5.1. Let $\psi_n = \phi_{n+1} - \phi_n$. Then $\psi_n \geq 0$ and $\sum_{n=1}^{m} \psi_n = \phi_{m+1} - \phi_1$. Hence $\sum_n \psi_n(t)$ diverges for $t \in E$. Also,

$$\sum_{n=1}^{m} \int |\psi_n| \, dt = \sum_{n=1}^{m} \int (\phi_{n+1} - \phi_n) \, dt = \int \phi_{m+1} \, dt - \int \phi_1 \, dt.$$

Hence,

$$\sum_n \int |\psi_n| \, dt = \lim_{n \to \infty} \int \phi_n \, dt - \int \phi_1 \, dt < \infty.$$

(c) $\Rightarrow$ (b). Obvious.

(b) $\Rightarrow$ (a). Let $\{\psi_n\}$ be a sequence in C_0, where $\sum_n \psi_n(t)$ diverges for each t in E and $\sum_n \int |\psi_n| dt < \infty$. Now let $\phi_m = \sum_{n=1}^{m} |\psi_n|$. It is easy to check that $\{\phi_n\}$ satisfies the conditions of Definition D.5.1, hence E is a null set. ∎

EXAMPLE 1. Let E consist of a single point $\{p\}$. Then E is a null set. To see this we simply let ψ_n be a function in C_0 that satisfies $\psi_n(p) = 1$ and $\int |\psi_n| dt = 1/n^2$, as shown in Figure D.5.1. Since $\sum_n \psi_n(p) = +\infty$ and $\sum_n \int |\psi_n| dt < \infty$, we see that $\{p\}$ is a null set, by statement (b) in the last theorem. ∎

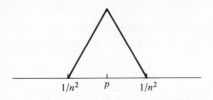

$1/n^2$　p　$1/n^2$

Figure D.5.1.

EXAMPLE 2.   Let $E_1, E_2, \ldots$ be a sequence of null sets. We claim that the union $E = \bigcup_{n=1}^{\infty} E_n$ is a null set. In order to see this we let $\{\phi_{nk}\}$ be a sequence in C_0 that satisfies

$$\sum_n |\phi_{nk}(t)| = +\infty, \qquad \text{for all } t \text{ in } E_k,$$

$$\sum_n \int |\phi_{nk}| \, dt = M_k < \infty.$$

Now let $\psi_{nk} = 2^{-k} M_k^{-1} \phi_{nk}$, which is a sequence in C_0. If $t \in E$, then $t \in E_k$ for some k and

$$\sum_{n,k} |\psi_{nk}(t)| = \sum_k 2^{-k} M_k^{-1} \sum_n |\phi_{nk}(t)| = +\infty.$$

Furthermore,

$$\sum_{n,k} \int |\psi_{nk}| \, dt = \sum_k 2^{-k} M_k^{-1} \sum_n \int |\phi_{nk}| \, dt = \sum_k 2^{-k} < \infty.$$

It follows that E is a null set.   ∎

By combining Examples 1 and 2 we see that any countable set is a null set. In particular, the set of rational numbers is a null set. In Section 9, we shall discuss the Cantor set. This is an example of an uncountable set which is a null set.

EXAMPLE 3.   Let us now give an example of a set that is not a null set. Let E be any nontrivial interval with end points $a < b$. Now choose α, β so that $a < \alpha < \beta < b$. Now let ξ be a nonnegative function in C_0 that satisfies

$$\xi(t) = \begin{cases} N, & t \text{ in } [\alpha, \beta], \\ 0, & \text{outside } [a, b], \end{cases}$$

as shown in Figure D.5.2. If E were a null set, then we could find an increasing sequence $\{\phi_n\}$ in C_0 such that $\phi_n(t) \to +\infty$ for $a < t < b$, and $\lim_{n \to \infty} \int \phi_n \, dt < \infty$.

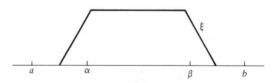

Figure D.5.2.

Now set

$$\psi_n = (\xi - \phi_n)^+.$$

It is easy to see that $\{\psi_n\}$ is a decreasing sequence and that $\lim_{n \to \infty} \psi_n(t) = 0$ for all t. Hence by property (I.4) we see that $\lim \int \psi_n \, dt = 0$. By using the fact that $(\xi - \phi_n) \leq (\xi - \phi_n)^+ = \psi_n$, we get

$$N(\beta - \alpha) - \int_\alpha^\beta \phi_1 \, dt = \int_\alpha^\beta (\xi - \phi_1) \, dt = \int_\alpha^\beta [(\xi - \phi_n) + (\phi_n - \phi_1)] \, dt$$

$$\leq \int_\alpha^\beta (\psi_n + \phi_n - \phi_1) \, dt$$

$$\leq \int (\psi_n + \phi_n - \phi_1) \, dt,$$

since $\phi_n - \phi_1 \geq 0$ and $\psi_n \geq 0$. Now let $n \to \infty$ and we get

$$N(\beta - \alpha) \leq \lim_{n \to \infty} \int \phi_n \, dt.$$

Since N is arbitrary, this implies that $\lim_{n \to \infty} \int \phi_n \, dt = +\infty$, which is a contradiction. Hence E is not a null set. ∎

EXERCISE

1. Let I_k, $k = 1, 2, \ldots$, be a sequence of intervals with end points $a_k - b_k$. Define the *total length* of the sequence $\{I_k\}$ by $\sum_{k=1}^\infty (b_k - a_k)$. A set $E \subset R$ is said to be a *set of measure zero* if for every $\varepsilon > 0$ there is a sequence $\{I_k\}$ of intervals with total length $\leq \varepsilon$ and such that $E \subset \bigcup_k I_k$. Show that a set $E \subset R$ is a null set if and only if it is a set of measure zero.

6.   CONVERGENCE ALMOST EVERYWHERE

We will see soon that from the point of view of integration theory, null sets can be ignored, that is, they are negligible. In order to make this precise, we need the following definitions.

D.6.1 DEFINITION.   If some property holds for all real numbers t outside some null set, then we shall say that this property holds *almost everywhere* (which is usually abbreviated "a.e."). For example, a sequence of functions $\{f_n\}$ is said to converge to f almost everywhere, that is,

$$\lim_{n \to \infty} f_n = f \quad \text{(a.e.)},$$

if the set of real numbers T, for which $\{f_n(t)\}$ fails to converge to $f(t)$, is a null set. Similarly we say that

$$\sum_n f_n(t) = F \quad \text{(a.e.)},$$

if the set of real numbers t, for which $\sum_n f_n(t) \neq F(t)$, is a null set. Also $f \leq g$ (a.e.) means that the set $\{t : f(t) > g(t)\}$ is a null set.

The first consequence of this concept is that we can prove an "almost everywhere" form of Dini's Theorem.

D.6.2 THEOREM.   *Let $\{\phi_n\}$ be a decreasing sequence of nonnegative functions in C_0 such that*

$$\lim_{n \to \infty} \phi_n = 0 \qquad (a.e.).$$

Then $\lim_{n \to \infty} \int \phi_n \, dt = 0$.

Proof: Let $\varepsilon > 0$ by given. We will now show that $\lim_{n \to \infty} \int \phi_n \, dt \leq \varepsilon$.

First we note that since $\{\phi_n\}$ is decreasing and $\phi_n \geq 0$, the limit $\lim_{n \to \infty} \phi_n(t)$ exists for all t. Next let E be the null set

$$E = \left\{ t : \lim_{n \to \infty} \phi_n(t) \neq 0 \right\}.$$

Now use the definition of a null set to construct an increasing sequence $\{\psi_n\}$ so that $\psi_n(t) \to + \infty$ for $t \in E$ and

$$\int \psi_n \, dt \leq K < \infty$$

for all n, where K is an appropriate constant.

Now set $\xi_n = \phi_n - \varepsilon K^{-1} \psi_n$. It is clear that $\{\xi_n\}$ is a decreasing sequence and therefore $\{\xi_n^{+}\}$ is a decreasing sequence. Furthermore $\lim_{n \to \infty} \xi_n^{+}(t) = 0$ for all t. Therefore, by property (I.4), we get

$$\lim_{n \to \infty} \int \xi_n^{+} \, dt = 0.$$

Since

$$\phi_n = \xi_n + \varepsilon K^{-1} \psi_n \leq \xi_n^{+} + \varepsilon K^{-1} \psi_n,$$

we get

$$\lim_{n \to \infty} \int \phi_n \, dt \leq \lim_{n \to \infty} \int \xi_n^{+} \, dt + \lim_{n \to \infty} \varepsilon K^{-1} \int \psi_n \, dt \leq \varepsilon.$$

Now let $\varepsilon \to 0$ and we get $\lim_{n \to \infty} \int \phi_n \, dt = 0$.   ∎

D.6.3 COROLLARY.   *Let $\psi \in C_0$ and let $\{\phi_n\}$ be a sequence of nonnegative functions in C_0. Assume that*

$$\sum_n \phi_n \geq \psi \qquad (a.e.).$$

Then $\sum_n \int \phi_n \, dt \geq \int \psi \, dt$.

Proof: Let $\xi_n = (\psi - \sum_{i=1}^{n} \phi_i)$. Then $\{\xi_n\}$ and $\{\xi_n^{+}\}$ are decreasing sequences. Furthermore,

$$\lim_{n \to \infty} \xi_n^{+} = 0 \qquad (a.e.)$$

which implies that $\lim_{n\to\infty} \int \xi_n^+ \, dt = 0$. Since

$$\psi = \xi_n + \sum_{i=1}^{n} \phi_i \leq \xi_n^+ + \sum_{i=1}^{n} \phi_i$$

we get

$$\int \psi \, dt \leq \int \xi_n^+ \, dt + \sum_{i=1}^{n} \int \phi_i \, dt.$$

By letting $n \to \infty$ we get the desired conclusion. ∎

D.6.4 COROLLARY. *Let $\{\phi_n\}$ be a sequence in C_0 such that $\sum_{n=1}^{\infty} \phi_n = 0$ (a.e.) and $\sum_{n=1}^{\infty} \int |\phi_n| dt < \infty$. Then*

$$\sum_{n=1}^{\infty} \int \phi_n \, dt = 0.$$

Proof: It follows from Theorem D.5.2 that the set

$$E = \left\{ t : \sum_{n=1}^{\infty} |\phi_n(t)| = +\infty \right\}$$

is a null set. Since $\phi_n^+ \leq |\phi_n|$ and $\phi_n^- \leq |\phi_n|$, the series $\sum_n \phi_n^+(t)$ and $\sum_n \phi_n^-(t)$ converge absolutely for $t \notin E$. Hence, for $t \notin E$ we have

$$\sum_{n=1}^{\infty} \phi_n^+(t) - \sum_{n=1}^{\infty} \phi_n^-(t) = \sum_{n=1}^{\infty} [\phi_n^+(t) - \phi_n^-(t)] = \sum_{n=1}^{\infty} \phi_n(t) = 0,$$

that is, $\sum_{n=1}^{\infty} \phi_n^+(t) = \sum_{n=1}^{\infty} \phi_n^-(t)$ for $t \notin E$. Now let m be fixed and set $\psi = \sum_{n=1}^{m} \phi_n^+$. One then has

$$\psi \leq \sum_{n=1}^{\infty} \phi_n^- \quad \text{(a.e.)}$$

and by the last corollary we get

$$\int \psi \, dt = \sum_{n=1}^{m} \int \phi_n^+ \, dt \leq \sum_{n=1}^{\infty} \int \phi_n^- \, dt.$$

Now let $m \to \infty$ and we get

$$\sum_{n=1}^{\infty} \int \phi_n^+ \, dt \leq \sum_{n=1}^{\infty} \int \phi_n^- \, dt. \tag{D.6.1}$$

By interchanging the role of ϕ_n^+ and ϕ_n^- we can prove that the inequality in (D.6.1) can be reversed. We thus have

$$0 = \sum_{n=1}^{\infty} \int \phi_n^+ \, dt - \sum_{n=1}^{\infty} \int \phi_n^- \, dt$$

$$= \sum_{n=1}^{\infty} \left(\int \phi_n^+ \, dt - \int \phi_n^- \, dt \right)$$

$$= \sum_{n=1}^{\infty} \int (\phi_n^+ - \phi_n^-) \, dt = \sum_{n=1}^{\infty} \int \phi_n \, dt. \quad ∎$$

D.6.5 COROLLARY. *Let $\{\phi_n\}$ and $\{\psi_n\}$ be two sequences from C_0 satisfying the following*:

$$\sum_n \phi_n = \sum_n \psi_n \quad \text{(a.e.)},$$

$$\sum_n \int |\phi_n|\, dt < \infty, \qquad \sum_n \int |\psi_n|\, dt < \infty.$$

Then

$$\sum_n \int \phi_n\, dt = \sum_n \int \psi_n\, dt.$$

Proof: Let $\xi_n = \phi_n - \psi_n$. It is a simple matter to show that $\{\xi_n\}$ satisfies the hypotheses of the last corollary. One then has

$$0 = \sum_n \int \xi_n\, dt = \sum_n \int \phi_n\, dt - \sum_n \int \psi_n\, dt. \quad \blacksquare$$

7.   THE LEBESGUE INTEGRAL

Let us summarize the conclusions of the last section. Suppose we are given a sequence $\{\phi_n\}$ in C_0 with the property that $\sum_n \int |\phi_n|dt < \infty$. Then we know two things:

(1) The series $\sum_n \phi_n$ converges almost everywhere, and
(2) The series $\sum_n \int \phi_n\, dt$ converges.

If f is a function that satisfies

$$f = \sum_n \phi_n \quad \text{(a.e.)},$$

then it seems natural to define the integral of f by

$$\int f\, dt = \sum_n \int \phi_n\, dt,$$

which is what we do.

D.7.1 DEFINITION.   Let $f: R \to R$ be a function that satisfies $f = \sum_n \phi_n$ (a.e.) for some sequence $\{\phi_n\}$ in C_0 with $\sum_n \int |\phi_n|dt < \infty$. Then f is said to be *Lebesgue integrable* and the (*Lebesgue*) *integral* of f is given by

$$\int f\, dt = \sum_n \int \phi_n\, dt.$$

We shall let $L(-\infty,\infty)$ or L denote the class of Lebesgue integrable functions.

This definition is well formulated since the value of $\int f\, dt$ does not depend on the approximating sequence $\{\phi_n\}$ (see Corollary D.6.5).

It follows that every function in C_0 is Lebesgue integrable, that is, $C_0 \subset L$ and the Lebesgue integral agrees with the Riemann integral. Indeed, if $f \in C_0$, let

$\phi_1 = f$, $\phi_n = 0$, $n = 2, 3, \ldots$. Then $f = \sum_n \phi_n$ and $\sum_n \int |\phi_n| dt = \int |\phi_1| dt < \infty$. So f is Lebesgue integrable. Finally

$$\int f \, dt = \sum_n \int \phi_n \, dt = \int \phi_1 \, dt,$$

that is, the two integrals agree.

It is convenient to have a characterization of L in terms of sequences.

D.7.2 THEOREM.   *Let* $f: R \to R$ *be given. Then* $f \in L$ *if and only if there is a sequence* $\{\psi_n\}$ *in* C_0 *that satisfies*

$$\lim_{n \to \infty} \psi_n = f \quad \text{(a.e.)}$$

$$\lim_{m, n \to \infty} \int |\psi_n - \psi_m| \, dt = 0. \tag{D.7.1}$$

In this case one has

$$\int f \, dt = \lim_{n \to \infty} \int \psi_n \, dt. \tag{D.7.2}$$

Proof: Let $f \in L$ and let $\{\phi_n\}$ be a sequence in C_0 that satisfies the conditions of Definition D.7.1. It is now easy to check that the sequence

$$\psi_m = \sum_{n=1}^{m} \phi_n$$

satisfies (D.7.1).

Conversely, if $\{\psi_n\}$ is a sequence that satisfies (D.7.1), then we can find a subsequence $\{\psi_{n_k}\}$ such that

$$\int |\psi_{n_{k+1}} - \psi_{n_k}| \, dt \le 2^{-k}.$$

If we set $\phi_1 = \psi_{n_1}$, and $\phi_k = \psi_{n_k} - \psi_{n_{k-1}}$ for $k = 2, 3, \ldots$, then it is easy to check that $\{\phi_k\}$ satisfies the conditions of Definition D.7.1. Hence $f \in L$.

We shall leave the proof of (D.7.2) as an exercise.   ∎

In the next theorem we shall list some of the elementary properties of the Lebesgue integral. While these properties are important they do not form our main objective. The most important properties of the Lebesgue integral will be discussed in the next section where we present the basic limit theorems.

D.7.3 THEOREM.   *Let* $f, g \in L$ *and let* α *and* β *be real numbers. Then the following statements are valid:*

(a) $\alpha f + \beta g$ *is in* L *and* $\int (\alpha f + \beta g) dt = \alpha \int f \, dt + \beta \int g \, dt$.

(b) $|f|$ *is in* L *and* $|\int f \, dt| \le \int |f| \, dt$.

(c) *If* $f \ge 0$, *then* $\int f \, dt \ge 0$.

(d) $f \wedge g$ *and* $f \vee g$ *are in* L.

Proof: By using Definition D.7.1 and Theorem D.7.2 one can find sequences $\{\phi_n\}$, $\{\psi_n\}$, and $\{\xi_n\}$ in C_0 such that $f = \sum_n \phi_n$ (a.e.), $g = \sum_n \psi_n$ (a.e.), $f = \lim \xi_n$ (a.e.), and where

$$\sum_n \int |\phi_n|\, dt < \infty, \quad \sum_n \int |\psi_n|\, dt < \infty, \quad \lim_{m,\,n\to\infty} \int |\xi_m - \xi_n|\, dt \to 0.$$

(a) Let E_1 and E_2 be null sets with the property that for $t \notin E_1$ one has $f(t) = \sum_n \phi_n(t)$, and for $t \notin E_2$ one has $g(t) = \sum_n \psi_n(t)$. Now define

$$E_3 = \left\{ t : \sum_n |\phi_n(t)| = +\infty \right\},$$

$$E_4 = \left\{ t : \sum_n |\psi_n(t)| = +\infty \right\}.$$

It follows that E_3 and E_4 are null sets and $E = \bigcup_{i=1}^{4} E_i$ is a null set. It is easy to see that for $t \notin E$ one has

$$\alpha f(t) + \beta g(t) = \sum_n \alpha \phi_n(t) + \beta \psi_n(t).$$

Furthermore,

$$\sum_n \int |\alpha \phi_n + \beta \psi_n|\, dt \leq \sum_n |\alpha| \int |\phi_n|\, dt + \sum_n |\beta| \int |\psi_n|\, dt < \infty.$$

Hence $\alpha f + \beta g$ is in L, and

$$\int (\alpha f + \beta g)\, dt = \sum_n \int (\alpha \phi_n + \beta \psi_n)\, dt$$

$$= \alpha \sum_n \int \phi_n\, dt + \beta \sum_n \int \psi_n\, dt$$

$$= \alpha \int f\, dt + \beta \int g\, dt.$$

(b) Since $|f| = \lim |\xi_n|$ (a.e.) and $||\xi_m| - |\xi_n|| \leq |\xi_m - \xi_n|$, we see that

$$\lim_{m,\,n\to\infty} \int ||\xi_m| - |\xi_n||\, dt \to 0.$$

Hence $|f|$ is in L by Theorem D.7.2. Furthermore,

$$\left| \int \xi_n\, dt \right| \leq \int |\xi_n|\, dt,$$

so

$$\left| \int f\, dt \right| = \lim_{n\to\infty} \left| \int \xi_n\, dt \right| \leq \lim_{n\to\infty} \int |\xi_n|\, dt = \int |f|\, dt.$$

(c) If $f \geq 0$, then $f = \lim |\xi_n|$ (a.e.) and by the last paragraph one has

$$\int f\, dt = \lim_{n\to\infty} \int |\xi_n|\, dt \geq 0.$$

(d) This follows from (a) and (b) once one observes that

$$f \wedge g = \tfrac{1}{2}(f + g) - \tfrac{1}{2}|f - g|$$
$$f \vee g = \tfrac{1}{2}(f + g) + \tfrac{1}{2}|f - g|. \quad \blacksquare$$

EXAMPLE 1.   We have shown that C_0 lies in L. Let us now show that certain piecewise continuous functions lie in L. To begin let $f: R \to R$ be a bounded function with the property that

$$f: (a,b) \to R \text{ is continuous, and}$$
$$f(t) = 0 \text{ for } t \notin [a,b],$$

where $a < b$. Let us show that $f \in L$ (see Figure D.7.1). Since f is bounded, there is an M such that $|f(t)| \leq M$ for all t. Choose monotone sequences $\{a_n\}$ and $\{b_n\}$ in

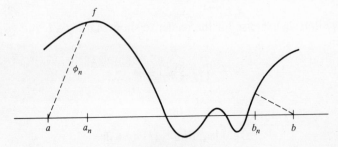

Figure D.7.1.

$[a,b]$ so that $a_n \leq b_n$, $a = \lim a_n$, and $b = \lim b_n$, for example, $a_n = a + 1/n$, $b_n = b - 1/n$. Define ϕ_n in C_0 so that $|\phi_n(t)| \leq M$ for $a \leq t \leq b$ and

$$\phi_n(t) = f(t), \quad a_n \leq t \leq b_n$$
$$= 0, \quad t \notin [a,b].$$

Then $f = \lim \phi_n$ (a.e.). Furthermore for $n \geq m$ one has

$$\int |\phi_n - \phi_m| dt \leq 2M(|a_m - a| + |b - b_m|) \to 0$$

as $n, m \to \infty$. It follows from Theorem D.7.2 that $f \in L$. Furthermore, it is easy to see that $\int f \, dt = \int_a^b f \, dt$, where $\int_a^b f \, dt$ is the Riemann integral of f.

Since a finite linear combination of functions in L is also in L, we see that any bounded, piecewise continuous function with compact support is in L. In particular any step function with compact support is in L. A *step function* is defined to be a piecewise constant function.[1] $\blacksquare$

We will need another characterization of L in the next section.

---

[1] We warn the reader that some authors use the term "step function" in a somewhat different manner.

If $f \in L$, then there is a sequence $\{\phi_n\}$ in C_0 with $f = \sum_{n=1}^{\infty} \phi_n$ (a.e.) and $\sum_{n=1}^{\infty} \int |\phi_n| dt < \infty$. Now let $m \geq 1$ and define $\psi_1 = \sum_{n=1}^{m+1} \phi_n$, and $\psi_k = \phi_{m+k}$. Then $f = \sum_{k=1}^{\infty} \psi_k$ and $\sum_{k=1}^{\infty} \int |\psi_k| dt < \infty$. Since $\sum_{k=2}^{\infty} \int |\psi_k| dt = \sum_{n=m+2}^{\infty} \int |\phi_n| dt \to 0$ as $m \to \infty$, we see that the series $\sum_{k=2}^{\infty} \int |\psi_k| dt$ can be made arbitrarily small by choosing m large. We thus have the following result.

D.7.4 THEOREM. *A function f is in L if and only if for every $\varepsilon > 0$ there is a sequence $\{\psi_n\}$ in C_0 that satisfies*

$$f = \sum_{n=1}^{\infty} \psi_n \quad \text{(a.e.)}, \tag{D.7.3}$$

and

$$\sum_{n=2}^{\infty} \int |\psi_n| \, dt \leq \varepsilon.$$

We leave it as an exercise for the reader to show that if $f \in L$ and $\{\psi_n\}$ satisfies (D.7.3), then

$$\int |f - \psi_1| \leq \varepsilon, \tag{D.7.4}$$

and

$$\sum_{n=1}^{\infty} \int |\psi_n| \, dt \leq \int |f| \, dt + 2\varepsilon. \tag{D.7.5}$$

EXERCISES

1. Prove Equation (D.7.2).
2. Show that (D.7.3) implies (D.7.4) and (D.7.5).
3. Let $f \in L$ with $\int |f| dt = 0$. Show that $f = 0$ (a.e.).

8.   LIMIT THEOREMS

In this section we come to the heart of the Lebesgue theory. The limit theorems we shall now discuss have an importance which is hard to overemphasize. We ask that the reader take particular note of them.

We are primarily interested in three types of limit theorems. Let $\{f_n\}$ be a sequence of functions in L and let $f = \lim f_n$. We will consider the following cases

    I.   MONOTONE CONVERGENCE $f_n \leq f_{n+1}$.

    II.   NONMONOTONE CONVERGENCE (FATOU'S THEOREM).

    III.   DOMINATED CONVERGENCE $|f_n(t)| \leq g(t)$.

One of course expects a stronger conclusion with a stronger hypothesis, and this will indeed be the case. Each of the theorems on sequences will have a counterpart for infinite series.

We begin this section with a discussion of the Levi Theorem for infinite series, $f = \sum_n f_n$. This theorem is a capsule version of the last three sections. The important thing to note is that the functions f_n now belong to L and not merely to C_0.

D.8.1 THEOREM. (LEVI.) *Let $\{f_n\}$ be a sequence in L with $\sum_{n=1}^{\infty} \int |f_n|\, dt < \infty$. Then the series $\sum_n f_n$ converges almost everywhere. Moreover, if $f = \sum_n f_n$ (a.e.), then $f \in L$ and*

$$\int f\, dt = \sum_n \int f_n\, dt.$$

Proof: It follows from Theorem D.7.4 and (D.7.5) that for each n, there is a sequence $\{\psi_{nk}\}$ in C_0 with $f_n = \sum_{k=1}^{\infty} \psi_{nk}$ (a.e.) and

$$\sum_{k=1}^{\infty} \int |\psi_{nk}|\, dt \le \int |f_n|\, dt + 2^{-n}.$$

Let E_n be the null set for which $f_n(t) = \sum_k \psi_{nk}(t)$ for $t \notin E_n$. Since

$$\sum_n \sum_k \int |\psi_{nk}|\, dt \le \sum_n \int |f_n|\, dt + \sum_n 2^{-n} < \infty,$$

it follows from the definition of a null set, that there is a null set E such that the series $\sum_n \sum_k \psi_{nk}(t)$ converges (absolutely) for $t \notin E$. Hence, if t is not in the null set $E \cup (\bigcup_n E_n)$, the series

$$\sum_n f_n(t) = \sum_n \sum_k \psi_{nk}(t)$$

converges, that is, this series converges almost everywhere.

If $f = \sum_n f_n$ (a.e.), then $f = \sum_n \sum_k \psi_{nk}$ (a.e.) and thus $f \in L$. Furthermore,

$$\int f\, dt = \sum_n \left(\sum_k \int \psi_{nk}\, dt \right) = \sum_n \int f_n\, dt. \quad \blacksquare$$

Next we turn to the Monotone Convergence Theorem, which is a simple corollary of the Levi Theorem. But before we do that, let us recall that a sequence of functions $\{f_n\}$ is said to be *increasing* if $f_n \le f_{n+1}$ for all n. If one has $f_n \ge f_{n+1}$, then the sequence is said to be *decreasing*.

D.8.2 THEOREM. (MONOTONE CONVERGENCE.) *Let $\{f_n\}$ be an increasing sequence in L with $\int f_n\, dt \le M < \infty$ for all n. Let $f = \lim f_n$, then $f \in L$ and*

$$\int f\, dt = \lim_{n \to \infty} \int f_n\, dt.$$

Proof: Let $g_n = f_n - f_{n-1}$, where $f_0 = 0$. Then $g_n \geq 0$ and

$$\sum_n \int |g_n|\, dt = \sum_n \int g_n\, dt = \lim_{n \to \infty} \int f_n\, dt < \infty,$$

and $f = \sum_n g_n$ (a.e.). Hence by Levi's Theorem $f \in L$ and

$$\int f\, dt = \sum_n \int g_n\, dt = \lim_{n \to \infty} \int f_n\, dt. \quad \blacksquare$$

One can obviously get the same conclusion if one assumes instead that $\{f_n\}$ is a decreasing sequence and $\int f_n\, dt$ is bounded below.

The next result is concerned with unrestricted convergence of a sequence $\{f_n\}$.

D.8.3 THEOREM. (FATOU.) *Let $\{f_n\}$ be a sequence of nonnegative functions in L with*

$$\liminf_{n \to \infty} \int f_n\, dt = \lim_{n \to \infty} \left[\inf_{n \leq k} \int f_k\, dt \right] < \infty$$

and assume that $f = \lim f_n$ (a.e.). Then $f \in L$ and

$$\int f\, dt \leq \liminf_{n \to \infty} \int f_n\, dt.$$

Proof: Let n be fixed and for each integer $k \geq 1$ let

$$h_{n,k} = f_n \wedge f_{n+1} \wedge \cdots \wedge f_{n+k}.$$

It follows from Theorem D.7.3 (d) that $h_{n,k} \in L$ and $0 \leq h_{n,k} \leq f_n$. Furthermore $\{h_{n,k}\}$ is decreasing in k. If we let

$$g_n = \lim_{k \to \infty} h_{n,k}$$

it follows from the Monotone Convergence Theorem that $g_n \in L$ and

$$\int g_n\, dt = \lim_{k \to \infty} \int h_{n,k}\, dt \leq \int f_n\, dt.$$

It is clear that g_n is increasing in n. Since $f = \lim g_n$ (a.e.) one has $f \in L$ and

$$\int f\, dt = \lim_{n \to \infty} \int g_n\, dt \leq \liminf_{n \to \infty} \int f_n\, dt. \quad \blacksquare$$

The Fatou Theorem can be reformulated for infinite series. Let $g_N = \sum_{n=1}^N f_n$ and assume that $g_N \geq 0$ for all N. Furthermore, assume that

$$\liminf_{N \to \infty} \sum_{n=1}^N \int f_n\, dt < \infty,$$

and that $g = \sum_{n=1}^\infty f_n$ (a.e.). Then $g \in L$ and

$$\int g\, dt = \int \sum_{n=1}^\infty f_n\, dt \leq \liminf_{N \to \infty} \sum_{n=1}^N \int f_n\, dt.$$

We now turn to the most important of the limit theorems.

D.8.4 THEOREM. (DOMINATED CONVERGENCE.) *Let $\{f_n\}$ be a sequence in L and assume that $f = \lim f_n$ (a.e.). If there is a $g \in L$ such that $|f_n| \le g$ (a.e.), then $f \in L$ and*

$$\int f\, dt = \lim_{n \to \infty} \int f_n\, dt. \tag{D.8.1}$$

Proof: Since $0 \le g + f_n \le 2g$ we know that

$$\liminf_{n \to \infty} \int (g + f_n)\, dt \le \int 2g\, dt < \infty.$$

Hence by Fatou's Theorem $g + f = \lim (g + f_n)$ (a.e.) is in L. Thus $f = (g + f) - g$ is in L. Furthermore,

$$\int (g + f)\, dt \le \liminf_{n \to \infty} \int (g + f_n)\, dt,$$

which implies that

$$\int f\, dt \le \liminf_{n \to \infty} \int f_n\, dt. \tag{D.8.2}$$

Similarly we have $0 \le g - f_n \le 2g$, so

$$\int (g - f)\, dt \le \liminf_{n \to \infty} \int (g - f_n)\, dt,$$

which implies that

$$-\int f\, dt \le \liminf_{n \to \infty} \int (-f_n)\, dt = -\limsup_{n \to \infty} \int f_n\, dt,$$

or

$$\limsup_{n \to \infty} \int f_n\, dt \le \int f\, dt. \tag{D.8.3}$$

By combining (D.8.2) and (D.8.3) we see that $\lim_{n \to \infty} \int f_n\, dt$ exists and (D.8.1) is valid. ∎

D.8.5 COROLLARY. *Let $f = \lim f_n$ (a.e.), where $\{f_n\}$ is a sequence in L. If $|f| \le g$ (a.e.), where $g \in L$, then $f \in L$.*

Proof: Define g_n by

$$g_n(t) = \begin{cases} g(t), & \text{if } g(t) < f_n(t), \\ f_n(t), & \text{if } -g(t) \le f_n(t) \le g(t), \\ -g(t), & \text{if } f_n(t) < g(t) \end{cases}$$

or equivalently

$$g_n = g \wedge [f_n \vee (-g)].$$

Hence $g_n \in L, |g_n| \leq g$ (a.e.) and $f = \lim g_n$ (a.e.). So by the Dominated Convergence Theorem we see that $f \in L$. ∎

D.8.6 COROLLARY.   *Let $f = \lim f_n$ (a.e.), where $\{f_n\}$ is a sequence in L. Then $f \in L$ if and only if $|f| \in L$.*

Proof: This follows from the last corollary (let $g = |f|$) and Theorem D.7.3 (b). ∎

The Dominated Convergence Theorem can also be reformulated for infinite series.

D.8.7 COROLLARY.   *Let $\{f_n\}$ be a sequence in L and assume that $f = \sum_{n=1}^{\infty} f_n$ (a.e.). If there is a $g \in L$ such that*

$$\left| \sum_{n=1}^{N} f_n \right| \leq |g| \quad \text{(a.e.)}$$

for all N, then $f \in L$ and

$$\int f \, dt = \int \sum_{n=1}^{\infty} f_n \, dt = \sum_{n=1}^{\infty} \int f_n \, dt.$$

EXERCISES

1. Extend Fatou's Theorem to the following case. Let $\{f_n\}$ be a sequence of non-negative functions, where

$$\liminf_{n \to \infty} \int f_n \, dt < \infty.$$

   Show that $\liminf_{n \to \infty} f_n(t)$ exists almost everywhere and if $f = \liminf_{n \to \infty} f_n(t)$, then

$$\int f \, dt \leq \liminf_{n \to \infty} \int f_n \, dt.$$

2. Consider the differential equation $X' = f(X,t)$ where $f: R \times R \to R$ is a continuous function. Assume that $|f(X,t)| \leq m(t)$ for all X in R and t in some interval $I = [a,b]$ where $m(t)$ is the Lebesgue integrable. Let $\{X_n(t)\}$ be a sequence of solutions of this equation on I and assume that $X_n(t) \to X_0(t)$ for all t in I. Show that X_0 is a solution. [*Hint*: Convert the differential equation into an integral equation $X(t) = X(a) + \int_a^t f(X(s), s) \, ds$. Now apply the appropriate limit theorems.]

3. A typical control theory problem can be formulated as follows: Consider the differential equation $X' = f(X,t,u)$, where f is continuous and u is a control parameter. One seeks a control $u(t)$ such that the solution $X(t)$ satisfies $X(0) = X_0$, and $X(1) = X_1$, where X_0 and X_1 are fixed. (If u does this we say that it is admissible.) Furthermore, we require u to be chosen so that the integral

$\int_0^1 |u(t)|\,dt$ assumes its minimum value. Assume that one has found a sequence of controllers $\{u_n\}$ such that

$$\int_0^1 |u_n(t)|\,dt \to \inf\left\{\int_0^1 |u(t)|\,dt : u \text{ is admissible}\right\}$$

$$u_0 = \lim u_n \quad (\text{a.e. on } [0,1]),$$

where u_0 is admissible. Show that

$$\int_0^1 |u_0(t)|\,dt = \inf\left\{\int_0^1 |u(t)|\,dt : u \text{ is admissible}\right\},$$

that is, u_0 is an "optimal" control.

4. First let $f_n \geq 0$ satisfy $\int |f_n|^p\,dt < \infty$, for some p with $1 \leq p < \infty$ and then let $\|f_n\|_p = \left(\int |f_n|^p\,dt\right)^{1/p}$. Assume that $\sum_n \|f_n\|_p < \infty$.
   (a) Show that the series $\sum_n f_n$ converges almost everywhere.
   (b) If $f = \sum_n f_n$ (a.e.) show that $\|f\|_p \leq \sum_n \|f_n\|_p$.
   [*Hint*: Let $g_n = \sum_{i=1}^n f_i$ and use the Minkowski Inequality to show that $\|g_n\|_p \leq \sum_{i=1}^n \|f_i\|_p$. Now apply the Monotone Convergence Theorem.]

5. Let f be a Riemann-integrable function on the interval $I = [a,b]$. Assume that $f(t) = 0$ for $t \notin I$. Show that f is in L and that the Riemann integral of f and the Lebesgue integral of f agree. [*Hint*: Use Theorem D.2.3 and the Monotone Convergence Theorem.]

6. Prove Corollary D.8.7.

9.   MISCELLANY

In this section we collect a number of refinements or extensions of the notion of the integral presented above.

Complex-Valued Functions

Let $f = u + iv$, where u and v are the real and imaginary parts of f. If both u and v belong to L, we say that $f \in L$ and define

$$\int f\,dt = \int u\,dt + i \int v\,dt.$$

We invite the reader to prove the Dominated Convergence Theorem for complex-valued functions. Another interesting (and useful) fact is to show that for a complex-valued function f one has $f \in L$ if and only if $|f| \in L$.

Vector-Valued Functions

If $f : R \to R^n$, then one has $f = (f_1, f_2, \ldots, f_n)$, where $f_i : R \to R$, $1 \leq i \leq n$. Now define $\int f\,dt$ to be the vector

$$\int f\,dt = \left(\int f_1\,dt, \int f_2\,dt, \ldots, \int f_n\,dt\right)$$

provided $f_i \in L$, $1 \le i \le n$. We ask the reader to show that

$$\left\| \int f \, dt \right\| \le \int \|f\| \, dt,$$

where $\| \cdot \|$ is any norm on R^n.

Characteristic Functions and Measure

If A is a set in R, then the characteristic function of A, χ_A, is given by

$$\chi_A(t) = \begin{cases} 1, & \text{if } t \in A, \\ 0, & \text{if } t \notin A. \end{cases}$$

If $\chi_A \in L$, then we shall say that A is *measurable* and define the *Lebesgue measure* of A by

$$m(A) = \int \chi_A \, dt.$$

More generally, if χ_A is a characteristic function and $A = \bigcup_{n=1}^{\infty} A_n$, where $\{A_n\}$ is a sequence of mutually disjoint sets with $\chi_{A_n} \in L$, then we shall say that A is *measurable* and define the *Lebesgue measure* A by

$$m(A) = \sum_{n=1}^{\infty} \int \chi_{A_n} \, dt,$$

which may be $+\infty$. One may ask whether $m(A)$ depends on the sequence $\{A_n\}$. The answer, as seen in Corollary D.6.5, is negative when $\sum_{n=1}^{\infty} m(A_n)$ is finite. Moreover, even when this sum is $+\infty$, the answer is the same.

What are some of the properties of the collection of measurable sets? We shall list them here and ask the reader to verify our assertions.

1. If A is measurable, then the complement A' is measurable.

2. If $\{A_1, A_2, \ldots\}$ is a sequence of measurable sets, then $\bigcup_{n=1}^{\infty} A_n$ and $\bigcap_{n=1}^{\infty} A_n$ are measurable.

3. Every interval is measurable.

4. Every open set and every closed set is measurable.

5. Let A be a measurable set with *measure zero*, that is, $m(A) = 0$. Then A is a null set.

6. Let $\{A_1, A_2, \ldots\}$ be a sequence of mutually disjoint measurable sets, then

$$m\left(\bigcup_{n=1}^{\infty} A_n \right) = \sum_{n=1}^{\infty} m(A_n).$$

7. Let A and B be measurable sets with $A \subseteq B$. Then $m(A) \le m(B)$.

Measurable Functions

A function $f: R \to R$ is said to be *measurable* if $f^{-1}(0)$ is a measurable set in R whenever 0 is an open set in R. (Compare this with the characterization of a continuous function in Theorem 3.9.7.) It is easy to see that f is a measurable function if and only if

$$\{t : f(t) < a\} \tag{D.9.1}$$

is a measurable set for every a in R. One can also replace the inequality $<$ in (D.9.1) by either $\leq$, or $>$, or $\geq$ without changing the collection of measurable functions.

Many of the properties of measurable functions are easy consequences of the above properties. For example,

1. Every continuous function is measurable.

2. If f and g are measurable, then $f \wedge g$, $f \vee g$, fg, and $f + g$ are measurable.

3. If $f = \lim f_n$ (a.e.), where the functions f_n are measurable, then f is measurable. (Hence the functions in L are measurable.)

4. If $\{f_n\}$ is a sequence of measurable functions, then $\sup\{f_1, f_2, \ldots\}$ and $\inf\{f_1, f_2, \ldots\}$ are measurable.

5. If f is a measurable function with $|f| \leq g$ (a.e.), where $g \in L$, then $f \in L$.

The Notation $\int_A f \, dt$

Let $f \in L$ and let A be a measurable set. One then defines $\int_A f \, dt$ by

$$\int_A f \, dt = \int f \cdot \chi_A \, dt.$$

If A is the interval $[a,b]$, then one oftentimes writes

$$\int_A f \, dt = \int_a^b f \, dt.$$

If $A = R$, then $\int_A f \, dt$ becomes $\int f \, dt$ which is sometimes written as

$$\int f \, dt = \int_{-\infty}^{\infty} f \, dt.$$

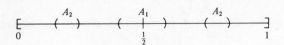

Figure D.9.1.

The Cantor Set

Let us now consider an example of a particularly interesting subset of $E = [0,1]$. We will now define a family $\{A_n\}$ of disjoint sets where each A_n is the union of a finite number of intervals. Each A_n will be open and $A = \bigcup_{n=1}^{\infty} A_n$ will be an open set in E. The complement $A_n' = E - A_n$ will, of course, be closed and each

A_n' will be nonempty. Furthermore, the sequence $C_N = \bigcap_{n=1}^{N} A_n'$ will be chosen to be a decreasing sequence of nonempty closed sets, and therefore the intersection

$$C = \bigcap_{n=1}^{\infty} A_n' = \bigcap_{N=1}^{\infty} C_N$$

is nonempty, closed, and compact. (Refer to Exercise 32, Section 3.17.) The set C will be the *Cantor set.* Now for the specifics.

Let A_1 be an open interval of length a, $0 < a < 1$, with center at $\frac{1}{2}$, that is, $A_1 = (1/2 - a/2, 1/2 + a/2)$. (We will be particularly interested in the case $a = \frac{1}{3}$, then $A_1 = (\frac{1}{3}, \frac{2}{3})$ is the (open) middle third of the interval $E = [0,1]$.) The complement A_1' is then the disjoint union of two closed intervals, I_1^1 and I_2^1 of equal length, $m(I_1^1) = m(I_2^1) = (1 - a)/2$. Now let r be a fixed real number satisfying $0 < r \le 1 - a$. See Fig. D.9.1.

A_2 is defined to be the union of two open intervals J_1^1 and J_2^1, each of length $ar/2$, where J_i^1 and I_i^1 ($i = 1,2$) have the same center. (In the special case $a = \frac{1}{3}$ and $r = \frac{2}{3}$, the intervals J_1^1 and J_2^1 form the middle third of the interval I_1^1 and I_2^1, that is, $A_2 = (\frac{1}{9}, \frac{2}{9}) \cup (\frac{7}{9}, \frac{8}{9})$.) The complement of $A_1 \cup A_2$ now consists of four disjoint closed intervals I_1^2, I_2^2, I_3^2, and I_4^2 of equal length, $(1 - a - ar)/4$.

Now assume that $A_1, A_2, \ldots, A_n$ have been chosen so that they are disjoint and open, and that each A_k, $k = 1, 2, \ldots, n$, consists of 2^{k-1} open intervals of equal length with $m(A_k) = a \cdot r^{k-1}$. Also the set $E - (A_1 \cup \cdots \cup A_k)$ consists of 2^k disjoint closed intervals $I_1^k, \ldots, I_{2^k}^k$ of equal length. It is easy to see that

$$m(I_i^k) = 2^{-k}(1 - a - ar - \cdots - ar^{k-1}), \qquad (i = 1, \ldots, 2^k).$$

We now define A_{n+1} as follows: A_{n+1} will be the union of 2^n open intervals $J_1^n, \ldots, J_{2^n}^n$, each of length $2^{-n}ar^n$, where J_i^n and I_i^n have the same centers. Then $m(A_{n+1}) = ar^n$ and the set $E - (A_1 \cup \cdots \cup A_{n+1})$ consists of 2^{n+1} disjoint closed intervals of equal length.

The open sets A_n have now been defined for all n, by mathematical induction. They are disjoint and $m(A_n) = ar^{n-1}$. Therefore, if $A = \bigcup_{n=1}^{\infty} A_n$, then A is open and measurable and by property (6) for the measure m one has

$$m(A) = \sum_{n=1}^{\infty} ar^{n-1} = \frac{a}{1 - r}.$$

The complement of A, $C = E - A$, is nonempty, closed, and measurable and

$$m(C) = m(E) - m(A) = \frac{1 - r - a}{1 - r}.$$

C is defined to be the *Cantor set.* The following theorem gives the pertinent properties of the Cantor set.

D.9.1 THEOREM.   *Let C be a Cantor set in $[0,1]$. Then C is nonempty, closed, compact, nowhere dense and $0 \le m(C) < 1$. Moreover, C contains uncountably many points.*

Proof: We have already shown that C is nonempty, closed, compact, and $0 \le m(C) < 1$. Let us now show that C is nowhere dense. This means that Int C, the interior of C, is empty. Let $x \in C$. We want to show that there does not exist a nontrivial open interval $B_\varepsilon(x) = \{y: |x - y| < \varepsilon\}$ which lies entirely in C. In other words, every nontrivial open interval containing x also contains points in the complement C'.

Let A_n be defined as above and let $C_n = E - (A_1 \cup \cdots \cup A_n)$. Then C_n consists of 2^n disjoint closed intervals, each of length

$$2^{-n}(1 - a - ar - \cdots - ar^{n-1}) \le 2^{-n}.$$

Therefore, if $y \in C_n$ and $\delta \ge 2^{-n}$, then it is also true that $B_\delta(y)$ does not lie entirely in C_n, since $m(B_\delta(y)) = 2 \cdot \delta > 2^{-n}$.

Now

$$C = \bigcap_{n=1}^{\infty} C_n,$$

so if $x \in C$, then $x \in C_n$ for every $n = 1, 2, \ldots$. Now let $\varepsilon > 0$ be given and choose n so that $\varepsilon \ge 2^{-n}$. Then $B_\varepsilon(x)$ does not lie entirely in C_n, that is, there is a y in $B_\varepsilon(x)$ such that $y \notin C_n$. Hence, $y \notin C$, so we have shown that C is nowhere dense.

We will leave the proof of the last statement, that C is uncountable, as an exercise.   ∎

In the special case where $a = \frac{1}{3}$ and $r = \frac{2}{3}$, we see that $m(C) = 0$, that is, the Cantor set has measure zero. This is true more generally when a is arbitrary $(0 < a < 1)$ and $r = 1 - a$. However, if $0 < r < 1 - a$, then $m(C) > 0$. That is, C is then a set which has positive measure, but it does not contain a nontrivial interval!

EXERCISES

1. By using the definition of the Riemann integral above show directly that χ_C is Riemann integrable if and only if $m(C) = 0$.

2. Let $f:[a,b] \to R$ be given and define

$$\text{Discont}(f) = \{t \in [a,b]: f \text{ is not continuous at } t\}.$$

It can be shown (see Riesz and Sz. Nagy [1, pp. 23–24]) that the function f is Riemann integrable if and only if Discont(f) is a set of measure zero. Let χ_C be the characteristic function of the Cantor set. Show that χ_C is Riemann integrable if and only if $m(C) = 0$.

3. Extend the concept of a measurable function to complex-valued functions. Do the same for vector-valued functions. Which, if any, of the properties of real-valued measurable functions are valid for vector-valued measurable functions?

10.   OTHER DEFINITIONS OF THE INTEGRAL

The construction or the definition of the (Lebesgue) integral we have presented is based on the Daniell approach. That is, we began with a space of functions C_0 satisfying properties (I.1), (I.2), (I.3), and (I.4) in Section 4. This approach can be generalized. What is really needed is that at the outset we be given a set $\mathscr{E}$ of functions, call them elementary functions, and an integral defined on $\mathscr{E}$ so that the four properties of Section 4 are satisfied. We then extend the integral so that it is defined on a superset of $\mathscr{E}$. In this section we shall point out two such generalizations and invite the reader to carry out the details of the construction of the integral.

EXAMPLE 1. (STEP FUNCTIONS.)   Let $\mathscr{E}$ denote the collection of all real-valued step functions defined on R and with compact support. A function $\phi \in \mathscr{E}$ can be described as follows: There are two sequences of real numbers $\{t_0, t_1, \ldots, t_n\}$ and $\{\alpha_1, \alpha_2, \ldots, \alpha_n\}$, which depend on ϕ, such that $t_0 < t_1 < \cdots < t_n$ and[2]

$$\phi(t) = \alpha_i, \qquad t_{i-1} < t < t_i, \qquad i = 1, 2, \ldots, n,$$
$$\phi(t) = 0, \qquad t < t_0 \text{ or } t_n < t. \tag{D.10.1}$$

Define the integral of ϕ by

$$\int \phi \, dt = \sum_{i=1}^{n} \alpha_i (t_i - t_{i-1}), \tag{D.10.2}$$

where ϕ satisfies (D.10.1). It is easy to see that this class of functions $\mathscr{E}$, together with the integral defined by (D.10.2), satisfy properties (I.1), (I.2), and (I.3) in Section 4. Property (I.4) is somewhat harder to prove, and this argument is outlined in the exercises.

From this point one can use the Daniell approach to define null sets, convergence almost everywhere, and finally the Lebesgue integral. In order to show that this definition of the Lebesgue integral agrees with that given in Section D.7 one need only prove that, in terms of the new integral, every function in C_0 is integrable and the integral agrees with the Riemann integral. We invite the reader to check these assertions. ∎

EXAMPLE 2. (MEASURE SPACES.)   Let X be a nonempty set and let $\mathscr{B}$ denote a σ-*field* of subsets of X. This means that the empty set $\varnothing$ is in $\mathscr{B}$ and whenever $\{E_1, E_2, \ldots\}$ belong to $\mathscr{B}$, then the complement $E_1' = X - E_1$ and $\bigcup_{n=1}^{\infty} E_n$ belong to $\mathscr{B}$. This also implies that $\bigcap_{n=1}^{\infty} E_n = \bigcup_{n=1}^{\infty} E_n'$ belongs to $\mathscr{B}$. Let μ be a *positive*[3] *measure* defined on $\mathscr{B}$. This means that μ maps $\mathscr{B}$ into the reals R in such a way that $\mu(\varnothing) = 0$, $\mu(E) \geq 0$ for all E in $\mathscr{B}$, and

$$\mu\left(\bigcup_{n=1}^{\infty} E_n\right) = \sum_{n=1}^{\infty} \mu(E_n)$$

---

[2] It is not necessary for our purposes to specify the values of ϕ at the points $\{t_0, t_1, \ldots, t_n\}$. (Why?)

[3] The adjective "positive" refers to the condition $\mu(E) \geq 0$. If we drop this condition we have an arbitrary measure. (See Section D.14.)

whenever $\{E_1, E_2, \ldots\}$ is a sequence of mutually disjoint sets in $\mathscr{B}$. Some specific examples of measure spaces are discussed in the exercises.

We now wish to define an integral on X. For this purpose we shall let $\mathscr{B}_0$ denote those sets E in $\mathscr{B}$ that satisfy $\mu(E) < \infty$. A function $\phi\colon X \to R$ is said to be a *simple function* if there are a finite number of disjoint sets $\{E_0, E_1, \ldots, E_n\}$ in $\mathscr{B}$, and real numbers $\{\alpha_0, \alpha_1, \ldots, \alpha_n\}$ such that $\alpha_0 = 0$, $\{E_1, \ldots, E_n\}$ belong to $\mathscr{B}_0$, $\bigcup_{i=0}^{\infty} E_i = X$, and $\phi(x) = \alpha_i$, if $x \in E_i$, $0 \le i \le n$. We allow the possibility that any of the sets E_i, including E_0, may be empty. We now define the integral of ϕ by

$$\int \phi \mu(dx) = \sum_{i=1}^{n} \alpha_i \mu(E_i). \tag{D.10.3}$$

It is easy to see that the class $\mathscr{F}$ of all simple functions together with the integral defined by (D.10.3) satisfies properties (I.1), (I.2), and (I.3) in Section 4. We ask the reader to check this. Let us prove here the generalization of Dini's Theorem, property (I.4).

In order to do this we will need the following property of the measure μ. "If $\{A_n\}$ is a decreasing sequence from $\mathscr{B}_0$ and $\bigcap_{n=1}^{\infty} A_n = \varnothing$, then $\mu(A_n) \to 0$ as $n \to \infty$."

Let $\{\phi_n\}$ be a decreasing sequence of nonnegative simple functions with $\lim \phi_n(x) = 0$ for all x in X. Let E denote the set in X for which $\phi_1(x) > 0$. Then $E \in \mathscr{B}_0$. Let $M < \infty$ satisfy $\phi_1(x) \le M$ for all x in X. Since $0 \le \phi_n \le \phi_1$ one has $\phi_n(x) = 0$ when $x \notin E$, and $\phi_n(x) \le M$ for all x in X.

Let $\varepsilon > 0$ be given and define A_n by

$$A_n = \{x \in E\colon \phi_n(x) > \varepsilon\}.$$

It follows that $\{A_n\}$ is a decreasing sequence in $\mathscr{B}_0$ with $\bigcap_{n=1}^{\infty} A_n = \varnothing$. Now let ψ_n be the simple function

$$\psi_n(x) = \begin{cases} M, & x \in A_n, \\ \varepsilon, & x \in E - A_n, \\ 0, & x \notin E. \end{cases}$$

Then $0 \le \phi_n \le \psi_n$, and

$$\int \phi_n \mu(dx) \le \int \psi_n \mu(dx) = \varepsilon \mu(E - A_n) + M \mu(A_n)$$
$$\le \varepsilon \mu(E) + M \mu(A_n).$$

Now choose N so that $\mu(A_n) \le \varepsilon$ whenever $n \ge N$. One then has

$$\int \phi_n \mu(dx) \le (\mu(E) + M)\varepsilon, \qquad n \ge N.$$

Thus $\lim_{n \to \infty} \int \phi_n \mu(dx) = 0$, which proves (I.4).

Starting from this point we can again use the Daniell approach to define null sets, convergence almost everywhere and the integral $\int f\mu(dx)$. A variation does occur when one defines a measurable set and a measurable function. Specifically a *measurable set* is simply a set in the σ-field $\mathscr{B}$. Then a real-valued function

$f: X \to R$ is said to be *measurable* if for every open set O in $R, f^{-1}(O)$ is a measurable set in X. (Also see Exercise 8.)

An important version of this example occurs when one is discussing probability spaces. In that case μ is a probability measure and $\mu(X) = 1$. We shall treat this case further in Appendix E. ∎

EXERCISES

1. Show that Properties (I.1), (I.2), and (I.3) hold for the integral in (D.10.2) defined on the step functions $\mathscr{E}$.

2. Show that Property (I.4) holds for the integral on the step functions $\mathscr{E}$.

3. The *Borel field* $\mathscr{B}$ in R is defined to be the smallest σ-field of sets from R that contains all open intervals (a,b). The sets in the Borel field are called Borel sets.
   (a) Show that every interval is a Borel set.
   (b) Show that every open set is a Borel set.
   (c) Show that every closed set is a Borel set.
   (d) If $I = (a,b)$, let $m(I) = b - a$. Show that there is one and only one measure μ defined on $\mathscr{B}$ that satisfies $\mu(I) = m(I)$ for every open interval I. (This measure is called the *Lebesgue measure*.)
   (e) Show that the definition of the integral in Section 7 agrees with the definition in Example 2 when the measure μ is the Lebesgue measure.

4. Let $f \in L$ and $f \geq 0$. Let $\mathscr{B}$ denote the Borel sets in R. For $E \in \mathscr{B}$ let
$$\mu(E) = \int f \cdot \chi_E \, dt.$$
   (a) Show that μ is a positive measure on R.
   (b) Show that if g is a simple function, then $\int g\mu(dx) = \int f \cdot g \, dt$.
   (c) What can you say about $\int g\mu(dx)$ when g is not a simple function?

5. Let $\mathscr{B}$ be the Borel field in R and define μ on $\mathscr{B}$ by
$$\mu(E) = \begin{cases} 1, & \text{if } 0 \in E, \\ 0, & \text{if } 0 \notin E. \end{cases}$$
   (a) Show that μ is a measure on R.
   (b) Compute $\int g\mu(dx)$ when g is a continuous function.

6. The *Borel field* on R^n is the smallest σ-field of sets from R^n that contains all open rectangles $\{(x_1,\ldots,x_n): a_i < x_i < b_i, i = 1,2,\ldots,n\}$. Define the *Lebesgue measure* on R^n by defining
$$m(\{(x_1,\ldots,x_n): a_i < x_i < b_i, i = 1,2,\ldots,n\}) = \prod_{i=1}^{n}(b_i - a_i)$$
for each rectangle and then extending m to $\mathscr{B}$. Construct the Lebesgue integral R^n.

7. Let L be the space of Lebesgue-integrable functions on R and let $\int$ denote the Lebesgue integral.
   (a) Show that L satisfies properties (I.1), (I.2), (I.3), and (I.4).

Use the Daniell approach starting with L to answer the following questions:

(b) Describe the null sets in terms of Section D.5.
(c) Describe convergence almost everywhere in terms of Section D.6.
(d) Describe the extended integral in terms of the Lebesgue integral. [*Hint*: Use Levi's Theorem.]

8. Use the definition of measurable function in Example 2 and the definition of Borel set in Exercise 3 to show that a function $f: X \to R$ is measurable if and only if $f^{-1}(B)$ is measurable in X for every Borel set $B \subset R$.

11. THE LEBESGUE SPACES L_p

The Lebesgue spaces L_p play a central role in the study of functional analysis. These spaces are employed extensively in the text. Here we shall limit our discussion to the definition and the proof of a basic theorem, which says that the Lebesgue spaces are complete.

D.11.1 DEFINITION. Let E be a measurable set in R and let p satisfy $1 \le p < \infty$. A function $f: E \to R$ is said to belong to $L_p(E)$ if

$$\int_E |f|^p \, dt < \infty,$$

or equivalently, if $|f \cdot \chi_E|^p$ belongs to L. Moreover, a norm is defined on $L_p(E)$ by $\|f\|_p = (\int_E |f|^p \, dt)^{1/p}$.

D.11.2 THEOREM. *Let $\{f_n\}$ be a sequence in $L_p(E), 1 \le p < \infty$, with the property that*

$$\lim_{m,n \to \infty} \int_E |f_m - f_n|^p \, dt = 0.$$

Then there is an $f \in L_p(E)$ that satisfies $f = \lim f_n$ (a.e.) and

$$\lim_{n \to \infty} \int_E |f - f_n|^p \, dt = 0. \qquad (D.11.1)$$

Proof: It will suffice to assume that $E = R$. If not, we could replace f_n by $\hat{f}_n$, where

$$\hat{f}_n(t) = \begin{cases} f_n(t), & t \in E, \\ 0, & t \notin E. \end{cases}$$

Then $\{\hat{f}_n\}$ satisfies

$$\int |\hat{f}_m - \hat{f}_n|^p \, dt = \int_E |f_m - f_n|^p \, dt.$$

Take the case $p = 1$ first. If

$$\lim_{m,n \to \infty} \int |f_m - f_n| \, dt = 0, \qquad (D.11.2)$$

then we can find a subsequence $\{f_{n_k}\}$ that satisfies

$$\int |f_{n_{k+1}} - f_{n_k}|\, dt \le 2^{-k}.$$

Now set $g_1 = f_{n_1}$, $g_k = f_{n_k} - f_{n_{k-1}}$, $k = 2, 3, \ldots$. Then the sequence $\{g_k\}$ satisfies $\sum_{k=1}^{\infty} \int |g_k|\, dt < \infty$. So by the Levi Theorem D.8.1 there is an f in L such that

$$= \sum_{k=1}^{\infty} g_k = \lim_{k \to \infty} f_{n_k} \qquad \text{(a.e.)}.$$

It follows from (D.11.2) that for every $\varepsilon > 0$ there is an M such that for $n_k, n \ge M$ one has

$$\int |f_{n_k} - f_n|\, dt \le \varepsilon.$$

By now letting $n_k \to \infty$ we get

$$\int |f - f_n|\, dt \le \varepsilon,$$

hence (D.11.1) is true for the case $p = 1$.

The proof for the general case $1 \le p < \infty$ is similar. First choose a subsequence $\{f_{n_k}\}$ so that

$$\left(\int |f_{n_{k+1}} - f_{n_k}|^p\, dt \right)^{1/p} \le 2^{-k}$$

and define $\{g_k\}$ as above. It follows from Exercise 4, Section D.8 that the series $\sum_{k=1}^{\infty} |g_k|$ converges almost everywhere, hence the series $\sum_{k=1}^{\infty} g_k$ converges almost everywhere. If we let

$$f = \sum_{k=1}^{\infty} g_k = \lim_{k \to \infty} f_{n_k} \qquad \text{(a.e.)},$$

then we can repeat the above argument to show that $f \in L_p$ and that (D.11.1) holds. ∎

There is also a Lebesgue space for $p = \infty$, which we now define.

D.11.3 DEFINITION.   Let E be a measurable set in R. Then $L_\infty(E)$ will denote the collection of all bounded measurable functions on E. The *essential supremum* of a function $f \in L_\infty(E)$ is given by

$$\operatorname*{ess.\,sup.}_{E} |f| = \inf\{B : |f| \le B \text{ (a.e.)}\}.$$

If f is continuous, it is easy to see that ess. sup. $|f|$ agrees with sup $|f|$.

There is a completeness theorem for L_∞. We leave the proof as an exercise.

D.11.4 THEOREM.   *Let $\{f_n\}$ be a sequence in $L_\infty(E)$ with the property that*

$$\lim_{m, n \to \infty} \left(\operatorname*{ess.\,sup.}_{E} |f_n - f_m| \right) = 0.$$

Then there is an f in $L_\infty(E)$ that satisfies $f = \lim f_n$ (a.e.) and

$$\lim_{n\to\infty} \left(\text{ess. sup.}_{E} |f - f_n| \right) = 0.$$

The concepts and results of this section can be extended to complex-valued functions and even vector-valued functions. Furthermore, these results are valid for any integral that is defined using the Daniell approach. In particular, these results extend to the integral defined on general measure spaces in Example 2, Section D.10. We shall discuss these extensions in the exercises.

EXERCISES

1. Prove Theorem D.11.4. [*Hint*: Show that the set

$$\{t : f_n(t) \text{ is not a Cauchy sequence}\}$$

   is a null set.]

2. Prove Theorems D.11.2 and D.11.4 for complex-valued functions.

3. Prove Theorems D.11.2. and D.11.4 for vector-valued functions.

4. Determine f where $f = \lim f_n$ (a.e.) and

$$f_n(t) = \begin{cases} n/2, & |t| \leq 1/n, \\ 0, & |t| > 1/n. \end{cases}$$

   Does $\int |f - f_n|^p \, dt \to 0$ as $n \to \infty$ for any $p, 1 \leq p < \infty$? Does ess. sup. $|f - f_n| \to 0$ as $m \to \infty$?

5. Converse of Hölder Inequality: Assume that $\left| \int_E f(t)g(t) \, dt \right| \leq K \left(\int_E |g(t)|^q \, dt \right)^{1/q}$ for every $g \in L_q(E)$. Show that $f \in L_p(E)$, where $p^{-1} + q^{-1} = 1$, and that $\left(\int_E |f(t)|^p \, dt \right)^{1/p} \leq K$.

12. DENSE SUBSPACES OF L_p, $1 \leq p < \infty$

In the study of certain linear operators, particularly unbounded operators, one is interested in knowing whether the domain, or the range, of an operator is a dense subspace of some normed linear space. In this section we wish to prove an omnibus theorem which asserts that the space $C_0^\infty = C_0^\infty(R,R)$, the space of infinitely differentiable functions with compact support, is dense in L_p, $1 \leq p < \infty$. It is immediate then that if any subspace $\mathscr{D}$ of L_p contains C_0^∞, then $\mathscr{D}$ is dense in L_p.

Consider $L_p = L_p(-\infty,\infty)$ with the usual norm

$$\|f\|_p = \left(\int |f|^p \, dt \right)^{1/p}.$$

It is a consequence of Theorem D.7.4 and Inequality (D.7.4) that the space C_0 is dense in $L_1 = L$. It is easy to show that C_0 is dense in L_p for all p, $1 \leq p < \infty$, as we shall now see. Let p be fixed with $1 \leq p < \infty$ and let $f \in L_p$. Define f_n by

$$f_n(t) = \begin{cases} n, & \text{if } f(t) > n \\ f(t), & \text{if } -n \leq f(t) \leq n \\ -n, & \text{if } f(t) < -n. \end{cases}$$

For $n = 1, 2, \ldots$ one has $|f_n|^p \leq |f|^p$, hence $f_n \in L_p$. Furthermore, $\lim f_n = f$ (a.e.) and

$$|f - f_n|^p \leq (|f| + |f_n|)^p \leq 2^p |f|^p.$$

By the Dominated Convergence Theorem, one has

$$\|f - f_n\|_p = (\textstyle\int |f - f_n|^p \, dt)^{1/p} \to 0, \qquad \text{as} \quad n \to \infty.$$

Conclusion: *The bounded functions are dense in L_p, $1 \leq p < \infty$.*

Let f be a bounded function in L_p and let $f_n = f\chi[-n,n]$, where $\chi[-n,n]$ is the characteristic function of the interval $[-n,n]$. One then has $|f_n|^p \leq |f|^p$ hence $f_n \in L_p$ and $f = \lim f_n$ (a.e.). By repeating the reasoning of the last paragraph we can show that $\|f - f_n\| \to 0$ as $n \to \infty$.

Conclusion: *The bounded functions with compact support are dense in L_p, $1 \leq p < \infty$.*

If f is a bounded function with compact support and $f \in L_p$, then we claim that $f \in L_1$. Indeed, if $|f| \leq M$ (a.e.), where M is a positive constant, and if f vanishes outside the interval $[a,b]$, then $|f| \leq M\chi[a,b]$ (a.e.), and by Corollary D.8.5 we see that $f \in L_1$.

D.12.1 THEOREM.   *The space C_0 is dense in L_p, $1 \leq p < \infty$.*

Proof: Let $f \in L_p$ and let $\varepsilon > 0$ be given. Choose a bounded function g in L_p, where g has compact support and $\|f - g\|_p \leq \varepsilon$. Say that $|g| \leq M\chi_I$, where $I = [a,b]$ is a bounded interval. Since $g \in L_1$, it follows that we can find a sequence $\{\phi_n\}$ in C_0 such that $g = \lim \phi_n$ (a.e.). Without any loss of generality we can assume that $|\phi_n| \leq M\chi_I$. Then

$$|g - \phi_n|^p \leq (|g| + |\phi_n|)^p \leq (2M)^p\chi_I.$$

It follows from the Dominated Convergence Theorem that

$$\|g - \phi_n\|_p \to 0 \quad \text{as} \quad n \to \infty.$$

Now choose n so that $\|g - \phi_n\|_p \leq \varepsilon$. It then follows from the Minkowski Inequality that $\|f - \phi_n\|_p \leq 2\varepsilon$.   ∎

Let us now prove the main theorem of this section. Let C_0^∞ denote the space of real-valued infinitely differentiable functions with compact support.

D.12.2 THEOREM.   *The space C_0^∞ is dense in L_p, $1 \leq p < \infty$.*

Proof: Let $f \in L_p$ and let $\varepsilon > 0$ be given. Choose $\phi \in C_0$ so that $\|f - \phi\|_p \leq \varepsilon$. For any $n = 1, 2, \ldots$, let $\psi_n \in C_0^\infty$ that satisfies $0 \leq \psi_n$, $\psi_n(t) = 0$ for $|t| \geq 1/n$ and $\int \psi_n \, dt = 1$. For example, we might take

$$\psi_n(t) = \begin{cases} 0, & \text{if } |t| \geq 1/n, \\[2mm] C \exp\left\{\dfrac{1}{|nt|^2 - 1}\right\}, & \text{if } |t| \leq 1/n, \end{cases}$$

where C is an appropriate constant. Now set

$$\phi_n(t) = \int_{-\infty}^{\infty} \phi(s)\psi_n(t-s)\,ds = \int_{-\infty}^{\infty} \phi(t-s)\psi_n(s)\,ds = \int_{-1/n}^{1/n} \phi(t-s)\psi_n(s)\,ds.$$

It follows from the first equality that ϕ_n is infinitely differentiable. Since ϕ has compact support we see that $\phi_n \in C_0^{\infty}$.

If $|\phi| \le M$, then

$$|\phi_n(t)| \le \int_{-1/n}^{1/n} |\phi(t-s)|\psi_n(s)\,ds \le M\int_{-1/n}^{1/n} \psi_n(s)\,ds = M.$$

Next we claim that $\phi = \lim \phi_n$. To prove this we shall compute

$$\phi(t) - \phi_n(t) = \phi(t) - \int_{-1/n}^{1/n} \phi(t-s)\psi_n(s)\,ds$$

$$= \int_{-1/n}^{1/n} [\phi(t) - \phi(t-s)]\psi_n(s)\,ds.$$

Since ϕ is continuous at t, for every $\varepsilon > 0$ we can find an N such that

$$|\phi(t) - \phi(t-s)| \le \varepsilon$$

whenever $|s| \le 1/N$. It follows, then, that for $n \ge N$ one has $|\phi(t) - \phi_n(t)| \le \varepsilon$.

It then follows from the Dominated Convergence Theorem that $\|\phi - \phi_n\|_p \to 0$ as $n \to \infty$. If we choose n so that $\|\phi - \phi_n\|_p \le \varepsilon$, then the Minkowski Inequality assures us that for this n we have $\|f - \phi_n\| \le 2\varepsilon$. ∎

EXERCISES

1. The space L_{∞} was omitted for good reason. Show that C_0 is not dense in L_{∞}.

2. Let $C_0^{\infty}[a,b]$ denote the space of all infinitely differentiable functions f with the property that f and all its derivatives vanish at a and b. Show that $C_0^{\infty}[a,b]$ is dense in $L_p[a,b]$ for $1 \le p < \infty$.

3. Let $C_0^{\infty}(R^n)$ denote the space of all infinitely differentiable real-valued functions $u(x_1,\ldots,x_n)$ with compact support. Show that $C_0^{\infty}(R^n)$ is dense in $L_p(R^n)$, $1 \le p < \infty$.

4. Let Ω denote an open set in R^n and let $C_0^{\infty}(\Omega)$ denote the space of all infinitely differentiable real-valued functions $u(x_1,\ldots,x_n)$ with compact support inside Ω. Show that $C_0^{\infty}(\Omega)$ is dense in $L_p(\Omega)$, $1 \le p < \infty$.

13. DIFFERENTIATION

In this section we shall study the relationship between F and f when F and f satisfy

$$F(t) = F(a) + \int_a^t f(s)\,ds, \qquad a \le t \le b, \tag{D.13.1}$$

and $f \in L_1[a,b]$. If f is continuous, then the Fundamental Theorem of Calculus tells us that

$$\frac{dF}{dt} = f \quad \text{on} \quad a \le t \le b. \tag{D.13.2}$$

Conversely, if F is any C^1-function that satisfies (D.13.2), then F is given by (D.13.1). Our interest here is in determining what happens when f is not continuous.

Our discussion will be a bit sketchy and we will leave a number of nontrivial details for the reader to check. The goal is to give a complete characterization of functions F that satisfy (D.13.1) for some f in $L_1[a,b]$. We shall see that F satisfies (D.13.1) if and only if F is absolutely continuous. Furthermore, we shall see that if F is absolutely continuous, then dF/dt exists almost everywhere, and if $dF/dt = f$ (a.e.), then F and f satisfy (D.13.1).

D.13.1 DEFINITION. Let I be a compact interval. A function F is said to be *absolutely continuous on I* if for every $\varepsilon > 0$ there is a $\delta > 0$ such that whenever $I_k = [a_k, b_k]$ are nonoverlapping intervals in I with $\sum_{k=1}^{n} |b_k - a_k| \le \delta$, one has $\sum_{k=1}^{n} |F(b_k) - F(a_k)| \le \varepsilon$. If I is an arbitrary interval, we say that F is *absolutely continuous on I* if F is absolutely continuous on every compact subinterval.

We can now make two observations. First, every absolutely continuous function is continuous, and second, every continuously differentiable function is absolutely continuous. That absolute continuity implies continuity follows directly from the definitions. The second statement is a special case of Theorem D.13.3 given below.

We will need the following lemma.

D.13.2 LEMMA. *Let F satisfy a Lipschitz condition on I (that is, there is a constant $K \ge 0$ such that $|F(t) - F(s)| \le K|t - s|$ for all t and s in I). Then F is absolutely continuous.*

Proof: Since

$$\sum_{k=1}^{n} |F(b_k) - F(a_k)| \le K \sum_{k=1}^{n} |b_k - a_k|,$$

we see that if

$$\sum_{k=1}^{n} |b_k - a_k| \le \varepsilon K^{-1} = \delta,$$

then

$$\sum_{k=1}^{n} |F(b_k) - F(a_k)| \le \varepsilon. \quad \blacksquare$$

This lemma can be generalized to the case where F satisfies a local Lipschitz condition on I, which means that for every compact set $J \subset I$, there is a $K \ge 0$ such that

$$|F(t) - F(s)| \le K|t - s|$$

for all t and s in J. (In this case, K gets larger as the interval J gets larger.) By the last lemma, we see that F is absolutely continuous on every compact interval $J \subset I$, so F is absolutely continuous on I.

The following theorem completely characterizes the class of absolutely continuous functions. It also gives the counterpart of the Fundamental Theorem of Calculus. We will not prove the entire theorem here since this would require techniques which are beyond the scope of this book. For a complete discussion of this theorem we refer the reader to Riesz and Sz. Nagy [1, pp. 50–54] and Royden [1, pp. 80–92].

D.13.3 THEOREM.   *A function F defined on I is absolutely continuous if and only if F satisfies (D.13.1) for some $f \in L_1[a,b]$. Moreover, under this condition, F' exists a.e. and $F' = f$ (a.e.).*

Proof: We will prove here that if F can be expressed by (D.13.1) then F is absolutely continuous. Let J be a compact interval in I. If f is bounded on J, say $|f(t)| \leq B$ (a.e.) on J, then F satisfies a Lipschitz condition on J, since for t and s in J with $t \geq s$, one has

$$|F(t) - F(s)| \leq \int_s^t |f(u)| du \leq B|t - s|.$$

Hence by Lemma D.13.2, F is absolutely continuous. If f is not bounded on J, then for every $\varepsilon > 0$ we can write $f = g + h$, where g is bounded on J (say that $|g| \leq n$ (a.e.)) and $\int_J |h| \, dt < \varepsilon/2$, by Section D.12. One then has

$$|F(t) - F(s)| \leq \int_s^t |g(u)| \, du + \int_s^t |h(u)| \, du$$

$$\leq n|t - s| + \frac{\varepsilon}{2}.$$

Now set $\delta = \varepsilon(2n)^{-1}$. If $I_1, \ldots, I_n$ are disjoint intervals in J with $\sum_{k=1}^n |b_k - a_k| \leq \delta$, then $\sum_{k=1}^n |f(b_k) - f(a_k)| \leq n\delta + \varepsilon/2 = \varepsilon$. Hence f is absolutely continuous. ∎

We omit the rest of the proof which can be found in the references cited above. However a word of caution to the reader is necessary. There do exist continuous functions F with the property that dF/dt exists almost everywhere but

$$F(t) \neq F(a) + \int_a^t \frac{dF}{ds}(s) \, ds.$$

In fact there does exist a *strictly* increasing continuous function F with the property that $dF/dt = 0$ (a.e.), see Riesz and Sz. Nagy [1, p. 48]. Needless to say such a function is not absolutely continuous.

EXERCISES

1. If $a \leq b$ we define $\int_b^a f(s) \, ds = -\int_a^b f(s) \, ds$. Show that if $F(t) = \int_t^b f(s) \, ds$ for $a \leq t \leq b$, where $f \in L_1[a,b]$, then $F' = -f$ (a.e.).

2. (Differential Equations) Let $f(x,t)$ be a measurable function that is continuous in x for each t. Assume there is an $m \in L_1[a,b]$ such that $|f(x,t)| \le m(t)$ for all x and t.

(a) Show that if $\phi(t)$, $a \le t \le b$, is an absolutely continuous function that satisfies

$$\phi'(t) = f(\phi(t),t) \quad \text{(a.e.)}, \quad \text{and} \quad \phi(a) = \phi_0, \tag{D.13.3}$$

then

$$\phi(t) = \phi_0 + \int_a^t f(\phi(s),s) \, ds, \ a \le t \le b. \tag{D.13.4}$$

(b) Conversely, show that if ϕ satisfies (D.13.4), then ϕ is an absolutely continuous function that satisfies (D.13.3).

14. THE RADON-NIKODYM THEOREM

The Radon-Nikodym Theorem is simply an extension of the results of the last section to arbitrary measures. We shall present this theorem here without proof. For our purpose, this theorem is used primarily to discuss the conditional expectation operator in Section E.5.

Let X be a nonempty set and let $\mathscr{F}$ be a σ-field of subsets in X, as in Example 2, Section 10. Recall that a *measure* on X is a mapping μ of $\mathscr{F}$ into the extended real numbers $\tilde{R} = R \cup \{+\infty\} \cup \{-\infty\}$ such that $\mu(\varnothing) = 0$ and

$$\mu\left(\bigcup_{n=1}^{\infty} E_n\right) = \sum_{n=1}^{\infty} \mu(E_n) \tag{D.14.1}$$

whenever $\{E_1, E_2, \ldots\}$ is a sequence of mutually disjoint sets in $\mathscr{F}$. Earlier we had considered only positive measures, but now we wish to consider measures μ of the form[4]

$$\mu = \mu^+ - \mu^-,$$

where μ^+ and μ^- are positive measures. The absolute value of μ is then the positive measure

$$|\mu| = \mu^+ + \mu^-.$$

We shall say that the measure μ is *σ-finite* on X if there is a sequence $\{E_1, E_2, \ldots\}$ of measurable sets that satisfy $|\mu|(E_n) < \infty$ and $\bigcup_{n=1}^{\infty} E_n = X$. If $|\mu|(X) < \infty$, then we say that μ is a *finite* measure.

We note that the Lebesgue measure is σ-finite on R and finite on any bounded interval.

Now let μ and ν be two measures on $(X, \mathscr{F})$. We shall say that ν is *absolutely continuous* with respect to μ if $|\nu|(E) = 0$ whenever $|\mu|(E) = 0$.

---

[4] One can show that every measure has this form, see Royden [1, pp. 202–206].

EXAMPLE 1.   Let $f \in L$ and let $\mathscr{F} = \mathscr{B}$ denote the Borel sets in R. Let μ denote the Lebesgue measure on R and for $E \in \mathscr{B}$ define $v(E)$ by

$$v(E) = \int_E f \, dt = \int f \chi_E \, dt. \qquad (D.14.2)$$

If $f = f^+ - f^-$, then we see that

$$v^+(E) = \int_E f^+ \, dt, \qquad v^-(E) = \int_E f^- \, dt.$$

Furthermore $v(\varnothing) = 0$. Also, if $\{E_1, E_2, \ldots\}$ is a sequence of mutually disjoint sets, then

$$\left| \sum_{n=1}^N f \chi_{E_n} \right| \le \sum_{n=1}^N |f| \chi_{E_n} \le |f|.$$

So by the Dominated Convergence Theorem one has

$$v \left(\bigcup_{n=1}^\infty E_n \right) = \int f \left(\sum_{n=1}^\infty \chi_{E_n} \right) dt = \sum_{n=1}^\infty \int f \chi_{E_n} \, dt$$

$$= \sum_{n=1}^\infty v(E_n).$$

Hence v is a measure on $\mathscr{B}$. Finally, we note that v is absolutely continuous with respect to μ.

The Radon-Nikodym Theorem, which we state next, tells us that every finite measure v that is absolutely continuous with respect to the Lebesgue measure μ must satisfy (D.14.2) for some f in L. ∎

D.14.1 THEOREM. (RADON-NIKODYM.)   *Let μ be a σ-finite measure on $(X, \mathscr{F})$ and let v be a measure that is absolutely continuous with respect to μ. Then there is a measurable function f such that*

$$v(E) = \int_E f(x) \mu(dx) \qquad (D.14.3)$$

for all $E \in \mathscr{F}$. Furthermore v is finite if and only if $f \in L_1(X, \mathscr{F}, \mu)$.

The function f appearing in (D.14.3) is called the *Radon-Nikodym derivative* of v with respect to μ.

EXAMPLE 2.   One can reduce the theory of the last section to a special case of the Radon-Nikodym Theorem. We shall illustrate this reduction by considering a monotone increasing function F that is absolutely continuous.

One can construct a measure v on the Borel sets $\mathscr{B}$ by using F. First, for each interval E with end points $\{a, b\}$, where $a < b$, we let

$$v(E) = F(b) - F(a).$$

Next we extend v to countable unions of intervals by using (D.14.1). Continuing this way, by repeated use of (D.14.1), we can extend v to all of $\mathscr{B}$.

Finally the fact that the measure v is absolutely continuous with respect to the Lebesgue measure μ is a direct consequence of the definition of absolute continuity for functions and the characterization of a null set appearing in Exercise 1, Section 5. ∎

15.   FUBINI THEOREM

We close this appendix with an abbreviated statement of the Fubini Theorem, which gives conditions under which one can interchange the order of integration.

For this we are interested in functions of the form $f(x,y)$ defined on $A \times B$ and we ask: When is it true that

$$\int_A \left[\int_B f(x,y)\, dy \right] dx = \int_B \left[\int_A f(x,y)\, dx \right] dy? \qquad (D.15.1)$$

One can show (Royden [1, pp. 233–234]) that if $f(x,y)$ is a measurable function that satisfies either one of the conditions:

(i) $f \in L_1(A \times B)$;
(ii) $f \geq 0$;

then (D.15.1) is valid.

SUGGESTED REFERENCES

Asplund and Bungart [1]   Royden [1]
Halmos [1]   Rudin [1]
Loomis [1]

Appendix E

Probability Spaces
and Stochastic Processes

1. PROBABILITY SPACES

Probability spaces are the basic mathematical models for random phenomena. Intuitively the situation is as follows. Suppose one is given a collection of random phenomena. One then pictures that off someplace in the background is a nonempty set Ω, which is called the *sample space*. On each experiment, the "forces that be" pick one of the *sample points* $\omega \in \Omega$. This sample point ω, in turn, determines the values associated with all the random phenomena in that particular experiment.

In addition one is also interested in certain *events*, which are merely subsets, or perhaps, distinguished subsets of Ω. The relationship $\omega \in A$ means that the event A "occurred" during the given experiment. The *probability* that the event A will occur is then a real number $P(A)$ satisfying $0 \leq P(A) \leq 1$.

Let us now make this intuitive picture more precise. For this purpose, we will assume that the reader is familiar with Appendix D, in general, and Section D.10, in particular.

A *probability space* is a triple $(\Omega, \mathscr{F}, P)$, where Ω is a nonempty set, $\mathscr{F}$ is a σ-field of subsets of Ω and P is a positive measure with $P(\Omega) = 1$. In this case P is sometimes called a *probability measure*.

Before we look at some examples of probability spaces let us list here a partial dictionary relating certain measure-theoretic and probability-theoretic terms

Probability Space	Measure Space
Sample Point	Point in the Space
Event	Measurable Set
Probability	Measure
Sure Event	Whole Space
Impossible Event	Empty Set
Event with Probability Zero	Set of Measure Zero
Almost Sure	Almost Everywhere
Random Variable	Measurable Function
Expectation	Integral

EXAMPLE 1.   Let Ω denote the collection of all possible initial moves for white in a chess game. Thus Ω consists of 20 points. The reader should easily think of several candidates for probability measures on Ω. ∎

EXAMPLE 2.   Let Ω denote the collection of all possible outcomes of an experiment involving 50 flips of a coin. A typical sample point, then, would be an ordered

50-tuple $(H,H,T,\ldots,H)$ consisting of heads (H) and tails (T). Let p denote the probability of getting H on any toss and $q = 1 - p$ the probability of getting T. Let A be the event consisting of all outcomes with n heads and $50 - n$ tails. It is well known that

$$P(A) = \frac{50!}{(50 - n)!\, n!}\, p^n q^{50-n}. \quad \blacksquare$$

EXAMPLE 3. Let $\Omega = R$ and let $f \in L_2(-\infty,\infty)$ with $\int_{-\infty}^{\infty} |f|^2\, dt = 1$. Then

$$P(A) = \int_A |f|^2\, dt$$

is a probability measure on Ω. This type of probability space occurs in the study of quantum mechanics. $\blacksquare$

EXAMPLE 4. Let $\Omega = R^2$ and using polar coordinates define $f: R^2 \to R$ by

$$f(r,\theta) = \left[\sqrt{2}\,\pi^{3/2} r \exp\left(\frac{r^2}{2}\right)\right]^{-1}.$$

Then

$$P(A) = \iint_A f\, dA$$

is a probability measure on Ω. $\blacksquare$

2. RANDOM VARIABLES AND DISTRIBUTION FUNCTIONS

A probability space is an abstraction that appears only implicitly in many applications. The concrete formulation of these problems is usually in terms of certain random variables and their distributions.

A *(real) random variable* is simply a real-valued measurable function defined on the sample space Ω. (See Example 2, Section D.10.) A *(complex) random variable* is a complex-valued function whose real and imaginary parts are real random variables.

Although the definition of a random variable is in terms of a function defined on the sample space, in many applications one is oftentimes only interested in the distribution function of a random variable.

Let X be a real random variable defined on a probability space $(\Omega,\mathscr{F},P)$. The *(probability) distribution function* of X, denoted by $F(x)$, is

$$F(x) = P[X(\omega) \le x].$$

That is, $F(x)$ is the probability of the event

$$A = \{\omega \in \Omega: X(\omega) \le x\}.$$

For complex random variables, both the real and imaginary parts have probability distribution functions.

EXAMPLE 1. Let Ω denote the 36 possible outcomes of rolling two dice. Assume that each outcome is equally likely, so that the probability measure is merely $P(A) = \text{card}(A)/36$. Now let X be the random variable that assigns to each outcome the total points on the two dice. The distribution function for x is illustrated in Figure E.2.1. ∎

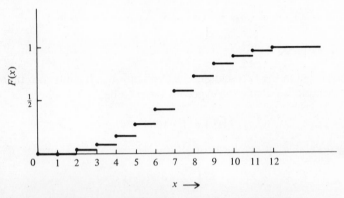

Figure E.2.1. Distribution Function for Example 1.

Let us note a few properties of the distribution function. First we note that $0 \leq F(x) \leq 1$ for all x and $F(-\infty) = 0$ and $F(\infty) = 1$. Also, if $x \leq y$, then

$$\{\omega \in \Omega: X(\omega) \leq x\} \subseteq \{\omega \in \Omega: X(\omega) \leq y\}.$$

Hence $F(x) \leq F(y)$, that is, F is monotone increasing. Also, if $x_1 < x_2$, then

$$P[x_1 < X(\omega) \leq x_2] = F(x_2) - F(x_1).$$

If, in addition, $F(x)$ is absolutely continuous (see Section D.13), then there is a function $f \in L_1(-\infty,\infty)$ such that

$$F(x) = \int_{-\infty}^{x} f(t)\, dt$$

for all x. In this case we say that f is the (*probability*) *density function* for the random variable X.

EXAMPLE 2. Consider $\Omega = R$. Let $g \in L_2(-\infty,\infty)$, with $\int_{-\infty}^{\infty} |g|^2\, dt = 1$, and let

$$P(A) = \int_{A} |g|^2\, dt$$

be the probability measure on Ω. Let $X(t) = t$ be a random variable on Ω. The distribution function for X is

$$F(x) = P[t \leq x] = \int_{-\infty}^{x} |g|^2\, dt$$

and the density function is $|g(x)|^2$. We ask the reader to compute the distribution and density functions for the random variable $X(t) = t^3$. ∎

Let X and Y be two real random variables on Ω. The *joint (probability) distribution function* is defined by

$$F(x,y) = P[X(\omega) \leq x \text{ and } Y(\omega) \leq y].$$

Similarly the *joint distribution function* for the real random variables $\{X_1,\ldots,X_n\}$ is given by

$$F(x_1,x_2,\ldots,x_n) = P[X_1 \leq x_1, X_2 \leq x_2,\ldots,X_n \leq x_n].$$

3.   EXPECTATION

As suggested in the first section, the expectation is the integral. Indeed, if X is a random variable we define the expected value to be $E[X]$, where[1]

$$E[X] = \int_\Omega X(\omega)P(d\omega).$$

The expected value may not exist for all random variables. However, if

$$E[|X|] = \int_\Omega |X(\omega)|P(d\omega) < \infty,$$

then $E[X]$ is finite and well defined. Thus, if X belongs to $L_1(\Omega,\mathscr{F},P)$, then X has a finite expected value.

A real random variable X in $L_2(\Omega,\mathscr{F},P)$, that is, X satisfies $E[|X|^2] < \infty$, is said to have a *finite second moment*. Since[2] $L_2(\Omega,\mathscr{F},P) \subset L_1(\Omega,\mathscr{F},P)$, we see that if X has a finite second moment, then it has a finite expected value. In this case, we define the *variance* $\sigma^2(X)$ by

$$\sigma^2(X) = E[|X - E[X]|^2]$$

and $\sigma(X)$ is called the *standard deviation* of X.

If X and Y are two real random variables in $L_2(\Omega,\mathscr{F},P)$, then the covariance of X and Y is defined by

$$\mathrm{Cov}(X,Y) = E[(X - \mu_x)(Y - \mu_y)],$$

where $\mu_x = E[X]$ and $\mu_y = E[Y]$. The *correlation coefficient* is, then,

$$\rho(X,Y) = \frac{\mathrm{Cov}(X,Y)}{\sigma_x \sigma_y},$$

where σ_x and σ_y denote the standard deviation of X and Y, respectively.

EXERCISES

1. Show that $L_2(\Omega,\mathscr{F},P) \subset L_1(\Omega,\mathscr{F},P)$.

2. Let X be a complex random variable in $L_2(\Omega,\mathscr{F},P)$. Show that

$$E[|X - E[X]|^2] = E[|X|^2] - |E[X]|^2.$$

---

[1] Again we assume familiarity with the integral as constructed in Appendix D, especially in Example 2, Section D.10.
[2] See Exercise 1.

3. Let X and Y be real random variables in $L_2(\Omega,\mathscr{F},P)$. Show that

$$\mathrm{Cov}(X,Y) = E[XY] - \mu_x\mu_y,$$

where $\mu_x = E[X]$ and $\mu_y = E[Y]$.

4. STOCHASTIC INDEPENDENCE

Let X and Y be real random variables on a probability space $(\Omega,\mathscr{F},P)$. We wish to introduce the concept of stochastic independence for these random variables. Roughly, this means the values for X should somehow be independent of the values for Y. This concept can be made precise by using the distribution functions. Specifically, we say that X and Y are (*stochastically*) *independent* if

$$P[X(\omega) \le x \text{ and } Y(\omega) \le y] = P[X(\omega) \le x] \cdot P[Y(\omega) \le y] \qquad (\text{E.4.1})$$

for all x and y. Similarly, we say that a finite collection of real random variables $\{X_1,\dots,X_n\}$ is (*stochastically*) *independent* if

$$P[X_1 \le x_1, \dots, X_n \le x_n] = P[X_1 \le x_1] \cdots P[X_n \le x_n]$$

for all $x_1, \dots, x_n$. In general, an arbitrary collection of real random variables is said to be (*stochastically*) *independent* if every finite subcollection is independent.

We say that two complex random variables X and Y are (*stochastically*) *independent* if each of the following four sets of real random variables are independent:

$$\{\mathrm{Re}\ X, \mathrm{Re}\ Y\}, \qquad \{\mathrm{Re}\ X, \mathrm{Im}\ Y\},$$
$$\{\mathrm{Im}\ X, \mathrm{Re}\ Y\}, \qquad \{\mathrm{Im}\ X, \mathrm{Im}\ Y\}.$$

In a similar way we define stochastic independence for finite and infinite collections of complex random variables.

There is one very important fact concerning independent random variables.

E.4.1 THEOREM. *Let X and Y be stochastically independent random variables in $L_2(\Omega,\mathscr{F},P)$. Then*

$$E[XY] = E[X]E[Y] \quad and \quad E[X\bar{Y}] = E[X]\overline{E[Y]}.$$

We shall not prove this theorem here since it involves concepts not fully developed in this book. However, a proof can be found in Loeve [1].

Before we look at some examples let us note that if X and Y are real random variables, then they are stochastically independent if and only if

$$P[X(\omega) > x \text{ and } Y(\omega) > y] = P[X(\omega) > x]P[Y(\omega) > y] \qquad (\text{E.4.2})$$

for all x and y.

EXAMPLE 1. Let A and B be two events in Ω, and let X_A and X_B denote their characteristic functions, that is,

$$X_A(\omega) = 1, \qquad \omega \in A$$
$$= 0, \qquad \omega \notin A.$$

We claim that the random variables X_A and X_B are independent if and only if

$$P(A \cap B) = P(A)P(B). \tag{E.4.3}$$

Indeed if (E.4.2) is valid for all x and y, then for $x < 1$ and $y < 1$ one has

$$P(A \cap B) = P[X_A(\omega) > x \text{ and } X_B(\omega) > y]$$
$$= P[X_A(\omega) > x]P[X_B(\omega) > y] = P(A)P(B).$$

Conversely, if (E.4.3) is valid, then it is easy to see that by reversing the above reasoning (E.4.2) holds whenever $x < 1$ and $y < 1$. However, the cases where $x \geq 1$ or $y \geq 1$ are trivially checked. ∎

EXAMPLE 2. Let X and Y be independent random variables and let u and v be two measurable functions. Then $u(X)$ and $v(Y)$ are independent random variables. ∎

We caution the reader not to confuse the concept of stochastic independence with the concept of linear independence introduced in Chapter 4.

EXERCISES

1. Let X and Y be nonzero stochastically independent random variables in $L_2(\Omega, \mathscr{F}, P)$.
   (a) Show that $\text{Cov}(X, Y) = 0$.
   (b) Show that if $E[X] = E[Y] = 0$, then X and Y are linearly independent.
   [*Hint*: Let $Z = aX + bY = 0$, then compute $\text{Cov}(Z, X)$ and $\text{Cov}(Z, Y)$.]
2. Construct two linearly independent random variables X and Y such that $E[X] = E[Y] = 0$ and where X and Y are not stochastically independent.
3. Let $\{X_1, \ldots, X_n\}$ be a collection of stochastically independent random variables. Let $U = g(X_1, \ldots, X_m)$ and $V = h(X_{m+1}, \ldots, X_n)$, where $1 \leq m \leq n$. Show that U and V are stochastically independent.

5. CONDITIONAL EXPECTATION OPERATOR

Let $(\Omega, \mathscr{F}, P)$ be a probability space and let $\mathscr{B}$ be a subcollection of $\mathscr{F}$. One says that $\mathscr{B}$ is a *sub-σ-field* if $\mathscr{B}$ itself is a σ-field. Let us now look at a few examples.

EXAMPLE 1. $\mathscr{B} = \{\varnothing, \Omega\}$. ∎

EXAMPLE 2. $\mathscr{B} = \mathscr{F}$. ∎

Of course, these two examples represent the extreme cases. All other sub-σ-fields lie in between.

EXAMPLE 3. Let $A \in \mathscr{F}$. Then $\mathscr{B} = \{\varnothing, A, A', \Omega\}$ is the σ-field generated by A. ∎

EXAMPLE 4. Let $A, B \in \mathcal{F}$, where $A \cap B = \emptyset$, and let

$$C = (A \cup B)' = A' \cap B'.$$

Then

$$\mathcal{B} = \{\emptyset, A, B, C, A', B', C', \Omega\}$$

is the sub-σ-field generated by A and B, that is, $\mathcal{B}$ is the smallest σ-field containing A and B. (Describe the sub-σ-field generated by A and B when one does not assume that $A \cap B = \emptyset$.) ∎

The sub-σ-fields we shall be primarily interested in are described in the following example.

EXAMPLE 5. Let Y be a real random variable on $(\Omega, \mathcal{F}, P)$. This means that for every Borel set A in R, the set $Y^{-1}(A) \in \mathcal{F}$. Now let $\mathcal{B}$ be the smallest σ-field in $\mathcal{F}$ that contains all events of the form $Y^{-1}(A)$, where A is a Borel set in R. Then $\mathcal{B}$ is said to be the *(sub)-σ-field generated* by Y.

For example, if $Y = X_A$ is the characteristic function of an event A, then $\mathcal{B}$ is as given in Example 3. Or, if $Y = \alpha X_A + \beta X_B$, where $0 < \alpha < \beta$ and $A \cap B = \emptyset$, then $\mathcal{B}$ is given in Example 4. ∎

We can now define conditional expectation. First let $\mathcal{B}$ be a sub-σ-field of $\mathcal{F}$ and let $X \in L_1(\Omega, \mathcal{F}, P)$. Then define $v(B)$ by

$$v(B) = \int_B X(\omega) P(d\omega), \qquad B \in \mathcal{B}.$$

It is easy to see that v is a measure on $\mathcal{B}$ and that it is absolutely continuous with respect to $P_\mathcal{B}$, which is the restriction of P to $\mathcal{B}$. So by that Radon-Nikodym Theorem there is a (unique) random variable $E^\mathcal{B}[X]$ that is measurable[3] with respect to $\mathcal{B}$ and that satisfies

$$v(B) = \int_B E^\mathcal{B}[X](\omega) P_\mathcal{B}(d\omega) = \int_B X(\omega) P(d\omega), \qquad \text{(E.5.1)}$$

or, as it is sometimes written,

$$\int_B E^\mathcal{B}[X]\, dP_\mathcal{B} = \int_B X\, dP.$$

This random variable $E^\mathcal{B}[X]$ is called the *conditional expectation of X with respect to $\mathcal{B}$*. If $\mathcal{B}$ is the σ-field generated by a random variable Y, then we shall denote $E^\mathcal{B}[X]$ by $E^Y[X]$ and call it the *conditional expectation of X with respect to Y*.

---

[3] There is subtlety here which should not be overlooked. The first random variable X is measurable with respect to $\mathcal{F}$, whereas $E^\mathcal{B}[X]$ is measurable with respect to $\mathcal{B}$. That is, $X^{-1}(A) \in \mathcal{F}$ and $E^\mathcal{B}[X]^{-1}(A) \in \mathcal{B}$ for all Borel sets in R. If X itself is measurable with respect to $\mathcal{B}$, then the uniqueness part of the Radon-Nikodym Theorem would tell us that $X = E^\mathcal{B}[X]$.

EXAMPLE 6.   If $\mathcal{B} = \{\varnothing, \Omega\}$, then $E^{\mathcal{B}}[X] = E[X]$, that is, $E^{\mathcal{B}}$ maps X onto the constant function $E[X]$.   ∎

EXAMPLE 7.   If $\mathcal{B} = \mathcal{F}$, then $E^{\mathcal{B}}[X] = X$.   ∎

EXAMPLE 8.   If $\mathcal{B} = \{\omega, A, A', \Omega\}$ and if $0 < P(A) < 1$, then

$$E^{\mathcal{B}}[X](\omega_0) = \frac{1}{P(A)} \int_A X(\omega) P(d\omega), \qquad \text{if } \omega_0 \in A,$$

$$= \frac{1}{P(A')} \int_{A'} X(\omega) P(d\omega), \qquad \text{if } \omega_0 \in A'. \quad ∎$$

EXERCISES

1. Let $X = c = $ constant (a.e.), that is, almost everywhere. Show that
$$E^{\mathcal{B}}[X] = c \text{ (a.e.).}$$

2. Show that $E^{\mathcal{B}}$ is a linear operator on $L_1(\Omega, \mathcal{F}, P)$.

3. Show that if $X \le Y$ (a.e.), then $E^{\mathcal{B}}[X] \le E^{\mathcal{B}}[Y]$ (a.e.).

4. Show that
$$-E^{\mathcal{B}}[|X|] \le E^{\mathcal{B}}[X] \le E^{\mathcal{B}}[|X|] \qquad \text{(a.e.).}$$

5. Show that $|E^{\mathcal{B}}[X]| \le E^{\mathcal{B}}[|X|]$ (a.e.) and hence, $E^{\mathcal{B}}$ maps $L_1(\Omega, \mathcal{F}, P)$ into $L_1(\Omega, \mathcal{B}, P)$.

6. Show that $E^{\mathcal{B}}$ is a projection on $L_1(\Omega, \mathcal{F}, P)$.

7. Show that $E^{\mathcal{B}}$ maps $L_2(\Omega, \mathcal{F}, P)$ into $L_2(\Omega, \mathcal{F}, P)$ and that E is a projection on this space.

8. (Dominated Convergence.) Assume that $\lim X_n = X$ (a.e.), where $|X_n| \le Y$ (a.e.). Show that if $Y \in L_1(\Omega, \mathcal{F}, P)$, then $\lim E^{\mathcal{B}}[X_n] = E^{\mathcal{B}}[X]$ (a.e.).

9. Show that
$$\int_\Omega E^{\mathcal{B}}[|X|^2] \, dP_{\mathcal{B}} = \int_\Omega |X|^2 \, dP \tag{E.5.2}$$

for all X in $L_2(\Omega, \mathcal{F}, P)$. [*Hint*: First show that (E.5.2) holds for simple functions and then pass to the limit.]

10. Let $X, Y \in L_2(\Omega, \mathcal{F}, P)$ and assume that X is measurable with respect to $\mathcal{B}$. Show that
$$E^{\mathcal{B}}[XY] = X E^{\mathcal{B}}[Y]. \tag{E.5.3}$$

[*Hint*: First show that (E.5.3) holds when X is a simple function and then pass to the limit.]

11. Let $X,\ Y \in L_2(\Omega, \mathscr{F}, P)$. Show that

$$E^{\mathscr{B}}[E^{\mathscr{B}}[X]\,Y] = E^{\mathscr{B}}[X]\,E^{\mathscr{B}}[Y].$$

12. Discuss $E^{\mathscr{B}}[X]$ when X is complex-valued. Show that the above properties are valid even in this case.

6. STOCHASTIC PROCESSES

A *stochastic process* is simply a family of random variables X_t, where t lies in some index set. If t belongs to some interval on the real line, this is referred to as a continuous process. If t ranges over some countable set, this is referred to as a discrete process. In the later case one often writes X_n in place of X_t.

EXAMPLE 1. For t in the interval I, let $X(t, \omega)$ be a random variable with finite second moment, that is

$$E[|X(t, \cdot)|^2] = \int_{\Omega} |X(t, \omega)|^2 P(d\omega) < \infty.$$

Let us abbreviate the notation and write $X(t)$ in place of $X(t, \omega)$.

One says that $X(t)$ is continuous in t if for every $\varepsilon > 0$ there is a $\delta > 0$ such that if $|h| \le \delta$, then

$$E[|X(t) - X(t + h)|^2] \le \varepsilon^2.$$

Assume that $E[X(t)] = 0$ for all t in I. Now define the covariance function by

$$r(t,s) = E[X(t)\overline{X(s)}].$$

We leave it to the reader to show that $r(t,s)$ is continuous in t and s when $X(t)$ is continuous in t. ∎

SUGGESTED REFERENCES

Doob [1]

Feller [1]

Halmos [1]

Kolmogorov [1]

Loeve [1]

References

S. AGMON
  [1] *Lectures on Elliptic Boundary Value Problems*. Princeton, N.J.: D. Van Nostrand Company, Inc., 1965.

N. I. AKHIEZER AND I. M. GLAZMAN
  [1] *Theory of Linear Operators in Hilbert Space*. New York: Frederick Ungar Publishing Co., 1961.

N. ARONSZAJN
  [1] "Approximation techniques for eigenvalues of completely continuous symmetric operators." *Proceedings of the Symposium on Spectral Theory and Differential Problems*. Stillwater: Oklahoma College, 1951, pp. 179–202.

R. B. ASH
  [1] *Information Theory*. New York: John Wiley & Sons, Inc., 1965.

E. ASPLUND AND L. BUNGART
  [1] *A First Course in Integration*. New York: Holt, Rinehart and Winston, Inc., 1966.

G. BACHMAN AND L. NARICI
  [1] *Functional Analysis*. New York: Academic Press, Inc., 1966.

S. BANACH
  [1] *Theorie des operations lineaires*. New York: Chelsea Publishing Company, 1955.

R. G. BARTLE
  [1] *The Elements of Real Analysis*. New York: John Wiley & Sons, Inc., 1964.

L. BERS, F. JOHN, AND M. SCHECHTER
  [1] *Partial Differential Equations*. New York: Interscience Publishers, 1964.

A. S. BESICOVITCH
  [1] *Almost Periodic Functions*. New York: Dover Publications, Inc., 1954.

F. BEUTLER
  [1] "Sampling theorems and bases in a Hilbert space," *Information and Control*, **4**, No. 2–3 (Sept. 1961), 97–117.

P. BILLINGSLEY
  [1] *Ergodic Theory and Information*. New York: John Wiley & Sons, Inc., 1965.
  [2] *Convergence of Probability Measures*. New York: John Wiley & Sons, Inc., 1968.

R. P. BOAS, JR.
  [1] *A Primer of Real Functions*. Washington, D.C.: Math. Assn. of America, 1960.

609

S. BOCHNER

[1] *Fourier Transforms*. Princeton, N.J.: Princeton University Press, 1949.

F. E. BROWDER

[1] "Nonlinear mappings of nonexpensive and accretive type in Banach spaces," *Bull. Amer. Math. Soc.*, **73** (1967), 875–882.

L. CESARI

[1] *Asymptotic Behavior and Stability Problems in Ordinary Differential Equations*. Berlin: Springer-Verlag, 1963.

E. CODDINGTON AND N. LEVINSON

[1] *Ordinary Differential Equations*. New York: McGraw-Hill, Inc., 1955.

R. COURANT AND D. HILBERT

[1] *Methods of Mathematical Physics*. Vol. I and II. New York: Interscience Publishers, 1953 and 1962.

R. COURANT AND H. ROBBINS

[1] *What is Mathematics?* London: Oxford University Press, 1941.

M. DAMBORG AND A. NAYLOR

[1] "The fundamental structure of input-output stability for feedback systems," *IEEE Transactions on System Science and Cybernetics*, Vol. SSC-6, No. 2 (April 1970).

M. M. DAY

[1] *Normed Linear Spaces*. Berlin: Springer-Verlag, 1962.

J. A. DIEUDONNÉ

[1] *Foundation of Modern Analysis*. New York: Academic Press, Inc., 1960.

J. L. DOOB

[1] *Stochastic Processes*. New York: John Wiley & Sons, Inc., 1953.

N. DUNFORD AND J. SCHWARTZ

[1] *Linear Operators*, Parts I and II. New York: Interscience Publishers, 1958 and 1963.

A. DVORETZKY AND C. ROGERS

[1] "Absolute and unconditional convergence in normed linear spaces," *Proc. Nat Acad. Sci. U.S.A.* **36** (1950), 192–197.

R. E. EDWARDS

[1] *Functional Analysis*. New York: Holt, Rinehart and Winston, Inc., 1965.

[2] *Fourier Series*, Vol. I and II. New York: Holt, Rinehart and Winston, Inc., 1967.

L. ENGEL

[1] *How to Buy Stocks*, 4th edition. New York: Bantam Books, 1967.

W. FELLER

[1] *An Introduction to Probability Theory and Its Applications*, Vol. I and II. New York: John Wiley & Sons, Inc., 1957 and 1966.

A. FRIEDMAN

[1] *Generalized Functions and Partial Differential Equations*. Englewood Cliffs, N.J.: Prentice-Hall, Inc., 1963.

[2] *Partial Differential Equations*. New York: Holt, Rinehart and Winston, Inc., 1969.

B. FUGLEDE
  [1] "A commutivity theorem for normal operators," *Proc. Nat. Acad. Sci.,*
  U.S.A. **36** (1950), 35–40.

P. R. GARABEDIAN
  [1] *Partial Differential Equations.* New York: John Wiley & Sons, Inc.,
  1964.

I. I. GIKHMAN AND A. V. SKOROKHOD
  [1] *Introduction to the Theory of Random Processes.* Philadelphia: W. B.
  Saunders Company, 1969.

A. M. GLEASON
  [1] "Measures on the closed subspaces of a Hilbert space," *J. Math Mech.*
  6 (1957), 885–893.

C. GOFFMAN AND G. PEDRICK
  [1] *First Course in Functional Analysis.* Englewood Cliffs, N.J.: Prentice Hall,
  Inc., 1965.

R. R. GOLDBERG
  [1] *Fourier Transforms.* New York: Cambridge University Press, 1962.

S. GOLDBERG
  [1] *Unbounded Linear Operators.* New York: McGraw-Hill, Inc., 1966.

P. HALMOS
  [1] *Measure Theory.* Princeton, N.J.: D. Van Nostrand Company, Inc., 1950.
  [2] *Lectures on Ergodic Theory.* New York: Chelsea Publishing Company, 1956.
  [3] *Introduction to Hilbert Space.* New York: Chelsea Publishing Company,
  1957.
  [4] *Finite Dimensional Vector Spaces.* Princeton, N.J.: D. Van Nostrand
  Company, Inc., 1958.
  [5] *Naive Set Theory.* Princeton, N. J.: D. Van Nostrand Company, Inc., 1960.

G. H. HARDY, J. E. LITTLEWOOD, AND G. POLYA
  [1] *Inequalities.* New York: Cambridge University Press, 1952.

F. HAUSDORFF
  [1] *Mengenlehre.* New York: Dover Publications, Inc., 1944.

G. HELLWIG
  [1] *Differential Operators of Mathematical Physics.* Reading, Mass.: Addison-
  Wesley Publishing Company, Inc., 1967.

E. HEWITT
  [1] "The role of compactness in analysis," *Amer. Math. Monthly* **67** (1960),
  499–516.

E. HEWITT AND K. STROMBERG
  [1] *Real and Abstract Analysis.* Berlin: Springer-Verlag, 1965.

D. HILBERT
  [1] *Grundzüge Einer Allgemeinen Theorie der Linearen Integralgleichungen.*
  New York: Chelsea Publishing Company, 1952.

E. HILLE AND R. S. PHILLIPS
  [1] *Functional Analysis and Semi-Groups.* Providence, R.I.: American Mathematical Society, 1957.

J. INDRITZ
   [1] *Methods in Analysis*. New York: The Macmillan Company, 1963.
J. M. JAUCH
   [1] "Theory of the scattering operator," *Helv. Phys. Acta* **31** (1958), 127–158 and 661–684.
   [2] *Foundations of Quantum Mechanics*. Reading, Mass.: Addison-Wesley Publishing Company, Inc., 1968.
T. KATO
   [1] "On the existence of solutions of the helium wave equations," *Trans. Amer. Math. Soc.*, **70** (1951), 212–218.
   [2] *Perturbation Theory for Linear Operators*. Berlin: Springer-Verlag, 1966.
J. KELLEY
   [1] *General Topology*. Princeton, N.J.: D. Van Nostrand Company, Inc., 1955.
A. N. KOLMOGOROV
   [1] *Foundations of the Theory of Probability*. New York: Chelsea Publishing Company, 1950.
A. N. KOLMOGOROV AND S. V. FOMIN
   [1] *Elements of the Theory of Functions and Functional Analysis*, Vol. I and II. Albany, N.Y.: Graylock, 1957 and 1961.
M. A. KRASNOSEL'SKII AND YA. B. RUTICKII
   [1] *Convex Functions and Orlicz Spaces*. Groningen, Netherlands: P. Noordhoff, N.V., 1961.
C. LANCZOS
   [1] *Linear Differential Operators*. Princeton, N.J.: D. Van Nostrand Company, Inc., 1961.
E. B. LEE AND L. MARKUS
   [1] *Foundations of Optimal Control Theory*. New York: John Wiley & Sons, Inc., 1967.
M. LOÈVE
   [1] *Probability Theory*. Princeton, N.J.: D. Van Nostrand Company, Inc., 1960.
L. H. LOOMIS
   [1] *An Introduction to Abstract Harmonic Analysis*. Princeton, N.J.: D. Van Nostrand Company, Inc., 1953.
E. R. LORCH
   [1] *Spectral Theory*. New York: Oxford University Press, 1962.
   [2] "The spectral theorem." in *Studies in Mathematics*, Vol. I. Math. Assn. Amer., 1962.
W. MAAK
   [1] *An Introduction to Modern Calculus*. New York: Holt, Rinehart and Winston, Inc., 1963.
N. MEYERS AND J. SERRIN
   [1] "$H = W$," *Proc. Nat. Acad. Sci. U.S.A.* **51** (1964), 1055–1056.
S. G. MIKHLIN (editor)
   [1] *Linear Equations of Mathematical Physics*. New York: Holt, Rinehart and Winston, Inc., 1967.

M. A. NAIMARK
   [1] *Normed Rings*. Groningen, Netherlands: P. Noordhoff, N.V., 1960.
Z. NEHARI
   [1] *Conformal Mapping*. New York: McGraw-Hill, Inc., 1952.
E. D. NERING
   [1] *Linear Algebra and Matrix Theory*. New York: John Wiley & Sons, Inc., 1963.
J. VON NEUMANN
   [1] *The Mathematical Foundations of Quantum Mechanics*. Princeton, N. J.: Princeton University Press, 1955.
L. NEUSTADT
   [1] "Minimum effort control," *J. SIAM Control*, **1** (1962), 16–31.
R. E. A. C. PALEY AND N. WIENER
   [1] *Fourier Transforms in the Complex Domain*. Providence, R.I.: American Mathematical Society, 1934.
W. A. PORTER
   [1] *Modern Foundations of Systems Engineering*. New York: The Macmillan Company, 1966.
YU. V. PROKHOROV AND M. FISZ
   [1] "A characterization of normal distributions in Hilbert space," *Theory Prob. and Its Appl.* **2** (1957), 468–469.
R. PROSSER AND W. ROOT
   [1] "The ε-entropy and ε-capacity of certain time-invariant channels," *J. Math. Anal. Appl.* **21** (1968), 233–241.
F. RIESZ AND B. SZ.-NAGY
   [1] *Functional Analysis*. New York: Frederick Ungar Publishing Co., 1955.
H. L. ROYDEN
   [1] *Real Analysis*. New York: The Macmillan Company, 1963.
W. RUDIN
   [1] *Principals of Mathematical Analysis*, 2nd edition. New York: McGraw-Hill, Inc., 1964.
L. SCHWARTZ
   [1] *Thèorie des Distributions*, Vol. I and II. Paris: Hermann & Cie, 1951.
G. R. SELL AND H. WEINBERGER
   [1] "Periodic behavior in a food chain," (to appear).
G. F. SIMMONS
   [1] *Introduction to Topology and Modern Analysis*. New York: McGraw-Hill, Inc., 1963.
S. L. SOBOLEV
   [1] *Applications of Functional Analysis in Mathematical Physics*. Providence, R.I.: American Mathematical Society, 1963.
M. H. STONE
   [1] *Linear Transformations in Hilbert Space*. Providence R.I.: American Mathematical Society, 1932.

A. E. TAYLOR

[1] *Advanced Calculus*. Waltham, Mass.: Blaisdell Publishing Company, 1955.
[2] *Introduction to Functional Analysis*. New York: John Wiley & Sons, Inc., 1958.

J. G. TRUXAL

[1] *Automatic Feedback Control System Synthesis*. New York: McGraw-Hill, Inc., 1955.

A. WILANSKY

[1] *Functional Analysis*. Waltham, Mass.: Blaisdell Publishing Company, 1964.

R. L. WILDER

[1] *Introduction to the Foundations of Mathematics*. New York: John Wiley & Sons, Inc., 1952.

D. YOULA, L. CASTRIOTA, AND H. CARLIN

[1] "Bounded real scattering matrices and foundations of linear passive network theory," *IRE Trans. on Circuit Theory* (March 1959).

A. C. ZAANEN

[1] *Linear Analysis*. Groningen, Netherlands: P. Noordhoff, N.V., 1953.

A. ZYGMUND

[1] *Trigonometric Series*, Vol. I and II. New York: Chelsea Publishing Company, 1952.

Index of Symbols

Index

617